REVIEW OF HUMAN DEVELOPMENT

REVIEW OF HUMAN DEVELOPMENT

Edited by

TIFFANY M. FIELD
University of Miami

ALETHA HUSTON
University of Kansas

HERBERT C. QUAY
University of Miami

LILLIAN TROLL
Rutgers University

GORDON E. FINLEY
Florida International University

A Wiley-Interscience Publication

JOHN WILEY & SONS

New York · **Chichester** · **Brisbane** · **Toronto** · **Singapore**

Copyright © 1982 by John Wiley & Sons, Inc.

All rights reserved. Published simultaneously in Canada.

Reproduction or translation of any part of this work beyond that permitted by Section 107 or 108 of the 1976 United States Copyright Act without the permission of the copyright owner is unlawful. Requests for permission or further information should be addressed to the Permissions Department, John Wiley & Sons, Inc.

Library of Congress Cataloging in Publication Data:

Main entry under title:
Review of human development.

   "A Wiley-Interscience publication."
   Includes index.
   1. Developmental psychology—Addresses, essays, lectures.   I. Field, Tiffany.

BF713.5.R48   155   81-21886
ISBN 0-471-08116-7   AACR2

Printed in the United States of America

10 9 8 7 6 5 4 3 2 1

CONTRIBUTORS

MARGARET M. BALTES
Department of Gerontopsychiatry
Free University
Berlin, Germany

PAUL B. BALTES
Max Planck Institute on Human
   Development and Education
Berlin, Germany

MARK A. BARNETT
Department of Psychology
Kansas State University
Manhattan, Kansas

ALBERT CARON
Laboratory of Developmental
   Psychology
National Institutes of Mental Health
Bethesda, Maryland

ROSE CARON
Laboratory of Developmental
   Psychology
National Institutes of Mental Health
Bethesda, Maryland

CATHERINE S. CHILMAN
School of Social Welfare
University of Wisconsin
Milwaukee, Wisconsin

PAUL COSTA
National Institute of Aging
Gerontology Research Center
Baltimore City Hospital
Baltimore, Maryland

NELSON COWAN
Department of Psychology
University of Wisconsin
Madison, Wisconsin

JOAN T. ERBER
Department of Psychology
Washington University
St. Louis, Missouri

JOSEPH FAGAN III
Perceptual Development Laboratory
Psychology Department
Case Western Reserve University
Cleveland, Ohio

BEVERLY I. FAGOT
Department of Psychology
University of Oregon
Eugene, Oregon

TIFFANY M. FIELD
Department of Pediatrics
Mailman Center for Child Development
University of Miami Medical School
Miami, Florida

GORDON FINLEY
Department of Psychology
Florida International University
Tamiami Campus
Miami, Florida

WYNDOL FURMAN
Department of Psychology
University of Denver
Denver, Colorado

RENEE GARFINKEL
Clinic for Intellectual Function
   in the Aging
Department of Psychiatry and
   Geropsychiatric Institute
The Graduate Hospital
Philadelphia, Pennsylvania

JACOB L. GEWIRTZ
Department of Psychology
Florida International University
Tamiami Campus
Miami, Florida

GUNHILD HAGESTAD
Department of Individual and Family
 Studies
College of Human Development
Pennsylvania State University
University Park, Pennsylvania

MARGARET HELLIE HUYCK
Psychology Department
Illinois Institute for Gerontology
Chicago, Illinois

NATHAN KOGAN
Department of Psychology
New School of Social Research
New York, New York

M. POWELL LAWTON
Behavioral Sciences Section
Philadelphia Geriatric Center
University of Minnesota
Philadelphia, Pennsylvania

LYNN S. LIBEN
Department of Psychology
University of Pittsburgh
Pittsburgh, Pennsylvania

GISELA LABOUVIE-VIEF
Psychology Department
Wayne State University
Detroit, Michigan

LEWIS P. LIPSITT
Child Study Center
Brown University
Providence, Rhode Island

JUDITH LIST
Institute of Child Development
University of Minnesota
Minneapolis, Minnesota

MICHAEL LEWIS
Rutgers Medical School
CMDNJ
New Brunswick, New Jersey

KENT A. MCCLELLAND
Department of Sociology
University of Miami
Coral Gables, Florida

ROBERT MCCRAE
Gerontology Research Center
National Institute of Aging
Baltimore City Hospital
Baltimore, Maryland

PHILIP A. MORSE
Department of Psychology
Waisman Center
University of Wisconsin
Madison, Wisconsin

DAVID MOSHMAN
Department of Psychology
University of Nebraska
Lincoln, Nebraska

EDITH NEIMARK
Department of Psychology
Douglass College
Rutgers University
New Brunswick, New Jersey

JUDITH L. NEWMAN
Department of Psychology
Temple University
Philadelphia, Pennsylvania

WILLIS F. OVERTON
Department of Psychology
Temple University
Philadelphia, Pennsylvania

HANUŠ PAPOUŠEK
Max Planck Institut fur Psychiatrie
Deutsche Forschungsantalt fur
 Psychiatrie
Monchen, West Germany

MECHTHILD PAPOUŠEK
Max Planck Institut fur Psychiatrie
Deutsche Forschungsantalt fur Psychiatrie
Monchen, West Germany

CHARLOTTE J. PATTERSON
Department of Psychology
University of Virginia
Charlottesville, Virginia

MARION PERLMUTTER
Institute of Child Development
University of Minnesota
Minneapolis, Minnesota

SLOBODAN B. PETROVICH
Psychology Department
University of Maryland
Baltimore, Maryland

HERBERT C. QUAY
Applied Social Sciences
University of Miami
Coral Gables, Florida

PATRICK RABBITT
Department of Experimental Psychology
Queen's College
University of Oxford
Oxford, England

MABEL RICE
Department of Human Development
University of Kansas
Lawrence, Kansas

JAIPAUL ROOPNARINE
Mailman Center for Child Development
University of Miami Medical School
Miami, Florida

HOLLY A. RUFF
Department of Pediatrics
Albert Einstein Medical College
Yeshiva University
Bronx, New York

JAMES SHERMAN
Department of Human Development
University of Kansas
Lawrence, Kansas

PATRICIA B. SUTKER
Department of Psychiatry and Behavioral Sciences
Medical University of South Carolina
Charleston, South Carolina

BARBARA TURNER
School of Education
University of Massachusetts
Amherst, Massachusetts

CASTELLANO B. TURNER
School of Education
University of Massachusetts
Amherst, Massachusetts

PREFACE

The purpose of this volume is to provide the researcher, teacher, student, and other persons interested in developmental psychology a current view of the "state of the art" research on topics at each stage of development. Thus a researcher or student of infancy who is teaching a child development course can become acquainted with research being conducted in the childhood and adolescent stages, or the researcher of aging who is tracking the life-span development of a process can become familiar with current work on that process at other stages.

Most developmental psychologists are identified with the study of given processes at given stages rather than multiple processes at a given stage or a single process across stages. The organization of this volume reflects that reality, reviews of process-oriented research within a given stage of development from infancy to aging. The contributors were selected because they are actively conducting research on topics of current interest at the different stages and are among the leading researchers in the areas they review. Thus their reviews cover their own research as well as that of their colleagues. Brief thumbnail sketches of the chapters, although they do not do justice to the material reviewed, will be provided as an overview of the volume, since the diversity of the volume discouraged an attempt at integration.

INFANCY

Gewirtz and Petrovich set the stage for the volume and the section on infancy with an evolutionary perspective on human development, providing animal and human examples of genetic change as the basis of organic evolution and behavior change as the basis of cultural evolution.

In the second chapter of the infancy section, I introduce the term *social perception* to describe a fast accumulating literature on the very young infant's unique, adaptive and precocious perception of social stimuli, such as human faces and voices, and discuss the special case that can be made for the perception of "humanness" from the start of life.

A very comprehensive review of infant auditory and speech perception is then provided by Morse and Cowan. Their exhaustive references and tables provide an exemplary resource for the student interested in auditory and speech perception.

Lipsitt presents the "new look" in the field of infant learning, which emphasizes contextual and ethological conditions. He reviews the infant habituation and conditioning literature and the learned adaptations of infants that "tend to promote the pleasures of sensation and circumvent annoying or painful stimulation."

Infant memory, an increasingly popular area of study because of the recent reports of a link between early visual recognition memory and later intelligence, is reviewed by Fagan. His discussion includes an elaboration of the paradigms used and the temporal and contextual factors which influence infant recognition memory.

A natural sequel is the chapter on the development of object perception in infancy by Ruff. Despite the importance of object perception for cognitive abilities, such as object permanence and categorization, this area of research has only recently become popular, and Ruff insightfully outlines some of the issues in need of further investigation.

The Carons introduce their chapter on cognitive development in early infancy with a section on the conceptual trends in adult cognition. They then present the Piagetian and Gibsonian perspectives that have guided the field. Rich in theoretical discussion and empirical detail, the Carons' chapter integrates a wealth of research material which basically suggests that the conceptual foundation of intellect is formed during early infancy and that infants, like adults, "ignore isolated details of experience in favor of the more abstract information that subordinates these details."

From perception and cognition, the section on infancy moves to social development. The Papoušeks provide a unique approach to the discussion of early social interactions by treating them in the context of evolution and general systems theory and then by reviewing the literature on developmental changes and failures in infant–adult interaction.

Roopnarine and I follow with a chapter on the very recently popular research on infant–peer interactions. Arguing that infants' interactions with peers and adults may develop in parallel, we challenge the view that infant–adult interactions are a necessary precursor to infant–peer interactions, a view which may have derived from studying adult and peer interactions at different stages of infancy and observing infants in strange settings with unfamiliar peers.

Finally, Lewis rounds out the infancy section with a global view of the infant's expanding social network and how the infant is socialized by learning to become a member of these different networks: family, school, and society.

The research reviewed in the infancy section contributes an overall picture of the infant as significantly more sophisticated perceptually, cognitively, and socially than had been thought a decade ago, particularly the 0- to 6-month-old infants who are the subject of most of these chapters and, indeed, most infancy research. Other general impressions of these reviews are that there is a growing awareness of the importance of ecological or contextual factors in research on infants and an increasing appreciation of the relationships between the development of perceptual, cognitive, and social processes in infancy as reflected in terms like *social perception* and *social cognition*.

CHILDHOOD

The opening chapter of this section by Overton and Newman presents the competence–activation/utilization approach to cognitive development, an integration of competence models, such as that of Piaget, and performance models. A basic competence is developed and environmental factors may activate competence. If factors enhance relatively adequate performance or are effective following the demonstration of mature competence, then the function of those factors is utilization. These authors offer paradigms and empirical data from the literature on children's cognitive performance in support of this model.

A behavioral approach to children's learning is then presented by Sherman and is contrasted with the structural approach to learning. The acquisition of response classes—through teaching, observational learning, and environmental consequences—and the defining properties of response classes and stimulus classes are described in this chapter.

Rice reviews, in the following chapter, the development of knowledge on child language from Chomsky's description of the formal structure of language, to theories that link syntax with semantic structures, to the inclusion of sociolinguistic knowledge as an area of child language study. She then discusses competing hypotheses on the role of cognition and the role of the environment in the acquisition of communicative competence.

In the next chapter in the childhood section Liben reviews empirical work on children's memory and suggests that development of memory is tied to cognitive advances in other areas such as logical operations, knowledge, representation, and metacognition. In addition, she emphasizes the importance of cross-cultural approaches to the study of memory development for determining the universality of the phenomena and the relative contribution of variables, which are confounded in our culture, such as chronological age and years of schooling.

The topic of self-control and self-regulation is discussed by Patterson, self-control being viewed as the "figure" and self-regulation as the "ground" of voluntary activity. In a chapter rich in language, metaphor, and philosophy, Patterson reviews research on the question of how young children can use their developing cognitive and linguistic skills to develop self-control, including the effects of imaginal transformations and self-instructional plans and the child's knowledge about self-control processes.

Fagot then reviews research on the influences of parents, teachers, and peers on early sex-role development. In suggesting new directions for research, she highlights the importance of specifying how the child's own characteristics contribute to adult and peer pressures and the varying influences of adults and peers at various stages of the child's sex-role development.

Research on empathy and prosocial behavior in children is reviewed by Barnett who suggests that the focus of this area needs to be broadened to include affective responses to the full range of emotions rather than the typically studied responses to the distress and sadness of others. Further, he emphasizes that naturalistic studies of the role of empathy in the child's evolving interpersonal relations, such as the ability to form friends, may expand our understanding of empathy and its function in the affective and social development of the child.

A natural sequel to the chapter on empathy is the chapter on children's friendships by Furman. In this review resides an example of how we stage-oriented developmental psychologists might benefit from viewing the development of phenomena such as friendship, in this case, or other phenomena such as love, work, and play, important to all of us at all stages, in a life-span context. Furman reviews data on friendship from infancy to adulthood as well as the trials and tribulations that go with friendships across development. He neglects, only, friendship among the aging, a time when making new friends may be critical given the loss of old friends.

ADOLESCENCE

The section on adolescence contains fewer chapters, reflecting the reality that there are fewer developmental psychologists studying adolescence—some say because when they were studying developmental psychology they themselves were "virtual" adolescents, making the subject "too close to home." Adolescence is also less well demarcated and has not generally been considered as a period in which basic cognitive and social behavior is rapidly developing. Another reason offered for the very few studies on adolescents is that they are less readily accessible for study, often resenting "interference" from adult researchers.

Moshman and Neimark open this section with a chapter on the development of the adolescent as logician, scientist, philosopher, and artist. The common developmental trends thought to occur across these four disparate areas include a "finer differentiation of components and their integration into a more flexibly organized whole" and "increasing detachment of thought from both context and personal viewpoint."

Problems highly visible at this stage, drug and alcohol behaviors, are reviewed by Sutker who presents data on drug use patterns, prevalence, and frequency estimates; variation of use by age, gender, and ethnicity; personal characteristics among drug abuse samples; and environmental factors.

Another problem associated with this stage, that of adolescent aggression, is reviewed by Quay. The literature on the etiology and treatment of two different dimensions of aggression, conduct disorder and socialized aggression, is reviewed, a literature which points to socialized aggression being less seriously pathological than conduct disorder.

McClelland, in a chapter entitled "The Adolescent Subculture in the Schools," discusses the research generated by the Coleman report on the "adolescent subculture that repudiated adult interests," a literature which largely repudiates Coleman. While agreeing that high school students' behavior may be somewhat distinctive, McClelland maintains that the patterns of difference are not so consistent to merit the term *subculture*.

Finally, Chilman reviews research on causes and consequences of adolescent childbearing, on contraceptive use, abortion, and births to unmarried teenagers.

Whereas the infant and the child, as presented in the earlier sections, are depicted as more sophisticated perceptually, socially, and cognitively than research of a decade ago suggested, the adolescent is portrayed in this section as fraught with problems. Adolescents may not be a subculture, but the frequent focus of research on the problems of adolescence and problem teenagers perpetuates a view of them as a subculture, essentially problematic and different from children and adults.

ADULTHOOD

Until recently the adulthood stage was infrequently studied, perhaps because of the tradition of developmental psychologists to research early development or perhaps because many of our stellar theorists, such as Piaget, maintained that development is completed by adulthood. Nonetheless, the reviews in this section on recent research in cognitive development, mental health, friendship, marriage, family, and work suggest that development does indeed continue through adulthood.

Neimark introduces this section with a review of cognitive development during adulthood. She discusses the factors that affect cognitive performance of adults, including content knowledge, structural learning, field factors, and individual difference factors of motivation and bias. Although Neimark suggests that there is a cognitive development unique to the adult stage, it is without a "directing" theory, and empirical data collecting may be hampered by adults being too busy with practical matters to serve as research subjects and by the difficulty of tapping representative adult thought under laboratory conditions.

Labouvie-Vief, like Neimark, discusses the somewhat static and negative view that has been taken of adult cognition, as "a state in which the organism achieves an equilibrium some time between the early rush of childhood changes and terminal deteriorative biological processes." She claims that the ability to tolerate and create instability and remain open to new information might be viewed as just as essential to survival as the ability to maintain stability. Logic during adulthood may become subordinated to "the social context of thought."

Turner and Turner review current literature on mental health during adulthood and how conceptions of positive mental health have revealed a social-class and gender bias. They suggest in their overview that cohort differences have been most apparent in gender comparisons and least apparent for social-class and race comparisons, reflecting the "failure of social systems rather than the shortcomings of individuals."

Friendship and marriage over the adult years are reviewed by Huyck. Literature on marriage, marriage styles, remarriage, same-sex and cross-sex friendships, and ethnic variations in these are reviewed. Sex and age differences in patterns of intimacy are clear, but whether they reflect social or historical effects of developmental change is unclear.

Hagestad, in her chapter on the family, presents literature from both those who have focused on young children and their parents and those who have more recently focused on aged parents and their middle-aged children, the split being referred to as the alpha and the omega of family research. She then discusses research strategies, the multigenerational view of family, and the reciprocal influence of parent and child as a lifelong process.

Finally, for this section, Garfinkel, in a chapter entitled "By the Sweat of Your Brow," reviews research on the role of work in adult development. Although virtually ignored as a research subject until very recently, as is reflected in the length of this chapter, Garfinkel suggests that "it may be the single most dominant influence on the life of an adult, touching everything from speech patterns to friendship networks."

AGING

Finley introduces the section on aging with a review of the theoretical and empirical literature on the effects of modernization on the aged. Finley begins with population profiles of the more and less developed regions of the world and then turns to the data which support different models, a curvilinear, or "cyclical," model of modernization and a linear, or "decline," model of modernization. For future research he suggests that modernization and aging need to be studied cross-culturally as multidimensional entities and distinctions need to be made between factors such as status and attitudes and the actual treatment accorded the aged.

Baltes and Baltes review research on variability and plasticity in aging with a focus on two research programs: operant-ecological research on elderly dependency in nursing homes and cognitive training research in the area of psychometric intelligence. The findings summarized indicate that dependency in older nursing home residents is not solely regulated by biological conditions; dependency is modifiable, and it is acquired and maintained "by the nature of the existing social contingencies with which the elderly live." In addition, there is much evidence for the elderly's ability to benefit from experience in cognitive practice.

In a chapter entitled "Breakdown of Control Processes in Old Age," Rabbitt suggests that recent models based on elementary control systems theory may provide better descriptions of change. He then reviews data showing breakdown of control of performance in a laboratory experiment and examples of changes in efficiency in everyday life situations such as clumsiness and loss of motor control and keeping track of everyday conversations.

Perlmutter and List subscribe to the view of learning as a lifelong activity rather than a function which declines among the aged. Although age-related declines have been noted in the "micromechanisms of learning" used in laboratory situations, effective learning has also been documented among older adults in the macrolearning styles used in everyday situations. Following a discussion of the nature of learning and learning in laboratory situations, these authors turn to learning in naturally occurring situations, focusing on physical skills and the literature on physical training and sensory adaptation; cognitive skills and the literature on media use and formal education; and social skills with particular attention to retirement adjustment and psychotherapy.

Erber, in a chapter on memory among the aged, discusses theoretical models of memory; memory stores including sensory, primary, secondary, and tertiary memory; and noncognitive factors affecting memory including motivation, anxiety, and cautiousness. She concludes that older people seem to have difficulty mostly in the realm of secondary memory which is "most apparent in laboratory-type episodic memory tasks when structure is not provided, when little or no retrieval support is available, and, in some cases, when several elements must be attended to simultaneously." When any or all of these are facilitated, performance markedly improves.

Kogan reviews research of the 1970s on cognitive styles in older adults, or the individual variation in modes of attending, perceiving, remembering, and thinking. These styles include reflection–impulsivity, field dependence–independence, and conceptualization styles. Kogan suggests an examination of demographic, health, ability, and personality correlates to better understand the considerable variability on cognitive styles among the old.

Stability rather than change is noted for personality development during adulthood and old age, according to McCrae and Costa. Following a review of alternative models of personality and aging, they suggest that instead of asking how personality is changed by aging, we might ask how the life course is changed by "enduring personality dispositions."

Finally, Lawton, in the last chapter of the volume, entitled "The Well-Being and Mental Health of the Aged," describes four domains of psychological well-being that can be conceptually and empirically differentiated, including happiness, positive affect, negative affect, and "generalized life satisfaction based on perceived consequences between desired and achieved life goals." He concludes with the question: "If stable, intrapersonal, temperamental factors contribute to the experience of positive and negative affect, how do external experiences modify such long-term propensities?"

Thus in the section on aging, where one might expect to find a focus on decline of developmental processes, instead several of the authors have noted these processes as being stable and ongoing. Discontinuities or declines noted at this stage might relate more to environmental factors, such as the treatment of the aged and roles accorded to the aged, which may, in turn, contribute to the considerable individual variability noted on several of the processes discussed.

SUMMARY

As one reader to another, there are some general impressions of the volume I would like to share. Throughout each of the stages reviewed, one is impressed with the increasingly sophisticated human being, whether infant or aged, that is being revealed by increasingly sophisticated research. Also apparent across all chapters is the developmental psychologist's growing awareness of the interrelatedness of perceptual, cognitive, and social developmental processes and the effects of contextual or ecological factors on these.

Finally, although developmental psychologists persist in studying these processes at different stages, perhaps depending upon the age of their favorite subjects, there is a continuity of processes throughout. While being an infant/child researcher, the descriptions of theory, behavior, and process at the later stages seemed familiar to me. Almost every contributor spoke of the importance of studying individuals and their differences across development for further enlightenment of the developmental process.

TIFFANY M. FIELD

Miami, Florida
March 1982

CONTENTS

INFANCY

1. Early Social and Attachment Learning in the Frame of Organic and Cultural Evolution — 3
   Jacob L. Gewirtz & Slobodan B. Petrovich

2. Social Perception and Responsivity in Early Infancy — 20
   Tiffany M. Field

3. Infant Auditory and Speech Perception — 32
   Philip A. Morse & Nelson Cowan

4. Infant Learning — 62
   Lewis P. Lipsitt

5. Infant Memory — 79
   Joseph F. Fagan III

6. The Development of Object Perception in Infancy — 93
   Holly A. Ruff

7. Cognitive Development in Early Infancy — 107
   Albert J. Caron & Rose F. Caron

8. Infant–Adult Social Interactions: Their Origins, Dimensions, and Failures — 148
   Hanuš Papoušek & Mechthild Papoušek

9. Infant–Peer Interactions — 164
   Tiffany M. Field & Jaipaul L. Roopnarine

10. The Social Network Systems Model: Toward a Theory of Social Development — 180
    Michael Lewis

CHILDHOOD

11. Cognitive Development: A Competence–Activation/Utilization Approach — 217
    Willis F. Overton & Judith L. Newman

12. Behavioral Approach to Children's Learning — 242
    James A. Sherman

13. Child Language: What Children Know and How — 253
    Mabel L. Rice

14.	**The Developmental Study of Children's Memory** *Lynn S. Liben*	269
15.	**Self-Control and Self-Regulation in Childhood** *Charlotte J. Patterson*	290
16.	**Adults As Socializing Agents** *Beverly I. Fagot*	304
17.	**Empathy and Prosocial Behavior in Children** *Mark A. Barnett*	316
18.	**Children's Friendships** *Wyndol Furman*	327

ADOLESCENCE

19.	**Four Aspects of Adolescent Cognitive Development** *David Moshman & Edith Neimark*	343
20.	**Adolescent Drug and Alcohol Behaviors** *Patricia B. Sutker*	356
21.	**Adolescent Aggression** *Herbert C. Quay*	381
22.	**Adolescent Subculture in the Schools** *Kent A. McClelland*	395
23.	**Adolescent Childbearing in the United States: Apparent Causes and Consequences** *Catherine S. Chilman*	418

ADULTHOOD

24.	**Cognitive Development in Adulthood: Using What You've Got** *Edith D. Neimark*	435
25.	**Discontinuities in Development from Childhood to Adulthood: A Cognitive-Developmental View** *Gisela Labouvie-Vief*	447
26.	**Mental Health in the Adult Years** *Barbara F. Turner & Castellano B. Turner*	456
27.	**From Gregariousness to Intimacy: Marriage and Friendship over the Adult Years** *Margaret Hellie Huyck*	471
28.	**Parent and Child: Generations in the Family** *Gunhild O. Hagestad*	485

| 29. | By the Sweat of Your Brow
Renee Garfinkel | 500 |

AGING

30.	Modernization and Aging *Gordon E. Finley*	511
31.	Microanalytical Research on Environmental Factors and Plasticity in Psychological Aging *Margaret M. Baltes & Paul B. Baltes*	524
32.	Breakdown of Control Processes in Old Age *Patrick Rabbitt*	540
33.	Learning in Later Adulthood *Marion Perlmutter & Judith A. List*	551
34.	Memory and Age *Joan T. Erber*	569
35.	Cognitive Styles in Older Adults *Nathan Kogan*	586
36.	Aging, the Life Course, and Models of Personality *Robert R. McCrae & Paul T. Costa, Jr.*	602
37.	The Well-Being and Mental Health of the Aged *M. Powell Lawton*	614

Author Index 629

Subject Index 655

REVIEW OF
HUMAN DEVELOPMENT

INFANCY

EARLY SOCIAL AND ATTACHMENT LEARNING IN THE FRAME OF ORGANIC AND CULTURAL EVOLUTION

Jacob L. Gewirtz
*State University of New York at Stony Brook
and Florida International University*

Slobodan B. Petrovich
University of Maryland Baltimore County

This chapter examines some mechanisms of early social learning and attempts to cast the empirical record in a broad conceptual framework that views *genetic change* and *learning* as the two major processes by which individuals or species progressively acquire structural and functional characteristics that allow them to interact and adapt to the environment of the ecosystem. The issues dealt with are thought to be generic for all theories of development, regardless of their orientation, heuristic flavor, or degree of formality. On such a basis, the investigation of early human social learning would proceed, in the same way as would the inquiries into the development of all classes of behavior, by analysis of the controlling variables, both of the present and the past. Aspects of development indexed by such terms as species-specific behaviors, prenatal and perinatal determinants, maturational constraints, critical periods, and the like would only qualify aspects of this model but would not change its essential features.

At the outset, our goal is to elucidate some of the phylogenetic and ontogenetic constraints on early learning and development. We begin our attempt by providing a paleoanthropological perspective, followed by examples of ecological adaptation of behavior. In the frame of cultural evolution and its outcomes, our treatment considers ecological determinants of infant social behavior and of learning and concludes with consideration of the plausible roles of learning in illustrative cases of early human social development. For these cases, the heuristic emphasis will be on conditioning paradigms and operant learning, in particular, for that paradigm exemplifies adaptation within the cultural-evolutionary frame. (There exist, to be sure, other paradigms that could be applied to some of these cases; e.g., Gewirtz, 1972a.)

A PALEOANTHROPOLOGICAL PERSPECTIVE

Perhaps as early as 5 million years ago, our hominid ancestors walked this planet. How did the ascent of *Homo sapiens* proceed from those animal beginnings to that phase in the 1850s when Charles Darwin and Alfred Russell Wallace independently proposed a "new" process of creation—evolution by natural selection? In their different ways, physical, natural, behavioral, and social science are each striving to provide an answer to that question. The contributions of Darwin and Wallace planted the seeds for the powerful scientific and intellectual conceptualization that is still unfolding. Evolution is becoming increasingly comprehensible as a process, inescapable as a fact, and all-embracing as a concept (Ghiselin, 1969; Huxley, 1960; Kennedy, 1980).

Normally, chimpanzees and other great apes possess 48 chromosomes, while *Homo sapiens* has 46. These differences between humans and their nearest primate relative—the chimpanzee—are thought largely due to chromosomal mutations. Even so, most human and chimpanzee chromosome pairs are so similar that there is no question about the empirical evidence supporting their homologous nature. The great apes have retained their quadrupedal posture, while evolution has moved humans in the direction of bipedalism, thereby freeing hands for manipulating objects and developing tools. It is the change in teeth that marks the separation of the primate line that leads to *Homo erectus*. The jaw of *Ramapithecus* found in Kenya and India dates back about 14 million years, with humanlike teeth but without the great canines of the anthropoid apes. With a larger brain (Jerison, 1973) and free hands, ancestors of humans began to develop tools, communicate, foster social organization, and conquer the use of fire (Bodmer & Cavalli-Sforza, 1976). For at least a million years after that, human ancestors lived animallike as predators, scavengers, foragers, and hunters. There are no significant fossil records to forecast the emergence of art and culture. Then, some 20,000-odd years ago, warmed by fire, humans began to communicate the hunter's awareness of the prey animal by rock paintings in caves like those of Altamira. Such cave paintings signaled an evolutionary emergent—the capacity of the human animal to affect and think about the environment beyond itself. Cultural evolution was occurring, with predictable concurrent features: plants and animals were domesticated; agricultural settlements were developed; tools were improved by discoveries of copper, bronze, and, much later, steel; trade centers grew and urbanization began; and communicative skills, including language and writing, developed. Once the evolutionary process produced a species with *culture,* the knowledge of nature was mastered at an ever-accelerating pace. As earlier noted, the hominid ancestors of the human walked this planet for several million years. And the cave paintings date back more than 20,000 years. Since that point in time, it took *Homo sapiens* about 25,000 years to acquire knowledge and retrospectively to examine its origins, as Darwin and Wallace did. A cumulative storehouse of cultural information provided the human species with an ever-increasing capacity to master, exploit, and understand nature and its place in it (Bronowski, 1973; Campbell, 1975; Campbell, B., 1979; Jaynes, 1976; Lasker, 1969; Murdy, 1975; Naroll & Naroll, 1973).

When the conceptual framework of evolution is applied to understanding human emergence and existence, it is evident that two very different processes have been at work. Under the process of *organic* evolution, the human organism has evolved as have other animal species, through a creative process wherein living matter has responded to changes in its environment, based on evolutionary principles of genetic variation and systematic selective retention. In contrast, the process of *cultural* evolution has involved features very different from those of organic evolution. Among these differences are (1) the mechanism of organic evolution is Mendelian, while the mechanism of cultural evolution is Lamarckian—in the sense that the information is passed from one individual to another and from one generation to the next via mechanisms of communication, learning, and skill acquisition rather than through genes; (2) cultural evolution is potentially much more rapid, and given language, computer language and retrieval processes, and communication modes, the storehouse of information has been retrieved and transmitted at ever-increasing speeds; (3) the basic elements of organic evolution are the genes and the gene-phenotypes—that is, the individual—while cultural evolution may include the emergence of a social unit as an entity in selection; (4) organic evolution is blind and opportunistic, while cultural evolution is heavily influenced by learning, tradition, foresight, and the ability to conceive of a "better way of life" for oneself and one's offspring; (5) through organic evolution, organisms respond to contingencies of survival and continue to adapt to ecological demands; in contrast, through cultural evolution, humans modify and shape the environment to fit their requirements (Campbell, 1975; Lorenz, 1969; Petrovich, 1978; Piaget, 1978).

The coevolutionary nature of the genetic and learning mechanisms is not always understood. There exists an innate storehouse of information—the gene pool. However, *Homo sapiens* is also capable of learning during ontogeny, thereby acquiring and adding new knowledge upon an ever-expanding base. Infinitely more than any other animal, by the quality of its biology, the human is a creature of socialization to a given cultural context and, as such, singularly a product of its learning (e.g., Campbell, 1979; Eibl-Eibesfeldt, 1975; Freud, 1953; Hinde, 1974; Lasker, 1969; Lorenz, 1965, 1969, 1973; Murdy, 1975; Piaget, 1971; Skinner, 1966; Stern, 1973).

BIOLOGICAL CONSTRAINTS ON BEHAVIORAL DEVELOPMENT AND THE ECOLOGICAL ADAPTATION OF BEHAVIOR

From a paleontological perspective, evolution by natural selection has not been a uniform, progressive or linear, or evenhanded process. Experimental approaches based on phylogenetic relatedness have proved heuristically inadequate in their handling of many of the complexities characteristic of behavioral universes (Atz, 1970; Hodos & Campbell, 1969). For example, the original taxonomic groups were generally differentiated on the basis of structural phenotypes that were often difficult, if not impossible, to relate and incorporate in the investigation of behavioral similarities. Even though there are *no* behavioral fossils, investigators were forced at times to make inferences and speculate about the behavior of extinct life forms based on the fossil record of structural phenotypes. In addition, the evidence stemming from a thorough comparative analysis within and across species and phyla has raised two provocative questions: (1) What is the evolutionary explanation for the remarkable behavioral similarity among phylogenetically unrelated or distantly related organisms? and (2) Why are there important behavior differences between phylogenetically related species?

The answers were simple but relatively slow in emerging. Behavioral outcomes are evolutionary reflections of demands exerted by ecological contingencies. Thus the study of the ecological adaptation of behavior—*teleonomy*—provides the evidence for the thesis that similar behaviors among unrelated, or distantly related, forms as well as different behaviors among phylogenetically closely related species result from differential survival values of behavioral adaptations to selection pressures characteristic of contingency demands of species-specific ecological niches (Pittendrigh, 1958; Lorenz, 1969; Tinbergen, 1972; Williams, 1966).

Tinbergen (e.g., 1972) and his students demonstrated convincingly the heuristic value of conceptualizations based on recognition of the importance of teleonomy. Cullen's work (1957) stands out in particular. She showed that the behavioral repertoire of a cliff-nesting gull (*kittiwake*) is similar to its ground-nesting-gull relative, except for many behavioral patterns closely associated with the ecological peculiarity of its nesting, which takes place on narrow ledges of cliff faces. For example, the cliff-nesting gull fails to learn the identity of its own young, whereas the ground-nesting gull readily does. Cliff-nesting-gull young do *not* wander about on the narrow cliff ledges, while ground-nesting-gull young move about and often intermingle. The learning ability involving a discrimination by the ground-nesting gull of its hatchlings seems to have evolved as a discrete unit associated with the care of the young in the context of specific adaptation to nesting habitats. In the cliff-nester, there has been no selection for this sort of learning; if hatchlings are not in the nest, searching would seem unproductive.

All organisms must adapt to contingencies of ecological selection, such as food, shelter, predation, and climatic changes. The importance of these factors in the proximal evolution of behavior can be investigated experimentally. For instance, the evolved survival value of egg pigmentation and shading is lost once the young hatch, exposing the inner white surface of the eggshell and thereby betraying a nest to potential predators. Thus Tinbergen (1963) was able to demonstrate the adaptive fitness of such behaviors as eggshell removal by gulls in the context of preserving nest camouflage. Similarly, in response to predation pressures, many species have evolved complex behavioral countermeasures, including removing cloacal droppings from the young and flying to dump them away from the nesting area.

Clinal (graduated) variations represent another clear manifestation of specialized adaptations along ecological gradients. For example, as latitude increased in the northerly or southerly direction, a number of general trends involving morphological, physiological, and behavioral adaptations became apparent: body size increases (Bergmann's rule); tails, ears, bills, and limbs become relatively short (Allen's rule); relative length of the hair increases; wings become more pointed; the relative size of the heart, pancreas, liver, kidney, stomach, and intestine increases; there is a reduction in pigments, phaeomelanins and eumelanins (Gloger's rule); relative oxygen consumption and metabolic needs decrease, and general activity decreases; migratory instincts become more manifest; larger and warmer "nests" are constructed (King's rule); home ranges become larger, with territorial behavior more pronounced; and photoperiodic rhythms become more evident. As Thiessen (1972) has noted, none of these "clinal laws" could have been predicted from theoretical and empirical approaches rooted in phylogenetic sophistication. They became apparent only when climatic demands of species-specific ecological niches were considered.

Ecologically oriented research has made it possible to show that "good sleepers" are generally predators, have secure sleeping places, or both, while "poor sleepers" tend to be subject to predation at any time (Allison & Van Twyver, 1970). A knowledge and understanding of the behavioral repertoire of some 300 species in their natural habitats was the best predictor of their exploration and curiosity as captive animals in the zoo (Glickman & Sroges, 1966). Many investigators have attempted to study the adaptive significance of different modes of social organization that exist in nature, and at present we are witnessing comprehensive and provocative attempts at such a synthesis (Chagnon & Irons, 1979; Wilson, 1975).

The teleonomic approach has refocused comparative studies of animal learning (Bolles, 1975; Hinde & Stevenson-Hinde, 1973; Seligman, 1970; Seligman & Hager, 1972). In the last decade, a clear departure from the classical orientations has occurred. The shift from hypotheses emphasizing the principle of phylogenetic relatedness (e.g., Bitterman, 1960, 1965) to those exploring ecological perspectives and biological constraints is unmistakable (Lockard, 1971; Mason & Lott, 1976). Two assumptions characterize current orientations: (1) learning is influenced by species-typical constraints and (2) because some forms of learning are particular modes of adaptation to specific ecological contingencies, based on their biological preparedness animals learn associations that are characteristic of the natural history of the species. For example, being a scavenger species, rats will learn quickly to associate taste CSs with stomach illness UCSs (prepared association). In this instance, an association will occur between taste quality and the toxic agent inducing illness, even when the delay between ingestion and the aversive condition is more than an hour. More extended training is required for rats to learn to associate exteroceptive CSs with shock UCSs (an example of unprepared association). In contrast, as an example of counterprepared association, rats will not learn to associate taste with shock or visual or auditory stimuli paired with the toxic agent inducing stomach illness (Garcia, Ervin, & Koelling, 1966; Garcia & Koelling, 1966). Further illustrations are found in other areas, ranging from acquisition of bird song to imprinting in precocial birds.

Acquisition of the Bird Song. A few examples can suggest the range of biological constraints affecting the ontogeny of the bird song. The parasitic cowbird lays its eggs in the nest of a variety of host species, thereby exposing its young to a broad scope of prenatal and postnatal experiences. Nevertheless, as adults the cowbirds have no difficulty singing or recognizing a conspecific song. On the other hand, male white-crowned sparrows are predisposed to learn, first, the species-typical song and, subsequently, the regional dialect. Failure of these sparrows to learn a local dialect may lead to their reproductive isolation. In this frame, Marler (1970) has pointed out that male adaptation to a particular habitat predicted better the importance of specific experiences on the acquisition of the adult song than did the phylogenetic relatedness of the species. Further, based on their reviews of the empirical record, Marler (1970, 1976) and Nottebohm (1970) have underscored the many parallels that are to be found between song acquisition and the development of language. Acoustical stimulation allows sounds to be heard and remembered; the organism hears itself and matches that against what has been learned previously or against a neural template. Auditory feedback is critical for development but does not appear to be as important subsequently. Relevant learning takes place during an optimal period early in the organism's life. Learning is involved in the transition from a subsong or babbling to the adult type of vocalization. Vocal learning is to some extent "stimulus free." There are specific predispositions to learning certain sounds over others, and in both cases—bird song and language ontogeny—there is lateralization of control in the central nervous system. Marler and Nottebohm each stress that these observations do not imply the existence or discovery of language in birds, but that there may be a set of rules that would evolve in nature in organisms engaging in vocal learning. Speech development and song development both proceed along well-defined biological boundaries.

IMPRINTING

Lorenz (1935) applied the German term *Prägung* to a process by which the newly born young animal of many precocial avian species (e.g., the chicken, Greylag goose, jackdaw, mallard, pheasant, partridge) forms a relatively rapid social *bond* or *attachment* to a biologically appropriate conspecific or, in its absence, to some available object, even one from the human species. *Prägung* ("stamping in," "coining") was translated into English as *imprinting*. This metaphoric bond or attachment is connoted by the close reliance of the behavior of the young animal on stimuli provided by its parental figure, and is often denoted by a variety of the young animal's cued-response patterns to parents. Thus imprinting is a label, a heuristic tool, an abstraction conceived to vary along a dimension of theoretical and analytical abstraction comprised of various levels of behavioral analysis—ranging from those on a microlevel (e.g., biochemical, physiological analyses) to those on a macrolevel (e.g., social, ecological analyses). The notions of preparedness and biological constraints are implicit in the imprinting conception. Thus Lorenz (1935) proposed that imprinting has several salient features: imprinting takes place in a highly limited maturation period early in the life of the organism; imprinting has lasting effects; resulting from natural selection, imprinting involves cues that are species-typical; only specific response patterns of the young animal are involved in the imprinting to the social objects; and imprinting could affect the later expression of behaviors (e.g., sexual) not yet performed by the young animal.

Lorenz (1935, 1969), Hess (1959, 1964, 1973), and other early, systematic investigators of the imprinting process in infant animals of precocial species assumed that a learning mechanism underlays imprinting, but one that is biologically constrained. Therefore, they emphasized more the biological boundaries of the acquisition process than its general features. Thus Hess (1959, 1964, 1973) concluded that "association learning" was not involved in imprinting because imprinting involved a distinct critical period during which outcomes associated with association learning did not occur: (a) the early learning that occurred was more influential than later, recent learning; (b) noxious stimulation facilitated rather than inhibited acquisition; (c) trial massing and effort facilitated rather than limited acquisition, and (d) such drugs as muscle relaxants interfered with attachment acquisition to a surrogate imprinting object but ordinarily did not affect learning. This emphasis by Hess and others on the biological constraints on the learning characterizing imprinting was taken by many to be equivalent to the position that if learning is at all involved in imprinting, it must be a form unique to the imprinting process (Bateson, 1971; Hinde, 1970; Rajecki, Lamb, & Obmascher, 1978). In this frame, the imprinting process was thought to be involved in the formation of primary social bond–relationships (Hess, 1962, 1973; Scott, 1960, 1968).

Concurrently, a number of investigators have approached imprinting as a process that involves routine learning paradigms and they have viewed social bond–relationships as the outcomes of that learning. Thus, Gewirtz (1961) approached the imprinting process in operant-learning terms, Bateson (1966, 1971) in generic learning terms, Cairns (1966) in terms of contiguity learning, and Hoffman and associates (Hoffman & DePaulo, 1977; Hoffman & Ratner, 1973) in terms of various learning paradigms. Like Scott (1960, 1968), the listed investigators have reiterated the similarities between the processes involved in imprinting in presocial avian species and those involved in early socialization and the formation of primary social bonds in dogs, monkeys, and humans.

The investigation of imprinting in avian and mammalian species has necessitated the reconsideration and updating of some of the claims originally made (Gottlieb, 1973, 1980; Hess, 1973; Hess & Petrovich, 1977; Roy, 1980). For instance, subsequent research revealed that different species exploit different sensory modalities for the development of the attachment bond, ranging from chemical to auditory, visual, tactile, and thermal. Even so, when some of these ecological, contextual, or species-specific differences were filtered out, the emerging picture supported the thesis that many animals "imprint" and that the development of such a "bond" promotes the survival of the individual and the species. In the next section we examine how the phenomenon of imprinting has been applied to the consequences of early human attachment experiences and subsequent behavioral development.

Imprinting as a Model for Human Attachment

The striking features of the imprinting process have appealed to many who sought an animal model for object fixation or for sexuality channeled in the direction of the biologically inappropriate object, and for whom the notion of the development of "bonding" is the key to individual and species survival. The experimental demonstration of a critical period for imprinting was seen as compatible with theoretical propositions advanced by such sequential-stage theorists as Freud (early), Erikson, and Piaget. However, while looking for similarities between the animal and the human condition, some important ecological and behavioral incompatibilities were overlooked. Lorenz (1935), Hess (1973), and others used precocial, rapidly maturing, avian species capable of complex perceptual-motor integration soon after hatching. It is highly adaptive for a mallard duckling to be able effectively to follow the hen and move in the water in the first 48 hours after hatching, for that behavioral repertoire mitigates against predation and provides access to food and shelter. The ecological contingencies and species-specific maturation processes are surely different for the human infant, for whom there is a much better ecological and behavioral analog in altricial-species development. Altricial young are helpless at birth, and attachment and often social learning requires extensive periods of their interaction with parents and, often also, siblings (Cairns, 1966; Gewirtz, 1961,

1969; Hinde, 1974; Klaus & Kennell, 1976; Roy, 1980). In this frame, also, Hess (1973) has argued that the imprinting process is characteristic of the socialization of precocial species and that its generalization to humans is unwarranted and may deflect attention from the unique features of imprinting. In particular, the fact that critical periods in human behavioral development have yet to be shown conclusively led Hess (e.g., 1973, p. 341) to conclude that human imprinting was yet to be demonstrated.

Like Hess, Bowlby (1969) focused on the comparative aspects of imprinting and early socialization in a number of species. However, Bowlby failed to take into account the ecological dimensions of the underlying processes. The emphasis of his analysis was on the existence of a "bond" that promotes individual and species survival. The apparent resemblances between the bonding process in infrahuman and human young provided the bases for Gray's (1958) and Bowlby's (1969) cases for human imprinting. Thus Bowlby stated:

> So far as is at present known, the way in which attachment behavior develops in the human infant and becomes focused on a discriminated figure is sufficiently like the way in which it develops in other mammals, and in birds, for it to be included legitimately under the heading of imprinting. (Bowlby, 1969, p. 223)

Bowlby's (1969) synthesis of the literature (and Ainsworth's derivative approach—e.g., 1972) has greatly influenced the current views of human attachment. In this frame, given the dyadic nature of socialization and the species-typical plasticity of *Homo sapiens,* it can be useful to differentiate between approaches to attachment that do not employ conditioning paradigms (e.g., Bowlby, 1958, 1969; Gray, 1958), and approaches that do (e.g., Cairns, 1966; Gewirtz, 1961, 1972b,c; Hoffman & Ratner, 1973). Bowlby's and Gray's conceptualizations stop short of specifying the role of learning in imprinting, with Bowlby employing instead an ad hoc goal-correction conception. In the context of our focus on cultural evolution via acquisition mechanisms, we will here emphasize such learning processes in our approach to the phenomena of imprinting/attachment and socialization in general.

As noted earlier, in counterpoint to the gap in the Bowlby (1958) and Gray (1958) approaches, Gewirtz in 1961 approached imprinting/attachment employing routine learning paradigms. He detailed an instrumental-conditioning model for the simultaneous acquisition of behavioral attachment of mother to infant and of infant to mother. This model assumed that both mother and infant respond differentially to the other's behaviors. With examples from diverse interaction situations, Gewirtz illustrated how the responses of each could concurrently influence (condition) the responses of the other. In this learning process, stimuli provided by the appearance and behavior of the mother might acquire discriminative and reinforcing control over the infant's responses, while stimuli from the appearance and behavior of the infant might acquire discriminative and reinforcing control over the mother's responses—and denote the acquisition by each of an attachment to the other. This conditioning approach was open to the operation of unconditioned stimuli (releasers) for species-specific responses, as identified in animal-behavior research lore to that time and as proposed by Bowlby (1958). Paralleling Bowlby's releaser-stimulus analysis, Gewirtz (1961) analyzed the learning processes that were potentially involved. He detailed how the infant's head turns, eye contacts, smiles, and cries, being reinforced potentially by a mother's systematic responding to them, in that very same context might be reinforcing other responses of the mother that the infant's head turns, eye contacts, smiles, cry cessations, and other social behaviors followed routinely. He noted also how often these concurrent conditioning processes might occur even without the mother's awareness that her responses were changing systematically (Gewirtz, 1969, 1972b, 1978). A series of paradigmatic experiments on these reciprocal-influence processes was reported (Gewirtz & Boyd, 1977b). Other analyses were noted earlier that are compatible with this theoretical approach, on potential conditioning mechanisms underlying infant attachment (e.g., Hoffman & DePaulo, 1977; Hoffman & Ratner, 1973) and on the influence, via a conditioning mechanism, of infant behavior on maternal responses (e.g., Gewirtz & Boyd, 1977b; Rheingold, 1969).

SOCIAL LEARNING IN INFANCY

Learning as Process

In the frame of our earlier discussion of "Lamarckian" cultural evolution and imprinting/attachment formation, the study of learning focuses on processes wherein environmental events modify behavior systematically. Consensually valid definitions may consider learning to be denoted by relatively permanent changes in behavior that occur as a result of practice (Kimble, 1961), as a process manifesting itself in adaptive changes in behavior resulting from specific experiences (Thorpe, 1963) or as a systematic change in behavior due to recurring environmental conditions (e.g., Gewirtz, 1969). Implicit in such definitions are two features: that the learning process is denoted by changes from one functional "state" to another and that adaptive change in behavior is normally involved in learning. From a comparative perspective, a consideration of learning mechanisms leads into the literature on sensitization, habituation, respondent (classical, associative, Pavlovian) and operant (instrumental, Thorndikian, trial-and-error) conditioning, imprinting, learning via imitation, and cognition. Learning via such mechanisms is ubiquitous in the animal world, can occur in earliest life, and is demonstrated readily in the human infant (e.g., Birrer, 1977; Brackbill, Fitzgerald, & Lintz, 1967; Emde & Robinson, 1976; Fitzgerald & Brackbill, 1976; Millar, 1976; Piaget, 1971; Sameroff & Cavanagh, 1979).

How early any such learning can occur is a secondary question. The primary question is the nature of the *process* whereby the learning occurs. Even so, we note that learning has been demonstrated in the first week of human infancy (e.g., Sameroff, 1968; Siqueland & Lipsitt, 1966). Thus the newborn human can detect changes in environmental events that are contingent on its own behavior and can adapt to such changes—that is, it can learn (Papoušek, 1961, 1977).

In this paper, we employ the paradigm of operant learning as exemplar of the learning process, for under the operant-learning paradigm, acquisition and its reversal via extinction or counterlearning epitomize adaptive processes within the frame of cultural evolution as earlier discussed. In operant learning, *acquisition* is denoted by a systematic increase in the probability (or some other measure, like amplitude) of a behavior effected by its environmental consequence (e.g., food or some visual or auditory display), and *extinction* is denoted by a systematic decrease in probability of the behavior effected by the deletion of the contingent event. Investigators have shown in human infants, children, and adults that diverse behaviors that ordinarily enter social interchanges can be conditioned readily under the operant paradigm. For instance, operant learning in infants has been demonstrated for head turns (Caron, 1967; Papoušek, 1961; Siqueland & Lipsitt, 1966), eye contacts (Etzel & Gewirtz, 1967), smiles (Brackbill, 1958; Etzel & Gewirtz, 1967), and cry cessations (Etzel & Gewirtz, 1967). A critical summary of such demonstrations has been supplied by Millar (1976).

The functional-behavioral approach emphasized here provides a useful system for heuristic analysis and for designing and working in new environmental settings. It affords a ready means of focusing simultaneously on the functional elements of the environment and of the child behavior affected by them. In the analysis of operant learning in any behavior arena, numerous definitions are possible for stimuli and responses. A functional analysis examines the relations, if any, between particular sets of definitions of the terms "stimulus" and "response," focusing on systematic changes in some attribute (e.g., rate) of the behavior and the environmental event contingent upon it, compared with when the environmental event is not so presented. The conditioning denoted confirms the functional utility of the category definitions developed and leads to the contingent event being termed the "reinforcing stimulus" (or reinforcer) for the behavioral event that is termed the "response." The reinforcement conception implies only that there exist events which, when made contingent upon responses, will change the rates of some of them systematically. Thus reinforcing stimuli need not exist under all conditions for every potential response; a contingent event that can reinforce one response need not serve as a reinforcer for that response under all other contextual-setting conditions or for another potential response; and the fact that an event functions as a reinforcing stimulus in one context does not preclude its functioning in different stimulus roles in other contexts (Catania, 1973; Gewirtz, 1971, 1972d).

Social Learning

The term "social learning" defines a category of adaptive behavior that involves stimuli provided by people, but otherwise follows the same principles as does nonsocial learning. Social learning is indispensable given the cultural basis of civilization. It results from, and is involved in, man's dealings with other humans and accounts for many of the distinctive behavioral qualities of humans. Thus, much cueing and reinforcement of behavior in complex human-learning situations is of a social nature; for example, discriminative settings for parental, teacher, or peer approval. If social stimuli have not acquired discriminative and reinforcing value for the child, his learned skills may not attain appropriate levels. Those who have a deficiency in the experience base underlying social learning are thus unable to fend for themselves in society, and may be labeled "autistic." For these reasons, the study of the social stimulus environment, social responses, and the underlying learning process is most important.

The social *environment* consists of those functional stimuli which are provided by people; and social *behaviors* are those under the actual or potential control of social *stimuli,* in their acquisition or maintenance, or both. Except that stimulus classes occurring in natural settings are likely to be more variable than stimuli in contrived laboratory experiments and that the term "social stimuli" usually denotes those occurring in interchange in natural settings, there is thought to be nothing intrinsically special about stimuli provided by people or about social settings as contexts for learning.

Learning Constraints and Facilitators

Both the infrahuman and the human infant come into the world with facilitators and constraints on their behavior—some unlearned/biological, others ecological (Gewirtz, 1961, 1972d; Hinde & Stevenson-Hinde, 1973). Thus physical and chemical events, as monitored by the infant's sensory receptors/processors, can either limit or expedite behavioral development; for example, for such reflexes as sucking and grasping. In this context, there are sometimes instances where at some point along an ordered temporal or contextual dimension, a behavior (or learning) fails to be manifested and where at a later (or second) point along that ordered dimension that behavior (or learning) is shown. The concept of "preparedness" (Seligman, 1970) has occasionally been used, on an ad hoc basis, in such instances to explain response occurrence after nonoccurrence. Where a dimension of time is involved, there is an implicit appeal in such usage to a mechanism of *maturation* to explain the differential. However, an explanation in terms of preparedness is incomplete, for that concept indicates little either about the phylogeny or the ontogeny of the response in question, and tells us even less about the underlying process than do ordinarily such setting-condition variables as are termed "deprivation–satiation," "set," or "state."

Social behavior and learning may be qualified by the physical and social ecology of a setting and can be facilitated or limited by the manipulation of ecological factors. The term "ecology" stands for the gross conditions of an environment that determine which events and behaviors can occur in a setting, and specifically whether or not a child can detect a stimulus, emit a behavior, or be reinforced contingent on responding. Such conditions are involved as the available space, the type and number of materials positioned in that space, and the type and number of peers and adults there. Similarly, the rules and regulations governing a setting represent social facilitators or constraints on behavior systems. Thus ecological conditions can insulate a child against, or cause the child to be exposed to, adults, other children, or their specific activities, and in that way can determine whether or not the child can emit particular responses or have responses maintained by particular consequences.

The Role of Early Learning: Crying as an Illustrative Case

In this section, selected data and interpretations are summarized in a focal arena of early interchange between infant and caregiver. Basic concepts are used to detail how the instrumental-crying behavior in the human infant's repertoire can be maintained, modified, or eliminated by the availability of reinforcement contingencies provided by caregiver responding; how it can come under discriminative control; and how, in the process of its differentiation, it can in turn come to control diverse concurrent maternal responses to the infant. Further, criterion indices of attachment have been provided by crying in particular contexts, as upon signs of a mother's departure or a stranger's approach (Schaffer & Emerson, 1964). In addition, crying has often denoted physical distress or emotional reaction to environmental change. Thus an elucidation of crying as instrumental, as cued, and as elicited can serve to exemplify how behavioral concepts can be used to order the acquisition of diverse behavior systems by the child in early life.

The Conditioning of Crying

From patterns of correlations across the four quarters of the first year of life derived from longitudinal observations of 26 mother–infant pairs, Bell and Ainsworth (1972) concluded that consistent and prompt maternal responding to infant crying was associated with reduced frequency and duration of infant crying in later time quarters. They contrasted this interpretive finding with their conception of the "popular belief" that under the operant-learning paradigm, contingent maternal responding should reinforce/increase infant crying and maternal failure to respond should extinguish/decrease that crying (pp. 1187–1188). Bell and Ainsworth did not separate elicited from instrumental crying, and their conclusion was shown to be unsupported by their published data on both conceptual and methodological grounds (Gewirtz & Boyd, 1977a). Even so, their unwarranted conclusion presents the occasion to sketch briefly how the routine variables of elementary reinforcement schedule and latency conditions provided by caregiver responding can determine infant instrumental-crying patterns. For simplicity, it is assumed that a particular class of infant-crying responses constitutes a functional unit and that there is present a reinforcing agent who will respond rapidly, consistently, and contingently to that class; and that the parental responses, according to some schedule, will provide reinforcing stimuli for the crying responses in that class. In this frame, potential classes of infant operant crying and stimuli provided by parental-response contingencies on that crying can be defined in great variety, including combinations of such crying attributes as latency, duration, and intensity. Thus crying of short or long duration or short but consistently intense crying may comprise the response class upon which reinforcing maternal behavior is consistently and immediately contingent. Alternatively, a mother may respond only to precursors of crying or to noncrying responses (as on a DRO schedule) but never to crying itself. At the same time, diverse definitions of maternal responses involving combinations of these same attributes may function to provide discriminative (cue) and/or reinforcing stimuli for diverse classes of infant-crying responses.

One responsive mother may effectively shape her infant's loud, lengthy cries by ignoring both short, low-intensity precursors of crying and short, low-intensity cries and by responding expeditiously only to high-intensity, long-duration cries. The inclination of this mother actually may be to ignore her infant's cries when his needs have been met, but she may find it impossible to ignore the crying when it has become sufficiently intense or loud. A second responsive mother may shape the short, low-intensity cries of her infant by ignoring both the precursors of crying and lengthy, loud cries, while she responds rapidly and decisively only to short, low-intensity cries. In yet another contrast, a third responsive mother may foster behavior incompatible with her infant's crying; she may respond with dispatch only to short, low-intensity precursors of her infant's crying and/or to noncrying responses, to rear a child who cries rarely (and then mainly when painful events elicit crying). At a gross level, each of these three interaction patterns (and diverse other patterns) could contribute to a positive correlation between infant crying and maternal ignoring of crying, much like the pattern Bell and Ainsworth reported. However, the process denoted by the functional relations between infant and mother behaviors would be very different for each case.

Operant-learning concepts also order the effects of diverse infant-crying patterns on maternal responding as well as the effects of a wide range of maternal-responding patterns on cued infant operant-crying patterns, often used as indicators of attachment (Gewirtz, 1977; Gewirtz & Boyd, 1977b).

SUMMARY

Genetic change is the basis of *organic evolution;* and systematic behavior change denoting *learning* is the basis of *cultural evolution*. The *co*evolutionary nature of genetic change and learning has not always been understood. In this chapter, these two change mechanisms have been viewed as the major processes whereby individuals and species acquire structural and functional characteristics that facilitate their interaction with, and adaptation to, their environments. Both types of evolution can facilitate or constrain learning. In the case of humans, the innate storehouse of information represented by the gene pool is augmented by learning during ontogeny. Infinitely more than any other species of animal, the quality of its biology makes the human a creature of socialization in a given cultural context and a singular product of its learning.

A *functional*-behavior approach and the *operant*-learning process have been emphasized in this chapter. This is because, in a paradigmatic way, operant learning focuses on environmental context and consequences that change or maintain behavior, thus epitomizing adaptive processes within the frame of cultural evolution. Using examples from the maternal–child interaction arena of infant crying (an arena that often has provided infant-attachment indices), emphasis was placed on the processes whereby infant (and parent) instrumental-learning occurs. These operant-learning processes reflect the *influence* that maternal behavior can have over infant responses and that infant behavior can have over maternal responses.

REFERENCES

Ainsworth, M. D. S. Attachment and dependency: A comparison. In J. L. Gewirtz (Ed.), *Attachment and dependency*. Washington, D.C.: Winston, 1972.

Allison, T., & Van Twyver, H. B. The evolution of sleep. *Natural History,* 1970, **79**, 56–65.

Atz, J. W. The application of the idea of homology to behavior. In L. R. Aronson, E. Tobach, D. S. Lehrman, & J. S. Rosenblatt (Eds.), *Development and evolution of behavior*. San Francisco: Freeman, 1970.

Bateson, P. P. G. The characteristics and context of imprinting. *Biological Review,* 1966, **41**, 177–220.

Bateson, P. P. G. Imprinting. In H. Moltz (Ed.), *The ontogeny of vertebrate behavior*. New York: Academic, 1971.

Bell, S. M., & Ainsworth, M. D. S. Infant crying and maternal responsiveness. *Child Development,* 1972, **43**, 1171–1190.

Birrer, C. Fundamental education: Towards a theory of learning in infancy. *Genetic Psychology Monographs,* 1977, **96**, 247–335.

Bitterman, M. E. Toward a comparative psychology of learning. *American Psychologist,* 1960, **15**, 704–712.

Bitterman, M. E. Phyletic differences in learning. *American Psychologist,* 1965, **20**, 396–410.

Bodmer, W. F., & Cavalli-Sforza, L. L. *Genetics, evolution and man.* San Francisco: Freeman, 1976.

Bolles, R. C. *Theory of motivation.* New York: Harper & Row, 1975.

Bowlby, J. The nature of the child's tie to his mother. *International Journal of Psychoanalysis,* 1958, **39**, 350–373.

Bowlby, J. *Attachment and loss.* Vol. 1: *Attachment.* New York: Basic Books, 1969.

Brackbill, Y. Extinction of the smiling response in infants as a function of reinforcement schedule. *Child Development,* 1958, **29**, 115–124.

Brackbill, Y., Fitzgerald, H. E., & Lintz, L. M. A developmental study of classical conditioning. *Monographs of the Society for Research in Child Development,* 1967, **32**.

Bronowski, J. *The ascent of man.* Boston: Little, Brown, 1973.

Cairns, R. B. Attachment behavior in mammals. *Psychological Review,* 1966, **3**, 409–426.

Campbell, B. *Humankind emerging* (2nd ed.). Boston: Little, Brown, 1979.

Campbell, D. T. On conflicts between biological and social evolution and between psychology and moral tradition. *American Psychologist,* 1975, **30**, 1103–1126.

Caron, R. F. Visual reinforcement of head-turning in young infants. *Journal of Experimental Child Psychology,* 1967, **5**, 489–511.

Catania, A. C. The nature of learning. In J. A. Nevin & G. S. Reynolds (Eds.), *The study of behavior.* Glenview, Ill.: Scott, Foresman, 1973.

Chagnon, N. A., & Irons, W. (Eds.), *Evolutionary biology and human social behavior: An anthropological perspective.* North Scituate, Mass.: Duxbury, 1979.

Cullen, E. Adaptations in the kittiwake to cliff-nesting. *Ibis,* 1957, **99**, 275–302.

Eibl-Eibesfeldt, I. *Ethology: The biology of behavior.* New York: Holt, Rinehart and Winston, 1975.

Emde, R. N., & Robinson, J. The first two months: Research in developmental psychobiology and the changing view of the newborn. In J. Noshpitz & J. Call (Eds.), *Basic handbook of child psychiatry.* New York: Basic Books, 1976.

Etzel, B. C., & Gewirtz, J. L. Experimental modification of caretaker-maintained high rate operant crying in a 6- and a 20-week old infant (*Infans tyrannotearus*): Extinction of crying with reinforcement of eye contact and smiling. *Journal of Experimental Child Psychology,* 1967, **5**, 303–317.

Fitzgerald, H. E., & Brackbill, Y. Classical conditioning in infancy: Development and constraints. *Psychological Bulletin,* 1976, **83**, 353–376.

Freud, S. *The standard edition of the complete psychological works of Sigmund Freud.* James Strachey (Ed.). London: Hogarth, 1953.

Garcia, J., Ervin, F., & Koelling, R. Learning with prolonged delay of reinforcement. *Psychonomic Science,* 1966, **5**, 121–122.

Garcia, J., & Koelling, R. Relation of cue to consequence in avoidance learning. *Psychonomic Science,* 1966, **4**, 123–124.

Gewirtz, J. L. A learning analysis of the effects of normal stimulation, privation, and deprivation on the acquisition of social motivation and attachment. In B. M. Foss (Ed.), *Determinants of infant behaviour.* London: Methuen (New York: Wiley), 1961, 213–299.

Gewirtz, J. L. Mechanisms of social learning: Some roles of stimulation and behavior in early human development. In D. A. Goslin (Ed.), *Handbook of socialization theory and research.* Chicago: Rand McNally, 1969, 57–212.

Gewirtz, J. L. (Ed.), *Attachment and dependency.* Washington, D.C.: Winston, 1972. (Distributed by Halsted Press Division of John Wiley & Sons, New York.) (a)

Gewirtz, J. L. Attachment, dependency, and a distinction in terms of stimulus control. In J. L. Gewirtz (Ed.), *Attachment and dependency.* Washington, D.C.: Winston, 1972, 139–177. (b)

Gewirtz, J. L. On the selection and use of attachment and dependence indices. In J. L. Gewirtz (Ed.), *Attachment and dependency.* Washington, D.C.: Winston, 1972, 179–215. (c)

Gewirtz, J. L. Some contextual determinants of stimulus potency. In R. D. Parke (Ed.), *Recent trends in social learning theory.* New York: Academic, 1972. (d)

Gewirtz, J. L. Maternal responding and the conditioning of infant crying: Directions of influence within the attachment–acquisition process. In B. C. Etzel, J. M. LeBlanc, & D. M. Baer (Eds.), *New developments in behavior research: Theories, methods, and application.* Hillsdale, N.J.: Erlbaum, 1977.

Gewirtz, J. L. Social learning in early human development. In A. C. Catania & T. A. Brigham (Eds.), *Handbook of applied behavior analysis.* New York: Irvington, 1978, 105–141.

Gewirtz, J. L. Continuity–discontinuity from the perspective of operant social-learning theory. Paper presented at the Fifth Biennial Conference of the International Society for the Study of Behavioral Development, Lund, Sweden, June 1979.

Gewirtz, J. L., & Boyd, E. F. Does maternal responding imply reduced infant crying? A critique of the 1972 Bell and Ainsworth report. *Child Development,* 1977, **48**, 1200–1207. (a)

Gewirtz, J. L., & Boyd, E. F. Experiments on mother–infant interaction underlying mutual attachment acquisition: The infant conditions the mother. In T. Alloway, P. Pliner, & L. Krames (Eds.), *Attachment behaviour. Advances in the study of communication and affect.* Vol. 3. New York and London: Plenum, 1977. (b)

Ghiselin, M. *The triumph of the Darwinian method.* Berkeley: University of California Press, 1969.

Glickman, S. E., & Sroges, W. R. Curiosity in zoo animals. *Behaviour,* 1966, **26**, 151–158.

Gottlieb, G. Neglected developmental variables in the study of species identification in birds. *Psychological Bulletin,* 1973, **79**, 362–372.

Gottlieb, G. Development of species identification in ducklings: VI. Specific embryonic experience required to maintain species-typical perception in Peking ducklings. *Journal of Comparative and Physiological Psychology,* 1980, **94**, 579–587.

Gray, P. H. Theory and evidence of imprinting in human infants. *Journal of Psychology,* 1958, **46**, 155–166.

Hess, E. H. Imprinting. *Science,* 1959, **130**, 133–141.

Hess, E. H. Imprinting and the critical period concept. In E. L. Bliss (Ed.), *Roots of behavior.* New York: Hoeber-Harper, 1962.

Hess, E. H. Imprinting in birds. *Science,* 1964, **146**, 1128–1139.

Hess, E. H. *Imprinting: Early experience and the developmental psychobiology of attachment.* New York: D. Van Nostrand, 1973.

Hess, E. H., & Petrovich, S. B. (Eds.), *Imprinting: Benchmark papers in animal behavior.* Stroudsburg, Pa.: Dowden, Hutchinson & Ross, 1977.

Hinde, R. A. *Animal behavior: A synthesis of ethology and comparative psychology* (2nd ed.). New York: McGraw-Hill, 1970.

Hinde, R. A. *Biological bases of human social behavior.* New York: McGraw-Hill, 1974.

Hinde, R. A., & Stevenson-Hinde, J. (Eds.), *Constraints on learning.* New York: Academic, 1973.

Hodos, W., & Campbell, C. B. G. Scale naturae: Why there is no theory in comparative psychology. *Psychological Review,* 1969, **76**, 337–350.

Hoffman, H. S., & DePaulo, P. Behavioral control by an imprinting stimulus. *American Scientist,* 1977, **65**, 58–66.

Hoffman, H. S., & Ratner, A. M. A reinforcement model of imprinting: Implications for socialization in monkeys and man. *Psychological Review,* 1973, **80,** 527–544.

Huxley, J. S. The emergence of Darwinism. In S. Tax (Ed.), *Evolution after Darwin.* Vol. I. *The evolution of life.* Chicago: The University of Chicago Press, 1960.

Jaynes, J. *The origin of consciousness in the breakdown of the bicameral mind.* Boston: Houghton-Mifflin, 1976.

Jerison, H. J. *Evolution of the brain and intelligence.* New York: Academic, 1973.

Kennedy, G. E. *Paleoanthropology.* New York: McGraw-Hill, 1980.

Kimble, K. A. *Hilgard and Marquis' conditioning and learning* (2nd ed.). New York: Appleton-Century-Crofts, 1961.

Klaus, M. H., & Kennel, H. J. *Maternal infant bonding.* St. Louis: Mosby, 1976.

Lasker, G. W. Human biological adaptability. *Science,* 1969, **166,** 1480–1486.

Lockard, R. B. Reflections on the fall of comparative psychology: Is there a message for us all? *American Psychologist,* 1971, **26,** 168–179.

Lorenz, K. Der Kumpan in der Umwelt des Vogels. *Journal für Ornithologie,* 1935, **83,** 137–213.

Lorenz, K. *Evolution and modification of behavior.* Chicago: University of Chicago Press, 1965.

Lorenz, K. Innate basis of learning. In K. Pribram (Ed.), *On the biology of learning.* New York: Harcourt Brace Jovanovich, 1969.

Lorenz, K. *Civilized man's eight deadly sins.* New York: Harcourt Brace Jovanovich, 1973.

Marler, P. A comparative approach to vocal development: Song learning in the white-crowned sparrow. *Journal of Comparative and Physiological Psychology,* 1970, **71,** No. 2, Part 2, 1–25.

Marler, P. Sensory templates in species-specific behavior. In J. C. Fentress (Ed.), *Simpler networks and behavior.* Sunderland, Mass.: Sinauer, 1976.

Mason, W. A., & Lott, F. D. Ethology and comparative psychology. *Annual Review of Psychology,* 1976, **27,** 129–154.

Millar, W. S. Operant acquisition of social behaviors in infancy: Basic problems and constraints. In H. W. Reese (Ed.), *Advances in child development and behavior.* Vol. 11. New York: Academic Press, 1976.

Murdy, W. H. Anthropocentrism: A modern version. *Science,* 1975, **187,** 1168–1172.

Naroll, R., & Naroll, F. *Main currents in cultural anthropology.* New York: Appleton-Century-Crofts, 1973.

Nottebohm, F. Ontogeny of bird song. *Science,* 1970, **167,** 950–956.

Papoušek, H. Conditioned head rotation reflexes in infants in the first months of life. *Acta Paediatrica,* 1961, **50,** 565–576.

Papoušek, H. Entwicklung der Lernfahigkeit im Sauglingsalter [The development of learning ability in infancy]. In G. Nissen (Ed.), *Intelligenz, Lernen und Lernstorungen.* Berlin: Springer-Verlag, 1977.

Petrovich, S. B. Adaptation and evolution of behavior. In G. U. Balis, L. Wurmser, E. McDaniel, & R. G. Grenell (Eds.), *Dimensions of behavior.* Boston: Butterworth, 1978.

Piaget, J. *Biology and knowledge.* Chicago: University of Chicago Press, 1971.

Piaget, J. *Behavior and evolution.* New York: Pantheon, 1978.

Pittendrigh, C. S. Adaptation, natural selection, and behavior. In A. Roe & G. G. Simpson (Eds.), *Behavior and evolution.* New Haven: Yale University Press, 1958.

Rajecki, D. W., Lamb, M. E., & Obmascher, P. Toward a general theory of infantile attachment: A comparative view of aspects of the social bond. *Behavioral and Brain Sciences,* 1978, **1,** 417–464.

Rheingold, H. L. The social and socializing infant. In D. A. Goslin (Ed.), *Handbook of socialization theory and research.* Chicago: Rand McNally, 1969.

Roy, M. A. (Ed.), *Species identity and attachment: A phylogenetic evaluation.* New York: Garland STPM Press, 1980.

Sameroff, A. J. The components of sucking in the human newborn. *Journal of Experimental Child Psychology,* 1968, **6**, 607–623.

Sameroff, A. J., & Cavanagh, J. P. Learning in infancy: A developmental perspective. In J. D. Osofsky (Ed.), *Handbook of infant development.* New York: Wiley, 1979.

Schaffer, H. R., & Emerson, P. The development of social attachment in infancy. *Monographs of the Society for Research in Child Development,* 1964, **29** (3, Whole No. 94).

Scott, J. P. Comparative social psychology. In R. H. Waters, D. A. Rethlingshafer, & W. E. Caldwell (Eds.), *Principles of comparative psychology.* New York: McGraw-Hill, 1960.

Scott, J. P. *Early experience and the organization of behavior.* Belmont, Calif.: Brooks/Cole, 1968.

Seligman, M. E. P. On the generality of laws of learning. *Psychological Review,* 1970, **77**, 406–418.

Seligman, M. E. P., & Hager, J. L. (Eds.), *Biological boundaries of learning.* New York: Appleton-Century-Crofts, 1972.

Siqueland, E. R., & Lipsitt, L. P. Conditioned head-turning behavior in newborns. *Journal of Experimental Child Psychology,* 1966, **3**, 356–376.

Skinner, B. F. The phylogeny and ontogeny of behavior. *Science,* 1966, **153**, 1205–1213.

Stern, C. *Principles of human genetics.* San Francisco: Freeman, 1973.

Thiessen, D. D. A move toward species-specific analyses in behavior genetics. *Behavior Genetics,* 1972, **2**, 115–126.

Thorpe, W. H. *Learning and instinct in animals.* London: Methuen, 1963.

Tinbergen, N. The shell menace. *Natural History,* 1963, **72**, 28–35.

Tinbergen, N. *The animal and its world.* London: Allen & Unwin, 1972.

Williams, G. C. *Adaptation and natural selection.* Princeton: Princeton University Press, 1966.

Wilson, E. O. *Sociobiology: The new synthesis.* Cambridge, Mass.: Belknap-Harvard University Press, 1975.

2

SOCIAL PERCEPTION AND RESPONSIVITY IN EARLY INFANCY

Tiffany M. Field
Mailman Center for Child Development
University of Miami Medical School

How is it that neonates can discriminate their mother's voice from that of a stranger, yet only later perceive single speech sound contrasts? How is it that neonates, who reputedly do not scan the inner features of the face and do not have the necessary contrast sensitivity to perceive facial features, discriminate and imitate facial expressions?

Social perception is not a recognized term like visual and auditory perception, but a term which could be used to describe a body of literature which is accumulating and suggesting a unique and perhaps precocious perception of social objects by the infant. While the visual and auditory perception literature suggests that the very young infant may be lacking the necessary perceptual equipment to discriminate complex, animate stimuli, such as faces and voices, the use of those stimuli reveals fairly sophisticated perceptual skills from the first days of life. This chapter will review some of the data on infant perception of faces and voices and offer some ideas about why the young infant may appear perceptually precocious when presented with these stimuli.

The question of why infants may perceive such stimuli as faces and voices more readily than other visual and auditory stimuli is complicated by stimulus features and methodological variables. While evolutionary theorists and nativists suggest that there may be innate social stimulus feature detectors or releasing mechanisms by which a very young infant perceives faces or voices, the possibility also remains that faces and voices are simply more interesting stimuli. Real speech sounds are more salient than computer-simulated, single speech contrasts, just as real three-dimensional faces are more interesting than photographed faces. The salience or attention-getting and attention-maintaining qualities of social stimuli undoubtedly contribute to the infant seeming more precocious in the mother's arms than in the experimenter's lab.

The more sophisticated responses to social stimuli, not predicted by responses to inanimate stimuli, may also relate to experimental conditions. For the presentation of social stimuli, infants are often placed in an upright position. For inanimate stimuli they have typically been placed in a supine position, a position notably less conducive to alerting (Gardner & Turkewitz, 1980; Korner & Thoman, 1970). The supine position is characteristically more soporific and conducive to involuntary tonic neck reflexes and limb movements which, in themselves, alter the infant's state. The experimental boxes often used for the presentation of inanimate stimuli or for corneal photography may introduce distracting sources of stimulation, for example, peepholes, mirrors, and camera lenses, looming stimulation or distressing claustrophobia (a condition not yet studied in the infant). The characteristically more distant placement of inanimate rather than social stimuli may also contribute to differential perceptual performance. Young infants, while capable of seeing beyond the optimal 7 to 10-in. range, may not be as captivated by more distal stimulation. In addition, the experimenter's usual attempt to standardize testing and schedule sessions at an optimal time for the infant—for example, prior to feeding (Pomerleau-Malcuit & Clifton, 1973), postfeeding (Gardner & Turkewitz, 1980), or midway between feedings (Brazelton, 1973)—may be problematic inasmuch as there are mixed findings regarding the optimal time, and the variability on this factor may be considerable across infants. In addition, inanimate stimuli are often presented in a standardized fashion with relatively fixed intertrial intervals, while social stimulation usually occurs at variable intervals and features intervening manipulations to alert the infant. Infants may require varying amounts of intertrial stimulation to maintain an alert state.

More abstract explanations for differential responding to social and inanimate stimulation have been raised by ethologists and nativists. The former group has argued that social stimuli are ecologically more meaningful and responses to those stimuli are like fixed action patterns which are adaptive in the evolutionary sense (Tinbergen, 1951). In a similar vein, the nativist argues that responses to social stimuli may be inborn, enabled by special stimulus feature detectors for speech (Eimas, 1975) or for faces (Bower & Wishart, 1980). Supramodal perception may be present from birth, enabling the integration of multimodal stimulation (Bower, 1974; Meltzoff & Moore, 1977; Meltzoff & Borton, 1979). Social stimuli may be more salient and more readily perceived because they are multimodal. A social stimulus which is typically an auditory-visual-tactile-olfactory stimulus contains more information, may impinge on more receptors, and may "fire more feature detector cells" or more complex cells.

From an ecological perspective, some have argued that the perceptual system is most attuned to those relationships in the environment which are important for survival. The expectation, according to this model, is that perceptual development proceeds from the more abstract to the simple (Bower & Wishart, 1980) or from higher-order to lower-order structures (Gibson, 1979). A real face or voice might be more readily perceived because it is a more abstract stimulus than a computer-simulated speech sound or a face pattern. A mother's and stranger's voice contain complex, abstract features and differences, while inanimate stimuli are usually varied on a specific and single dimension. A specific change in a unidimensional stimulus may be more difficult to detect than a more complex change in a multidimensional stimulus. Association areas of the brain, prior to their development of neuronal specificity, may be more receptive to more abstract, multimodal stimulation, and hypercomplex cells may be more readily activated than simple or complex cortical cells during early development (Hubel & Wiesel, 1968). Supramodal perception, whatever the underlying mechanism, may precede the perception of more specific changes within a single modality.

VOICE PERCEPTION

Several researchers have noted neonates' preference for auditory stimuli in the frequency band of the human voice (Hutt et al., 1968; Kearsley, 1973). Condon and Sander (1974) presented tapes of human voices to neonates and found that they moved their limbs in synchrony with the speech segments. When each frame of the film was superimposed on the audio transcript, distinct limb movements corresponded to the onsets of separate speech segments. A replication attempt with additional controls is currently in progress (Dowd, 1980), since Dowd has speculated that the infants' limb movements may reflect self-synchrony or movements synchronous with speech rhythms rather than speech segments. Nonetheless, the neonates' synchronous movements to speech tapes and the absence of similar movements to tapping sounds suggest the greater salience of the more complex speech tract.

A very large literature has accumulated over the last decade on the young infants' discrimination of speech contrasts (see Chapter 3). The categorical perception of speech sounds—like *bas* and *pas, das* and *gas*—has been investigated using a number of different methodologies including heart-rate changes (Morse, 1972), high-amplitude sucking (Eimas, 1975), and a reinforcement paradigm labeled "VRISD" (visually reinforced infant speech discrimination) in which a dancing, musical bear reinforces head turning in the direction of speech sound changes (Eilers, Wilson, & Moore, 1977). As noted in Chapter 3, the age at which infants discriminate different speech contrasts varies according to the complexity of the contrast and the type of paradigm used. While infants perform these discriminations at earlier ages when more interesting paradigms are used, such as the dancing-bear reinforcement procedure (Eilers et al., 1977), these discriminations seem less impressive in light of the neonate's ability to discriminate his mother's from a stranger's voice.

In a study which demonstrated neonates' preference for their mother's voice (DeCasper & Fifer, 1980), neonates were presented with audiotapes of their mothers and other mothers reading Dr. Suess's *And to Think That I Saw It on Mulberry Street* (certainly an interesting stimulus). Infants sucked on a nonnutritive nipple connected to a pressure transducer which then triggered, via a programmer, the mother's or other mother's audiotape as a function of the length of the interval between sucking bursts. Depending on their random assignment to groups, infants had to shorten (as compared to baseline) or lengthen their interburst intervals to produce their mother's voice. The effects of response requirements and voice characteristics were controlled by requiring half the infants to respond after short, and half after long, intervals and by having each maternal voice serve as the nonmaternal voice for another infant. Additional controls included reversing the response requirements for half the infants and using a different discriminative stimulus (a 400-Hz tone) for the tone and no-tone periods during which a sucking burst might produce the mother's reading of Dr. Suess. This well-controlled study suggests that infants prefer their mother's voice and learn to produce her voice by varying their nonnutritive-sucking behavior.

FACE PERCEPTION

Two schools of thought with somewhat different views and data present the infant as a perceiver of lower-order structures (Salapatek, 1975; Salapatek & Kessen, 1966) and a perceiver of higher-order structures (Gibson, 1979). The lower-order-structure school presents visual-scanning research which suggests that young infants do not scan an entire stimulus array but generally attend to only segregated portions of patterns such as outer contours. The higher-order-structure proponents present data suggesting that infants are able to perceive higher-order relationships in visual arrays (Treiber & Wilcox, 1980).

Several investigators have used corneal photography to document the scanning patterns of infants typically 1 to 3 months old when presented with a mother's and stranger's mirrored reflections (Haith, Bergman, & Moore, 1977; Maurer & Salapatek, 1976), or a real face versus schematic face photos (Hainline, 1978). A consistent finding across these studies was that the infants, particularly the younger infants, fixated more frequently on the outer contours of the face stimulus than the facial features. Some methodological factors which may have contributed to this finding include the supine position of the infant, the distance of the mirrored reflection, the moving stimulus, and the corneal photography of one eye. While the infant's head was cradled in a supine position, the tonic neck reflex of 1-month-old infants may have interfered with midline looking. In addition, in two out of three of the conditions, the face was moving side to side which, in itself, might contribute to more fixation on the contours than the central features of the face. Further, the considerable distance of the face stimulus from the infant (19 in.) may have contributed to the facial features appearing less distinctive than the light–dark contrast of the facial contours. Finally, Slater and Findlay (1972) have reported marked and consistent disparities between the center of the pupil and the fixation point with the direction of error being opposite in the two eyes. They suggested a 10° correction factor for neonates. This correction factor was not used by the above authors, although that amount of disparity may have accounted for the off-target looking they reported.

Despite these methodological problems, the consistent reporting of contour versus facial-feature scanning makes the discrimination of facial features by the very young infant seem unlikely. In addition, research by Banks and colleagues (Banks & Souther, 1979) suggests that infants as young as 1 month have insufficient acuity and contrast sensitivity to discriminate facial features. Banks and Souther (1979) used linear systems analysis and a contrast sensitivity function to predict the infant's contrast sensitivity to photographs of the facial expressions *happy, sad,* and *surprise*. The photographs were reduced by a Fourier transform projection laser and lenses to the amount of contrast sensitivity 1- and 3-month-old infants would be expected to have, given their contrast sensitivity to vertical gratings as measured by a visual preference technique in a previous study (Banks & Salapatek, 1978). For the 1-month-old infants, the photographed faces were indiscernible with very faint information in the region of the eyes and mouth. The photographs of faces the 3-month-old infants were predicted to perceive had more discernible features, with the *surprise* face being more distinctive than the other two expressions. Based on the predictions of Banks and Souther (1979), infants less than 3 months of age would not be able to detect internal features of the face.

Since there are several studies which suggest that even neonates perceive and discriminate facial features, there may be problems associated with generalizing from linear gradient perception data to perception of live faces. First, the visual preference technique used in the linear gradient study of Banks and Salapatek (1978) may not be the most effective assessment of discrimination in the very young infant, as has been discussed at some length by a number of researchers (Cohen & Gelber, 1975; Nelson, Morse, & Leavitt, 1979). Second, linear gratings may be intrinsically uninteresting to the infant, as is suggested by their preference for curvilinear versus rectilinear patterns (Fantz & Miranda, 1975). The hypercomplex cells which are reputedly activated by curved lines, as opposed to the simple and complex cortical cells activated by straight lines, may be more functional during early infancy (Hubel & Wiesel, 1968). Third, the use of linear systems analysis may treat low- and high-frequency information equivalently, leading to a reduction of both low-frequency information (which is apparently perceived more readily by the young infant) and high-frequency information. Facial features may contain more low- than high-frequency information, information which would have been artifactually reduced in the photographs using the linear systems analysis technique. Finally, even if the reduction of information by this technique yields very faint photographs, it is unclear whether the three-dimensional, live face would be as indiscernible as a photograph. Infants may, in fact, prefer live faces to photographs because the features are less faint than those of a photograph. In any case, the empirical data on early perception and responsivity to faces suggest that a prediction made from contrast sensitivity of linear gradients to facial features may not be ecologically meaningful or accurate.

Nine-minute-old newborns, for example, have been noted to discriminate face patterns in which eyes, mouth, and nose features have been arranged as in a real face from those patterns in which facial features have been partially or generally scrambled (Goren, Sarty, & Wu, 1975). Tracking was most reliably elicited by regular as opposed to scrambled face patterns, suggesting a preference for the regular face pattern.

A number of investigators have demonstrated an even more precocious skill of the neonate and very young infants which implies perception of facial features, namely the imitation of facial movements (Dunkeld, 1978; Field, Woodson, Greenberg, & Cohen, 1980; Gardner & Gardner, 1970; Maratos, 1973; Meltzoff & Moore, 1977). Meltzoff and Moore (1977) modeled behaviors already in the infant's repertoire; for example, lip protrusion, mouth opening, and tongue thrusting for infants 12 to 17 days old. A number of control procedures, including rating of the videotaped imitations by naive coders, was used to avoid potential experimenter biases. They then compared the incidence of the infant's behaviors, for example, tongue protrusion, across all the expressions being modeled (lip protrusion, mouth opening, and tongue protrusion). This base of comparisons was used to determine whether the modeled behavior, such as tongue protrusion, occurred more frequently during the modeling of tongue protrusion than the other behaviors modeled, independent of variable baseline and arousal differences in responses. The expected neonatal behaviors occurred more frequently during the models of those behaviors, suggesting a rudimentary form of imitation. While some have argued that these responses are merely fixed action patterns (Masters, 1979)—particularly inasmuch as Jacobson (1979) reported that infant tongue protrusions are elicited not only by a tongue protrusion model but also, to a slightly lesser degree by a moving pen (Jacobson, 1979)—the data, nonetheless, suggest that neonates are perceiving and discriminating movements of facial features such as the mouth.

A more recent study in our lab provides supportive data for the neonate's ability to perceive, discriminate, and differentially respond to facial features (Field et al., 1980). Using a trials-to-criterion visual habituation paradigm, 1-day-old neonates were presented a live happy, sad, or surprised face in a counterbalanced order. The model held the neonate in a face-to-face position so the infant's face was approximately 9 in from the model's face. The model provided kinesthetic and auditory stimulation (two bobs up and down and two clicks of the tongue) to elicit the infant's attention. As soon as the infant looked at the model's face, the model emitted a happy, sad, or surprised face, and sustained the expression until the infant looked away. The same facial expression was repeated until the trial in which the infant looked at the model's face for less than 2 seconds (the criterion trial). The model then emitted the next facial expression for a series of trials to criterion.

The neonate showed habituation of the expressions or a decrease of looking time over trials and dishabituated or showed increased looking to the new expression, suggesting that he was able to discriminate the different facial expressions. In addition, his scanning patterns varied as a function of the type of expression. For example, his scanning of the mouth and eye regions was evenly distributed for the surprise face which featured both mouth and eye widening. However, he looked primarily at the mouth region, which is the salient feature of change, in the case of the happy and sad faces. This observation suggests that neonates do scan the internal features of the face, unlike the findings of outer contour scanning discussed earlier. These discrepant findings may relate to the closer distance of the model's face in this study or the salient feature changes of the expressions.

Finally, the infant's mouth, eye, and brow behaviors were analyzed across the modeled expressions, as in the Meltzoff and Moore study, in order to assess their incidence independent of variable baseline arousal and response levels. The results support the findings of Meltzoff and Moore (1977) inasmuch as there was mimicking of components of these facial expressions by the neonates. For example, eye widening occurred more frequently during the eye-widening, surprise expression than the other expressions. Similarly, lip protrusion occurred more frequently during the lip-protruding, sad expression, mouth opening during the mouth-opening, surprised expression. In addition, a relaxed brow occurred more frequently during the happy expression, a knit brow during the sad expression, and a raised brow during the surprise expression. Finally, the coder, who was blind to the expression being modeled, guessed the modeled expression at greater than chance probability simply by observing the face of the neonate. These data suggest, then, that neonates not only discriminate particular features of different facial expressions but also respond differentially to them with imitative behaviors or behaviors of similar appearance to those of the model. (See Figure 2.1.)

FACES AND VOICES

The face and voice together may be more salient than the face or voice alone because of its multimodal or multidimensional nature. The talking face might also be considered a more abstract, complex stimulus, and as such, the proponents of ear-

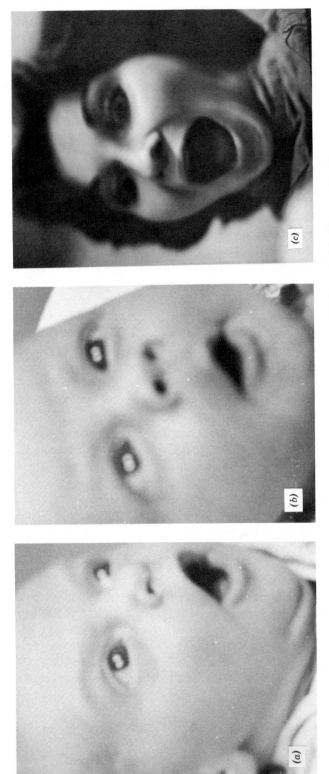

Figure 2.1. Examples of facial responses of neonates (a and b) to a model of surprised facial expression (c).

lier perception of higher-order structures would predict that it would be more readily perceived than the face or voice alone. While the particular study required to answer this question has not yet been conducted, there are a number of studies suggesting sophisticated discriminations of talking faces by very young infants.

First, the neonate is noted to track more readily the talking face than the face alone (the quiet face) during the Brazelton Neonatal Behavior Assessment procedure (Brazelton, 1973). Another demonstration of the salience and perception of joint face-voice stimulus is that slightly older infants, at 3 weeks of age, are noted to show "disturbed" responses if the mother's voice emanates from a different direction than her face (Aronson & Rosenbloom, 1971). These investigators positioned the mother face to face with her infant located behind a soundproof screen. The volumes of the loudspeakers were then made unequal so that the location of the voice shifted toward the louder speaker and away from the seen location of the mouth. Disturbed behaviors of the 3-week-old infant were interpreted as responses to a violation of the expectation that the mother's voice emanates from the direction of her mouth, suggesting that face and voice are perceived as an integrated stimulus.

The perception of specific faces and voices has been documented for the very young infant, aged 1 week to 2 months, in a series of studies reviewed in Carpenter (1974). Carpenter et al. (Carpenter, Tecce, Stechler, & Friedman, 1970) presented a mother's face, a manikin's face, and an abstract "facial" form (a kitchen collander bent to an oval shape and painted flesh color with three colored knobs attached to create "features") to infants ranging from 1 to 8 weeks of age. The authors were surprised to find the least amount of looking at the mother's face. However, gaze averting and fussy behavior of the infants in the presence of the mother suggested to the investigators that the infants were avoiding the mother's face possibly because she constituted a violation of expectancy. The mother's face was silent rather than talkative.

In a second study of infants, age 2 to 7 weeks, Carpenter (1974) presented a mother's silent face, a mother's face and voice, a mother's and stranger's voice, and parallel conditions for the stranger's face. The faces were live, the voices were tape recorded, and the "speaker" mimed the words. Infants as young as 2 weeks differentiated the mother's from the stranger's face, irrespective of the voice conditions. Each face was looked at longer if accompanied by a voice, but the mother's face received more looking than the stranger's face with either voice. More immediate gaze aversion of wider angles was noted when faces were presented with mismatched voices, reminiscent of the findings of Aronson and Rosenbloom (1971) and suggestive of the infant having formulated an association between faces and voices and particularly the face and voice of his mother.

Differential responding to different individuals' faces and voices and to face and voice changes within individuals has been demonstrated for very young infants by a number of face-to-face interaction studies. Infants as young as 4 weeks of age have been noted to show different behaviors when positioned en face with an infant peer and with his mother (Fogel, 1980). The infant fixedly stares and moves his limbs in a jerky manner in the presence of a peer, while multiple brow movements and smooth limb movements predominate in the mother's presence.

The slightly older 4-month-old infant responds differentially to his co-twin and a mirror image of himself by showing more gazing at the mirror but more smiling, vocalizing, and reaching toward his co-twin (Field, 1979a). Similarly, the 4-month-old infant responds differentially to a talking, head-nodding, infant-sized raggedy ann doll and his animated mother, showing more gazing at the doll but more social behaviors when face to face with his mother (Field, 1979b). Finally, the face and voice of the infant's mother and father are clearly discriminated as manifested by more smiling and laughing to the father and more gazing at the mother (Field, 1981). While the faces and voices of these social objects clearly differ on a number of variables, such as frequency, intensity, and contingent responsivity, the very young infant's differential responses suggest discrimination of the variations in these very complex face and voice stimuli.

Perhaps even more impressive are the infant's discriminations and differential responding to subtle changes of the face and voice made by a single individual. Dramatic differences are noted in the young infant's behavior when a mother, in the middle of a spontaneous interaction, becomes silent or still-faced (Fogel, 1980; Stoller & Field, 1981; Tronick, Als, Adamson, Wise, & Brazelton, 1978).

Similarly, when the mother continues to look at the infant but shows voice changes as she speaks to an adult (whom she is now seeing instead of her infant by an arrangement of mirror reflections), the infant's behavior is also altered, as if he is aware that she is no longer talking to him (Trevarthen, 1974). Finally, as the mother's change in activity level or quantity of facial-vocal stimulation occurs when she shifts to being "imitative of her infant's behaviors" or to attempting to "keep her infant's attention," her infant's gaze and affective behaviors change dramatically (Field, 1977).

These subtle and not so subtle changes of faces and voices either as the interaction partner is varied or the behavior of a particular interaction partner is modified are responded to immediately by the infant. Changes in the infant's attentive, affective, and cardiac activity suggest that the infant has discriminated the different faces and voices.

SUPRAMODAL INTEGRATION

How is it, then, that very young infants who show somewhat limited abilities to discriminate simple face patterns and single speech contrast stimuli also show relatively sophisticated discriminations of social stimuli such as faces and voices alone and faces and voices together? Bower and Wishart (1979) suggest:

> [The] baby's first acquisition, possibly an intrauterine acquisition, is the very high level idea that the world has a three-dimensional structure and a temporal structure. This structural idea is then specified in terms of whatever variables are available. It is possible that the form of the variables are specified but not their modalities. (p. 80)

Faces and voices which have both a three-dimensional and a temporal structure appear to be more readily perceived than more specific face patterns and two-dimensional face photographs or single speech contrasts. A commonly cited example of this general or abstract to the specific course of development is the perception of sounds of all human languages at a very early age followed by an inability to differentiate those at an older age.

Supramodal or intermodal perceptual abilities may precede perception of unimodal, unidimensional stimuli just as neurones are thought to develop only gradually from a very abstract, uncommitted form following from experiences with the environment (Jacobson, 1970). This early plasticity may enable the infant to perceive multidimensional information prior to information from single modalities. Examples of intermodal or supramodal perception include blind infants seeing by ultrasonic devices (Bower, 1974) and oral-tactile perception of a stimulus following only visual experience with that stimulus (Meltzoff & Borton, 1979).

Bower and Wishart (1979) suggest that the child is born with a very abstract awareness of "humanness" plus some very general strategies for signaling that awareness to others, strategies like imitation and interactional synchrony. Both the perceptions and strategies may become more specific with experience. Photographs and face patterns do not have the three-dimensional, abstract quality of the "human" face. Their specificity and unidimensionality may be discriminated only at a later stage, as has been demonstrated for photographed faces. The infant may discriminate a very abstract stimulus such as his mother's voice reading *And to Think that I Saw It on Mulberry Street* but fail to recognize contrasts of her single speech sounds. While her speech record is more abstract than her speech sounds, hers is a specific voice. The neonate may have acquired sufficient experience in utero to perceive her specific voice (as opposed to a stranger's or even the father's voice). Whether multidimensional complex social stimuli are more salient, whether hypercomplex cells are more developed, or supramodal perception more precocious, the very young infant appears more able to perceive and respond to social stimulation. Despite the complexities of the study of social perception, the very young infant's precocity in that area suggests the importance of further investigation.

REFERENCES

Aronson, E., & Rosenbloom, S. Space perception in early infancy: Perception within a common auditory–visual space. *Science,* 1971, **172,** 1161–1163.

Banks, M. S., & Salapatek, P. Activity and contrast sensitivity in 1-, 2-, and 3-month-old human infants. *Investigative Ophthalmology and Visual Science,* 1978, **17,** 361–365.

Bower, T. G. R. *Development in infancy.* San Francisco: Freeman, 1974.

Bower, T. G. R., & Wishart, J. G. Towards a unitary theory of development. In E. B. Thoman (Ed.), *Origins of the infant's social responsiveness.* Hillsdale, N.J.: Erlbaum, 1979.

Brazelton, T. B. *Neonatal behavioral assessment scale.* London: Spastic International Medical Publications, 1973.

Carpenter, G. C. Visual regard of moving and stationary faces in early infancy. *Merrill-Palmer Quarterly,* 1974, **20,** 181–194.

Carpenter, G. C., Tecce, J. J., Stechler, G., & Friedman, S. Differential visual behavior to human and humanoid faces in early infancy. *Merrill-Palmer Quarterly,* 1970, **16,** 91–108.

Cohen, L., & Gelber, E. Visual habituation and memory in infancy. In L. B. Cohen & P. M. Salapatek (Eds.), *Infant perception and cognition.* New York: Academic, 1975.

Condon, W. S., & Sander, L. W. Synchrony demonstrated between movements of the neonate and adult speech. *Child Development,* 1974, **45,** 456–462.

DeCasper, A. J., & Fifer, W. P. Of human bonding: Newborns prefer their mothers' voices. *Science,* 1980, **208,** 1174–1176.

Dowd, J. The temporal organization of spontaneous movements in the human infant. Ph.D. dissertation, University of Massachusetts, 1980.

Dunkeld, J. The development of imitation in infancy. Doctoral dissertation, University of Edinburgh, 1978.

Eilers, R. E., Wilson, W. R., & Moore, J. M. Developmental changes in speech discrimination in infants. *Journal of Speech and Hearing Research,* 1977, **20,** 766–780.

Eimas, P. Speech perception in early infancy. In L. B. Cohen & P. H. Salapatek (Eds.), *Infant perception and cognition.* New York: Academic, 1975.

Fantz, R. L., & Miranda, S. B. Newborn infant attention to form of contour. *Child Development,* 1975, **46,** 224–228.

Field, T. Effects of early separation, interactive deficits and experimental manipulations on infant–mother face-to-face interaction. *Child Development,* 1977, **48,** 763–771.

Field, T. Differential behavioral and cardiac responses of 3-month-old infants to a mirror and peer. *Infant Behavior and Development,* 1979, **2,** 179–184. (a)

Field, T. Visual and cardiac responses to inanimate faces by young term and preterm infants. *Child Development,* 1979, **50,** 188–194. (b)

Field, T. Gaze behavior of normal and high-risk infants during early interactions. *Journal of the American Academy of Child Psychology,* 1981, **20,** 308–317.

Field, T., Woodson, R., Greenberg, R., & Cohen, D. Discrimination and imitation of facial expressions by neonates. Unpublished manuscript, University of Miami, 1980.

Fogel, A. Peer versus mother directed behavior in one- to three-month-old infants. *Infant Behavior and Development,* 1980, **2,** 215–226.

Gardner, J., & Gardner, H. A note on selective imitation by a six-week-old infant. *Child Development,* 1970, **41,** 1209–1213.

Gardner, J. M., & Turkewitz, G. The effect of arousal level on visual preferences in preterm infants. Unpublished paper, Albert Einstein College of Medicine, 1980.

Gibson, E. J., Owsley, C. J., & Johnston, J. Perception of invariants by five-month-old infants: Differentiation of two types of motion. *Developmental Psychology,* 1978, **14,** 407–415.

Goren, C. C., Sarty, M., & Wu, P. K. Visual following and pattern discrimination of face-like stimuli by newborn infants. *Pediatrics,* 1975, **56,** 544–549.

Hainline, L. Developmental changes in visual scanning face and non-face patterns by infants. *Journal of Experimental Child Psychology,* 1978, **25,** 90–115.

Haith, M. M., Bergman, T., & Moore, M. J. Eye contact and face scanning in early infancy. *Science,* 1977, **198,** 853–855.

Hubel, D. H., & Wiesel, T. N. Receptive fields and functional architecture of monkey striate cortex. *Journal of Physiology,* 1968, **195,** 215.

Hutt, S. J., Hutt, C., Lenard, H. G., Bermuth, H. V., & Muntjewerff, W. J. Auditory responsivity in the human neonate. *Nature,* 1968, **218,** 888–890.

Jacobson, M. Development of central connections and formation of neuronal circuits. *Developmental Neurobiology,* 1970, **4,** 294–344.

Jacobson, S. W. Matching behavior in the young infant. *Child Development,* 1979, **50,** 425–430.

Kearsley, R. B. The newborn's response to auditory stimulation: A demonstration of orienting and defensive behavior. *Child Development,* 1973, **44,** 582–590.

Korner, A. F., & Thoman, E. B. Visual alertness in neonates as evoked by maternal care. *Journal of Experimental Child Psychology,* 1970, **10,** 67–78.

Maratos, O. The origin and development of imitation in the first six months of life. Doctoral dissertation, University of Geneva, 1973.

Masters, J. C. Interpreting "Imitative" responses in early infancy. *Science,* 1979, **205,** 215.

Maurer, D., & Salapatek, P. Developmental changes in the scanning of faces by young infants. *Child Development,* 1976, **47,** 523.

Meltzoff, A. N., & Borton, R. W. Intermodal matching by human neonates. *Nature,* 1979, **282,** 403–404.

Meltzoff, A. N., & Moore, M. K. Imitation of facial and manual gestures by human neonates. *Science,* 1977, **198,** 75–78.

Morse, P. A. The discrimination of speech and nonspeech stimuli in early infancy. *Journal of Experimental Child Psychology,* 1972, **3**, 477–492.

Nelson, C. A., Morse, P. A., & Leavitt, L. A. Recognition of facial expressions by seven-month-old infants. *Child Development,* 1979, **50**, 1239–1242.

Pomerleau-Malcuit, A., & Clifton, R. K. Neonatal heart rate response to tactile, auditory and vestibular stimulation in different states. *Child Development,* 1973, **44**, 485–496.

Salapatek, P. Pattern perception in early infancy. In L. Cohen & P. Salapatek (Eds.), *Infant perceptions: From sensation to cognition* (Vol. 1). *Basic Visual Processes.* New York: Academic, 1975.

Salapatek, P., & Kessen, W. Visual scanning of triangles by the human newborn. *Journal of Experimental Child Psychology,* 1966, **3**, 155–167.

Slater, A. M., & Findlay, J. M. The measurement of fixation positions in the newborn baby. *Journal of Experimental Child Psychology,* 1972, **14**, 349–364.

Souther, A. F., & Banks, M. S. The human face: A view from the infant's eye. Paper presented at the meetings of the Society for Research in Child Development, San Francisco, Calif., March 1979.

Stoller, S., & Field, T. Alteration of mother and infant behavior and heart rate during a still-face perturbation of face-to-face interaction. In T. Field & A. Fogel (Eds.), *Emotions and interactions.* Hillsdale, N.J.: Erlbaum, 1981.

Tinbergen, N. *A study of instinct.* New York: Oxford University Press, 1951.

Treiber, F., & Wilcox, S. Perception of a "subjective" contour by infants. *Child Development,* 1980, **51**, 915–917.

Trevarthen, C. Conversations with a two-month-old. *New Scientist,* May 1974, pp. 230–233.

Tronick, E., Als, H., Adamson, L., Wise, S., & Brazelton, T. B. The infant's response to entrapment between contradictory messages in face-to-face interaction. *Journal of Child Psychiatry,* 1978, **17**, 1–13.

3

INFANT AUDITORY AND SPEECH PERCEPTION

Philip A. Morse
Nelson Cowan
Department of Psychology
University of Wisconsin at Madison

Speech communication is a process that is at once technical, social, and cognitive. Thus it is reasonable that the infant's perception of speech draws upon capabilities that range from sensations mediated by specific neurological structures to holistic perceptions that integrate acquired knowledge. Although several reviews of the infant speech perception literature have documented impressive *phonetic* perceptual abilities (Eilers, 1980; Eimas, 1978; Eimas & Tartter, 1979; Jusczyk, in press; Kuhl, 1979; Morse, 1978, 1979; Trehub, 1979; Trehub, Bull, & Schneider, in press), the full range of perceptual capabilities necessary for the early development of language has not been given much attention.

This chapter begins by considering two aspects of auditory development fundamental to the infant's perception of speech sounds: the ability to hear sounds and the processes underlying auditory perception. The section entitled "Sensation" focuses on the methods employed in evaluating threshold responses in infants and how infants' sensitivity to different frequencies compares to adult listeners'. The perception of speech sounds also depends upon the ability to process the information in auditory stimuli once they are detected. Accordingly, the section entitled "Auditory Perception" considers processes involved in the infant's extraction and retention of auditory information. The subsequent section entitled "Phonetic Perception" examines studies of infants' processing of information in human speech. In particular, we explore the ability to make simple auditory discriminations between speech sounds, to make categorical discriminations, and even to exhibit "perceptual constancy" for various categories of vowels and consonants. However, most studies of phonetic perception have investigated the infant's discrimination of isolated syllables, and the understanding of human verbal communication involves much more than this. Therefore, in the following section entitled "From Phonetics to Words," we discuss the infant's phonetic perception of multisyllabic utterances, the perception of voices and suprasegmental information, and some of the issues involved in the eventual recognition of complex auditory patterns such as "names" or "words." Finally, in the closing section of this review, "Issues in Infant Speech Perception," three major problem areas are addressed that have attracted the attention of researchers in this field: methodological issues, brain–behavior relationships, and the speech perception abilities that may be species-specific.

SENSATION

The most basic capability underlying speech perception is of course the ability to hear. Infants' hearing abilities generally have been underestimated due to infants' refusal to give overt signs of having heard. Researchers have been resourceful in seeking evidence of infants' early auditory abilities, and have found that even fetuses respond physiologically to auditory stimulation (Murphy & Smyth, 1962; Johansson, Wedenberg, & Westin, 1964). Consistent with this conclusion is the anatomical evidence indicating that the inner and middle ears reach adult size and shape before birth (Elliott & Elliott, 1964). Finally, there is no lack of prenatal auditory stimulation: the sound level at the fetus's head, due mainly to the mother's pulse, has been measured at 72–85 dB (Bench, 1968; Walker, Grimwade, & Wood, 1971), which approximates the intensity of adult conversation. Furthermore, external noises of low frequencies may reach the fetus with as little as 20 dB of attenuation (Walker et al., 1971).

Although the peripheral mechanism for hearing is largely mature at birth, the neurological components that process auditory signals develop substantially after birth. One way that the neurological development of audition can be examined is through the components of electrical brain responses to auditory signals. When electrical potentials are recorded from the scalp and averaged across many stimulus presentations, an "auditory evoked potential" (AEP) results. Characteristic waves in the first 10 msec of the AEP have been attributed to the transmission of the signal through the auditory nerve and through the brainstem nuclei involved in audition (Jewett, Romano, & Williston, 1970; Picton, Hillyard, Krang, & Galambos, 1974). An examination of the latency and size of AEP components in infants suggests that transmission of the signal through the auditory nerve is adultlike by the sixth postnatal week, but that transmission through the brainstem continues to mature throughout the first 2 years of life (Salamy, McKean, & Buda, 1975; Salamy & McKean, 1976; Hecox & Galambos, 1974). This sequence of neurological development is also supported by studies indicating that the relatively peripheral structures (e.g., the eighth nerve, superior olive, and lateral lemniscus) are fully myelinated by the ninth fetal month (Yakovlev & Lecours, 1967), but that the inferior colliculus, a brain structure further along in the chain of auditory transmission, is not well myelinated at birth (Rorke & Riggs, 1969).

Despite these data from developmental neurology, the acuity of infants' hearing relative to adults' is at best a matter of debate. One problem in assessing the infant's hearing is that most available response measures do not have sensitivities that are equivalent in adults to *voluntary* thresholds. Furthermore, different measures yield discrepant estimates of developmental changes in sensitivity. The wide variety of response measures that have been employed includes reflex and overt motor reactions, heart rate change, nonnutritive sucking, early, middle, and late components of the average evoked brain potential, electromyographic potentials, cochlear potentials, and measurements of the middle ear reflex. The results of tests conducted with these measures differ in the observed hearing sensitivity of infants relative to adults, the responses to specific sound frequencies, and the types of developmental change. Responses also interact with the behavioral state of the infant. For example, Eisenberg (1976) reported that neonates responded to a high-pitched whistle (with motor movements, eye movements, or behavioral arousal or quieting) much more frequently when awake or in a light sleep than in a deep sleep. In contrast, low-pitched sounds elicited about the same number of responses in every state. Similarly, Bench and Mentz (1975) found that neonates respond better to a 90-dB white noise when asleep, but respond better to a 60-dB noise when awake. At present, the correct explanation for such stimulus–state interactions is unknown.

One way to resolve this discrepancy between response measures is to assume that the reasonably sensitive measures yielding the smallest estimates of infant–adult differences provide the most accurate assessments of infant hearing capabilities. With this criterion, the early component of the AEP, termed a "brainstem evoked response," has proven important (Schulman-Galambos & Galambos, 1970; Galambos & Hecox, 1978). Normal newborns display sensitivity to clicks about 10 dB above the adult threshold. An even smaller estimate of the newborn–adult discrepancy has been obtained with respiration as the response (Bradford, 1975).

A second approach to the assessment of hearing in infancy is to select the procedure that most closely resembles adult voluntary threshold procedures. With infants over 6 months of age, the Visual Reinforcement Audiometry (VRA) procedure can be used to measure voluntary responses to sound (Moore, Thompson, & Thompson, 1975; Trehub, Schneider, & Endman, 1980; Schneider, Trehub, & Bull, 1980). In this procedure, infants are presented with auditory stimuli and reinforced with a colorful, moving display for a headturn in response to the stimulus. Differential (left vs. right) responding is required and the response interval is unlimited, so that the test is in essence one of forced choice and is appropriate for a signal detection analysis. In a recent review of the infant audiological data, Schneider, Trehub, and Bull (1979) compared the VRA procedure to other tests of infants' hearing abilities. They suggested that most audiological tests have found infants to be less sensitive to sounds than adults by 30 dB or more. Furthermore, the tests have found either no difference in sensitivity at different frequencies or else a lack of sensitivity in the high frequencies relative to adults. In contrast, the VRA results suggest that infants' hearing threshold is only about 15–20 dB above the adult threshold and that infants have a differential deficit in the *low* rather than the high frequencies.

Schneider et al. also suggest that unlike the VRA, most measures of infants' hearing are contaminated by motivational and state factors. Motor and physiological responding to a sound may not always occur when the infant detects it; that is, the response may be dependent on the sound's arousal value, salience, or significance to the subject. Low-frequency sounds are more soothing and rewarding to infants than high-frequency sounds (Bench, 1969; Birns, Blank, Bridger, & Escalona, 1965), whereas high-frequency sounds are more disturbing; that is, result in fewer motor reflexes and more orienting reactions (Eisenberg, 1976). Thus some types of unconditioned responding may be elicited by only a subset of the sounds an infant is able to detect.

Schneider et al. (1979) point out that the VRA procedure provides the infant with a new, extrinsic reinforcement for responding to the stimulus, so that one need not depend on the sound's intrinsic motivating qualities as in previous tests of hearing. Therefore, they conclude that the VRA procedure most accurately reflects infant hearing thresholds. However, most past investigations of infants' sound frequency sensitivities employed simple or complex tone stimuli or clicks. In contrast, Schneider et al. (1980) and Trehub et al. (1980) employed noise-band stimuli. This stimulus difference could account for differences in the results. There is an important reason why sensitivity functions for tones and noise-band stimuli might differ. The bandwidths of octave-band stimuli are frequency-dependent; for example, the lowest band frequency employed by Trehub et al. (1980), centered at 200 Hz, would have a bandwidth of about 140 Hz. In contrast, the highest band, centered at 10,000 Hz, would have a bandwidth of roughly 7070 Hz. Consequently, at equal sound pressure levels these two stimuli would have unequal "spectrum levels," or sound pressure levels per cycle (cf. Hirsch, 1952; Yost & Nielsen, 1977). The experienced loudness and the detection threshold of a sound depend upon this distribution of energy across frequencies. A specific sound pressure is experienced as louder the more its energy is spread across frequencies and extends beyond a single critical bandwidth (Scharf, 1970). A difference between infants and adults in critical bandwidths as a function of frequency could lead to substantially different developmental trends in detection and experienced loudness of tones versus noise bands.

Thus our present knowledge of infants' basic hearing capabilities is sparse, but several paradigms promise a wealth of knowledge. For example, brainstem-evoked response studies demonstrate sensitivity to clicks near the adult threshold in newborns. Furthermore, the application of signal detection procedures to infant hearing tests, especially with the VRA procedure, may soon yield unbiased estimates of tone thresholds in infants over 6 months of age. Regardless of whether infants prove to be more sensitive in the high- or low-frequency regions, the data using various procedures clearly indicate that infants have good hearing in the intermediate-frequency range and thus come well equipped to listen to the sounds of human speech. A few authors (Eisenberg, 1976; Lenard, Bernuth, & Hutt, 1969) have taken this thesis further, proposing that there is a sensitivity maximum in the speech carrier frequencies. However, at present there is insufficient evidence to support this stronger claim (although cf. Weir, 1979).

Stimulus detection is not the only important issue related to sensation. Little is currently known about infants' psychophysical scaling of stimuli, although infant researchers are close to having a procedure for such investigations. Some classic stimulus scaling procedures used with adults measure "just noticeable differences" (jnd) between stimuli along a continuum (e.g., tone frequency), and these jnds enable one to match physical and psychophysical scale values (e.g., frequency and pitch). An infant procedure shown capable of measuring absolute thresholds might be adaptable to the measurement of jnds as well. The minimal stimulus change along a dimension (e.g., pitch) at which discrimination performance differs from chance might be taken to reflect one jnd (see Aslin, Hennesy, Pisoni, & Percy, 1979, for an example of this approach with a speech continuum using a variation of the VRA procedure.)

AUDITORY PERCEPTION

A speech stimulus, once detected, remains relatively useless to the developing infant until its qualities have been further resolved and can be stored and retrieved. This additional processing of the stimulus depends upon central brain structures which, unlike some of the more peripheral structures discussed above, are far from mature at birth (see Hecox, 1975, and Berg & Berg, 1980, for reviews). For example, the inferior colliculus and medial geniculate are not as well myelinated at birth as lower auditory centers are, and the auditory cortex and corpus callosum may be even less mature. Responses of these immature central structures are reflected in the later components of the average evoked potential. Unlike components within the first few msec, these middle and late components (up to 500 msec) change dramatically in the first year of life (e.g., Ohlrich, Barnet, Weiss, & Shanks, 1978). Further, studies of heart rate and of visual and motor reflex activity suggest that at 2 to 3 months of age, there is a shift of control from subcortical to cortical centers of the brain (Graham, Strock, & Zeigler, 1980; Bronson, 1974; Karmel & Maisel, 1975).

Although adult levels of neurological maturity do not appear to be prerequisite for the ability to discriminate auditory stimuli, it remains an open question if infants are neurologically equipped to process sounds with the same precision and resolution as adults. For example, in one generally accepted view of adult auditory processing, incoming sounds are held in a buffer memory, or "echoic storage," for about 250 msec, and within that time the information within the sound leading to its identification is extracted (Massaro, 1975). This time period is well-suited to the perception of syllables within speech. However, little is known about this time period of sound recognition in early infancy. It is possible that infants do not extract physical information from sounds as quickly as adults. This might contribute to a "blooming, buzzing confusion" within infant auditory and speech perception. It may also be the case that the echoic storage of a sound lingers on in the infant longer than in adults, so that if the incoming signals are slowed sufficiently (as is often the case in maternal speech to infants), adequate perception becomes possible.

There is in fact some indirect evidence that infants do not process information in the same time period as adults. Little (1971) found that classical eyelid conditioning can be obtained in infants under 1 month of age with a CS–UCS interval of 1500 msec, but not with shorter intervals. In contrast, the optimal interval in adults is 250–500 msec (Gormezano & Moore, 1969). Additional evidence that adults and infants differ in the time of processing comes from research on prepulse modification of the startle response (Graham et al., 1980). A loud, sudden sound typically results in a startle reaction that can be indexed by an eyeblink response. However, this startle reaction is partially inhibited in adults by a brief, moderately intense stimulus preceding the startle pulse by as little as 16 msec and up to 300 msec. In infants 6 to 9 weeks of age, however, startle inhibition was not observed with prepulse intervals of 75 or 175 msec. Yet substantial startle inhibition *was* obtained with intervals of 225 and 275 msec. Thus processing of the prepulse stimulus may be slower in the infant. Graham et al. interpreted this result along with other evidence to suggest that infants may be selectively immature in neurological components whose response to stimulation involves a short time constant. Other work also indicates that prestimulation facilitates the newborn heart rate response to a square-wave tone when it precedes the tone by 4000 or 250 msec, but not 100 msec (VerHoeve, Leavitt, & Morse, 1980).

In our nonnutritive-sucking laboratory, we have begun to explore methods for the investigation of echoic storage in infancy (Cowan, 1978; Cowan, Suomi, & Morse, 1981), using a modification of the recognition-masking paradigm employed to study echoic storage in adults (see Massaro, 1975). In the adult paradigm, two brief sounds are presented with a short, variable stimulus onset asynchrony (SOA), and the subject is required to identify either the first sound (in the backward-masking procedure) or the second sound (in the forward-masking procedure). Inasmuch as the second stimulus interferes with the echoic storage of the first stimulus, recognition of the first stimulus is impeded when the SOA is too short; that is, when it is 250 msec or less. Comparable interference with recognition of the second stimulus in the pair does not occur, because echoic storage for that stimulus is uninterrupted. In our infant procedure, there are multiple presentations of a stimulus pair in a trial. Either the first stimulus of the pair changes within the trial (backward masking), the second stimulus changes (forward masking), or neither change (control). Cowan et al. employed a modified version of the traditional nonnutritive-sucking discrimination procedure (e.g., Eimas, Siqueland, Jusczyk, & Vigorito, 1971). Infants 8 to 9 weeks of age were presented with pairs of 50-msec, synthetic vowels ([a] and [ɛ]), with a 0-msec intra-pair ISI (i.e., with an SOA = 50 msec) and with about 1 sec between pairs. In the backward-masking condition, infants received a contrast in which the first member of the pair changed (e.g., [a–a] vs. [ɛ–a]), whereas in the forward-masking condition the second member of the pair changed (e.g., [a–a] vs. [a–ɛ]). Finally, control subjects received the same pair (e.g., [a–a]) throughout the test procedure. The results revealed that only infants in the forward-masking condition demonstrated discrimination of the vowel change. This pattern of results suggests that infants (like adults) *are* able to discriminate vowel differences with this brief interval within pairs. However, when echoic storage for the change in vowel is impeded (i.e., by the second vowel in each pair in the backward-masking condition), infants do not readily exhibit discrimination of these vowel contrasts. Additional studies are needed to examine further the pattern of information extraction from echoic storage in early infancy and its maturational course.

Auditory information processing does not stop with the identification of the features of a sound from echoic storage. In order to compare one sound with another one to determine if it is "familiar" or "novel," as infants are required to do in speech and auditory discrimination paradigms, they must retain sounds in short-term memory (STM) long enough to make this comparison successfully. Recently several studies have begun to examine more explicitly the role of STM in different paradigms used to study infant speech perception (Swoboda, Kass, Morse, & Leavitt, 1978; Leavitt, Brown, Morse, & Graham, 1976). However, many more systematic studies of infant auditory short-term memory are needed to describe such fundamental STM features as the role of decay, interference, and the size of the infant's STM store. Finally, if the infant is going to learn to recognize a series of "familiar" sounds as being names of objects and people that are important to him/her, the infant must be able to retain these sound patterns (in long-term memory) across hours and days. We know virtually nothing about this aspect of the infant's memory for speech sounds, although informal observations of infants who have been trained to recognize a particular contrast in the head-turning paradigm suggest that infants can remember some familiar speech sounds up to at least a week (Kuhl, personal communication).

PHONETIC PERCEPTION

The primary approach taken by most researchers in studying the development of the infant's speech perception abilities has been to investigate the perception of specific phonetic contrasts that have previously been well-studied in adult listeners. Accordingly, researchers have focused primarily on three different levels of phonetic perception: auditory discrimination, categorical discrimination, and perceptual constancy.

Studies at the *auditory* level examine the infant's abilities to discriminate differences in the major acoustic cues that signal consonant and vowel contrasts. The study of infant speech perception would not be very complicated if for every phonetic consonant and vowel contrast in adult speech there existed a single, corresponding acoustic difference in the waveform. If this were the case, we would only need to verify at what age infants discriminate these specific acoustic differences and study how the adult category names for these different acoustic patterns are acquired, presumably according to the principles of paired-associate learning. However, most perceived phonetic categories and features are *not* related in a simple 1 : 1 manner to cues in the acoustic waveform, but instead rely upon a complex code for their perception (e.g., Liberman, Cooper, Shankweiler, & Studdert-Kennedy, 1967; Liberman, 1970). The speech code contains several different levels or aspects of complexity relating phonetic and acoustic information together that the developing infant must eventually master. One aspect of the speech code is the phenomenon known as "categorical perception," which refers to the pattern of adult discrimination and identification observed when listeners are only able to discriminate acoustic differences to the extent that they can label them differentially. In other words, listeners discriminate well between category boundaries, but poorly (near chance levels) within phonetic categories. In contrast to categorical perception, in which certain variations along a single acoustic *dimension* are ignored (categorized), in perceptual constancy the listener treats a wide variety of acoustic stimuli as belonging to the same phonetic category. A consonant category may contain variations in speaker, vowel environment, intonation, position in the syllable, and/or the relevant acoustic cue. For example, the acoustic cues distinguishing [p] from other phonetic categories include (1) a difference in voice-onset-time in the context of [ba] versus [pa] (Abramson & Lisker, 1970), (2) a difference in vowel duration in the context of [mab] versus [map] (Raphael, 1972), and (3) the presence or absence of a silent gap between [s] and [l] in the contrast [slɪt] versus [splɪt] (Dorman, Raphael, & Liberman, 1979). At this more complex level of the speech code, the relationship of phonetic categories to acoustic cues may be described as more abstract, although it is possible that the relationship can nevertheless be specified (e.g., Stevens & Blumstein, 1978).

Infant Testing Paradigms

Studies of infant speech perception have relied primarily on three paradigms and their variations for assessing discrimination: nonnutritive sucking, heart rate, and head turning. In all these procedures an individual infant is presented with a familiar or background speech stimulus for a period of time and then his/her reaction is measured to a series of novel or change stimuli at the appropriate time in the experiment. In the classical version of the high-amplitude sucking (HAS) paradigm (Eimas, Siqueland, Juscyzk, & Vigorito, 1971) following a no-sound baseline, a speech sound is played through a loudspeaker to the infant contingent upon the infant's rate of high-amplitude sucking. The more the infant sucks on a nonnutritive nipple, the more repetitions of that sound the infant receives. As the infant discovers this contingency, the rate of high-amplitude sucking increases (acquisition). Eventually, the infant begins to satiate or habituate to the sound, and the frequency of high-amplitude sucking decreases. If sucking decreases at a rate that falls within a predetermined range, the sound is changed, and a subsequent increase in responding relative to a no-change control condition is interpreted as evidence of discrimination of the two different sounds. Eimas and Miller (1980) have recently introduced a variant of this HAS paradigm in which following baseline they present a speech sound contingent upon high-amplitude sucking for 7 minutes (no matter how much the infant sucks) and then change to a novel sound. As before, discrimination is inferred from an increase in sucking rate relative to a no-change control. We have recently developed a very different modification of the HAS paradigm (Cowan et al., 1981), in which the infant receives speech stimuli contingent upon high-amplitude sucking for 10 minutes following the 1 minute baseline but in which the "familiar" and "novel" stimuli change every 30 seconds in the experimental condition (e.g., from [ba] to [pa] and back to [ba]). Discrimination is indexed by an increase in sucking above baseline by the experimental group relative to that of a no-change control group. These HAS paradigms are of use only with infants between birth and 4 months of age.

In contrast to the HAS paradigm, the presentation of sounds in the heart rate (HR) procedure first introduced by Moffitt (1971) is entirely under the control of the experimenter. If a speech sound or string of speech sounds is played at a moderate intensity, infants will generally respond to the sound onset with heart rate deceleration (orienting). With repeated trials of these same sounds, this initial orienting response (OR) tends to habituate. After a fixed number of trials (to permit the habituation of the cardiac OR), a novel sound is presented for a series of trials, and if recovery of the OR (cardiac deceleration) occurs to the new sounds, discrimination of the stimulus contrast is inferred. Drawbacks in the use of HR as a measure of speech discrimination are the developmental changes in the cardiac OR and the sensitivity of HR activity to state changes.

Finally, in the head-turning (HT) procedure developed by Eilers, Wilson, and Moore (1977), the infant is seated on a parent's lap and entertained at midline by an assistant playing with an interesting toy. From a speaker located in front and to the side of the infant (about 45°) is presented a constant speech sound as a background stimulus. After listening to this background sound for awhile, the presentation of a novel stimulus periodically interrupts the background stimulation. If the infants turns his/her head toward the speaker when this change occurs, a visual reinforcer (e.g., an animated monkey or dancing bear) is illuminated within a darkened box for the infant to view. If the infant produces a head turn more frequently on those trials in which a change in sounds occurs compared to no-change control trials, then discrimination of the contrast is inferred. This procedure and its many variants have been found to be primarily applicable to listeners 6 months of age and older.

Infants' Discrimination of Consonant and Vowel Contrasts

Armed with these paradigms, investigators have explored the infant's (1) simple auditory discrimination of various acoustic cues in speech sounds, (2) categorical discrimination of phonetic contrasts, and (3) perceptual constancy; that is, more abstract phonetic organization of speech sounds. As can be seen in Table 3.1, the cumulative results of these studies indicate an impressive set of auditory and phonetic speech perception abilities. The sections that follow present the evidence for the discrimination (for each of these three levels) of place of articulation, voicing, fricative, manner of articulation, liquid, and vowel contrasts, respectively.

The first study of infants' *auditory* discrimination of critical acoustic cues in speech was Moffitt's (1971) investigation of place of articulation using an HR paradigm in 4- to 5-month-olds. He found that infants could discriminate formant transitions that differentiate the syllables [ba] and [ga]. Evidence of similar discriminative abilities was subsequently obtained for younger infants (Morse, 1972; Leavitt, Brown, Morse, & Graham, 1976). Miller, Morse, and Dorman (1977) have further demonstrated that infants can discriminate a place contrast differing in only the initial 8–32 msec release burst cue (rather than in formant transition cues). Other studies have evidenced discrimination of place cues in the *final* position (Juscyzk, 1977) and in a *medial* position in bisyllabic stimuli (Williams, 1977; Juscyzk & Thompson, 1978).

Evidence that infants have categorical discrimination for place of articulation was first demonstrated in two studies by Eimas (1974a). He found that infants 2 to 3 months of age discriminated the place continuum [bae–dae–gae] categorically, exhibiting discrimination of between-category pairs but *not* of within-category contrasts along his formant transition continuum. Further, he found that infants, like adults, demonstrated noncategorical discrimination of these same critical cues when they were presented in isolation as nonspeech sounds. Replications of this finding of categorical discrimination for place of articulation have been successfully carried out with different stimuli using an HR paradigm (Miller & Morse, 1976; Till, 1976).

The investigation of perceptual constancy or more abstract phonetic categorization for place of articulation has proved much more difficult. Fodor, Garrett, and Brill (1975) employed a head-turn conditioning procedure with syllables such as [pi], [ka], and [pu]. They reported that infants conditioned more quickly when syllables with the same initial consonant (e.g., [pi] and [pu] vs. [ka]) were paired together than when different ones were reinforced (e.g., [pi] and [ka] vs. [pu]). These results were interpreted as evidence that infants have perceptual constancy for consonants that occur in syllables with varying vowel contexts. However, since natural speech stimuli were employed in this study, it is possible that infants were responding to a similar initial release burst in syllables with the same consonant rather than correctly classifying varied acoustic cues across vowel contexts. In our own laboratory we unsuccessfully tried to demonstrate phonetic constancy for the syllables [di], [du], and [gɛ] using an operant head-turning paradigm that required an apparently Herculean feat of the infant; that is, differential head turning at 4 to 4½ months of age (Morse, Leavitt, Donovan, Kolton,

Miller, & Judd-Engel, 1975). More recently, we studied the problem of perceptual constancy using the HAS paradigm (Suomi & Morse, 1979) by familiarizing infants with randomized tokens from the stimulus set [bi, bo, bae, bu, ba, bɛ] and examining their discrimination of this set from the corresponding velar set [gi, go, gae, gu, ga, gɛ]. However, infants consistently failed to meet our satiation criteria required for a stimulus change in the test paradigm. Perhaps these variations in vowel environment simply proved too interesting for infants at this age, thus preventing us from examining their perceptual constancy for place of articulation with this paradigm. Although these results are generally not encouraging, they may indicate limitations of the paradigms rather than the infant's lack of abstract phonetic categories for place of articulation.

Studies of the infant's discrimination of other phonetic features/categories may be similarly organized into those of auditory discrimination, categorical discrimination, and perceptual constancy (see Table 3.1). In experiments of the discrimination of voicing contrasts, that is, [ba] versus [pa] cued by differences in voice-onset-time (VOT), investigators have demonstrated auditory discrimination in syllable-initial position (Trehub & Rabinovitch, 1972; Eilers, Morse, Gavin, & Oller, 1981), in intervocalic position (Trehub, 1973c), and in syllable-final position, cued by vowel duration differences (Eilers, 1977).

The first study that demonstrated *categorical* discrimination of a speech contrast by infants was one of VOT conducted by Eimas, Siqueland, Jusczyk, and Vigorito (1971) in infants 1 and 4 months of age with the HAS paradigm. Infants discriminated a 20-msec VOT difference in the syllables [ba] versus [pa] when it straddled the adult phonetic boundary (between [ba]: +20-msec VOT and [pa]: +40-msec VOT) but failed to discriminate this difference within the [ba] category (−20-msec VOT vs. 0-msec VOT) or the [pa] category (+60-msec VOT vs. +80-msec VOT). This finding has subsequently been confirmed in other studies with bilabial (Miller, 1974) and alveolar (Eimas, 1976b) contrasts. Unfortunately, no work has been done examining perceptual constancy or abstract phonetic categorization for voicing in early infancy.

In addition to the many studies of the infant's perception of place of articulation and voicing contrasts, considerable effort has been invested in examining the discrimination of *fricative* contrasts in the first year of life. These contrasts are particularly interesting since (1) young children appear to have difficulty perceiving them (Abbs & Minifie, 1969; Eilers & Oller, 1976) and (2) some of these contrasts contain high-frequency cues that infants may have difficulty hearing. Eilers and Minifie (1975) found that infants between 4 and 17 weeks of age could discriminate such fricative contrasts as [sa] versus [va] and [sa] versus [ʃa] (as in "sha"). Subsequent studies revealed evidence of infant discrimination for the fricative pairs [sa] versus [za], [as] versus [az], [fi] versus [θi], [fa] versus [θa], and [af] versus [aθ] (Eilers, 1977; Eilers, Wilson, & Moore, 1977; Holmberg, Morgan, & Kuhl, 1977). Recent work by Jusczyk, Murray, and Bayley (1979) with synthetic and natural speech tokens of [fa] versus [θa] further demonstrated that infants are able to discriminate fricative contrasts cued by differences in either frication or formant frequencies. However, infants do not appear to demonstrate the pattern of categorical-like discrimination observed in adult listeners in the fricative pair [fa] versus [θa] (Carden, Levitt, & Jusczyk, 1979; Jusczyk et al., 1979; Jusczyk, in press). Finally, Kuhl and her colleagues (Holmberg, Morgan, & Kuhl, 1977; Kuhl, 1980) have dem-

TABLE 3.1. Studies of Infant Speech Perception: Auditory, Categorical Discrimination, and Perceptual Constancy

Phonetic Categories	Level	Stimulus Contrasts	Ages	Paradigm	Reference
Place of articulation	Auditory	[ba] vs. [ga]	4–5 mo	HR	Moffitt (1971)
		[ba] vs. [ga]	6 wk	HAS	Morse (1972)
		[ba] vs. [ga]	6 wk	HR	Leavitt, Brown, Morse, & Graham (1976)
		[bu] vs. [gu]–bursts	3–4 mo	HR	Miller, Morse, & Dorman (1977)
		[d] vs. [g]–initial and final position	2 mo	HAS	Jusczyk (1977)
		[d] vs. [g]–medial position	2–3 mo	HAS	Williams (1977)
		[d] vs. [g]–initial and medial positions	2 mo	HAS	Jusczyk & Thompson (1978)
	Categorical discrimination	[dae] vs. [gae]	2–3 mo	HAS	Eimas (1974a)
		[dae] vs. [gae]	3–4 mo	HR	Miller & Morse (1976)
		[ba] vs. [da]	4 mo	HR	Till (1976)
	Perceptual constancy	[pi], [ka], [pu]	3½–4½ mo	conditioned HT	Fodor, Garrett, & Brill (1975)
		[di], [du], [gɛ]	4–4½ mo	conditioned HT	Morse, Leavitt, Donovan, Kolton, Miller, & Judd-Engel (1975)
Voicing	Auditory	[b] vs. [g]	2 mo	HAS	Suomi & Morse (1979)
		[ba] vs. [pa], [da] vs. [ta]	4–17 wk	HAS	Trehub & Rabinovitch (1972)
		[ba] vs. [pa], [du] vs. [tu]	6–8 mo	HT	Eilers, Morse, Gavin, & Oller (1981)
		[aba] vs. [apa], [ataba] vs. [atapa]	4–17 wk	HAS	Trehub (1973a)
		[t] vs. [d]–final position	6–14 wk	HAS	Eilers (1977)
	Categorical discrimination	[ba] vs. [pa]	1 & 4 mo	HAS	Eimas, Siqueland, Jusczyk, & Vigorito (1971)
		[ba] vs. [pa]	6–8 wk	HAS	Miller (1974)

Category	Subcategory	Contrast	Age	Method	Reference
Fricatives	Auditory	[da] vs. [ta]	2–3 mo	HAS	Eimas (1975b)
		[ba] vs. [pa]	6 mo	HT	Eilers, Wilson, & Moore (1979)
		[sa] vs. [va], [sa] vs. [ʃa], [sa] vs. [za], [as] vs. [az]	4–17 wk	HAS	Eilers & Minifie (1969)
		[sa] vs. [za], [as] vs. [az]	5–19 wk	HAS	Eilers (1977)
		[fi] vs. [θi], [fa] vs. [θa], [sa] vs. [ʃa], [sa] vs. [va]	6 & 12 mo	HT	Eilers, Wilson, & Moore (1977)
	Categorical discrimination	[fa] vs. [θa], [ða] vs. [va]	2 mo	HAS	Jusczyk, Murray, & Bayley (1979)
	Perceptual constancy	[f] vs. [θ] and [s] vs. [ʃ] in initial and final positions	6 mo	HT	Holmberg, Morgan, & Kuhl (1977); Kuhl (1980); Kuhl (1980)
Manner	Auditory	[bɛ] vs. [wɛ] vs. [uɛ]	6–8 mo	HAS	Hillenbrand, Minifie, & Edwards (1977)
	Categorical discrimination	[ba] vs. [ma]	2–4 mo	HAS	Eimas & Miller (in press)
	Perceptual constancy	[ba] vs. [wa]	2–4 mo	HAS	Eimas & Miller (1980)
		[m,n,ŋ] vs. [b,d,g]	6 mo	HT	Hillenbrand (1980)
Liquids	Categorical discrimination	[ra] vs. [la]	2–3 mo	HAS	Eimas (1975a)
Silence	Auditory	[slɪt] vs [splɪt]	6–8 mo	HT	Morse, Eilers, & Gavin (in press)
Vowels	Auditory	[i] vs [u], [i] vs. [a]	4–17 wk	HAS	Trehub (1973b)
	Categorical discrimination	[i] vs. [ɪ] 240 msec	2 mo	HAS	Swoboda, Morse, & Leavitt (1976)
		[i] vs. [ɪ] 60 msec	2 mo	HAS	Swoboda, Kass, Morse, & Leavitt (1978)
	Perceptual constancy	[i] vs. [a]	1–4 mo	HAS	Kuhl & Miller (in press)
		[a] vs. [ɔ]	6 mo	HT	Kuhl (1977)
		[a] vs. [i]	5½–6½ mo	HT	Kuhl (1979)

43

onstrated that infants exhibit *perceptual constancy* for fricative contrasts. In this HT study infants are first trained to discriminate a single contrast; for example, [sa] versus [ʃa]. In subsequent sessions variations in individual tokens, vowel context, and speaker are added in stages until the infant has been trained to discriminate the categories [s] versus [ʃ] across these varying contexts.

The few studies of the infant's perception of manner of articulation have yielded interesting results. First, a study by Hillenbrand, Minifie, and Edwards (1977) demonstrated that infants could discriminate distinctions between the stop consonant [bɛ], the semivowel [wɛ], and the diphthong [uɛ] when cued by differences in the rate and duration of the formant transitions (with [bɛ] containing the shortest duration and most rapid rate and [uɛ] the longest duration and slowest rate of transitional change). Eimas and Miller (in press) studying the categorical discrimination of the oral/nasal distinction ([ba] vs. [ma]) found that infants discriminated the between-category contrast ([ba] vs. [ma]), but discrimination of the within- and between-category contrasts did not differ reliably. Thus although infants are able to discriminate this manner difference auditorily, they do *not* evidence the categoricallike discrimination observed in adult listeners. In a recent article, Eimas and Miller (1980) examined the categorical discrimination of the stop [ba] versus the semivowel [wa]. Not only did they demonstrate that infants discriminated the underlying duration of formant transition cues categorically but they also showed very elegantly that infants' perception of this cue is context-dependent. More specifically, the category boundary in discrimination was found to shift dramatically as a function of the total duration of the syllable (80 vs. 296 msec), thereby following the pattern observed in adult listeners (Miller & Liberman, 1979). Thus infants' categorical discrimination of this cue, like adults', is not absolute but is relative to other contextual information. Finally, evidence of perceptual constancy for manner was observed with nasals ([m], [n], [ŋ]) vs. plosives ([b], [d], [g]) in varying vowel and speaker contexts (Hillenbrand, 1980).

The infant's discrimination of liquids has been investigated in a single study by Eimas (1975a). In this experiment infants discriminated categorically differences in the starting frequency of the third formant when it cued the contrast [ra] versus [la], but their discrimination of this same cue was more continuous when it was presented in a nonspeech context (i.e., in isolation).

The final study of infant consonant discrimination concerns the use of silence in cueing a phonetic contrast. Morse, Eilers, and Gavin (in press) studied the infant's perception of silence as a cue in the contrast "slit" versus "split." They observed that infants discriminated small amounts of silence spliced between the [s] and [l] of "slit" and their data indicated that infants did not treat these hybrid "slit plus silence" stimuli differently from "split" stimuli, which raises the interesting possibility that in this context infants (like adults) may perceive silence as [p]-like.

Turning to studies of *vowel* perception in early infancy, the pattern of discrimination results observed is both similar and dissimilar to that found for consonants. First, Trehub (1973b) has demonstrated auditory discrimination of two different vowel contrasts. Unlike the evidence of categorical discrimination generally observed for consonants, infants and adults discriminate vowels more continuously. Swoboda, Morse, and Leavitt (1976) presented infants with 240-msec tokens of the vowels [i] versus [ɪ] and observed that infants discriminated both between- *and* within-category pairs. Further, when the duration of these same stimuli was reduced to 60 msec, discrimination of this vowel contrast was more categoricallike and within-category discrimination was less evident (Swoboda, Kass, Morse, & Leavitt, 1978). This pattern of continuous versus categorical discrimination is consistent with a model of vowel perception that assumes that for relatively shorter vowels auditory short-term memory is less available, forcing the listener to rely upon phonetic short-term memory and resulting in more categoricallike performance (Fujisaki & Kawashima, 1970; Pisoni, 1973, 1975). Finally, Kuhl and her colleagues have examined perceptual constancy for vowel contrasts. Using the HAS paradigm, Kuhl and Miller (in press) found that 2-month-old infants could discriminate the vowel contrast [i] versus [a] which contained variations in intonation contour (but could not similarly discriminate differences in intonation with variations in vowel quality). In a series of HT studies (Kuhl, 1977, 1979) similar perceptual constancy for vowels was observed for the vowel contrasts [a] versus [i] and [a] versus [ɔ] with variations in intonation *and* speaker.

In summarizing the findings in infant phonetic perception reviewed above, (1) there is impressive evidence of auditory discrimination of a wide range of phonetic cues in consonant and vowel contrasts; (2) with only a few exceptions, infants exhibit the pattern of categorical discrimination of consonants and continuous discrimination of vowels that characterizes adult listeners' performance; and (3) in several instances the infant's discrimination of speech cues exhibits aspects of perceptual constancy with phonetic categories that subsume varying contexts, such as vowel, intonation, and speaker environment (Kuhl, 1980), or are sensitive to syllable duration (Eimas & Miller, 1980). Thus the young infant has been shown to possess in both quantity and quality many of the phonetic perceptual abilities of the adult listener. Although the developmental course of the acquisition of these abilities has not been extensively investigated due to procedural problems, it may be that some or perhaps most of these phonetic abilities are available at birth or soon after.

FROM PHONETICS TO WORDS

Although the evidence of the infant's auditory and phonetic perceptual abilities is impressive, it is generally limited to studies of isolated syllables. The development of the early stages of receptive language, for example, the ability to learn names for objects, requires several additional speech perceptual skills of the infant. To begin with, since words do not generally consist of single syllables, the infant must be able to recognize phonetic differences in multisyllabic tokens as well. Further, the infant must acquire perceptual knowledge of several of the organizational aspects of speech.

Knowledge of the organizational properties of speech at an intrasyllabic level of analysis was examined in a study by Eimas and Miller (1976), who concluded that 2- to 3-month-old infants can discriminate differences both in single consonant *features* and in the *rearrangement* of these features in pairs of syllables. In addition to the possibility that infants may perceive syllables as consisting of features that may be rearranged or recombined, the work of Bertoncini and Mehler (in press) suggests that infants may perceive the syllable as a basic unit of speech perception.

Beyond an appreciation of the structure of the syllable and the discrimination of contrasts in isolated syllables lie the challenges of perceiving speech in multisyllabic strings and even sentences. Several investigations have demonstrated that infants in the first few months can discriminate syllable contrasts in bisyllabic contexts. For example, Jusczyk and Thompson (1978) observed that infants discriminated [bada] versus [baga] as well as [daba] versus [gaba], and Williams (1977) found that a place contrast in medial position in vowel–consonant–vowel tokens was discriminated when a brief silent closure period preceded the consonant. Similarly, Trehub (1973a) observed that infants could discriminate voicing differences ([b] vs. [p]) in bisyllabic contexts, but not in trisyllabic contexts (e.g., [atapa] vs. [ataba]). Trehub (1976b) has suggested that the relatively shorter duration of the syllables in the trisyllabic contexts may have made the discrimination of voicing more difficult.

The infant eventually needs to be able to recognize familiar syllabic patterns in the stream of running speech. We have recently begun to study some of the factors that might play a role for the infant in this task (Goodsitt, Morse, Cowan, & VerHoeve, 1981). Using an HT paradigm, we first trained 7-month-old infants to discriminate the contrast [ba] versus [du]. Then in subsequent sessions we tested these infants on the same target syllables within three-syllable strings with 50 msec between syllables. Two factors were of interest to us: serial position in the sequence and constancy of the context syllables. In the *constant* background condition infants received each [ba] or [du] target embedded in either the initial, medial, or final position of a trisyllabic sequence that included two identical syllables (either [ti] or [ko]), e.g., [tibati], [kokoba], or [dukoko]). In the *mixed* background condition infants received each target in a trisyllabic sequence with both [ko] *and* [ti]; for example, [kobati], [tikoba], or [dukoti]. Infants did not find it easier to recognize the familiar contrast in any particular position in the sequence of syllables, but they were able to recognize the target contrast better when the background was a constant as opposed to a mixed multisyllabic context. Additional work along this line might (1) examine if infants can learn to recognize such a contrast without prior training with it, (2) explore streams of speech context longer than trisyllabic ones, and (3) investigate the recognition of *multisyllabic* targets in running speech. It is this last task that comes the closest to researching the infant's ability to recognize familiar combinations of syllables that might serve as names for objects and people. Although many studies have been done exploring the *production* of first words in infants, little is known about the early perceptual recognition of words.

Yet it is not enough that the infant be able to recognize a "word" or set of familiar syllables in a string of speech sounds uttered by some well-meaning adult. The infant must also come to associate particular sound patterns with objects, events, and even emotions. Perhaps the most important "objects" in the infant's life are his/her caregivers. Thus it is of interest to inquire about the infant's ability to discriminate different voices, to categorize voices into male and female, and to categorize individual voices as belonging to mother or father. Several studies have documented the infant's discrimination of two different voices (Kaplan, 1969; Turnure, 1971), as well as that of samples of the same voice (Culp & Boyd, 1975; Culp & Gallas, 1975). In a recent series of experiments in our own laboratory (Miller, Younger, & Morse, in press) we investigated the 7-month-old infant's categorization of male versus female voices using a modified HT paradigm. In one experiment infants were presented with a set of female voices saying "hi" on one channel of the tape recorder and a set of male "hi" tokens on the other channel. Infants readily learned to discriminate this contrast. Evidence that this discrimination involved male/female categorization was obtained in a control experiment in which infants were unable to learn to discriminate these same voices when they were randomly organized into two "categories," each containing both male and female voices. A third experiment indicated that although pitch (i.e., the perception of fundamental frequency) may serve as one cue to which infants are attending when classifying these voices, it could not account fully for this ability. Finally, a recent study of male and female voice categorization suggests that 6-month-olds exhibit more categorical discrimination of male versus female voices than 2-month-olds (C. Miller, personal communication).

Studies of the infant's differential sensitivity to the maternal voice have been carried out in older infants (Mills & Melhuish, 1974; Mehler, Bertonicini, Baurière, & Jassik-Gershenfeld, 1978; Miller, 1979), as well as most recently in newborns (DeCasper & Fifer, 1980). According to DeCasper and Fifer, a preference for the voice of the infant's mother appears to be evident within the first few days after birth as shown in a modified (preferential) sucking procedure in which changes in the infants' sucking responses resulted in presentation of their mother's voice more often than that of another female. Evidence that infants can integrate auditory and visual information in identifying objects and events has been presented by several investigators (Cohen, 1973; Lewis & Hurowitz, 1977; Spelke, 1979; Spelke & Owsley, 1979). Of particular interest is the work of Spelke and Owsley (1979) who demonstrated that the 7-month-old and possibly the 4-month-old tend to look to the correct speaking parent (mother vs. father) even when their voices were tape-recorded and not spatially or temporally coincident with the visual stimuli. These results suggest that infants at these ages have bimodal or intermodal knowledge of their parents' identity.

Finally, turning to the infant's perception of other qualities in the speech of adults, studies have revealed that infants in the first few months of life can detect differences in the stress placed on a syllable (Spring & Dale, 1977) and in the intonation pattern imposed on a syllable (Morse, 1972; Kuhl & Miller, in press). However, little if anything is known about the infant's discrimination or categorization of intonation contours over longer strings of syllables, such as words or sentences. This is true not only for intonational cues that signal sentence type (e.g., declarative statement vs. question), but it is also the case for the suprasegmental features that characterize the many emotions conveyed by the adult speaker's voice.

Thus to summarize our knowledge to date of the infant's speech perceptual abilities: (1) considerable evidence indicates that infants discriminate auditorily and categorically, as well as exhibit perceptual constancy for a number of phonetic contrasts; (2) the data suggest that the infant's knowledge of the organization of speech includes some information about the syllable and its discrimination/recognition in multisyllabic sequences, but that we still have much to learn about the infant's ability to recognize familiar syllabic patterns in running speech and his/her ability to recognize "names" of objects or people in the first year of life; (3) infants appear to have some very impressive knowledge of the voices of males versus females as well as particular individuals (especially mother), although the earliest age at which voices and faces are bimodally integrated needs further study; (4) very little is known about the infant's understanding of the *complex* patterns of intonation and other suprasegmental information employed by adult speakers in signaling sentence type, pragmatic information, and various emotions; and finally (5) virtually nothing is known regarding the relationships of these various speech perception abilities in the developing infant to language development milestones, for example, in production.

ISSUES IN INFANT SPEECH PERCEPTION

In its first decade the field of infant speech perception has not only documented a wide variety of auditory and phonetic abilities in the young infant but also has generated a number of important methodological and theoretical issues, some of which have been met by considerable disagreement (cf. Doty, 1972; Butterfield & Cairns, 1974; Morse, 1978; Eilers, Gavin, & Wilson, 1979, 1980; Aslin & Pisoni, 1980; Jusczyk, in press). In this section we address three of the major topics that have interested investigators in this field: (1) methodological issues, (2) brain–behavior relationships, and (3) the "special" nature and origins of the infant's speech perception abilities.

Methodological Issues

Several methodological issues emerged early in the research on infant speech perception. For example, some investigators felt that studies of infant speech perception should employ natural speech stimuli to ensure ecological validity, whereas other researchers argued for the use of synthetic stimuli that permitted careful control over acoustic cues. Today, most investigators would probably agree that each type of stimulus is useful, but for different types of experimental questions. Synthetic speech sounds are generally necessary when the investigator wishes to know to what acoustic cues the infant can respond, and control of these cues cannot be accomplished with natural tokens. On the other hand, natural speech stimuli may be preferable in studies in which it is difficult to synthesize the appropriate tokens, in which the researcher wishes to establish evidence of the discrimination before proceeding to determine the specific discriminative cues employed by the infant, or in which the critical manipulations address differences in paradigms, cross-language populations, ages, multisyllabic environments, and so on.

Another issue of concern is the importance of such state changes as fussiness, crying, or sleep during the experimental session. In studies of infant heart rate, state changes produce substantial effects (Berg, Berg, & Graham, 1971; Eisenberg, 1976; Rewey, 1973). In nonnutritive HAS studies, opinions differ (cf. Trehub, 1975; Morse, 1978). In contrast, state factors appear to be of less concern in the HT paradigm. This is probably due to the older age of the infant coupled with the role of the assistant in keeping an infant entertained during the session. Whatever the reasons, the subject attrition rate appears to be the lowest in this paradigm, making it an easier procedure to use in repeated testing of the same infant.

Investigators have also questioned the processes that underlie the infant's performance in a particular testing paradigm. For example, the processes underlying "acquisition" in the HAS paradigm have been explored (Butterfield & Cairns, 1974; Swoboda, Morse, & Leavitt, 1976; Trehub & Chang, 1977; Williams & Golenski,

1978; Miller & Morse, 1979). The results of this work reveal that as earlier work assumed, the contingency between the sound presentation and high-amplitude sucking is an important component of this paradigm (cf. Siqueland & DeLucia, 1969). Further, the study by Miller and Morse (1979) suggests that the role of adaptation of feature detectors in the acquisition phase of the HAS paradigm may also have an effect. Other studies investigating processes underlying the infant's test performance have examined the role of memory in discriminative responding. For example, in the HR paradigm with its long intertrial intervals between trains of stimuli (during which the listeners must remember the stimuli presented), younger infants may fail to exhibit discrimination of a speech contrast. However, when a modified HR paradigm is employed in which the intertrial intervals are eliminated, infants do demonstrate discrimination of this same contrast (Leavitt et al., 1976; Miller et al., 1977). The role of memory in the HAS paradigm has been explored in two different ways. First, the relative shift from continuous to more categoricallike discrimination of long versus short vowels (Swoboda et al., 1976; Swoboda et al., 1978) is consistent with an account of vowel discrimination that emphasizes the role of phonetic *and* auditory short-term memory (STM) for long vowels but the reduced availability of auditory STM in short vowels, resulting in more categorical-like discrimination (cf. Fujisaki & Kawashima, 1970; Pisoni, 1973, 1975). Second, Swoboda et al. (1978) also observed that discrimination of vowel contrasts in the HAS paradigm is inversely related to the length of the interval between the last familiar stimulus and the first novel syllable (the longer the interval, the poorer the discrimination).

Another area of interest to some researchers has been the question of individual differences in infant speech perception. Miller (1979) has explored this issue using HR procedures and Swoboda et al. (1976, 1978) have examined differences between normal and at-risk infants using the HAS paradigm. In addition, the HT procedure, and its many variations, holds considerable promise for investigating questions of individual differences, since it is particularly well suited for the collection of multiple trials on the same infant.

Brain–Behavior Relationships

The second topic in infant speech perception that has generated considerable interest in the past decade is the relationship of these speech perception behavioral responses to the underlying physiology of the infant's auditory system. One dominant theme in much of this interest has been the role of feature detectors in speech perception. In studies with adult listeners, Eimas and his coworkers (e.g., Eimas & Corbit, 1973; Eimas, Cooper, & Corbit, 1973; Eimas & Tartter, 1979) have been able to shift phonetic category boundaries by repeatedly presenting tokens from one end of a phonetic continuum and, in order to account for this adaptation, have suggested that the human cortex contains feature detector mechanisms that respond to phonetic or auditory aspects of the sounds of human speech. This suggestion is consistent with evidence of neurons that serve as acoustic feature detectors in the auditory systems of other species (Ades, 1976; Eimas & Tartter, 1979). The infant's ability to discriminate speech contrasts both auditorily and categorically at such an early age could similarly be based on the functioning of some types of feature detectors. Moreover, the adaptation of feature detectors through repeated stimulus presentation and recovery through a stimulus change might serve to enhance discrimination in infant paradigms. The work of Miller and Morse (1979) indicates that adaptation effects can occur at least for *adults* who receive the same short sequences of stimuli as infants in the HAS paradigm. However, although a feature detector model provides an apparently biological interpretation of the infant speech perception data, other research with the adaptation paradigm has questioned the mechanisms involved in speech adaptation studies; for example, whether the detectors operate on acoustic or phonetic features and whether they involve perceptual shifts or response biases (Ades, 1976).

Another approach to the biological/physiological bases of infant speech perception has been the exploration of hemispheric laterality effects for speech and nonspeech sounds. In most adults, speech perception is lateralized in the left hemisphere, whereas tonal or melodic perception is lateralized in the right hemisphere. These asymmetries may be seen in behavioral tests of dichotic listening, involving the simultaneous presentation of different stimuli to the two ears, which yield superior perception of speech sounds in the right ear (right-ear advantage, or REA) and better performance with musical timbre in the left ear (left-ear advantage, or LEA). In addition, electrophysiological recordings from the surface of the scalp also exhibit differential activity over the left hemisphere (LH) for many speech sounds and over the right hemisphere (RH) for different musical stimuli.

Studies with place of articulation contrasts using behavioral techniques have revealed that infants discriminated these speech contrasts better in the right ear (REA) and nonspeech (music) better in the left ear (LEA). Entus (1977) observed this result using the HAS paradigm with infants 3 to 20 weeks of age, although a study by Vargha-Khadem and Corballis (1977) failed to replicate this finding. Nevertheless, Best and her coworkers, using an HR paradigm, did observe an REA for place discrimination in 3- and 4-month-olds and an LEA for music in infants 2, 3, and 4 months of age (Glanville, Best, & Levinson, 1977; Best, Hoffman, & Glanville, 1979).

The first few studies to investigate the infant's pattern of differential electrical activity over the two hemispheres for speech and nonspeech stimuli using averaged evoked potential (AEP) techniques found that newborns, 5-month-old infants, 1-year-old children, and adults responded to speech sounds and words with a greater AEP amplitude over the LH and responded to nonspeech sounds with a greater AEP amplitude over the RH (Molfese, 1972, 1977; Molfese, Freeman, & Palermo, 1975). More recent AEP studies of place of articulation contrasts by Molfese and his coworkers have revealed that a component of the AEP of adults (Molfese, 1978a), infants as young as 30 hours of age (Molfese & Molfese, 1979a), and even preterm infants (Molfese & Molfese, in press) characteristic only of the LH was found to vary systematically as a function of changes in formant transition. Further, Molfese (1978a, 1980a) has also observed evidence of a component of the adult and infant AEP that reveals the ability of only the LH to differentiate the consonants [b] and [g] independent of vowel context, suggesting the presence of perceptual constancy. Finally, some of Molfese's results have indicated that these effects may depend on the speech (normal formant bandwidth) rather than nonspeech (sinewave bandwidth) structure of the stimuli (see Molfese, in press, for a discussion of these factors). In contrast to the AEP studies of place of articulation perception, voicing contrasts yield an RH component of the AEP that relates to categorical differences in VOT. This result has been observed in adults (Molfese, 1978a), 4-year-old children (Molfese & Hess, 1978), and 2-month-old infants (Molfese & Molfese, 1979b), but not in newborns (Molfese & Molfese, 1979b). Finally, a similar RH component related to nonspeech differences in the relative time of onset of two tones (tone-onset-time) has been observed in adults and children (Molfese, 1980b, Molfese, Erwin, & Deen, 1980). In sum, these infant studies of hemispheric laterality indicate that infants exhibit patterns of hemispheric differences in processing speech and nonspeech stimuli generally similar to adult listeners. Furthermore, hemispheric responses to speech contrasts may depend on the particular acoustic–phonetic features (e.g., place of articulation vs. voicing).

The "Special" Nature of Infant Speech Perception

The third and perhaps most provocative (as well as anthropocentric) question that has captured the interest of researchers in infant speech perception is whether or not there exist perceptual abilities in humans that are specific to speech in addition to those of general auditory perceptual organization. In the research on hemispheric laterality this issue has surfaced in perceptual comparisons of (1) normal formant bandwidth versus otherwise identical nonspeech sinewave bandwidth stimuli, (2) the perception of tone-onset-time (TOT) and VOT contrasts, and (3) speech versus music perception. Experiments have also been carried out using a variety of other nonspeech stimuli with some speechlike properties. For example, Jusczyk, Rosner, Cutting, Foard, and Smith (1977), examining the discrimination of "pluck" versus "bow" sinewave stimuli differing in rise time, observed that infants discriminated these sounds categorically, as adults had been shown to do. More recently, Jusczyk, Pisoni, Walley, and Murray (1980) tested infants on their categorical discrimination of a TOT continuum previously employed with adult listeners. However, the infant's pattern of discrimination in this study did *not* mirror the categorical boundaries of adults. Finally, Eimas (1974a; 1975b) has reported that infants respond differently to formant transitional cues in a nonspeech context compared to a speech context, a pattern similar to that observed in adult listeners. In general, the studies of nonspeech stimuli reveal that infants treat these stimuli in a manner similar to adults (one exception being Jusczyk et al., 1980).

The question of the special nature of speech perception is also related to the innateness of these abilities. If infants can be shown to discriminate speech contrasts (especially categorically) without previous exposure to these contrasts, this would suggest that those abilities are innate and perhaps even limited to humans. We know that many of the speech perceptual abilities of the infant antedate the infant's differential production of these same speech contrasts. This has led several investigators to question the role of early listening experience in the infant's perception of speech sound differences by presenting nonnative language contrasts to infants for discriminative testing. Some of these results are consistent with a model of early experience that posits that infants may begin with the ability to perceive (or very quickly learn to perceive) contrasts that are *not* native to the language environment in which they live. Furthermore, with continued exposure to a limited language environment, unused discriminative abilities may disappear or decline. For example, Trehub (1976a, 1978) reported that infants, but not English-speaking adults, could discriminate the foreign Czech contrast, [ža] versus [řa]. Similarly, Eimas (1975) showed that infants, like English-speaking adults but unlike Japanese adults (Miyakawa, Strange, Verbrugge, Liberman, Jenkins, & Fujimura, 1975), can discriminate categorically the English contrast [ra] versus [la]. Finally, 7-month-old, English-learning infants have been shown capable of discriminating two Hindi contrasts that Hindi adults could and English-speaking adults could not discriminate (Werker, Gilbert, Humphrey, & Tees, 1981).

A somewhat different picture has emerged from cross-language studies of voicing (VOT). Adult studies of voicing (Lisker & Abramson, 1964; Abramson & Lisker, 1970) have revealed that most languages divide the VOT continuum into two or three categories, with one boundary for bilabial stops falling in the +30-msec VOT region (English /b/ vs. /p/) and one in the prevoiced region (approximately −20-msec VOT: Spanish /b/ vs. /p/). Studies of infant VOT discrimination have consistently revealed that infants discriminate contrasts across the English, +30-msec VOT boundary, even if such a contrast is not in their native language environment. This result was obtained for Kikuyu-learning infants in Kenya by Streeter (1976), for Guatemalan Spanish-learning infants by Lasky, Syrdal-Lasky, and Klein (1975), and for infants learning Cuban Spanish (Eilers, Gavin, & Wilson, 1979). In contrast, discrimination across the prevoiced boundary has not been easy to document in infants for whom this is a foreign contrast, whereas infants who hear it in their native language discriminate this contrast well. Evidence of discrimination in this prevoiced region in English-learning infants was observed in very few subjects or for very large differences in VOT (Eimas, 1975b; Eilers, Wilson, & Moore, 1979). In contrast, discrimination across the prevoiced boundary did occur in infants exposed to this contrast through their native language (e.g., Streeter, 1976). The composite pattern of discrimination for these two boundaries is illustrated in a study by Eilers, Gavin, and Wilson (1979) that compared English-learning and Spanish-learning infants' discrimination across the English and Spanish boundaries. Both groups discriminated the English contrast, but only the Spanish-learning infants performed well on the Spanish contrast. These results suggest that discrimination of this *prevoiced* VOT contrast appears to benefit from linguistic experience in the first few months after birth. In contrast, discrimination of the English contrast does not appear to require such experience. However, recent evidence from a study of Eilers, Morse, Gavin, and Oller (1981) suggests that the relative salience of the English boundary for infants may be due to multiple acoustic cues as suggested by Stevens and Klatt (1974). When infants were tested with the English contrast [du] versus [tu] cued *solely* by onset of voicing, discrimination was conspicuously absent.

Thus, the cross-linguistic results indicate that some contrasts are generally salient for all infants and/or that language-specific experience is *not* necessary for the infant to be able to discriminate these contrasts. In some cases, adults may lose this discriminative ability due to the lack of exposure to these nonnative contrasts, although there is some debate regarding the comparison of infant and adult data and testing paradigms. The discrimination of other speech contrasts does appear to benefit from early language experience. Clearly, only a handful of cross-linguistic contrasts and studies have been carried out to date, but already the results indicate that a model of the role of early experience in the development of infant speech perception will need to consider the degree of salience of different acoustic cues and permit language-specific experience to be necessary for some cues and unnecessary for others in order for infants to demonstrate discriminative behavior (Aslin & Pisoni, 1980; Eilers, Gavin, & Oller, 1980). Finally, a recent study by Eilers and her coworkers (Eilers, Gavin, & Oller, in press) further indicates that Cuban Spanish-learning infants who live in the bilingual Miami community do *better* than English-learning infants (who hear little Spanish) at discriminating a contrast (Czech [ža] vs. [řa]) that is foreign to both English- and Spanish-learning infants. This suggests that a *varied* language experience may facilitate the discrimination of contrasts in languages to which the infant is not exposed.

The third approach to the special nature of speech perception in the infant is to investigate similar abilities in nonhuman primates and other species. The results of these studies have generally suggested that many of the auditory and categorical discriminative abilities of the infant are also shared by other species as well. For example, Morse and Snowdon (1975) demonstrated that rhesus monkeys also exhibit between-category discrimination for place of articulation differences that is superior to within-category performance (see also Sinnott, Beecher, Moody, & Stebbins, 1976; Waters & Wilson, 1976; Kuhl & Padden, under review b). Research with chinchillas has further demonstrated that categorical discrimination of VOT may be found in these nonprimate mammals (Kuhl & Miller, 1975, 1978; Kuhl, 1976, 1979), and similar results have recently been reported in the monkey (Kuhl & Padden, under review a). More recently, Molfese, Morse, Taylor, and Erwin (1981) have demonstrated that infant rhesus monkeys exhibit a component of the AEP that reflects perceptual constancy for /b/ versus /g/ (independent of vowel context). Taken together, these animal findings indicate that many aspects of what we have examined earlier under the subheading of "Phonetic Perception" may turn out to be accounted for by the psychoacoustic-processing abilities of the mammalian auditory system (Kuhl, 1979).

The prospect that much of what we have studied to date in infant speech perception is *not* unique to our species prompts us to ponder what aspects of speech perception *are* likely to be special to humans and to human infants. Several interesting possibilities lie before us: first, much of what is referred to as phonetic categorization in adult and infant humans, as well as in nonhumans, may indeed reflect invariant properties of speech sounds, albeit complex ones (e.g., Stevens & Blumstein, 1978). Perhaps in these more complex categories, we shall find that the infant differs from nonhuman species, although this level of speech perceptual ability may require considerable linguistic experience. One example, mentioned earlier in this chapter, is the category /p/ with acoustic cues that include silence ([slɪt] vs. [splɪt]), voice-onset-time ([bɪt] vs. [pɪt]), and vowel duration ([mab] vs. [map]). The usefulness of this avenue of research will probably depend upon a finer understanding and organization of the more complex levels of phonetic categorization and the acoustic invariants that may underlie them. A second (although less likely) possibility is that there are no qualitative differences in the types of complexity of phonetic categories available to the infant versus the animal, and that speech categorization abilities are characteristic of the mammalian or primate auditory subcortical or cortical systems. Third, and perhaps most intriguing, is the possibility that infants differ from other animals in the ability to organize temporally or cross modally their auditory–linguistic knowledge. This may become evident in their recognition of complex combinations of phonemes in running speech (eventually in the forms of words and sentences). Alternatively, infants may be especially good at learning to associate auditory patterns or names with familiar objects, a skill that has been difficult to demonstrate beyond simple commands in nonhuman species (Gilman, 1921; Warden & Warner, 1928; Geschwind, personal communication) and may require species-specific cortical structures for language and/or cognition.

REFERENCES

Abbs, J., & Minifie, F. The effects of acoustic cues in fricatives on perceptual confusions in preschool children. *Journal of the Acoustical Society of America,* 1969, **45**, 1535–1542.

Abramson, A., & Lisker, L. Discriminability along the voicing continuum: Cross-language tests. In *Proceedings of the Sixth International Congress of Phonetic Sciences, Prague, 1967.* Prague: Academia, 1970, pp. 569–573.

Ades, A. E. Adapting the feature detectors for speech perception. In E. C. T. Walker & R. J. Wales (Eds.), *New approaches to language mechanisms.* The Hague: North Holland, 1976.

Aslin, R., Hennessy, B., Pisoni, D., & Perry, A. Individual infants' discrimination of voice onset time: Evidence for three modes of voicing. Paper presented at the biennial meeting of the Society for Research in Child Development, San Francisco, Calif., 1979.

Aslin, R., & Pisoni, D. Effects of early linguistic experience on speech discrimination by infants: A critique of Eilers, Gavin, and Wilson, 1979. *Child Development,* 1980, **51**, 107–112.

Barnet, A. B., Olrich, E. S., Weiss, I. P., & Shanks, B. Auditory evoked potentials during sleep in normal children from ten days to three years of age. *Electroencephalography and Clinical Neurophysiology,* 1975, **39**, 29–41.

Bench, R. J. Sound transmission to the human foetus through the maternal abdominal wall. *J. Genet. Psychol.,* 1968, **113**, 85–87.

Bench, J. Some effects of audio-frequency stimulation on the crying baby. *Journal of Aud. Research,* 1969, **9**, 112–128.

Bench, J., & Mentz, L. Stimulus complexity, state, and infants' auditory behavioral responses. *British Journal of Disorders of Communication,* 1975, **10**, 52–60.

Berg, K., Berg, W., & Graham, F. Infant heart rate response as a function of stimulus state. *Psychophysiology,* 1971, **8**, 30–44.

Berg, W. K., & Berg, K. M. Psychophysiological development in infancy: State, sensory function and attention. In J. Osofsky (Ed.), *Handbook of infant development.* New York: Wiley, 1980.

Bertoncini, J., & Mehler, J. Syllables as units in infant speech perception. *Infant behavior and development,* in press.

Best, C., Hoffman, H., & Glanville, B. Brain lateralization in 2-, 3-, and 4-month-olds for phonetic and musical timbre discriminations under memory load. *Haskins Laboratories Status Report on Speech Research,* 1979, **SR-59/60**, 1–30.

Birns, B., Blank, M., Bridger, W. H., & Escalona, S. K. Behavioral inhibition in neonates produced by auditory stimuli. *Child Development,* 1965, **36**, 639–645.

Bradford, L. J. Respiratory audiometry. In L. J. Bradford (Ed.), *Physiological measures of the audio-vestibular system.* New York: Academic, 1975.

Bronson, G. The postnatal growth of visual capacity. *Child Development,* 1974, **45**, 873–890.

Butterfield, E., & Cairns, G. Discussion summary—Infant reception research. In R. Schiefelbusch & L. Lloyd (Eds.), *Language perspectives—Acquisition, retardation, and intervention.* Baltimore: University Park Press, 1974, pp. 75–102.

Carden, G., Levitt, A., & Jusczyk, P. Manner judgments affect the locus of the labial/dental boundary for stops and fricatives. Paper presented at the 97th meeting of the Acoustical Society of America, Cambridge, Mass., June 1979.

Cohen, S. Infant attentional behavior to face–voice incongruity. Paper presented at the biennial meeting of the Society for Research in Child Development, Philadelphia, Pa., 1973.

Cowan, N. The processing of brief vowels in infancy. *Univ. of Wisconsin Infant Development Laboratory Research Status Report,* 1978, **2**, 217–274.

Cowan, N., Suomi, K., & Morse, P. Echoic storage in infant perception. To be presented to the Society for Research in Child Development, 1981.

Culp, R., & Boyd, E. Visual fixation and the effect of voice quality and context differences in 2-month-old infants. In F. Horowitz (Ed.), Visual attention, auditory stimulation, and language discrimination in young infants. *Monographs of the Society for Research in Child Development,* 1975, **39**, 78–91.

Culp, R., Gallas, H. Discrimination of male voice quality by 8- and 9-week-old infants. Paper presented at the biennial meeting of the Society for Research in Child Development, Denver, Colo., 1975.

DeCasper, A., & Fifer, W. Of human bonding: Newborns prefer their mother's voices. *Science,* 1980, **208**, 1174–1176.

Dorman, M., Raphael, L., & Liberman, A. Some experiments on the sound of silence in phonetic perception. *Journal of the Acoustical Society of America,* 1979, **65**, 1518–1532.

Doty, D. Infant speech perception: Report of a conference held at the University of Minnesota, June 20–22, 1972. *Human Development,* 1974, **17**, 74–80.

Eilers, R. Context-sensitive perception of naturally produced stop and fricative consonants by infants. *Journal of the Acoustical Society of America,* 1977, **61**, 1321–1336.

Eilers, R. Infant speech perception: History and mystery. In G. Yeni-Komshian, J. Kavanagh, & C. Ferguson (Eds.), *Child phonology: Perception and production.* New York: Academic Press, 1980.

Eilers, R., Gavin, W., & Oller, D. K. Cross-linguistic perception in infancy: Early effects of linguistic experience. *Journal of Child Language,* in press.

Eilers, R., Gavin, W., & Wilson, W. Linguistic experience and phonemic perception in infancy: A cross-linguistic study. *Child Development,* 1979, **50**, 14–18.

Eilers, R., Gavin, W., & Wilson, W. Effects of early linguistic experience on speech discrimination by infants: A reply. *Child Development,* 1980, **51**, 113–117.

Eilers, R., & Minifie, F. Fricative discrimination in early infancy. *Journal of Speech and Hearing Research,* 1975, **18**, 158–167.

Eilers, R., Morse, P., Gavin, W., & Oller, D. K. Discrimination of voice onset time in infancy. *Journal of the Acoustical Society of America,* 1981, **70**, 955–965.

Eilers, R., & Oller, D. K. The role of speech discrimination in developmental sound substitutions. *Journal of Child Language,* 1976, **3**, 319–329.

Eilers, R., Wilson, W., & Moore, J. Developmental changes in speech discrimination in infants. *Journal of Speech and Hearing Research,* 1977, **20**, 766–780.

Eilers, R., Wilson, W., & Moore, J. Speech discrimination in the language-innocent and language-wise: A study in the perception of voice-onset-time. *Journal of Child Language,* 1979, **6**, 1–18.

Eimas, P. Auditory and linguistic processing of cues for place of articulation by infants. *Perception and Psychophysics,* 1974, **16**, 513–521. (a)

Eimas, P. Auditory and phonetic coding of the cues for speech: Discrimination of the [r–l] distinction by young infants. *Perception and Psychophysics,* 1975, **18**, 341–347. (a)

Eimas, P. Speech perception in infancy. In L. Cohen & P. Salapatek (Eds.), *Infant perception: From sensation to cognition* (Vol. II). New York: Academic Press, 1975, pp. 193–231. (b)

Eimas, P., Cooper, W., & Corbit, J. Some properties of linguistic feature detectors. *Perception and Psychophysics,* 1973, **13**, 247–252.

Eimas, P., & Corbit, J. Selective adaptation of linguistic feature detectors. *Cognitive Psychology,* 1973, **4,** 99–109.

Eimas, P., & Miller, J. Contextual effects in infant speech perception. *Science,* 1980, **209,** 1140–1141.

Eimas, P., & Miller, J. Discrimination of the information for manner of articulation by young infants. *Infant Behavior and Development,* in press.

Eimas, P., Siqueland, E., Jusczyk, P., & Vigorito, J. Speech perception in infants. *Science,* 1971, **171,** 303–306.

Eimas, P. D., & Tartter, V. C. On the development of speech perception: Mechanisms and analogies. In H. W. Reese & L. P. Lipsitt (Eds.), *Advances in child development and behavior* (Vol. 13). New York: Academic, 1979, pp. 155–193.

Eisenberg, R. *Auditory competence in early life: The roots of communicative behavior.* Baltimore: University Park Press, 1976.

Elliott, G. B., & Elliott, K. A. Some pathological, radiological and clinical implications of the precocious development of the human ear. *Laryngoscope,* 1964, **74,** 1160–1171.

Entus, A. Hemispheric asymmetry in processing of dichotically presented speech and nonspeech stimuli by infants. In S. Segalowitz & F. Gruber (Eds.), *Language development and neurological theory.* New York: Academic, 1977, pp. 63–73.

Fodor, J., Garrett, M., Brill, S. Pi ka pu: The perception of speech sounds by prelinguistic infants. *Perception and Psychophysics,* 1975, **18,** 74–78.

Fujisaki, H., & Kawashima, T. Some experiments on speech perception and a model for the perceptual mechanism. *Annual Report of the Engineering Research Institute, Faculty of Engineering, University of Tokyo, Tokyo,* 1970, **29,** 207–214.

Fulton, R., Gorzycki, P., & Hull, W. Hearing assessment with young children. *Journal of Speech and Hearing Disorders,* 1975, **40,** 397–404.

Galambos, R., & Hecox, K. E. Clinical applications of the auditory brain stem response. *The Otolaryngologic Clinics of North America,* 1978, **11,** 709–722.

Gilman, E. A dog's diary. *Journal of Comparative Psychology,* 1921, **1,** 309–315.

Glanville, B., Best, C., & Levenson, R. A cardiac measure of cerebral asymmetries in infant auditory perception. *Developmental Psychology,* 1977, **13,** 54–59.

Goodsitt, J., Morse, P., Cowan, N., & VerHoeve, J. Infant speech perception in a multisyllabic world. Paper presented at the biennial Meetings of the Society for Research in Child Development, Boston, Mass., 1981.

Gormezano, I., & Moore, J. W. Classical conditioning. In M. H. Marx (Ed.), *Learning: Processes.* New York: MacMillan, 1969.

Graham, F. K., Strock, B. D., & Zeigler, B. L. Excitatory and inhibitory influences on reflex responsiveness. In W. A. Collins (Ed.), *Minnesota symposium on child psychology* (Vol. 14). Hillsdale, N.J.: Erlbaum, 1980.

Hecox, K. Electrophysiological correlates of human auditory development. In L. B. Cohen & P. Salapatek (Eds.), *Infant perception: From sensation to cognition* (Vol. 2). New York: Academic, 1975.

Hecox, K., & Galambos, R. Brainstem auditory evoked responses in human infants. *Arch. Otolaryngol.,* 1974, **99,** 30–33.

Hillenbrand, J. Perceptual organization of speech sounds by young infants. Unpublished Ph.D. dissertation, University of Washington, Seattle, Wash., 1980.

Hillenbrand, J., Minifie, F., & Edwards, T. Tempo of frequency change as a cue in speech sound discrimination by infants. Paper presented at the biennial meeting of the Society for Research in Child Development, New Orleans, La., 1977.

Hirsh, I. J. *The measurement of hearing.* New York: McGraw-Hill, 1952.

Holmberg, T., Morgan, K., & Kuhl, P. Speech perception in early infancy: Discrimination of fricative contrasts. *Journal of Acoustical Society of America,* 1977, **62** (Suppl. 1), S99(A).

Jewett, D. L., Romano, M. N., & Williston, J. S. Human auditory evoked potentials: Possible brainstem components detected on the scalp. *Science,* 1970, **167,** 1517–1518.

Johansson, B., Wedenberg, E., & Westin, B. Measurement of tone response by the human fetus. *Acta Otolaryngol.* 1964, **57**, 188–192.

Jusczyk, P. Perception of syllable-final stop consonants by 2-month-old infants. *Perception and Psychophysics,* 1977, **21**, 450–454.

Jusczyk, P. Infant speech perception: A critical appraisal. In P. Eimas & J. Miller (Eds.), *Perspectives on the study of speech.* Hillsdale, N.J.: Erlbaum, in press.

Jusczyk, P., Murray, J., & Bayley, J. Perception of place of articulation in fricatives and stops by infants. Paper presented at the Biennial Meeting of the Society for Research in Child Development, San Francisco, Calif., 1979.

Jusczyk, P., Pisoni, D., Walley, A., & Murray, J. Discrimination of relative onset-time of two-component tones by infants. *Journal of the Acoustical Society of America,* 1980, **67**, 262–270.

Jusczyk, P., Rosner, B., Cutting, J., Foard, C., & Smith, L. Categorial perception of nonspeech sounds in 2-month-old infants. *Perception and Psychophysics,* 1977, **21**, 50–54.

Jusczyk, P., & Thompson, E. Perception of a phonetic contrast in multisyllabic utterances by two-month-old infants. *Perception and Psychophysics,* 1978, **23**, 105–109.

Kaplan, E. The role of intonation in the acquisition of language. Unpublished Ph.D. dissertation, Cornell University, Ithaca, N.Y., 1969.

Karmel, B. Z., & Maisel, E. B. A neuronal activity model for infant visual attention. In L. B. Cohen & P. Salapatek (Eds.), *Infant perception: From sensation to cognition* (Vol. I). New York: Academic, 1975.

Kuhl, P. Speech perception by the chinchilla: Categorical perception of synthetic alveolar plosive consonants. *Journal of the Acoustical Society of America,* 1976, **60** (Suppl. 1), S81(A).

Kuhl, P. Speech perception in early infancy: Perceptual constancy for the vowel categories /a/ and /ɔ/. *Journal of the Acoustical Society of America,* 1977, **61** (Suppl. 1), S39(A).

Kuhl, P. Models and mechanisms in speech perception. *Brain, Behavior, and Evolution,* 1979, **16**, 374–408. (a)

Kuhl, P. Speech perception in early infancy: Perceptual constancy for spectrally dissimilar vowel categories. *Journal of the Acoustical Society of America,* 1979, **66**, 1668–1679. (b)

Kuhl, P. Perceptual constancy for speech sound categories. In G. Yeni-Komshian, J. Kavanagh, & C. Ferguson (Eds.), *Child phonology: Perception and production.* New York: Academic, 1980.

Kuhl, P., & Miller, J. Speech perception by the chinchilla: Voiced–voiceless distinction in alveolar plosive consonants. *Science,* 1975, **190**, 69–72.

Kuhl, P., & Miller, J. Speech perception by the chinchilla: Identification functions for synthetic VOT stimuli. *Journal of the Acoustical Society of America,* 1978, **63**, 905–917.

Kuhl, P., & Miller, J. Discrimination of auditory target dimensions in the presence or absence of orthogonal variation in a second dimension: Implications for developmental speech perception and development of attention and memory. *Perception and Psychophysics,* in press.

Kuhl, P., & Padden, D. Speech discrimination by macaques: Auditory constraints on the evolution of language. Under review. (a)

Kuhl, P., & Padden, D. Speech perception by macaques: Enhanced discrimination at the phonetic boundaries between speech–sound categories. Under review. (b)

Lasky, R., Syrdal-Lasky, A., & Klein, R. VOT discrimination by four and six and a half month old infants from Spanish environments. *Journal of Experimental Child Psychology,* 1975, **20**, 215–225.

Leavitt, L., Brown, J., Morse, P., & Graham, F. Cardiac orienting and auditory discrimination in 6-week infants. *Developmental Psychology,* 1976, **12**, 514–523.

Lenard, H. G., Bernuth, M., & Hutt, S. J. Acoustic evoked responses in newborn infants. The influence of pitch and complexity of the stimulus. *Electroencephalography and Clinical Neurophysiology,* 1969, **27**, 121–127.

Lewis, M., & Hurowitz, L. Intermodal person schema in infancy: Perception within a common auditory–visual space. Paper presented at the meeting of the Eastern Psychological Association, Boston, Mass., 1977.

Liberman, A. The grammars of speech and language. *Cognitive Psychology,* 1970, **1**, 301–323.

Liberman, A., Cooper, F., Shankweiler, D., & Studdert-Kennedy, M. Perception of the speech code. *Psychological Review,* 1967, **74**, 431–461.

Lisker, L., & Abramson, A. A cross-language study of voicing in initial stops: Acoustical measurements. *Word,* 1964, **20**, 384–422.

Little, A. M. Eyelid conditioning in the human infant as a function of the ISI. Paper presented at the biennial meeting of the Society for Research in Child Development. Minneapolis, 1971.

Marquis, D. Can conditioned response be established in the newborn infant? *Journal of Genetic Psychology,* 1931, **39**, 479–492.

Massaro, D. W. *Experimental psychology and information processing.* Chicago: Rand McNally, 175.

Mehler, J., Bertoncini, J., Barrière, M., & Jassik-Gershenfeld, D. Infant recognition of mother's voice. *Perception,* 1978, **7**, 491–497.

Miller, C. Individual differences in infant speech perception: A method of assessment. In H. D. Kimmel, E. H. van Olst, & J. F. Orlbeke (Eds.), *The orienting reflex in humans.* New York: Erlbaum, 1978, pp. 625–632.

Miller, C. Voice recognition and categorization in infants. *Univ. of Wisconsin Infant Development Laboratory Research Status Report,* 1979, **3**, 95–178.

Miller, C., & Morse, P. The "heart" of categorical speech discrimination in young infants. *Journal of Speech and Hearing Research,* 1976, **19**, 578–589.

Miller, C., & Morse, P. Selective adaptation effects in infant speech perception paradigms. *Journal of Acoustical Society of America,* 1979, **65**, 789–798.

Miller, C., Morse, P., & Dorman, M. Cardiac indices of infant speech perception: Orienting and burst discrimination. *Quarterly Journal of Experimental Psychology,* 1977, **29**, 533–545.

Miller, C., Younger, B., & Morse, P. The categorization of male and female voices in infancy. *Infant Behavior and Development,* in press.

Miller, J. Phonetic determination of infant speech perception. Unpublished Ph.D. dissertation. University of Minnesota, Minneapolis, Minn., 1974.

Miller, J., & Eimas, P., Organization in infant speech perception. *Canadian Journal of Psychology,* 1979, **33**, 353–367.

Miller, J., & Liberman, A. Some effects of later-occurring information on the perception of stop consonant and semivowel. *Perception and Psychophysics,* 1979, **25**, 457–465.

Mills, M., & Melhuish, E. Recognition of mother's voice in early infancy. *Nature,* 1974, **252**, 123–124.

Miyawaki, K., Strange, W., Verbrugge, R., Liberman, A., Jenkins, J., & Fijimura, O. An effect of linguistic experience: The discrimination of /r/ and /l/ by native speakers of Japanese and English. *Perception and Psychophysics,* 1975, **18**, 331–340.

Moffitt, A. Consonant cue perception by twenty- to twenty-four-week-old infants. *Child Development,* 1971, **42**, 717–731.

Molfese, D. Cerebral asymmetry in infants, children and adults: Auditory evoked responses to speech and music stimuli. Unpublished Ph.D. dissertation, Pennsylvania State University, University Park, 1972.

Molfese, D. Infant cerebral asymmetry. In S. Segalowicz & F. Gruber (Eds.), *Language Development and Neurological Theory.* New York: Academic, 1977.

Molfese, D. Electrophysiological correlates of categorical speech perception in adults. *Brain and Language,* 1978, **5**, 25–35. (a)

Molfese, D. Left and right hemispheric involvement in speech perception: Electrophysiological correlates. *Perception and Psychophysics,* 1978, **23**, 237–243. (b)

Molfese, D. The phoneme and the engram: Electrophysiological evidence for the acoustic invariant in stop consonants. *Brain and Language,* 1980, **9,** 372–376. (a)

Molfese, D. Hemispheric specialization for temporal information: implications for the processing of voicing cues during speech perception. *Brain and Language,* 1980, **11,** 285–299. (b)

Molfese, D. Neural mechanisms underlying the processing of speech information in infants and adults: Suggestions of differences in development and structure from electrophysiological research. In U. Kirk (Ed.), *Neuropsychology of language, reading, and spelling.* New York: Academic, in press.

Molfese, D. Erwin, R., & Deen, M. Hemispheric discrimination of tone onset times by preschool children. Paper presented at the annual meeting of the Psychonomic Society, St. Louis, Mo., 1980.

Molfese, D., Freeman, R., Jr., & Palermo, D. The ontogeny of lateralization for speech and nonspeech stimuli. *Language and Brain,* 1975, **2,** 356–368.

Molfese, D., & Hess, R. Speech perception in nursery school age children: Sex and hemispheric differences. *Journal of Experimental Child Psychology,* 1978, **26,** 71–84.

Molfese, D., & Molfese, V. Hemisphere and stimulus differences as reflected in the cortical responses of newborn infants to speech stimuli. *Developmental Psychology,* 1979, **15,** 505–511. (a)

Molfese, D., & Molfese, V. Infant speech perception: Learned or innate. In H. A. Whitaker & H. Whitaker (Eds.), *Advances in neurolinguistics* (Vol. 4). New York: Academic, 1979. (b)

Molfese, D., & Molfese, V. Cortical responses of preterm infants to phonetic and nonphonetic speech stimuli. *Developmental Psychology,* in press.

Molfese, D., Morse, P., Taylor, J., & Erwin, R. Cortical responses of infant rhesus monkeys to consonant invariant cues: Evidence for an innate basis for language development. Paper presented at the International Neuropsychological Society, New Orleans, La., 1981.

Moore, J. M., Thompson, G., & Thompson, M. Auditory localization of infants as a function of reinforcement conditions. *Journal of Speech and Hearing Disorders,* 1975, **40,** 29–34.

Morse, P. The discrimination of speech and nonspeech stimuli in early infancy. *Journal of Experimental Child Psychology,* 1972, **14,** 477–492.

Morse, P. Infant speech perception: Origins, processes, and *Alpha Centauri.* In F. Minifie & L. Lloyd (Eds.), *Communicative and cognitive abilities—Early behavioral assessment.* Baltimore: University Park Press, 1978, pp. 195–227.

Morse, P. The infancy of infant speech perception: The first decade of research. *Brain, Behavior, and Evolution,* 1979, **16,** 331–373.

Morse, P., Eilers, R., & Gavin, W. The perception of the sound of silence in early infancy. *Child Development,* in press.

Morse, P., Leavitt, L., Donovan, W., Kolton, S., Miller, C., & Judd-Engel, N. Headturning to speech: Explorations beyond the heart and the pacifier. *Univ. of Wisconsin Infant Development Laboratory Research Status Report,* 1975, **1,** 347–351.

Morse, P., & Snowdon, C. An investigation of categorical speech discrimination by rhesus monkeys. *Perception and Psychophysics,* 1975, **17,** 9–16.

Morse, P., & Suomi, K. Probing perceptual constancy for consonants with the nonnutritive sucking paradigm. *Univ. of Wisconsin Infant Development Laboratory Research Status Report,* 1979, **3,** 311–319.

Murphy, K. P., & Smyth, C. N. Response of foetus to auditory stimulation. *The Lancet,* 1962, **1,** 972–973.

Ohlrich, E. S., Barnet, A. B., Weiss, I. P., & Shanks, B. L. Auditory evoked potential development in early childhood: A longitudinal study. *Electroencephalography and Clinical Neurophysiology,* 1978, **44,** 411–423.

Picton, T. W., Hillyard, S. A., Krang, H. I., & Galambos, R. Human auditory evoked potentials. I. Evaluation of components. *Electroenceph. Clin. Neurophysiol.,* 1974, **36,** 179–190.

Pisoni, D. Auditory and phonetic memory codes in the discrimination of consonants and vowels. *Perception and Psychophysics,* 1973, **13,** 253–260.

Pisoni, D. Auditory and phonetic short-term memory and vowel perception. *Memory and Cognition,* 1975, **3,** 7–18.

Raphael, L. Preceding vowel duration as a cue to the voicing characteristics of word-final consonants in English. *Journal of the Acoustical Society of America,* 1972, **51,** 1296–1303.

Rewey, H. Developmental change in infant heart rate response during sleeping and waking states. *Developmental Psychology,* 1973, **8,** 35–41.

Rorke, L. B., & Riggs, H. E. *Myelination of the brain in the newborn.* Philadelphia: J. B. Lippincott, 1969.

Salamy, A., & McKean, C. M. Postnatal development of human brainstem potentials during the first year of life. *Electroencephalography and Clinical Neurophysiology,* 1976, **40,** 418–426.

Salamy, A., McKean, C. M., & Buda, F. Maturational changes in auditory transmission as reflected in human brainstem potentials. *Brain Research,* 1975, **96,** 361–366.

Schraf, B. Critical bands. In J. V. Tobias (Ed.), *Foundations of modern auditory theory* (Vol. 1). New York: Academic, 1970.

Schneider, B., Trehub, S., & Bull, D. The development of basic auditory processes in infants. *Canadian Journal of Psychology,* 1979, **33,** 306–319.

Schneider, B., Trehub, S. E., & Bull, D. High-frequency sensitivity in infants. *Science,* 1980, **207,** 1003–1004.

Schulman-Galambos, C., & Galambos, R. Brain-stem evoked response audiometry in newborn hearing screening. *Archives of Otolaryngology,* 1979, **105,** 86–90.

Sinnott, J., Beecher, M., Moody, D., & Stebbins, W. Speech sound discrimination by humans and monkeys. *Journal of the Acoustical Society of America,* 1976, **60,** 687–695.

Siqueland, E., & DeLucia, C. Visual reinforcement of non-nutritive sucking in human infants. *Science,* 1969, **165,** 1144–1146.

Spelke, E. Perceiving bimodally specified events in infancy. *Developmental Psychology,* 1979, **15,** 626–636.

Spelke, E., & Owsley, C. Intermodal exploration and knowledge in infancy. *Infant Behavior and Development,* 1979, **2,** 13–27.

Spock, B. *Baby and Child Care* (4th ed.). New York: Pocket Books, 1968.

Spring, D., & Dale, P. The discrimination of linguistic stress in early infancy. *Journal of Speech and Hearing Research,* 1977, **20,** 224–231.

Stevens, K., & Blumstein, S. Invariant cues for place-of-articulation in stop consonants. *Journal of the Acoustical Society of America,* 1978, **64,** 1358–1368.

Stevens, K., & Klatt, D. Role of format transitions in the voiced-voiceless distinctions for stops. *Journal of the Acoustical Society of America,* 1974, **55,** 653–659.

Streeter, L. Language perception of 2-month-old infants shows effects of both innate mechanisms and experience. *Nature,* 1976, **259,** 39–41.

Swoboda, P., Kass, J., Morse, P., & Leavitt, L. Memory factors in infant vowel discrimination of normal and at-risk infants. *Child Development,* 1978, **49,** 332–339.

Swoboda, P., Morse, P., & Leavitt, L. Continuous vowel discrimination in normal and at-risk infants. *Child Development,* 1976, **47,** 459–465.

Till, J. Infants' discrimination of speech and nonspeech stimuli. Unpublished Ph.D. Dissertation, University of Iowa, Iowa City, Iowa, 1976.

Trehub, S. Auditory–linguistic sensitivity in infants. Unpublished Ph.D. dissertation, McGill University, Montreal, Quebec, Canada, 1973. (a)

Trehub, S. Infants' sensitivity to vowel and tonal contrasts. *Developmental Psychology,* 1973, **9,** 81–96. (b)

Trehub, S. The problem of state in infant speech discrimination studies. *Developmental Psychology,* 1975, **11,** 116.

Trehub, S. The discrimination of foreign speech contrasts by infants and adults. *Child Development,* 1976, **47,** 466–472. (a)

Trehub, S. Infants' discrimination of multisyllabic stimuli: The role of temporal factors. Paper presented at the annual meeting of the American Speech and Hearing Association, Houston, Tex., 1976. (b)

Trehub, S. Reflections on the development of speech perception. *Canadian Journal of Psychology,* 1979, **33,** 368–381.

Trehub, S., Bull, D., & Schneider, B. Infant speech and nonspeech perception: A review and reevaluation. In R. Schiefelbusch & D. Bricker (Eds.), *Early language: Acquisition and intervention.* Baltimore: University Park Press, in press.

Trehub, S., & Chang, H. Speech as reinforcing stimulation for infants. *Developmental Psychology,* 1977, **13,** 170–171.

Trehub, S., & Rabinovitch, S. Auditory–linguistic sensitivity in early infancy. *Developmental Psychology,* 1972, **6,** 74–77.

Trehub, S. E., Schneider, B. A., and Endman, M. Developmental changes in infants' sensitivity to octave-band noises. *Journal of Experimental Child Psychology,* 1980, **29,** 282–293.

Turnure, C. Response to voice of mother and stranger by babies in the first year. *Developmental Psychology,* 1971, **4,** 182–190.

Vargha-Khadem, F., & Corballis, M. Cerebral asymmetry in infants. Paper presented at the biennial meeting of the Society for Research in Child Development, New Orleans, La., 1977.

VerHoeve, J. N., Leavitt, L. A., & Morse, P. A. Modulation of neonatal heart rate response by prestimulation. *Pediatrics Research,* 1980, **14,** 439.

Walker, D., Grimwade, J., & Wood, C. Intrauterine noise: A component of the fetal environment. *American Journal of Obstetric Gynecology,* 1971, 91–95.

Warden, C., & Warner, L. The sensory capacities and intelligence of dogs, with a report on the ability of the noted dog "Fellow" to respond to verbal stimuli. *The Quarterly Review of Biology,* 1928, **3,** 1–28.

Waters, R., & Wilson, W. Speech perception by rhesus monkeys: The voicing distinction in synthesized labial and velar stop consonants. *Perception and Psychophysics,* 1976, **19,** 285–289.

Weir, C. Auditory frequency sensitivity of human newborns: Some data with improved acoustic and behavioral controls. *Perception & Psychophysics,* 1979, **26,** 287–294.

Werker, J., Gilbert, J., Humphrey, K., & Tees, R. Developmental aspects of cross-language speech perception. *Child Development,* 1981, **52,** 349–355.

Williams, L. The effects of phonetic environment and stress placement on infant discrimination of place of stop articulation. Paper presented at the Second Annual Boston University Conference on Language Development, Boston, Mass., 1977.

Williams, L., & Golenski, J. Infant speech sound discrimination: The effects of contingent versus non-contingent stimulus presentation. *Child Development,* 1978, **49,** 213–217.

Yakovlev, P. I., & Lecours, A. The myelogenetic cycles of regional maturation of the brain. In A. Minkowski (Ed.), *Regional development of the brain in early life.* Philadelphia: Davis, 1967.

Yost, W. A., & Nielsen, D. W. *Fundamentals of hearing.* New York: Holt, Rinehart, and Winston, 1977.

4

INFANT LEARNING

Lewis P. Lipsitt
Brown University

Infant learning processes constitute some of the earliest manifestations of adaptation to environmental exigencies. Although not often framed in this way, the biological requirements for survival include learned adaptations which tend to promote the pleasures of sensation and circumvent annoying or painful stimulation. The hedonic substructure, which mediates approach and avoidance responses, dates ontogenetically to very early embryonic stages and guides the organism through a myriad of experiences that can be hazardous or can promote well-being, survival, and perpetuation of the species. In general, pleasant hedonic stimulation facilitates tissue growth and maintenance, and annoying hedonic stimulation threatens tissue damage and inanition. It is because of this remarkable confluence of interaction involving the hedonic roots of behavior, on the one hand, and the availability of a species-specific repertoire of approach and avoidance responses, on the other, that Thorndike was able to proclaim the most evident and most celebrated law of learning present in all cognitive creatures: Those responses that are followed by a satisfying state of affairs will tend to be repeated when similar stimulating conditions present themselves again, and those followed by annoying circumstances will not.

More studies of infant learning processes have been carried out in the past 20 years than occurred in the entire previous history of humanity. To encapsulate the progress made and the advances in knowledge accumulated in the few pages of this chapter will remain, at the end, an unmet goal. The best we can do is present categorical descriptions of methods used, with the rubrics that have evolved to organize these methods, and offer some exemplary findings that will, we hope, reflect some of the intellectual excitement of the field and the directions that the field may take in the near future.

It is not without design that the opening paragraph was given a pronounced biological flavor, for the new look in the field of infant learning is one which emphasizes the contextual and ethological conditions within which organisms survive and

---

Preliminary work on this chapter was done while the author was a Fellow of the Center for Advanced Study in the Behavioral Sciences at Stanford with support from the Spencer Foundation and the James McKeen Cattell Foundation. Thanks are also due the Harris Foundation and the March of Dimes Birth Defects Foundation for support of various aspects of the author's research reported here.

thrive. Indeed, we will see that the capacity of the infant to search for and find sustenance (called "rooting behavior" in the newborn) and to defend against threatening stimulation (such as in thwarting threats to adequate respiration) are already present, on an unconditioned, unlearned basis, in the normal newborn virtually as a gift of the species. This biological presence of survival-promoting behaviors is what makes learning possible. Learned behaviors will indeed supplant those constitutionally given behaviors eventually; as the infant matures, the cerebral cortex comes into increasing control of psychomotor activities and socialization rituals proliferate. Without a basic unconditioned response repertoire there can be no learning. Learning is a biological process, usually adaptive but, as with all things organic, sometimes maladaptive as well.

NEONATAL LEARNING

Classical Conditioning

The great interest in the conditionability of the human infant has followed naturally from the arising awareness (through well-designed experiments on early sensory processes) that the human newborn enters the world with all sensory systems functioning. If the newborn can taste and smell and feel touch, then it is quite possible that perceptual experiences involving these functioning sensory modalities will be lasting ones. Although the central nervous system of the newborn is clearly not fully functional, it may take only a modicum of stimulation to sufficiently affect the appropriate nervous system pathways to generate a memory trace. Memory traces are the mediating or substantive structure of which learning effects are made.

Given that the pioneering work on classical conditioning was done in the Soviet Union, it is of more than passing interest that the earliest studies of classical conditioning in children were done by Russians but were, in fact, failures. After a number of attempts to obtain conditioned responding to previously neutral stimuli, paired with unconditioned stimulation, Krasnogorskii (1913) came to the conclusion that classical conditioning was precluded by cortical immaturity in the first 6 months of human postnatal life. Although it is now quite clear, on the basis of many studies, that cerebral "permissions" for conditioning to occur are present from the earliest moments of life in normal infants, there remain subtleties of methodology and classification to be worked out. Reviews of the infant learning literature have concluded that classical conditioning in the newborn is possible and present (Lipsitt, 1963; Horowitz, 1969; Hulsebus, 1974), but one must admit that there is lingering doubt in some quarters as to the appropriateness of the classical-conditioning terminology for the experientially induced effects obtained in the so-called classical-conditioning studies (Sameroff, 1972).

Classical conditioning involves the recruitment of response to a previously neutral stimulus, due to the simultaneous presentation of that neutral stimulus with some other stimulus which is effective at the outset in eliciting the relevant response. In one of the earliest studies, Denisova and Figurin (1929) demonstrated that infants will engage in anticipatory sucking movements when they are placed in their accustomed feeding position, after they have been treated to a number of trials wherein the opportunity to suck has immediately succeeded placement of the infant in the feeding position. Ineffective in the elicitation of sucking behavior at the outset, the feeding position is the conditioning stimulus. Documentation of a classical-conditioning effect necessitates that other events are ruled out that could produce a conditioninglike effect. It has to be demonstrated that (1) the presumably learned behavior was actually a response to the conditioning stimulus rather than the unconditioned stimulus and (2) the response enhancement observed cannot be attributed to maturational or age change in the infant's response repertoire.

Figure 4.1 shows a 2-month-old infant equipped for recording of response in a classical eyelid-conditioning study.

Both the eye-blink response and sucking behavior have been successfully conditioned to tactile stimuli. Lipsitt, Kaye, and Bosack (1966) demonstrated enhancement of sucking on a nonoptimal tube through the pairing with it of a known effective elicitor of sucking. The infant was presented with a 5% dextrose solution associated with presentation of the tube. Control infants did not receive pairing of the conditioning (tube) and unconditioned (dextrose) stimuli but did receive un-

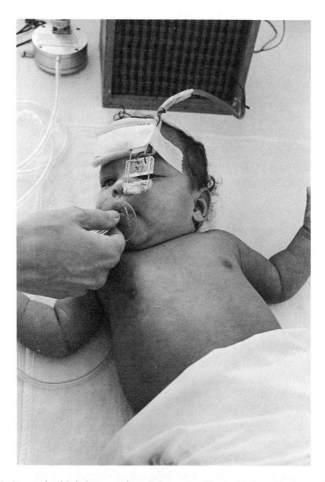

Figure 4.1. A 2-month-old infant equipped for recording of response in a classical eyelid-conditioning study.

paired presentations of the same stimuli. The infants receiving paired stimulation came to respond anticipatorily to the presentation of the tube alone with enhanced sucking behavior.

The so-called Babkin reflex involves several components, including gaping, sucking, turning the head to midline, raising the head, and in some instances eyelid closure and forearm flexion. This congenital pattern of behavior is elicited by exerting pressure on the infant's palms. Kaye (1965) successfully conditioned this response by raising the infant's arms from its sides up to the head, just before application of the palm pressure. The infants demonstrated conditioned responding by eventually gaping and turning their heads to midline in response only to the arm motion, a finding replicated by Connolly and Stratton (1969) who were able to demonstrate this type of conditioning to both auditory and kinesthetic stimulation.

Time can be used as a conditioning stimulus. Temporal conditioning has been studied in connection with feeding schedules. When an infant is put on a clock-controlled feeding schedule, such as a 3-hour interval, some of the behavioral perturbations occurring upon elimination of the feeding might be attributed to a temporal-conditioning phenomenon. A group of newborns was switched from a fixed 3-hour schedule to a 4-hour schedule by Marquis (1941). This group's behavior was compared with that of a control group which was on a 4-hour schedule throughout. Infants with changing schedules engaged in more body activity and behavioral disruption than the control group during the fourth hour when the feed would ordinarily have been given. Although infants on a 3-hour schedule may not eat as much as those on a 4-hour schedule, and the effect might therefore be due partly to greater hunger in the switched group (Lipsitt, 1963), it is a reasonable assumption that over the course of a day's feeding on either a 3- or 4-hour schedule, infants consume approximately equal quantities of milk. Changes of feeding schedules have been shown to induce anticipatory increases not only in body activity (Bystroletova, 1954) but in blood leukocyte count as well (Krachkovskaia, 1959).

Few studies have involved visual stimuli in conditioning work with infants, perhaps because tactual and auditory stimulation are essentially inescapable when administered, while visual stimulation requires an alert and attentive infant. In a study by Kasatkin and Levikova (1935), 6 infants from 14 to 48 days of age acquired stable conditioned visual responses. A colored light was presented with the feeding bottle, and after numerous trials, a response occurred to the light. The response consisted of turning the head and making mouth movements typical of feeding. After the infants were conditioned to a specific color, generalized responses were noted on tests with other colors on the continuum near to the conditioning stimulus. Response differentiation ultimately took place, such that the infants continued their conditioned behavior in response to the reinforced color, but not to the nonreinforced colors.

Auditory stimuli have been used successfully in classical conditioning with newborns; for example, Marquis (1941) used the sound of a buzzer preceding the offering of a bottle nipple during the first 10 days of the infants' lives. Within 5 training days, the 10 infants studied were making sucking responses to the buzzer sound. Kasatkin and Levikova (1935) also demonstrated conditioned differentiation of tones by the time infants in their study were 2 to 3 months old. Lipsitt and Kaye (1964) presented a low-frequency loud tone (93 dB) with the insertion of a nipple in the mouth of 3- and 4-day-old infants. A control group involved the noncontiguous presentation of the same tone and nipple. The tone was presented alone as a test for conditioning on every fifth trial, and following training, all babies received extinction trials with the test tone alone. Classical appetitive conditioning was confirmed, although the training did not manifest an effect until the extinction condition was administered.

Operant Conditioning

Whereas classical conditioning involves learning a new association between a neutral and an initially effective stimulus, operant learning results from the efficacy of reward and punishment, delivered contingently upon execution of some specific response. There is much remaining to be learned concerning the underlying processes of operant learning in infants, but considerable progress has been made in recent years.

The components of sucking patterns of newborns can be altered by specific environmental conditions, as demonstrated by Sameroff (1972) who showed that if the infant is fed contingently upon the negative-pressure component of sucking behavior, the infant will use more negative pressure than when this contingency is not in effect. Sameroff showed this by differentially reinforcing the two major components of the sucking response, positive pressure exerted by gum action, and negative pressure in the oral cavity.

Another study in which operant sucking was controlled by contingent feeding was that of Brown (1972). She based her study on the hypothesis of Premack (1965) that a response higher in the habit hierarchy (i.e., a response with higher strength) may be used to reinforce a weaker response in the hierarchy. Brown presented 2- and 3-day-old infants with a nipple, either a regular commercial nipple or a blunt variation of it, followed by access to another nipple, either regular or blunt. On the second day of conditioning, the rate of sucking on the regular nipple in the "nipple followed by blunt" group was reliably lower than the rate of sucking on the regular nipple in the "nipple followed by nipple" group. Sucking rates for the blunt oral stimulus in the "blunt followed by nipple" group, moreover, were significantly higher than such sucking in the "blunt followed by blunt" group. Besides confirming Premack's hypothesis, the study is an unusual but interesting demonstration of operant learning by infants.

Figure 4.2 shows a procedure for recording head turning in a newborn.

Elaborations of traditional operant-conditioning paradigms have yielded successful learning demonstrations in newborns. Siqueland and Lipsitt (1966) adapted the conditioned head-turning techniques first reported by Papoušek (1959, 1960); they studied the effect of contingent reinforcement on ipsilateral head movements. Instead of awaiting the natural or spontaneous occurrence of head turning, as is the customary style in operant conditioning, the head turns in their study were elicited by a touch at the side of the infant's mouth. This stimulus produced head turning on about 30% of the preconditioning trials; that is, the base rate was .30. In three methodological variations of the procedure, nutritive reinforcement was administered contingent upon the elicited head-turning movements, and the general result was that head-turning responses were enhanced by the reinforcement condition. In the first of the studies, each infant served as its own control to guard against the

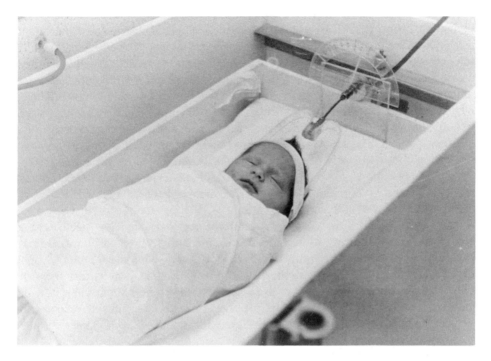

Figure 4.2. A procedure for recording head turning in a newborn. A plastic harness is gently attached with a Velcro band to infant's head and, on the other side, to a rotary transducer that connects with a polygraph, for recording of amplitude of head turns in conjunction with other events; for example, an auditory stimulus.

possibility that any effect obtained could be attributed to changes in state rather than in a memorial or learning process. Two different auditory stimuli were used. On alternate trials the stimulus for head turning was a different tone–touch combination. On positive trials, the baby received one tone–touch combination to the left cheek, and on the other half of the trials, a different tone–touch combination to the same cheek served as the negative stimulus. On positive trials, ipsilateral head turns were reinforced by opportunity to suck for dextrose, while on negative trials no such opportunity occurred. The results showed that the positive stimulus combination facilitated increased responding, from 30 to 83%, whereas on negative trials the response probability remained essentially the same throughout. The total training period took 30 minutes, following which the positive and negative stimulus conditions were reversed. In this second half of the study, the previous stimulus combination was no longer reinforced and the previous negative stimulus was made positive. A behavioral shift then took place, in the second 30-minute session, accommodating to the changed stimulus conditions. Thus the contingent application of incentive conditions, such as the presentation of food, may alter the probabilities of reflexive responding typical of infants within the first days of life.

High-amplitude sucking behavior has been used to study operant learning in infants, in association with visual reinforcement. Studies carried out by Siqueland and colleagues (Milewski & Siqueland, 1975; Siqueland & DeLucia, 1969; Werner & Siqueland, 1978) have involved babies from the preterm period to 4 months of age by using a screen to show slides of varying illuminations, with the level of illumination being controlled by the sucking behavior (an operant) of the infant. Using this technique of "conjugate reinforcement," infants showed typical acquisition and extinction effects, increasing their sucking rate to receive visual incentives, and decreasing sucking either when the screen no longer provided contingently presented visual stimulation or after many trials of the same stimulus had produced habituation (and therefore diminished sucking behavior) to it. Occurrence of habituation to a frequently presented stimulus is essentially a "familiarization phenomenon," which makes it possible to record recovery of sucking behavior to novel stimulation and to quantify perceived degree of novelty in terms of level of recovery or dishabituation. A study by Milewski and Siqueland (1975) showed that infants as young as 1 month of age can discriminate between familiar and novel stimuli on the basis of both color and pattern.

Sucking behavior in the contingent reinforcement paradigm has been used to study the effect of auditory stimuli as well. In one such study with infants, Eimas, Siqueland, Jusczyk, and Vigorito (1971) showed that as with visual stimuli, young infants suck to hear auditory stimuli for a period of time, whereupon their interest lags and sucking wanes or even ceases. At that point a slight change in the auditory stimulus, such as from "ba" to "ga," causes a significant recovery of the sucking response relative to control infants for whom no such change in auditory stimulus has taken place.

Thus we see that operant learning takes place in the very young infant and that the operant paradigm can be used as a means to the study of sensory and discriminative capacities of the young infant as well as of the rudiments of cognitive differentiation.

Habituation as a Learning Process

The condition of the organism, in classical and operant learning, is presumed altered by the conditioning experience, and the residual of the experience is thought to be mediated by brain structures. The effect of the conditioning is essentially permanent, although the effect can be superseded by other processes, such as extinction or the superimposition of counterconditioning. In classical conditioning, a new neurally mediated association is effected between a neutral and an effective stimulus, and in operant learning, a neurally mediated association between response-produced stimulation and a reward stimulus takes place. Although habituation does involve gradual behavioral changes as a consequence of repetitive stimulation and does produce behavioral alterations which last for a time, as in learning processes, it does not involve the clear development of new, or enhancement of old, associations as in classical and operant conditioning.

Human infants will frequently respond vigorously to a new or sudden stimulus which, if repeated on several subsequent occasions, will lose the capacity for evoking the response which occurred initially. Although the diminution of behavior is sometimes very rapid, the waning of response is typically gradual (Clifton & Nelson, 1976; Engen & Lipsitt, 1965; Kessen, Haith, & Salapatek, 1970). If the stimulus is not inordinately powerful, response strength with repetitive presentations may reduce to zero, even though first stimulus presentations may have produced a marked startle response and highly accelerated heart rates. The phenomenon of habituation has come to be of great importance in the study of infant behavior and development, because it presents an excellent tool for the investigation of stimulus discriminability by inarticulate organisms. Following habituation to one stimulus, a different stimulus can then be administered which, if sufficiently distinctive from the first, will yield a "recovery response," representing a dishabituation process. The occurrence of dishabituation reveals that the infant recognizes the difference between (i.e., responds differentially to) the two stimuli. Thus the discriminative capabilities of infants in any sensory modality can be documented. Among the first such studies of this type were those of Bartoshuk (1962a, 1962b), in which he was able to show, in newborns, that cardiac acceleration to auditory stimuli will diminish with repeated presentations, and that newborns will show recovery of response with presentation of a novel auditory stimulus, even if this novel stimulus involves a backward glissando, or reverse slide, of a stimulus gradually changing in frequency.

Reviewers of infant habituation studies (e.g., Kessen, Haith, & Salapatek, 1970) have noted that the process is one which enables the study of developing memory, that the phenomenon is one which reveals adequacy of cortical functioning, and that central nervous system deficits may be detected through documentation of habituation deficits.

Pioneering work on visual habituation of infants was carried out by Fantz, whose early studies showed that gaze duration can be objectively measured and used over repetitive trials to assess the baby's waning interest in visual stimuli (Fantz, 1961, 1964). His earliest studies showed that newborns have a habituation deficit relative to infants of 10 weeks of age and beyond.

Because habituation does relate to memory function mediated by the brain, its assessment is sometimes included in tests of individual differences in newborns (Brazelton, 1973). Lewis (1967) has, in fact, shown that impaired habituation of infants relates to delivery-room scores (the Apgar categories) and other measures of perinatal distress and brain damage. Some data exist to the effect that anencephalic and hydranencephalic infants show no habituation (Brackbill, 1971; Wolff, 1969), but two recent studies are exceptions (Graham, Leavitt, & Strock, 1978; Tuber, Berntson, Bachman, & Allen, 1980).

Habituation using the sucking response has been used to document deficits of infants born under conditions of unusual cranial pressure or with the umbilicus around the neck. Either of these eventualities may produce anoxia or asphyxia; a deficit in habituation has also been demonstrated by Soviet investigators, using a very interesting paradigm (Bronshtein, Antonova, Kamenetskaya, Luppova, & Sytova, 1958). Infants in their studies were presented with a stimulus, such as sound, during sucking episodes. Infants of normal birth manifested sucking interruption when they were presented with such a stimulus, but then diminished in their response of sucking suppression with repetitive presentations of that stimulus. For the infants born at risk, however, many more stimulus presentations were required to habituate the sucking-suppression response, and the most seriously at-risk babies did not habituate at all.

Finally, it may be said that infants with Down's syndrome have shown less habituation than normal infants of the same age (Barnet, Olrich, & Shanks, 1974), and quite definite effects of obstetrical anesthesia have also been obtained. Bowes, Brackbill, Conway, and Steinschneider (1970) examined infants at 2 and 5 days of age, some of these being babies whose mothers had received high dosages of anesthesia, and others, low dosages or none at all. Infants of the former category required as many as 4 times more trials to habituate than did those receiving little medication. It appears that the habituation paradigm is successful in providing a reasonably simple and noninvasive assessment of the infant's capacity for processing information and for discriminating modest differences in stimulus inputs.

Conditioning as a Case of Prepared Responding

As Rovee-Collier and Lipsitt (1982) have pointed out, the distinctions between classical and operant learning are far from clear-cut. Both paradigms frequently yield behaviors containing components of each. In classical eyelid conditioning, for example, the conditioned response is typically learned in such a way as to anticipate the noxious unconditioned stimulus so as to avoid it (Kimble, 1961). Indeed, Sameroff (1971, 1972) suggests that classical conditioning of newborns is, in fact, an instance of instrumental conditioning. Again, the important question relates to how the newborn capitalizes upon general evironmental characteristics to develop specific behaviors appropriate to it. Only recently (cf. Hinde, 1979) has there been adequate attention paid by researchers to the ecological niche and ethological character of the organism in connection with these learning processes. Most studies thus far have involved, for example, study of the infant in isolation from its usual habitat. In our zeal to discover whether newborns classically and operantly condition, we have instrumented experimental environments and stimulus situations for them which we presumed would support and facilitate those conditioning processes if the infants were capable of them. In so doing, we often tended to forget that in the "natural state," which after all is the normal biologic condition, infants are found with, around, and among other persons. It is a relatively recent development that we are studying babies in association with their parents or other caretaking individuals; we are using natural foods and odors as reinforcers of behavior; we are talking in the infant's ears instead of using audio oscillators to generate highly specifiable (but nonhuman) sounds; and we are looking at the noise and light cycles of neonatal intensive-care units as unusual developmental environments which have "evolved" too quickly to be seen as "true" biological adaptations.

Data available on infant conditioning reveal that conditioning proceeds most readily and quickest when the environmental conditioners capitalize best upon responses which are biologically primed for promotion (Rovee-Collier & Lipsitt, 1982). It is easier to teach a dog to jump in the air for food when hungry than when satiated, and conditioning goes fastest when the food reward administered is one which is of high incentive value for the species and individual being trained. Animal lovers and trainers have been aware of this since the beginning of bestial domesticity, but the assertion of the importance of biological roots of behavior is quite new (cf. Breland & Breland, 1961; Seligman & Hager, 1972). The natural habitat of the organism and the stimulating conditions indigenous to it are probably critical factors in the enhancement of conditioning behavior of infants.

Miscellany of Research Advances in Infant Learning

It is impossible in such brief space as this to cover more than the most salient, and perhaps the best supported, findings to date on aspects of infant learning. The reader should appreciate that this is a field of inquiry in which very rapid advances have been made in recent years. There has been a great upsurge of interest in infantile learning processes by animal researchers (e.g., Brake, 1981; Hall, 1979), some of whom have taken their methodological leads from students and investigators of human infant learning processes. This interest has been sparked in part by the newfound realization that certain perceptual–experiential processes of infants, such as those involving psychophysiological effects of early visual experience, are quite clearly dependent upon a kind of critical period existing early in the animal's life (see St. James-Roberts, 1979).

Another area of real progress has been in the study of so-called conjugate reinforcement and its effects on infants. On the basis of numerous studies reviewed by Rovee-Collier and Gekoski (1979), it is apparent that conjugate reinforcement is a powerful tool for facilitating the onset and strengthening of learning in infants and that the learned responses acquired under these conditions possibly have greater transfer potential to new situations and to later test periods. Conjugate reinforcement differs from the usual operant reinforcement procedure in that the rate or amplitude of response directly determines the intensity of the consequences. Response rate and strength directly modulate the flow of reinforcement, rather than discrete responses having discrete reward consequents as, for example, in the receipt of one food pellet for one criterional response. Advances in this area are imminent and bear watching (see Rovee-Collier, Sullivan, Enright, Lucas, & Fagen, 1980).

THE SPECIAL CASE OF 6 WEEKS AND BEYOND

Rovee-Collier and Gekoski (1979) have suggested that there is a significant developmental shift in humans at around 4 weeks of age. While the neonate is quite competent in modifying and controlling its own environment in certain respects, it can be concluded from the successes and failures in infant-conditioning studies that the newborn is highly sensitive to response cost and behaves in a manner which maximizes the benefits of a given activity in relation to the required energy expenditure for its execution. This viewpoint is compatible with that of Skinner (1938) who suggested that organisms tend to adopt easy, efficient response topologies in the absence of response-contingent incentives. There is, so to speak, a trade-off between required response energy expenditure and the value of the consequent response. For whatever reason, experimenters have had little success in conditioning high-energy behaviors such as foot kicking (Solkoff & Cotton, 1975) until human infants are over 1 month of age. It may or may not be important that this is the approximate age at which the human infant achieves a modicum of physiological control over its own body temperature. Regardless, it would appear that at around 6 weeks of age there is a critical transition in sensory sensitivity, attentiveness, deliberative control of psychomotor activities, and seeming pleasure in the self-delivery of exteroceptive stimulation. Not so surprisingly, this period coincides with Piaget's "secondary circular reaction," in which the infant first comes to appreciate that his or her own behavior has something to do with the resulting stimulus feedback. In a real sense the normal human infant has become a biofeedback organism. At this time, the baby seems capable of a rudimentary understanding of cause–effect relationships with respect to self-generated environmental perturbations.

There are numerous responses, like the grasp reflex of the newborn, which are strong at birth and begin to weaken soon thereafter. These are completely gone by 5 months of age, but even by 6 weeks of age a considerable response diminution has taken place. Turning the head to a touch near the mouth, grasping at a tactual stimulus in the palm of the hand, and swimming movements when the infant is placed in water all go through observable and fairly quantifiable changes with the passage of time. Response progressions in all these modalities suggest that the behaviors are obligatory at the outset but gradually become more deliberative or more "voluntary" as time and experience progress (McGraw, 1943). McGraw's careful observations of reflex ontogeny in the first year of life have shown that the transitional period between a reflexive and a voluntary phase of the response usually occurs between 2 and 5 months. McGraw suggested that the transitional phases of these patterns of response, as in aquatic behavior, can be characterized as "disorganized behavior" or "struggling activity." On the basis of gross brain development data (Dobbing, 1974) and data relating to the enormous rate of dendritic proliferation occurring within the first 2 months of life (Purpura, 1974), it appears that neural ontogenesis and behavioral maturation constitute a symbiotic course of development such that the accretion of brain tissues permits certain increasingly complex behaviors to occur. But experience, too, can generate changes in rate of neuronal growth and achievement of function (Hubel & Wiesel, 1970). By the same token, if the infant does not make a smooth transition from the earlier stages of reflex functioning to the more mature voluntary and deliberative stages, as described by McGraw, the infant may enter a stage of jeopardy with regard to its ability to defend against threatening stimulation.

Behavioral Transitions, Learning Processes, and Developmental Jeopardy

The period between 2 and 4 months seems to represent a major transitional phase for the learned embellishment and voluntary expression of many reflexes which had been previously elicited only under highly circumscribed conditions (Lipsitt, 1976; McGraw, 1943). The classic work of McGraw on progressions in aquatic behavior with age show that the reflex patterns are supplanted eventually by more deliberative, probably learned, patterns of behavior with organizations of their own. Interestingly, this is precisely the period of greatest risk of crib death. Although McGraw did not study or allude to the sudden infant death phenomenon, she did describe the transitional period as one involving neurobehavioral disorganization and disarray. She described three distinct phases of reflex development and transition: (1) the waning of the basic reflex, (2) the onset of a transitional period of response confusion, and (3) the eventual predominance of a more "mature" learned response. McGraw supposed that the transition is one essentially from subcortical to cortical responding.

Few studies have explored the role of practice in facilitating the transition from obligatory reflexive responding to what might be regarded as operant responding. Ambulatory behavior is apparently related, to an extent, to opportunities for exercise of the reflexive components of the eventual voluntary behaviors (Konner, 1973; Zelazo, 1976; Zelazo, Zelazo, & Kolb, 1972), but certainly not enough is known about the processes of transition or the mechanisms whereby this practice has its effect. The adaptive value of practicing motor skills, particularly those involving extensive coordination, has been shown in animals (e.g., Kavanau, 1967). Differential opportunity to exercise these skills and normal individual variation in reflex characteristics may separate those infants during developmental transitional periods whose behavior patterns are inadequate from those who are competent at moments critical to survival.

Variation in opportunities to exercise reflexes and the emerging skills which capitalize upon the prior presence of those congenital reflexes could have some bearing on whether and how uneventfully the infant will pass through the transitional period of behavioral confusion or psychomotor disarray. Similarly, if the quality of the initial reflex upon which later behavior will ultimately be based is compromised by birth defect, perinatal hazards, or environmental insufficiencies, the practice required for the transition through the period of confusion and into a period of adaptive cortically mediated learning might be less effective. An ineffective transition may well lead to some condition of developmental jeopardy and perhaps even a life crisis in the first year. The *process of maturation* involving the interaction of constitutional with experiential determinants may be more important than the achievement of maturational milestones. Experiences endured in the course of maturation may be precisely those responsible for the kinds of learning that endure most profoundly. Similarly, infants born at risk or in developmental jeopardy within the first year of life might be just those requiring special experiential interventions to obviate the developmental obstacles inherent in the periods of neurobehavioral transition.

Crib Death and Infantile Behavior Processes

Nearly 8000 babies die in the United States each year, most of them between 2 and 4 months of age, with no evidence of any diagnosable problem prior to death. "Crib death" is the most common cause of death in the first year, excluding the especially hazardous first few days of life. Some of these seem, on close examination of perinatal history or through an incisive autopsy, to have had predisposing deficits, but most die of causes yet to be elucidated. A plausible suggestion is that constitutional deficits conspire with environmental insufficiencies in some cases (Lipsitt, 1971, 1976). Crib death shows some seasonal variation, with peaks in the colder months. More cases of sudden infant death syndrome (SIDS) occur during sleep; the infants are usually found in the morning after no apparent sign of struggle. They simply stop breathing, with little evidence of any agonal experience. A mild upper-respiratory infection is found in 40–50% of the cases, with parents reporting a runny nose or raspy breathing, but pathologists seldom find evidence of a lethal respiratory disorder, which in fact would change the diagnosis. With no sign of a febrile condition, the baby has been regarded in the few days immediately preceding death as essentially normal (Valdes-Dapena, 1980).

Several recent publications (e.g., Anderson & Rosenblith, 1971; Lipsitt, 1976, 1978; Lipsitt, Sturner, & Burke, 1979; Naeye, Ladis, & Drage, 1976; Protestos, Carpenter, McWeeny, & Emery, 1973; Steinschneider, 1976; Thoman, Miano, & Freese, 1977) suggest that babies who succumb to crib death are as a group (with exceptions, of course) the products of a difficult, but not *seriously* aberrant, gestation and birth. A number of indices of physiological and behavioral insufficiency are present in the earliest days and weeks of life. They are more frequently premature or small for gestational age, more often require delivery of oxygen at birth, and more frequently show indications of respiratory distress in the few days immediately following birth. Most infants who become cases of crib death are discharged from hospital as normal, and most have been asymptomatic up to the moment of death, but statistical examination of birth and developmental records reveals that, contrasted with control cases, they have often harbored signs which taken as a constellation might constitute a pattern of psychobiologic jeopardy. The histories of SIDS cases often reveal hazardous signs of respiratory instability, behavioral lethargy, and inability to engage in appropriate defensive maneuvers in response to aversive or annoying stimulation (Lipsitt, 1976). As a result of perinatal adversities or early developmental lethargy, often involving periodic apneic spells and associated anoxia, eventual SIDS victims often have generally subdued activity. They suck weakly, move less, respond with higher thresholds to noxious events, and, in general, engage their environment less. Such infants *may* subject themselves to fewer opportunities for learning than "normal" infants do.

Although considerable medical attention has been devoted to the problem of crib death, only recently have the infant's behavioral contributions to its own fitness (Lipsitt, 1976, 1979) been considered seriously. Both Naeye (Naeye, Messmer, Specht, & Merritt, 1976) and Valdes-Dapena (1980), two prominent American pathologists studying crib death, have recently documented or acknowledged behavioral insufficiencies or idiosyncrasies as relevant and worthy of further research.

The notion of a stimulation and learning aberration contributing to infantile death is not novel, for both Spitz (1965) and Goldfarb (1945) have spoken convincingly of the debilitating effects on infants, including death, of certain types of adverse environments. Psychobiological researchers have for some time (e.g., Cannon, 1932; Richter, 1958; Seligman, 1975) called attention to the phenomenon of death in adults and animals from psychological causes—for example, soldiers giving up and dying on the battlefield without apparent cause, spouses dying shortly after the deaths of their partners, and animals giving up and dying in apparent desperation and despondency when confronted with life stresses that seem inescapable. Seligman (1975) described the conditioning processes that conspire to create helplessness and hopelessness that can lead to stupor and death.

Some neonatal congenital responses have significance in promoting survival and adaptation to the biological condition of the species (Emde & Robinson, 1976). When the infant is touched on the side of the mouth, for example, the head turns ipsilaterally, whereupon mouth opening occurs and is further promoted by more touch. The infant's mouth opens, the lips close around the nipple, a pressure seal is created, and regular rhythmic sucking occurs depending on the shape of the oral stimulus and quality, or "incentive value," of the fluid which the infant receives. The sucking response is, in fact, modulated in several of its dimensions by the sweetness of the fluid; sucking "incentives" are operative from the earliest moments after birth (Lipsitt, 1976).

Strikingly apparent at birth and soon after are several congenital response systems, such as the grasping reflex, the Babkin response, primitive reaching responses, and obligatory visual attention. These diminish in frequency and intensity with time, especially during the period 2 to 4 months of age (McGraw, 1943; Paine, 1976). The grasp reflex becomes a slow exploratory mode of response to pressure on the palm of the hand. Some of these responses are displaced or complemented by learned responses mediated by higher cortical centers. These eventual "voluntary" responses contrast with the brain-stem control that existed before experience could superimpose itself upon the lower-brain function. Shortly after birth and particularly in the first 2 months of human life (Dobbing, 1975; Purpura, 1975), there is rapid development of neural function that is probably critical for the experiential accretion of learned responses. Some behavioral patterns apparently *must* be learned during this time. The relevant age may be the time by which the unlearned protective reflexes have diminished to an ineffective level; otherwise, the organism might not have been prepared adequately for survival.

The normal neonate responds to respiratory occlusion, or even to the *threat* of occlusion, with a series of defensive actions (Brazelton, 1973; Graham, 1956; Anderson & Rosenblith, 1971). Gunther (1955, 1961) observed struggling of newborns as they suckled at the breast, a response that may be likened to an enraged response which escalates as the stimulus is prolonged. In the intact neonate, the behavior pattern is essentially failsafe, culminating in crying and ultimately freeing the respiratory passages. If the newborn does not have a strong head-and-hand defensive response to respiration threats or to head restraint, the defensive behaviors which must ultimately supplant this congenital response by 2 to 4 months of age may not become learned.

SIDS cases have as a group begun life with some organismic deficits associated with perinatal stresses. Respiratory occlusion, failure of appropriate defensive behavior, and inadequate compensatory mouth-breathing when the nostrils are clogged could conceivably lead to anoxia and a comatose state and, ultimately, to the infant's death, particularly when all these conditions converge, as in sleeping infants with a cold and a history of insufficient response to threatening stimulation. A hypothesis worthy of serious investigation, then, is that infants will be in particular jeopardy at the vulnerable age range for crib death if they have not learned to engage in responses necessary for clearing of the respiratory passages or clearing the way *to* those passages when threatened with occlusion.

SUMMARY

Various procedures for the study of the mechanisms of infant learning have been reviewed briefly. Exemplary findings have been reported, but an exhaustive synthesis has not been attempted. It is quite clear that human newborns are importantly affected by experiential inputs, through the processes of habituation and of classical and operant learning. As technology and the experimental methodologies are refined for the study of behavioral change with cumulative stimulation, infants seem increasingly competent. Application of the techniques in comparative investigations reveals that high-risk babies are frequently compromised in their behavioral capacities relative to physiologically normal, full-term infants.

REFERENCES

Anderson, R. B., & Rosenblith, J. F. Sudden unexpected death syndrome: Early indicators. *Biologia Neonatorum,* 1971, **18**, 395–406.

Barnet, A. B., Olrich, E. S., & Shanks, B. L. EEG evoked responses to repetitive stimulation in normal and Down's syndrome infants. *Developmental Medicine and Child Neurology,* 1974, **5**, 612–619.

Bartoshuk, A. K. Human neonatal cardiac acceleration to sound: Habituation and dishabituation. *Perceptual and Motor Skills,* 1962, **15**, 15–27. (a)

Bartoshuk, A. K. Response decrement with repeated elicitation of human neonatal cardiac acceleration to sound. *Journal of Comparative and Physiological Psychology,* 1962, **55**, 9–13. (b)

Bowes, W., Brackbill, Y., Conway, E., & Steinschneider, A. The effects of obstetrical medication on fetus and infant. *Monographs of the Society for Research in Child Development,* 1970, **35**, 3–25.

Brackbill, Y. The role of the cortex in orienting: Orienting reflex in an anencephalic human infant. *Developmental Psychology,* 1971, 5, 195–201.

Brake, S. C. Suckling infant rats learn a preference for a novel olfactory stimulus paired with milk delivery. *Science,* 1981, **211**, 506–508.

Brazelton, T. B. *Neonatal behavioral assessment scale.* Philadelphia: William Heinemann Medical Books, 1973.

Breland, K., & Breland, M. The misbehavior of organisms. *American Psychologist,* 1961, **61**, 681–684.

Bronshtein, A. T., Antonova, T. G., Kamenetskaya, N. H., Luppova, V. A., & Sytova, V. A. On the development of the functions of analyzers in infants and some animals at the early stage of ontogenesis. In: *Problems of evolution of physiological functions,* 1958. U.S.S.R.: Academy of Science (U.S. Department of H.E.W., Translation Service, 1960.)

Brown, J. Instrumental control of the sucking response in human newborns. *Journal of Experimental Child Psychology,* 1972, **14**, 66–80.

Bystroletova, G. N. The formation in neonates of a conditioned reflex to time in connection with daily feeding rhythm. *Zhurnal Vysshei Veyatel'Nosti Imeni I.P. Pavlova,* 1954, **4**, 601.

Cannon, W. B. *The wisdom of the body.* New York: W. W. Norton, 1972.

Clifton, R. K., & Newson, M. N. Developmental study of habituation in infants: The importance of paradigm, response system, and state. In T. J. Tighe & R. N. Leaton (Eds.), *Habituation.* Hillsdale, N.J.: Erlbaum, 1976.

Connolly, K., & Stratton, T. An exploration of some parameters affecting classical conditioning in the neonate. *Child Development,* 1969, **40**, 431–441.

Denisova, M. P., & Figurin, N. L. K voprosu o pervykh sochetatelnykh pishchevykh refleksakh u grudnykh detei. *Vop. genet. Refleks. Pedol.,* 1929, **1**, 81–88.

Dobbing, J. Human brain development and its vulnerability. *Mead Johnson symposium on perinatal and developmental medicine: Biologic and clinical aspects of brain development,* 1974, No. 6.

Eimas, P. D., Siqueland, E. R., Jusczyk, P., & Vigorito, J. Speech perception in infants. *Science,* 1971, **171**, 303–306.

Emde, R. N., & Robinson, J. The first two months: Recent research in developmental psychobiology and the changing view of the newborn. In J. Noshpitz & J. Call (Eds.), *Basic handbook of child psychiatry.* New York: Basic Books, 1976.

Engen, T., & Lipsitt, L. P. Decrement and recovery of responses to olfactory stimuli in the human neonate. *Journal of Comparative and Physiological Psychology,* 1965, **59**, 312–316.

Fantz, R. L. The origin of form perception. *Scientific American,* 1961, **204**, 66–72.

Fantz, R. L. Visual experiences in infants: Decreased attention to familiar patterns relative to novel ones. *Science,* 1964, **146**, 668–670.

Goldfarb, W. Psychological privation in infancy and subsequent adjustment. *American Journal of Orthopsychiatry,* 1945, **15**, 247–255.

Graham, F. K., Leavitt, L. A., Strock, B. D., & Brown, J. Precocious cardiac orienting in a human anencephalic infant. *Science,* 1978, **199**, 322–324.

Gunther, M. Instinct and the nursing couple. *Lancet,* 1955, **1**, 575.

Gunther, M. Infant behavior at the breast. In B. Foss (Ed.), *Determinants of infant behavior.* London: Methuen, 1961.

Hall, W. G. Feeding and behavioral activation in infant rats. *Science,* 1979, **205**, 206–208.

Hinde, R. A. *Biological bases of human social behavior.* New York: McGraw-Hill, 1974.

Horowitz, F. D. Learning, developmental research, and individual differences. In L. P. Lipsitt & H. W. Reese (Eds.), *Advances in child development and behavior* (Vol. 4). New York: Academic, 1969.

Hubel, D. H., & Wiesel, T. N. The period of susceptibility to the physiological effects of unilateral eye closure in kittens. *Journal of Physiology,* 1970, **206**, 419–436.

Hulsebus, R. C. Operant conditioning of infant behavior: A review. In H. W. Reese (Ed.), *Advances in child development and behavior* (Vol. 8). New York: Academic, 1973.

Kasatkin, N. I., & Levikova, A. M. On the development of early conditional reflexes and differentiations of auditory stimuli in infants. *Journal of Experimental Psychology,* 1935, **18**, 1–19.

Kavanau, J. L. Behavior of captive white-footed mice. *Science,* 1967, **155**, 1623–1639.

Kaye, H. The conditioned Babkin reflex in human newborns. *Psychonomic Science,* 1965, **2**, 287–288.

Kessen, W., Haith, M. M., & Salapatek, P. H. Human infancy: A bibliography and guide. In P. H. Mussen (Ed.), *Carmichael's manual of child psychology* (Vol. 1). New York: Wiley, 1970.

Kimble, G. A. *Hilgard and Marquis' conditioning and learning* (2nd ed.). New York: Appleton-Century-Crofts, 1961.

Konner, M. Newborn walking: Additional data. *Science,* 1973, **178**, 307.

Krachkovskaia, M. V. Reflex changes in the leukocyte count of newborn infants in relation to food intake. *Pavlov Journal of Higher Nervous Activity,* 1959, **9**, 193–199.

Krasnogorskii, N. I. Über die grundmechanismen der arbeit der grosshernrunde bei Kindern. Jahrbuch für Kinderheilkunde, 1913, **78**, 373–389.

Lewis, M. The meaning of a response, or why researchers in infant behavior should be oriental metaphysicians. *Merrill-Palmer Quarterly,* 1967, **13**, 7–18.

Lipsitt, L. P. Learning in the first year of life. In L. P. Lipsitt & C. C. Spiker (Eds.), *Advances in child development and behavior* (Vol. 1). New York: Academic, 1963.

Lipsitt, L. P. Infant anger: Toward an understanding of the ontogenesis of human aggression. Unpublished paper presented at the Department of Psychiatry, The Center for the Health Sciences, University of California at Los Angeles, March 4, 1971.

Lipsitt, L. P. Developmental psychobiology comes of age: A discussion. In L. P. Lipsitt (Ed.), *Developmental psychobiology: The significance of infancy.* Hillsdale, N.J.: Erlbaum, 1976.

Lipsitt, L. P. Perinatal indicators and psychophysiological precursors of crib death. In F. D. Horowitz (Ed.), *Early developmental hazards: Predictors and precautions.* Boulder, Colo.: Westview Press, 1978.

Lipsitt, L. P., & Kaye, H. Conditioned sucking in the human newborn. *Psychonomic Science,* 1964, **1**, 29–30.

Lipsitt, L. P., Kaye, H., & Bosack, T. N. Enhancement of neonatal sucking through reinforcement. *Journal of Experimental Child Psychology,* 1966, **4**, 163–168.

Lipsitt, L. P., Sturner, W. Q., & Burke, P. Perinatal indicators and subsequent crib death. *Infant Behavior and Development,* 1979, **2**, 325–328.

McGraw, M. B. *The neuromuscular maturation of the human infant.* New York: Hafner, 1943.

Marquis, D. P. Learning in the neonate: The modification of behavior under three feeding schedules. *Journal of Experimental Psychology,* 1941, **29**, 263–282.

Milewski, A. E., & Siqueland, E. R. Discrimination of color and pattern novelty in 1-month human infants. *Journal of Experimental Child Psychology,* 1975, **19**, 122–136.

Naeye, R., Ladis, B., & Drage, J. S. SIDS: A prospective study. *American Journal of Diseases of Children,* 1976, **130**, 1207–1210.

Naeye, R., Messmer, J., III, Specht, T., & Merritt, F. Sudden infant death syndrome temperament before death. *Journal of Pediatrics,* 1976, **88**, 511–515.

Papoušek, H. A method of studying conditioned food reflexes in young children up to the age of 6 months. *Pavlov Journal of Higher Nervous Activity,* 1959, **9**, 136–140.

Papoušek, H. Conditioned motor digestive reflexes in infants. II. A new experimental method for the investigation. *Cesk. Pediatrics,* 1960, **15**, 981–988.

Premack, D. Reinforcement theory. In D. Levin (Ed.), *Nebraska symposium on motivation.* Lincoln: University of Nebraska Press, 1965.

Protestos, C., Carpenter, R., McWeeny, P., & Emery, J. Obstetric and perinatal histories of children who died unexpectedly (cot death). *Archives of Disease in Childhood,* 1973, **48**, 835–841.

Purpura, D. P. Neuronal migration and dendritic differentiation: Normal and aberrant development of human cerebral cortex. In *Mead Johnson symposium on perinatal and developmental medicine,* 1974, No. 6.

Richter, C. On the phenomenon of sudden death in animals and man. *Psychosomatic Medicine,* 1957, **19**, 191–198.

Rovee-Collier, C. K., & Gekoski, M. J. The economics of infancy: A review of conjugate reinforcement. In H. W. Reese & L. P. Lipsitt (Eds.), *Advances in child development and behavior* (Vol. 13). New York: Academic, 1979.

Rovee-Collier, C. K., & Lipsitt, L. P. Learning, adaptation, and memory in the newborn. In P. Stratton (Ed.), *Psychology of the human newborn.* Chichester, England: Wiley, 1982.

Rovee-Collier, C. K., Sullivan, M. W., Enright, M., Lucas, D., & Fagen, J. W. Reactivation of infant memory. *Science,* 1980, **208**, 1159–1161.

St. James-Roberts, I. Neurological plasticity, recovery from brain insult, and child development. In H. W. Reese & L. P. Lipsitt (Eds.), *Advances in child development and behavior* (Vol. 14). New York: Academic, 1979.

Sameroff, A. J. Learning and adaptation in infancy: A comparison of models. In H. W. Reese (Ed.), *Advances in child development and behavior* (Vol. 7). New York: Academic, 1972.

Seligman, M. E. P. *Helplessness.* San Francisco: Freeman, 1975.

Seligman, M. E. P., & Hager, J. L., (Eds.). *Biological boundaries of learning.* New York: Appleton-Century-Crofts, 1972.

Siqueland, E. R., & DeLucia, C. A. Visual reinforcement of non-nutritive sucking in human infants. *Science,* 1969, **165**, 1144–1146.

Siqueland, E. R., & Lipsitt, L. P. Conditioned head-turning behavior in newborns. *Journal of Experimental Child Psychology,* 1966, **3**, 356–376.

Skinner, B. F. *The behavior of organisms.* New York: Appleton-Century, 1938.

Solkoff, N., & Cotton, C. Contingency awareness in premature infants. *Perceptual and Motor Skills,* 1975, **41**, 709–710.

Spitz, R. Hospitalism: A follow-up report. In: *The psychoanalytic study of the child* (Vol. 2). New York: International Universities Press, 1946.

Steinschneider, A. Implications of the sudden infant death syndrome for the study of sleep in infancy. In A. D. Pick (Ed.), *Minnesota symposium on child psychology* (Vol. 9). Minneapolis: University of Minnesota Press, 1975.

Thoman, E. B., Miano, V. N., & Freese, M. P. The role of respiratory instability in the sudden infant death syndrome. *Developmental Medicine and Child Neurology,* 1977, **19**, 729–738.

Valdes-Dapena, M. Sudden infant death syndrome: A review of the medical literature 1974–1979. *Pediatrics,* 1980, **66**, 597–614.

Werner, J. S., & Siqueland, E. R. Visual recognition memory in the preterm infant. *Infant Behavior and Development,* 1978, **1**, 79–94.

Wolff, P. H. What we must and must not teach our young children from what we know about early cognitive development. Reprinted from *Planning for better learning,* S.I.M.P./William Heinemann Medical Books, 1969.

Zelazo, P. R. From reflexive to instrumental behavior. In L. P. Lipsitt (Ed.), *Developmental psychobiology: The significance of infancy.* Hillsdale, N.J.: Erlbaum, 1976.

Zelazo, P., Zelazo, N. A., & Kolb, S. Newborn walking. *Science,* 1972, **177**, 1057–1060.

5

INFANT MEMORY

Joseph F. Fagan III
Professor of Psychology
Case Western Reserve University

The purpose of this chapter is to provide a summary of the major findings on visual recognition memory for infants ranging in age from birth to the seventh month of life. More detailed reviews of infant visual recognition memory are available (Olson, 1976; Werner & Perlmutter, 1979). The first aim is to present operational definitions of infant visual perception and recognition. A basic measure of visual perception is the tendency of the infant to devote more fixation to some stimuli than to others. One such naturally occurring preference is the infant's devotion of more attention to a novel than to a previously seen target. Novelty preferences may be taken to indicate recognition memory. Following a summary of paradigms employed to test the infant's differential response to novel and previously seen targets, a discussion of the origins and development of visual recognition is presented. The discussion emphasizes the fact that while recognition is possible at any age, the kind of information that is encoded by the infant varies with age. The focus of the chapter then shifts to a consideration of temporal and contextual factors which influence infant recognition memory. Specifically, consideration is given to the effects of study time on recognition; evidence for long-term memory and for forgetting is noted; and the role which study plays in facilitation of recognition is discussed. Recent data indicating a link between early recognition memory and later intelligence are then examined. In the final section of the chapter the focus is on how recognition testing has been used to study the infant's perceptual world and also on practical implications arising from the study of infant visual recognition.

OPERATIONAL DEFINITIONS

Three major techniques are currently used to assess the visual recognition capabilities of infants. All three techniques are based on the assumption that recognition memory is indicated by differential responsiveness to a novel stimulus and a previously exposed stimulus. Two of the paradigms employ direct measurement of the infant's visual interest or differential looking. The third provides an indirect estimate

---

The preparation of this chapter was supported by Major Research Project Grant HD-11089 from the National Institute of Child Health and Human Development.

80   Infant Memory

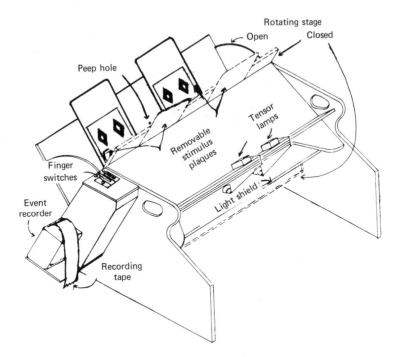

Figure 5.1. Schematic view of visual interest test apparatus.

of visual interest by measuring, instead, the infant's rate of sucking where sucking is employed as an instrumental response by the infant to produce visual stimulation. A test of the infant's visual interest was developed by Fantz in 1956 and rests on the assumption that if an infant looks more at one stimulus than at another, the infant must be able to differentiate between the two targets. The procedure for determining an infant's visual fixation is simple. The infant is placed in front of a "stage" on which targets are secured (see Figure 5.1). With the targets in place, an observer, looking through a peephole centered between the targets, observes the corneal reflection of a target over the pupils of the infant's eyes and records the length of fixation paid to each stimulus.

Based on the visual interest test, three paradigms have been developed to test infant recognition. In one procedure (e.g., Caron & Caron, 1968), a stimulus is presented for a number of trials and then a new stimulus is introduced. Typically, the infant's response to the repeatedly exposed target declines, or "habituates," over trials but returns to its initial level, or "dishabituates," when a novel target is introduced. Presumably, this habituation–dishabituation sequence indicates that the infant has stored some information about the repeatedly exposed stimulus. If, following the initial decline and recovery of response, the old target is reintroduced and the infant's response again declines, there is some indication of delayed recognition.

A second procedure (e.g., Fantz, 1964; Fagan, 1970) based on the visual interest test is to expose the infant to a target for a certain period of time, usually from 1 to 2 minutes, and then to present him with the recently exposed and novel target simultaneously. Infants typically devote the greater part of their visual fixation to the novel target when tested with this paired-comparison approach. Delayed recognition memory is easily tested by varying the time that elapses between the end of the study period and the presentation of the test pairing.

In the third paradigm used to test visual recognition (Siqueland & DeLucia, 1969), a visual stimulus is brought into focus as a contingent reinforcement for high-amplitude sucking. As the infant habituates to the repeated target, sucking declines, and after the infant reaches some criterion of habituation, a new target is introduced. If the sucking response increases in the presence of the novel target, recognition is inferred.

ORIGINS AND DEVELOPMENT

Visual recognition memory may be demonstrated at any age during infancy depending upon the discriminability of the previously exposed and novel targets with which the infant is faced. Generally, targets differing along many dimensions are differentiated on a recognition test at an earlier age than are pairs of stimuli with fewer between-target differences. Moreover, total maturation level as indexed by age from conception appears to play a more important role in determining when preferences for novelty will appear than does length of visual experience as defined by age from birth.

Memory for the previously exposed member of a pair of easily scannable targets differing along a variety of dimensions is demonstrable during the first month of life (Friedman, 1972; Friedman, Bruno, & Vietze, 1974; Milewski & Siqueland, 1975; Milewski, 1978). A demonstration of early recognition is provided in a study by Werner and Siqueland (1978) who employed the high-amplitude sucking paradigm to test 6-day-old infants who had been born 5 weeks prior to term. Following at least 5 minutes of study, the neonates in Werner and Siqueland's experiment were able to differentiate between novel and previously exposed checkerboards that varied in size and number of pattern elements as well as in hue and brightness.

While differentiation between grossly different visual targets may be made on recognition testing during the first week of life, stimuli differing only in orientation, form, or patterning may not be differentiated on a recognition test until about the third or fourth month (e.g., Fantz, Fagan, & Miranda, 1975). Even more difficult are subtle pattern distinctions, such as between achromatic photos of two unfamiliar faces, distinctions that are not likely to be revealed on recognition tests until 5 months (Fagan, 1979). The fact that novelty preferences vary with the nature of the stimulus differences between previously exposed and novel stimuli has been used by investigators to chart theoretically interesting instances of early perceptual development. For example, in a recent study, Fagan and Shepherd (1979) explored the development of 4- to 6-month-old infants' ability to recognize the orientation of a face. Recognition was inferred from the infant's preference for a novel target. The particular orientations, which infants were asked to identify on a recognition test, were chosen to test Braine's (in press) theory of the development of orientation perception.

Braine assumes that the identification of the orientation of a face develops in three stages: the initial distinction is between the face in its upright position and the face positioned either upside down or sideways; distinctions among nonupright orientations, such as upside down or sideways, occur in the second stage; and the third stage is characterized by the discrimination of left from right. Combining their recognition test results with those of others (e.g., Watson, 1966; McGurk, 1970; Fagan, 1972), Fagan and Shepherd provide a summary of the distinctions among facial orientations which are or are not accomplished from 4 to 6 months of age, noting the correspondence of the empirical evidence to Braine's theory. In brief, the results of Fagan and Shepherd, taken together with earlier work, confirm Braine's assumption that distinctions among orientations of a face develop in a sequence in which a differentiation of upright from nonupright precedes distinctions among nonuprights which, in turn, are solved earlier than are left–right discriminations. A detailed discussion of orientation perception is beyond the scope of this chapter. The point of the illustration is that recognition testing may be used to provide a description of the kinds of information which infants encode over age and that such description may aid in evaluating theories of perceptual development.

The final point to be noted with regard to the origins and development of visual recognition is that maturational status, rather than length of visual experience, seems to be the more important determinant of when recognition memory is exhibited. An experiment by Fagan, Fantz, and Miranda (1971) provided pertinent data on the question of whether emergence of novelty preferences, given a particular pair of targets, is more influenced by the length of the infant's experience in the visual world or by the infant's general maturational level. Fagan et al. selected samples of infants born at varying lengths of gestation (i.e., infants at varying maturational levels) and tested their recognition memory for abstract patterning at various postnatal ages (i.e., after differing lengths of visual experience). When responsiveness to novelty was plotted as a function of postnatal age, infants born at term gestation exhibited a reliable preference for novelty about 4 weeks sooner than infants born preterm. Since postnatal age but not gestational age was equated for the two groups in such an analysis, it was obvious that sheer length of visual experience was not determining responsiveness to novelty.

A second analysis that compared the performances of term and preterm infants as a function of conceptional age (i.e., gestational age plus postnatal age) resulted in virtually identical novelty preferences over conceptional age for the two groups. Plotting by conceptional age, in effect, matched the term and preterm infants for total maturational level. The conclusion drawn by Fagan et al. was that total maturational level plays a more important role than length of visual experience in determining when preference for novelty will appear. Subsequent studies by Sigman and Parmelee (1974) and by Rose (1980), in which term and preterm infants have been compared, have confirmed the fact that degree of responsiveness to visual novelty is more similar for term and preterm infants when the infants are matched on conceptional rather than on postnatal age.

STUDY TIME

Infants do not seem to require lengthy study of a target in order to later recognize that target. An experiment by Fagan (1974), for example, asked whether brief amounts of study exposure prior to recognition testing would be effective in allowing novelty preferences to emerge for 5-month-olds. A related question was whether longer study was needed for more difficult discriminations. Difficulty of discrimination was defined by the age at which novelty preferences are first shown for a particular discrimination. Among the tasks included in the Fagan (1974) study were pairs of novel and previously exposed targets which involved discriminations among abstract stimuli varying along a number of dimensions, among abstract targets varying only in pattern arrangement, and among photos of faces (see Figure 5.2). The amount of study time necessary to elicit a novelty preference on recognition testing varied over tasks. As little as 4 seconds of study was needed to differentiate a novel from a previously seen target when the targets varied widely (top row, Figure 5.2). A similar level of novelty preference was not reached for pairs of targets differing solely on patterning (middle row, Figure 5.2) unless 17 seconds had been spent studying the to-be-familiar target. Distinctions among faces (bottom row, Figure 5.2) required 20 to 30 seconds of prior study. The order of task difficulty corresponded to age-related differences in ease of discrimination, with tasks requiring little study also being the tasks solved at an earlier age.

Since the publication of the Fagan (1974) study, similar findings for infants tested at 5 to 6 months of age, with regard to the efficacy of brief study time and the interaction of study time with target discriminability, have been reported by a number of investigators. Specifically, studies by Cornell (1979) and by Rose (1980) have confirmed the fact that widely varying abstract patterns may be differentiated following as little as 5 to 10 seconds of study time. As was the case in the Fagan (1974) study, infants tested by Fagan (1977a), by Cornell (1979), or by Lasky (1980) on abstract patterns which differed only in arrangement of elements required between 15 and 20 seconds of study before one pattern was distinguished from the other. Finally, novelty preference for photos of faces, as in the Fagan (1974) report, emerged after 20 to 30 seconds of familiarization for infants tested by Cornell (1979), Lasky (1980), and Rose (1980).

In short, infants at 5 to 6 months of age are able to recognize a target following a relatively brief exposure to that stimulus. The more similar the to-be-remembered stimulus is to the novel target, the more study time is required to elicit a novelty preference. Moreover, it is also possible that the order of emergence of novelty preferences for particular tasks over age may be recaptured at a single age by varying study time. The latter conclusion indicates that manipulation of both study time and task difficulty at a later age may provide investigators with a simple empirical means for generating and checking hypotheses about the course of information processing over earlier months.

The fact that brief periods of study are effective in revealing infants' recognition memory also has an interesting methodological implication. All the studies demonstrating the utility of brief study time for infant recognition memory have employed the paired-comparison procedure where the infant is exposed to a target for a period of time and then is presented with the previously exposed and novel target simultaneously. A second procedure, as noted in our discussion of operational definitions of infant memory, is to present the same stimulus for a number of trials and then to introduce a new stimulus. In most cases, the infant's response to the repeatedly exposed target declines or habituates over trials but returns to a high level or dishabituates upon introduction of a novel target. In a variant of this procedure, infants are exposed to the same target until a criterion of habituation has been reached and then the novel target is introduced. The experiments on study

Figure 5.2. Pairs of stimuli whose members were differentiated from each other by infants at 5 months following various lengths of study time.

time noted here, however, indicate that recognition may be accomplished long before any reliable decline over trials or any habituation to criterion can be demonstrated. In effect, such findings indicate that the 5- to 6-month-old infant's memory capabilities may be underestimated when a habituation–dishabituation sequence serves as the operational definition of recognition memory.

LONG-TERM MEMORY

The infant's visual recognition memory, at least by 5 months of age, is long lasting and not easily disrupted. A study by Fagan (1970, Experiment I) provided some indirect support for the existence of 24- to 48-hour retention of information gained from exposure to abstract patterns on the part of 5-month-old infants. A direct test of long-term memory for abstract patterns was made in a study by Fagan (1973) in which infants 5 to 6 months of age recognized, after 2 days, which member of a pair of targets they had originally studied, even when the stimuli differed only in patterning. A second experiment in the Fagan (1973) study examined the 5-month-old infant's delayed recognition for photos of faces at intervals of 3 hours, 1, 2, 7, and 14 days. Retention of the information conveyed by a face photo was demonstrated at each interval.

Confirmation of the main findings from the 1970 and 1973 studies by Fagan has been provided by other investigators. Martin (1975) and Strauss and Cohen (1980) demonstrated 24-hour retention of the information conveyed in abstract forms on the part of 5-month-olds and 1 to 2 week retention for abstract patterns at 7 months of age has been reported by Topinka and Steinberg (1978). Finally, infants at 5 to 6 months of age in a study by Cornell (1979) showed 48-hour memory for abstract patterns and face photos when examined for a "savings" effect. Specifically, Cornell (1979) found that a brief study period, sufficient to produce immediate recognition, did not lead to 48-hour retention. However, when allowed a second exposure to a to-be-remembered target just prior to a 48-hour recognition test, an exposure too brief to allow even immediate recognition, infants demonstrated reliable delayed recognition. Cornell's study not only provides a demonstration of long-term recognition but also shows that infants can profit from temporally separate exposures to a target.

Long-term recognition memory is not only present in the early months of life, it is also not easily disrupted. In a study by Fagan (1971), for example, both immediate and delayed recognition tests were made for each of three sets of abstract black-and-white patterns administered during a single test session. The 5-month-old infants in the Fagan (1971) study demonstrated reliably more attention to novel than to familiar targets on immediate and delayed recognition tests for each of the three problems. Degree of differential fixation to novel targets exhibited no decline from immediate to delayed testing and was not altered by the serial order which the problem occupied during testing. In other words, there was no evidence for disruption of memory due to proactive or retroactive interference.

In addition, a series of experiments by Fagan (1977b) sought to induce 5-month-old infants to forget which face photo they had seen before by providing the infants with interference from other face photos or line drawings of faces following initial study of a to-be-remembered face. The general results of the Fagan (1977b) experiments were that highly similar intervening targets could, if presented soon after study, lead to loss of recognition. The deleterious effects of such intervention were quite limited, however. Recovery of recognition occurred after a 1-minute rest, and memory loss was easily prevented by further, brief exposure to the previously studied target. In short, the Fagan (1977b) results indicate that particular amnesiac effects in infant memory could either be prevented or shown to be short-lived. Findings similar to Fagan's (1977b) with regard to the infant's resistance to interference have been reported by Bornstein (1976) for color memory, by McCall, Kennedy, and Dodds (1977) using form-color patterns, and by Cohen, Deloache, and Pearl (1977) for faces.

In summary, various studies confirm the existence of long-term recognition memory on the order of days and even weeks on the part of 5- to 7-month-old infants. Moreover, such memory is quite robust with forgetting occurring only under very circumscribed conditions.

FACILITATION OF RECOGNITION

Certain conditions of study are more apt to result in recognition of a target than are others. Facilitation of recognition may be accomplished either by altering the stimulus or the temporal context of study. Specifically, recognition may be facilitated by allowing the infant to study both the to-be-remembered target and a related target prior to recognition testing. The initial demonstration of the fact that provision of related instances of a target during study may aid later recognition was provided in a study by Fagan (1976). Fagan (1976, Experiments 4 & 5) asked whether an infant would choose a female face as novel on pairings of a male and female face even though a different male face had been shown for study. The 7-month-old infants in the Fagan (1976) study did identify the male face as familiar but only when at least two other males had been shown prior to recognition testing. In a second study, Fagan (1978a) tested the replicability of his original finding (Fagan, 1976) that exposing the infant to related instances of a face during study aids later recognition of that face. In confirmation of the earlier result, Fagan (1978a) found that the 7-month-old infant's recognition memory for a man's face (when that man was to be differentiated from another man) may be improved by allowing the infant prior study of various poses of the to-be-remembered man.

Moreover, the Fagan (1978a) study (Experiment 3) along with another experiment (Fagan, 1978b) found that the provision of related instances of a target during study facilitates not only facial recognition but the recognition of abstract patterns as well. Fagan (1978a) also explored particular stimulus presentation conditions during study which might affect facilitation of recognition. In the Fagan (1978a) series, it was necessary, for example, to present only one associated stimulus along with a to-be-remembered target during study in order to aid recognition. Such facilitation appeared to be most dependent upon the kind of related target shown for study and was more likely to be demonstrated when simultaneous rather than successive exposure to related instances was allowed.

Aside from the studies by Fagan (1976, 1978a, 1978b), instances of facilitation of infants' recognition due to prior study of related targets are contained in reports by Ruff (1978), Olson (1979), and Nelson, Morse, and Leavitt (1979). In the

Ruff (1978) experiment, 9-month-old infants' recognition of the form of an object was enhanced by allowing the infant to study paired instances of an object which were identical in form but which varied in color, size, and orientation. Olson (1979) found memory for abstract designs to be facilitated when 4- to 5-month-old infants were shown multiple items from the same category of patterns prior to recognition testing. Finally, Nelson et al. found that 7-month-old infants could demonstrate generalized recognition of facial expression (e.g., recognizing a happy or a fearful expression across different faces) only when the infant had been exposed to the to-be-remembered expression posed by at least two different models during study. In summary, permitting the infant to see related instances of a to-be-remembered target during study may facilitate the later recognition of that particular target whether it is a face (Fagan, 1978a) or an abstract pattern (Fagan, 1978a, 1978b; Olson, 1979). Moreover, such study conditions may also facilitate the recognition of an invariant feature of a target, such as the sex of a face (Fagan, 1976), the expression of a face (Nelson et al., 1979), or the form of an object (Ruff, 1978).

The temporal context of study may also facilitate later recognition of a target. In a recent experiment by Cornell (1980), 6-month-old infants were asked to distinguish between a male and a female face after retention intervals of 5 seconds, 1 minute, 5 minutes, and 1 hour. All infants were allowed exactly 20 seconds over four 5-second periods to study the to-be-remembered face. For half of the infants, study was "massed" with only 3 seconds separating one 5-second study period from another. The remaining infants were given "distributed" practice in that a 1-minute intertrial interval was imposed between study periods. All infants demonstrated immediate (i.e., 5-second delay) recognition memory. Only infants given distributed study, however, displayed recognition memory at the remaining retention intervals. Cornell's demonstration that distributed study facilitates infants' delayed recognition memory indicates that the temporal as well as the stimulus context of study may alter the infant's subsequent recognition of a target.

In brief, there is evidence to indicate that certain stimulus and temporal conditions of study may facilitate recognition. The exploration of the parameters surrounding the effect of study context on recognition and the suggestion of possible mechanisms which may account for the effect remain as key empirical and theoretical issues. It is likely that the study of how recognition may be facilitated will receive increasing attention from investigators of infant memory since a great deal of both theoretical and practical benefit can come from knowing how to ensure that particular information will be encoded by an infant.

INFANT RECOGNITION MEMORY AND LATER INTELLIGENCE

Infant recognition ability, as measured by extent of visual preferences for novel targets, appears to be related to later cognitive functioning. Indirect support for a link between early recognition memory and later intelligence is provided by studies where groups of infants expected to differ in intelligence later in life have been found to differ in the ability to recognize a familiar visual stimulus. Fantz and Nevis (1967) compared home-reared offspring of highly intelligent parents with institution-reared offspring of women of average intelligence. Miranda and Fantz (1974) compared Down syndrome with normal infants. In each study, the infants in groups expected to be more intelligent later in life were also superior in visual recognition during infancy.

A direct test of the assumption that variations in infant recognition memory represent individual differences in intelligence has been made by Fagan and McGrath (1981). Specifically, 93 infants who had participated in earlier studies by Fagan (1971, 1973, 1976, 1977b), in which they had been given multiple-paired presentations of novel and previously seen stimuli, were revisited after approximately 4 to 7 years and given vocabulary tests of intelligence. Fagan and McGrath found that tests of infant visual recognition memory are valid predictors of later intelligence. Significant correlations between infants' preferential visual fixation to novel stimuli and later intelligence of .37 and of .57 for the 4- and 7-year samples were obtained. The correlations for the males (.31 and .49) and for the females (.40 and .67) at 4 and 7 years, respectively, did not depart markedly from the overall coefficients at each age or, within ages, from one another. Substantial correlations between infant memory and later vocabulary were also reported at each age when variation due to parental educational level was factored out of the original correlations. In short, Fagan and McGrath obtained statistically significant relationships between infant recognition scores and later vocabulary tests. Moreover, the values did not vary by sex and could not be accounted for on the basis of variations in socioeconomic status.

It is important to note that the correlations of .37 and .57 reported by Fagan and McGrath very likely underestimate the predictive validity of infant memory tests due to the restricted range of intelligence within which their predictions had to be made and due to the relatively low reliability of the infant memory scores they employed. Peabody IQ scores, for example, for their subjects ranged from 90 to 130. The low reliability of the infant memory scores was a simple function of the small number of items (i.e., usually three novelty tests) upon which the memory scores were based. In effect, higher predictive validity coefficients than those obtained by Fagan and McGrath would be expected given a wider range of cognitive functioning in a population or when infant visual recognition scores may be based on a greater number of recognition tests.

In short, there is support for the assumption that variations in early visual recognition memory reflect and predict variations in intelligent functioning. Practically, the development of refined tests of infant memory may accomplish what years of research on infant sensorimotor abilities has not done; that is, predict later intellectual functioning. Theoretically, the demonstration of continuity between early visual recognition memory and later intelligence raises the question of the basis for the continuity and calls for the specification of similarities in the informational character and the solution processes between early visual recognition tasks and later intelligence tests. Our present purpose, however, is simply to note that the demonstration of a link between early visual memory and later intelligence raises the issue of the basis of such a continuity, an issue sure to be of importance for the further study and analysis of infant visual recognition.

SUMMARY AND DISCUSSION

The infant's tendency to devote more fixation to a novel than to a previously exposed target is an operational definition of visual recognition memory. Infants, from birth, are able to differentiate among highly discriminable targets on a recognition test. Successively finer distinctions are made with increasing maturational level. The infant, at least from 5 months of age, requires relatively little study of a target for subsequent recognition with more or less study needed to encode different kinds of information. Also by 5 months of age, the infant's recognition memory is long lasting and is not easily disrupted. The stimulus or temporal context in which a stimulus is presented for study may facilitate the later recognition of that stimulus. Finally, it appears that variations in early recognition memory may be predictive of later intellectual functioning.

Perhaps the most important implication to draw from the study of infant recognition memory is that such tests of recognition provide information on the perceptual–cognitive world of the infant. For example, by controlling the manner in which novel and previously exposed targets vary, investigators have explored the development of the infant's perception of color (Bornstein, 1978), shape (Schwartz & Day, 1979), and orientation (Fagan & Shepherd, 1979). Further, the study of responsiveness to novelty has also made it possible to discover if an infant, on recognition testing, realizes that a previously seen stimulus has, in some respect, remained the same. In other words, it is possible to discover whether specific features of a target are perceived as invariant by an infant.

In a study by Fagan (1977a), for example, infants were allowed to study a form–color compound and then were presented with a pair of targets with a familiar and a novel cue along one dimension and the same two novel cues along the other. To test whether color could serve as a basis for recognition, the infant might study a red diamond and then be tested on the pairing red square versus green square. In this example, the only dimension containing a familiar and a novel cue is color; hence a reliable preference for novelty would indicate that color had been encoded as an invariant feature during study. Fagan (1977a) found that infants were able to encode either the invariant form or color of a target as a basis for later recognition. Many experiments have demonstrated that infants are able to detect features of a stimulus which remain invariant from study to test.

McGurk (1972), Cornell (1975), and Fagan (1979) found that infants between 4 and 6 months of age recognize a pattern even though its orientation has been changed, and studies by Fagan (1976), Cohen and Strauss (1979), Fagan and Singer (1979), Nelson et al. (1979), and Strauss (1979) indicate that infants from 7 months of age recognize invariant aspects of faces. Infants, by 5 months of age, are also able to recognize information common to an object and to a picture of that object (Dirks & Gibson, 1977; Rose, 1977; Deloache, Strauss, & Maynard, 1979) and, by 1 year, can employ information gained by study in one modality to solve a recognition test given in another modality (Gottfried, Rose, & Bridger, 1977, 1978; Rose, Gottfried, & Bridger, 1978). While additional examples might be given of the kinds of information that infants encode, the point is simply that tests of visual recognition tell us about the infant's developing ability to perceive, to abstract, to categorize, and to transfer information.

The last implication to be noted here is that the study of infant visual recognition may result in the measurement of early intelligence. Such an outcome would represent a clear example of the derivation of a clinical tool from basic research. Practically, the results noted earlier, which demonstrate links between early visual memory and later intelligence, encourage continued research on the development of more extensive tests of infant visual memory as an estimate of infant intelligence.

Such tests, when developed, would have a number of uses. They would have immediate application as screening devices for infants who, because of various prenatal or perinatal factors, are to be at risk for intellectual retardation. Another important use of tests of infant memory would be to determine the short-term effects of intervention programs. Finally, memory tests developed for infants might be employed to measure cognitive functioning in populations such as the profoundly retarded for whom nonverbal tests are required, a possibility currently being explored by Shepherd and Fagan (1981).

REFERENCES

Bornstein, M. H. Infants' recognition memory for hue. *Developmental Psychology,* 1976, **12,** 185–191.

Bornstein, M. H. Chromatic vision in infancy. In H. W. Reese & L. P. Lipsitt (Eds.), *Advances in child development and behavior* (Vol. 12). New York: Academic Press, 1978.

Braine, L. G. Early stages in the perception of orientation. In M. Bortner (Ed.), *Cognitive development and growth,* in press.

Caron, R. F., & Caron, A. J. The effects of repeated exposure and stimulus complexity on visual fixation in infants. *Psychonomic Science,* 1968, **10,** 207–208.

Cohen, L. B., Deloache, J. S., & Pearl, R. A. An examination of interference effects in infants' memory for faces. *Child Development,* 1977, **48,** 88–96.

Cohen, L. B., & Strauss, M. S. Concept acquisition in the human infant. *Child Development,* 1979, **50,** 419–424.

Cornell, E. H. Infants' visual attention to pattern arrangement and orientation. *Child Development,* 1975, **46,** 229–232.

Cornell, E. H. Infants' recognition memory, forgetting, and savings. *Journal of Experimental Child Psychology,* 1979, **28,** 359–374.

Cornell, E. H. Distributed study facilitates infants' delayed recognition memory. Paper presented at the International Conference on Infant Studies, New Haven, Conn., 1980.

Deloache, J. S., Strauss, M. S. & Maynard, J. Picture perception in infancy. *Infant Behavior and Development,* 1979, **2,** 77–89.

Dirks, J., & Gibson, E. J. Infants' perception of similarity between live people and their photographs. *Child Development,* 1977, **48,** 124–130.

Fagan, J. F. Memory in the infant. *Journal of Experimental Child Psychology,* 1970, **9,** 217–226.

Fagan, J. F. Infants' recognition memory for a series of visual stimuli. *Journal of Experimental Child Psychology,* 1971, **11,** 244–250.

Fagan, J. F. Infants' recognition memory for faces. *Journal of Experimental Child Psychology,* 1972, **14,** 453–476.

Fagan, J. F. Infants' delayed recognition memory and forgetting. *Journal of Experimental Child Psychology,* 1973, **16,** 424–450.

Fagan, J. F. Infant recognition memory: The effects of length of familiarization and type of discrimination task. *Child Development,* 1974, **45,** 351–356.

Fagan, J. F. Infants' recognition of invariant features of faces. *Child Development,* 1976, **47,** 627–638.

Fagan, J. F. An attention model of infant recognition. *Child Development,* 1977, **48,** 345–359. (a)

Fagan, J. F. Infant recognition memory: Studies in forgetting. *Child Development,* 1977, **48,** 68–78. (b)

Fagan, J. F. Facilitation of infants' recognition memory. *Child Development,* 1978, **49,** 1066–1075. (a)

Fagan, J. F. Infant recognition memory and early cognitive ability: Empirical, theoretical, and remedial considerations. In F. D. Minifie & L. L. Lloyd (Eds.), *Communicative and cognitive abilities—Early behavioral assessment*. Baltimore: University Park Press, 1978. (b)

Fagan, J. F. The origins of facial pattern recognition. In M. Bornstein & W. Kessen (Eds.), *Psychological development from infancy*. Hillsdale, N.J.: Erlbaum, 1979.

Fagan, J. F., Fantz, R. L., & Miranda, S. B. Infants' attention to novel stimuli as a function of postnatal and conceptional age. Paper presented at Society for Research in Child Development Meeting, April 4, 1971.

Fagan, J. F., & McGrath, S. K. Infant recognition memory and later intelligence. *Intelligence*, 1981, **5**, 121–130.

Fagan, J. F., & Shepherd, P. A. Infants' perception of face orientation. *Infant Behavior and Development*, 1979, **2**, 227–234.

Fagan, J. F., & Singer, L. T. The role of simple feature differences in infants' recognition of faces. *Infant Behavior and Development*, 1979, **2**, 39–45.

Fantz, R. L. A method for studying early visual development. *Perceptual and Motor Skills*, 1956, **6**, 13–15.

Fantz, R. L. Visual experience in infants: Decreased attention to familiar patterns relative to novel ones. *Science*, 1964, **146**, 668–670.

Fantz, R. L., Fagan, J. F., & Miranda, S. B. Early perceptual development as shown by visual discrimination, selectivity, and memory with varying stimulus and population parameters. In L. Cohen & P. Salapatek (Eds.), *Infant perception: From sensation to cognition: Basic visual processes* (Vol. 1). New York: Academic Press, 1975.

Fantz, R. L., & Nevis, S. The predictive value of changes in visual preferences in early infancy. In J. Hellmuth (Ed.), *The exceptional infant* (Vol. 1). Seattle: Special Child Publications, 1967.

Friedman, S. B. Habituation and recovery of visual response in the alert human newborn. *Journal of Experimental Child Psychology*, 1972, **13**, 339–349.

Friedman, S. B., Bruno, L. A., & Vietze, P. Newborn habituation to visual stimuli: A sex difference in novelty detection. *Journal of Experimental Child Psychology*, 1974, **18**, 242–251.

Gottfried, A. W., Rose, S. A., & Bridger, W. H. Cross-modal transfer in human infants. *Child Development*, 1977, **48**, 118–123.

Gottfried, A. W., Rose, S. A., & Bridger, W. H. Effects of visual, haptic, and manipulatory experiences on infants' visual recognition memory of objects. *Developmental Psychology*, 1978, **14**, 305–312.

Lasky, R. E. Length of familiarization and preference for novel and familiar stimuli. *Infant Behavior and Development*, 1980, **3**, 15–28.

McCall, R. B., Kennedy, C. B., & Dodds, C. The interfering effects of distracting stimuli on the infant's memory. *Child Development*, 1977, **48**, 79–87.

McGurk, H. The role of object orientation in infant perception. *Journal of Experimental Child Psychology*, 1970, **9**, 363–373.

McGurk, H. Infant discrimination of orientation. *Journal of Experimental Child Psychology*, 1972, **14**, 151–164.

Martin, R. M. Effects of familiar and complex stimuli on infant attention. *Developmental Psychology*, 1975, **11**, 178–185.

Milewski, A. E. Young infants' visual processing of internal and adjacent shapes. *Infant Behavior and Development*, 1978, **1**, 359–371.

Milewski, A. E., & Siqueland, E. R. Discrimination of color and pattern novelty in one-month infants. *Journal of Experimental Child Psychology*, 1975, **19**, 122–136.

Miranda, S. B., & Fantz, R. L. Recognition memory in Down's syndrome and normal infants. *Child Development*, 1974, **45**, 651–660.

Nelson, C. A., Morse, P. A., & Leavitt, L. A. Recognition of facial expressions by seven-month-old infants. *Child Development*, 1979, **50**, 1239–1242.

Olson, G. M. An information processing analysis of visual memory and habituation in infants.

In T. J. Tighe & N. Leaton (Eds.), *Habituation: Perspectives from child development, animal behavior, and neurophysiology.* Hillsdale, N.J.: Erlbaum, 1976.

Olson, G. M. Infant recognition memory for briefly presented visual stimuli. *Infant Behavior and Development,* 1979, **2**, 123–134.

Rose, S. A. Infants' transfer of response between two-dimensional and three-dimensional stimuli. *Child Development,* 1977, **48**, 1086–1091.

Rose, S. A. Enhancing visual recognition memory in preterm infants. *Developmental Psychology,* 1980, **16**, 85–92.

Rose, S. A., Gottfried, A. W., & Bridger, W. H. Cross-modal transfer in infants: Relationship to prematurity and socioeconomic background. *Developmental Psychology,* 1978, **14**, 643–652.

Ruff, H. A. Infant recognition of the invariant form of objects. *Child Development,* 1978, **49**, 293–306.

Schwartz, M., & Day, R. H. Visual shape perception in early infancy. *Monographs of the Society for Research in Child Development,* 1979, 182.

Shepherd, P. A., & Fagan, J. F. Visual pattern detection and recognition memory in children with profound mental retardation. In N. R. Ellis (Ed.), *International review of research in mental retardation,* New York: Academic, 1981.

Sigman, M., & Parmalee, A. H. Visual preferences of four-month-old premature and fullterm infants. *Child Development,* 1974, **45**, 959–965.

Siqueland, E. R., & DeLucia, C. A. Visual reinforcement of non-nutritive sucking in human infants. *Science,* 1969, **165**, 1144–1146.

Strauss, M. S. Abstraction of prototypical information by adults and 10-month-old infants. *Journal of Experimental Psychology: Human Language and Memory,* 1979, **5**, 618–632.

Strauss, M. S., & Cohen, L. B. Infant immediate and delayed memory for perceptual dimensions. Paper presented at the International Conference on Infant Studies, New Haven, Conn., April 1980.

Topinka, C. V., & Steinberg, B. Visual recognition memory in 3½ and 7½ month-old infants. Paper presented at the International Conference on Infant Studies, Providence, R.I.: March 1978.

Watson, J. S. Perception of object orientation in infants. *Merrill-Palmer Quarterly,* 1966, **12**, 73–94.

Werner, J. S., & Perlmutter, M. Development of visual memory in infants. In H. W. Reese & L. P. Lipsitt (Eds.), *Advances in child development and behavior* (Vol. 14). New York: Academic, 1979.

Werner, J. S., & Siqueland, E. R. Visual recognition memory in the preterm infant. *Infant Behavior and Development,* 1978, **1**, 79–94.

6

THE DEVELOPMENT OF OBJECT PERCEPTION IN INFANCY

Holly A. Ruff
Department of Pediatrics
Albert Einstein College of Medicine

The perception of animate and inanimate objects is important for the infant's emerging social skills, for his ability to move around and manipulate his environment, and for the development of cognitive abilities, particularly object permanence and object categorization. Despite the burgeoning of studies in the area of infant perception, however, relatively few studies have directly investigated the perception of objects; as a consequence, a number of important aspects of infant perception have been neglected. The purpose of this chapter, therefore, is to discuss the development of the perception required for the infant's active commerce with the three-dimensional world, to consider issues in need of investigation, as well as to review what is already known. To keep the topic within manageable proportions, the discussion will be limited to development in the first 6 months.

BACKGROUND

Objects are enclosed substances with surfaces independent of the surrounding surfaces. As such, they are free to move or be moved. Since object perception must occur in a world in which both the observer and the objects around him move, an adequate theory of object perception must be based, at least in part, on the changing stimulation which results from such motion. For this reason, I have found James Gibson's view of perception (1966, 1979) to be particularly helpful. He maintained that there is information in the changing optic array for both change and nonchange in the objective world.[1] The perception of a stable world is possible be-

---

[1] The optic array refers to the structured light that reaches the eye; this structure is imposed on the unstructured illumination by the physical structure of the environment from which the light is reflected. The resulting discontinuities of color and brightness provide information about the environment's structure.

I would like to thank Richard Aslin, Katharine Lawson, Gerald Turkewitz, and Herbert Vaughan for commenting critically on an earlier draft of this chapter.

cause, through all the changes, something in the optic array remains the same. That which does not change is referred to as invariance and is what specifies the enduring objective structure of the environment. Although not everyone would agree (see Braunstein, 1976), perception is considered here to be the detection of invariance.

An object's structural properties are critical for perception and recognition; since these properties can be analyzed in terms of surface shape and slant and the relationship of one surface to another, object perception is based on principles which underlie perception in general. Surfaces of objects are textured, and essential to perception is the ability to detect brightness margins between texture elements. Object perception then requires that we be able to detect edges, that is, the place at which two surfaces meet. According to Gibson (1979), the most important source of information for an object's edge is the accretion and deletion of the background texture that occurs when the object moves or when the observer moves while watching it. The accretion and deletion reverse when either the object or the observer reverses direction; Gibson calls this general event a "reversible occlusion." The locus along which texture appears and disappears is the edge of an object in front of its background. The regularity and ubiquity of this phenomenon make it a powerful source of information about the separation of the object from the rest of the environment. The information from accretion and deletion is correlated with other kinds of information that can also be used in the perception of edges. For example, stereopsis, the binocular perception of depth resulting from retinal disparity, allows the observer to see an edge separated from the background. Other information for edges includes texture differences between the object and the background, and frequently, brightness and color contrast as well. Edges and boundaries, however, are not marked unambiguously by contrast, for such contrast also occurs on flat surfaces where there is no structural discontinuity. On the other hand, accretion and deletion of background texture can specify an occluding edge where there is no contrast in texture, brightness, or color (Gibson, Kaplan, Reynolds, & Wheeler, 1969). Nevertheless, perception of edges involves occlusions, binocular disparity, *and* the various kinds of contrast often enough that they are highly correlated.

In addition to detecting edges, the observer must also be able to accurately perceive the shape and slant of the object's surfaces. For this, gradients of texture are important (Gibson, 1950; Braunstein, 1976). The material composition of most objects is relatively homogeneous and results in a uniform pattern of color or brightness differences. The texture of a surface becomes optically more dense as the surface recedes from the observer; the steeper the angle, the faster the increase in density. Texture gradients can also specify the kind of surface; thus the texture density is maximum at the center of a concave surface, while it is maximum at the edges of a convex one. Adults are able to use density gradients for perception of a slanting surface though there is not a one-to-one relationship between the gradient and the degree of perceived slant (Gibson, 1950a; Flock, 1964). There are, however, other sources of information about slant.

Additional information about the slant of surfaces comes from the systematic projective change that an object's outline undergoes as it moves. For example, a surface which rotates away from the frontoparallel plane to a tilt of 45° with reference to the line of sight undergoes progressive compression in one direction, or foreshortening; the same is true for all texture elements on the surface. Information from progressive foreshortening of texture elements does not require uniform distribution of elements or even regularity in size.

A final aspect of object perception is the ability to see the relationship of one part of the object to another. Again, powerful sources of information are available during motion of the object or observer, although binocular perception may make some of the same information available to the stationary observer of a stationary object. Just as occlusions make it possible to see the object's position relative to the rest of the environment, they also provide information about the relative positions of parts of a complex object. For example, as an object moves laterally away from the observer, nearer portions of the object will occlude those surfaces which are further from the observer. During rotation, one side of the object will gradually be hidden from view as the "back" side gradually comes into view. The nature of the change will help to specify the type of edge as well as the arrangement of parts.

As an object moves, the projected shape of the object changes; the projected outline provides some information about the shape of contour but is ambiguous in that several possible objects could project in the same way. For rigid objects, the family of perspective changes can specify the structure of the object (Hay, 1966) in a way that a single perspective cannot. Johansson (1977) has suggested that the visual system analyzes the stimulus flow according to the principles of projective geometry. To support this suggestion, he provides evidence that the human observer is predisposed to interpret even movement in two dimensions as rigid motion of a surface in depth rather than as deformation in the frontoparallel plane.

The visual system is, however, more flexible and complex since it must also extract invariants from nonrigid as well as from rigid motion. Investigation shows that adults can perceive the expression on a face (Bassili, 1978) and the sex of a person walking (Cutting, Profitt, & Koslowzki, 1978) from nothing more than spots of light attached to them. Some of this ability involves extracting invariants from the underlying rigid structure, but it also involves detection of particular kinds of motion, for example, the movement of the face when smiling. There is much that needs to be understood about the means by which adults use projective transformations, but the combination of such information with that from occlusions and texture gradients make perception highly veridical.

Moving from summarizing the sources of information adults can and probably do use in their perception of objects, the rest of the chapter concentrates on development during infancy. How do infants come to use occlusions, binocular disparity, texture gradients, and projective transformations to perceive the structure of objects? It would be arrogant to propose to answer this question in any definitive way, but I would like to suggest an outline that is consistent with the general theoretical framework discussed above.

DEVELOPMENT DURING INFANCY

Eleanor Gibson (1969) suggests that development proceeds because of "perceptual learning," a term that refers to the gradual differentiation of perception that takes place with continued exposure to an object. Since perception of objects involves the detection of invariance, differentiation, in this case, is the detection of more and more specific invariants. As the specificity in the perception of a single object increases, so does the observer's ability to discriminate that object from other similar ones. It is assumed that the newborn infant starts with some capacity to perceive regularities in the flow of stimulation and therefore to detect some very general invariants; perceptual development involves increasing specificity. Also, since the detection of invariants in a given object should facilitate detection of invariants in other objects that share some of the same characteristics, one would expect the rate of perceptual learning to increase with the infant's increased experience. One constraint on early perception, then, is simply lack of experience. The infant's development also depends on the maturation of the system, and anatomical and physiological immaturities in this system impose another constraint on the infant's ability to perceive objects. Finally, perception, as it is being discussed here, is an active process involving exploration by the observer; thus, development during infancy almost certainly involves increasing efficiency in the process of exploration as well as the development of new modes of exploration.

The Starting Point

One way to examine the origins of object perception is to look at what the newborn is capable of. The development of the visual system is so rapid that infant sensory capacities by 6 months approximate those of the adult; for example, acuity (Dobson & Teller, 1978), accommodative capacity (Banks, 1980), and visual evoked potential (Sokol & Jones, 1979) are close to adult values at that time. Yet it must be said that the newborn starts life with an immature and imperfect system. He cannot resolve fine detail because his acuity is poor; it has been estimated to be around 20/800 at best (Dobson & Teller, 1978). He has some accommodative capacity, but his depth of focus is large enough that there must be considerable blur (Banks, 1980). The poor acuity and accommodation mean that the newborn will be sensitive mainly to low spatial frequencies. In addition, the newborn's contrast sensitivity is considerably inferior to that of the adult (Atkinson, Braddick, & Braddick, 1974; Banks & Salapatek, 1978), so that contrast must be fairly high before the infant can detect the difference between light and dark parts of the stimulus. There is a relationship between spatial frequency and contrast, so that the higher the contrast, the smaller the frequency that can potentially be seen. Most experiments with newborns capitalize on this by presenting infants with high-contrast black-and-white displays.

Another aspect of visual functioning that is immature at birth is binocular vision. Aslin and Dumais (1980) report that bifoveal fixation and convergence are unreliable until 2 or 3 months of age. Both deficiencies potentially affect the system's ability to use the discrepancy in the views from the two eyes to perceive an object in depth. Bifoveal fixation and convergence are not sufficient, however, because stereopsis depends upon the detection of binocular disparity, and infants do not seem to be sensitive to binocular disparity until 3½ months of age (Fox, Aslin, Shea, & Dumais, 1980).

How do these limitations of the system affect object perception? Because brightness contrast makes the perception of margins between texture elements possible and because perception of texture elements is important for seeing occluding edges, the newborn infant will see occlusions only of the larger and more contrasting elements of the background. Perception of the nature and slant of surfaces usually depends upon the perception of finer texture elements, so discrimination of flat and curved surfaces, angular and rounded corners, and frontal and slanted surfaces may not be possible until around 2 months of age. The fineness and accuracy of the infant's ability to perceive the location of objects and their edges may be limited by poor convergence and the lack of stereopsis, although the infant who does not move much or reach for objects does not require much information about object solidity and relative placement.

The important point, however, is that the immaturities of the visual system do not make object perception based on monocular information impossible, even for the youngest infant. Banks and Salapatek (1978) write "since the intensity distribution of any two-dimensional, time-invariant stimulus can be specified in terms of its spatial frequency content, the CSF (contrast sensitivity function) provides a useful estimate of the form information to which an observer is sensitive" (p. 361). On this basis, however, the abilities of the newborn could be underestimated. The normal stimulus for perception is neither two-dimensional nor time-invariant. The stimulation from the changing optic array means that much potential information is there even for an immature and inefficient system. For example, edges that are invisible when the object is stationary, that is, not of sufficient contrast, may become visible during movement, because of the accretion and deletion of the background. The results of experiments on infant perception, therefore, may overestimate the functional consequences of the system's immaturity by not simulating the richness of the infant's normal experience.

Although the newborn's sensory capacities do not preclude object perception at some level, other kinds of studies are necessary to show whether infants, in fact, are sensitive to information specifying structural properties of objects. There are many studies which demonstrate the newborn's discriminative ability; Fantz, Fagan, and Miranda (1975) report, for example, that newborns prefer (and, therefore, must discriminate) patterned over nonpatterned stimuli, curved lines over straight ones, and patterns with more and larger elements over those with fewer and smaller ones. These studies are important in that they show the infant's differential sensitivity to patterned stimulation and, possibly, their ability to discriminate between some projected contours. The studies cannot, however, provide a complete picture of object perception because the stimuli used are two-dimensional and do not allow us to test for the perception of occlusions, the perception of shape and slant through gradients of texture density, or sensitivity to perspective transformations. Studies with moving three-dimensional displays are necessary to determine what discriminations between objects can be made; then there will be a need for studies which separate experimentally the different possible sources of information in order to determine the basis for the discriminations. The task will be a difficult one.

The Detection of Invariance

There seem to be no data that contradict the idea that the newborn is capable of detecting invariance, but neither do we have data that demonstrate this ability. There is, on the other hand, useful data with older infants and an examination of these studies follows.

First, a brief digression into methodological issues seems desirable. Discrimination is demonstrated whenever an infant responds in some differential manner to two displays. Simple preference techniques have been very useful because an infant's significantly longer looking at one stimulus in a pair of stimuli means that he must be discriminating between them on some basis. Since no conclusions can be drawn from lack of preference, experimenters turned to techniques which expose the baby to one display and then test his response to the previously exposed display versus his response to a new one. Two versions of this technique have been developed. One (Fagan, 1970) involves familiarizing the infant with one display for a given length of time and then presenting the infant with the familiarized display and a novel one. Any significant difference in response to the two shows discrimination. The second technique (Jeffrey & Cohen, 1971) involves exposing the infant to a given display until he habituates; that is, shows a response decrement. Then the previously exposed display and a new one are presented on successive trials. If the infant shows an increase in responsiveness to the new display and continues to show a response decrement to the old one, discrimination is shown. Positive results from both techniques also demonstrate the infant's recognition memory of the familiarized, or habituated, display (see Chapter 5 for further discussion of this aspect). Although negative results again are hard to interpret, the technique presumably provides some motivation for looking at one rather than another of two displays that may, initially, have been equally attractive. In this sense, it represents an alternative to more laborious conditioning techniques. The novel display can also be varied in many ways to determine what the basis of the discrimination and recognition is. Both techniques can be modified to test for the infant's ability to detect invariance in objects (1) by varying the object during familiarization or habituation and/or (2) by varying the familiarized object during the test. In either case, positive results show that the infant is not simply discriminating or recognizing exactly the same display but has indeed detected some invariant aspect of the object. For the purposes of this chapter, these techniques are important for what they tell us about the detection of invariance and not for what they tell about recognition memory as such.

Several studies are relevant to the question of whether infants can detect structural invariance in objects. Bower (1966) conditioned 2-month-old infants to turn their heads in response to a flat rectangle and found that they generalized the conditioned response to the same rectangle when both the orientation and the projected shape of the object had changed. Generalization to conditions in which only the projected shape or only the orientation changed was much less. His data suggest that the infants were attending to the real shape of the rectangle; they must, then, have detected some invariant of the shape.

Day and McKenzie (1973) report confirming data. They recorded the looking times of 6- to 16-week-old infants under three conditions: (1) eight 20-second presentations of a stationary cube in the same orientation on each trial; (2) eight 20-second presentations of a stationary cube in a different orientation on each trial; and (3) eight 20-second presentations of cutout photographs of the condition in which the orientation varied. They found that infants habituated equally rapidly to the first two conditions involving the real cube, suggesting that they were attending to some invariant property of the cube regardless of change in orientation. Further, they found no habituation in the condition with the photographs; this result raises the possibility that the infants, in the first two conditions, were using binocular perception or information resulting from head motion. While these data are suggestive, the investigators did not incorporate any test to determine what specific aspect of the three-dimensional displays were being habituated to; the infants may, therefore, have been responding to invariants specifying a general characteristic, such as straight edges, rather than invariants specifying the particular relationship of surfaces in a cube. Such an interpretation is supported by a study by Cook, Field, and Griffiths (1978) who found that infants of the same age showed an equally rapid response decrement to (1) a cube that differed in orientation from trial to trial for eight trials and (2) a cube and a wedge alternating from trial to trial and also differing in orientation.

Caron, Caron, and Carlson (1979) provide data, however, that can be interpreted unequivocally as showing that infants respond to a particular relationship of edges.[2] They repeatedly presented 12-week-old infants with either a square or an equal-sized trapezoid and varied the slant, and, therefore, the projected shape, from trial to trial. When each infant reached a criterion of habituation, he was presented with the habituated shape in a novel orientation and the other, novel, shape. The infants continued to show response decrement to the old shape in a novel orientation but showed response recovery to the novel shape. While these results show response to structural invariants, similar studies with solid forms would provide a better basis for determining what information infants use in their pickup of invariants in situations where they are presented with stationary objects.

The studies by Bower, Day and McKenzie, and Caron et al. were all done in the classic shape constancy tradition,[3] and they show clearly that infants respond to the real shape, or structure, and not to a particular projection on the retina. In a framework which emphasizes the stimulus flow, however, the classic problem of how the observer perceives the real shape from a static retinal image no longer exists (Johansson, 1977). Instead, the problem becomes one of how the invariant structure of the object is detected in changing stimulation. In fact, changing stimulation should make such detection easier by providing more information. It is reasonable to suppose, therefore, that infants younger than 2 to 3 months could perceive the structural invariance of the simple objects used in the above studies and perhaps perceive more complex ones if they had an opportunity to view them during movement. Experiments addressing this particular possibility have not been done, but E. Gibson, Owsley, Walker, and Megaw-Nyce (1979) have shown that detecting invariance in a moving object is at least as easy; that is, 3-month-old infants who are habituated to a geometric form as it moves respond to that form as familiar, even when it undergoes a novel motion. In addition, Owsley (1980) found that rotation facilitated the monocular perception and recognition of a wedge for 4-month-olds. That is, infants habituated to the rotating wedge generalized habituation to the stationary wedge in two different positions and recovered responsiveness to a cube, while infants habituated to the stationary wedge failed to show differentiation on the test.

Ruff (1980) also found that 6-month-olds' recognition of a fairly complex abstract structure appeared to be facilitated by some simple movements; that is, the infants showed better discrimination between novel and familiar objects after a moving familiarization than after an essentially stationary familiarization, even after equivalent periods of looking. However, some movements (i.e., rotation around one or more axes) seemed to make the task more difficult than other movements (i.e.,

---

[2] Since they used a flat cutout form rather than a solid form, it is not possible to speak of the relationship of surfaces.
[3] See Day and McKenzie (1977) for a review of the development of the constancies during infancy.

displacement without rotation). This differential effect suggests that the complexity of the optical transformation affects the ease with which information about structure can be extracted. The stimulus flow will vary in its complexity according to the complexity of both the object that is moving and the motion itself. The young infant's experience may mean that he is more likely to be affected by these variations than an adult who has already extracted the invariants of both motion and structure in many objects and events.

A similar picture emerges for infants' perception of the human face, an example of an animate, elastic object. Barrera and Maurer (in press) and Maurer and Heroux (1980) provide evidence for 3-month-old infants' ability to discriminate between photographs of their mother's face and a strange female's face and to discriminate between photographs of the father's face and a strange male's face. The transfer between the live face and the photograph suggests that infants have learned to detect some invariants in the faces of their parents by three months; given the amount of experience with their parents' faces prior to three months, this ability is not surprising. Detecting invariants in and discriminating among strange faces, however, seems to be more difficult. Dirks and Gibson (1977) found that 5-month-olds transferred an habituated response from the live face of a previously unknown person to his photograph and recovered response to the photograph of a complete stranger. This discrimination occurred, however, only when there were gross differences between the two faces. Again, however, there is indirect evidence that object motion is facilitating. Spelke (1975) reports results which strongly suggest that 4-month-old infants can extract from moving pictures the structural invariance of an individual face and recognize that face when it is engaged in a novel activity, even when the faces to be discriminated on the recognition test were very similar.

The above work suggests that, by 2 to 3 months, infants can detect some structural invariants in rigid geometrical forms and in animate forms, and that their detection may well be facilitated by object motion and the consequent changes in the optic array. Clearly, much more detailed work is necessary to determine the exact nature of the invariants detected. One would expect, on the basis of the principle of differentiation, that the invariants the infant detects first will be of a general nature; but the detection of general invariants still make it possible for him to discriminate among many classes of objects. For example, two- to three-month-old infants show a clear discrimination between curved and straight patterns by their strong preference for the curves (Fantz & Nevis, 1967; Ruff & Birch, 1974) and such discrimination may occur even earlier (Fantz & Miranda, 1975). To see when the same behavior occurs with curved and straight edges of three-dimensional objects would not be difficult, but the problem has not yet been investigated. Another distinction is between elastic and rigid substances. Gibson, Owsley, & Johnston (1978) found that 5 month-old infants habituated to an object which made only rigid motions would recover to the object undergoing elastic deformation, but not to the object making a novel rigid motion. Since the object had spots on it, the authors suggest that the discrimination was made on the basis of the fact that the adjacency relationships among texture elements remain the same under rigid motion but alter under elastic motion. Their explanation is not definitive since the contour of elastic objects also changes with motion, but the study represents an important area of discrimination and recognition to be pursued. Both of the mentioned discriminations are important for the general differentiation between animate and inanimate objects. Another important basis on which classes of objects can be discriminated is whether their edges or surfaces are convex or concave. Yonas (1973) found some evidence for discrimination of these characteristics before 6 months using visual fixation as a measure. Ruff (1981) found that 6-month-olds, during looking and manipulation could differentiate cubes with small round protusions from ones with small round indentations.

A more subtle kind of discrimination that infants learn to make is between different configurations of the same elements. Kagan (1970) reports that 4-month-old infants, given 3 weeks experience with a mobile of three geometric forms, responded to the novelty of a mobile with a different spatial arrangement of the three forms. These results suggest recognition of the familiar configuration. On the other hand, Ruff (1978) found that 6-month-olds, in two separate problems with 3 minutes of familiarization time, did not differentiate between objects that varied only in arrangement of parts. The studies mentioned here are too few to elucidate the sequence in which these differentiations may develop. Studies which test several possibilities at different ages would be helpful. The results from long term familiarization as seen in the Kagan study would be particularly helpful since infants' perceptual learning as defined by E. Gibson requires time and continued exposure.

Although the general course of development is probably from the detection of general invariants to detection of more specific ones, an important factor that is hard to pinpoint is motivation. Motivation should play a role in the eventual detection of the structural uniqueness of a given object. In many situations there is no need to detect anything but the most general characteristics of an object. On the occasions when it is important to perceive more specifically, as with recognition of parents, the necessary differentiation is more likely to take place. Also, the need, or motivation, to detect certain invariants will probably change from situation to situation, with the relevance of particular characteristics changing with the child's activity (Ruff, 1980). The role of motivational factors in the detection of structural invariance would be difficult to assess experimentally, but our understanding of object perception and differentiation will eventually require experimental attention to these factors.

Exploration of Objects

The broadest definition of the visual system includes the various capabilities of the observer for clarifying, exploring, and searching for information in the world around him (Gibson, 1966). We need, therefore, to consider the infant's developing ability to use his own movement in the service of perception. Eye movements are exploratory in that they make it possible for different parts of the same object to be fixated. One would expect scanning to change as a reflection of the infant's increasing efficiency in taking in information. The scanning of figures, for example, increases from two fixations per second at birth to four fixations per second in early childhood and the corresponding length of fixation decreases (Kurtzberg & Vaughan, 1978). Some investigators (Leahy, 1976; Salapatek, 1975) have noted an increase during the first few months in the extent to which the contours of a figure are scanned; it has also been noted that, at about two months, infants begin to scan the internal elements of a figure or object as well as the external contours (Maurer & Salapatek, 1976; Salapatek, 1975), though movement of the internal elements increases their salience for younger infants (Bushnell, 1979; Girton, 1979). An interesting area of investigation would be to look at possible changes in the infant's eye movements with increasing familiarization. Leahy (1976) found that, with time, infants showed decreasing fixations to the most salient feature of the figure, that is, the one that initially elicited the most fixations. There may be changes, as well, in the pattern of fixations—for example, toward a more stereotyped sequence. Although scanning continues to become more efficient into the school years (Day, 1975), it appears that important changes occur during the infant's rapid development.

The infant combines eye movements and head movement in order to follow moving targets. This following is important for two reasons: one is that it keeps the most sensitive portion of the retina, the fovea, directed toward the object; the second reason is that following allows the observer to experience the transformations and changes so important for detecting invariants of structure. Newborn infants are capable of following targets moving in an arc around their heads (Barten, Birns, & Ronch, 1971; Kremenitzer, Vaughan, Kurtzberg, & Dowling, 1979). In the first two months, infants increase markedly the amount of time they spend following objects moving at right angles to the line of sight (Ruff, Lawson, & Daum, in preparation). The quality of their visual following seems to change from one that is predominately saccadic, or "jerky," to a following that involves more smooth pursuit (Aslin, 1981; Dayton & Jones, 1964), though the use of transformations in the perception of structure could theoretically occur during either kind of following. Although we still need to determine what information is attended to and picked up during visual following, it is clear that, from birth, infants are, to some extent, capable of the movements necessary for such detection.

Head movements are critical for the perception of occlusions. All observers, including infants, must be able to obtain information about a static world; that is, there are times when the observer is the only moving part of the immediate environment. A head movement to the right will uncover background texture at the right edge of an object and cover texture at the left edge; the return movement to the left will cause the reverse. As discussed before, these reversible occlusions show the observer where the edges of the objects are and what objects lie behind it. For an adult observer, slight head movements may be all that are necessary to provide him with this unambiguous information about the structure of the environment and of objects. The extent to which infants use head movement to provide themselves with information is a completely unexplored area, and, for the moment, it can only be hypothesized that older infants show better recognition of structural properties partly because they have learned better how to use systematic head movements.

Movement of the rest of the body is relevant also. While infants in the first few months do not locomote, they are carried from place to place and therefore have an opportunity to experience a greater degree of change than can be experienced from head movement alone. Held and his colleagues (Held & Bauer, 1966; Held & Hein, 1963) have demonstrated that young cats and monkeys who do not have control over their own movements are impaired in the development of perceptual-motor coordination. There is no evidence, however, that visual perception as such is interfered with, and Walk (1979) cites evidence that some visual observation of moving objects may compensate for the lack of self-produced motion in the development of such perceptual-motor activities such as avoiding the visual cliff. Normally, the infant experiences both self-produced and other-produced movement, and the two may be reciprocally facilitative. The infant may be able to generalize from his experience with his own head motion to experience with motion that is not under his control. On the other hand, the more sweeping changes experienced while being carried may make the invariants in the environment and in objects easier to detect for the inexperienced infant and may facilitate his pick up of information from the more restricted changes involved in head movement.

Although the focus of this chapter has been on the first six months, it should be noted that a very important form of exploration in the first year is the manipulation of objects. During manipulation, the infant of six months and older can use his hands to move objects in ways that produce different views, perspective changes, and occlusions, and thereby provide himself information about the object. The infant younger than six months also has an opportunity to move objects with his hands and to observe the motion, but his motor control is limited so that the resulting movements cannot be as systematic as those of the older infant. On the other hand, the younger infant does have control over the movements of his own hand and spends considerable time in observing it; from such experience, he may pick up information about both objects and motions.

SUMMARY

There are many ways in which the infant is an observer who can use his own activity to extract relevant information about his environment. The kinds of information become more and more specific as the infant gains more experience and as his visual system becomes more finely tuned. A tremendous amount of perceptual learning takes place in the first six months and infants are soon able to recognize many objects and aspects of the world around them.

The view of perceptual development that has been proposed here suggests that the changes with age are not qualitative. That is, development of object perception is not a matter of stages but involves more gradual transitions in terms of increasing efficiency of exploration and increasing specificity of recognition (E. Gibson, 1969, 1977). The assumption is that the newborn is ready for the detection of invariants, though only at the most general level. Our further understanding of perceptual development will depend heavily upon the precision with which we can analyze the complex stimulation facing the active infant in the three-dimensional world. Only then will it be possible to specify what information infants use at different times and in different situations and to determine how they do so. It is an exciting challenge.

REFERENCES

Aslin, R. N. Development of smooth pursuit in human infants. In D. F. Fisher, R. A. Monty, & J. W. Senders (Eds.), *Eye movements: Cognition and visual perception.* Hillsdale, N.J.: Erlbaum, 1981.

Aslin, R. N., & Dumais, S. T. Binocular vision in infants: A review and a theoretical framework. In L. Lipsitt & H. Reese (Eds.), *Advances in Child Development and Behavior* (Vol. 15). New York: Academic, 1980.

Atkinson, J., Braddick, O., & Braddick, F. Acuity and contrast sensitivity of infant vision. *Nature,* 1974, **247**, 403.

Banks, M. S. The development of visual accommodation during early infancy. *Child Development,* 1980, **51,** 646–666.

Banks, M. S., & Salapatek, P. Acuity and contrast sensitivity in 1, 2, and 3 month old human infants. *Investigative Ophthalmology and Visual Sciences,* 1978, **17,** 361–365.

Barrera, M., & Maurer, D. Recognition of mother's photographed face by the three-month-old infant. *Child Development,* in press.

Barten, S., Birns, B., & Ronch, J. Individual differences in the visual pursuit behavior of neonates. *Child Development,* 1971, **42,** 313–319.

Bassili, J. N. Facial motion in the perception of faces and of emotional expression. *Journal of Experimental Psychology: Human Perception and Performance,* 1978, **4,** 373–379.

Bower, T. G. R. Slant perception and shape constancy in infants. *Science,* 1966, **151,** 832–834.

Braunstein, M. L. *Depth perception through motion.* New York: Academic, 1976.

Bushnell, I. W. R. Modification of the externality effect in young infants. *Journal of Experimental Child Psychology,* 1979, **28,** 211–229.

Caron, A. J., Caron, R. F., & Carlson, V. R. Infant perception of the invariant shape of objects in slant. *Child Development,* 1979, **50,** 716–721.

Cook, M., Field, J., & Griffiths, K. The perception of solid form in early infancy. *Child Development,* 1978, **49,** 866–869.

Cutting, J. E., Proffitt, D. R., Kozlowski, L. T. A biomechanical invariant for gait perception. *Journal of Experimental Psychology: Human Perception and Performance,* 1978, **4,** 357–372.

Day, M. C. Developmental trends in visual scanning. *Advances in Child Development and Behavior,* 1975, **10,** 154–195.

Day, R. H., & McKenzie, B. E. Perceptual shape constancy in early infancy. *Perception,* 1973, **2,** 315–320.

Day, R. H., & McKenzie, B. E. Constancies in the perceptual world of the infant. In W. Epstein (Ed.), *Stability and constancy in visual perception.* New York: Wiley, 1977.

Dayton, G. O., & Jones, M. H. Analysis of characteristics of fixation reflex in infants by use of direct current electrooculography. *Neurology,* 1964, **14,** 1152–1156.

Dirks, J., & Gibson, E. Infants' perception of similarity between live people and their photographs. *Child Development,* 1977, **48,** 124–130.

Dobson, V., & Teller, D. Visual acuity in human infants: A review and comparison of behavioral and electrophysiological studies. *Vision Research,* 1978, **18,** 1469–1483.

Fagan, J. F. Memory in the infant. *Journal of Experimental Child Psychology,* 1970, **9,** 217–226.

Fantz, R. L., Fagan, J. F., & Miranda, S. B. Early visual selectivity. In L. B. Cohen & P. Salapatek (Eds.), *Infant perception: From sensation to cognition.* New York: Academic, 1975.

Fantz, R. L., & Miranda, S. B. Newborn attention to form of contour. *Child Development,* 1975, **46,** 224–228.

Fantz, R. L., & Nevis, S. Pattern preferences and perceptual–cognitive development in early infancy. *Merrill-Palmer Quarterly,* 1967, **13,** 77–108.

Flock, H. R. A possible optical basis for monocular slant perception. *Psychological Review,* 1964, **71,** 380–391.

Fox, R., Aslin, R. N., Shea, S. L., & Dumais, S. T. Stereopsis in human infants. *Science,* 1980, **207,** 323–324.

Gibson, E. J. *Principles of perceptual learning and development.* New York: Appleton-Century-Crofts, 1969.

Gibson, E. J. How perception really develops: A view from outside the network. In D. LaBerge & S. J. Samuels. *Basic Processes in reading: Perception and comprehension.* Hillsdale, N.J.: Erlbaum, 1977.

Gibson, E. J., Owsley, C. J., & Johnston, J. Perception of invariants by five-month-old infants: Differentiation of two types of motion. *Developmental Psychology*, 1978, **14**, 407–415.

Gibson, E. J., Owsley, C. J., Walker, A., & Megaw-Nyce, J. Development of the perception of invariants: substance and shape. *Perception*, 1979, **8**, 609–619.

Gibson, J. J. *Perception of the visual world*. Boston: Houghton-Mifflin, 1950.

Gibson, J. J. The perception of visual surfaces. *American Journal of Psychology*, 1950, **63**, 367–384.

Gibson, J. J. *The senses considered as perceptual systems*. Boston: Houghton-Mifflin, 1966.

Gibson, J. J. *An ecological approach to visual perception*. Boston: Houghton-Mifflin, 1979.

Gibson, J. J., Kaplan, G. A., Reynolds, H. N., & Wheeler, K. The change from visual to invisible: A study of optical transitions. *Perception and Psychophysics*, 1969, **5**, 113–116.

Girton, M. R. Infants' attention to intrastimulus motion. *Journal of Experimental Child Psychology*, 1979, **28**, 416–423.

Hay, J. C. Optical motions and space perception. *Psychological Review*, 1966, **73**, 550–565.

Held, R., & Bauer, J. A. Visually guided reaching in infant monkeys after restricted rearing. *Science*, 1967, **155**, 718–720.

Held, R., & Hein, A. Movement-produced stimulation in the development of visually guided behavior. *Journal of Comparative and Physiological Psychology*, 1963, **56**, 872–876.

Jeffrey, W. E., & Cohen, L. B. Habituation in the human infant. *Advances in Child Development and Behavior*, 1971, **6**, 63–97.

Johansson, G. Spatial constancy and motion in visual perception. In W. Epstein (Ed.), *Stability and constancy in visual perception*. New York: Wiley, 1977.

Kagan, J. Attention and psychological change in the young child. *Science*, 1970, **170**, 826–832.

Kremenitzer, J. P., Vaughan, H. G., Kurtzberg, D., & Dowling, K. Smooth-pursuit eye movements in the newborn infant. *Child Development*, 1979, **50**, 442–448.

Kurtzberg, D., & Vaughan, H. G. Maturation and task specificity of cortical potentials associated with visual scanning. In D. Lehmann & E. Callaway (Eds.), *Human evoked potentials*. New York: Plenum, 1979.

Leahy, R. L. Development of preferences and processes of visual scanning in the human infant during the first 3 months of life. *Developmental Psychology*, 1976, **12**, 250–254.

Maurer, D., & Heroux, L. The perception of faces by three-month-old infants. Paper presented at the International Conference on Infant Studies, New Haven, April 1980.

Maurer, D., & Salapatek, P. Developmental changes in the scanning of faces by young infants. *Child Development*, 1976, **47**, 523–527.

Ruff, H. A. Infant recognition of the invariant form of objects. *Child Development*, 1978, **49**, 293–306.

Ruff, H. A. The development of perception and recognition of objects. *Child Development*, 1980, **51**, 981–992.

Ruff, H. A. The effect of context on infants' responses to novel objects. *Developmental Psychology*, 1981, **17**, 87–89.

Ruff, H. A. The effect of object movement on infants' detection of object structure. *Developmental Psychology*, in press.

Ruff, H. A., & Birch, H. G. Visual fixation in three-month-old infants: The effect of concentricity, curvilinearity and number of directions. *Journal of Experimental Child Psychology*, 1974, **17**, 460–473.

Ruff, H. A., Lawson, K. R., & Daum, C. Visual following of moving objects by full-term and preterm infants. Manuscript in preparation.

Salapatek, P. Pattern perception in early infancy. In L. B. Cohen & P. Salapatek (Eds.), *Infant perception: From sensation to cognition*. New York: Academic Press, 1975.

Sokol, S., & Jones, K. Implicit time of pattern evoked potentials in infants; An index of maturation of spatial vision. *Vision Research*, 1979, **19**, 747–755.

Spelke, E. Infant perception of faces in motion. Paper presented at the meeting of the American Psychological Association, Chicago, September 1975.

Walk, R. D. Depth perception and a laughing heaven. In A. D. Pick (Ed.), *Perception and its development: A tribute to Eleanor J. Gibson.* Hillsdale, N.J.: Erlbaum, 1979.

Yonas, A. Development of spatial reference systems in perception of shading information for depth. Paper presented at the biennial meeting of the Society for Research in Child Development, Philadelphia, March 1973.

7

COGNITIVE DEVELOPMENT IN EARLY INFANCY

Albert J. Caron
Rose F. Caron
The National Institute of Mental Health

Cognitive development in its most general sense is concerned with how we acquire and use knowledge in adapting to the vicissitudes of the world. To talk of knowledge is at once to consider its organization, because knowledge, when unorganized, is literally the absence of knowledge. The ability of humans to structure information is, to say the least, prodigious. We differentiate the external world of objects and creatures from the inner world of thoughts and feelings. External things, in turn, tend to be linked in a variety of ways. One prominent form of organization is categorical grouping: four-legged, furry things that bark and wag their tails are all dogs; tables, lamps, and chairs are all furniture. Categories themselves can be arranged hierarchically: all terriers are dogs, all dogs are mammals, and all mammals are animals. In addition to categorization we also combine objects and events spatially, temporally, contextually, and causally. We know that ceilings are above floors, that autumn follows summer, that lions live in jungles, and that heat melts ice. Some of us know even grander things: the theory of evolution or the second law of thermodynamics. And in addition to this immense knowledge of "what," we have an equally vast knowledge of "how," which, though often less conscious than the former, is no less organized (as Lashley, 1951, was the first to discern). Thus we know how to move efficiently in space, to manipulate objects, to speak, read, write, play musical instruments, and so forth. The capacity to organize information and skills has given humans a formidable evolutionary advantage: it allows them to reduce the diversity of the world and to render it coherent and predictable so that goals can be achieved with minimum expenditure of effort.

Our concern in this chapter is with the rudiments of knowledge formation in infancy, a period when intellect seems so patently bereft of the characteristics described above that it is hard to imagine it as related in any substantive way to the complex information-processing capacities of adults or, for that matter, of children.

The preparation of this chapter was partially supported by grants from the March of Dimes-Birth Defects Foundation and the W. T. Grant Foundation.

As Flavell (1977) has noted, most of us would probably judge the intellectual gap between neonate and toddler as much greater than that between toddler and adult. Indeed, research on infant cognition was largely a matter of documenting the emergence of adultlike behaviors until well after midcentury when interest turned inward to the structures and processes that might underlie observed behavioral change.

The shift in emphasis was due to a number of developments. First and foremost was the impact of Piagetian theory (Piaget, 1951, 1952, 1954), which attributed the formation of the basic categories of mind—space, time, objects, causality, symbols—to the internalization of sensorimotor activity during infancy. While Piaget's observations tended to stress the shortcomings of the very young infant, a second influence, the incorporation into infancy research of sophisticated methodological techniques, began to draw attention to the presence of unexpected perceptual, mnemonic, and intellectual competencies in early life. This trend received additional reinforcement from James Gibson's (1966, 1979) theory of ecological optics as disseminated by Eleanor Gibson (1969) and T. G. R. Bower (1974). According to this view, the sensory systems of infants are attuned through evolution to the significant properties of the environment, so that the path to knowledge formation leads directly through perception, unabetted by motoric activity. Finally, there was the input of developmental psycholinguists who, finding themselves released from the constraint of conceptualizing language purely in syntactic terms, came to suspect that the emergence of rudimentary speech in the second year of life might depend on the prior buildup of prelinguistic concepts in the first year.

Although each of these developments helped establish the legitimacy of infant cognition, inquiry remained fragmented, both theoretically and empirically, well into the last decade. Researchers working within the Piagetian frame, for example, tended to focus on the critical transition periods of later infancy (8 to 18 months) when presumably the infant first becomes aware of permanent objects differentiated from the self (cf. Gratch, 1975). Gibsonian theorists, on the other hand, examined the discriminative capacities of younger infants (0 to 6 months) in an effort to demonstrate that the eye perceives objectively without benefit of the hand or of movement in space (cf. Bower, 1966b). Investigators of a more empirical bent sought, by and large, to demonstrate the applicability to the infant of conventional experimental paradigms—habituation, classical and operant conditioning—and in so doing generated ad hoc theories of attention, discrimination, and recognition memory to account for their data (Cohen & Gelber, 1975; Fagan, 1977; Jeffrey, 1968; Kagan, 1970; McCall, 1971; Watson, 1967). Psycholinguists, meanwhile, focused on the nature of the older infant's first concepts and on his strategies for linking concepts with words (Anglin, 1978; Clark, 1973, 1975; Nelson, 1973, 1974).

In recent years, confrontation in theory and research between these points of view has accelerated. In particular, new findings generated by perceptually oriented theorists are being brought to bear on long-held Piagetian conceptions (e.g., Gibson, Owsley, & Johnston, 1978; Ruff, 1980), and experimentalists are moving beyond basic perceptual research to form the empirical base for a viable cognitive psychology of infancy that Piagetians as well as Gibsonians will ultimately have to address (see, for example, Cohen, 1979). Likewise, the search for the origins of communicative behavior and social cognition has prompted speculation that infants are peculiarly sensitive to the animate and interactive properties of persons, and that detection of these properties may precede awareness of static features of objects (Bruner, 1977; Field, chapter 2, this volume; Trevarthen, 1977). And overriding these trends, important themes have begun to crystallize in adult cognition that offer challenging perspectives for developmentalists of all persuasions. We have at hand, consequently, a new and provocative body of research and thinking whose implications for infant cognition have barely begun to be absorbed. In the balance of this chapter, we examine the shifting Zeitgeist and what it portends for our understanding of infant mentality. We begin with a brief overview of the adult trends, and then survey current developmental theories as they dovetail with issues in the adult literature. Finally, we review the empirical evidence in the light of these issues.

CONCEPTUAL TRENDS IN ADULT COGNITION

Perhaps the most significant development in cognitive psychology within the last decade has been the recognition that almost every aspect of human mentation (perception, memory, comprehension, language, etc.) is vitally dependent on one's organized knowledge of the real world. As psycholinguistic research became increasingly semantic in orientation and as memory research began to incorporate more meaningful stimulus material—sentences, stories, and so forth—it quickly became apparent that general knowledge other than that in the input itself was essential to comprehension and retention (for recent reviews, see Lachman, Lachman, & Butterfield, 1979; Puff, 1979; Weimer & Palermo, 1974). Given this insight, it remained for theory and research to address the monumental task of delineating the nature and structure of the permanent knowledge base.

Permanent Knowledge

An initial attempt to characterize permanent knowledge was provided by Tulving (1972) who sought to distinguish the memory that underlies general knowledge (e.g., our knowledge that canaries are birds) from memory for immediate experience (e.g., "what happened when my canary got out of its cage last week"). The latter he regarded as an "episodic" memory store that preserves the autobiographical, spatiotemporal particulars of experience. The former, termed "semantic" memory, he viewed as a permanent record of one's general definitional knowledge of words and symbols, and their meanings.

As originally conceived, semantic memory was intended to account for our ability to answer questions about aspects of general knowledge, such as "Name a bird beginning with the letter C," "Is a canary an animal?," "Does it eat?," and the like. The relative speed with which adults could perform such tasks suggested to early investigators that knowledge is organized in categorical units (animals, furniture, clothing, etc.) and that its overall structure is hierarchical (canary is stored under bird, bird under animal, and so on). At each of these categorical nodes, moreover, there was presumably an associated list of the features most defining for that level (e.g., "sing" for canary, "fly" for bird, "eat" for animal). It is not our purpose here to go into the details of such models, but rather to point out that their underlying premise—that hierarchical taxonomies constitute the sole or major form of mental organization—has been called into question by recent research and analysis. Thus

Rosch (1973, 1975) has demonstrated that natural categories are rarely organized in tight, logical fashion where all instances possessing the critical attributes have equal status. On the contrary, some exemplars are more *prototypic* or typical of a concept than others because they share nondefining as well as defining features with the category (e.g., since canaries, like many birds, sing and fly, they are considered more representative of the bird category than ostriches). Concept membership, in other words, is not absolute but a matter of degree, and Rosch suggests that categories may be organized in terms of bestfitting prototypes rather than strictly defining features. A further critique followed from the demonstration that the defining features of concepts are not fixed but flexible, and are affected by context. For example, drawings of cups that are not quite prototypic are perceived as *cups* when filled with coffee, but are seen as *pots* when filled with flowers (Labov, 1973).

But perhaps most damaging to the view of semantic memory as exclusively taxonomic was the observation that permanent knowledge is not limited to individual lexical items, but rather contains combinations of words that convey complex relational ideas. Underlying these combinations and accounting for their meaning is a vast, alinguistic conceptual base consisting not simply of compendiums of features but of abstract contextual frames in which persons, things, and their many possible interactions are embedded (Bransford & McCarrell, 1974; Kintsch, 1974; Miller & Johnson-Laird, 1976; Schank & Abelson, 1977). According to Schank and Abelson (1977), moreover, these frames are essentially episodic or eventlike in nature (analogous to a theatrical script) and derive from personal experience. A script, for Schank and Abelson, is a generalized sequence of events organized around a common goal such as "going to a restaurant" or "attending a birthday party," where there are a number of prescribed roles (such as customer, waiter, chef, etc.) and where individual events occur in coherent, causal relationship. Investigators working on computer comprehension of natural discourse realized that one had to build some such knowledge into the mechanism in order for it to be able to decode the highly elliptical forms of ordinary conversation. In typical discourse, one might say "We ate out last night; the service was terrible." A computer would have difficulty interpreting that statement, unless it had access to some sort of restaurant script.

Event Contexts

The notion of abstract event contexts arrived almost simultaneously on the psychological scene from diverse sources under a variety of labels: "frames" (Goffman, 1974; Minsky, 1975), "schemas" (Neisser, 1976; Rumelhart & Ortony, 1977), "schematic memory" (Mandler, 1979), "conceptual frameworks" (Miller & Johnson-Laird, 1976), and "scripts" (Schank & Abelson, 1977). While the perspectives and assumptions of the individual investigators differ, they share a fundamental proposition: knowledge of event frameworks renders the things and happenings of the world intelligible. The case has been most cogently put by Bransford & McCarrell (1974), who maintain that perception of isolated objects carries no meaning at all, and that entities become meaningful only when we know their role in events, that is, their relationships to other things, their functions, actions, how they originate, and so forth. As John Dewey (1960) noted a half century ago, "The brute thing, the thing without meaning to us, is something whose relations are not grasped" (p. 135). Bransford and his colleagues have extended this theme to memory, which they view not as a storage of infinite particulars but as a process of recreating the contexts in which particulars were originally experienced (Bransford, McCarrell, Franks, & Nitsch, 1977). In sum, there is a growing consensus that perception, memory, and comprehension of perceived objects and events, and of the words and sentences that symbolize them, are based on our cognitive ability to generate event frames that reveal their complex interrelationships.

Frames, or schemas, are direct descendants of Bartlett's (1932) "schemata," and like schemata they are essentially structural or holistic concepts; that is, they embody not simply aggregations of elements but the generalized relationships holding between elements (Anderson, 1977). In contemporary schema theory such generalized schemata are held to represent both the structure of objects (e.g., the configuration of the human face) and of the scenes, events, and sequences of events in which objects participate (Rumelhart & Ortony, 1977). At each of these levels the basic organizational components are relational concepts: spatial–configural and part–whole in the case of objects and scenes; temporal, functional, actional, and causal in the case of events.

Summary and Developmental Implications

Permanent knowledge, then, is now thought to have a dual conceptual structure: a basic schematic organization in which the particulars of experience are linked relationally and a categorical organization in which linkage occurs by similarity, most likely prototypic (Mandler, 1979). The former presumably underlies our ability to perceive the structure of objects and to grasp their overall significance within the context of events. The latter allows us to identify previously unseen objects as members of a familiar class and thus to anticipate their significance prior to experience. Both forms of conceptual organization, finally, provide a basis for lexical knowledge: schemata accounting for our understanding of the *intensional* or connotative meaning of words, categories for comprehension of their *extensional* or referential meaning (Miller & Johnson-Laird, 1976).

To date, the work on memory organization has focused almost exclusively on adult representational systems, without regard to developmental considerations. For students of early development, however, it raises potentially fruitful hypotheses, in particular: (1) sensitivity to dynamic aspects of the environment plays a major role in cognitive development, (2) detection of relational properties of objects and events takes precedence over detection of elemental features, and (3) the ability to form concepts or categories, relational and otherwise, precedes the development of language. The central issues, of course, are from what source, by what means, and when do infants come to cognize real world information. Specifically, do relational concepts originate in the environment or are they supplied by the organism in an effort to interpret sensory input that is basically devoid of meaning? Similarly, are object categories mere abstractions of the features perceived to be common to a set of things or are they more idealized constructions of mind that bear no precise correspondence to any individual item? Those familiar with philosophical speculation on the origin of human cognition will recognize these questions as the cutting edge of historical debate between empiricists and rationalists. For empiricists, the world is the ultimate source of meaningful ideas that gradually accrue through association of contiguous sensations and abstraction of perceived invariant features. For rationalists, meaning is a product of inborn relational categories of mind (space, time, causality, etc.) that impose structure on sensory experience but require no recourse to it. Variants of these arguments, as we shall see, persist on the contemporary developmental scene where they form the principal bone of contention between Gibsonians and Piagetians.

Closely linked to the question of the origin of mental organization is the question of the method of its acquisition. Obviously, if relational categories are products of mind, they should not be affected by experience. If, on the other hand, they are extracted from the environment, how are they in fact extracted? By perceptual observation alone? Or does perception play a secondary role to action? Further, if relational knowledge derives from experience (perceptual or actional), at what point in development does it emerge? Are young infants unable to appreciate the real structure of the world or is the connectivity of things apprehended from birth and simply expanded in scope during the course of development? In the following section we survey prevailing developmental theories with an eye to how they address these issues and to their possible points of convergence with hypotheses suggested by the adult theories. We begin as we must with the seminal contributions of Piaget.

EARLY COGNITIVE DEVELOPMENT: THEORETICAL PERSPECTIVES

Piaget

Piaget's view of the origin of intellect occupies a middle ground between traditional empiricist and rationalist positions. In the spirit of rationalism, he asserts that reality is never known directly, but is interpreted through generalized schemata that are constructions of the organism. In accord with empiricism, on the other hand, he holds that schemata are not innate, but are gradually formed as a consequence of experience (Piaget, 1951, 1952, 1954, 1969, 1971). The experience that matters, however, that takes the infant from an empty, objectless world to a world of permanent objects bound by space, time, and causal dependency, is not perceptual but actional, not based on passive observation of things but on predispositions to respond to them. Thus Piaget's infant schemes are schemes of doing, of knowing how things can be operated on or transformed, rather than schemes of being, of knowing the states and properties of things. Schemes of being—"figurative" schemes in Piaget's terminology—represent the static physical appearance of objects and hence, he maintains, can mediate only recognition or imagery. By contrast, schemes of doing, or "operative" schemes, provide the framework for interpreting percepts, which otherwise remain fragmentary and meaningless.

Operative schemes are thought to emerge over a protracted period through the progressive intercoordination and internalization of discrete sensorimotor sequences that are repeatedly carried out to specific classes of stimuli. Beginning with built-in reflexes, such as sucking, looking, and the palmar grasp, the infant passes gradually from external to intentional control of behavior through six invariantly ordered stages: (1) simple reflex reactions (0 to 1 month); (2) coordination of one reflex scheme with another, such as bringing thumb to mouth for sucking (1 to 4 months); (3) use of coordinated schemes to produce interesting environmental effects (e.g., repetitive shaking of a rattle to generate sounds—4 to 8 months); (4) higher-order coordination of schemes into intentional means-ends behavior (e.g., pushing aside one object to obtain another—8 to 12 months); (5) purposeful, trial and error exploration to discover new means, and the first use of tools (e.g., pulling a string to secure an out-of-reach object—12 to 18 months); and finally (6) the invention of new means through internally represented schemes (18 to 24 months). Accounting for this progression are the complementary adaptational processes of *assimilation* and *accommodation*. Assimilation is the process whereby events are construed in terms of presently available schematic structures, whereas accommodation is the reciprocal function in which existing structures are modified to take account of novel aspects of events. Through the combined influence of schemata on input and input on schemata, the infant's knowledge system is steadily revised.

The notion that things have permanent existence even when not observed and act as sources of causality independent of the self—concepts basic to the understanding of object and event relationships—are also held to develop slowly and not to be fully realized until Stage 6 (18 to 24 months). As to object permanence, prior to 8 months an object that is completely occluded from view is held to be out of existence, since infants generally will not attempt to retrieve it. At this time, objects exist egocentrically, that is, only to the extent that they engage perception and action. Between 8 and 12 months (Stage 4), infants will retrieve a hidden object, but confine their search to the place of initial disappearance, despite observing the object transferred to a second hiding place (the Stage 4 place error). Although Piaget believes that infants come to perceive objects as having constant size and shape during Stage 4, he regards the place error as evidence that they have not yet dissociated objects from their own action schemes and hence cannot yet represent things in an objective spatial framework. In Stage 5, they no longer commit the place error, but according to Piaget, still have only a partially objectified sense of space, because if an object is invisibly conveyed from one place to another (e.g., inside one's hand) and then the conveyor (the hand) is shown to be empty, they cannot infer that it has been surreptitiously released at the new location but persist in searching for it inside the visibly empty conveyor. Not until Stage 6 do infants become able to infer invisible displacements of hidden objects and thereby reveal a full-fledged understanding of objects existing independently in a common space.

The question of object representation, unlike action representation, poses a dilemma for Piaget, for how can an object be symbolized by schemata that contain only dispositions to respond to it. How, in short, can a response represent the thing responded to? Piaget resolves this problem by basing object representation on a unique class of responses; those that *imitate* the actions of objects. The ability to construct internal models of external objects is held to be a by-product of the developing capacity to imitate. According to Piaget, genuine imitation does not appear until Stage 3 (4 to 8 months), but here the child imitates only those behaviors that he can see or hear himself produce (e.g., manual gestures or vocalizations but not facial expressions). In Stage 4, he becomes able to imitate actions that he cannot actually see or hear himself perform, and in Stage 5, he imitates movements of both animate and inanimate objects that he has never before performed himself. By Stage 6, he becomes capable of *deferred imitation,* in which actions previously witnessed are reproduced at a later time (indicating that all events are now capable of internal representation).

As to the development of causality, Piaget (1954) sees the infant passing through three general periods marked by a gradual differentiation of an objectified and spatialized causality from an initial matrix in which causality is confounded with the subject's own actions. Thus, in the first period (0 to 4 months), the infant experiences inchoate feelings of effort in movement but has no understanding that his actions are the source of his subsequent percepts (which remain inseparable from action). In the second period (4 to 8 months), he begins to dissociate his behaviors from their consequences and means from ends but ignores the necessary external connections that link actions with results. At this time he believes his own movements bring about events in a "magical–phenomenalistic" way. Finally, between 8 and 18 months, cause and effect relations are slowly externalized. Initially in this span, objects are seen as causal centers but only in limited circumstances (e.g., Jacqueline's recognition that her father's hand produces wiggling). Eventually (between 12 and 18 months), there is full acknowledgement of external independent centers of causality and of the self as an object under the causal control of others.

In summary, through much of the first year of life, the Piagetian infant lives in an unstable world of fleeting, impermanent images haphazardly related to one another and to his own activity. His sensory systems, initially discrete and uncorrelated, register only static fragments of reality that do not become coherent until the second year when coordinated action reveals the spatiotemporal–causal connectivity of things. As we shall see below, recent theory and research have begun to challenge this view.

The Gibsons

Piaget's constructivist account of infant cognitive development stands in sharp contrast to the radical realism of James Gibson (1950, 1966, 1979). In line with empiricist approaches, Gibson sees the world as the source of meaningful information. However, the world is so copious in information that sensations require no associative embellishment to impart meaning (e.g., retinal images need not be linked with kinesthetic sensations of reaching or crawling to convey depth information); rather, valid knowledge is directly available to each of the senses as invariant patterns in the changing stimulus array (e.g., gradients of texture density that provide direct visual information for depth). Moreover, contrary to Piaget, the senses are not dissociated at birth, but the same information can be registered by different senses amodally. Given these conditions, Gibson concludes that information about the solidity and permanence of objects, their interrelationships (causal and otherwise) in space and time, their affordable functions, and so forth, is directly detectable from birth, without benefit of action or its mental derivatives.

Although Gibson maintains that valid information is accessible from birth, perception is not unaffected by experience. Experience serves to enhance the efficiency with which information is extracted from the environment and, more importantly, to "attune" the perceptual system to increasingly subtle and differentiated aspects of stimulation. Eleanor Gibson (1969) initially conceptualized this differentiation process as the detection of distinctive features of sets of things. The infant discriminates first those features that distinguish one set from another (e.g., the presence of eyes, mouth, etc., in the case of faces) and then the features that differentiate individual members within the set (e.g., eye color, thickness of lips, shape of the nose). However, Gibson has since become disenchanted with this notion, claiming instead that distinctive features may be useful for distinguishing two-dimensional stimuli such as pictures or letters but are less important for distinguishing "real things happening over time in three dimensions" (Gibson & Levin, 1979, p. 243). She argues, for example, that one seldom recognizes faces by eye color or nose shape but rather by some general structural property that remains invariant over changing perspectives. Thus, Gibson now gives priority to what she originally assumed to be a later phase of the differentiation process—detection of invariant structures—and here again development proceeds from less to greater differentiation.

But how exactly do infants become attuned to invariant structures? E. Gibson (1969) postulates three general mechanisms governing this process: abstraction, filtering, and attention. "Abstraction" refers to the fact that invariant properties of stimuli become perceptually salient when presented with concomitantly varying properties. For example, the constant shape of an object begins to stand out as it is viewed across varying perspectives. "Filtering" designates the converse process whereby continually varying properties come to be ignored or "tuned out." Thus, the varying orientations of the object gradually diminish in perceptual potency. Finally, "attention" refers to the fact that perceptual systems are not passive, but actively seek out information—not in the service of extrinsic reinforcement, but because it is in the nature of such systems to strive to "reduce uncertainty."

In a thoughtful essay, Ruff (1980) has attempted to specify in Gibsonian terms the precise conditions that foster abstraction of the structure of individual objects. The major contributing condition is held to be the continuous movement of objects in relation to one another and to the infant (or the movement of the infant in relation to objects). When an object moves so as to occlude parts of the background, textural elements of the background are deleted at the forward edge and incremented at the trailing edge, thus serving to differentiate the edges of the object. Likewise, perspective transformations such as rotation or non-rotational, displacement movements reveal the spatial relationships of the various parts of an object, because of the covering or uncovering of one part by another. Finally, movements that alter or fail to alter surface texture tell the observer whether an object is elastic or rigid. Subsequently, as infants become able to grasp or manipulate objects, the resulting visual changes as well as the tactual/kinaesthetic input itself disclose more refined aspects of object structure. Still later, broader activities in which objects participate establish functional invariants. Thus, a child who has the experience of drinking milk from a cup extracts invariants specifying cup as something to drink from and milk as fluid and ingestible. This will also draw attention reciprocally to finer aspects of structure so that the infant can now begin to detect differences and similarities across sets of things and thereby form object categories. Finally, as the child sees and interacts with the same object in a variety of event contexts, many more details of its structure and function become salient so that it now comes to exist as a unique entity.

The development of object awareness, then, is patently different in the Gibsonian and Piagetian schemes. Whereas for Piaget the invariant object is a mental construction derived from acting upon the environment, for the Gibsons it is detected *in* the environment in the context of change. Although action plays a role in both theories, the Gibsons emphasize the perceptual consequences of activity, while

Piaget stresses activity itself or, more precisely, the transformations it produces in the states of things. In both systems, awareness develops gradually, but whereas for Piaget the concept of permanently existing objects is apprehended during the second year, for the Gibsons recognition of unique objects should occur sometime during the latter half of the first year. The problem of existence and related conceptual issues that Piaget views as impervious to perceptual analysis the Gibsons see as aspects of event perception, a topic to which we now turn.

The extent to which Gibsonian theory regards perception as a metaphor for mind may be gleaned from its treatment of events. For Piaget, as we have seen, events are inevitably misinterpreted by infants until they develop an objectified sense of space, time, and causality. The disappearance of an object wipes out its existence for the Stage 3 infant. Likewise, the transfer of an object from place A to place B is not regarded by the Stage 4 child as excluding its presence in place A. For the Gibsons, on the other hand, events are objectively perceived as unified happenings over time. A ball that rolls toward the infant is not seen as a series of balls changing in size but as one ball changing in position in space. Constancy and the perception of an event are thus reciprocal phenomena (E. Gibson, 1969). This applies as well to existence constancy, information for which is conveyed by the nature of disappearance events: gradual occlusion of an object specifies its continued existence, whereas fading or imploding indicates distintegration (Gibson & Gibson, 1955). Gibson insists, too, that the invariant dynamic properties constituting an event are simultaneously available to a number of senses, and hence will be perceived as amodal information. For example, the lip movements of a speaker overlap in tempo with the vocalization being uttered so that vision and audition access a single invariant temporal pattern abstracted from sensory specifics. Because they are multiply detectable, moreover, events are assumed to be discriminable early in life.

Not only is the structure of simple events directly perceived but so, too, are the complex relationships (functional, causal, etc.) embodied in events. For Piaget, the development of functional and causal understanding is interrelated; that is, the ability to comprehend the functional role of objects depends ultimately on an understanding of causal relations. The infant may perceive that a cup holds liquid, but Piaget would argue that this is a local percept involving no appreciation of the containing function of the walls of the cup. For Gibson, on the other hand, the functions or "affordances" of objects are directly detectable as invariant properties of stimulation (J. Gibson, 1966, 1977). The affordance of an object is the relationship between its substance and surface properties, on the one hand, and the behavioral capacities of an organism, on the other. Thus for an animal with hands a small solid object affords grasping; for an animal having no hands it does not have this affordance and is not perceived as such. In this sense, "perception of affordance is inseparable from proprioception of one's own body" (J. Gibson, 1977, p. 79). While the invariant information specifying affordances is thought to be given in ambient light, experience may be required to direct attention to it, particularly, as we have seen, in the case of manufactured objects. More natural affordances, however, such as the support function of solid, extended surfaces, may be perceived at birth, and others, such as the animate quality of living things, are assumed to be so unique that infants learn "almost immediately" to differentiate them.

For Gibson, causality, too, is directly perceivable. The invariant information that specifies causal transactions may be complex and not readily extracted, but Gibson concurs with Michotte (1963) that the impression of cause, at least in mechanical collisions (the "billiard ball" effect), should be detectable at birth. While neither Gibson nor Michotte are explicit as to what the invariant transactional pattern underlying causality might be, Gestalt psychologists have speculated that causal impression depends on perceived transfer of attributes from instigating to effected event: the ball that is hit launches off with velocity and direction commensurate with the movement of the initial ball (Koffka, 1935). Both events must also obey various spatial and temporal constraints: a ball that is not hit should not be launched nor should the impacted ball fail to move almost immediately.

In sum, the Gibsons' treatment of events, no less than that of objects, could hardly be more disparate from that of Piaget. What stands out particularly in their overall approach is the steadfast refusal to invoke mnemonic constructs to account for the infant's interpretation of the complex happenings of the world. In their view, the infant does not become less stimulus bound with increasing experience, but rather becomes more adept at extracting information from stimuli. Mental processes, they insist, do not enhance the validity of the information so detected. Certainly questions may be raised about this position. One problem that would appear to pose difficulties for the Gibsons is the detection of novelty. For example, if an infant has become attuned to a particular invariant across a number of transformations, how does he subsequently recognize a deviation from this invariant? The radio receiver metaphor would appear to break down here, because a receiver simply fails to register information to which it is not attuned, and yet infants do recognize novelty, because they pay more attention to novel than to previously familiarized events. (Indeed, this phenomenon has provided the operational basis for the bulk of recent advances in infancy research.) The ability to respond to novelty suggests that something has been carried over from past experience which serves as a standard of comparison for incoming stimulation, and that the discrepancy between incoming and stored information elicits the heightened attention. It is precisely this observation that has generated the theoretical views to which we now turn.

Neo-Constructivism

Although many developmentalists concur with the Gibsonian premise that perception rather than motor activity is the principal avenue to cognition, they are less inclined to reduce cognition to perception and are more willing to entertain the possibility that representational constructs mediate human mentation. While a number of theorists subscribe to this view, it has been most explicitly set forth by Kagan (1971, 1976, 1979). Two themes are central to Kagan's position: attention is guided by internal structures and cognitive development is under maturational control. Like Piaget, Kagan holds that the infant's first cognitive structure is a schema but a schema conceived as an abstract representation of an object, not as a representation of action taken on the object. A schema for an entity, therefore, can be established merely by attending to it, and static as well as dynamic visual stimuli can generate schemas. The schema, too, is essentially structural, preserving the relations among features rather than simply the features themselves.

Schemata are assumed to form in the second or third month, and once established they govern the infant's deployment of attention to new stimuli. Specifically, stimuli that are moderately different from an established schema will recruit maximum attention, whereas those that are identical to it or extremely different will recruit minimum attention (a principle generally known as the "discrepancy hypothesis"). The creation of "schematic prototypes," or categories representing classes of stimuli, constitutes a second step in schema formation. In contrast with the Gibsons, Kagan regards this process not as one of simple detection of invariants but rather as an active mental construction, although he does not specify what this process might entail. Once formed, however, schematic categories permit recognition of new events as instances or noninstances of a class through reference to the prototype. A final step in cognitive development involves the formation in the second year of "symbolic categories" where representation is mediated not by image-like templates but by arbitrary units such as words.

Kagan (1976) contends that maturational factors produce a quantal shift in cognition sometime in the last quarter of the first year. The simultaneous appearance in that period of rudimentary object permanence, increased babbling in response to speech, heightened attention to discrepant events, and stranger anxiety are believed to signal the appearance of a new competence called "activation of relational structures." The infant now does not simply respond to ongoing events, but can generate representations of past and future events with which to compare representations of the momentary situation. This permits higher-order representation of possible causes and sequelae of experience so that, unlike the younger infant who merely recognizes an event as familiar, the one-year-old can interpret that event within a broader framework. Thus for Kagan, as for Piaget, knowledge of meaningful relationships between objects begins to occur late in the first year and is indirect—mediated by extended event schemata.

Kagan's overall theory is representative of constructivist positions held by many developmentalists who may otherwise deviate from a number of his particular assumptions. For example, there are probably few who would accept his maturational bias and its implied rejection of continuity in development. However, all are in one way or another committed to the proposition that mnemonic structures underlie perception and categorization.

Identity Theory

Still another group of theorists—notably Bower (1974, 1975, 1979), Moore (1975; Moore & Meltzoff, 1978), and Harris (in press)—take a position which, while more Gibsonian in tone than Kagan's, also recognizes the need to provide the infant with mental representations. These serve, however, not to direct attention but to interpret the displacement and disappearance events that Piaget saw as problematic for the infant. While the views of this group vary in specifics, all regard the resolution of the object problem as fundamentally one of establishing the identity of objects.

Bower takes an extreme stand relative to Piaget. In light of J. Gibson's (1966) analysis of object permanence as an early appearing perceptual constancy, and given his own research purporting to show that 2-month-olds expect occluded objects to reappear (Bower, 1967) and that 5-month-olds search for objects that disappear in darkness (Bower & Wishart, 1972), he maintains that infants have a sense of object existence at least by 5 months of age (Bower, 1974). Accordingly, he sees the infant's problem not as having to discover the permanence of objects but as having to locate them when they shift in position. Moore (1975) also reduces permanence to a search issue, although he believes that permanence per se does not develop until about 9 months of age (Moore & Meltzoff, 1978). Both Bower and Moore take the problem of search an additional step, claiming that it hinges on an even more fundamental process—the developing rules of object identity—because until one can identify a displaced object as identical to the original, there is no need to initiate search.

Moore and Meltzoff (1978) see identity rules developing in three phases. Prior to 5 months of age, infants identify objects separately in terms of place and path of movement, so that different objects that occupy the same place and that move in the same trajectory are seen as the same object, and, conversely, the same object in two different places or when moving and stationary is seen as two different objects. Between 5 and 8 months they begin to coordinate place and movement and to shift to a "feature rule" (identification of objects by constant features). At this time, an object is perceived as the same object whether it moves, stops, or changes direction. Finally, at about 9 months of age infants begin to maintain the identity of objects that disappear from view, initially following visible displacement and eventually following invisible displacement. Once an object that disappears and reappears is identified as the same object, the infant comes to conclude that disappearing objects continue to exist while occluded.

Harris (in press) also places greater emphasis on search and identity than on permanence as the young infant's major conceptual problem. However, in contrast with Moore and Meltzoff's account, which he feels is inconsistent with existing data, he stresses two alternative confusions that beset the infant: (1) how to encode position and (2) how to distinguish between duplicates (which can appear simultaneously in two places) and uniques (which can exist at one time in only one place). Harris agrees with Butterworth (1975, 1978) and Bremner (1978, Bremner & Bryant, 1977) who claim that young infants have difficulty specifying the position of an object. At first they define position *egocentrically,* relative to the self, and, according to Butterworth, will regard an object that occupies a stable position in relation to the self as the same object. Gradually, with the onset of crawling, they come to define position *allocentrically,* in relation to an external framework, so that they now see objects that maintain a constant position, or even move, within a stable frame as the same (e.g., "still in that room"). Harris argues, however, that these notions beg a more fundamental question: how does the infant know, when operating with either code, whether he is confronting the same object on different sightings or merely a similar object; that is, whether he is dealing with successive appearances of a single object or with different instances of a class of objects? Harris claims that all of the evidence on infant search (from visual tracking to the place error) implies that the infant behaves as though he is dealing with various instances of a class of objects (where the movement of an entity does not negate its existence in a prior place) rather than with a unique object. Thus as opposed to Bower and Moore, who believe the infant resolves the identity problem when he begins to define objects by similarity of features, Harris maintains that this is not sufficient and that the infant must also discover that some similars can be in two different places simultaneously, whereas others (uniques) can be there only successively. Prolonged experience with displacement and manipulation of objects is thought to establish this discrimination.

Overall, identity theory differs from that of Piaget in shifting the conceptual problem of infancy from object permanence and egocentricity of action to object identity and egocentricity of perception. In contrast with Gibson, it stresses that information available in the environment is not sufficient to preserve the infant from error about the location and identity of objects. On the other hand, it assumes with Gibson that error is eventually resolved through feedback from the environment.

Psycholinguistic Speculation

Theories of early language development provide still another perspective on the cognitive growth of the young infant. We know that the child understands and produces his first words at about 11 to 12 months and begins to combine words into rudimentary sentences at approximately 18 months. A number of implications follow from these observations. For one thing, since many first words are labels not for particular items but for classes of objects (e.g., "doggie"), infants must have the capacity to categorize events prior to naming them (MacNamara, 1972). Initially, they may "overextend" (Clark, 1973) or "underextend" (Anglin, 1978) the referential domain of labels (e.g., calling a horse "doggie" or refusing to apply the term to a Great Dane), but however imperfect their performance, it must rest on a preverbal ability to conceptualize. Secondly, the ability to comprehend strings of words suggests that the infant must not only be able to segment the environment into distinctive object classes but must also be able to integrate them in terms of various abstract relational concepts—agency of an action, recipient, possession, location, temporal sequence, etc.—concepts encoded by the syntactic conventions of language. Inasmuch as some psycholinguists (e.g., Greenfield & Smith, 1976) believe that early one word utterances represent not simply objects but more elaborate propositions, then conceptualization in the first year would be expected to include such relational concepts.

Recent linguistic speculation has been directed to the origin and nature of preverbal conceptualization. Nelson, following Schank and Abelson (1977), holds that concepts of the natural world are derived from, and defined by, a context or system in which they are embedded. In early life, such contexts are episodic, consisting initially of specific event structures that contain "particular and idiosyncratic information regarding possible actors, actions, locations, results of action, etc." (Nelson, 1977, p. 222). With further experience, these structures become more generalized into broader event sequences, or "miniscripts" that relate to a single goal, for example, "eating time," "going for a walk time," and so forth. Ultimately, the child may embed concepts within either an episodic system or a context-free, categorical system where logical rather than physical–spatiotemporal relationships provide the organizational basis.

Since object and event concepts arise interdependently, the core meaning of object concepts for Nelson is thought to reside in their dynamic properties—their characteristic actions, or how they are acted upon or used. Perceptual properties (shape, color, size, etc.) serve merely to identify a new object as a probable class member before one has had a chance to observe its defining dynamic qualities. Thus for a young child the concept of "dog" would consist principally of its barking, running, and tail-wagging traits and secondarily of its four-leggedness or furriness. Nelson's hypothesis is based on her observation that the bulk of children's first words are names for objects that move, change, or are manipulable (e.g., pets, food, people, shoes, etc.) as opposed to objects that are rarely transformed (tables, sofas, trees). Support for this view has been equivocal, in that infants sometimes generalize words to novel referents that are perceptually similar even though they know them to have different functions (Clark, 1975). However, this might reflect confusion about the basis for applying names—whether by form or function—rather than an indictment of Nelson's basic position. It may be the case, too, that infants utter some words whose meanings they comprehend only in a denotative way and that they come to understand their broader intensional meanings at a later time. In the last analysis, however, if words are to be used coherently rather than by rote, the role of objects in events must eventually be grasped.

EMPIRICS

In the span of a few years, research on infant cognition has become a thriving cottage industry whose products extend from perception to language across a broad range of methodologies and perspectives. Since it is hardly possible to examine all of this literature here, we have tried to confine our review to the first year of life and specifically to the organizational themes suggested by the adult trends. Thus, we focus on the infant's ability to detect relational or patterned information, first in space, then in time, and finally between objects in extended event contexts. At the same time, we examine the capacity of infants to categorize information, paying particular attention to the possible role of representation in categorization. Some important topics—notably the detailed work on space perception and on object permanence have been slighted but these have received extensive coverage elsewhere (see Bower, 1975; Gratch, 1975; Harris, 1975, in press; Moore & Meltzoff, 1978; Yonas & Pick, 1975; Yonas, 1979).

Configurations in Space

Contrary to Piagetian doctrine, young infants seem able to detect various aspects of space and to do so with some degree of intersensory coordination. For example, newborns have been shown to turn their eyes toward a peripheral sound (Wertheimer, 1961), and 4-month-olds, to reach more quickly and more frequently for targets that are within reach than out of reach (Field, 1976; Rezba, 1977), although they perform with limited accuracy until about 7½ months (Gordon, Lamson, & Yonas, 1978). Likewise, infants of 4½ months (but not 3½ months) try to avoid a virtual object that appears on a collision course with the face (Yonas, Oberg, & Norcia, 1978), indicating that optical cues provide direct information for depth prior to extensive experience with feedback from reaching, as Gibson maintains. Our primary concern here, however, is not with the development of space perception per se but with the ability of infants to interrelate information within space, that is, to detect spatial configurations. Theoretically speaking, Piaget, Gibson, and Kagan would all concur that infants should be able to recognize two-dimensional spatial patterns, although Piaget would argue that such configurations are figural impressions having no objective significance. While Gibson and Kagan would also grant young infants the ability to detect configurations in three-dimensional space (e.g., the structure of solid objects), Piaget would disagree because depth in near space presumably does not become organized until the occurrence of reaching and touching, and in far space until the onset of crawling.

Before examining the evidence, it is important to be clear about what is meant by a configuration and how it differs from other stimulus attributes. In this regard, some distinctions proposed by Garner (1978) provide a helpful guide. Garner distinguishes between *component* and *wholistic* properties of stimuli. Component properties consist of either dimensions or features, dimensions being components that vary across mutually exclusive levels (e.g., a face may have a long or short nose, but not both) and that are either continuous (e.g., loudness) or discrete (e.g., form). Features, on the other hand, are components that either exist or do not exist in relation to a stimulus (e.g., a dog may have spots or not have spots). Thus, whether a component is regarded as a feature or a dimension depends on whether existence or level is its characteristic mode of variation. Wholistic properties are of three types: simple wholes, templates, and configurations. Simple wholes are no more than the sum of their parts. Templates, while also in a sense equal to the sum of their parts, are prototypes of a set of stimuli having modal or average values on the most relevant attributes of the set, with the values of irrelevant attributes being averaged out to zero. A prototypic Cocker Spaniel, for example, may be conceived as an assemblage of the average values of the relevant dimensions (body size, ear length, etc.) of all Cocker Spaniels. However, the templates of real objects usually cannot be considered independently of their configurations. Configural wholes are particularly important because they are emergent properties, that is, more than the sum of their parts. They involve such relations between components as symmetry, spatial arrangements, repetition, etc., and are of special interest to developmentalists because the components can be varied along many levels without affecting the relation (e.g., nose length or eye shape can be changed without altering the face configuration).

One further issue to be kept in mind before turning to the literature bears on methodology. As previously noted, if infants look longer at a novel than at a previously familiarized or habituated stimulus, we may conclude that they can discriminate the two targets. However, this standard procedure does not tell us whether discrimination was based on perceived configural differences between the two stimuli or simply on perceived differences between componential properties (e.g., change in position of an individual element). To control for the effects of element change, the familiar stimulus may be varied at test along one or more component dimensions (e.g., a familiarized face might be shown in a new pose). If habituation generalizes to the altered familiar stimulus but not to the similarly altered novel stimulus (e.g., a new face in the new pose), configural discrimination can be inferred. One problem with this approach is that should the infant respond equivalently to both changes, one cannot conclude that he is unable to process configural information. To get around this difficulty, investigators have begun to introduce componential variation during habituation (e.g., showing the same face in a number of poses). Configural discrimination is then inferred from generalization of habituation to a further instance of the componential change (the same face in a totally different pose) and from recovery of responding to the novel configuration in the altered state (a different face in the new pose). This invariance or categorization procedure, because it habituates responding to irrelevant stimulus change, provides the most definitive test of the infant's discriminative capabilities.

Two-Dimensional Configurations. Considerable evidence is now at hand which suggests that infants are sensitive at an early age to spatial relationships between the same simple elements presented in two-dimensional visual displays. Schwartz and Day (1979), for example, found that infants of 2 to 4 months who were visually habituated to two lines forming a 90° angle subsequently yielded stronger recovery to a new angle in the original orientation than to the familiarized angle rotated 90°, indicating that they were more attuned to the angular relationships than to the orientation of the individual lines. Likewise, a number of investigators have demonstrated that 4- and 5-month-olds, following habituation to configurations of identical small shapes, pay more attention to a new configuration composed of the same shapes than to the same configuration composed of novel shapes (Cornell, 1975; Fagan, 1977; Vurpillot, Ruel, & Castrec, 1977). Using an invariance procedure, Milewski (1979) has since shown that 3-month-olds can recognize such patterns not only across shifts in element shape but also across discriminable changes in element density. In a recent study, Caron and Caron (1981) found that an invariant configuration of nonidentical elements could also be abstracted across discriminable changes in element shape. Their 4-month-old infants were habituated to four stimuli involving the same "little above big" relationship (e.g., $\overset{\circ}{O}, \overset{\diamond}{\diamond}, \overset{\triangle}{\triangle}$), whereupon they were tested with new shapes either in the same relationship ($\overset{x}{X}$) or in an altered relationship ($\overset{X}{x}$). The infants generalized habituation (yielded minimal recovery) to the familiar relationship and recovered significantly to the novel relationship, whereas control subjects who were habituated to a single exemplar recovered equally to both test stimuli. That the propensity to detect relationships between similar elements also applies to nonspatial configurations has been demonstrated in studies involving repetition or numerosity patterns. Infants from 4 to 7 months were shown to abstract the quantity of small sets (two vs. three) that varied in density, color, size, brightness, and shape (Starkey & Cooper, 1980; Starkey, & Gelman, 1980), and very recently neonates have been found to do the same across changes in density (Antell & Keating, unpublished manuscript).

The ability of infants to detect two-dimensional shapes composed of continuous contour lines has been examined in studies where the figures have been rotated both in the fronto-parallel plane and in the third dimension. Thus in the previously cited study by Schwartz and Day (1979) infants of 2 to 4 months who were habituated to a square were subsequently tested with a 45° rotation of the square that appeared as a diamond and with a rhomboid figure of same side length as the square. Dishabituation was stronger to the rhombus. In still another experiment, following habituation to a rectangle, infants were shown a 90° rotation of the rectangle and a square. Dishabituation was again greater to the new form. Unfortunately, clearcut interpretation of these results in terms of shape discrimination is not possible, because discrimination of angles could have accounted for the first set of results, and of side length, the second set. Two further studies, however, provide less equivocal data. Bower (1966a) conditioned 2-month-olds to turn their heads to a thin wooden rectangle and found that they generalized responding to the same rectangle at a new slant but not to a trapezoidal stimulus that projected the same retinal shape as the conditioned stimulus. Caron, Caron, and Carlson (1979), using an invariance procedure, habituated 3-month-olds to either a flat square or an equal-sized flat trapezoid that varied in slant, and hence in projective shape, from trial to trial. At test, both the familiar and novel stimuli were shown for the first time in the fronto-parallel plane. Generalization of habituation occurred to the familiar shape in the novel orientation, whereas attention recovered to the novel shape. The results of both studies are consistent with the interpretation that 2- to 3-month-old infants can detect complete, two-dimensional forms.

A possible exception to the above finding comes from a study of illusory forms (Bertenthal, Haith, & Tucker, 1980). Black circles with cut-out 90° wedges (e.g., ◖) may be arranged in groups of four to give an impression of a square; when arranged randomly the illusion is eliminated. When 3-, 5-, and 7-month-old infants were habituated to the illusory arrangement, only the 7-month-olds recovered to a subsequently presented random arrangement. Since these infants could not discriminate two different random arrangements, it is apparent that the response of the experimental group reflected configurational rather than featural sensitivity. The inability of 3- and 5-month-olds to detect the configuration under these conditions may be due to many factors, not the least of which may be the lack of firmly established schemata for intact shapes. While assimilation of stimuli to stable schemata may be necessary to create illusory effects, schemata are yet to be formed in the case of de novo perception. Be that as it may, the overall evidence indicates that prior to 4 months infants are more disposed to detect configural than componential information in two-dimensional displays composed of similar elements.

Discrimination of two-dimensional patterns composed of entirely different elements appears to occur sometime after 4 months. Bower (1966c), for example, conditioned infants to turn their heads in the presence of a figure composed of a cross circumscribed by a circle with a small dot on either side of the cross. In a subsequent test phase, he presented the original stimulus and each component separately (circle, cross, and dots). He reasoned that if the infants had detected the figure as a compound, responding would be greater to the whole than to the sum of the parts. Infants of 2 to 5 months were tested, but not until 5 months did the predicted summation pattern emerge. Similar results have been obtained in a number of habituation studies (Cornell & Strauss, 1973; Miller, 1972; Miller, Ryan, Sinnott & Wilson, 1976). Apparently, infants, like adults, find it easier to establish Gestalts for similar than dissimilar elements, a principle that also applies to three-dimensional objects such as the human face, as we shall presently see.

Three-Dimensional Configurations: Objects. The infant's ability to detect three-dimensional structure has been examined with both artificial and natural objects, the human face being the most conspicuous example of the latter. Consistent with J. Gibson's (1966) contention that such invariants become salient in the context of continuous movement, E. Gibson and her colleagues (Gibson, Owsley, Walker, & Megaw-Nyce, 1979) found that 3-month-olds, following habituation to a three-dimentional geometric object undergoing two different rotational motions, subsequently paid more attention to a novel than to the familiar object, whether both were in motion or motionless. In another study, Owsley (1980) reported that 4-month-olds tested under monocular viewing conditions were better able to discriminate a cube from a wedge, if the wedge had been habituated under rotation than in either a single stationary position (where it projected as a cube) or in multiple stationary positions. Similarly, Ruff (1978, 1980) showed that 6-month-olds could differentiate a novel from a familiar complex form after viewing the latter undergoing displacement movement, but not after it had been seen in a variety of static orientations. Given the variety of local percepts produced by these different movements, it would be hard to rationalize any of the preceding results in terms of discrimination of components. That self-produced movement, where infants have some control over the transformations they perceive, may also contribute to these effects was evidenced in a recent experiment by Bertenthal, Campos, and Benson (1980). When locomotor and prelocomotor 7-month-olds were familiarized with a solid form shown in a variety of static orientations, only the locomotor infants could then discriminate a novel from the familiar object. Whether locomotion itself or some correlated factor such as conceptual maturity is responsible for these results may be better determined by training procedures. Overall, these studies provide strong evidence that structural information about objects is differentiated in the context of events, and that simple structures can be detected prior to extensive manipulative experience.

The previous experiments were concerned with perception of a single object across perspective transformations. It has also been shown that infants can detect configural invariants common to sets of different objects, that is, that they can form object categories. For example, Cohen and Caputo (1978) found that 7-month-olds, following habituation to photographs of a number of stuffed animals, could generalize to new instances of the category and at the same time discriminate a member of a new category (a rattle). No younger ages were tested. It is unfortunate that the novel category was not better matched to the original in such component dimensions as texture and brightness, so that a configural invariant could have been definitively identified as the basis for categorization. In a second experiment, this condition was met for a 12 month group who were able to differentiate photographs of dogs from antelopes. In still another study, Ross (1980), using plastic and styrofoam models of men, animals, food, and furniture, obtained appropriate categorical responding in each case at 12 months. The food and furniture categories are of particular interest, because they involve abstract functional invariants rather than configural commonality per se, and thus bear more directly on the question of affordances, an issue yet to be addressed.

Faces. Studies of face perception are more difficult to interpret because much of the work has used photographs or schematic drawings rather than actual faces. In addition, investigators have often not been clear about the level of configural invariant they were asking the infant to discriminate: (1) the general structure of the face, (2) categorical groupings of types of faces such as males versus females or happy versus sad expressions, or (3) configurations specifying unique faces. Given the Gibsonian premise that perceptual development proceeds from general to more specific levels of differentiation, one might expect to find that facial structure is discriminated before categories and categories before uniques.

As regards the structure of the face, early research suggested that infants can differentiate naturally arranged schematic faces from scrambled faces at about 4 months of age (Haaf & Bell, 1967; McCall & Kagan, 1967). However, evidence from a recent study using more sophisticated procedures (Maurer & Barrera, 1981) indicated that 2-month-olds (but not 1-month-olds) can discriminate regular faces from both scrambled symmetric and scrambled asymmetric arrangements. This finding, of course, cannot tell us whether any aspects of the face were actually configured, since as the authors themselves point out, discrimination could have been based on noticing that a single feature was not in its proper place. Because young infants, when scanning the face, look longer at the eyes than at any other feature (Hainline, 1978; Haith, Bergman, & Moore, 1977) and because the eyes involve identical elements, which may be easier to process, one might expect the eye pairing to be among the first aspects of the face to become configured. There is some indication that this is the case and that it occurs at least by 4 months of age. Thus, Caron, Caron, Caldwell, and Weiss (1973) found that when 4-month-olds were habituated to schematic faces with either one or both eyes scrambled, they recovered fixation to a regularly arranged face, indicating that it was perceived as different from the scrambled versions. On the other hand, when just the nose and mouth were scrambled, the regular face was not perceived as dissimilar. That the pairing of the eyes rather than their location was the critical invariant was suggested by the fact that the regular face was seen as similar to a face with proper head orientation but inverted internal features. Five-month-olds, by contrast, perceived the regular face as distinct from all three of the above distortions, indicating that by that age the entire face had become configured. The use of static, two-dimensional faces obviously constrains the generalizability of these findings, and it would therefore seem advisable to apply the techniques Ruff (1978) employed with objects to this problem as well. For example, infants might be shown a series of three-dimensional, life-like faces, each moving in various ways, and might then be tested with a novel face paired with various scrambled versions of that face.

One implication of the findings by Caron et al. (1973) is that neither facial categories nor unique faces should be recognized prior to 5 months of age. With regard to categories, there is indeed no hard evidence indicating that infants can detect invariants specifying sex, age, or facial expression prior to 5 months. For example, Cornell (1974) found that 5-month-olds were unable to generalize habituation to a photograph of a new male or female following familiarization with multiple faces of either sex. Six-month-olds, on the other hand, could make this transfer although not robustly. Fagan (1976), using a similar technique, obtained stronger effects with a 7-month sample (the only age group tested), and, more to the point, Cohen and Strauss (1979) showed that neither 4- nor 6-month-olds, but only 7-month-olds, could generalize to a photograph of a new female face following habituation to different females in varying poses. Unfortunately, Cohen and Strauss did not use a male face at test, so one can't be certain that the obtained categorization at 7 months was sex-based. The important point, however, is that their younger groups were unable to categorize the faces on any basis. In the only study known to us that has examined categorization of age in infancy, Fagan and Singer (1979) found that 5½-month-olds could discriminate two rather similar looking, but age disparate, photographs (a bald-headed man and a baby) yet were unable to differentiate the same adult from a very different appearing male. Although the authors attribute their results to the abstraction of an age-related invariant, the use of only two faces and the failure to test for generalization to novel categorical exemplars qualifies this interpretation. Turning finally to facial expressions, while there is evidence that 3- and 4-month-olds can discriminate photographs of various primary emotions (Barrera & Maurer, 1981a; LaBarbera, Izard, Vietze, & Parisi, 1976;

Young-Browne, Rosenfeld, & Horowitz, 1977), these data, too, cannot be interpreted as a categorization phenomenon because of failure to test for generalization. On the other hand Caron, Caron, and Myers (in press) have demonstrated that among 4-, 6-, and 7-month groups, only the last could generalize an invariant expression viewed across different people to a new person displaying the same expression.

Again, it is possible that the preceding studies are underestimating the infant's capacity to categorize faces, because in no instance were infants shown real, dynamic stimuli. Interestingly, Walker (1980) has shown that 5-month-olds, following habituation to films of happy facial and vocal expressions posed by different people, dishabituated more strongly to a new person in an altered facial and vocal expression than to the same new person in the familiar expression. Of course, visual and auditory inputs were confounded here, but the results nevertheless suggest that our current estimates of when infants begin to categorize faces may not be conclusive.

The ability of infants to perceive unique faces has been investigated with both static and dynamic stimuli. Studies using photographs indicate that discrimination of individual faces does not occur later than categorization, but at about the same time. Fagan (1976) and Cohen and Strauss (1979) have shown that 7-month-olds exposed to a person presented in various orientations can then differentiate that person from a novel individual of the same sex when each is shown in a new orientation. Cohen and Strauss, moreover, demonstrated that neither 4- nor 6-month-olds could detect this invariant, both groups responding instead to the shift in orientation. Spelke (1975), on the other hand, found that infants as young as 4 months, following exposure to a female face performing six different filmed activities, could then recognize her performing two new activities and differentiate her from a very similar woman performing the new activities. This study, in contrast with the two preceding experiments, constitutes prima facie evidence for the centrality of movement in articulating facial invariants. It is regrettable, however, that motionless faces were not also presented at test so that one could have determined whether discrimination had been based on movement invariants (tempo, etc.) or on structural invariants. Nevertheless, the results do suggest that infants might be capable at a very early age of discriminating still photographs of very familiar persons, that is, those they have already seen in numerous and varied action situations. Indeed, there is evidence that 3-month-old infants can differentiate photographs of their mother's face from a strange female's face and of their father's face from a strange male's face (Barrera & Maurer, 1981b, 1981c; Maurer & Heroux, 1980). However, without benefit of an invariance procedure (all presentations were frontal), we cannot be certain that discrimination did not occur on the basis of differences in components (e.g., lip thickness, hair style, etc.) or of differences in partial invariants such as eye separation. It would be of more than passing interest to repeat this study showing the mother in front view during habituation and in full profile at test paired with a stranger in full profile. If this discrimination could be made at 3 months, everything thus far known about infant face recognition would have to be recast.

Spatial Configurations in Vision and Touch. Information about the structure of an object is available to both vision and touch, an observation that prompted Gibson to speculate that such patterns can be detected in a supramodal manner and that recognition once achieved in one modality should be readily transferable to the other. For Piaget, on the other hand, cross-modal integration would depend on coordination of the action schemes for touching and looking, an achievement not expected to occur until after four months of age.

Attempts to demonstrate transfer from vision to touch at an early age have yielded mixed results. Although Bower, Broughton, and Moore (1970) claimed that neonates showed appropriate hand shaping as they reached toward an object, Gordon and Yonas (1976) found no evidence for adaptive responding in 5½-month-olds. On the other hand, attempts to assess transfer from touch to vision have been more successful. Meltzoff and Borton (1979) found that 4-week-old infants who were allowed to suck on a pacifier with either a plain or nubbed surface subsequently looked longer at the pacifier they had sucked. Since sucking and looking are not easily coordinated action schemes, the findings challenge the Piagetian position. It is not entirely clear, however, that a structural property was detected here. The presence or absence of surface protuberances is a featural rather than a configural property; the nubs were not arranged in any particular pattern nor were the infants required to discriminate different shapes of pacifiers. As to shape per se, the available evidence comes from manual manipulation, where cross-modal recognition of geometric forms (spheres and cubes) has been demonstrated at 6 months of age (Ruff & Kohler, 1978). Comparable effects have also been found with 9-month-olds (Bryant, Jones, Claxton, & Perkins, 1972) and with 12-month-olds (Gottfried, Rose, & Bridger, 1977), but of course combined visual and tactual experience could account for all these results. Since coordinated manual manipulation of objects does not emerge much before 5 months, integration of manual and visual input cannot be used to provide a test of amodal perception of object structure at younger ages. Such a test will have to come from studies employing oral manipulation along the lines of the Meltzoff and Borton (1979) experiment, but thus far none have appeared.

Spatial Relations: Summary. Clearly, infants seem able to detect configurations in both two and three dimensions prior to 5 months of age, a period considerably earlier than Piaget would allow in the case of three-dimensional structures. In addition, simple geometric forms seem to be differentiated earlier than more complex structures involving unlike components, as in the human face and in the objects employed by Ruff (1980). Movement appears to be the major factor contributing to the infant's ability to extract invariant forms, and for this reason the structure of animate objects should be recognizable sooner than those of inanimate entities. This supposition gains support from the fact that the structure of the face could be discriminated at 5 months in static presentations (Caron et al., 1973), whereas Ruff's somewhat simpler inanimate forms could not be detected until 6 months, and then only in the context of movement. Exactly when facial structure becomes discriminable is moot at this point, because assessments have not been conducted under the most favorable conditions (three dimensional models shown in movement). Finally, although from a Gibsonian perspective one would expect infants to be able to abstract general facial invariants such as those specifying age and sex sooner than the invariants associated with unique faces, the evidence does not bear this out. Whereas facial categories do not appear to be discriminably prior to 7 months, there is some indication that very familiar faces may be differentiated considerably earlier than 7 months.

Coding of Invariants

Given that infants can detect an invariant object as well as general invariants corresponding to classes of objects, a question arises as to how they accomplish this feat. Do they, as Gibson claims, simply extract the invariant properties that are actually present in the stimulus array? Or do they, as Kagan speculates, form a representational prototype or schema that contains configural properties common to all members of a stimulus set but is much less specific as to the details of individual components. From the adult categorization literature we get an inkling of what this prototyping process might involve. Specifically, when adults are shown a set of objects such as faces they form a schematic representation that contains either average values of the various component attributes (e.g. head size, nose length, etc.) or modal values of components (e.g., if most instances have curly hair, then the prototype would have curly hair). Typically, adults abstract modal values for easily discriminable, discontinuous dimensions such as hair type and average values for harder-to-discriminate, continuous dimensions such as nose length, although they sometimes abstract modal values for these as well (Goldman & Homa, 1977; Neumann, 1977). The important point, however, is that the prototype does not correspond in its component values to any specific object of the set. This implies that if a new exemplar of the category were shown that approximated the average or modal values of the previously shown series, it would be seen as more similar to what had gone before than an actual instance contained in the series.

In an innovative study, Strauss (1979) put this prediction to test as a way of determining whether infants, too, might form prototypes rather than extract invariants. He first familiarized his 10-month-old subjects with a number of schematic faces in which various component values—eye separation, nose width, nose length, and face length—varied from trial to trial. The infants saw many more extreme examples of each feature (e.g., very wide and very narrow noses) than average examples. When tested with a face containing nothing but extreme values and one with nothing but average values, the infants generalized more to the average face, and looked longer at the extreme faces, even though they had seen these average values much less frequently. Likewise, they looked longer at a novel than at the average face at test, but failed to discriminate the extremes from the novel face. In a second study, Strauss (1981) found that infants perceived an average face never shown during familiarization as more familiar than an actually familiarized face.

Strauss concluded that infants may be more constructive processors of their visual environments than the Gibsons believe and that invariant information is stored in abstract prototype form rather than simply being detected in stimuli. Taken at face value, Strauss' findings are hard to rationalize in Gibsonian terms, but how much of a challenge they pose would appear to depend on the level of abstraction being considered. In particular, recognition of both very broad categories (e.g. faceness) as well as unique objects would seem to involve detection of the actual invariant information presented in stimulus arrays, as the Gibsons claim. For example, faceness is specified by the general structural arrangement of components, and such arrangements are invariant in all faces (eyes are never found below the nose). Likewise, recognition of a unique face requires detection of the specific values of component dimensions (hair color, nose length, etc.) which again are available in the stimulus array corresponding to an individual face. On the other hand, when we consider intermediate-level categories (e.g. Oriental versus Occidental faces) Gibsonian theory falters, for it is here that averaging or schematizing of componential values seems to take place (e.g. Oriental faces are represented by prototypical eye shape, skin color, hair texture, etc. that probably bear little direct correspondence to any specific Oriental ever encountered). Nevertheless, if the Gibsons were to concede that the invariant of a set of varying feature values is an average or mode, they could accommodate Strauss's data.

A more fundamental problem raised by Strauss's experiment, and also touched on earlier in our discussion of the detection of novelty, revolves about the issue of representation. The Gibsons maintain that perception is direct and requires no mediation by mnemonic structures. It was pointed out, however, that if perceptual analyzers are merely attuned to invariant properties, there is no way to account for detection of novelty, since novel stimuli should simply be passed over just as a radio receiver fails to register frequencies to which it is not tuned. The fact that infants do not ignore stimuli that deviate from a previously exposed set of objects suggests rather that the invariants of the set have been stored and are activated during examination of subsequent incoming stimulation.

Harris (in press), in recognition of this problem, has proposed that storage involves a dual process: the properties of a newly encountered stimulus are registered to the extent that their values depart from a previous prototype, and the properties whose values show zero departure from the prototype are added incrementally to the prototype. Such a "double storage system" accounts for the ability to detect invariance, because randomly varying attributes (e.g., orientation changes of a particular face) are averaged out, and also for the ability to detect difference, because a new variation that has not been averaged out (e.g., a new face) will elicit further inspection.

Harris's hypothesis allows us to determine whether a newly encountered stimulus is assimilated to a previously formed prototype or is perceived de novo. If it is assimilated, then those component values of the stimulus that are irrelevant to the prototype (e.g., shape of eyes in the case of the face schema) should register weakly, and subsequent changes in these values (e.g., a new eye shape) should be ignored relative to changes in relevant components (e.g., addition of a third eye). However, if a stimulus is not perceived as an instance of a prototype, subsequent changes in every componential value should readily be detected. A case in point is an experiment by Kagan and his colleagues (Kagan, Linn, Mount, Reznick, & Hiatt, 1979), though they do not interpret their data in these terms. Five-month-old infants, previously shown to be able to discriminate two horizontally arranged black dots from two similarly configured black triangles, were subsequently found to be unable to make this discrimination when the two figures were enclosed in a circular frame so that the dots resembled eyes. On the other hand, they were able under these circumstances to discriminate the two horizontal dots from two *vertical* dots and (marginally) from *three* horizontal dots, though not from two larger or two smaller horizontal dots. The results are precisely what would have been expected if the enclosed figures had indeed been perceived as facial exemplars, for change in orientation and in number would be more disruptive of an eye pattern than change in shape or size. Thus Harris's double storage mechanism provides an explanation of the attentional effects of stimulus change that appears to be more satisfactory than the originally formulated discrepancy hypothesis. The latter assumed that small and large deviations from a previously formed schema would command little attention relative to an intermediate change. By defining change as relevant or irrelevant to a prototype, Harris avoids the ambiguity inherent in specifying degree of change, a problem that has always plagued the discrepancy hypothesis.

Dynamic Configurations

As became evident in our survey of the adult cognition literature, the contexts that predominate in establishing the meaning of things are dynamic or eventlike in nature. Likewise, our examination of the infant literature pointed to the informative role of events in differentiating the structure of objects. Both considerations raise the inevitable question as to when infants come to perceive the structure of events themselves. The problem is of particular interest, because most of the significant things in the environment are capable of a variety of self-generated movements with important adaptational consequences. Again, according to Gibsonian theory, such patterns should be detectable early in life on an amodal basis. According to Piagetian theory, on the other hand, an infant under six months should have difficulty perceiving invariant events, because (1) perception is assumed to be figurative in this period, a sequence of unintegrated sensations, and (2) the different senses are presumably not yet fully coordinated.

Unimodally Perceived Events. Evidence that infants under 6 months of age can discriminate complex event information, both unimodally and multimodally presented, has accumulated steadily in recent years. With regard to purely visible events, Gibson et al. (1978) have shown that 5-month-olds, following visual habituation to a sponge rubber object undergoing varying rigid motions (30° rotations around different axes), recovered significantly to a new deformation motion (squeezing) but not to a new rigid motion. While the authors rightly concluded that the infants had detected the invariant structural property of rigidity, it is also clear that they had discriminated two different types of motion (non-deforming versus deforming). In a subsequent study (Walker, Owsley, & Gibson, 1977), the ability to abstract a single rigid motion across different shapes, and, again, to discriminate it from a deforming motion, was shown to be within the capability of 3-month-olds. Whether discrimination was based on the contour changes associated with deformation or on the density changes of the speckled surface of the deformed object cannot be determined, but it is nevertheless apparent that some dynamic invariant had been abstracted.

Audible aspects of events abound with configural properties. Musical sounds have melodic and rhythmic patterns that can be detected across changes in pitch, loudness, instrument, and voice. Natural speech is rich in intonational, stress, rate, morphemic, and syntactic patterns that can be extracted by the human ear across differences in many acoustic parameters. Evidence that infants can also discriminate some of these invariants has begun to accumulate. For example, at the phonemic level of speech, children of 4 months or younger have been shown to discriminate different consonants on a categorical basis (Eimas, 1975). At this same age they can also abstract invariant vowels across changes in pitch contour (Kuhl & Miller, 1975) and at 6 months, constant vowels across different speakers (Kuhl, 1976). While these properties are more featural than configural, other evidence suggests that infants are also able to detect patterned qualities of speech. Thus,

2-month-olds appear to be sensitive to the rhythmic and rhyming aspects of speech (Horowitz, 1974) and 1-month-olds have been found to recognize the voice of a parent (Mehler, Bertoncini, Barriere, & Jassik-Gerschenfeld, 1978). In a particularly well-controlled study, DeCasper and Fifer (1980) found that newborns would readily alter their sucking patterns to hear their mother's voice, but made significantly fewer adjustments to hear the voice of another infant's mother. Although the precise characteristics that underlay this discrimination cannot be determined, it is hard to imagine that they involve only featural components. Perhaps more compelling with regard to configural discrimination is a recent demonstration that 7-month-olds could not only generalize across male and female voices, but could distinguish between these categories independent of differences in vocal frequency (Miller, Younger, & Morse, 1980). Finally, definitive evidence of discrimination of temporal configurations is provided by a study using nonvocal stimuli (Chang & Trehub, 1977). Five-month-olds were first habituated to a six-tone melody and were then found to generalize habituation to a transposition of the pattern five semitones above or below the original but to dishabituate to a control pattern composed of the same novel tones in scrambled order. Comparable generalization and discrimination effects were demonstrated for a 2–4 tonal beat as opposed to a 4–2 beat. Clearly, infants had detected the overall event configuration rather than the absolute pitch of the individual tones.

Bimodally Specified Events. Earlier, attention was drawn to the fact that some sights and sounds such as rhythmic hand movement and clapping noises are temporally synchronized and specify the same event. As noted, these events are of particular theoretical interest because from a Gibsonian perspective infants should be able to match information across these modalities independently of prior associative experience. Conversely, the occurrence of sensory correlations at early ages would be hard for Piagetians to explain because there are no obvious actions that infants might undertake to provide a basis for schema formation. Lastly, such findings would constitute evidence that very young infants can perceive highly abstract temporal relationships.

A dramatic demonstration of the infant's ability to detect synchrony between different sensory events comes from a study normally not considered in this context (Meltzoff & Moore, 1977). Two- to 3-week-olds were shown to be able to selectively imitate a number of adult gestures (tongue protrusion, mouth opening, lip protrusion, and sequential finger movement). Given the variety of gestures, the effect cannot reasonably be attributed to general arousal produced by moving stimuli, although Jacobson (1979) has since found that tongue protrusion will also occur to in and out movements of a pen and less so to a ball. Thus, although imitation may not have been confined to anatomically identical but to similarly shaped moving objects, the possibility remains that infants were matching a visual event to kinaethetic sensations associated with their own body movement. Moreover, since there generally was a delay between termination of the stimulus and onset of the infant's response, the authors' conclusion that this behavior was mediated by some sort of supramodal representation cannot be dismissed.

The ability of infants to detect temporal synchronization between simultaneously occurring sounds and sights has been examined by Spelke (1979a) in a series of experiments. In an initial study (Spelke, 1976), 4-month-olds were shown two films side by side on a split screen and simultaneously heard a soundtrack matched to one of the films played through a centrally located speaker. The films were a woman playing peekaboo and a musical sequence played on toy percussion instruments. The infants watched the sound-appropriate event a significant proportion of the time. A second experiment (Spelke, 1978) indicated that the effect occurred only in the percussion sequence. Bahrick, Walker, and Neisser (1978) reported similar results with 4-month-olds using somewhat less familiar events (hands playing pat-a-cake, a slinky toy being opened and closed with two hands, and two colored sticks striking a toy xylophone in time with a nursery tune). Because these events were all somewhat familiar and hence might have become associated through past experience, Spelke (1979b) extended her research to artificially synchronized events: toy animals moving up and down with soundbursts accompanying the visible impact of the ground. Although no clear preference was shown by her 4-month-old subjects for the acoustically specified event, sensitivity to intermodal correspondence was demonstrated by a method involving visual search (direction of first look following brief periods of sound presentation alone). Subsequent experiments demonstrated that the effect could be produced (1) when tempo was intermodally related but the sounds and impacts were not synchronized and (2) when sounds and impacts were synchronized but tempo was constant in the two films.

In still another study (reported in Spelke & Cortelyou, 1981), 4 month infants were shown side-by-side films of two unfamiliar women speaking with exaggerated expressions typical of speech to infants. The sound track of each woman was played in turn, the voices being synchronized with the facial movements of the speaker. Infants looked significantly more at whichever woman was "speaking," and during a subsequent search test they looked first and more often toward the woman whose voice they heard. Given that these were totally unfamiliar persons whose voices and faces had never been associated, the results indicate that infants can detect the invariant temporal relationship between the sight of a speaking person and the sound of his or her voice, even when they are spatially dissociated. Similar findings were reported by Dodd (1979), who showed that infants will look more at an unfamiliar face when it moves in synchrony with the voice than when it is nonsynchronous.

Perhaps the most impressive evidence of the infant's ability to detect bimodally specified events comes from a recent study by Walker, Bahrick, and Neisser (1980). Four-month-olds were shown two *superimposed* naturalistic films on the same screen and were "biased" to attend to one of them by simultaneously playing its soundtrack. The soundtrack was periodically turned off as the film images were separated, thereby permitting a test of preference for the nonsynchronized ("novel") film. Infants significantly preferred the novel film, and to the same extent as control infants who had previously seen a single film. Both groups yielded significantly greater preference than a second control group who had viewed an out-of-focus version of the overlapped films. Thus, amodal information from the soundtrack alone could not have produced the results; rather, the infants had apparently been able to selectively differentiate the acoustically-specified event from the superimposed complex.

These studies, taken together, provide strong evidence that young human infants are sensitive to the abstract structural aspects of events and that they pick up this information multimodally without benefit of prior sensory association. Empiricists and Piagetians, alike, would be hard pressed to account for these data.

Cross-Modal Association. Some sights and sounds can be linked only through learned association because they share no common configural properties. Although mouth movements may be synchronized with speech, the face of a person has no intrinsic correlation with his or her voice quality. We raise this issue at this point not because it tells us anything about configural discrimination but because it bears on the question of identity. If infants quickly learn to associate sounds (or other sensory input) with sights, they should be better able to specify the uniqueness of things. Cookies and spoons may exist in two places simultaneously, but a mother with distinctive voice and expressive qualities (to say nothing of smell and touch) is never so located.

A number of studies have examined the infant's ability to connect sights and sounds. Lyons-Ruth (1977) found that infants of 4 months will look more at a toy when its previously familiarized sound is produced than at a toy whose sound has never been heard. With regard to face-voice relationships, Lewis and Hurowitz (1977) reported that infants of 1 and 4 months looked away more often from mismatched pairings of mothers' and strangers' face and voice than from matched pairings, even though face and voice always emanated from the same location. Since face and voice were synchronized in matched pairings, but were unsynchronized in mismatched pairings, the results could be a function of either perception of synchrony or of learned association. These factors were not confounded in a further study by Spelke and Owsley (1979) where infants from 3½ to 7½ months saw their two parents sitting motionless side by side and heard the voices of each in turn through a central speaker. Despite the absence of synchrony, infants over the entire age range tended to look more often at the voice-appropriate parent. Interestingly, when this procedure was repeated with the mother and a female stranger, infants tended to look more often at the inappropriate person. Since the original results with mother and father were replicated, it is possible that in the presence of unfamiliar stimuli, infants seek to assimilate the unknown to the known. Be that as it may, it appears that by 4 months of age infants have learned to associate some properties of the face and voice of a parent.

Object to Object Relations

Objects in Relation to Themselves. Our concern in this section is with the infant's understanding of objects that shift position in space, that is, with the classic issue of object permanence. Since this problem has received extensive treatment in several excellent reviews, we shall not attempt to retrace this terrain, but instead will examine in a very general way the evidence bearing on the question of object identity, that is, whether infants see objects that are displaced in space as one and the same object. As noted in our theoretical overview, both Bower (1974) and Moore and Meltzoff (1978) would expect infants to preserve the identity of continuously moving objects at 5 months of age and of objects that appear in different locations at 9 months of age. The evidence pertaining to object recognition suggests, as we have seen, that by about 7 months infants can discriminate between different people on a purely visual basis. If one also takes into account the demonstrated ability of infants to coordinate visual and auditory information as well as their extensive experience with their own caretakers, they may well be able to recognize familiar persons prior to 7 months. Of course, the critical question, as stated by Harris (in press), is whether they see the same person who appears at two separate locations as the identical person or merely as two very similar instances of a class of objects. Harris insists that infants until late in Stage 4 (11 to 12 months) confront a world filled with similars and may even regard successive appearances of their own parents as duplication events.

Harris cites various lines of evidence to buttress his position. First, various experiments indicate that when an object moves behind a screen and reemerges at the other side, infants of 9 months or younger show little understanding that they are witnessing one and the same object on a continuous trajectory. Thus, after watching an object enter and exit from behind a screen a number of times, they rarely look ahead on later trials to anticipate its exit, implying that they believe two separate events are involved (Meicler & Gratch, 1980; Nelson, 1971). If later the object is stopped behind the screen, again they fail to look toward the exit point (Meicler & Gratch, 1980). Finally, if a different object emerges from behind the screen, they exhibit little evidence of surprise (Goldberg, 1976; Meicler & Gratch, 1980; Muller & Aslin, 1978). Contrary to this last observation, however, Moore, Borton, and Darby (1978) found that both 5- and 9-month-old infants showed more looking back or looking away behavior when identity was changed during occlusion than when it was preserved. As Harris himself points out, this may be due to the fact that Moore et al. stopped their object after it emerged from behind the screen, whereas in each of the other cases the objects were in continuous movement across trials and thus might have captured attention and thereby suppressed search. Support for this interpretation, and for the possible existence-enhancing effects of a highly familiar object, comes from a study that also used an intertrial stop procedure (Boomer, 1977). Eight-month-old infants (but not 5-month-olds) evidenced both surprise and search behavior when their own mother walked behind a screen and a new female emerged in her place.

Evidence related to the place error is also not entirely consistent with Harris's view. Infants do sometimes continue to search for an object in its old hiding place even when it remains fully visible at the new location (Butterworth, 1975; Harris, 1974), which seems to imply that they believe there are two objects: one at the new location and one at the old. However, in a study where the infant's own mother disappeared first through one door and then through another (Corter, Zucker, & Calligan, 1980), 9-month-olds searched perseveratively at the initial door following displacement. Corter et al. concluded that, since by 9 months of age infants can identify their own mother, the place error cannot be explained by inability to identify objects in two different places as the same object. Harris (in press), on the other hand, interprets their data as evidence for the possibility that the infant believes he has more than one mother who can each be simultaneously in two places. However, Corter et al.'s additional finding that infants showed signs of distress when confronted with a change of hiding place would appear to be incompatible with the hypothesis that they thought a mother might exist behind either door and more consistent with the possibility that they were confused about her location. Acredolo (1978, 1979) has found that infants do not begin to use external landmarks to locate objects until sometime between 9 and 11 months, and that even at 9 months they react egocentrically in an unfamiliar laboratory setting.

In sum, the evidence is equivocal with regard to identity theory. Harris's version of the theory is an attractive one, because it appears to resolve some anomalous findings, but at this point we simply do not have conclusive data as to whether infants regard mothers as they do spoons: as reproducible carbon copies.

Event Roles. If infants under 6 months can perceive the invariant structure of events and, reciprocally, the invariant structure of objects that participate in events, soon thereafter, according to Gibsonian theory, they should begin to recognize the appropriate roles of objects within familiar event contexts. By "roles" is meant the functions or affordances of inanimate objects and the actional capacities of animate objects, such as the ability to initiate behavior, to communicate, to use objects instrumentally, and to produce effects on things and people. From a Piagetian perspective, infants would not be expected to distinguish appropriate from inappropriate roles until they had developed object permanence, rudimentary tool use, and the externalization of object relations (sometime after 12 months).

Unfortunately, little systematic work has been done on this problem in the crucial 6- to 12-month period. As noted, Nelson (1977) has argued that the meaning core of a child's first concepts (even prior to labeling) is functional/dynamic rather than perceptual. She found that 8- and 10-month-olds pay more attention to a previously explored object undergoing a new motion than to either a new object undergoing the same new motion or a new object undergoing the old motion (Nelson, 1979), thereby implying that the dynamic property had been more salient during initial exposure. Since the objects used in this experiment were relatively unfamiliar and the motions were arbitrarily related to each, the study tells us that the motion of an object may command attentional priority but not whether it is perceived as more affordable by the object than another motion. More to the point would be the demonstration that an 8-month-old could recognize, for example, that the spontaneous movement of a person is appropriate, whereas the same motion of a table is anomalous, or, to take a functional case, that a cup can be "poured into" but a toy block cannot.

There has been some recent work with older infants that does bear tangentially on this issue. Using habituation procedures with filmed events, Golinkoff (1975; Golinkoff & Kerr, 1978) attempted to determine whether infants of 14 to 24 months understand the concept of animacy. The subjects were habituated to sequences of either a person pushing another person or a person pushing an inanimate object (table or chair), whereupon the roles were reversed. Although infants at all ages could detect the role reversals, they did so equally for both films, suggesting that they had not perceived the agency of the inanimate object as anomalous. A similar finding was reported by Gilmore, Suci, and Chan (1974) for 18-month-olds. Golinkoff has since questioned the implications of these studies on the grounds that infants see many inanimate objects in agentive roles on TV. Accordingly, in a further experiment (Golinkoff & Harding, 1980) she created a real life anomaly—a spontaneously moving real chair. Ratings of infants' emotional responses revealed that 64 percent of 16-month-olds and 83 percent of 24-month-olds now considered the event a violation. Unfortunately, since younger infants were not used and in view of the previous negative findings, no firm conclusions can be drawn as to when infants come to link animacy with specific objects. It might be noted, however, that in research where a more subtle social role violation was examined—the sight of adults crawling on the floor or sucking a baby bottle—the occurrence of laughter reached significant proportions at 12 months of age

(Sroufe & Wunsch, 1972; Cicchetti & Sroufe, 1976). Also Ross's (1980) previously cited evidence that 12-month-old infants can categorize food and furniture suggests that functional similarity between very disparate items was detected at this age. Be that as it may, more extensive research with younger infants using much simpler types of object affordances is sorely needed.

Causality. Although predictions from Piagetian and Gibsonian theory are most clearly differentiated with regard to early causal development, there have been only a handful of experimental infant studies directed to this issue. Using Piaget's experimental situations, Uzgiris and Hunt (1975) developed a number of sensorimotor assessment scales, including an operational causality scale, that yielded a sequence of development from magical to objectified causality very similar to Piaget's. Evidence of attribution of causal power to an external agent appeared initially at 5 months when infants touched the examiner's hands, ostensibly to induce him to repeat an interesting action. More definitive evidence of externalization occurred at 12 to 15 months when, for example, infants would return a mechanical toy to the adult after it had stopped moving. However, subjects showed no recognition of the direct ways to activate this toy until about 18 months and then only after prior demonstration. Though consistent with Piaget's observations, these results are not necessarily inconsistent with J. Gibson's speculations. They suggest that more naturalistic forms of causality (as in the case of the hands) may be recognized much earlier than causal situations where the source of potency (the key in the case of the windup toy) is arbitrary from the infant's point of view. Gibson (1977) has noted than an arbitrary combination of properties invented by an experimenter (e.g., that a blue triangle on a panel specifies a banana behind the panel) should be harder for an animal to detect than a natural invariant combination (e.g., that a long, rounded yellow surface specifies a banana behind its skin).

Only two experiments have focused directly on the Michotte-Gibson hypothesis that perception of mechanical causality is innate. Ball (1973) habituated infants, ranging in age from 2½ months to 2½ years, to a sequence in which a moving red object disappeared behind a screen and a white object subsequently emerged from the other side. The relative velocities were those reported by Michotte (1963) to have produced maximal causal impression in adults. After 10 repetitions the screen was removed. One half of the subjects then saw 10 trials in which the red object collided with the white one, while the other half saw motion of the same two objects without collision. The noncollision subjects spent significantly more time tracking the display than the collison subjects, whereas control groups exposed only to the critical trials showed no difference in attention to the two displays. Although Ball interpreted his results as indicating that the screen trials had set up an expectation of collision as an innate perceptual basis for causality, his conclusion appears stronger than the data warrant. For one thing, Ball's older infants had had ample visual and tactual-kinaesthetic experience which could have affected their expectations. More important, however, there is no way to determine whether increased attention to the noncontact event was due to the absence of collision (a violation of causality) or simply to the stopping of the red object (a violation of identity). If, as Moore (1975) has conjectured, infants under 5 months identify objects by trajectory rather than by features, then during habituation the red and white objects might actually have appeared to the younger subjects as a single entity on a continuous path of motion. Olum (1958), who repeated Michotte's experiments with 7-year-olds, reported that many of her subjects saw the display as the continuous movement of a single object and often did not even see a second object. In that case, the contact trials in Ball's test series would have appeared as further repetitions of the familiar continuous movement sequence, but the cessation of movement of the red object in the noncontact trials could well have been viewed as the sudden introduction of a second (stationary) object rather than as a causal anomaly.

Borton (1979) subsequently modified Ball's experiment with a homogeneous group of 3-month-olds who watched a moving object collide with a stationary one and launch it off, and also saw the same event without collision. There were no prior habituation trials. The fact that under the noncontact condition infants exhibited more visual search behavior (looking back, looking ahead, looking away) led him to conclude that the infants understood that physical contact was necessary for one object to transmit its energy of motion to another. In one of his experiments Borton took pains to disconfound violation of identity and of causality (by having the second object hesitate and thus continue in the projected trajectory of the first object even though there had been no contact). Since his results in this condition were essentially the same as those in his other noncontact conditions, violation of identity appears to have been eliminated as a factor contributing to his findings. If this manipulation can be accepted as an adequate control, then Borton's study provides provisional support for Gibson's ecological theory of physical causality.

There have been no systematic attempts to investigate other parameters besides physical contact as conditions of causal impression, although some studies in the learning literature bear on the role of spatiotemporal and intensity concordances. Thus Millar and Schaffer (1972, 1973) found that 6-month-olds could acquire an operant response if the spatial displacement between feedback source and manipulandum were no greater than 5°. A 60° displacement failed to produce learning at this age, though it did so in 9- and 12-month-olds. In addition, the fact that very young infants can be easily conditioned by means of conjugate reinforcement (cf. Rovee-Collier & Gekoski, 1979) where intensity and frequency relations between response and feedback are closely matched, suggests that such factors enhance impressions of causal efficacy. Watson (1977) has also found that three-month-old infants make sophisticated use of temporal information to evaluate the "necessity" and "sufficiency" of behavior in an instrumental response situation where reinforcers are sometimes free and sometimes contingent on responding. Overall, these studies are consistent with Testa's (1974) thesis that real world causal relations are central to all learning, at both the animal and human levels.

A last word is in order regarding comprehension of social casuality. Here, too, the Piagetian and Gibsonian positions deviate in clear-cut fashion. Piaget makes no attempt to distinguish between knowledge of animate and inanimate sources of causality. Jacqueline, between 12 and 18 months, recognizes both that her father is the originator of the force that blows her hair and that a ball placed on top of an incline will roll down. Whether she differentiates between these two causal events, or between her own respective roles in initiating them, is not addressed. However, it is possible that while infants may see mechanical causality in terms of such parameters as spatial contiguity and transfer of attributes from cause to effect, these factors are not necessary conditions for recognizing social causality (where influence may occur at a distance and cause and effect need not share qualitative features). In fact, as Carlson-Luden (1979) has observed, what Piaget described as magical behavior in Stage 3—behaving to make interesting things happen—may be an appropriate mechanism for influencing animate objects. Gibson, as noted previously, feels that the responsive affordance of animate objects to one's own behavior are perceived early in life, although he doesn't specify what the basis of these percepts might be.

Carlson-Luden (1979) sought to determine whether 10-month-old infants understood social and physical causality in the same terms. She devised a situation where either a physical or a social (human) event could be activated when a knob was pushed either toward or away from the event. The infants used the knob significantly more for the physical than the social reinforcer, and while they pushed significantly more toward than away from the physical event, they exhibited no difference in direction for the social event. The results indicated that the use of the knob in the "toward" direction was deemed appropriate for the physical consequence but was seen as inappropriate for the human consequence. These results, taken together with Bates (1976) finding that the onset of communicative intention in infants occurs at 9 to 10 months of age, indicate that understanding of some aspects of social causality may occur much earlier than Piaget suspected.

In sum, rudimentary awareness of both physical and social causality may emerge well before 12 months of age, and at least in the case of physical causality, there is some reason to believe that it is based on the extraction of an event invariant, as J. Gibson has proposed. However, the data are more provocative than definitive, thus ensuring that this will be a fertile research area in coming years.

CODA

If evidence were needed that the conceptual foundation of intellect is laid early in life, it abounds in the preceding pages. One clear generalization that emerges from these data is that the human infant, like his adult counterpart, ignores the isolated details of experience in favor of the more abstract information that organizes these details. Well before 6 months of age he can detect the invariant patternings within objects and events regardless of variation in their components. During this period, too, he can categorize otherwise discriminable stimuli in terms of their general similarities. And, finally, although the evidence is less firm, he seems to be sensitive during the first year to relationships between objects in event contexts.

To accomplish these organizational feats, the infant has available a more formidable sensory armamentarium than would have been suspected from Piagetian theory. As Gibson clearly recognized, objects and events are perceived through multiple sensory channels that register the same information supramodally, so that infants need not depend on experience to bring about coordination of different modalities. Moreover, vision provides information that is as veridical as that originating from touch, kinaesthesis, or audition, and again each modality seems to have direct access to the structural properties of things.

Does this mean, then, that the infant confronts the world with no more than a sophisticated detection system? In barest terms, does he simply extract information as it exists in the environment? In all probability, the answer is no. While both the structure and details of things seem to be directly detectable, the fact that under some circumstances details will be modalized or averaged suggests that input may also be transformed during the perceptual process. More important, the infant's ability to recognize similarities and differences between events and to correlate information across modalities appears to depend on the capacity to represent information in memory. Likewise, young babies are prone to make more errors, particularly about the location and identity of things, than a simple detection system would allow. This implies that they perceive events within incipient frames and must learn to modify such frames in order to undo error.

This observation brings us full circle to the question of frames that initiated the present essay. As Piaget insisted, comprehension is not solely a matter of perception and, as we have seen from our survey of the adult literature, events are interpreted and reinterpreted in the light of internalized schemes. These may not necessarily arise from the interiorization of action, as Piaget claimed, but his metaphor of the young infant as scientist, who continually revises his interpretive frames in the light of new experience and who assimilates ongoing events to existing frames, is nonetheless apt. As Weimer (1977) has observed, "every scientific revolutionary has been an assimilator and Einstein was not just more skilled in looking than Newton" (p. 484). Humans do a lot of guessing about things they can't see directly, and this process involves several levels of abstraction from perceived invariances. Thus, perception may take the infant a long way toward knowing, but whether it takes him the whole way seems questionable.

REFERENCES

Acredelo, L. P. Development of spatial orientation in infancy. *Developmental Psychology,* 1978, **14**, 224–234.

Acredelo, L. P. Laboratory versus home: The effect of environment on the 9-month-old infant's choice of spatial reference system. *Developmental Psychology,* 1979, **15**, 585–593.

Anderson, R. C. The notion of schemata and the education enterprise. In R. C. Anderson, R. J. Spiro, & W. E. Montague (Eds.), *Schooling and the acquisition of knowledge.* Hillsdale, N.J.: Erlbaum, 1977.

Anglin, J. From reference to meaning. *Child Development,* 1978, **49**, 969–976.

Antell, S., & Keating, D. Perception of numerical invariance by neonates. Unpublished manuscript.

Bahrick, L., Walker, A., & Neisser, U. Infants' perception of multimodal information in novel events. Paper presented at the meeting of the Eastern Psychological Association, Washington, D.C., March 1978.

Ball, W. N. The perception of causality in the infant. Unpublished paper, Report #37, Developmental Program, Department of Psychology, University of Michigan, 1973.

Barrera, M. E., & Maurer, D. The perception of facial expressions by the three-month-old. *Child Development,* 1981, **52**, 203–207. (a)

Barrera, M. E., & Maurer, D. Discrimination of strangers by the three-month-old. *Child Development,* 1981, **52**, 558–564. (b)

Barrera, M. E., & Maurer, D. Recognition of mother's photographed face by the three-month-old infant. *Child Development,* 1981, **52**, 714–717. (c)

Bartlett, F. C. *Remembering: An experimental and social study.* Cambridge, England: Cambridge University Press, 1932.

Bates, E. Pragmatics and Sociolinguistics in child language. In D. Morehead & A. Morehead (Eds.), *Normal and deficient child language*. Baltimore: University Park Press, 1976.

Bertenthal, B. I., Campos, J. J., & Benson, N. Self produced locomotion and the perception of invariant relations: Is there a connection? Paper presented to the International Conference on Infant Studies, New Haven, April 1980.

Bertenthal, B. I., Haith, M. K., & Tucker, P. *Infants' sensitivity to subjective contours.* Paper presented to the International Conference on Infant Studies, New Haven, April, 1980.

Boomer, M. Object permanence development in the infant-mother relationship: An investigation of the cognitive aspect of attachment. Unpublished doctoral dissertation, University of California, Berkeley, 1977.

Borton, R. W. The perception of causality in infants. Unpublished doctoral dissertation, 1979, University of Washington.

Bower, T. G. R. Slant perception and shape constancy in infants. *Science,* 1966, **151**, 832–834. (a)

Bower, T. G. R. The visual world of infants. *Scientific American,* 1966, **215**, 80–92. (b)

Bower, T. G. R. Heterogeneous summation in human infants. *Animal Behavior,* 1966, **14**, 395–398. (c)

Bower, T. G. R. The development of object permanence. Some studies of existence constancy. *Perception and Psychophysics,* 1967, **2**, 411–418.

Bower, T. G. R. *Development in infancy*. San Francisco: Freeman, 1974.

Bower, T. G. R. Infant perception of the third dimension and object concept development. In L. B. Cohen & P. Salapatek (Eds.), Infant perception: From sensation to cognition, Vol. II, New York: Academic, 1975.

Bower, T. G. R. *Human Development*. San Francisco: Freeman, 1979.

Bower, T. G. R. & Wishart, J. G. The effects of motor skill on object permanence. *Cognition,* 1972, **1**, 165–171.

Bower, T. G. R., Broughton, J. M., & Moore, M. K. Demonstration of intention in the reaching behavior of neonate humans. *Nature,* 1970, **228** (No. 5272).

Bransford, J. D., & McCarrell, N. S. A sketch of a cognitive approach to comprehension. In W. Weimer and D. S. Palermo (Eds.), *Cognition and the symbolic processes*. Hillsdale, N.J.: Erlbaum, 1974.

Bransford, J., McCarrell, N., Franks, J., & Nitsch, K. Toward unexplaining memory. In R. Shaw & J. Bransford (Eds.), *Perceiving, acting, and knowing*. Hillsdale, N.J.: Erlbaum, 1977.

Bremner, J. G. Egocentric versus allocentric spatial coding in nine-month-old infants: Factors influencing the choice of code. *Developmental Psychology,* 1978, **14**, 346–366.

Bremner, J. G., & Bryant, P. E. Place versus response as the basis of spatial errors made by young infants. *Journal of Experimental Child Psychology,* 1977, **23**, 162–171.

Bruner, J. S. Early social interaction and language acquisition. In H. R. Schaffer (Ed.), *Studies of mother-infant interaction*. New York: Academic, 1977.

Bryant, P. E., Jones, P., Claxton, V., & Perkins, J. Recognition of shapes across modalities by infants. *Nature,* 1972, **240**, 303–304.

Butterworth, G. Object identity in infancy: The interaction of spatial location codes in determining search errors. *Child Development,* 1975, **46**, 866–870.

Butterworth, G. Thought and things: Piaget's theory. In A. Burton and J. Radford (Eds.), *Perspectives on thinking*. London: Methuen, 1978.

Carlson-Luden, V. Causal understanding in the ten-month-old. Unpublished doctoral dissertation, 1979, University of Colorado.

Caron, A. J., & Caron, R. F. Processing of relational information as an index of infant risk. In Friedman, S. L. & Sigman, M. (Eds.), *Preterm birth and psychological development*. New York: Academic, 1981.

Caron, A. J., Caron, R. F., & Carlson, V. R. Infant perception of the invariant shape of an object varying in slant. *Child Development,* 1979, **50**, 716–721.

Caron, A. J., Caron, R. F., Caldwell, R., & Weiss, S. Infant perception of the structural properties of the face. *Developmental Psychology,* 1973, **9,** 385–399.

Caron, R. F., Caron, A. J., & Myers, R. S. Abstraction of invariant face expressions in infancy. *Child Development,* in press.

Chang, N., & Trehub, S. Auditory processing of relational information by young infants. *Journal of Experimental Child Psychology,* 1977, **24,** 324–331.

Cicchetti, D., & Sroufe, L. A. The relationship between affective and cognitive development in Down's syndrome infants. *Child Development,* 1976, **47,** 920–929.

Clark, E. V. What's in a word? On the child's acquisition of semantics in his first language. In T. Moore (Ed.), *Cognitive development and the acquisition of language.* New York: Academic, 1973.

Clark, E. V. Knowledge, context, and strategy in the acquisition of meaning. In D. Dato (Ed.), *Developmental psycholinguistics: Theory and applications.* 26th Annual Georgetown University Round Table. Washington, D.C.: Georgetown University Press, 1975.

Cohen, L. Our developing knowledge of infant perception and cognition. *American Psychologist,* 1979, **34,** 894–899.

Cohen, L., & Caputo, N. Instructing infants to respond to perceptual categories. Paper presented at the meeting of the International Conference on Infancy Studies, Providence, Rhode Island, March 1978.

Cohen, L. B., & Gelber, E. R. Infant visual memory. In L. B. Cohen & P. Salapatek (Eds.), *Infant perception: From sensation to cognition.* Vol. I. New York: Academic, 1975.

Cohen, L. B., & Strauss, M. S. Concept acquisition in the human infant. *Child Development,* 1979, **50,** 419–424.

Cornell, E. Infants' discrimination of photographs of faces following redundant presentations. *Journal of Experimental Child Psychology,* 1974, **18,** 98–106.

Cornell, E. H. Infant's visual attention to pattern arrangement and orientation. *Child Development,* 1975, **46,** 229–232.

Cornell, E. H., & Strauss, M. S. Infants' responsiveness to compounds of habituated visual stimuli. *Developmental Psychology,* 1973, **9,** 73–78.

Corter, C. M., Zucker, K. J., & Calligan, R. F. Patterns in the infant's search for mother during brief separation. *Developmental Psychology,* 1980, **16,** 62–69.

DeCasper, A. J., & Fifer, W. P. Of human bonding: Newborns prefer their mothers' voices. *Science,* 1980, **208,** 1174–1176.

Dewey, J. *The quest for certainty.* New York: Capricorn, 1960.

Dodd, B. Lipreading in infants: Attention to speech presented in- and out-of-synchrony. *Cognitive Psychology,* 1979, 11, 478–484.

Eimas, P. Speech perception in early infancy. In L. B. Cohen & P. Salapatek (Eds.), *Infant Perception,* Vol. II. New York: Academic, 1975.

Fagan, J. F. Infant's recognition of invariant features of faces. *Child Development,* 1976, **47,** 627–638.

Fagan, J. F. An attention model of infant recognition. *Child Development,* 1977, **48,** 345–359.

Fagan, J. F. Origins of facial pattern recognition. In M. Bornstein & W. Kessen (Eds.), *Psychological development from infancy: Image to intention.* Hillsdale, N.J.: Erlbaum, 1979.

Fagan, J. F., & Singer, L. T. The role of simple feature differences in infant's recognition of faces. *Infant Behavior and Development,* 1979, **2,** 39–48.

Field, J. Relation of young infants' reaching behaviour to stimulus distance and solidity. *Developmental Psychology,* 1976, **12,** 444–448.

Flavell, J. H. *Cognitive development.* Englewood Cliffs, N.J.: Prentice-Hall, 1977.

Garner, W. R. Aspects of a stimulus: Features, dimensions, and configurations. In E. Rosch & B. B. Lloyd (Eds.), *Cognition and categorization.* Hillsdale, N.J.: Erlbaum, 1978.

Gibson, E. J. *Principles of perceptual learning and development.* New York: Appleton, 1969.

Gibson, E. J., & Levin, H. Afterword. In A. D. Pick (Ed.), *Perception and its development.* Hillsdale, N.J.: Erlbaum, 1979.

Gibson, E. J., Owsley, C. J., & Johnston, J. Perception of invariants by five-month-old infants: Differentiation of two types of motion. *Developmental Psychology*, 1978, **14**, 407–415.

Gibson, E. J., Owsley, C. J., Walker, A., & Megaw-Nyce, J. Development of the perception of invariants: Substance and shape. *Perception*, 1979, **8**, 609–619.

Gibson, J. J. *The perception of the visual world*. Boston: Houghton-Mifflin, 1950.

Gibson, J. J. *The senses considered as perceptual systems*. Boston: Houghton-Mifflin, 1966.

Gibson, J. J. The theory of affordances. In R. Shaw & J. Bransford (Eds.), *Perceiving, acting, and knowing*. Hillsdale, N.J.: Erlbaum, 1977.

Gibson, J. J. *An ecological approach to visual perception*. Boston: Houghton-Mifflin, 1979.

Gibson, J. J., & Gibson, E. J. Perceptual learning: Differentiation or enrichment? *Psychological Review*, 1955, **62**, 32–41.

Gilmore, L. M., Suci, G., & Chan, S. Heart rate deceleration as a function of viewing complex visual events in 18-month-old infants. Paper presented at the meetings of the American Psychological Association, New Orleans, 1974.

Goffman, E. *Frame analysis*. New York: Harper & Row, 1974.

Goldberg, S. Visual tracking and existence constancy in 5-month-old infants. *Journal of Experimental Child Psychology*, 1976, **22**, 478–491.

Goldman, D., & Homa, D. Integrative and metric properties of abstracted information as a function of category discriminability, instance variability, and experience. *Journal of Experimental Psychology: Human Learning and Memory*, 1977, **3**, 375–385.

Golinkoff, R. M. Semantic development in infants: The concept of agent and recipient. *Merrill-Palmer Quarterly*, 1975, **21**, 181–193.

Golinkoff, R. M., & Harding, C. G. The development of causality: The distinction between animates and inanimates. Paper presented to the International Conference on Infant Studies, New Haven, 1980.

Golinkoff, R. M., & Kerr, J. C. Infant's perception of semantically defined action role changes in filmed events. *Merrill Palmer Quarterly*, 1978, **24**, 53–61.

Gordon, F. R., & Yonas, A. Sensitivity to binocular depth information. *Journal of Experimental Child Psychology*, 1976, **22**, 413–422.

Gordon, F. R., Lamson, G., & Yonas, A. *Reaching to a virtual object*. Unpublished manuscript, University of Minnesota, 1978.

Gottfried, A. W., Rose, S. A., & Bridger, W. H. Cross modal transfer in human infants. *Child Development*, 1977, **48**, 118–123.

Gratch, G. Recent studies based on Piaget's view of object concept development. In L. Cohen and P. Salapatek, *Infant perception: From sensation to cognition*. New York: Academic, 1975.

Greenfield, P., & Smith, J. H. *The structure of communication in early language development*. New York: Academic, 1975.

Haaf, R., & Bell, R. A facial dimension in visual discrimination by human infants. *Child Development*, 1967, **38**, 893–899.

Hainline, L. Developmental changes in visual scanning of face and non-face patterns by infants. *Journal of Experimental Child Psychology*, 1978, **25**, 90–115.

Haith, M. M., Bergman, T., & Moore, M. J. Eye contact and face scanning in early infancy. *Science*, 1977, **198**, 853–855.

Harris, P. L. Perseverative search at a visibly empty place by young infants. *Journal of Experimental Child Psychology*, 1974, **18**, 535–542.

Harris, P. L. Development of search and object permanence during infancy. *Psychological Bulletin*, 1975, **82**, 332–344.

Harris, P. L. Infant Cognition. In J. J. Campos & M. M. Haith (Eds.), *Handbook of Child Psychology*, Vol. I. New York: in press.

Horowitz, F. D. Visual attention, auditory stimulation, and learning discrimination in young infants. Monographs of the Society for Research in Child Development, 1974, **39**, (5–6, Serial No. 158).

Jacobson, S. Matching behavior in the young infant. *Child Development,* 1979, **50**, 425–430.

Jeffrey, W. The orienting reflex and attention in cognitive development. *Psychological Review,* 1968, **75**, 323–334.

Kagan, J. Attention and psychological change in the young child. *Science,* 1970, **170**, 826–832.

Kagan, J. *Change and continuity in infancy.* New York: Wiley, 1971.

Kagan, J. Emergent themes in human development, *American Scientist,* 1976, **64**, 186–196.

Kagan, J. Structure and process in the human infant: The ontogeny of mental representation. In M. Bornstein and W. Kessen (Eds.), *Psychological development from infancy: Image to intention.* Hillsdale, N.J.: Erlbaum, 1979.

Kagan, J., Linn, S., Mount, R., Reznick, J., & Hiatt, S. Asymmetry of influence in the dishabituation paradigm. *Canadian Journal of Psychology,* 1979, **33**, 288–305.

Kintsch, W. *The representation of meaning in memory.* Hillsdale, N.J.: Erlbaum, 1974.

Koffka, K. *Principles of Gestalt psychology.* New York: Harcourt, Brace, 1935.

Kuhl, P. K. Speech perception in early infancy: Perceptual constancy for vowel categories. *Journal of the Acoustical Society of America,* 1976, **60** (Suppl. 1), S90 (A).

Kuhl, P. K., & Miller, J. D. Speech perception in early infancy: Discrimination of speech sound categories. *Journal of the Acoustical Society of America,* 1975, **58** (Suppl. 1), S56 (A).

LaBarbera, J. D., Izard, C. E., Vietze, P., & Parisi, S. A. Four- and six-month-old infants' visual responses to joy, anger, and neutral expressions. *Child Development,* 1976, **47**, 535–538.

Labov, W. The boundaries of words and their meanings. In C. J. Bailey and R. W. Shuy (Eds.), *New ways of analyzing variation in English.* Washington, D.C.: Georgetown University Press, 1973.

Lachman, R., Lachman, J. L., & Butterfield, E. C. *Cognitive psychology and information processing: An introduction.* Hillsdale, N.J.: Erlbaum, 1979.

Lashley, K. S. The problem of serial order in behavior. In L. A. Jeffress (Ed.), *Cerebral mechanisms in behavior.* New York: Wiley, 1951.

Lewis, M., & Horowitz, L. Intermodal person schema in infancy: Perception within a common auditory-visual space. Paper presented at the meeting of the Eastern Psychological Association, Boston, April 1977.

Lyons-Ruth, K. Bimodal perception in infancy: Response to auditory-visual incongruity. *Child Development,* 1977, **13**, 492–500.

MacNamara, J. The cognitive basis of language learning in infants. *Psychological Review,* 1972, **79**, 1–13.

Mandler, J. Categorical and schematic organization in memory. In C. R. Puff (Ed.), *Memory organization and structure.* New York: Academic, 1979.

Maurer, D., & Barrera, M. Infants' perception of natural and distorted arrangements of a schematic face. *Child Development,* 1981, **52**, 196–203.

Maurer, D., & Heroux, L. The perception of faces by three-month-old infants. Paper presented at the International Conference on Infant Studies, New Haven, April 1980.

McCall, R. B. Attention in the infant: Avenue to the study of cognitive development. In D. Walcher, & D. Peters (Eds.), *Early childhood: The development of self-regulatory mechanisms.* New York: Academic, 1971.

McCall, R., & Kagan, J. Attention in infancy: Effects of complexity, contour, perimeter, and familiarity. *Child Development,* 1967, **38**, 939–952.

McKenzie, B. E. Infant perception of the invariant size of approaching and receding objects. Paper presented to the meeting of the International Conference on Infant Studies, New Haven, April 1980.

Mehler, J., Bertoncini, J., Barriere, M. & Jassik-Gerschenfeld, D. Infant recognition of the mother's voice. *Perception,* 1978, **7**, 491–497.

Meicler, M., & Gratch, G. Do 5-month-olds show object conception in Piaget's sense? *Infant Behavior and Development,* 1980, **3,** 265–282.

Meltzoff, A., & Borton, R. W. Intermodal matching by human neonates. *Nature,* 1979, **282,** 403–404.

Meltzoff, A. N., & Moore, K. M. Imitation of facial and manual gestures by human neonates. *Science,* 1977, **198,** 75–78.

Michotte, A. *The perception of causality.* London: Methuen, 1963.

Milewski, A. E. Visual discrimination and detection of configurational invariance in 3-month infants. *Developmental Psychology,* 1979, **15,** 357–363.

Millar, W. S., & Schaffer, H. R. The influence of spatially displaced feedback on infant operant conditioning. *Journal of Experimental Child Psychology.* 1972, **14,** 442–453.

Millar, W. S., & Schaffer, H. R. Visual-manipulative response strategies in infant operant conditioning with spatially displaced feedback. *British Journal of Psychology,* 1973, **64** (4), 545–552.

Miller, C. L., & Horowitz, F. D. Integration of auditory and visual cues in speaker classification by infants. Paper presented at the International Conference on Infant Studies, New Haven, April 1980.

Miller, C. L., Younger, B. A., & Morse, P. A. The categorization of male and female voices in infancy. Paper presented at the International Conference on Infant Studies, New Haven, April 1980.

Miller, D. J. Visual habituation in the human infant. *Child Development,* 1972, **43,** 481–493.

Miller, D. J., Ryan, E. B., Sinnott, J. P., & Wilson, M. A. Serial habituation in two-, three- and four month old infants. *Child Development,* 1976, **47,** 341–349.

Miller, G. A., & Johnson-Laird, P. N. *Language and perception.* Cambridge, Mass.: Harvard University Press, 1976.

Minsky, M. L. A framework for representing knowledge. In P. Winston (Ed.), *The psychology of computer vision.* New York: McGraw-Hill, 1975.

Moore, M. K. Object permanence and object identity: A stage-developmental model. Paper presented at the meeting of the Society for Research in Child Development, Denver, Colorado, 1975.

Moore, M. K., Borton, R., & Darby, B. L. Visual tracking in young infants: Evidence for object identity or object permanence? *Journal of Experimental Child Psychology,* 1978, **25,** 183–197.

Moore, M. K., & Meltzoff, A. N. Object permanence, imitation, and language development in infancy: Toward a neo-Piagetian perspective on communicative and cognitive development. In Minifie, F. P. and Lloyd, L. L. (Eds.), *Communicative and cognitive abilities-early behavioral assessment.* Baltimore: University Park Press, 1978, pp. 151–185.

Muller, A. A., & Aslin, R. N. Visual tracking as an index of the object concept. *Infant Behavior and Development,* 1978, **1,** 309–319.

Neisser, U. *Cognition and reality.* San Francisco: Freeman, 1976.

Nelson, K. Structure and strategy in learning to talk. *Monographs of the Society for Research in Child Development.* 1973, **38,** (1–27), Serial No. 149.

Nelson, K. Concept, word, and sentence: Interrelations in acquisition and development. *Psychological Review,* 1974, **81,** 267–285.

Nelson, K. Cognitive development and the acquisition of concepts. In R. C. Anderson, R. J. Spiro, & W. E. Montague (Eds.), *Schooling and the acquisition of knowledge.* Hillsdale, N.J.: Erlbaum, 1977.

Nelson, K. Explorations in the Development of a Functional Semantic System. In W. A. Collins (Ed.), *Children's language and communication.* Hillsdale, N.J.: Erlbaum, 1979.

Nelson, K. E. Accommodation of visual tracking patterns in human infants to object movement patterns. *Journal of Experimental Child Psychology,* 1971, **12,** 182–196.

Neumann, P. G. Visual prototype formation with discontinuous representation of dimensions of variability. *Memory and Cognition,* 1977, **5**, 187–197.

Olum, V. Developmental differences in the perception of causality under conditions of specific instructions. *Human Development,* 1958, 191–203.

Owsley, C. J., *Perceiving solid shape in early infancy: The role of kinetic information.* Paper presented to the International Conference on Infant Studies, New Haven, April 1980.

Piaget, J. *Play, dreams, and imitation in childhood.* New York: Norton, 1951.

Piaget, J. *The origins of intelligence in children.* New York: Norton, 1952.

Piaget, J. *The construction of reality in the child.* New York: Basic Books, 1954.

Piaget, J. *The mechanisms of perception.* London: Routledge and Kegan Paul, 1969.

Piaget, J. *Perception and knowledge.* Edinburgh, Scotland: University of Edinburgh Press, 1971.

Puff, R. C. (Ed.), *Memory organization and structure.* New York: Academic, 1979.

Rezba, C. *A study of infant binocular depth perception.* Unpublished undergraduate honors thesis, University of Minnesota, 1977.

Rosch, E. On the internal structure of perceptual and semantic categories. In T. E. Moore (Ed.), *Cognitive development and the acquisition of language.* New York: Academic, 1973.

Rosch, E. Cognitive representations of semantic categories. *Journal of Experimental Psychology: General.* 1975, **104**, 192–233.

Ross, G. S. Categorization in 1- to 2-year-olds. *Developmental Psychology,* 1980, **16**, 391–396.

Rovee-Collier, C. K. & Gekoski, M. J. The economics of infancy. In H. W. Reese and L. P. Lipsitt (Eds.), *Advances in child development and behavior* (Vol. 13), New York: Academic, 1979.

Ruff, H. A. Infant recognition of the invariant form of objects. *Child Development,* 1978, **49**, 293–306.

Ruff, H. A. The development of perception and recognition of objects. *Child Development,* 1980, **51**, 981–992.

Ruff, H. A., & Kohler, C. J. Tactual-visual transfer in six-month-old infants. *Infant Behavior and Development,* 1978, **1**, 259–264.

Rummelhart, D. E., & Ortony, A. The representation of knowledge in memory. In R. C. Anderson, R. J. Spiro & W. E. Montague (Eds.), *Schooling and the acquisition of knowledge.* Hillsdale, N. J.: Erlbaum, 1977.

Schank, R., & Abelson, R. *Scripts, plans, goals, and understanding.* Hillsdale, N.J.: Erlbaum, 1977.

Schwartz, M., & Day, R. H. Visual shape perception in early infancy. *Monographs of the Society for Research in Child Development,* 1979, **44** (7, Serial No. 182).

Spelke, E. Infant perception of faces in motion. Paper presented at the meeting of the American Psychological Association, Chicago, September, 1975.

Spelke, E. Infants' intermodal perception of events. *Cognitive Psychology,* 1976, **8**, 553–560.

Spelke, E. Intermodal exploration by four-month-old infants: Perception and knowledge of auditory-visual events. Unpublished doctoral dissertation. Cornell University, 1978.

Spelke, E. S. Exploring audible and visible events in infancy. In A. D. Pick (Ed.), *Perception and its development: A tribute to Eleanor J. Gibson,* Hillsdale, N.J.: Lawrence Erlbaum, 1979. (a)

Spelke, E. S. Perceiving bimodally specified events in infancy. *Developmental Psychology,* 1979, **15**, 626–636. (b)

Spelke, E. S., & Cortelyou, A. Perceptual aspects of social knowing: Looking and listening in infancy. In M. E. Lamb and L. R. Sherrod (Eds.), *Infant Social Cognition,* Hillsdale, N.J.: Erlbaum, 1981.

Spelke, E. S., & Owsley, C. Intermodal exploration and knowledge in infancy. *Infant Behaviour and Development,* 1979, **2**, 13–27.

Sroufe, L. A., & Wunsch, J. The development of laughter in the first year of life. *Child Development,* 1972, **43,** 1326–1344.

Starkey, P., & Cooper, R. G. Perception of numbers by human infants. *Science,* 1980, **210,** 1033–1034.

Starkey, P., Spelke, E., & Gelman, R. Number competence in infants. Paper presented at the International Conference on Infant Studies, New Haven, April 1980.

Strauss, M. S. Abstraction of prototypical information by adults and 10-month-old infants. *Journal of Experimental Psychology,* 1979, 5, 618–631.

Strauss, M. S. Infant memory of prototypical information. Paper presented at the meeting of the Society for Research in Child Development, April 1981.

Testa, T. Causal relationships and the acquisitions of avoidance response. *Psychology Review,* 1974, **51,** 491–505.

Trevarthen, C. Descriptive analyses of infant communicative behavior. In H. R. Schaffer (Ed.), *Studies in mother-infant interaction.* New York: Academic, 1977.

Tulving, E. Episodic and semantic memory. In E. Tulving & W. Donaldson (Eds.), *Organization of memory.* New York: Academic, 1972.

Uzgiris, I. C., & Hunt, J. McV. *Assessment in infancy: Ordinal scale of psychological development.* Urbana: University of Illinois Press, 1975.

Vurpillot, E., Ruel, J., & Castrec, A. L'organisation perceptive chez le nourrisson: Response au tout ou' a ses'elements. *Bulletin de Psychologie,* 1977, **327,** 396–405.

Walker, A. S. Perception of expressive behavior in infants. Unpublished doctoral dissertation, Cornell University, 1980.

Walker, A. S., Bahrick, L. E., & Neisser, U. Selective looking to multimodal events by infants. Paper presented at the meeting of the International Conference on Infant Studies, New Haven, April 1980.

Walker, A., Owsley, C., & Gibson, E. J. Differentiation of motions and shapes in motion by human infants. Paper presented at the Meeting of the Eastern Psychological Association, Boston, April 1977.

Watson, J. S. Memory and "Contingency Analysis" in infant development. *Merrill-Palmer Quarterly,* 1967, **13,** 55–76.

Watson, J. S. Infant perception of necessity and sufficiency in instrumental behavior. Paper presented to the biennial meeting of the Society for Research in Child Development, New Orleans, March 1977.

Weimer, W. B. Cognition: Forwards or backwards? (Review of cognition and reality: Principles and implications.) *Contemporary Psychology,* 1977, **22,** 483–484.

Weimer, W. B., & Palermo, D. S. (Eds.), *Cognition and the symbolic processes.* Potomac, Md.: Erlbaum, 1974.

Wertheimer, M. Psycho-motor coordination of auditory-visual space at birth. *Science,* 1961, **134,** 1692–1696.

Yonas, A. Studies of spatial perception in infancy. In A. D. Pick (Ed.): *Perception and Its Development.* Hillsdale, N.J.: Erlbaum, 1979.

Yonas, A., & Pick, H. L., Jr. An approach to the study of infant space perception. In L. B. Cohen & P. Salapatek (Eds.), *Infant perception: From sensation to cognition* (Vol. 2), New York: Academic, 1975.

Yonas, A., Oberg, C., & Norcia, A. Development of sensitivity to binocular information for the approach of an object. *Developmental Psychology,* 1978, **14,** 147–152.

Young-Browne, G., Rosenfeld, H. M., & Degen-Horowitz, F. Infant discrimination of facial expressions. *Child Development,* 1977, **48,** 555–562.

8

INFANT-ADULT SOCIAL INTERACTIONS
Their Origins, Dimensions, and Failures

Hanuš Papoušek
Professor of Developmental Psychobiology
Max Planck Institute for Psychiatry
Munich, F.R. Germany

Mechthild Papoušek
Research Psychiatrist and Neurologist in Developmental Psychobiology
Max Planck Institute for Psychiatry
Munich, F.R. Germany

A continually increasing number of students from different disciplines have become interested in the social development of human infants. Only in the last 5 years have extensive surveys documented this increase from multifaceted perspectives (Bell & Harper, 1977; Bullowa, 1979; Emde, Gaensbauer, & Harmon, 1976; Haith & Campos, 1977; Hunt, 1979; Klaus & Kennell, 1976; Lamb, 1976; Lewis & Brooks-Gunn, 1979; Lewis & Rosenblum, 1979; Osofsky, 1979; Schaffer, 1977; Simmel, 1980; Thoman, 1979; Thomas & Chess, 1980).

The rapid spread of multidisciplinary research on early social development has not only fertilized this area of inquiry but has also loosened the borderlines between former narrow zones of individual disciplines, has merged together theoretical concepts (Cairns, 1979; von Cranach, Foppe, Lepenies, & Ploog, 1979; Hinde, 1974; Leiderman, Tulkin, & Rosenfeld, 1977; Richards, M. P. M., 1974; Wilson, 1975), and, inevitably, has also brought a confusing debris of nonunifiable labels and speculative definitions.

Following foundations have kindly supported our research: Die Deutsche Forschungsgemeinschaft, Die Stiftung Volkswagenwerk, and Der Fond der Deutschen Bank beim Stifterverband für die Deutsche Wissenschaft. We owe special thanks to Prof. Detlev Ploog, the Director of Max-Planck Institute for Psychiatry, who has significantly influenced the biological and developmental approaches to the problem of psychiatry. Our modest contribution to this book is dedicated to Detlev Ploog's 60th birthday.

We also thank Peter Mangione, Ph.D., for helpful comments, and Dagmar Ellgring and Veronika Stroh for assistance in the preparation of this manuscript.

Moreover, the same behavioral units appear to have been interpreted differently across neighboring disciplines, although differential criteria remain undefined. For instance, smiling can be considered an independent emotional response, a social or communicative behavior, or an observable signal of a successful cognitive operation such as recognition.

The provocative complexity of the phenomenon of early social interaction has certainly nourished the interest of scientists as well as the introduction of innovative methodology (Lamb, Suomi, & Stephenson, 1979). Some puzzles like the intrinsic motivation for integrative needs must be solved by neuroscientists first. Only then shall we understand interrelationships between cognitive, social, and emotional behaviors. Conversely, recent technical advances have made available exact audiovisual records that can be reproduced ad libitum and analyzed step by step up to the level of microelements. Social relevance is another effective impulse stimulating the study of early social interaction at a time of increasing need to prevent more effectively the incidence of abnormalities in social development (Bruner, 1974; Newman & Newman, 1978; O'Connor, 1975; Parke & Collmer, 1975; Schaffer, 1971).

The rapid growth of human population represents an increasing burden on social adaptation as well as interindividual coexistence. Crises in the social adjustment to sexual life, parenthood, and child rearing may reflect the consequences of that burden (McCandless & Coop, 1979; Osofsky & Osofsky, 1972). The public, aware of the long-term effects of early social experience, keeps raising burning questions on infant rearing that researchers cannot ignore.

AUDIOVISUAL REPRODUCTION AND ANALYSIS OF INFANT–ADULT INTERACTION

Modern film and television techniques have facilitated the study of social interaction in important directions. Both methods have specific advantages and disadvantages; they can be combined, but neither substitute the other completely. For instance, remote control of filming can be facilitated with a television viewer. Filming provides better prints for publication or multiple copying. Videotaping enables immediate display of records as well as electronic operations such as combining records of two or more cameras or inserting time signals and polygraphic curves into pictures. Both techniques provide exact synchronization of picture and sound and, in addition, economical records of long-lasting observations with time-lapse technique. Slow-motion filming allows the analysis of very fast events.

Behavioral analysis has thus exceeded the former limits of simple observations in several ways including microanalysis, infrared recording with minimal illumination, and consecutive reproductions of observations whenever additional parameters are to be analyzed or the validity of analysis rechecked. New memory systems for instant storing, enlarging, and scanning of individual videopictures have been applied to analyze interactions (Papoušek & Papoušek, 1979a). The study of discriminative capacities of infants in preferential designs of experiments (e.g., Papoušek & Papoušek, 1974) is possible if several videorecorders can be combined for simultaneous stimulation with two different playbacks and recording the responses to such stimulation.

The former paper-and-pencil type of behavioral records forced a priori categorization and interpretative labeling of observed behaviors. Control of such classifications used to be very problematic even when complex event-recorders were introduced. Modern audiovisual recording techniques allow the postponement of interpretative attempts and the completion of detailed objective description first.

INTERACTION IN SOCIAL SYSTEMS

Given the complexity of human social development and the difficulty with its categorization, it may help to view it first from the abstract and "psychophysically neutral" level of the general systems theory (v. Bertalanffy, 1968). A system is a set of units with relationships among them, and to belong to a set means to have some common properties. Units in a system stand in a dynamic interaction since, in general, their relationships are neither unidirectional nor time-independent (steady). This is particularly true about living systems, differing from other systems in most general aspects such as those of thermodynamics.

Living and Nonliving Systems

General systems theory applies not only to preestablished physical or cybernetic systems, often representing closed systems, but also to organic structures expressed always in more or less unpredictable processes and bearing the features of open systems. Whereas physical events show a trend toward most probable states, that is, maximum entropy and progressive destruction of differentiation and order, living systems, both in ontogeny and phylogeny, develop toward a more improbable state, exhibiting increased differentiation and higher organization of matter. Living systems can produce materials carrying high free energy, and thus they can work against entropy and transport entropy.

Living systems contain genetic material and a nervous system, the latter being an essential subsystem that controls and integrates all subsystems (v. Bertalanffy, 1968; Miller, 1969). As open systems, living systems can maintain states distant from the thermodynamic equilibrium and spend existing potentials either in primary spontaneous activities or on releasing stimuli. Even under constant external conditions and in the absence of external stimuli, the living organism is basically an active system. Its subsystems are integrated together to form actively self-regulating, developing, and reproducing unitary systems with purposes and goals.

The functions of the nervous system encompass various degrees of complexity including a specifically high capacity for symbolism in man. The recognition of spontaneity and symbolism is crucial for understanding human behavior in general and social interaction in particular. In relation to human social behavior, symbolism allows communication with absent or distant partners and integration of experience across ages, continents, or even planets.

Social Interaction

Every living organism can survive only within a relatively narrow range of environmental conditions within which it interacts with its environment as a subsystem. Its interaction may be determined by an exposure to purely physical changes and accordingly correspond to a simple stimulus–response model. A categorically different situation is obtained if various species share the same environment and interact among themselves, in some cases as predators and preys and at other times in a symbiotic or neutral coexistence. Interaction among members of the same species acquires the character of a social interaction.

In trying to define specific properties of social interaction, we meet a difficult problem of classification. Classifying is a basic human activity, and our scientific observation also starts with descriptive classification, but the pigeonholes we use are often artificial and do not quite fit nature (Hinde, 1974). Hinde gives clear examples of major difficulties that arise in attempting to distinguish social behaviors from feeding, grooming, sexual reproduction, parenting, and so forth. In order to overcome the definitional ambiguity, we recommend, therefore, to view any conspecific interaction as social and any behavioral unit as indexing social behavior inasmuch as it occurs in the context of social interchange. Within the broadly defined category of social interaction, for descriptive purposes, it is still possible to distinguish lower-order subcategories, such as communicative, feeding, sexual, or parenting ones. Such definitional flexibility appears particularly advantageous in studying human social interaction, since due to his symbolic capacities, man can use almost any observable behavioral unit, for example, a word, a number, any movement, or even a pause in movements, as a means of social communication.

Social Communication

The symbolic capacity has placed human communication in a dominant position within social interaction. It has led to the development of languages and cultures, to the integration of experience across great distance and time, and to the emergence of self-consciousness. It has allowed people to construct theories or realms of belief reaching beyond concrete experience. Unfortunately, it has also contributed to the violation of the laws of nature for the sake of such beliefs (v. Bertalanffy, 1968).

Human communication exemplifies the programmatic structure of social interaction. Interaction serves some purposes, and its recognizable behavioral units are integrated hierarchically and performed in successive steps (Scheflen, 1969). Interaction programs are context- and culture-specific. They are influenced by genetic factors and also shaped into molds that are mutually recognizable and predictable by the members of a given culture.

Scheflen (1969) describes three orders in the behavioral integration of communication: (1) simple coordination of activities, whether consciously or not; (2) use of integrational signals helping to avoid ambiguities, regulate speed of performance, identify roles, states of attention or moods, monitor deviations in states, and refer to context; and (3) metacommunication, that is, communication about communication, which unlike the preceding signals can change the structure of programs in both innovative, or creative, and destructive ways. Within such a scheme, one function seems to us fundamental to all other ones, namely, the integration of experience (Papoušek & Papoušek, 1975; 1979b).

The Fundamental System of Adaptive Responses

Most observers of infants have been struck by the close interrelationship among emotional, social, and cognitive aspects of development (Ainsworth, Bell, & Stayton, 1974; Emde, Gaensbauer, & Harmon, 1976; Sroufe, 1979). These interrelationships are difficult to unravel, and as stated earlier, neuroscientific explanations are unsatisfactory; we continue to search for best fitting models.

In human infancy research, Papoušek and Papoušek (1975; 1979b) attribute the fundamental role of behavioral regulation to integrative nervous processes implicated in the processing of perceived sensations and in the organization of adaptive responses. Innate programs of integration with which the newborn responds to structural situational stimulations, for instance, programs for orienting, habituation, associative or instrumental learning, categorization, and detection of familiarity and regularity, seem to follow paths of hierarchical growth. With these programs human organisms respond not only to environmental but also to intrinsic signals and produce further intrinsic signals marking the results of integrative processes.

The operation of integrative processes is accompanied by observable changes in behavioral states, general mobility, facial expressions, and vocalizations of the infant. The caregiver can process these changes as communicative, social, or affective signals; the further unfolding of interaction and the caregiver's behavior confirm to what degree these signals have such functional value. This model of a fundamental adaptive response system, which mirrors contemporary ideas on the interrelation between actions, thoughts, and emotions (e.g., Bruner, 1974; 1975; Piaget & Inhelder, 1969), has lended itself to experimental analyses.

In nonsocial laboratory situations, Papoušek (1967) observed that affective signs of displeasure and pleasure accompany learning processes with increasing expressivity during the first 2 trimesters of infancy. Papoušek's observations indicate the presence of a motivational subsystem of intrinsic rewards resulting directly from integrative processes in the nervous system. Facial and vocal signs of displeasure appear during the infant's confrontation with too much novelty or with difficult learning or problem-solving situations. Signs of pleasure replace them simultaneously with successful outcomes of integrative processes, such as identification of a familiar object, fulfillment of prediction, or successful mastery of control over a contingent event. Within this context, pleasure is viewed as an index of intrinsic reward mechanisms linked with integrative operations and possibly related to the substrates of intracranial self-stimulation (Olds & Milner, 1954) that have been considered as motivational substrates in neurophysiology (Ploog, 1980). Moreover, this kind of intrinsic reward, which motivates the acquisition of knowledge, can exceed "gratifications of the other needs." As suggested by Papoušeks (1979b), the infant is as strongly pulled by the expectation of pleasure resulting from successful integration as it is pushed by unpleasant feelings connected with incongruency, dissonance, or failure in adjustment (Hebb, 1949; Hunt, 1966).

In this conceptualization, the acquisition of knowledge is viewed as a movement from "unknown" to "known" directed by the integrative processes in order to avoid, on one hand, stress from novelty and, on the other, stress from boredom (Papoušek & Papoušek, 1978). Novelty or incongruity compel the system to accumulate information and integrate a close unambiguous concept of the "unknown." However, an even higher level of integration can be used to reopen seemingly definite concepts, to take a nontraditional look at the known and make it unknown again. This tendency helps to avoid boredom with the known through playful or creative action (see Figure 8.1).

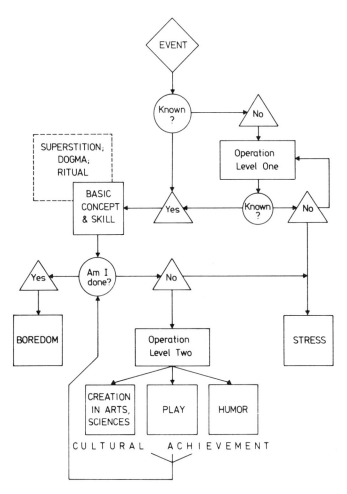

Figure 8.1. *Levels of knowledge acquisition:* Flowchart of integrative operations leading to higher levels of knowledge and protecting against stress from the unknown or against boredom from trivial knowledge. Operation level one includes learning basic concepts, rules, and skills and tends to create closed concepts. Operation level two tends to reopen concepts created at level one, view events from nontrivial standpoints, and create and solve new problems.

Overcoming the tendency to protect oneself against the distressing unknown with a closed system of simple concepts or often superficial, superstitious, or dogmatic beliefs and replacing them with more comprehensive and creative ones is essential for contributions to human culture from play, humor, arts, and sciences. This interpretation brings to mind Huizinga's (1962) concept of "homo ludens" and can be traced back to the seventeenth century to Comenius's ideas of pansophistic education through play in *Schola Materni Gremii* and in *Schola Ludus*.

The fundamental system of adaptive responses seems particularly relevant for the infant's social development for three reasons: (1) the fundamental role of integrative processes indicates the presence of biologically determined factors in evolution that favor the extensive development of integrative capacities in humans; (2) direct interrelationships among integrative processes, intrinsic motivation, and associated facial, gestural, and vocal behaviors allow the social environment to read and to influence the progressive operation of integrative processes in the infant; and (3) given a hierarchical order in integrative capacities, the more fundamental ones (including the detection of regularity, familiarity, or contingency) and the main forms of learning, imitation, and concept formation should developmentally precede the more complex forms of playful or creative processes, which in turn should be followed by the most complex functions of human abstraction, symbol use, and verbal communication. Such a developmental sequence assumes the functioning of mental processes, no matter how simple in form, during the preverbal period although originally psychologists considered these processes only in forms typical for subjects capable of conscious and linguistically competent mental operations.

Infant–Adult Interaction

The unusually long dependence of human progeny on more experienced caregivers, parental or nonparental, draws attention to the specific features of the infant–adult interaction. In the view of general systems theory, infant–adult interaction is a typical case of a dynamic interaction between two or more conspecific living systems. However, it also represents a specific interacting system in which both sides differ critically in the amount of integrated experience, social awareness, and knowledge. This difference gives the infant–adult interaction the character of a didactic system (Papoušek & Papoušek, 1978; 1981a) and thus points to new directions in both human and comparative research.

The forms of didactic activities performed by human social institutions at the level of conscious cultural interventions have been matters of scientific interest for centuries. Conversely, little is known about nonconscious didactic activities directed at very young infants and precursors in animal parenting. A teacher has to know how to get the pupil's attention, has to adjust his verbal communication to the pupil's level, and pass the preselected piece of information in a dosage corresponding to the pupil's capacity. Adequate programs helping the teacher are well known. Much less is known about programs used during infant–adult interactions in man or in animals.

PARENTAL BEHAVIOR

Among various types of infant–adult interaction, those with the infant's own parents are most important theoretically and practically. Biological parents can provide their children with experience most relevant to biological and cultural adjustment ranging from immunological resistance and ecological and climatic adjustment to experience resulting from social status. Parents represent the most typical union of both sexes as it has proved effective in reproduction within the given culture.

Both nature and culture assigned very different roles to the two parents with respect to reproduction and care of progeny. These differences are displayed by parents to progeny from an early age in different culture-dependent forms and influence the assimilation of cultural heritage (Bruner, 1971; Mead, 1946; Whiting & Whiting, 1975). Not enough attention has been paid to biologically determined universal features characterizing human parenting (Konner, 1977; Wolff, 1977), though some of these aspects merit attention from the view of the infant's social integration.

Biological Protection of Progeny

As regards the intrauterine protection of the fetus, only the neurohumoral regulation of maternal functions is relatively well known. Less is known about the psychological reintegration of maternal attitudes, interests, and preferences, the so-called primary maternal preoccupation (Winnicott, 1958) resulting in a state of heightened sensitivity to the infant's need. At its peak at the end of pregnancy, this state may be perceived almost as an illness, but tends to become repressed in memory a few weeks later. According to Winnicott's rich clinical experience, women with a strong male identification seldom reach this state and may "produce a child, but having missed the boat at the earliest stage." This indicates the role of psychobiological determinants in early infant–mother interaction and in its further successful development. Next to deviations in maternal sex identification, currently studied in connection with parental androgyny (Kelly & Worell, 1976), prenatal maternal stress can influence directly the emotionality if not the vitality of the fetus (Carlson & LaBarba, 1979; Matějček, Dytrych & Schüller, 1979).

The postpartum protection of the newborn and infant has been usurped by technology within modern cultures to such an extent that intuitive behaviors, and probably also researchers' interest in them, have been outdated. Behaviors facilitating nursing, such as massaging the breasts, stimulation of nipple erection before breast-feeding, and holding the infant in an appropriate position, have elicited limited interest. Other tendencies, such as touching the newborn with open palms, directing breath toward him, and giving him maximum bodily proximity, may be regarded as relics of basic pretechnological protection (Papoušek & Papoušek, 1979a); notwithstanding, they have been seen as signs of emotional attachment or communicative behavior in other studies (Klaus & Kennell, 1976; Stern, 1977).

Even less is known about the biology of the father's protective care. It may be determined by global sex differences, hormonal effects on the thresholds of parental behaviors, or by degree of androgyny. Evolution has provided an abundant distribution, as well as an early ontogenetic development, of paternal behaviors serving reproduction and care for progeny in animals. Thus it would be astonishing if natural selection had made the father a mere outsider in the initial care of the human infant. And if sociocultural trends have delegated him to the outsider status, could it be then that the father may produce a child but miss the boat?

Didactic Programs in Infant–Adult Interaction

No matter how well the bodily needs of the infant are met by nature or culture, its social integration can still suffer in the case of parental deprivation. The reasons for this neither are easy to identify and define nor can they be verified experimentally. Whatever may underlie "parental love" or "emotional attachment" seems to be of multifaceted structure, regulated by multidirectional interrelationships. There is no reason, however, for denying the fundamental role of integrative processes in this complex problem, both with respect to the infant's integrative capacity and the parent's propensity and ability to support the infant's integrative capacity, whether consciously or nonconsciously.

Conscious parental care, as documented in questionnaires or cultural-historical accounts, bears no direct relationship with the infant's integrative capacity (Bell &

Harper, 1977). On the contrary, a variety of cultural attitudes counteracts the cross-cultural regularity of social integration and acquisition of language. Consequently, it has become necessary to consider determinants of social integration that have escaped conscious awareness of both parents and observers (Papoušek & Papoušek, 1978). Here, the audiovisual microanalyses of interaction have facilitated a breakthrough in knowledge.

The present research pays increasing attention to behavioral sequences in interaction that facilitate communication between two dramatically different systems and helps the adult to adjust to the perceptual and integrative simplicity of the infant. These sequences are best understood when represented as a biological model of primary didactic care for the integrative growth of the infant (Papoušek & Papoušek, 1978). The evolutionary advantage of their nonconscious character can be seen in their weak dependence on sociocultural influence, cross-cultural universality, and the ease with which they can be carried out as compared to behaviors requiring rational conscious decisions (Papoušek & Papoušek, 1979a; 1981a). A better understanding of nonconscious parenting may also support the attempts to conceptualize the evolution of language and intelligence (Parker & Gibson, 1979) and the evolution of parental care (Rosenblatt, 1978).

Within this framework, the adult caretaker intuitively creates an abundance of learning situations supporting the growth of such capabilities as identification processes, control of environmental events, imitation, concept formation, acquisition of nonverbal and verbal communication, and so forth. The amount and quality of didactic situations exceed any set of experiences the nonliving environment can offer, and yet does not overtax the caretaker.

Unlike infants, whose intentions escape our knowledge, parents are motivated by intentions, expectancies, and goals. They support infantile behaviors that they find useful or pleasant and try to avoid noxious and nonadaptive behaviors while being guided by the expectancy of parental effectiveness among others. Following White (1959), Goldberg (1977) has suggested an efficacy model for parent–infant interaction. She has stressed that just as the infant derives feelings of efficacy from parent-provided contingency experience, parents derive feelings of efficacy from infant-provided contingency experience. The infant's responsivity, readability, and predictability enhance parental efficacy (Thomas & Chess, 1980).

Forms of Communication

To meet the demands of immature infants, the adult caregiver has to respond to a limited repertoire of infantile behaviors and to alter the adult mode of communication accordingly. Initial learning in human infants requires the infant's optimal waking state, a simple structure of stimuli and learning trials, and a relatively large number of repetitions of stimuli (Papoušek & Papoušek, 1975). In like manner, the following requisite can be asserted for the initial human communication: consistency in the performance of communicative elements (Messer, 1980), simultaneous engagement of several modalities (Turkewitz & McGuire, 1978), presence of eye-to-eye contact (Fogel, 1977; Robson, 1967), and support of imitative behaviors (Ryan, 1973; Papoušek & Papoušek, 1977).

Several components of nonconscious parenting fulfill the above conditions par excellence. The infant's behavioral state is checked regularly, for instance, through muscle tone (Papoušek & Papoušek, 1977), and manipulated so as to avoid ambiguous states and support either optimal waking or optimal sleep (Papoušek & Papoušek, 1981a; Thoman, Korner, & Beason-Williams, 1977). The adult tries to maintain the face-to-face position (Fogel, 1977; Stern, 1971), rewards the infant with vivid greeting responses for the achievement of eye-to-eye contact, and keeps the eye-to-eye distance during dialogues relatively constant around 22.5 cm (Papoušek & Papoušek, 1977; 1979a; Schoetzau & Papoušek, 1977).

The speech addressed to the infant illustrates the adult's adjustment particularly well. Its pitch is higher than in adult–adult conversation, the prosodic features are strikingly overexaggerated, and the structure is simplified and repetitive, often including imitation of the infant's vocal sounds (Snow, 1977). In most cases, communicative efforts by adults are not only visual and vocal; tactile, kinesthetic, and vestibular stimulations are also associated with the most relevant situational contexts in rather consistent combinations.

A particular significance has been attributed to nonverbal communication (Hinde, 1972; Mayo & LaFrance, 1978), especially in the case of emotional states or experiences. The classification (Sroufe, 1979), description (Blurton-Jones, 1974; Ekman & Oster, 1979), and development of emotions (Emde, Gaensbauer, & Harmon, 1976; Lewis & Rosenblum, 1978; Yarrow, 1979), as well as their recognition in infants and readability by parents (Contole, & Over, 1979), still pose more questions than they answer.

Adjustment to the Infant's Individuality

The adult's interactive adjustment shows an additional dynamic feature of specific didactic importance. The adult modulates the advancement of his communicative and didactic skills according to the development of the individual infant's capacity such that the newly emerging capacities receive the most supportive attention. Sometimes the infant's progress elicits a new strategy from the adult. For instance, the first repetitive syllables of the infant stimulate the adult to use them as potential protowords and derive some meaning from them (Papoušek & Papoušek, 1977). Likewise, various changes are elicited by the infant's locomotor progress (Green, Gustafson, & West, 1980).

This type of adjustment becomes particularly clear in cases in which the infant exhibits atypical development, such as in prematures (Field, Hallock, Dempsey, & Shuman, 1978; Goldberg, 1979), in retarded infants (Thoman, Becker, & Freese, 1978), and in Down's syndrome infants (Jones, 1977).

CHANGES IN SOCIAL INTERACTION DURING INFANCY

As regards the initial postpartum interaction, both the newborn's integrative competence and the adult's intuitive parenting create predispositions for the beginning of fundamental communication. However, such communication does not start entirely smoothly. Mothers typically complain about too little relatedness with babies for up to 4 or 6 weeks (Robson, 1979), showing a "maternal need to establish relatedness" that is satiated only by the infant's ability to respond with visual contact or a smile during the second month. The smile appears to accompany successful integrative processes in learning and cognition at about 1 month of age in nonsocial learning situations (Papoušek, 1967) and could thus result primarily from the infant's underlying "integrative needs." The acquisition of motor skills, such as grasping, pointing, or locomotion, also functions as triggering signals and unlocks novel conscious or nonconscious strategies in parenting. Some signals may fulfill parental expectations and bring about effective intrinsic rewards; others sharpen the awareness of new risks and increase anxieties.

The role of integrative processes becomes particularly evident in the development of preverbal vocal communication. During the gradual acquisition of the prerequisite skills, for example, control of expiration, vocal registers, and prosodic or musical elements, vocalization appears to be the most universal "educational toy" in the infant's learning repertoire and, at the same time, an object of finely graduated didactic routines possibly supporting not only the future acquisition of language but, more generally, imitative capacities, memory functions, concept formation, abstraction, and creative and symbolic capacities in play or song (Papoušek & Papoušek, 1981b). As an aside, it is interesting that among infrahuman primates only gibbons sing. However, their territorial singing follows prescribed scores with minimal plasticity and broadcasts precise information on the species and sex of individuals, the area occupied by the parents, and the presence of a junior ready to form a new pair (Marshall & Marshall, 1976).

From the perspective of the primary role of integrative processes, one may be tempted to compile a list of milestones to describe developmental changes in social interaction. Mead's (1934) model of three levels of communication—from momentary mutual adjustments followed by the conversation of gestures and extending to the use of words for significant symbols and for the reconstruction of self—could aid in the interactional research as the starting point in the hierarchical classification of communication within the framework of general systems theory (Scheflen, 1969). Similarly, Piaget's stages of intelligence (Piaget, 1952) could be built into such a framework, since they have already been utilized to organize and interpret a substantial portion of infancy research.

However, research on infant–adult interaction indicates an urgent need for more explanatory information rather than for developmental assessment. The construction of developmental scales should not inhibit longitudinal analysis of each individual interaction since its developmental course depends on the communication and coaction between mutually interdependent individuals who tend to increase the improbability of their behaviors over time. The adaptive flexibility of the adult compensates for great individual differences and even pathological deviations in the rate of integrative growth.

FAILURES IN INFANT–ADULT INTERACTION

Behavioral plasticity and compensatory adjustment within social interchanges can function as a buffer protecting both interacting partners against a variety of noxious factors of genetic or environmental origins. For instance, the risk of an unwanted pregnancy, deficit in the infant's capacities, or illness may still be overcome if a mutually rewarding infant–adult interaction can be established. Fraiberg (1971; 1974) has described increased tactile and kinesthetic stimulation as a compensation for handicaps in visual communication in caregivers of blind infants.

Conversely, similar deviations can disturb parenting to such an extent as to catalyze secondary deviations in the infant's behavior and to link reciprocal failures into a vicious circle that eventually leads to serious clinical problems (Papoušek & Papoušek, 1980). Next to the syndrome of social deprivation, the "maternal rejection" causing emotional disturbances and psychosomatic dwarfism has been added recently to the list of infancy disorders (MacCarthy, 1974). Maltreatment of infants by adults, the biological parents not excluded, is another category of clinical problems for which the common pathogenetic denominator may often be an interactional failure (Burgess, 1979).

An abnormal deviation in infant–adult interactions can have various genetic or environmental origins, which may be difficult to determine given current knowledge and diagnostic tools. Very little is known about the individual variability and clinical significance of nonconscious, intuitive parenting. The role of primary maternal preoccupation was mentioned earlier in this chapter. Fathers who cannot utilize the first weeks or months of fatherhood to regularly interact with their infant and make the fine grain interactional adjustments have only recently become targets of investigation. Their reduced engagement in early interchanges may attenuate the effectiveness of their interactions with infants; however, it may also reduce the risk of a recursively formed interactional failure. These questions certainly require more attention in future studies.

The infant's contribution to interactional failures is exemplified in preterm infants. They figure in interactional failures more frequently and are overrepresented among abused children (Elmer & Gregg, 1967; Klein & Stern, 1971) and in the population of divorced families (Leiderman & Seashore, 1975). Preterm infants, independently of the occurrence of respiratory distress, have also been found to be slower learners as compared to controls matched either according to chronological or gestational age (Papoušek & Papoušek, 1980). Maternal attitude can undergo critical fluctuations depending on the newborn's health in general (Klaus & Kennell, 1975; Lipsitt, 1979).

Combinations of biological and sociocultural factors presumably cause the majority of interactional failures. By reducing the effects of improved medical care for progeny, underscoring important theoretical problems, and representing natural experiments in a field characterized by limited systematic experimentation, interaction failures may soon attract increasing attention in different disciplines.

REFERENCES

Ainsworth, M. D. S., Bell, S., & Stayton, D. J. Infant–mother attachment and social development: "Socialisation" as a product of reciprocal responsiveness to signals. In M. P. M. Richards (Ed.), *The integration of a child into a social world*. Cambridge: Cambridge University Press, 1974, pp. 99–135.

Bell, R. A., & Harper, L. V. *Child effects on adults*. Hillsdale, N.J.: Erlbaum, 1977.

Blurton-Jones, N. G. (Ed.) *Ethological studies of child behavior*. Cambridge, England: Cambridge University Press, 1974.

Bruner, J. S. *The relevance of education*. New York: Norton, 1971.

Bruner, J. S. The organisation of early skilled action. In M. P. M. Richards (Ed.), *The integration of a child into a social world*. London: Cambridge University Press, 1974, pp. 167–184.

Bruner, J. S. The ontogenesis of speech acts. *Journal of Child Language,* 1975, **2**, 1–19.

Bullowa, M. (Ed.) *Before speech: The beginning of interpersonal communication.* Cambridge, England: Cambridge University Press, 1979.

Burgess, R. L. Child abuse: A social interactional analysis. In B. B. Lahey, & A. E. Kazdin (Eds.), *Advances in clinical child psychology* (Vol. 2). 1979, pp. 141–172.

Cairns, R. B. *Social development: The origins and plasticity of interchanges.* San Francisco: Freeman, 1979.

Carlson, D. B., & LaBarba, R. C. Maternal emotionality during pregnancy and reproductive outcome: A review of the literature. *International Journal of Behavioral Development,* 1979, **2**, 343–376.

Contole, J., & Over, R. Signal detection of infant social behavior. *Infant Behavior and Development,* 1979, **2**, 189–200.

Ekman, P., & Oster, H. Facial expressions of emotion. *Annual Review of Psychology,* 1979, **30**, 527–554.

Elmer, E., & Gregg, G. Developmental characteristics of abused children. *Pediatrics,* 1967, **40**, 596–602.

Emde, R., Gaensbauer, T., & Harmon, R. Emotional expression in infancy: A biobehavioral study. *Psychological Issues Monograph Series,* **10** (Monogr. No. 37), 1976.

Field, T., Hallock, N., Dempsey, J., & Shuman, H. H. Mothers' assessments of term infants and pre-term infants with respiratory distress syndrome: Reliability and predictive validity. *Child Psychiatry and Human Development,* 1978, **9**, 75–85.

Fogel, A. Temporal organization in mother–infant face-to-face interaction. In H. R. Schaffer (Ed.), *Studies in mother–infant interaction.* London: Academic, 1977, pp. 119–151.

Fontana, V. J. Further reflections on maltreatment of children. *Pediatrics,* 1973, **51**, 780–782.

Fraiberg, S. Intervention in infancy: A program for blind infants. *Journal of American Academy of Child Psychiatry,* 1971, **10**, 381–405.

Goldberg, S. Social competence in infancy: A model of parent–infant interaction. *Merrill-Palmer Quarterly,* 1977, **23**, 163–177.

Goldberg, S. Premature birth: Consequences for the parent–infant relationship. *American Scientist,* 1979, **67**, 214–220.

Green, J. A., Gustafson, G. E., & West, M. J. Effects of infant development on mother–infant interactions. *Child Development,* 1980, **51**, 199–207.

Haith, M. M., & Campos, J. J. Human infancy. *Annual Review of Psychology,* 1977, **28**, 251–293.

Hebb, D. O. *The Organization of Behavior.* New York: Wiley, 1949.

Hess, E. H. Ethology. In A. M. Freedman & H. I. Kaplan (Eds.), *Comprehensive textbook of psychiatry.* Baltimore: Williams & Wilkins, 1967, pp. 180–189.

Hinde, R. A. *Non-verbal communication.* London: University of Cambridge Press, 1972.

Hinde, R. A. *Biological bases of human social behavior.* New York: McGraw-Hill, 1974.

Huizinga, J. *Homo ludens. Vom Ursprung der Kultur im Spiel.* Hamburg: Reinbek, 1962.

Hunt, J. McV. The epigenesis of intrinsic motivation and early cognitive learning. In R. N. Haber (Ed.), *Current research in motivation.* New York: Holt, Rinehart, and Winston, 1966.

Hunt, J. McV. Psychological development: Early experience. *Annual Review of Psychology,* 1979, **30**, 103–143.

Jones, O. H. M. Mother–child communication with pre-linguistic Down's syndrome and normal infants. In H. R. Schaffer (Ed.), *Studies in mother–infant interactions.* London: Academic, 1977, pp. 379–401.

Kelly, J. A., & Worell, L. Parent behaviors related to masculine, feminine, and androgynous sex roles orientations. *Journal of Consulting and Clinical Psychology,* 1976, **44**, 843–851.

Klaus, M. H., & Kennell, J. H. *Maternal–infant bonding.* Saint Louis: Mosby, 1976.

Klein, M., & Stern, L. Low birth weight and the battered child syndrome. *American Journal of Diseases of Childhood,* 1971, **122**, 15–18.

Konner, M. Evolution of human behavior development. In P. H. Leiderman, S. R. Tulkin, & A. Rosenfeld (Eds.), *Culture and infancy.* New York: Academic, 1977, pp. 69–109.

Lamb, M. E. (Ed.) *The role of the father in child development.* New York: Wiley, 1976.

Lamb, M. E., Suomi, S. J., & Stephenson, G. R. (Eds.) *Social interaction analysis: Methodological issues.* Madison, Wis.: The University of Wisconsin Press, 1979.

Leiderman, P. H., & Seashore, M. J. Mother–infant neonatal separation: Some delayed consequences. In M. O'Connor (Ed.), *Parent–infant interaction.* CIBA Foundation Symposium 33 (New Series). Amsterdam: Elsevier, Excerpta Medica, North-Holland, 1975, pp. 213–232.

Leiderman, P. H., Tulkin, S. R., & Rosenfeld, A. (Eds.) *Culture and infancy: Variations in the human experience.* New York: Academic, 1977.

Lewis, M., & Brooks-Gunn, J. *Social cognition and the acquisition of self,* Plenum, New York, 1979.

Lewis, M., & Rosenblum, L. A. (Eds.) *The development of affect.* New York: Plenum, 1978.

Lewis, M., & Rosenblum, L. A. (Eds.) *The child and its family.* New York: Plenum, 1979.

Lipsitt, L. P. The newborn as informant. In R. B. Kearsley & I. E. Sigel (Eds.), *Infants at risk:* Assessment of cognitive functioning. Hillsdale, N.J.: Erlbaum, 1979, pp. 1–2.

McCandless, B. R., & Coop, R. H. *Adolescents: Behavior and development* (2nd ed.). New York: Holt, Rinehart and Winston, 1979.

MacCarthy, D. Effects of emotional disturbance and deprivation (maternal rejection) on somatic growth. In J. A. Davis & J. Dobbing (Eds.), *Scientific foundations of paediatrics,* London: Heinemann, 1974, pp. 55–67.

Marshall, J. T., Jr., & Marshall, E. R. Gibbons and their territorial songs. *Science,* 1976, **199,** 235–237.

Matějček, Z., Dytrych, Z., & Schüller, V. The Prague study of children born from unwanted pregnancies. *International Journal of Mental Health,* 1979, **7,** 63–77.

Mayo, C., & La France, M. On the acquisition of nonverbal communication: A review. *Merrill-Palmer Quarterly,* 1978, **24,** 213–228.

Mead, G. H. *Mind, self and society.* Chicago: University of Chicago Press, 1934.

Mead, M. Research in primitive children. In L. Carmichael (Ed.), *Manual of child psychology.* New York: Wiley, 1946, pp. 735–789.

Messer, D. J. The episodic structure of maternal speech to young children. *Journal of Child Language,* 1980, **7,** 29–40.

Miller, J. G. Living systems: Basic concepts. In W. Gray, F. J. Duhl, & N. D. Rizzo (Eds.), *General systems theory and psychiatry,* Boston: Little, Brown, 1969, pp. 51–133.

Newman, B. M., & Newman, P. R. *Infancy & childhood: Development & its contexts.* New York: Wiley, 1978.

O'Connor, M. (Ed.) *Parent–infant interaction.* CIBA Foundation Symposium 33 (New Series). Amsterdam: Elsevier, Excerpta Medica, North-Holland, 1975.

Olds, J., & Milner, P. Positive reinforcement produced by electrical stimulation of septal area and other regions of rat brain. *Journal of Comparative and Physiological Psychology,* 1954, **47,** 419–427.

Osofsky, J. D. (Ed.) *Handbook of infant development.* New York: Wiley, 1979.

Osofsky, J. D., & Osofsky, H. J. Androgyny as a life style. *Family Coordinator,* 1972, **21,** 411–418.

Papoušek, H. Experimental studies of appetitional behavior in human newborns and infants. In H. W. Stevenson, E. H. Hess, & H. L. Rheingold (Eds.), *Early behavior: Comparative and developmental approaches.* New York: Wiley, 1967, pp. 249–277.

Papoušek, H., & Papoušek, M. Cognitive aspects of preverbal social interaction between hu-

Papoušek, H., & Papoušek, M. Mirror image and self-recognition in young human infants: A new method of experimental analysis. *Developmental Psychobiology,* 1974, **7,** 149–157.

man infants and adults. In M. O'Connor (Ed.), *Parent–infant interaction.* CIBA Foundation Symposium 33 (New Series), 1975, pp. 241–260.

Papoušek, H., & Papoušek, M. Mothering and cognitive head-start: Psychobiological considerations. In H. R. Schaffer (Ed.), *Studies in mother–infant interaction.* London: Academic, 1977, pp. 63–85.

Papoušek, H., & Papoušek, M. Interdisciplinary parallels in studies of early human behavior: From physical to cognitive needs, from attachment to dyadic education. *International Journal of Behavioral Development,* 1978, **1,** 37–49.

Papoušek, H., & Papoušek, M. Early ontogeny of human social interaction: Its biological roots and social dimensions. In M. von Cranach, K. Foppa, W. Lepenies, & D. Plogg (Eds.), *Human ethology: Claims and limits of a new discipline.* Cambridge, England: Cambridge University Press, 1979, pp. 456–478. (a)

Papoušek, H., & Papoušek, M. The infant's fundamental adaptive response system in social interaction. In E. Thoman (Ed.), *Origins of the infant's social responsiveness.* Hillsdale, N.J.: Erlbaum, 1979, pp. 175–208. (b)

Papoušek, H., & Papoušek, M. Care for the normal and high-risk newborn: A psychobiological view of parental behavior. In S. Harel (Ed.), *The at risk infant,* International Congress Series No. 42. Amsterdam: Excerpta Medica, 1980, pp. 368–371.

Papoušek, H., & Papoušek, M. How human is the human newborn, and what else is to be done? In K. Bloom (Ed.), *Prospective issues in infant research.* Hillsdale, N.J.: Erlbaum, 1981, pp. 137–155. (a)

Papoušek, M., & Papoušek, H. Musical elements in the infant's vocalization: Their significance for communication, cognition, and creativity. In L. P. Lipsitt (Ed.), *Advances in infancy research* (Vol. 1). Norwood, N.J.: Ablex, 1981, pp. 163–224. (b)

Parke, R. D., & Collmer, C. W. Child abuse: An interdisciplinary perspective. In E. M. Hetherington (Ed.), *Review of child development research* (Vol. 5). Chicago: University of Chicago Press, 1975, pp. 504–590.

Parker, S. T., & Gibson, K. R. A developmental model for the evolution of language and intelligence in early hominids. *The Behavioral and Brain Sciences,* 1979, **2,** 367–408.

Piaget, J. *The origins of intelligence in children.* New York: International Universities Press, 1952.

Piaget, J., & Inhelder, B. *The Psychology of the Child.* New York: Basic Books, 1969.

Ploog, D. Soziobiologie der Primaten. In K. P. Kisker, J.-E. Meyer, C. Müller, & E. Strömgren (Eds.), *Psychiatrie der Gegenwart* (Vol. I, 2nd ed.). Berlin: Springer Verlag, 1980, pp. 379–544.

Richards, M. P. M. (Ed.) *The integration of a child into a social world.* London: Cambridge University Press, 1974.

Robson, K. The role of eye-to-eye contact in maternal-infant attachment. *Journal of Child Psychology and Psychiatry,* 1967, **8,** 13–25.

Robson, K. S. Development of the human infant from two to six months. In J. D. Noshpitz, J. D. Call, R. L. Cohen, & I. N. Berlin (Eds.), *Basic handbook of child psychiatry* (Vol. 1). New York: Basic Books, 1979, pp. 106–112.

Rosenblatt, J. S. Evolutionary background of human maternal behavior: Animal models. *Birth and the Family Journal,* 1978, **5,** 195–199.

Ryan, J. F. (1973). Interpretation and imitation in early language development. In R. A. Hinde & J. Stevenson-Hinde (Eds.), *Constraints on learning: Limitations and predispositions.* London/New York: Academic, 1973, pp. 427–443.

Schaffer, H. R. (Ed.) *The origins of human social relations.* New York: Academic, 1972.

Schaffer, H. R. (Ed.) *Studies in mother–infant interaction.* London: Academic, 1977.

Scheflen, A. E. Behavioral programs in human communication. In W. Gray, F. J. Duhl, & N. D. Rizzo (Eds.), *General systems theory and psychiatry.* Boston: Little, Brown, 1969, pp. 209–228.

Schoetzau, A., & Papoušek, H. Mütterliches Verhalten bei der Aufnahme von Blickkontakt mit dem Neugeborenen. *Zeitschrift für Entwicklungspsychologie und pädagogische Psychologie,* 1977, **9**, 231–239.

Simmel, E. C. (Ed.) *Early experience and early behavior: Implications for social development.* New York: Academic, 1980.

Snow, C. E. The development of conversation between mothers and babies. *Journal of Child Language,* 1977, **4**, 1–22.

Sroufe, L. A. Socioemotional development. In J. D. Osofsky (Ed.), *Handbook of infant development.* New York: Wiley, 1979, pp. 463–516.

Stern, D. N. A micro-analysis of mother–infant interaction: Behavior regulating social contact between a mother and her 3½-month-old twins. *Journal of American Academy of Child Psychiatry,* 1971, **10**, 501–517.

Stern, D. *The first relationship: Infant and mother.* The Developing Child Series, J. Bruner, M. Cole, & B. Lloyd (Eds.). London: Fontana Open Books, 1977.

Thoman, E. B. (Ed.) *Origins of the infant's social responsiveness.* Hillsdale, N.J.: Erlbaum, 1979.

Thoman, E. B., Becker, P. T., & Freese, M. P. Individual patterns of mother–infant interaction. In G. P. Sackett (Ed.), *Observing behaviors: I. Theory and applications in mental retardation.* Baltimore: University Park Press, 1978, pp. 95–114.

Thoman, E. B., Korner, A. F., & Beason-Williams, L. Modification of responsiveness to maternal vocalization in the neonate. *Child Development,* 1977, **48**, 563–569.

Thomas, A., & Chess, S. *The dynamics of psychological development.* New York: Brunner/Mazel, 1980.

Turkewitz, G., & McGuire, I. Intersensory functioning during early development. *International Journal of Mental Health,* 1978, **7**, 165–182.

von Bertalanffy, L. *Organismic psychology theory.* Barre, Mass.: Clark University Press with Barre Publishers, 1968.

von Cranach, M., Foppa, K., Lepenies, W., & Ploog, D. (Eds.) *Human ethology: Claims and limits of a new discipline.* Cambridge, England: Cambridge University Press, 1979.

White, R. W. Motivation reconsidered: The concept of competence. *Psychological Review,* 1959, **66**, 297–333.

Whiting, J. W. M., & Whiting, B. B. *Children of six cultures: A psycho–cultural analysis.* Cambridge, Mass.: Harvard University Press, 1975.

Wilson, E. O. *Sociobiology: The new synthesis.* Cambridge, Mass.: Harvard University Press, 1975.

Winnicott, D. W. Primary maternal preoccupation. In *Collected papers: Through paediatrics to psycho-analysis.* London and Tavistock, N.Y.: Basic Books, 1958, pp. 300–305.

Wolff, P. H. Biological variations and cultural diversity: An exploratory study. In P. H. Leiderman, S. R. Tulkin, & A. Rosenfeld (Eds.), *Culture and infancy.* New York: Academic, 1977, pp. 357–381.

Yarrow, L. J. Emotional development. *American Psychologist,* 1979, **34**, 951–957.

9
INFANT-PEER INTERACTIONS

Tiffany M. Field
Jaipaul L. Roopnarine
*Mailman Center for Child Development
University of Miami Medical School*

Increasing numbers of very young infants are attending nursery and day-care programs where their peers outnumber adults and become significant others in their early social development. Accordingly, infant–peer interactions have become a popular topic of research.

Almost every paper on infant–peer interaction during the last decade has commenced with a discussion of possible reasons for the long hiatus between research on infant–peer interactions in the 1930s and the 1970s. Prominent among the explanations offered is that the psychoanalytic view of the mother as primary social object dominated the literature during those intervening decades. An orientation that may have evolved from this view is that mother–infant (or father–infant) interactions precede and provide the foundation for infant–peer interactions (Eckerman, Whatley, & Kutz, 1975; Mueller & Vandell, 1979). Mueller and Vandell (1979), for example, in their recent review of the literature, refer to this precursor view repeatedly:

> Parents afford interactive relationships before anything else, including even the infants' own hands. Yet peers afford very little, while nonsocial objects are intermediate, being easier to control than are peers for many months. (p. 613)

> Therefore, purely on the basis of relative affordances, it is possible to conclude that normal parent–infant engagement will occur prior to normal peer–peer engagement. (p. 612)

Infants are viewed as sociable with their parents from the first weeks of life, but they are generally thought to develop peer sociability and interactions only later in infancy after having had interactive experiences with parents and objects.

Infants' interactions with mothers have generally been studied at an earlier developmental stage (the first few months) than infant–peer interactions which have typically been studied only later in the first year of infancy and toddlerhood. The precursor or sequential model may be an artifact of our having investigated mother–infant interactions at an earlier stage than infant–peer interactions. Studies of infant–peer dyads at this earlier stage might reveal a parallel rather than a sequential development of sociability to parents and peers.

Moreover, the sociability noted toward peers during the latter part of infancy is less than the sociability toward parents in the first months of infancy; this may relate to the infant having different developmental agendas during those two stages. Face-to-face social behaviors may be more salient in the infants' developmental agenda and repertoire during the first few months while nonsocial toy-directed behaviors, frequently noted during peer interactions, may be more consistent with developing fine motor skills at the end of the first year. The infant, at this stage, may be less preoccupied with social than object relations. Thus he would look relatively less social at this stage than during the first few months of infancy. Investigations of mother–infant interactions during the later stage may also be characterized as less social or more toy-oriented. In this way, the lesser sociability noted in the presence of peers may be an artifact of the developmental stage at which the infants are studied.

The lesser sociability observed with peers may also relate to factors introduced by the research situation; for example, the study of infants with no previous experience with peers, their play with unfamiliar peers, the presence of their mother as well as the unfamiliar mother of the peer, and the provision of toys. This chapter reviews some of the literature on infant–peer interactions in the context of these issues.

MOTHER–INFANT VERSUS INFANT–PEER INTERACTIONS

Many of the recent studies on infant–peer interactions have employed methodologies which are similar to those used in mother–infant interaction research. For example, infant–peer interactions are typically studied in a dyadic context and in the laboratory, and similar social behaviors have been coded and treated as criteria of the infant's sociability in both infant–infant and mother–infant interactions. Namely, infant–infant dyads are viewed as sociable or socially interactive if they engage in face-to-face behaviors, such as looking, smiling, vocalizing, and laughing at peers, and later, as they become more mobile, if they touch their peers, offer and accept toys, coordinate their toy play, and imitate their peers.

Mother–infant interactions have typically been staged in the laboratory in situations as ecologically meaningful as possible. For example, over the first few months, since the infant's repertoire and natural activities are limited to feeding and face-to-face behaviors (looking, smiling, cooing), mother–infant interaction studies have centered around feeding and face-to-face interactions (Bakeman & Brown, 1980; Brazelton, Koslowski, & Main, 1974; DiVitto & Goldberg, 1979; Field, 1977a,b, 1979a; Stern, 1974). From about age 5 to 12 months, mother–infant interactions are rarely studied because the infant appears to be less interested in interacting face to face than in playing with his hands or manipulating objects (Trevarthen, 1974). At 12 months, the researcher of mother–infant interactions typically stages a floor-play interaction in which the infant, mother, and toys are grouped for an interaction (Brachfeld, Goldberg, & Sloman, 1980; Field, 1979c). Investigators of mother–infant interactions have taken cues from the infant that feeding and face-to-face play are meaningful interactive contexts during the first few months, that the infant is not very interested in those activities over the next few months, and finally, that he will only cooperate at 12 months if toys are also featured in the interaction situation. The infant is very sociable at 3 months as manifested by looking, smiling, and cooing, but are we to assume that the infant is no longer sociable or socially interactive with his mother after 3 months because he is less interested in face-to-face interactions or at 12 months because he engages in more toy-directed behavior than the earlier looking, smiling, or cooing behaviors?

The infant–peer interaction researcher enters later in the first year and notes that in the presence of toys the infant shows less attention to peers (Eckerman & Whatley, 1977) and that without toys the infant treats his peer much like a toy (Bridges, 1933; Maudry & Nekula, 1939). Longitudinally, from that time through toddlerhood, only minimal changes are reported for social behaviors such as looking, vocalizing, smiling, and laughing, but involvement with the same play materials and coordinated play with toys usually increase. Are we again to conclude that infants are becoming less social? Are peer interactions of a different type than the interactions of mothers and infants because infants direct more looking, smiling, and cooing to mothers at the earlier stage and more toy-related behaviors in the presence of peers at the later stage? Observations at different stages of development confound these comparisons. The data on mother–infant interactions are derived from observations conducted in the first few months and the data for peer interactions from observations conducted in the latter part of the first year. The infant may relate to objects—social and nonsocial—according to a developmental scheme which depends on the facilitations and limitations of his behavioral repertoire and developmental agenda at the time of the observation.

The very few comparisons of interactions with mothers and peers during the first few months, in fact, suggest that the infant relates socially in similar ways to mothers and peers, namely looking, smiling, and cooing (Field, 1981; Fogel, 1980). At 12 months, when the infant's agenda is manipulation of objects and learning to navigate, he appears less social and more toy-directed in the presence of both peers and mothers. We would argue that the infant's repertoire of socially directed behaviors is basically very similar in the presence of mothers or peers. Peers and mothers are both treated as social objects, in a very different manner than nonsocial objects or toys. Any differences noted may relate more to the mother being not only a model of social behavior but also more predictable and contingently responsive than the peer. Her modeling and contingent responsivity may facilitate a smoother, more sociable-looking interaction, while the challenge and frustration of the peer's unpredictability may contribute to a less sociable interaction. However, the behaviors directed by the infant to its mother and peer are notably and similarly social, especially when contrasted to the behaviors directed toward objects. Some recent studies comparing the behaviors of very young infants directed to mother, peers, and objects support this view.

Young Infants' Behaviors in the Presence of Mother, Peers, and Objects

Peer- versus mother-directed behaviors of 1- to 3-month-old infants were recently investigated by Fogel (1980). In his study, infant behaviors were recorded when the infant was face to face with his mother or stranger peer and when the infant was alone. While socially directed behaviors were observed during both the mother and peer interactions, in contrast to the alone condition, there were some qualitative differences in behaviors emitted in the presence of the infant's mother and peer. Infants' behavior to the stranger peer had an intense, unbroken quality. Eyebrows were typically relaxed with a minimum of facial movement. Limb and body movement was rare, but when it occurred, it had an abrupt or jerky tempo. The peer condition evoked intense staring with occasional "strain forward" movements of the head, "apparently to get a closer look at the other infant." To their mothers, the infants' behavior involved multiple changes in brow expressions. Gazing was also accompanied by a variety of mouth, limb, hand, and finger movements, which were displayed in a smooth pattern. These latter movements, using fine motor coordination, were more frequently displayed to the mother in contrast to the more gross motor movements in the presence of the peer. Fogel reported that "not much went on during the 'alone' condition except the maintenance of a relaxed or receptive manner."

The qualitative difference between mother- and peer-directed behaviors, namely, the smoother or less abrupt quality of the facial and limb movements in the presence of the mother, are attributed by Fogel (1980) to the familiarity and the contingent responsivity of the mother. Smoother infant behaviors might be expected during mother–infant interactions due to her containing behaviors and her sensitivity to the infant's cues in accelerating and decelerating the stimulation she provides (Brazelton et al., 1974; Stern, 1974).

Despite the qualitative difference in smoothness of movements, the types of behaviors which were shown by the largest proportion of infants were the same in both peer and mother condition, namely, relaxed brow (72% in mother condition, 61% in peer condition), moderately raised brow (56% in both conditions), and mouth wide open (50% for mother, 44% for peer). Smiling, however, occurred among a greater proportion of infants in the presence of mothers rather than peers, perhaps because mothers are known to model smiling behavior and to contingently respond to several types of infant behaviors with smiling. Unfortunately, vocalizations (cooing and distress) were not coded in this study. Nonetheless, the results suggest a similar repertoire of social behaviors in the presence of mothers and peers and a relative infrequency of those behaviors when social objects are absent or the infant is alone.

Another recent study illustrates differences in the behavior of infants when face to face with peers and when presented with an object (Field, 1979a). Field presented a peer and a mirror sequentially and then simultaneously to 3-month-old infants. Their behaviors were videotaped and heart rate was monitored. Infants looked longer at the mirror, but smiled, vocalized, reached toward, and squirmed more in the presence of an infant peer. In addition, heart rate was elevated during the peer situation, perhaps because the infant was more aroused or more active. While the mirror image of the infant was similar in size and facial features to the live infant peer, thus providing a similar visual stimulus, the more sociable behaviors (smiling and vocalizing) directed to the peer suggest an awareness on the part of the infant of the greater potential for social interaction with another infant than with a mirror image of itself.

A replication of this study with a sample of twin infants yielded the same differences in behavior in the presence of the peer, in this case a familiar peer, and a mirror (Field & Ignatoff, 1980). However, a greater incidence of social behaviors (smiling, vocalizing, and reaching toward the peer) occurred among co-twins than among the unfamiliar infants of the previous study, suggesting that peer familiarity facilitated social behavior.

Subsequently, using a similar face-to-face methodology we compared 4-month-old infants' behaviors in the presence of the mother, a peer, a mirror, and an infant-sized raggedy ann doll (Field, 1981). The infants spent more time looking at the mirror and raggedy ann doll than the infant peer but less time looking at the peer than the mother. We interpreted these differences as being a function of the information-processing demands placed on the infant, with gaze aversion occurring more frequently in the presence of the peer, rather than the mirror or doll, because the peer is a multimodal, social stimulus and somewhat unpredictable. While the mother is also a social stimulus, her behaviors are more familiar, predictable, and contingent, and thus may require less information processing and arousal modulation as manifested by less gaze aversion in her presence than in the presence of the peer. While gaze behavior was the focus of this study, the data on other social behaviors (smiling and vocalizing) suggested that the infants, although they smiled and vocalized less to the peer than the mother, showed significantly more of these behaviors in the presence of the peer than the lifelike mirror image or raggedy ann doll.

Taken together, these studies suggest that behaviors toward peers are distinctly social at a very early stage in infancy. While smiling and vocalizing may occur less frequently in the presence of the peer than the mother, these behaviors occur more frequently with the peer than with less social objects such as a mirror and raggedy ann doll. The quantitative differences in social behaviors directed to peers and mothers reported by Field and colleagues and the qualitative differences in smoothness of movements noted by Fogel may relate to the mother being more familiar, predictable, and contingently responsive than the peer. That twin infants are more sociable together than infants and stranger peers suggests that familiarity and experience with peers may be a significant factor.

Further research is clearly needed to determine the role that such factors as familiarity, predictability, and contingency play in differentiating these early social interactions. Nonetheless, these data suggest that the infant is sociable in the presence of peers at a very early stage. The similar behaviors in the presence of mother and peer from the first weeks of life suggest that mother- and peer-directed social behaviors may develop in parallel rather than sequentially. If interactive experiences with the mother were requisite to the infant's peer-directed social behavior, different types of behaviors might be expected during mother and peer interactions. It is, of course, conceivable that the infant's prior interactive experiences with the mother during the first month of life provide that foundation. While these studies cannot resolve that question, they do suggest that by the end of the first month mother and peer are both effective elicitors of social behaviors. These data do not support the claims made by the Mueller and Vandell review (1979) when contrasting mothers with peers:

> Their "mother love" made them into almost perfect engagement-affording creatures to their largely incompetent offspring. . . . The situation vis-à-vis infant peers is very different. Even if an *en face* position could be sustained, which is doubtful given weakness of neck musculature, the biological rhythms of attention and withdrawal would prevent all but random synchronies in their gaze. Therefore, purely

on the basis of relative affordances, it is possible to conclude that normal parent–infant engagement will occur prior to normal peer–peer engagement. (p. 612)

Of course, most infants have more opportunities to interact with parents prior to peers. When given an early opportunity to interact with peers, however, infants appear to behave as sociably with peers as with parents. Interactions with adults do not seem to be a necessary foundation for interactions with peers. The very young infant smiles and coos at parents and peers, and the older infant, who has been the subject of most peer interaction research and the stimulus for the precedence theory, is not only less social with peers but also less social with parents, perhaps because smiles and coos are less salient activities in his repertoire at this time than are activities with toys. To advance a precursor model or a theory that infant–adult and infant–peer interactions develop in parallel requires that interactions between infants and parents and infants and peers be studied at similar stages of development.

Older Infants' Behaviors in the Presence of Mothers, Peers, and Objects

Following the first few months of infancy and until the latter part of the first year, there are very few studies on mother–infant and infant–peer interactions. The infant, during the second quarter of the first year, appears to be less interested in face-to-face interactions with his mother (Trevarthen, 1974). This probably extends to the en face interactions with peers and may explain the dearth of mother–infant and infant–peer studies during this period. Having discovered his hands, the infant may be less social at this time and more oriented to objects. Many of the social behaviors observed during this period appear to be mediated by objects.

It is only in the latter half of the first year that direct comparisons have been made between peer- and mother-directed social behaviors when both mother and peer are available in the same play situation. Vandell (1980) observed 6-, 9-, and 12-month-old infants during a peer–mother session conducted in a play laboratory. This paradigm, the observation of two mother–infant dyads playing in the laboratory, has been particularly popular for comparisons between peer- and mother-directed social behaviors. In addition, as in a number of other studies, the dyads were observed playing in a toys and a no-toys condition. Mothers were asked to place their infants together in the center of the room and to situate themselves along the wall and "to feel free to talk to one another, but not to initiate interactions with their infants. They were instructed to intervene only if they felt their babies needed them" (p. 356). The toys provided were four books, six small blocks, a rattle, a ball, some pop beads, and plastic keys.

For a comparison of the frequency of behaviors directed to peer and mother, Vandell (1980) reported the following: infants spent more time looking and vocalizing to their peers and more time touching their mothers. While it would appear from their table of mean frequencies that some of these differences did not occur for all three ages, no interaction effects were reported. For example, vocalizing to peers and mother is approximately equivalent at 6 months and at 12 months, but more frequently directed to peers at 9 months. Main effects for age were increases in vocalizing and looking at peer and mother, and infant-initiated proximity with the peer. Vandell (1980) noted that at 6 months no behavior was directed first to mother and only later to peer. In addition, positive correlations were found between the frequency of infants' vocalizations to mother and peer and smiling at mother and peer.

Cross-lagged correlations performed to determine whether infants generalized their behaviors from mother to peer or vice versa revealed no relationships. Vandell (1980) noted, "Much more apparent were the similarities in infants' behavior to mother and peer" (p. 359). As might be expected from earlier studies (Eckerman & Whatley, 1977; Ramey, Finkelstein, & O'Brien, 1976) and another study by Vandell (Vandell, Wilson, & Buchanan, 1980), the presence of toys depressed the likelihood of both peer- and mother-directed behaviors. While Vandell (1980) concluded that there is an underlying sociability in both infant–peer and infant–mother interactions despite some differences, she added the caveat that the infants may have been observed after the generalizations from mothers to peers had occurred or that other variables, such as an interaction pace, might yield evidence of a transfer from mother–infant to infant–infant interaction skills. One of the problems with this type of paradigm is that the peers were strangers. In addition, the mothers' presence may depress the amount of behavior directed toward peers in the same way that toys are noted to depress peers' sociability.

A comparison of peer- and mother-directed behaviors, made in a parent's cooperative nursery school setting, addresses this point (Field, 1979b). In this study, twelve 10- to 14-month-old infants and their mothers were observed over the course of a semester. Since the mothers rotated being in and out of the nursery room, comparisons could be made between the social behaviors occurring in a mother's absence and in her presence. As in most nursery school settings, many toys were available. When the infants' mothers were in the room, the infants vocalized more to their mothers than to peers. They also moved toward her, touched her, and offered her toys more often than they did to infant peers. However, they spent more time smiling at their peers and equivalent amounts of time looking at their peers and their mothers. When the infants' mothers were out of the room versus in the room, infants spent more time looking, smiling, vocalizing, and offering toys to their peers. In addition, they showed fewer negative behaviors such as taking toys from their peers and crying. Finally, as in the Vandell (1980) study, several peer-directed behaviors increased over the 4-month period, including smiling, moving toward, touching, and offering toys to their peers, while negative behaviors, such as toy snatching and crying, decreased. Behaviors directed to the peers' mothers occurred very infrequently, as has been noted by a number of other investigators (Eckerman & Whatley, 1977; Lewis et al., 1975).

These results suggest that peer contacts may have been inhibited by the presence of mothers in studies reporting fewer peer-directed and more negative social behaviors (Bronson, 1975) and may have been enhanced by increasing familiarity with their peers (Lewis et al., 1975; Mueller & Rich, 1976). The percentages of time that social behaviors were directed toward peers when mother was in the room versus out of the room in this study were very similar to those reported by Eckerman & Whatley (1977) for periods of toys being in the room versus out of the room. Mothers and toys seem to be strong competitors for peer-directed behaviors. That mothers and toys may be more salient stimuli is not surprising given that most infants have a longer history of relations with them than with peers. The fewer behaviors directed toward peers during the initial period of the study suggests that familiarity may be a critical variable in peer interaction studies. Familiarity with the peers' mothers, however, did not lead to an increase in those behaviors. The low frequency of behaviors directed to the peers' mothers suggests that the adult per se is not necessarily a more salient stimulus than a peer for infant social behaviors. In addition, this finding suggests that infants do not necessarily direct more behaviors to their mothers because as adults they initiate more interactions or are more contingently responsive, since peers' mothers were also presumably more contingently responsive than peers.

This study on the presence and absence of mothers and those studies on the presence and absence of toys during peer interactions suggest that the mother and toys are strong competitors for the attentions of an infant and may contribute to the relatively low incidence of socially directed behaviors toward peers. The increase of socially directed behavior over the course of a 4-month period may relate to increasing familiarity of the peers or a gradual diminution of toy-directed activity as infants become more agile at manipulating toys. Nonetheless, the infants are less social at this stage than during early infancy if behaviors such as smiling and vocalizing are used as criteria. Although these 10- to 14-month-old infants showed similar amounts of looking at their peers as the 4-month-old infants in the Field (1979a) peer–mirror study (41% for both age groups), the 4-month-old infants of the peer–mirror study smiled for 15% of the interaction time and the older 10- to 14-month-old infants smiled only 9% of the time. The younger infants also vocalized more frequently than the older infants (30% of the interaction time versus 12%). But the repertoire of the younger infants is limited to behaviors such as looking, smiling, and vocalizing, while the older infants were not only capable of but appeared more interested in crawling, walking, and manipulating toys. Being in an infant seat without toys in the younger age situation may facilitate early peer interactions. But it is not clear that a similar situation for the older infant (for example, being placed in a high chair without toys in an en face position with a peer) would necessarily facilitate social behaviors. Gross and fine motor activities appear to be more potent at this stage, and confinement to a high chair may elicit protest and fussy behavior instead.

During a motor milestone stage, for example, the onset of walking, social behaviors may occur less frequently. Fischer (1973), for example, noted that the infant who was the most mobile was the least social and verbal and, conversely, the infant who could not yet crawl or walk exhibited the greatest number of smiles and vocalizations.

Variability in the frequency of social behaviors across different stages of infancy is also suggested by the Eckerman and Whatley (1977) study. In this comparison of 10- to 12- and 22- to 24-month-old infants, equivalent amounts of peer watching occurred. Looking at peers seems to be the only high-frequency social behavior over the first 2 years, with Field (1979a) having reported 41% peer-directed looking for 3-month-olds and Eckerman & Whatley (1977) reporting 50 and 53% respectively for 10- to 12- and 22- to 24-month-old infants. Distal social behaviors such as smiling, vocalizing, and gesturing consistently occurred more frequently among the younger and older infants. We have suggested that these behaviors are more frequent at the younger age because the infant's repertoire is limited to these behaviors. The dip in social behaviors at 1 year relates to the prepotency of developing gross and fine motor skills. The increase in social behaviors again at 24 months and particularly those directed to peers may relate to an increase in cognitive and attention skills, enabling infants to relate simultaneously to objects and the bearer of objects, namely, the peer.

Eckerman et al. (Eckerman, Whatley, & Kutz, 1975) reported that "direct involvement in play" behaviors, such as imitating, showing and offering toys, and coordinated play, increased over the 10- to 24-month period. While involvement in play was less frequent at 10 to 12 months with both peers and mothers, by 22 to 24 months, play with the mother had declined in frequency while direct play with the peer increased markedly. An increase in cognitive skills and attentional capacity over this period may facilitate the infants' ability to relate simultaneously to peers and toys or to integrate peer and toy play, so that they are no longer competing for the attention of the infants. As Eckerman et al. (1975) point out, "Smiling and vocalizing to persons or contacting them may be prominent early behaviors, but other forms of social interaction require the child's integration of activities with things and people" (p. 48).

In addition to the effects of varying repertoires and developmental agendas and the influences of competitive sources of stimulation, such as mothers and toys, infant–peer play appears to be affected by peer experience and familiarity.

PEER EXPERIENCE AND FAMILIARITY

Familiarity seems to foster proximity and proximal behaviors, such as touching, hugging, and kissing. Just as mothers, who are more familiar to their infants than peers, are the recipients of more proximal behaviors, these behaviors are also directed toward familiar peers more frequently than unfamiliar peers.

A number of investigators have observed the peer-directed behaviors of infants playing with acquaintances versus strangers (Becker, 1977; Lewis et al., 1975; Mueller & Brenner, 1977; Young & Lewis, 1979). Becker (1977) compared peer-directed behaviors of infants who played together in the infants' homes for 10 sessions versus those of a control group who were observed at an initial session and at a time corresponding to the completion of the 10 play sessions of the experimental infants. She noted that the experimental infants directed more of their behavior to peers than to their mothers or toys, and their peer-directed behaviors increased over the 10 play sessions in contrast to the control infants whose behaviors did not change over the same time period.

Mueller and Brenner (1977) observed slightly older infants during dyadic play. A comparison between 16-month-old infants who had experienced 4 months of nursery school play with each other and 17-month-old infants with no experience revealed more coordinated social-directed behaviors and more sustained interactions among the familiar peers.

Lewis and colleagues (Lewis et al., 1975; Young & Lewis, 1979) attempted to determine whether familiarity per se or peer experience in general was the contributing factor to the greater amount of interaction observed among familiar peers. They observed the same infants in two circumstances, once with their acquainted peers and once with unfamiliar infants who had also had experience with other infants. More proximal activities (proximity, body contact, and touching) occurred among acquaintances than strangers. While there were no reliable differences in looking or vocalization as a function of peer familiarity, imitative behaviors and positive affect occurred more frequently among acquaintances. Thus familiarity or prior experience with a particular infant appeared to facilitate socially directed behaviors above and beyond the effects of general experience with peers. Familiarity and previous peer experience seem to be important variables in the study of early peer interactions (see Figures 9.1 and 9.2).

There appear to be variations in peer-directed behaviors as a function of a number of variables of this kind. Interactions in one's own home differ from those occurring in another infant's home, with infants engaging in more peer-directed behaviors in their own home (Becker, 1977). Similarly, observations of infants in open-field floor play with a number of toys (Fischer, 1973) versus playpen play with a limited number of toys (Maudry & Nekula, 1939) yield qualitative differ-

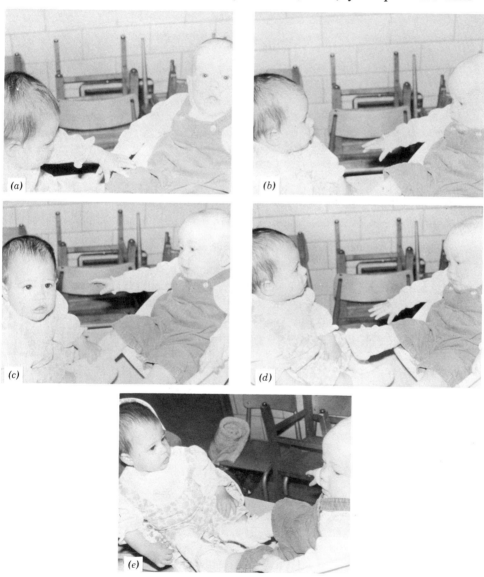

Figure 9.1. Unfamiliar peers in a laboratory setting alternately attending to each other.

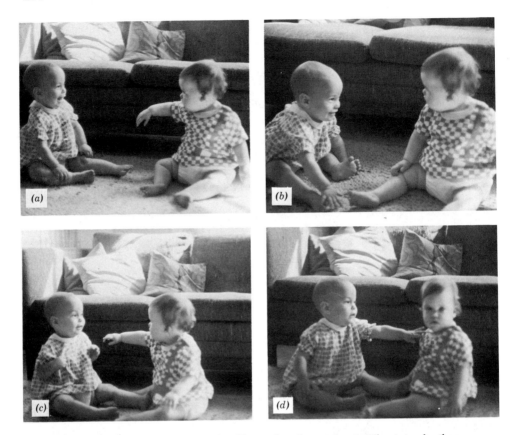

Figure 9.2. Familiar peers in a familiar setting alternately attending to each other.

ences, such as more positive play in the floor-play situation. These data, coupled with differences already noted in the studies of acquaintances versus stranger peers, mother-absence versus presence, and toy versus no-toy conditions, point to the importance of considering ecological variables. Most of the recent studies on early peer interactions may not be representative of the situation in which infants and their peers normally play. The peers have usually been unacquainted and have met only briefly in the presence of their mothers to engage in a dyadic or small group, laboratory situation. Since increasing numbers of infants and toddlers are being exposed to peers in group-care situations, such as day-care centers, family-care groups, and nursery schools, which feature multiple, familiar adults (teachers) and children, a multitude of toys, and the relative absence of parents, there is a need for studies of peer interactions in those settings.

SOCIAL BEHAVIORS OF INFANTS AND TODDLERS ATTENDING NURSERY SCHOOL

As noted above, there have been few observational studies of infants interacting in their natural settings. Accordingly, we designed a study in which infants and toddlers were observed interacting with their peers during free-play periods in their

TABLE 9.1. Mean Percentage of Time Children Engaged in Behaviors with Peers, Changes, and Significant Differences on the Behavioral Measures for the Two Groups

Behavior (Reliability)	First Observation		Second Observation		Longitudinal Changes	
	Infants	Toddlers	Infants	Toddlers	Infants	Toddlers
Watch (88)	20.57	17.18	39.57	34.55	**	**
Distal social behaviors (94)	5.00**	17.00	12.43*	31.09	*	**
Vocalize	3.29·	13.18	5.71	21.72		
Smile	.57	1.73	3.86	3.82		
Laugh	.00	.45	.43	2.64		
Fuss	1.14	1.64	2.43	2.91		
Physical contact (98)	5.85	3.90	12.00	13.18	*	*
Touch	5.14	3.54	10.43	12.18		
Strike	.71	.36	1.57	1.00		
Same play material (97)	3.42	4.36	7.28	10.36		
Proximity (88)	34.85*	46.81	52.14	57.63	**	*
Approach (98)	3.57	3.18	11.86*	5.27	**	*
Direct involvement in play (99)	4.28	7.27	9.00	14.46		.06
Imitate	.14	.73		2.82		
Offer toy	.14	.82		1.27		
Accept toy	.00	.45	.14	.64		
Take toy	1.14	.55	1.29	1.00		
Take over toy	.57	.45	1.00	.64		
Struggle	.71	1.36	3.14	3.27		
Coordinate	1.57	2.91	.57	4.82		

* $p < .05$
** $p < .01$

nursery classes (Roopnarine & Field, 1981): 11 infants and 11 toddlers were observed for a month at the beginning and a month at the end of the semester. We recorded the percentage of time that children engaged in the following behaviors with peers: distal social interaction, direct involvement in play, physical contact, approach, proximity, watching, and playing with the same materials (see Table 9.1). Our observations enabled us to examine the (1) differences between infants' and toddlers' social interaction with peers and (2) changes in children's interactions with peers over the four-month period.

In addition, relationships were assessed between developmental quotients and interaction behaviors and between extraversion–introversion ratings and interactive behaviors. These ratings were made according to the children's version of the Buck Affect Expression Rating Scale (Buck, 1977), which is comprised of 37 items on the child's extraversion–introversion and is rated on a Likert-type scale by independent observers.

There were no significant relationships between developmental quotients and behavior measures. By contrast, extraversion ratings were negatively related to watching and positively related to distal social interaction and proximity.

During the first month of observations, the toddlers directed more distal social behaviors toward their peers and were more likely to be proximal to them than were the infants. At the end of the semester, the toddlers again directed more distal social behaviors toward their peers than did the infants, but the infants were more likely to approach their peers than toddlers.

Comparisons of the time-one and time-two data indicated that the behaviors of both the infants and toddlers changed dramatically over the semester. Both the infants and toddlers showed more distal social behaviors, were more likely to monitor, were more likely to approach, be within proximity, and make physical contact with their peers. The toddlers also became more involved in direct play over the semester. Finally, comparisons of infants' interaction at time two and toddlers' interactions at time one revealed that the infants were more likely to monitor, approach their peers, and make physical contact with them than the toddlers.

These analyses of peer-directed behaviors during nursery school free play revealed that infants and toddlers directed significantly more positive than negative behaviors toward their peers, as has been reported by several other investigators. In addition, at both observation periods toddlers directed more distal social behaviors, particularly vocalizations and laughing, toward their peers than did the infants. Finally, over the one-semester period dramatic increases occurred for both groups on all behavior categories except for same-play materials for which no changes were noted and for direct involvement in play which only increased for toddlers.

Thus there were many more changes across the semester than there were age-group differences in these data. The amount of watching, distal social behaviors, physical contact, proximity, approach, and direct involvement in play increased across one semester, while differences between the age groups, despite their 15-month age difference, were considerably fewer. The toddlers showed more distal social behaviors at both ages, which were accounted for by vocalizing and laughing; smiling and fussing were approximately equivalent by the end of the semester for both groups. Although the infants were less proximal than the toddlers at the beginning of the semester, they were showing more approach behaviors than the toddlers by the end of the semester. On almost all the behaviors, then, except for the distal social behaviors (vocalizing and laughing) and direct involvement in play (coordinate play), the infants, despite being 15-months younger, looked approximately equivalent to the toddlers by the end of the semester, suggesting that they had experienced more change than the toddlers. Generally, these results suggest that familiarity with peers may effect more dramatic changes than developmental change per se, except for behaviors such as vocalizing, laughing, and coordinated toy play which may be more age-dependent.

Additional support for this interpretation is provided by a comparison between the data on these infants at the end of the semester ($M = 14$ months) and the 22- to 24-month old infants of the Eckerman et al. (1975) study, who had experienced only minimal peer contacts. At the end of the semester of observation, our infants, who then averaged 14 months, were watching their peers as frequently as the 22- to 24-month-old infants of the Eckerman et al. (1975) study (40 and 39%, respectively), engaging in distal social behaviors as frequently (13% for both groups), and making far more frequent physical contact (13%) than the 22- to 24-month-olds of Eckerman et al. (1975). However, the 14-month-olds of our study showed far less play with the same material (7 vs. 35%) and far less direct involvement in play (8 vs. 25%), particularly in the subcategory coordinated play (57 vs. 6%), than the 22- to 24-month-olds of Eckerman et al. (1975).

These comparisons coupled with data reported by Vandell et al. (1980) and by Young and Lewis (1977) on the effects of peer familiarity suggest that physical contact and proximity are examples of peer-directed behaviors which are most affected by familiarity. In those studies, approach, contact, and proximal behaviors were more frequent among familiar infants, but distal social behaviors and object-related behaviors did not differ as a function of familiarity. In the Eckerman et al. (1975) study on the effects of age differences across dyads of unfamiliar peers, physical-contact behaviors did not differ as a function of age, but object-related play (same-play materials and direct involvement in play) significantly increased across age. Thus there are some consistent findings in the literature suggesting that proximity–contact behaviors may be more affected by peer familiarity and that vocalizing and laughing, as well as object-related play such as coordinated play with the same material, are more age-dependent behaviors.

Comparisons across studies are tenuous at best, particularly inasmuch as these involved two different contexts: laboratory dyadic play and nursery group play. However, age and time (or familiarity) comparisons within this study suggest that familiarity with peers and social skill development across age may affect different peer-directed behaviors. That a greater number of those behaviors were affected by peer familiarity than by age differences per se highlights the salience of early peer contact.

SUMMARY

The small but growing literature on infant–peer interactions suggests, then, that infants are sociable with their peers from a very early age. Although some have proposed a precursor model in which infant sociability with peers is predicated on parent–infant interaction, that position was derived from a data base in which parent–infant interactions had been studied at an earlier age than infant–peer interactions. Despite some qualitative differences between early infant–parent and infant–peer interactions, there are many more similarities between these interactions as contrasted with ways in which infants relate to objects. Comparisons of very early infant–parent and infant–peer interactions suggest that sociability with parents and peers may develop in parallel.

An equally important problem with the infant–peer interaction literature has been that of ecological validity. Typically, infants have been observed in laboratory situations in the presence of their mothers and of stranger peers and their mothers. Several studies suggest that previous experience with peers, peer familiarity, the presence of the mother, and the environment—its familiarity and the presence of toys—significantly affect the infant's peer-directed behaviors. Observations of infants in more naturalistic settings, such as nursery schools and neighborhood play groups, may provide a clearer picture of the infant's developing interactions with peers.

REFERENCES

Bakeman, R., & Brown, J. Early interaction: Consequences for social and mental development at three years. *Child Development,* 1980, **51**, 437–447.

Becker, J. A learning analysis of the development of peer-oriented behavior in nine-month-old infants. *Developmental Psychology,* 1977, **13**, 481–491.

Brachfeld, S., Goldberg, S., & Sloman, J. Parent-infant interaction in free play at 8 and 12 months: Effects of prematurity and immaturity. *Infant Behavior and Development,* 1980, **4**, 289–306.

Brazelton, T. B., Koslowski, B., & Main, M. The origins of reciprocity: The early mother–infant interaction. In M. Lewis & L. Rosenblum (Eds.), *The effect of the infant on its caregiver.* New York: Wiley, 1974.

Bridges, I. A study of social development in early infancy. *Child Development,* 1933, **4**, 36–49.

Bronson, W. C. Developments in behavior with age-mates during the second year of life. In M. Lewis & L. A. Rosenblum (Eds.), *Friendship and peer relations.* New York: Wiley, 1975.

DiVitto, B., & Goldberg, S. The effects of newborn medical status on early parent–infant interaction. In T. Field, A. Sostek, S. Goldberg, & H. H. Shuman (Eds.), *Infants born at risk.* New York: Spectrum, 1979.

Eckerman, C. O., & Whatley, J. L. Toys and social interaction between infant peers. *Child Development,* 1977, **48**, 1645–1656.

Eckerman, C. O., Whatley, J. L., & Kutz, S. L. Growth of social play with peers during the second year of life. *Developmental Psychology,* 1975, **11**, 42–49.

Field, T. Effects of early separation, interactive deficits and experimental manipulations on infant–mother, face-to-face interaction. *Child Development,* 1977, **48**, 763–771. (a)

Field, T. Maternal stimulation during infant feeding. *Developmental Psychology,* 1977, **13**, 539–540. (b)

Field, T. Differential behavioral and cardiac response of 3-month-old infants to a mirror and peer. *Infant behavior and development,* 1979, **2**, 179–184. (a)

Field, T. Infant behaviors directed toward peers and adults in the presence and absence of mother. *Infant Behavior and Development,* 1979, **2**, 47–54. (b)

Field, T. Interaction patterns of high-risk and normal infants. In T. Field, A. Sostek, S. Goldberg, & H. H. Shuman (Eds.), *Infants born at risk.* New York: Spectrum Publications, 1979. (c)

Field, T. Gaze behavior of normal and high-risk infants during early interactions. *Journal of the American Academy of Child Psychiatry,* 1981, **20**, 308–317.

Field, T., & Ignatoff, E. Interactions of twins and their mothers. Unpublished paper, University of Miami, 1980.

Fischer, T. Rhythms of infant play. Unpublished master's thesis, Tufts University, 1973.

Fogel, A. Peer vs. mother-directed behavior in one-to-three-month-old infants. *Infant Behavior and Development,* 1980, **2**, 215–226.

Lewis, M., Young, G., Brooks, J., & Michalson, L. The beginning of friendship. In M. Lewis & L. Rosenblum (Eds.), *Friendship and peer relations.* New York: Wiley, 1975.

Maudry, M., & Nekula, M. Social relations between children of the same age during the first two years of life. *Journal of Genetic Psychology,* 1939, **54**, 193–215.

Mueller, E., & Brenner, J. The origins of social skills and interacting among playgroup toddlers. *Child Development,* 1977, **48**, 854–861.

Mueller, E., & Rich, A. Clustering and socially-directed behaviors in a play group of one-year-old boys. *Journal of Child Psychology and Psychiatry,* 1976, **17**, 315–322.

Mueller, E., & Vandell, D. Infant–infant interaction. In J. D. Osofsky (Ed.), *Handbook of infant development.* New York: Wiley, 1979.

Ramey, C. J., Finkelstein, N. W., & O'Brien, C. Toys and infant behavior in the first year of life. *Journal of Genetic Psychology,* 1976, **129,** 341–342.

Roopnarine, J., & Field, T. Peer-directed behaviors of infants and toddlers during nursery school play. In preparation, 1981.

Stern, D. N. Mother and infant at play. In M. Lewis & L. Rosenblum (Eds.), *The effect of the infant on its caregiver.* New York: Wiley, 1974.

Trevarthen, C. Conversations with a 2-month-old *New Scientist,* 1974, **22,** 230–235.

Vandell, D. Sociability with peers and mothers in the first year. *Developmental Psychology,* 1980, **16,** 355–361.

Vandell, D., Wilson, K., & Buchanan, N. Peer interaction in the first year of life: An examination of its structure, content, and sensitivity to toys. *Child Development,* 1980, **51,** 481–488.

Young, G., & Lewis, M. Effects of familiarity and maternal attention on infant peer relations. *Merrill-Palmer Quarterly,* 1979, **2,** 105–120.

10

THE SOCIAL NETWORK SYSTEMS MODEL
Toward a Theory of Social Development

Michael Lewis
Rutgers Medical School
CMDNJ
New Brunswick, N.J.

A network is an interconnection between events or people, and it is to the interconnection between social beings that this chapter is addressed. The newborn human infant enters into a world filled with networks, the most immediate and important of which is the family. Soon the infant's network will include significant others beyond family including friends, teachers, and eventually mates. Moreover, this principal network is embedded in other networks that form larger reference groups such as a clan, social class, or religious group. Still larger networks can be imagined that are made up of geographic regions (such as eastern or southern America) and countries or cultures (such as the Western Judeo-Christian culture). These networks are interconnected and exert influence on each other. They form the ecology of development (Bronfenbrenner, 1977). Each of these separate networks, as well as the entire system of networks, operate under general systems principles and, therefore, possess the characteristics of systems (Von Bertalanffy, 1967).

It is into this complex interdependent set of networks, into a changing array of people and institutions, behaviors and goals, that the infant is born, and it is within this array that the development and socialization of the child takes place. As has been pointed out by Lewis and Feiring (1978), the socialization of the infant is the process of learning to become a member of these different networks. The task of the socializer(s), then, is to teach the child the rules of membership. This teaching can be carried out through a variety of procedures, some of which are direct, as in didactic information exchange, and some of which involve more indirect interaction, such as modeling, imitation, and referencing (Bandura, 1969; Lewis, 1979b; Lewis & Feiring, 1981; Campos & Stenberg, 1981).

Portions of this paper were presented at an invited fiftieth annual address of the Eastern Psychological Association, Philadelphia, Pa., April 1979. Research reported was supported by NICHD Grant #NO 1-HD-82849.

The study of social development has received considerable attention, whereas less attention has been given to the study of dyadic interactions, relationships, social behavior, and the acquisition of social knowledge. Nevertheless, except for the work of attachment theorists, an outgrowth of the psychoanalytic theory, little effort has been taken to construct a comprehensive theory of social development. It would be presumptuous to assume that such a theory is readily available. What is possible is to outline some of the major issues confronting aspects of a theory of social development. Toward such an attempt this chapter is directed.

Social development has as its main themes the development of social knowledge, social behavior, and relationships. In a recent paper, Lewis (1980b) has attempted to articulate some of the concerns found in the study of social knowledge. In particular, knowledge of self, of others, and of the relationships between self and others and between others appears to be the appropriate domain of inquiry for this problem. In the study of social relationship, several problems seem important: the difference between interactions and relationships, as well as the development of relationships from interactions; the possible types of social relationships that exist in the adults' world and that appear to have their origin in the child's experience; and the interconnections that may exist between relationships. At this point, epigenetic and social network systems models will be discussed as well as the connection between relationships and needs.

INTERACTIONS AND RELATIONSHIPS

Although the study of social development has much to do with the establishment, maintenance, and use of social relationships, relatively little work has been directed to the study of relationships. In fact, most studies have implied that social interactions and relationships are synonymous. This may be due to the fact that interactions are relatively easily measured, while relationships are much more difficult to define and measure.

Lewis (Weinraub, Brooks, & Lewis, 1977; Lewis & Weinraub, 1976) has tried to make the distinction between interactions and relationships. For us, interactions are specifiable behaviors or sets of behaviors that are observable and, therefore, measurable. Relationships are inferred from interactions but are difficult to specify and are not easily measurable. For example, the behavior between two people constitutes an interaction, and one might be able to specify the relationship from this interaction. Likewise, knowing a relationship might enable us to predict an interaction. Thus, for example, a female nursing a baby may indicate a mother–child relationship, or a mother–child relationship might dictate a particular interaction, like nursing. However, as we can easily see, the inference of a relationship from an interaction or a set of interactions may be difficult. Likewise, specifying interactions given information about the relationship also is not obvious. Thus females other than the mother can nurse the child (wet nurses), and the mother–child relationship consists of complex and varied interactions including not only nursing and kissing but sometimes even hitting and spanking.

Lewis, Feiring, and Weinraub (1981) have argued that even though mother–child and father–child interactions are different, there is no reason to assume that the parent–child relationships are also different. In fact, although one important difference between the mother–child and father–child interactions is that they are predicated on different degrees of biological interactions, the relationships may be the same; for example, both may be equally viable attachment relationships.

> While the father is undistinguished from other male adults, the mother of the child is biologically obvious . . . the mother's interactions with the infant clearly distinguish her from all other women. Her biological interaction with the child through pregnancy and childbirth and then nursing makes known to her, to all other individuals, and to the child that she is "mother." . . . Fathers present a different case. Knowing that a man is father of a child tells us very little about that man's interactions with the child. (p. 12)

This lack of a predictable, biological, predisposed interaction pattern for fathers not only differentiates the father–infant interaction from that of the mother–infant interaction but also makes the nature of the father–infant relationship more difficult to discern. The loose connection between relationships and interactions serves to alert us to the need to study relationships as well as interactions.

Recently, Hinde (1976) has undertaken such an analysis in an attempt to develop a system for characterizing relationships. For Hinde, there are several dimensions that can be used to specify relationships. (1) Content of component interactions include what we have elsewhere called function or goal structures (Lewis & Feiring, 1978, 1979; Lewis & Weinraub, 1976). Here we refer to types of interactions, such as nursing, grooming, play, and education. (2) The diversity of interactions and the structure of the different possible types are a second dimension. As we have indicated, relationships are manifested in various interactions, and it may be the diversity and patterning of these interactions that are critical for understanding relationships. (3) Reciprocity versus complementarity refers to the status or power aspects of the interactions between people. In other words, interactions involve status dimensions. (4) Interactions also have quality dimensions or components, such as meshing of the dyadic members and turn taking. One in particular, the contingent nature of the interaction, has received much attention (see Lewis & Goldberg, 1969; Lewis & Coates, 1980; Goldberg, 1977). (5) The relative frequency and patterning of interactions constitutes a fifth dimension. Of specific importance for this aspect of relationships is the patterning of sets of interactions rather than the patterning of a simple interaction. (6) Multidimensional qualities is a general category used to consider a set of structural qualities of interactions as, for example, the length of a particular interaction and its short- and long-term duration. (7) Cognitive factors, the constructs of the mind, involve those mental processes that allow members of an interaction to think of the other member as well as of themselves. This aspect of interactions and relationships has received very little attention. Moreover, both the extent of the mental processes and the degree of social knowledge necessary to form relationships have been underestimated. (8) Penetration, or the intimacy of a relationship, is defined by Hinde as the "extent to which the personality of each is penetrated by the other" (p. 15).

Although this list of the dimensions of relationships could be added to or shortened, Hinde's analysis allows us to consider the complexities involved in understanding relationships. Furthermore, it is clear that any attempt to argue for a direct correspondence between interaction and relationship is not possible. This is especially so when the interactions we observe are limited in both quantity and duration.

FROM INTERACTIONS TO RELATIONSHIPS

At the heart of the issue of social development is the connection between interactions and relationships. In the first months of life, the infant's world is occupied primarily by interactions with other people and with objects. In contrast, the adult caregiver experiences both a relationship with the infant and interactions. The exploration of this asymmetry between adult and child is necessary in order to understand the growth of the child's social relations from these interactions.

Elsewhere, Lewis and Brooks-Gunn (1979) and Lewis (1979a) have argued that interactions are the material from which elaborate cognitive structures derive. If we hold that relationships are predicated, in large measure, on social cognitive structures, then it follows that interactions lead to relationships through the mediation of these structures. Such a view of the child's social development, that relationships derive from interactions, forces us to consider several issues. Before doing so, it must be noted that this view is implicit in research on attachment relationships (see Bowlby, 1969; Ainsworth, 1969) where early person–child interactions that have particular qualities, such as harmony, responsivity, positive reactions to distress, and high stimulation levels, to mention only a few, are believed to result in various types of attachment relationships. Although there is some research on this topic, more is necessary for us to understand the specific role of early interactions in attachment relationships. Nevertheless, a model of social interactions generating social relationships is an important aspect of children's social development.

Missing from the accounts of the attachment theorists, but necessary to a theory of social development, is a specification of the nature of the mediating social-cognitive structures required for developing relationships, since interactions and relationships do not and cannot have a one-to-one correspondence. One particular aspect of the social-cognitive structures needed for the development of relationships is the growth of the self.

The structure of the self has two components. The first is the existential self; that is, the distinction of a self separate from other people and objects. This self–other distinction occurs before the second component of the self becomes established. This process probably becomes stable around 8 or 9 months of age and manifests itself in self-permanence. Self-permanence, like object permanence, constitutes one of the first conservation achievements of the child: the ability to maintain an identity independent of setting or interaction. This self-permanence milestone allows for the development of the second aspect of the self, something we have called the "categorical self" (Lewis & Brooks-Gunn, 1979). The categorical self is that aspect of self-identity which consists of the ways we think about ourselves. Such common and early categories appear to include gender, age, competence, and value (good or bad) but may include others.

The formation of self-identity may be the crucial mediating social-cognitive variable since it permits consideration of both self and other. In particular, the development of self fosters more complex social-cognitive (even affective) processes, such as empathy. Empathy, the ability to place the self in the role of the other, is critical for the development of relationships (Lewis & Brooks-Gunn, 1979). Thus the development of the self and the capacity for empathy are the basis for the growth of relationships. The development of self may be a consequence of the child's very early interactions and thus may be the mediating variable between interactions and relationships.

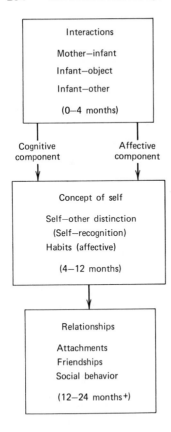

Figure 10.1. From interactions to self-concept to relationships.

Figure 10.1 presents a representation of this process. Lewis and Brooks-Gunn (1979) have shown that the development of the self may have its beginnings in the child's earliest interactions. However, it is not until around 8 months of age, when the infant has a sense of both self and object permanence, that a concept of self can be considered to be established. While some self-knowledge may be acquired earlier, it is not until the child has a unique sense of self, independent of the nature of the interaction, that a self-concept can be utilized by the child in forming relationships. In Figure 10.1, the child's interactions with objects and people are seen to lead to the creation of a concept of self. Such interactions probably also lead to the formation of what we call "habits." Habits are both cognitive and affective structures that derive from the child's associations of people with feeling states which are the consequence of interactions. Being fed by one's mother is an example of such a habit. These habits, together with the sense of self, enable the child to form relationships. Relationships may be possible only through the recognition of the existence of two selves—oneself and another's self, each having a separate identity and separate set of needs. Such a view of relationships as the negotiation of two separate selves is a common theme and is to be found in Sullivan's work (1953, 1954) as well as Youniss's (1980) notion of friendship.

As indicated in Figure 10.1, interactions may help to create a self which in turn forms the basis for relationships. From the data on self-recognition (Lewis & Brooks-Gunn, 1979) and the research on mother–infant interactions and attachment relationships, we can suggest proximate times for this process. In the first 3 to

4 months of life the infant's interactions with the environment result in a sense of self and a self–other distinction. What controls these early interactions is unknown but they are possibly due to several factors. Several investigators, Lipsitt (1980) and Papoušek and Papoušek (1980), as well as others, have viewed these early interactions in terms of complex social-reflexive behavior. Bower (1975) also has argued that rather than thinking of the newborn as possessing only a limited and simple set of initial schemata or structures (a view suggested by Piaget, 1963, as well), we must view the child as possessing a set of complex innate reflexes or schemata some of which seem to be interactive in nature. Early imitation (cf. Meltzoff & Moore, 1977) may be one set of these complex reflexes which serve to put the child in interaction with its social environment. Although the imitation phenomenon needs more study before we can be sure it is not an epiphenomenon (Lewis, 1979c; Waite & Lewis, 1979), such types of reflexes may be present and influential in other social-interactive behaviors.

Another possibility is that although the child affects the adult's behavior (Lewis & Rosenblum, 1974), it is still the adult who controls the interactions. Thus the major source of the interaction flow comes not from the infant but from the adult. We know that adults attribute to infants a wide set of abilities, including feeling, thinking, and knowing, long before children are capable of these activities (see Waite & Lewis, 1978, regarding the influence of parental beliefs on infant imitation; see also Pannabecker, Emde, Johnson, Stenberg, & Davis, 1980, on mothers' perception of infants' emotions). In this case it is the parents' behavior and attribution which sustain the early interactions, and like the "clever Hans" phenomenon, the infant's interactions are almost totally controlled by the adult. Of course, it is possible that our understanding of the timing of the child's cognitive ability is wrong and that children acquire complex cognitive structures much earlier. These structures then guide their social interaction.

The causes of these early interactions are speculative. Even so, the interactions occur and may provide the wellspring for the child's creation of important cognitive structures. To reiterate, in the creation of a sense of self and a set of habits, these early interactions provide the material for the building of relationships. From an ontogenetic point of view, the mediating structure between interactions and relationships is the development of self. Even beyond its developmental role, the sense of self may play an important role between interactions and relationships. Relationships are based on both direct and indirect interactions but require a sense of self and an integration of a self with other (empathy) to give them meaning. This is particularly true when we think of the recursive nature of thought: not only can I think about how I behave to you and how you behave to me, but I can think about how I think you think I behave toward you, and you to me. The complex recursive acts have been captured in Laing's *Knots* (1970) and have been discussed as particularly important in social cognition (Lewis, 1980b).

One final aspect of the interplay between interactions and relationships is important. Although interactions must vary in accord with various goals, settings, and other factors, their major function is to maintain the relationship. Thus it can be said that relationships can remain constant, while interactions change. For instance, various different interactions, particularly as a function of the age of the child, can be subsumed under an attachment relationship, a point made by Lewis and Ban (1971) and more recently by Sroufe (1979).

Although relationships rather than interactions should be our focus, almost all the work on early social development is restricted to describing interactions. Even the research on attachment relationships uses interactive measures to determine their nature. The restriction of our interest to interactions, rather than relationships, is due to our inability (or refusal) to consider what a person thinks or feels about the other and our overreliance on behavioral indices. In the subsequent sections of this chapter, it will be necessary to blur this distinction since not enough information exists to allow us to explore relationships. Even so, we need to be aware of the limitations imposed by such data.

TYPES OF RELATIONSHIPS

One important aspect of social development concerns the type and the nature of a social relationship that can be formed by the child. Some attention has been given to the attachment relationship, and implicit in the discussion of this type of relationship is that children are either attached or not attached to certain people. Moreover, within an attachment relationship two major distinctions—securely and insecurely attached—are usually made (Ainsworth et al., 1978). With the exception of this formulation and Lewis and Rosenblum's (1975, 1977) attempt to specify the various social objects that occupy the child's social space, little consideration has been given to other types of relationships.

Let us consider this problem from a broader perspective in order to explore the range of possible relationships that may exist in human experience. While this analysis relies on an adult perspective, evidence that children may also experience these different relationships appears to exist, evidence that helps support the analysis. At least three types of relationships can be conceived of, and it is likely that even more could be found. All of us, during our lives, have love relationships, friend relationships, and acquaintance relationships, each with a set of rules governing our actions and goals and each associated with different feeling states. Even within these categories, however, relationships vary and are quite complex.

Consider *love relationships* that take place both within the family as well as outside it. Love relationships may be of two kinds: those that are attachment relationships (that is, those which provide a secure base) and those that are not, for it is not clear that a secure base is a necessary part of all love relationships. For example, parents love their children, but children do not offer a secure base for parents. Love relationships may also be divided along another dimension as well; for instance, in some love relationships sexuality will play an important role, such as with a spouse, while in others, such as with mothers, fathers, and children, sexuality is absent. Thus within love relationships the dimensions of secure base (attachment) and sexuality form a complex structure for a variety of different "love relationships," all which exist in our adult experience. The four cells created by an attachment and sexuality matrix might contain mates as an example of attached and sexual; parents as an example of attached and nonsexual; boyfriend or girlfriend as nonattached and sexual; and parent to child as nonattached and nonsexual. Whether such a complex set of love relationships exists for children is unknown, and although a role for sexuality during childhood has been suggested (Freud, 1959), it is difficult to study. However, children can and probably do have strong love relationships that both have and do not have an attachment dimension. For example, children love both their parents and their younger siblings, but they are probably only attached to the former and not the latter.

Friendship relationships are different from love relationships, and although this difference may be difficult to describe, it is indicated by the language itself. Friendship relationships also vary along different dimensions and at times may merge with love relationships. Like love, friendship may or may not involve sexual behavior. Friendship relationships tend to vary with the age of the participants; thus they may involve same-aged peers or they may exist between older and younger persons, such as between a teacher and student. Like love, friendship relationships can be enduring and can exist even without extended interactions.

Acquaintance relationships are those relationships that tend to be the least enduring and the most specific to the individual interactions that bring them into existence. They usually occur as a consequence of particular social exchanges, such as with a postal worker or bank clerk. These relationships vary along a dimension of familiarity, from those in which the members recognize one another, know each other's names, and exchange information (such as between employers and employees or between a shop owner and a customer) to those less familiar interactions with the ticket collector on a train or with people whom we greet casually in passing on the street.

Although people do not have relationships with strangers, our analysis requires that this category of nonrelationships be included, especially since so much attention has been paid to children's social interactions with strangers. Strangers are by definition those people with whom we have no relationship and with whom we are unfamiliar. Yet even in this category of nonrelationships there are variations which may be of some importance to our analysis. For example, strangers who possess particular characteristics may elicit different interactions than strangers without those characteristics. Thus strangers of the same sex or racial background as the child are likely to evoke different interactions than strangers of the opposite sex or of another racial background (Lewis, 1980a).

In the study of the child's social development, the full array of possible relationships has not been explored or even considered. Attachment relationships, especially to the mother, have been investigated, and some studies of children's friendship patterns have been undertaken (Lewis & Rosenblum, 1975). Unfortunately, very few of these studies have looked at the growth of friendship. Studies of children's reactions to strangers have received attention (Lewis & Rosenblum, 1975); however, these studies usually have been designed to elicit fear rather than to study children's social behavior with strangers. By removing the mother, leaving the child alone, and requiring strangers to act in artificial ways, such as walking slowly and directly toward the child without talking or altering facial expression, most studies of children's behavior to strangers have been distorted.

For any complete study of social development, it is necessary to recognize that children do have relationships with people other than their parents and to trace the development of these relationships from the child's early social interactions. A complex array of relationships probably exists early in the life of the child, but without delineating the full range of these relationships, it is not possible to explore their development. There is significant evidence that young children are able to establish all three types of major relationships: love (both attachment and nonattachment), friendship, and acquaintance. The developmental sequence and the interconnection between different relationships is an issue to be addressed next.

INTERCONNECTION BETWEEN RELATIONSHIPS

The interconnection between relationships can be considered from both a static and a developmental point of view. In the former case, the study of the interconnection centers around the study of different social systems. For example, one might be interested in the number of friends a child has and try to relate the number of friends to such family dimensions as birth order or family size. In this regard, a recent study by Lewis may be of some interest. The question studied was whether the nature of the attachment relationship to the mother is related to the child's birth order; 174 children at 1 year of age were classified according to Ainsworth's attachment categories (Ainsworth et al., 1978), and it was found that 30 infants, or 17% were avoidant (type A), 125, or 72%, were secure (type B), and 19, or 11%, were resistant (type C), a distribution reported by others. However, when birth order and gender were considered, it was found that (1) firstborns were more securely attached than children in other birth positions and (2) second-borns were least securely attached, especially second-born males (only 52% were type B's). That attachment relationships are affected by the sex of the infant and by birth order suggests that the nature of one type of relationship may be dependent on other relationships, a point that will be stressed repeatedly in the analysis of social relationships within social systems.

From a developmental perspective, the study of interconnections is more articulated. The developmental perspective is primarily concerned with the effects of one set of relationships upon another set over time. Two views of the interconnection between relationships have been discussed—the epigenetic and social network systems models—and it is toward these that we now turn.

EPIGENETIC MODEL

The epigenetic model is the most accepted and utilized theory of the development of social relationships. As its central thesis is the assumption that one set of social experiences is directly connected to the next. More specifically, this model argues for a linear relationship such that the infant first adapts to one relationship and from this primary relationship all subsequent ones follow. This model is characterized by three features: (1) its fixed sequence, (2) its determinism, and (3) its trait or structural quality.

Fixed Sequence

The epigenetic theory argues for a linear progression in which the infant first adapts to one person within the family network, usually the mother, and from this basic adaptation all subsequent social relationships follow. While Bowlby (1969) originally believed in the unique biological relation between mother and infant as the origin of subsequent social adaptation, Ainsworth et al. (1978) subscribe to a more general formulation that specifies that attachment is a natural consequence of the infant's early interactions with others and that multiple attachments are possible. Thus the model of adaptation which stressed that the child first developed a relationship with the mother, then with the father and siblings, and then with others (for example, peers) is somewhat modified. Although a sequence still exists, it may not be linear; that is, the child's initial attachment relationship can be viewed as plural, with there being a number of others to whom the infant may form an attachment relationship. Even so, a fixed sequence argument still holds in that all subsequent relationships, such as those with peers, will follow and be a consequence of the earlier attachment relationship(s).

Unfortunately, the number and nature of the other people to whom the infant may be attached have not been studied. Except for the child's mother and father, few other attachment objects have been considered. This lack of consideration must be the function of a Western cultural bias since significant others within an extended family could include older siblings, grandparents, uncles, aunts, and cousins. Cross-cultural traditions concerning uncles, aunts, and grandparents should alert us to the fact that, at least for some cultures, these people might also be included in a list of possible attachment objects (Whiting & Whiting, 1975). In some sense, then, the number and nature of attachment figures is dependent on the structure and values of any particular culture. While the infant cannot survive without at least one adult figure caring for it, the nature and number of others involved in the child's life seem to be a function of the values of the larger social network. This issue needs to be examined since it obviously affects the sequential nature of social relationships. It may be a somewhat different model with potentially different consequences to have maternal attachment from which all other relationships follow rather than a multiple-attachment system from which other relationships follow.

Determinism

Not only do the theories of Freud and Bowlby postulate a fixed sequence in the development of relationships, but they also assume that later social experiences are determined by early ones, particularly the relationship between mother and child. This view is still widely held by others who see the mother–infant attachment determining the child's later peer relationships (Arend, Gove, & Sroufe, 1979; Matas et al., 1978). While much has been made of the deterministic nature of early relationships on subsequent social relationships, the data are rather sparse. While Bowlby (1951) points out that without a proper mother–infant relationship the child is "at risk" for a wide set of developmental dysfunctions, including death, the longitudinal nature of the proof necessary to back up this claim has made the demonstration of this aspect difficult.

Harlow's (1969) demonstration that motherless monkeys are at risk (1) for dysfunctional peer relationships—that is, they have difficulty mating—and (2) for mothering—that is, they make poor mothers who maltreat and even kill their young—is impressive evidence for the deterministic nature of early relationships. However, such a deterministic view has been challenged by a variety of other data, most of which pertains to the mother–child and child–child relationships.

As has been pointed out repeatedly, the effects on motherless monkeys, as reported by Harlow and Harlow (1965), were due not to the lack of a mother–child relationship but rather to the *complete* social isolation in which the monkeys were raised. Baby monkeys raised with other babies showed few of the effects originally reported (Harlow, 1969). This finding not only addresses the deterministic property of relationships but also demonstrates the multiple-path, as opposed to the fixed-path, aspects of social development. Lewis, Young, Brooks, and Michalson (1975) in discussing peer development point out that it is possible that peer relationships, rather than being determined in a linear fashion, evolve in parallel to other relationships. This view of a multiple-effect system is also offered by Harlow and Harlow (1965) and finds support in a variety of other studies. Thus rather than viewing peer relationships as determined by the mother–child or father–child attachment relationship, we can regard them as developing in tandem with the parent–child relationship. For example, a study by Cowen, Pedersen, Balijian, Izzo, and Trost (1973) suggests that there is a greater risk for adult dysfunctional social behavior in individuals with poor *peer* interactions in childhood. Harlow (1969) also looked at infant monkeys raised by mothers but without peer contact and another group raised with peers but without contact with the mother. Although there are some problems with the design (a mother and infant alone may have disturbed the mother's normal behavior), the general findings indicate that peer attachment in the absence of maternal attachment does not lead to subsequent peer dysfunction, while the absence of peer attachment results in serious disturbances. Being reared by a mother alone may have negative consequences for human infants as well; peer experience may be important for peer relationships! These results taken in toto indicate that there is no reason to suppose that maternal attachment per se will determine successful peer relationships in the absence of peer experience. Moreover, the data seem to indicate that peer experiences (peers and therapists) can overcome poor parental relationships and can influence future peer behavior. For example, Suomi and Harlow (1978) and Novak and Harlow (1975) have shown that the effects of being reared in social isolation (which results in poor peer relationships) can be counteracted through the use of peer therapists, especially if they are younger peers.

Further evidence of the importance of peers can be found in the Furman, Rake, and Hartup study (1979) which located 24 socially withdrawn preschool children. One-third of the children were assigned to a socialization experience with another child who was 15 months younger, another one-third were assigned a child within 3 months of their age, while the remaining one-third received no treatment. Play sessions outside the regular school classroom constituted the intervention. The follow-up to this intervention revealed that social activity of the "younger" therapists' group increased the most, followed by the "same age" therapists' group, with the "no treatment" group showing the least social activity. The authors report that the "younger" therapists' group appeared most similar to the general school population. While it is not known for certain, it can be assumed that the maternal attachment in these isolates was inadequate; nevertheless, successful peer experiences produced subsequent adequate peer interactions. This study and ones like it argue against a strict deterministic interconnection between relationships of the kind depicted by the epigenetic model. Hartup (1979) has concluded that "peer interaction is central in childhood socialization and the growth of social competence, and the suspicion grows that such competencies are direct derivations of early experiences in the peer culture" (p. 163).

A more direct approach to the issue of determinism would be to look at groups of children with inadequate attachment relationships to their mothers, who at the same time have adequate peer relations. Such a study could offer direct support against or for the deterministic nature of one relationship on another. Lewis and Schaeffer (1981) studied a group of abused and neglected infants and were concerned with the question of whether abused and neglected children—children who have insecure attachment relations with their mothers—would have good peer relations if they were placed in a day-care setting for 8 hours a day and given peer experience. In this study, two groups of infants aged 6 to 33 months were observed. Both groups were poor inner-city children and were similar in most respects, except that one group had been placed in a day-care center by the state's youth and family service agency because they were maltreated. After 4 months in the day-care setting, gross observation could not distinguish the abused children from the nonabused group. Their behavior toward peers, their free-play behavior, and their social interactions with caregivers indicated no obvious and discernible differences between these two groups of children. In order to obtain a more accurate measure of the infants' behavior, twelve 5-minute observations of each child were obtained by an observer who did not know what group each child belonged to. These observations were randomly scheduled over 3 weeks, and the only requirement was that the child could not be asleep or feeding. Thus 1 hour of interactive playtime behavior was observed for each child, which indicated almost no difference between these groups. The results of this study, along with those of Furman, Rake, and Hartup (1979), confirm the belief that poor infant–maternal relationships do not have poor peer relationships as a necessary consequence.

Although evidence of the deterministic nature of one relationship on another exists (Arend et al., 1979; Matas et al., 1978; Waters et al., 1979), the majority of the data leaves this question open. This is not to imply that one relationship cannot exert an effect on another. Indeed, in our social systems model just such an effect is envisioned. The difference between the positions concerns the deterministic nature of the effect. This will be explored under the section on trait, or character, effect.

Traits

The issue and controversy surrounding the nature of a trait, or an enduring aspect of personality, have dominated much contemporary thinking (Pervin, 1975). The issue is raised again when considering the effect of one relationship on another, and accounts for the major difference between the epigenetic and social network systems models. The clearest example of the difference between the epigenetic and social network system can be observed by considering the multiple attachment issue. That any individual child can have multiple attachments and that some of them may be secure while others are not, suggests that attachment refers to relationships not to the quality of a particular baby (A versus B babies, for example). Thus by considering multiple attachments we move from a consideration of a trait notion (a quality located in the child) to a systems notion in which the various relationship *vis à vis* a system influence the child. Simple attachment relationships give rise to exogenous qualities within the child thus an epigenetic approach, whereas multiple relationships give rise to endogenous forces such as the nature of the social network.

In the epigenetic view, the mother–infant, or earliest attachment, relationship endows the infant with a trait-, or characteristic-like structure, which is located within the organism. This trait or its absence then determines subsequent relationships. The nature of the trait has not been clarified. Sroufe (1979) and Block and Block (1979) have associated it with ego skills; however, it could be a trait such as self-esteem, self-efficacy, or some combination of the two. Whatever its nature, it is the presence or absence of the trait-like structure that influences other relationships. An often-used metaphor is that the child is like an empty vessel which needs filling. Once filled, the child can move on to new relationships. If the child is not filled, movement will be inhibited or the new relationships will differ in their nature or degree from those of the filled child. The task of the earliest attachment relationship is to fill the vessel or as sometimes thought to give the child those skills necessary to develop normally. Thus the explanation for the establishment of relationships rests with a mechanism that resides in the organism.

Also related to this trait issue is the notion of a critical period. Most theories, although it is seldom made explicit, assume that the attachment relationship in the opening months of life is critical and if this relationship is inadequate, then subsequent relationships, no matter how adequate, are not sufficient to alter the destructive effect. Thus there is a critical period with a beginning and an end after which experience has little effect. From a psychoanalytic point of view, these earlier relationships cannot be altered without therapeutic intervention. More recently, this critical-period concept has been adopted by Klaus and Kennell (1976) who argue for a very early critical period of bonding between mother and infant, which plays an important role in subsequent attachment relationships.

The notion of a trait provides a mechanism for the deterministic nature of the epigenetic model. One relationship can affect another through the creation of a trait in the child. The child then brings this trait to bear in its next relationship. Moreover, this trait or its absence, based on the outcome of the first relationship, is not easily affected by experience. The question remains, then, How do the data hold up against such trait notions? Certainly from an experiential point of view, early relationships appear to exhibit a strong influence on later ones, and a personality trait appears to be a reasonable mechanism for linking these experiences.

There are data that do not fit this model. Interestingly, the motherless-monkey literature again provides important information. Recall that the poor mothering behavior of motherless monkeys is used as an example of the consequences of the lack of some trait associated with having a good mothering experience. In other words, these monkeys, because of the lack of mothering, become poor mothers. What is not often attended to is the fact that after their first children, these motherless monkeys appear to be quite normal in their mothering of their second children (Rupenthal et al., 1976; Suomi, 1978). The fact is that the more time a mother spends with the first child, even though she maltreats it, the better mother she appears to be with the second child.

A similar analysis can be applied to the data on peer behavior of maltreated or abused human infants (Lewis & Schaeffer, 1981). If maltreated or abused infants have poor attachment relationships with their mothers and if this leads to a deficiency in a trait, then one would expect this missing trait to result in poor peer relationships. That this effect is eliminated by giving the infants peer experience again suggests that the concept of a trait as the deterministic factor may be in error. Lewis and Schaeffer (1981) have suggested two other ways in which maternal attachment may be related to peer relationships: (1) peer availability and (2) generalized fear.

Mothers who are inadequate in terms of "mothering" and who are unable to facilitate a secure attachment relationship may also prevent the infant from having peer experiences. Given that it is the mother who in our culture must facilitate early peer contact, her failure to do so will result in the absence of contact and, as a consequence, the development of inadequate peer-interaction skills. In this case, poor attachment and poor peer relationships are related, although the failure of the attachment relationship to the mother is not the cause of poor peer relationships, but rather, the cause is a lack of experience with peers.

Poor "mothering" resulting in a poor attachment relationship also may produce general fearfulness which has the effect of inhibiting contact with peers. It is this lack of contact, caused by fear, which affects peer relationships. This may explain why younger rather than same-aged or older peer therapists are better able to help depressed children; the younger peers may be less frightening to the "patient" since their skills are less developed. Notice, in both cases, it is not the absence of a trait caused by a poor interpersonal relationship with the mother that leads to poor subsequent relationships, but rather, the effect is more indirect and has to do with the interconnectedness of the social system.

In a study by Leiderman (1977) a secure attachment relationship and previous peer relationships led to good subsequent peer relationships. The data, however, left open the possibility that either kind of early relationship might have led to similar results. Unless we can separate the effects of early peer contact from secure maternal attachment relationships, one cannot logically exclude either possibility. This is also true for those studies which show that secure maternal attachments lead to subsequent peer adjustment (Matas et al., 1978; Waters et al., 1979). As we have suggested, in the social network system there may be a relationship between mothers who provide secure attachment and mothers who provide early peer relationships. Relationships are interdependent. However, this interdependence is a function of the social system, and in order to understand the interconnectedness, we must look at this system.

SOCIAL NETWORK SYSTEMS MODEL

The epigenetic model relies on features and mechanisms which are developmental in nature. As an example, the trait characteristic provides for a change in the internal structure of the child, a change brought about by a particular set of experiences. The social network systems model is less a developmental and more a social model. The causes of social behavior are found in the structure of the social system. The forces influencing social relationships and behaviors exist in the structure of the system, and when the system is altered, behavior and relations change.

Perhaps an example of the interplay between an individual's or dyad's behavior and a social network will underscore the importance of these issues. Rosenblum and Kaufman (1968) studied both bonnet and pigtail macaque monkeys. Bonnet macaque cluster together in matriarchal groups, so that a mother, her sisters, and her adult daughters and their babies might all be found huddled in close proximity.

The pigtail macaque, on the other hand, are much more isolate: there are no groups, just the adult female and her baby. Thus in the former case the baby is in close proximity to a large number and varied set of others besides its mother, while in the latter case the baby and mother are alone. The bonnet baby interacts daily and forms relationships with mother, aunts, grandmother, and cousins, while the pigtail baby only interacts and forms relationships with its mother.

When separated from its mother, the pigtail baby first shows marked distress, calling and moving about, and this is followed by a state of deep depression. The bonnet baby shows a markedly different pattern. It, too, is distressed by the withdrawal of the mother; however, it soon recovers as it is "adopted" and cared for by familiar others. This demonstration by Rosenblum and Kaufman (1968) illuminates the effects of the social network. In a system where there is only the mother, her loss constitutes an enormously significant event, one from which the infant is not likely to recover. In Bowlby's (1951) terms, the loss of the mother is a life-threatening event. In a system where there are other people, the loss of the mother, while significant, no longer becomes a life-threatening event and, in fact, given the proper care from others, appears to cause little subsequent disruption. Similar findings are reported by Robertson and Robertson (1971), Rutter (1979), and Tizard and Tizard (1971) with human infants. These findings require us to consider a social systems rather than an epigenetic analysis. Toward such a task we now turn our attention.

Systems in general and social systems in particular can be characterized by a number of features. In the discussion of the social network system, five features will be considered: (1) elements; (2) elements are related; (3) elements are nonadditive; the sum of the elements does not equal the total system; (4) elements operate under a steady state principle such that the elements have the ability to change and yet maintain the system; (5) systems are goal-oriented. Feiring and Lewis (1978) and Lewis, Feiring and Weinraub (1981) have outlined specific features of this system. The family social systems at any level—either of the cultural or of a family—can be characterized by these features.

Elements

Systems are composed of sets of elements. Within the family an element represents each individual member—mother, father, child—or it can represent dyads—mother–father, mother–child, father–child—or even triads when there is more than one child in the family. When considering units larger than the family, families themselves can become the elements of the system. There is no constraint on the number of elements or even the stability of the elements themselves. Thus within the family, elements can be both an individual as well as a dyad. In a family of 4 (2 children) there are a total of 4 simple elements, 6 dyads, and 4 triads, or a total of 14 possible elements. The potential array of different social elements that infants experience and that can influence them and are influenced by them is large. Unfortunately, there are almost no data on the number of different social contacts and their frequency and intensity as a function of the child's age. In one study of infant friendships, Lewis et al. (1975) found that infants had little opportunity to engage in peer contact. Only 20% of the mothers (mostly upper-middle- to middle-class Northeasterners) surveyed provided their 1-year-old infants with consistent peer exposure. Peer contact, however, can be quite variable; children in infant day-care programs have a great deal of peer contact. Peer contact may vary as a function of social class as well as the age of the child. At least for black ghetto children in the inner cities, peer contact is made early and is a chief social contact of even young children (Brown, 1971). Unfortunately, there are no large sets of data concerning the peer contact of infants.

This scarcity of data also applies to other potentially important figures. For example, there are no data as to the number of grandparent, aunt, or uncle experiences nor any real count of sibling contacts. Information regarding these relationships is also missing. Clearly, the number of others the child interacts with needs study, especially in light of changing child-rearing practices. Lewis and Feiring have collected data on the social networks of over 117 three-year-olds. These children live in the Princeton-Trenton, New Jersey, area and are all from mother–father intact homes. Demographically, they represent SES levels I–IV (Hollingshead, 1957) and are equally divided into first, second, third, and fourth birth orders. The children in our sample have approximately 5.8 friends (with a range from 2 to 12). In addition to mother, father, and siblings, the children come in contact with an average of 9.6 relatives (ranging from 4 to 26 relatives) including grandparents, aunts, uncles, and cousins. Mothers report that the children in our sample come in contact with almost all the people the mother would include as important in her social network, including her friends. The number of adults other than family that the child makes contact with is 7.1. On the average, 3.2, 4.4, and 4.4 relatives, friends, and adults, other than parents, are seen at least once a week. More work is necessary for us to determine the nature of the social network and the relationships that exist within this network. Moreover, it must be recognized that for some infants the only significant relationship may be between the mother and infant, while for others it may include fathers, caregivers (as in day care), siblings, peers, and relatives such as grandparents. Having considered multiple significant relationships, it is now possible to ask which ones exist, how they differ, and what their potential consequences.

In any discussion of the influence and significance of multiple social objects, it must be kept in mind that this cannot be determined apart from the socialization experience itself. Thus, for example, to study infants' peer behavior with infants who have had no previous peer experience fails to answer the question of the value or need of peer experience and its potential consequences. Only through studying varying networks can the relationship between the various social objects be understood.

Mothers. The role of mothers in young children's lives need not be discussed in detail. Mothers are the primary caregivers. Even when they do not care for the child directly, they assume the role of executor vis-à-vis the child. It is mothers who buy clothes for their children, and it is mothers who take their children to the pediatrician's office. When mothers put their children in infant day care or any alternative care program, it is almost always *other* mothers (women) who care for their children. Most change in the structure of the care of the very young in this society has been the substitution of someone else's mother for the biological mother.

Fathers. In the last 5 years, a wide number of studies and books have shown that fathers play a more important role in the young child's life than originally thought (Lamb, 1976; Pedersen, 1980, for example). Fathers can no longer be considered the forgotten parent. Rather than studying the amount of time the father spends with the infant, researchers have looked at the nature of the father–infant interaction. It is clear that fathers do not spend as much time with their infants as mothers. Even when home, the degree of fathers' interactions with their infants is relatively low (Ban & Lewis, 1974; Clarke-Stewart, 1978; Lewis & Weinraub, 1976; Parke & Sawin, 1980; Patterson, in press; Rebelsky & Hanks, 1971). Nonetheless, fathers can care for their young, and when they do, they do it as well as mothers (Parke & O'Leary, 1975).

Unfortunately, there are relatively little data on the amount of time and the nature of the interactions between fathers and infants in their everyday world or even in the Ainsworth et al. attachment situation. Although receiving some attention, exceptional families in which the father is the principal caregiver need to be studied in order to determine the effects of fathers as caregivers to the very young. Considerable data now exist that father–infant interactions are different than mother–infant interactions, the major difference being that fathers tend to play more and to play differently with their infants than mothers who tend to engage in more caregiver and punishing interactions (Lamb, 1976; Parke & Sawin, 1980; Lewis & Weinraub, 1976; Patterson, in press, Pedersen et al., 1979). That the interactions are different between infants and their mothers and fathers may have important consequences on their relationships. Unfortunately, too little is known to conclude more than that infants can be attached to fathers as well as mothers (Lamb, 1980).

Siblings. The limited study of sibling interactions and relationships is surprising in light of the fact that siblings live in the same house, often sleep in the same room, share the same parents and experiences, and live together for many years. Only recently have there been several studies of siblings. Siblings' interactions can be studied around communication, social behavior, and teaching. Dunn and Kendrick (1979; 1980) studied each of these types of interactions and found siblings, even young siblings (14 months of age and younger) capable of turn taking as well as empathic and teaching behaviors. Zajonc and Markus (1975) have shown the effect of siblings on intelligence test scores, and Dunn and Kendrick (1979) and Feiring and Lewis (1980b) have explored the effect of the birth of the second child on the first. Nonetheless, there are relatively little data on the subject of relationships between siblings except for anecdotal accounts of the socialization of one sibling by the other, the accounts of identical twins (Zazzo, 1948), and the effects of siblings left to themselves (Luria & Yudovich, 1971). In these accounts, it is clear that the siblings form intense and significant relationships. That infants can be upset by the loss of a sibling has been documented and suggests that in addition to the father and mother they should be included in the multiple-attachment category, although whether or not siblings form attachment relationships with each other may be a function of the amount of sibling contact allowed and the age spacing between them. In one laboratory study, Lamb (1978) has shown that older siblings (20 to 40 months older) show little attachment; however, this was not found for the younger sibling who was "inordinately" interested and influenced by the older sibling. Thus the data that do exist suggest infants form attachment relationships with siblings, especially in the case of younger infants. Moreover, infants close in age are likely to be treated as a unit (Lewis & Kreitzberg, 1979) and, therefore, are likely to have more interactions than siblings many years apart. Moreover, depending on the role assigned to the older sibling, an older sibling may be one of the primary caregivers of the child and, as such, may be the focus of an attachment (Whiting & Whiting, 1975). Like other attachment relationships, the degree to which siblings are attached may be a function of the values of the family as well as such factors as age spacing, sex of siblings, and number of sibling, to mention but a few. Certainly some siblings show a strong attachment toward one another that lasts a lifetime.

Other Relatives. Besides parents and siblings, other important relationships and interactions take place with family members who do not necessarily live in the same household. These relatives include grandparents as well as uncles, aunts, and cousins of the infant, and they have varying degrees of contact with the infant. They need to be considered as possible elements in the child's social system given their important connections to the parents of the child. Unfortunately, almost nothing is known about the interactions and relationships with these people.

There is information to allow us to speculate that for some cultures these relatives play an important role in the child's life. For example, some evidence exists that children are raised by their maternal grandmothers; this is especially true for children of teenage, inner-city, poverty mothers. The role of grandfathers or uncles and aunts, not to mention cousins, remains unknown. We do know that for some cultures uncles, especially the father's brother, have specific roles vis-à-vis the child. Thus at the death of the father, the protection and care of the infant become the uncle's responsibility (Frazer, 1915). Anecdotal reports indicate that these other relatives play important roles and have significant relationships in the lives of children. However, without more information, at least on the rate of contact, little can be said about them.

Peers. The study of peer relationships has an unusual history. Considerable work was undertaken in the 1930s, but almost no studies were conducted again until the 1970s (Lewis & Rosenblum, 1975; 1977). Successful peer interactions and significant relationships, even attachment relationships, appear to be a function of the social structure and cultural values. Work by Lewis et al. (1975), Mueller and Vandell (1979), and others indicate that infants are capable of empathic behavior and role taking as well as turn taking and teaching behaviors. Thus infants seem to have those skills necessary to form and maintain friendship relationships with peers. Lewis et al. (1975), for example, showed that 1-year-old infants can form at least some type of friendship as reflected by their different behavior toward familiar as opposed to unfamiliar peers. That older children are able to form significant relationships with peers is clear (Hartup, 1970), but depending on the social structure, these significant relationships may occur even earlier. Given the opportunity, very young children may form peer attachments; however, this phenomenon is very much dependent on the situation. For example, Freud and Dann (1951) and Gyomroi-Ludowyk (1963) have shown that peer attachments are possible and likely when infants are reared together, especially in the absence of significant other adult contact. Brown (1971) in his book on growing up in a ghetto also underscores the impact of the peer culture on 2- to 3-year-olds! Peer interactions and relationships have generally been underrated by the adult culture even though the more recent research indicates the constraints limiting contact between young children.

In short, the influences of the social system elements in the young infant's life are not necessarily limited to the mother. The research literature of the past 5 to 10 years indicates that the infant is embedded in a complex system of elements including mother, father, siblings, and other relatives especially grandparents, uncles and aunts, and peers. Moreover, the infant is capable of interacting with these others and in forming relationships with them. Some of these relationships may be attachments, some love without attachment, and some friendships. The constraints on these relationships probably rest on cultural factors. Thus the first requirement of a social network systems analysis is met: there are a multitude of possible elements.

Interconnection of Elements

Not only are systems characterized by sets of elements but by a set of interrelated elements; that is, by elements that are influenced by each other (Monane, 1967). Within the family the interaction of elements can be at several levels. At the simplest level the infant can affect its parents (Lewis & Rosenblum, 1974), the parents affect the infant, and the parents affect each other. Such effects come about through the direct interactions between family members. Recall that elements are not only individuals but may be dyads or even larger units. When larger elements are considered, the study of the interrelation of elements becomes more complex. For example, a child can affect not only each parent separately but also the parental interaction. The research on family size (number of children) and age of the child shows that children affect marital satisfaction (Rollins & Galligan, 1978). Likewise, the father can influence the mother and child individually as well as the mother–child interaction. Many different effects of this complex nature have been observed. Pedersen (1975) and Feiring (1975) observed the parental relationship as it affects the mother–child interactions, and several investigators have studied the effect of the father on the mother–child relationship (Clarke-Stewart, 1978; Lamb, 1978; Pedersen et al., 1977; Parke, 1978; Lewis & Weinraub, 1976). Circirelli (1975) looked at the influence of siblings on the mother–child relationships, and Dunn and Kendrick (1979) studied the effect of a newborn child (sibling) on the older-child–mother relationship (see Lewis & Rosenblum, 1977; Pedersen et al., 1979, for a review of such studies).

Lewis (Lewis & Weinraub, 1976; Lewis & Feiring, in press), Bronfenbrenner (1977), Lamb (1978), and Parke et al. (1979) have discussed some of the various ways that elements within a system may influence each other. Direct and indirect effects have been identified as two major classes (Lewis & Weinraub, 1976; Lewis & Feiring, 1981). Direct effects are those interactions which represent the influence of one person on the behavior of another when both are engaged in mutual interaction. Direct effects are usually observed in dyadic interactions but could involve more than two people as when a teacher instructs a class of students. Direct effects involve information gathered from participation in an interaction with another person or object and always involve the target person as one of the focused participants in the interaction. Direct effects have been studied for each member of the infant's family: mother–infant (Lewis, 1972; Stern, 1974; Brazelton, Koslowski, & Mains, 1974), father–infant (Brazelton, Yogman, Als, & Tronick, 1979), and sibling–infant (Greenbaum & Landau, 1977).

Indirect effects refer to two classes of interactions or influence. In the first, indirect effects are those sets of interactions which affect the target person but which occur in the absence of that person. While these sets of interactions affect the target person, they may be best described as influences which play their role indirectly in development by affecting direct effects or interactions. For example, the father by supporting the mother (both emotionally and physically) affects the mother–infant relationship. Several studies have demonstrated this effect (Barry, 1970; Heath, 1976; Feiring & Taylor, 1977).

Another category of indirect effects are those that occur in the presence of the target person even though the interaction is not directed toward or does not involve that person. These effects are based on information that is gathered from sources other than the direct interaction with another person and may be the result of (1) observation of another's interaction with persons or objects or (2) information gathered from another about the attitudes, behaviors, traits, or actions of a third person. Such indirect influences have been studied under a wide rubric including identification, modeling, imitation, and incidental, vicarious, and observational learning (see Lewis & Feiring, 1981). In the case of young infants and their interactions with others, several recent studies by Clarke-Stewart (1978), Feinman and Lewis (1981), and Feiring, Lewis, and Starr (1981) point out how the mother's interaction with others can affect the child's behavior toward others, particularly strangers. Thus, for example, if the mother acts positively toward a stranger, the child's behavior toward that stranger will be more positive than if the mother shows no interest at all (Feiring et al., 1981).

In short, the infant must adapt to a social system containing elements which vary in number and complexity and which influence the infant both directly through their interactions with the child and indirectly through their interactions with each other. The child establishes relationships within a network of already existing relationships.

Nonadditivity

Social systems also possess the quality of nonadditivity; that is, to know everything about the elements that comprise a system will not reveal everything about the operation of the whole. Any set of elements behaves quite differently within the system from the way they do in isolation. This rule holds for simple elements as well as for more complex ones. Within the family how an individual person behaves alone can be quite different from how that person behaves in the presence of another. Because observation of a person alone is nearly impossible (and unethical), the study of individual behavior alone is difficult. More common is the observation of a dyad alone or in the presence of a third member. For example, Clarke-Stewart (1978) observed mothers, fathers, and children in dyadic (parent–child) and triadic (mother–father–child) interactions and showed that the quantity and quality of behavior in the mother–child subsystem in isolation is changed when the mother–child subsystem is embedded in the mother–father–child subsystem. The mother initiated less talk, played with the child less, and was also less engaging, reinforcing, directive, and responsive in her child-directed interactions when the father was present than when she was alone with the child. Pedersen, Anderson, and Cain (1977) report that while the father-husband frequently divides his behavior between child and spouse in a three-person subsystem, the mother-wife spends much more time in dyadic interaction with the child than with her husband. The child also exerts influence on the parent system. Rosenblatt (1974) found that the presence of one or more children reduced adult–adult touching, talking, and smiling in selected public places, such as the zoo, park, and shopping centers. Lewis and Feiring (in press), looking at interactions at the dinner table, found that the number of children at the table affects the amount of mother–father verbal interaction and amount of positive affect exhibited between the two. As the number of children in the family increases, the amount of positive affect between parents decreases. A study by Parke and O'Leary (1975) illustrates both the qualities of interdependence and nonadditivity in the family system. When mothers were with their husbands, they were more likely to explore the child's body and smile at the child than when they were alone with the child. In addition, when mothers were with their husbands, they tended to touch their sons more than their daughters, whereas this difference was not evident when the mother and child were alone.

This rule of nonadditivity presents an obstacle for the study of interactions and relationships. There is some comfort in believing that the interactions one observes between two people remain the same or are invariant across situations or across changing social structures. The rule of nonadditivity suggests that this is not the case. The evidence that the relationships between elements are dependent on the set or subset of elements present suggests the relative nature of interactions. The effect on relationship is not known although it might well be the case that while interactions may vary, relationships do not. The relative nature of dyadic interactions makes the study of social behavior more difficult. Indeed, if the rule of nonadditivity is correct, the very act of observation (which introduces another element, the observer) must alter the behavior of those observed.

Steady State

Social systems are characterized by steady states. The term "steady state" describes the process whereby a system maintains itself while always changing to some degree. A steady state is characterized by the interplay of flexibility and stability by which a system endeavors to maintain a viable relationship among its elements and its environment. Social systems are defined as goal-oriented, and steady state processes are directed toward goal achievement. However, the same general goal may be served by different patterns of behavior as the system changes to adapt to its environment. Within the family such processes appear essential since one major change that occurs is the development of the child. Since children's skills, knowledge, and behavior continually change, these changes must be accounted for in maintaining stability. For example, Lewis and Ban (1971) have shown that the nature of the interactions between child and parent change over the first 2 years of the child's life. The amount of physical proximity, especially touching, decreases while the amount of distal contact, for example, looking and talking, increases. Thus the particular behaviors serving the relationship undergo change, yet the relationship probably remains the same.

The child's functioning in the family is described by adaptation (steady state) and by development of new behavior patterns (flexibility) in the service of prior goals (stability). For example, Dunn and Kendrick (1979) found that firstborn children's independent behavior increased upon the birth of a sibling. Although independent behavior occurred and was encouraged by the mother prior to the birth of the second child and was a developmental goal, the amount, kind, and opportunity for independent behavior changed as the family system changed to include a new member. This consistency in the midst of change has to be an important factor in development that bears on relationships and feelings. This principle of steady states allows for continuity within a changing structure, as well as within a changing developmental system, where new behaviors and new situations are observed (Lewis, 1980a,b).

Goals

Social systems are also characterized by their purposeful quality. The family system is generally thought to exist in order to perform certain functions or goal-oriented activities that are necessary both to the survival of its members and to the perpetuation of the specific culture and society. Family functions are often enumerated as procreation and child rearing, which suggests that the family is the principal agent of these societal goals. Beyond this level of generality are numerous other ways of describing and defining family functions (Lewis & Feiring, 1978; Parsons & Bales, 1955).

The epigenetic model of social development not only has restricted the nature of the family relationships available to the young child by focusing more or less exclusively on the mother, but it also has limited the types of activities or goals engaged in by the other family members. If only caregiving functions or goals are considered, then it makes some sense to study the mother as the most important (and only) element. However, other functions in the child's life, including, for example, play and teaching, may involve family members other than the mother.

The issue of functions, needs, and goals is quite complex. The study of the functions, needs, or goals of the young child is beyond the scope of this chapter. In order to create a list of functions or goals, one could utilize Wilson's (1975) adaptive functions or Murray's (1938) need functions. Lewis and Feiring (1979) have also defined some of the functions important for the infant, such as protection, caregiving, nurturance, play, exploration/learning, and social control. *Protection* includes protection from potential sources of danger, including both inanimate sources (e.g., falling off trees or being burned in fires) and animate (e.g., eaten by a predator, taken by a non-kin, or attacked by another). *Caregiving* includes feeding and cleaning at the least, and refers to a set of activities that center around biological needs relating to bodily activities. *Nurturance* is the function of love, or attachment, as specified by Bowlby (1969). *Play* refers to activities with no immediately obvious goal. These activities are engaged in for their own sake. *Exploration/learning* involves the social activity of finding out about the environment through either watching others, asking for information, or engaging in information acquisition with others. *Social control* represents the restriction of the infant's behavior because it interferes with the behavior of others or because it is useful in teaching specific rules. It differs from exploration/learning in that its major function is to restrict the infant's ongoing behavior.

More functions can be considered, but the discussion will be limited to these in order to proceed to the next topic of concern in social development: the connection between relationships and needs.

RELATIONSHIPS AND NEEDS

Once we accept the importance of multiple people in the infant's life, it becomes necessary to explore the connections between "what people do which things" and the possible outcomes given that certain people do certain things.

Any analysis of the family must consider the range of functions and the nature of the different members who satisfy these functions. Although different family members are generally characterized by particular social functions, it is often the case that persons and functions are only partially related (Lewis & Feiring, 1978, 1979; Lewis & Weinraub, 1976; Lewis, Feiring, & Weinraub, 1981). Consequently, the identity of the family member does not necessarily define the type or range of its social functions. Whereas caregiving (a function) and mother (a person) have been considered to be highly related, recent work indicates that fathers are equally adequate in performing this function. Play, another function, appears to be engaged in more by siblings and fathers than by mothers (Ban & Lewis, 1974; Clarke-Stewart, 1978; Lamb, 1976; Parke & Sawin, 1980).

As indicated previously, a family member may perform several functions within the context of the family system although different family members may be more associated with certain functions than others. Thus parents may do more caregiving and teaching, but children do more play. The relative amount of interaction time spent in different activities would certainly be expected to influence the child's development. For example, a child in a family where teaching was the most frequent function (regardless of to whom the teaching was being directed—child, sibling, or parent) would be expected to be different than a child in a family where caregiving was the most frequent activity. In addition, the different amount of various activities directed by the same individual to the child might influence the child's development. For example, a child whose mother engaged in more caregiving activity than play would be expected to be different from a child whose mother engaged in both equally.

In order to explore this problem a general systems analysis needs to be undertaken. In particular, it is necessary to explore the matrix constructed when multiple social objects and needs are related. Such a matrix has been presented in detail by Lewis and Feiring (1978; 1979) and discussed by Dunn and Kendrick (1979) and by Suomi (1979).

Social Matrix as a Measure of the Social Network: The Relation between People and Needs

Given the varied functions and the likelihood of multiple relationships in the life of the child, we need to develop a model that describes how these multiple people and functions operate. Moreover, having such a model enables us to explore how different matrixes produce different developmental outcomes. Such a model must focus on the relationship between people and functions.

		Social functions					
		F_1	F_2	F_3	F_4	F_5	F_n
		Protection $B_{11} B_{12} B_{13}$	Caregiving $B_{21} B_{22} B_{23}$	Nurturance $B_{31} B_{32} \ldots$	Play	Exploration/learning	B_n
			Feeding, changing	Rock, kiss			
P_1	Self						
P_2	Mother						
P_3	Father						
P_4	Peer						
P_5	Sibling						
P_6	Grandparent						
P_7	Aunt						
·	·						
·	·						
·	·						
P_n							

Figure 10.2. The relationship between social objects and functions.

In the model depicted in Figure 10.2, social objects form the y-axis, while social functions form the x-axis. Together they form a matrix of objects and functions (Lewis & Feiring, 1979). In this figure $P_1 \ldots P_n$ represents the set of social objects and includes self, mother, father, peer, sibling, grandparent, aunt, plus any other persons to be considered. The array of objects is determined by a variety of factors to be considered in turn. The functions labeled $F_1 \ldots F_n$ consist of protection, caregiving, nurturance, and so forth. The subset of behaviors $B_{11}, B_{12} \ldots, B_n$ represents the set of behaviors that can be measured and that are subsumed under the broad heading of a particular function. For example, under *caregiving*, feeding and changing could be included as different behaviors, while under *nurturance*, emotional behaviors such as kissing or responsivity might be considered. The general form of this model provides a framework for considering the complete array of social objects and functions that describe the child's social network. By examining the matrix, one can determine which social objects are present, what functions are being satisfied, and most important, which functions are achieved by which social object. By examining the horizontal axis of the matrix, one can obtain an idea of what functions are characteristic of a particular social object within this particular network. Thus this axis will inform us whether mothers are predominantly concerned with caregiving or fathers with play, a finding that is repeatedly reported. On the other hand, by examining the vertical axis of the matrix, one can study the functions characterizing the network of a particular child. Thus while all the listed functions are presumed to be important, it might be that for some children the caregiving function predominates, while for others caregiving, play, and learning are equally important. Likewise, a child whose network is predominantly concerned with protection will be quite different from one raised in a network concerned with nurturance. However, in order to determine the relationship of object to functions, we must turn our attention to the individual cells of the matrix.

The object–function relationship raises the general question of whether some object–function relationships are more important for the child's development than others. For example, what is the consequence of having an older sibling as caregiver as well as the mother instead of just the mother? Or, what is the consequence of having a caregiver as provider of care (as in day care) in addition to the mother instead of just the mother?

The answer to questions such as these requires that the long-term and higher-order goal structure of the family or society be considered. Thus the value assigned to the consequences of particular outcomes must ultimately play a role in any discussion of social development. For example, if successful peer social adjustment resulting in long-term stable and loving relationships is the goal, then one can examine which object–function best produces this outcome. On the other hand, if a socially isolated but artistically successful outcome is desired, then one asks, What produces this? In a word, any analysis of a consequence of an object–function matrix must depend on the specification of goals.

Several studies, some of which have already been discussed, have described particular object–function connections. As we have stated, these object–functions represent the Western cultural values as they now exist and do not necessarily reflect either an appropriateness across different cultures or an evaluation concerning which object–function represents the best outcome given a particular goal.

The little data that exist suggest that in the general American culture mothers mostly engage in caregiving, protective, and nurturance functions, while fathers, siblings, and peers engage in play. Teaching appears to be by parents and older siblings and peers. Similar findings across other cultures have been reported by Whiting and Whiting (1975), although important differences, especially in sibling and peer functions, have been found.

In an attempt to determine whether such an object–function matrix is applicable to young children, Edwards and Lewis (1979) explored how young children perceive the distribution of functions and persons. Persons were represented by dolls of infants, peers, older children, and parent-aged adults. Social functions were created for the child in the form of a story. In the two studies to be discussed, the children, ranging in age from 3 to 5 years, were asked to choose which person they wanted to interact with in a specific social activity. In all the studies, the functions were (1) being helped, (2) teaching about a toy, (3) sharing, and (4) play. For *help,* the adults were selected, while for someone to *play* with, children chose adults and infants last and peers and older children first. For *teaching* how to use a toy, older children were selected, while for *sharing* there were no preferences among social objects. In a second study, children were asked to point to photographs of infants, peers, older children, adults (parent-age), and grandparents in response to the social function questions. The results indicated that older persons were chosen for *help*. The *teaching* function supported the results of the first study in that older peers, rather than adults, were chosen. For *play,* peers and older children were preferred, and again there was no difference in the social object selected for *sharing* a replication of the earlier findings. Interestingly, the data on grandparent-age adults are most similar to that for parent-age adults.

How early the creation of an object–function connection is made is not known, however, a study by Lewis et al. (1975) indicates that by 1 year of age, infants play with peers more than with strange adult females or mothers and mothers are sought after for protection and nurturance more than strange females and peers. Such findings suggest that the discrimination of persons and functions begins very early in life.

The results of the Edwards and Lewis (1979) study also suggest that the persons and functions that describe a child's social system will be influenced by the situation in which the person performs the functions. One important aspect of a social system is the environment in which the system exists. Consequently, the family system, its members, and the functions performed by its members may be different depending on the situation or environment in which the family is interacting. The family system becomes an element in the larger system of the subculture or culture. Thus the quality of the elements can be considered to exist in order to perform certain functions or goal-oriented activities necessary to the survival of its members; families, as elements, perform similar activities for the perpetuation of the specific culture and society.

Some Variables Affecting the Social Matrix

Although individual differences in this matrix exist (e.g., each family may have idiosyncratic features), there are several constraints on the matrix that have to do with both ontogenetic and cultural differences. An examination of these constraints is useful in exploring social development.

Ontogeny. By examining the matrix at different points in a child's life, it should be possible to delineate developmental patterns. One might expect changes both in the composition of the social objects as well as in the nature of the functions. For example, in our society as children get older, they are likely to come into contact with more people, for example, peers and teachers. Likewise, new functions will emerge as caregiving and nurturance give way to teaching and control. Such changes can be observed in Figure 10.3 which shows a possible matrix for a 3-month-old infant and one for the same child at 6 years. Although the values are fictitious, they demonstrate the point. For the sake of simplicity, the same functions have been maintained, but their values change. When the infant is 3 months old, caregiving predominates over play and mother engages in the most caregiving activity. As already reported, fathers may play more with their children than mothers. By the time the child is 6 years old, new objects are added, caregiving gives way to play, and both father and mother play less with the child than do peers or siblings. Such a model is useful in charting the course of social development. Moreover, the use of this model can be used for prediction: Does a particular network at one age predict a later network? For example, what is the outcome for play when early caregiving is shared between several adults rather than one?

Cultural Differences. Cultural differences can also be characterized by differences in the matrix, and the usefulness of this model for this purpose should not be neglected. Figure 10.4 demonstrates differences between an Israeli 3-month-old child living in a kibbutz as compared to an American child of the same age. Again there are both object and function differences. Israeli children raised in a kibbutz when compared to American children have many more peers, as well as additional adults, with whom they interact. This results in a redistribution of the functions with the mother, father, and metapelet all involved in *caregiving* in one culture and only the mother and father in the other. Moreover, there is a radical change in both the distribution and amount of *play* when peers are introduced into the matrix. To be

206 The Social Network Systems Model

	F_1 Caregiving		F_2 Play		F_n
	B_{11}	B_{12} Feeding, dressing	B_{21}	B_{22} Hide & seek	B_n
P_1 Self		0		0	
P_2 Mother		30		5	
P_3 Father		10		10	
P_4 Grandmother		5		2	
•					
•					
•					
P_n					
		27		45	

a

	F_1 Caregiving		F_2 Play		F_n
	B_{11}	B_{12} Feeding, dressing	B_{21}	B_{22} Hide & seek	B_n
P_1 Self		10		25	
P_2 Mother		5		5	
P_3 Father		5		15	
P_4 Peer		1		25	
P_5 Teacher		5		15	
•					
•					
•					
P_n					

b

Figure 10.3. Ontogenetic changes in the social network matrix. The numbers shown in this figure do not represent real data and are for illustrative purposes only. Social network matrix: (*a*) target child at 3 months of age; (*b*) target child at 6 years of age.

noted in such cultural comparisons is the reduction in the caregiving role of the mother (and father) with the appearance of other adults; for example, the metapelet in the case of the kibbutz or other relatives in the case of cultures with extended families. These conclusions are supported by cross-cultural studies of Edwards and Whiting (1977) who found older siblings more involved in caregiving in nontechnological cultures. Konner (1975) also found similar results for ! Kung San infants as early as 1 to 13 weeks of age where nonmothers accounted for 20% of all physical contact recorded. Although there is little direct evidence, this phenomenon is probably less likely to be observed in the traditional American family where the child is rarely cared for or touched by other adults during this time period.

In a recent study of functions and objects within the family, Lewis and Feiring (in press) examined the functions of nurturance, caregiving, and information exchange. Although not all functions were observed, it was found that for the 3-year-old target-child, mothers engaged in more caregiving and nurturance than fathers,

	F_1 Caregiving		F_2 Play	F_n
	B_{11}	B_{12}		B_n
	Feeding, dressing			
P_1 Self	0			
P_2 Metapelet	25		10	
	5		5	
P_3 Father	2		2	
P_4 Peer	0		15	
P_6 Peer	0		20	
$\vdots$				
P_n				

a

	F_1 Caregiving		F_2 Play	F_n
	B_{11}	B_{12}		B_n
	Feeding, dressing			
P_1 Self	0		0	
P_2 Mother	25		10	
P_3 Father	15		10	
$\vdots$				
P_n				

b

Figure 10.4. The effect of culture on the social network matrix. The numbers shown in this figure do not represent real data and are for illustrative purposes only. (*a*) Social matrix of Israeli child at 3 months of age; (*b*) social matrix of American child at 3 months of age.

while both engaged in equal amounts of information exchange. The target-child exhibited equal amounts of nurturance to their mothers and fathers although they engaged in more information seeking with their mothers. Interestingly, when family size was considered, that is, when the number of children in the family increased, the object–function nature was altered. In particular, fathers increased their amount of caregiving for the target-child as the number of siblings increased; however, the total amount of nurturance for both mother and father decreased. Thus demographic variables, such as family size, sex of child, and possibly others, should affect this matrix. The fact that these variables can be shown to exert influence on the matrix provides some support for the validity of our formulation.

The construction of the social object–function matrix provides us with a model for understanding the child's social network. The variety of people and functions that comprise this network alters at least as a function of the child's age and culture. Until recently, the focus of most research has been restricted to examining particular cells of the matrix—such as the mother and caregiving functions. At the same time, certain functions and persons have been assumed to be synonymous. Moreover, by using the epigenetic model, researchers have assumed that the lack of a particular object–function leads to harmful consequences. If we consider the entire matrix, however, it may be possible to determine whether there is specificity in particular object–functions, such as whether mother and "mothering" are synonymous. In addition, it can be determined whether certain functions can be safely substituted by others; for example, whether a peer can be substituted for an adult in the functions of *nurturance* or *teaching*.

The social network systems model may be best for characterizing the child's social world as well as reflecting ontogenetic, cultural, and idiographic differences. In the child's life there exists a variety of significant social objects. Which social objects constitute the important relationships for the child depends on the structure and values of the particular culture. We view the social network as a system. The network is established by the culture for the transmission of cultural values. The composition of the network, the nature of the social objects, the functions they fulfill, and the relationships between objects and functions are the parts of the vehicle through which the cultural values are determined. In fact, this structure may be as important as the specific information conveyed; indeed, it may constitute the information itself. Finally, these comments should alert us to the dangers of assuming that the care of children represents some unchanging and absolute process. The care of children is the primary social activity of any society since it represents the single activity wherein the values of the society are preserved. Child care must reflect the values of the culture at large; therefore, one should expect it to change as these values change. The issue of social values must be addressed in order for us to deal with the ideal relationship (if any) between objects and functions and their interactions. The research task is to characterize various networks and to determine the outcome of various object–functions in terms of a set of goals and values rather than to argue for some ideal biological state.

SUMMARY

In the study of social development the development of social relationships must be a central topic. In this chapter several particular aspects of relationships were explored: the connection between interactions and relationships; various types of relationships; the interconnections between relationships which involve the two models, the epigenetic and social network system; and finally, the connection between relationships and needs. Left unconsidered were the elaborate cognitive underpinnings that not only mediate between interactions and relationships but also allow the infant to maintain and form relationships. Likewise for the adults, cognitive processes (including attribution) influence their behavior toward children.

In the section on the difference between interactions and relationships, we discussed the development of the self as one important aspect necessary for the ontogenesis of relationships. The nature and types of interactions that facilitate this development need to be explored, as well as the way in which the formation of relationships affects the developing concept of self.

While alternatives to the epigenetic model have been proposed, the test of the power of these alternatives is still to be done. Until sufficient data are gathered, it is impossible to reject either the epigenetic or the social network systems models. We believe that the social network systems model provides a broader framework into which the epigenetic model may fit. The interconnection between relationships and needs, at least as viewed from the social matrix perspective, depends upon the goals or values articulated by the society. In some sense, however, the model is goal free and provides only a structure by which we may view outcomes as they are related to object–functions. Thus in the final analysis a theory of social development rests in part on the ability to articulate and specify a set of goals.

REFERENCES

Arend, R., Gove, F., & Sroufe, L. Continuity of individual adaptation from infancy to kindergarten: A predictive study of ego-resiliency and curiosity in preschoolers. *Child Development,* 1979, **50**(4), 950–959.

Ainsworth, M. D. S. Object relations, dependency and attachment: A theoretical review of the infant-mother relationship. *Child Development,* 1969, **40**, 969–1026.

Ainsworth, M. D. S., Belhar, M. C., Waters, E., & Wall, S. *Patterns of attachment: A psychological study of the strange situation.* Hillsdale, N.J.: Erlbaum, 1978.

Ban, P., & Lewis, M. Mothers and fathers, girls and boys: Attachment behavior in the one-year-old. *Merrill-Palmer Quarterly,* 1974, **20**(3), 195–204.

Bandura, A. *Principles of behavior modification.* New York: Holt, Rinehart, and Winston, 1969.

Barry, W. A. Marriage research and conflict: An integrative review. *Psychological Bulletin,* 1970, **73**(1), 41–54.

Block, J., & Block, J. The role of ego control and ego-resiliency in the organization of behavior. In W. A. Collins (Ed.), *Minnesota symposia on child psychology* (Vol. 13). Hillsdale, N.J.: Erlbaum, 1979.

Bower, T. G. R. Infant perception of the third dimension and object concept development. In L. B. Cohen & P. Salapatek (Eds.), *Infant perception: From sensation to cognition* (Vol. II). New York: Academic Press, 1975.

Bowlby, J. *Maternal care and mental health.* Geneva: WHO, 1951.

Bowlby, J. *Attachment and loss: Attachment* (Vol. 1). New York: Basic Books, 1969.

Brazelton, T. B., Koslowski, B., & Main, M. The origins of reciprocity: The early mother–infant interaction. In M. Lewis & L. Rosenblum (Eds.), *The effect of the infant on its caregiver: The origins of behavior* (Vol. 1). New York: Wiley, 1974.

Brazelton, T. B., Yogman, M., Als, H., & Tronick, E. The infant as a focus for family reciprocity. In M. Lewis & L. Rosenblum (Eds.), *The child and its family: The genesis of behavior* (Vol. 2). New York: Plenum, 1979, 29–43.

Bronfenbrenner, U. Toward an experimental ecology of human development. *American Psychologist,* 1977, **32**, 513–531.

Brown, C. *Manchild in the promised land.* New York: Signet, 1971.

Campos, J. J., & Stenberg, C. R. Perception, appraisal and emotion: The onset of social referencing. In M. Lamb & L. Sheerod (Eds.), *Infant social cognition: Empirical and theoretical considerations.* Hillsdale, N.J.: Erlbaum, 1981.

Cicirelli, V. G. Effects of mother and older sibling on the problem solving behavior of the younger child. *Developmental Psychology,* 1975, **11**, 749–756.

Clarke-Stewart, K. A. And daddy makes three: The father's impact on mother and young child. *Child Development,* 1978, **49**(2), 466–478.

Cowen, E., Pedersen, A., Balijian, H., Izzo, L. D., & Trost, M. A. Long term follow-up of early detected vulnerable children. *Journal of Consulting and Clinical Psychology,* 1973, **41**, 438–446.

Dunn, J., & Kendrick, C. Interaction between young siblings in the context of family relationships. In M. Lewis & L. Rosenblum (Eds.), *The child and its family: The genesis of behavior* (Vol. 2). New York: Plenum, 1979.

Dunn, J., & Kendrick, C. The arrival of a sibling: Changes in interaction between mother and first-born child. *Journal of Child Psychology and Psychiatry,* 1980, **21**, 119–132.

Edwards, C. P., & Lewis, M. Young children's concepts of social relations: Social functions and social objects. In M. Lewis & L. Rosenblum (Eds.), *The child and its family: The genesis of behavior* (Vol. 2). New York: Plenum, 1979, 245–266.

Edwards, C. P., & Whiting, B. B. Dependency in dyadic context: New meaning for an old construct. Paper presented at the annual meeting of the Eastern Psychological Association, New York, April 1976.

Feinman, S., & Lewis, M. Maternal effects on infants' responses to strangers. Paper presented at the Society for Research in Child Development meetings, Boston, April 1981.

Feiring, C. The influence of the child and secondary parent on maternal behavior: Toward a social systems view of early infant–mother attachment. Doctoral dissertation, University of Pittsburgh, 1975.

Feiring, C., & Lewis, M. The child as a member of the family system. *Behavioral Science,* 1978, **23**, 225–233.

Feiring, C., & Lewis, M. Middle class differences in the mother–child interaction and the child's cognitive development. In T. Field (Ed.), *Culture and early interactions.* Hillsdale, N. J.: Erlbaum, 1980. (a)

Feiring, C., & Lewis, M. Children, parents and siblings: Possible sources of variance in the behavior of first born and only children. Paper presented at a Symposium for Cognitive and Social Consequences of Growing up without Siblings. American Psychological Association Convention, Montreal, Canada, 1980. (b)

Feiring, C., Lewis, M., & Starr, M. Indirect and direct effects on children's reactions to unfamiliar adults. Submitted for publication 1981.

Feiring, C., & Taylor, J. The influences of the infant and secondary parent on maternal behavior: Toward a social systems view of infant attachment. Unpublished manuscript, University of Pittsburgh, 1977.

Frazer, J. *The golden bough.* New York: Macmillan, 1915.

Freud, S. Instincts and their vicissitudes. *Collected papers.* New York: Basic Books, 1959. (Originally published, 1915.)

Freud, A., & Dann, S. An experiment in group upbringing. *Psychoanalytic Study of the Child,* 1951, **6**, 127–168.

Furman, W., Rake, D. F., & Hartup, W. W. Rehabilitation of socially-withdrawn children through mixed-age and same-age socialization. *Child Development,* 1979, **50**, 915–922.

Goldberg, S. Social competence in infancy: A model of parent–infant interaction. *Merrill-Palmer Quarterly,* 1977, **23**, 163–177.

Greenbaum, C. W., & Landau, R. In P. H. Leiderman, S. R. Tulkin, & A. Rosenfeld (Eds.), *Culture and infancy: Variations in the human experience.* New York: Academic, 1977, 245–270.

Gyomroi-Ludowyk, E. The analysis of a young concentration camp victim. *Psychoanalytic Study of the Child,* 1963, **18**, 484–510.

Harlow, H. F. Age-mate or peer affectional system. In D. S. Lehrman, R. A. Hinde, & E. Shaw (Eds.), *Advances in the study of behavior* (Vol. 2). New York: Academic, 1969.

Harlow, H. F., & Harlow, M. D. The affectionate systems. In A. M. Schrier, H. F. Harlow, & F. Stollnitz (Eds.), *Behavior of nonhuman primates* (Vol. 2). New York: Academic, 1965.

Hartup, W. W. Peer interaction and social organization. In P. Mussen (Ed.), *Carmichael's manual of child psychology.* New York: Wiley, 1970.

Hartup, W. W. Two social worlds: Family relations and peer relations. In M. Rutter (Ed.), *Scientific foundations of developmental psychiatry.* London: Heinemann, 1979.

Heath, D. H. Competent fathers: Their personality and marriage. *Human Development,* 1976, **19**, 26–39.

Hinde, R. A. Interactions, relationships, and social structure. *Man,* 1976, **11**, 1–17.

Hollingshead, A. B. *Two-factor index of social position.* New Haven, Conn.: Author, 1957.

Klaus, M., & Kennell, J. *Mother–infant bonding.* St. Louis: Mosby, 1976.

Konner, M. Relations among infants and juveniles in comparative perspective. In M. Lewis & L. Rosenblum (Eds.), *Friendship and peer relations: The origins of behavior* (Vol. IV). New York: Wiley, 1975.

Laing, R. D. *Knots.* New York: Pantheon, 1970.

Lamb, M. E. (Ed.) *The role of the father in child development.* New York: Wiley, 1976.

Lamb, M. E. Effects of stress and cohort on mother– and father–infant interaction. *Developmental Psychology,* 1976, **12**, 425–443.

Lamb, M. E. The father's role in the infant's social world. In J. H. Stevens & M. Matthews (Eds.), *Mother/child, father/child relationships.* Washington, D.C.: National Association for Education of Young Children, 1978.

Lamb, M. E. The development of parent-infant attachments in the first two years of life. In F. Pedersen (Ed.), *The father–infant relationship: Observational studies in the family settings.* New York: Praeger, 1980.

Leiderman, A. F. Preschoolers' competence with a peer: Relations with attachment and peer experience. *Child Development,* 1977, **48**, 1277–1287.

Lewis, M. State as an infant–environment interaction: An analysis of mother–infant interaction as a function of sex. *Merrill-Palmer Quarterly,* 1972, **18**, 95–121.

Lewis, M. The self as a developmental concept. *Human Development,* 1979, **22**, 416–419. (a)

Lewis, M. The social network: Toward a theory of social development. Fiftieth Anniversary invited address, Eastern Psychological Association meetings, Philadelphia, April 1979. (b)

Lewis, M. Issues in the study of imitation. Paper presented in a Symposium, Imitation in Infancy: What, When and How, at the Society for Research in Child Development Meetings, San Francisco, March 1979. Also in ERIC's Resources in Education (RIE), #ED 171 394, October 1979. (c)

Lewis, M. Issues in the development of fear. In I. L. Kutash & L. B. Schlesinger (Eds.), *Pressure point: Perspectives on stress and anxiety.* San Francisco: Jossey-Bass, 1980. (a)

Lewis, M. Newton, Einstein, Piaget and the concept of self. Invited plenary address before the Jean Piaget Society, Philadelphia, May–June 1980. (b)

Lewis, M., & Ban, P. Stability of attachment behavior: A transformational analysis. Paper presented at a symposium on attachment: Studies in stability and change, at the society for research in child development meetings, Minneapolis, April 1971.

Lewis, M., & Brooks-Gunn, J. *Social cognition and the acquisition of self.* New York: Plenum, 1979.

Lewis, M., & Coates, D. L. Mother-infant interactions and cognitive development in twelve-week-old infants. *Infant Behavior and Development,* 1980, **3**, 95–105.

Lewis, M., & Feiring, C. The child's social world. In R. M. Lerner & G. D. Spanier (Eds.), *Child influences on marital and family interaction: A life-span perspective.* New York: Academic, 1978.

Lewis, M., & Feiring, C. The child's social network: Social object, social functions and their relationship. In M. Lewis & L. Rosenblum (Eds.), *The child and its family: The genesis of behavior* (Vol. 2). New York: Plenum, 1979, 9–27.

Lewis, M., & Feiring, C. Direct and indirect interactions in social relationships. In L. Lipsitt (Ed.), *Advances in infancy research* (Vol. 1). New York: Ablex, 1981.

Lewis, M., & Feiring, C. Some American families at dinner. In L. Laosa & I. Sigel (Eds.), *The family as a learning environment* (Vol. 1). New York: Plenum, in press.

Lewis, M., Feiring, C., & Weinraub, M. The father as a member of the child's social network. In M. Lamb (Ed.), *The role of the father in child development* (2nd ed.). New York: Wiley, 1981.

Lewis, M., & Goldberg, S. Perceptual–cognitive development in infancy: A generalized expectancy model as a function of the mother–infant interaction. *Merrill-Palmer Quarterly,* 1969, **15**(1), 81–100.

Lewis, M., & Kreitzberg, V. The effects of birth order and spacing on mother–infant interactions. *Developmental Psychology,* 1979, **15**(6), 617–625.

Lewis, M., & Rosenblum, L. (Eds.) *The origins of fear: The origins of behavior* (Vol. 2). New York: Wiley, 1974.

Lewis, M., & Rosenblum, L. (Eds.) *Friendship and peer relations: The origins of behavior* (Vol. 4). New York: Wiley, 1975.

Lewis, M., & Rosenblum, L. (Eds.) *Interaction, conversation, and the development of language: The origins of behavior* (Vol. 5). New York: Wiley, 1977.

Lewis, M., & Rosenblum, L. (Eds.) *The development of affect: The genesis of behavior* (Vol. 1). New York: Plenum, 1978.

Lewis, M., & Schaeffer, S. Peer behavior and mother–infant interaction in maltreated children. In M. Lewis & L. Rosenblum (Eds.), *The uncommon child: The genesis of behavior* (Vol. 3). New York: Plenum, 1981.

Lewis, M., & Weinraub, M. The father's role in the child's social network. In M. E. Lamb (Ed.), *Role of the father in child development.* New York: Wiley, 1976, 157–184.

Lewis, M., Young, G., Brooks, J., & Michalson, L. The beginning of friendship. In M. Lewis & L. Rosenblum (Eds.), *Friendship and peer relations: The origins of behavior* (Vol. 4). New York: Wiley, 1975, 27–66.

Lipsitt, L. The enduring significance of reflexes in human infancy: Developmental shifts in the first month of life. Paper presented at the meetings of the Eastern Psychological Association, Hartford, April 1980.

Luria, A., & Yudovich, J. *Speech and the development of mental processes in the child.* London: Penguin, 1971.

Matas, L., Arend, R. A., & Sroufe, L. A. Continuity of adaptation in the second year: The relationship between quality of attachment and later competence. *Child Development,* 1978, **49**, 547–556.

Meltzoff, A. N., & Moore, M. K. Imitation of facial and manual gestures by human neonates. *Science,* 1977, **198**, 75–78.

Monane, J. H. *A sociology of human systems.* New York: Appleton-Century-Crofts, 1967, 33–44.

Mueller, E., & Vandell, D. Infant–infant interaction. In J. Osofsky (Ed.), *Handbook of infant development.* New York: Wiley, 1979.

Murray, H. A. *Explorations in personality.* New York: Oxford University Press, 1938.

Novak, M. A., & Harlow, H. F. Social recovery of monkeys isolated for the first year of life: Rehabilitation and therapy. *Developmental Psychology,* 1975, **11**, 453–465.

Pannabecker, B. J., Emde, R. N., Johnson, W., Stenberg, C., & Davis, M. Maternal perceptions of infant emotions from birth to 18 months: A preliminary report. Paper presented at the International Conference of Infant Studies meetings, New Haven, Conn., 1980.

Papoušek, H., & Papoušek, M. The common in the uncommon children: Comments on the child's integrative capacities and on initiative parenting. In M. Lewis & L. Rosenblum (Eds.), *The uncommon child: The genesis of behavior* (Vol. 3). New York: Plenum, 1980.

Parke, F. Perspectives in father–infant interaction. In J. Osofsky (Ed.), *Handbook of infancy.* New York: Wiley, 1978.

Parke, R. D., & O'Leary, S. Father–mother–infant interaction in the newborn period: Some findings, some observations, and some unresolved issues. In K. Riegel & J. Meacham (Eds.), *The developing individual in a changing world: Social and environmental issues* (Vol. 2). The Hugue: Mouton, 1975.

Parke, R. D., Power, T. G., & Gottman, J. M. Conceptualizing and quantifying influence patterns in the family triad. In M. E. Lamb, S. J. Suomi, & G. R. Stephenson (Eds.), *Social interaction analysis.* Madison, Wisc.: University of Wisconsin, 1979.

Parke, R. D., & Sawin, D. B. The family in early infancy: Social interactional and attitudinal analyses. In F. A. Pedersen (Ed.), *The father–infant relationship: Observational studies in the family setting.* New York: Praeger Special Studies, 1980.

Parsons, T., & Bales, R. F. *Family socialization and interaction process.* Glencoe, Ill.: Free Press, 1955.

Patterson, G. R. Mothers: The unacknowledged victims. *Monographs of the Society for Research in Child Development,* in press.

Pedersen, F. A. Mother, father and infant as an interactive system. Paper presented at the meetings of the American Psychological Association, Chicago, September 1975.

Pedersen, F. A. (Ed.) *The father–infant relationship: Observational studies in the family setting.* New York: Praeger, 1980.

Pedersen, F. A., Anderson, B. J., & Cain, R. L. An approach to understanding link-ups between the parent–infant and spouse relationship. Paper presented at the Society for Research in Child Development meetings, New Orleans, 1977.

Pedersen, F. A., Yarrow, L. J., Anderson, B. J., & Cain, R. L. Conceptualization of father influences in the infancy period. In M. Lewis & L. Rosenblum (Eds.), *The child and its family: The genesis of behavior* (Vol. 2). New York: Plenum, 1979, 45–66.

Pervin, L. A. *Personality: Theory, assessment, and research* (2nd ed.). New York: Wiley, 1975.

Piaget, J. *The origins of intelligence in children.* Translated by M. Cook. New York: W. W. Norton, 1963. (Originally published, 1936.)

Rebelsky, F., & Hanks, C. Fathers' verbal interaction with infants in the first three months of life. *Child Development,* 1971, **43**, 63–68.

Robertson, J., & Robertson, J. Young children in brief separation: A fresh look. *Psychoanalytic Study of the Child,* 1971, **26**, 264–315.

Rollins, B. C., & Galligan, R. The developing child and marital satisfaction of parents. In R. Lerner & G. Spanier (Eds.), *Child influences on marital and family interaction.* New York: Academic, 1978.

Rosenblatt, P. C. Behavior in public places: Comparisons of couples accompanied and unaccompanied by children. *Journal of Marriage and the Family,* 1974, **36**, 750–755.

Rosenblum, L. A., & Kaufman, I. C. Variations in infant development and response to maternal loss in monkeys. *American Journal of Orthopsychiatry,* 1968, **38**, 418–426.

Rupenthal, G. C., Arling, G. L., Harlow, H. F., Sackett, G. P., & Suomi, S. J. 10-year perspective of motherless mother monkey behavior. *Journal of Abnormal Psychology,* 1976, **85**, 341–349.

Rutter, M. Maternal deprivation, 1972–1978: New findings, new concepts, new approaches. *Child Development,* 1979, **50**, 283–305.

Sroufe, L. The coherence of individual development. *American Psychologist,* 1979, **34**, 834–841.

Stern, D. N. The goal and structure of mother–infant play. *Journal of the American Academy of Child Psychiatry,* 1974, **13**, 402–421.

Sullivan, H. *The interpersonal theory of psychiatry.* New York: Norton, 1953.

Sullivan, H. *The psychiatric interview.* New York: Norton, 1954.

Suomi, S. J. Maternal behavior by socially incompetent monkeys: Neglect and abuse of offspring. *Journal of Pediatric Psychology,* 1978, **3**, 28–34.

Suomi, S. J. Differential development in various social relationships by rhesus monkey infants. In M. Lewis & L. Rosenblum (Eds.), *The child and its family: The genesis of behavior* (Vol. 2). New York: Plenum, 1979, 219–244.

Suomi, S. J., & Harlow, H. F. Early experience and social development in rhesus monkeys. In M. E. Lamb (Ed.), *Social and personality development.* New York: Holt, Rinehart, & Winston, 1978.

Tizard, J., & Tizard, B. The social development of two-year-old children in residential nurseries. In R. Schaffer (Ed.), *The origins of human social relations.* New York: Academic, 1971.

Von Bertalanffy, L. *Robots, men and minds.* New York: Brazilles, 1967.

Waite, L. H., & Lewis, M. Maternal report on social imitation development during the first year. Paper presented at the Eighth Annual Symposium of the Piaget Society, Philadelphia, Pa., May 1978.

Waite, L. H., & Lewis, M. Early imitation with several models: An example of socio-affective development. Paper presented at the Society for Research in Child Development meetings, San Francisco, March 1979.

Waters, E., Wippman, J. & Sroufe, L. A. Attachment, positive affect, and competence in the peer group: Two studies in construct validation. *Child Development,* 1979, **50**(3), 821–829.

Weinraub, M., Brooks, J., & Lewis, M. The social network: A reconsideration of the concept of attachment. *Human Development,* 1977, **20**, 31–47.

Whiting, B. B., & Whiting, J. W. M. *Children of six cultures: A psychocultural analysis.* Cambridge, Mass.: Harvard University Press, 1975.

Wilson, E. O. *Sociobiology.* Cambridge: The Belknap Press of Harvard University Press, 1975.

Youniss, J. *Parents and peers in social development: A Sullivan-Piaget perspective.* Chicago: University of Chicago Press, 1980.

Zajonc, R. B., & Markus, G. B. Birth order and intellectual development, *Psychological Review,* 1975, **82**, 74–88.

Zazzo, R. Images du corp et conscience du soi. *Enfance,* 1948, **1**, 29–43.

CHILDHOOD

11

COGNITIVE DEVELOPMENT
A Competence-Activation/Utilization Approach

Willis F. Overton
Professor of Psychology
Temple University

Judith L. Newman
Assistant Professor of Human Development
The Pennsylvania State University

A dilemma that has obstructed the pursuit of satisfactory explanations of human behavior is that, on the one hand, there appears to be sufficient constancy of behavior across variable environmental conditions to warrant the inference that explanation requires the introduction of dispositional concepts (e.g., schemes, operations, structures, rules); on the other hand, behavior is also variable among individuals to the point that explanation requires the specification of environmental determinants. Frequently, investigators have sought to resolve the dilemma by denying or trivializing one or the other of its components. For example, contemporary structuralists sometimes appear to argue that universal structures, rules, or central-processing mechanisms are the sine qua non of explanation and treat individual differences and environmental effects as epiphenomena. Approaching from the opposite extreme, situationists propose that universal structures are at best temporary expedients and that ultimately these epiphenomena, as well as the variability of behavior, will be explainable by environmental determinants (Bowers, 1973).

Over the past several years, a relatively novel approach to the problem has been evolving in the field of cognitive development. This approach is characterized by the explicit assumption that both universal formal features and contingent features must be introduced as coequals for a complete understanding of behavior. From this perspective, the question is not which type of explanatory form ultimately explains behavior, but how are the two forms best interrelated to produce a more complete explanation. The approach itself has been generated by what has been termed the "competence–performance distinction" (Flavell & Wohlwill, 1969), but because this terminology has often led to some confusion, we will generally refer to the competence–activation/utilization model (Overton, 1976).

The aim of this chapter is to sketch several important features of this approach and to examine some of the empirical literature relevant to it. In pursuit of this aim we will first examine the competence–performance distinction. More specifically, we will examine the acquisition of competence, the research strategies generated by the distinction, and the criticisms raised against it. Next, issues and research entailing the activation of competence by various environmental contingencies will be explored, and finally, similar consideration will be given to factors that determine the utilization of competence.

THE COMPETENCE–PERFORMANCE DISTINCTION

This position maintains that a complete psychological theory would require two distinct components. The competence component or model entails idealized abstract systems of the form or structure of the individual's knowledge in various cognitive domains. The systems might alternatively be represented as rules, strategies, or logical systems. The performance, or activation/utilization, component or model involves the psychological processes along with task and situational factors that determine the application of competence in actual thought or behavior. Activation/utilization features might include memory, cognitive styles, task familiarity, stimulus features of the task, and so forth.

Because Piaget's theory of cognitive development has generated most of the research on this topic and because this theory presents the most clearly articulated competence model available, we will consider this theory throughout the remainder of this chapter both for illustrative purposes and for purposes of reviewing issues and research. However, competence–performance distinctions have been described in many areas of psychology. The current version was imported from psycholinguistics (Chomsky, 1965) by Flavell and Wohlwill (1969). In earlier years Werner (1937) made a similar distinction in cognitive development and termed it the process–achievement distinction. Hull (1943) employed the learning/habit–performance distinction, Lewin (1951) the genotype–phenotype distinction. More recently, in the area of verbal learning, Postman (1968) suggested relating "association" as a dispositional competence factor to performance factors such as drive and instructional effects. In the memory literature, Tulving and Pearlstone (1966) distinguished between the availability and accessibility of information, while Reese's (1962) concept of mediational deficiency and Flavell's (1970) concept of production deficiency focused on the absence of competence versus the failure of performance factors to elicit competence. Finally, in the area of social behavior, both Bandura (1977) and Mischel (1973) distinguish between an acquisition phase (competence) and a performance phase.

Piaget's theory is a theory of competence. As numerous investigators have noted, Piaget paid little systematic attention to the role of individual differences or contingent determinants of behavior. This should not be taken as a criticism, for Piaget chose an epistemologically distinct strategy from one that requires explanation of behavior in terms of contingent events. Piaget's strategy for understanding the cognitive domain, and Chomsky's for understanding language, requires that the theorist answer the following Kantian question: What must one *necessarily* assume about the nature of the organism in order for it to have the behaviors which it does exhibit? The method for answering this question is that of observing a sufficient subset of behaviors in the domain in question and constructing a competence model that best captures the universal features of the subset. If the model is valid and powerful, it will prove through further empirical demonstrations to be applicable to a much wider array of behaviors in the domain. The particular form of such a competence model is also guided by issues of parsimony, simplicity, internal coherence, and aesthetics (Miller, 1975; Stone & Day, 1980).

For Piaget, the construction of a model (and hence a *formal* explanation) of the universal features of cognitive skills was complicated by a developmental perspective which led to the view that there are progressive and qualitative changes in the nature of these features. This meant that several distinct but interrelated competence models were required. Piaget's solution was to construct four distinct models as the "best" representation of seemingly discontinuous cognitive skills that develop between infancy and adolescence. These are (1) the action schemes as integrated in structures d'ensemble at the sensory motor period; (2) the nascent operations of the preoperational period; (3) operatory structures as organized in logical groupings at the concrete operational period; and (4) operatory structures organized as a logical group at the period of formal operations. The first two of these competence models are presented in straightforward verbal format in the theory, while the competence models for the concrete operational and formal operational periods are presented in a logical format.

Two points are of general importance here. First, the format of the competence model is relatively independent of its validity (although it may be argued that logical–mathematical models are more powerful than other formats). Thus competence models may be presented as rules, logical–mathematical systems, strategies, structures, central-processing mechanisms, or in any number of other formats. The second point is that the failure of a particular competence model to demonstrate its adequacy through empirical tests is not a valid argument against the employment of competence models as necessary explanations of the domain under examination. This point may seem banal, but one hears arguments of the form "Since Piaget's INRC group is not a good model of adolescent thought, we should abandon a structural approach" with sufficient frequency to warrant this caveat.

The Acquisition of Competence

Before turning to an appraisal of competence models as psychological theory and consequently to the need for the introduction of activation/utilization processes, a few remarks are necessary concerning the developmental acquisition of the universal features represented by the competence model. These comments are made to note that contingent events do not enter as causal determinants with respect to competence, and hence their role must lie in other areas. Piaget's theory is again illustrative. As noted, the description of competence is provided by the structural and stage components of the theory. The answer to the question of how competence develops is provided by the functional component of the theory (see Overton, 1972). Here it is proposed that the infant begins with two defining characteristics: (1) a biological competence (actually a fifth model of competence but biological rather than psychological) and (2) an inherent activity. The latter is further specified by that primary phase of the activity which modifies the environment in conformity to the demands of the competence or structure (assimilation) and that phase which modifies the structures in conformity with the assimilated demands of the environment (accommodation). The theory assumes an inherent necessary movement from a state in which these twin activities are out of equilibrium to a state in which they are in equilibrium (i.e., match the ideal competence model of that period). The theory further assumes that following the development of competence at any given level, there is an inherent and necessary movement (reflective abstraction) to the next level (toward the next match) and the process repeats itself at each higher plane (the equilibration process). Two points should be made with respect to this developmental model. First, although the Genevans have recently tended to discuss the features of this model in process terms, it is basically a model that provides the necessary and universal features of development (Overton & Reese, 1981). That is, the developmental model is itself essentially a competence model which organizes the other competence models and thus provides an explanation (again formal) for their order and direction of development. The second and related point is that as a formal model, specific contingent events, such as environmental effects, do not enter as causal explanations of development. Contingent events are excluded by the logic of the position, and environmental effects enter only as they are themselves constituted by the formally defined processes. For this reason it has been maintained that contingent events can affect only the rate and terminal level of development and not the natural course of development (Overton & Reese, 1981).

The Competence–Activation/Utilization Approach

The ideal aim of any theory is to provide a complete explanation of the field under consideration. As described, Piaget's theory or any other competence model provides formal explanations; that is, explanation of the necessary and universal features of cognitive development. Thus competence theories explain the constancy of behavior—both momentary constancy and constancy of development—across variable environmental contingencies and across variable individuals. Such theories, although necessary for complete explanation, are not sufficient because they do not address the variability that also appears in behavior. In the past, such a deficiency in scope has often been interpreted as an inherent flaw in approach, and investigators who have been impressed with the variability of behavior have urged the abandonment of competence theories in favor of explanations based totally on contingent events. The unfortunate consequence of this latter approach is that as it is followed, it eventually becomes evident that explanations based on contingent events are not sufficient to account for the constancy of behavior (e.g., Cronbach, 1975; Gergen, in press). This in turn leads back to formal competence models, and the cycle repeats itself again and again.

The competence–activation/utilization approach is an attempt to break this vicious cycle. Conceptually, the approach involves the marriage of two distinct epistemologies. The marriage metaphor is here taken seriously because as in any good marriage the individual partners must maintain their individual identities and develop their own individual talents while cooperating and dividing their labors to achieve long-term aims and goals. The partners in the competence–activation/ utilization marriage are those who have pursued a rationalist understanding of science and those who have pursued an understanding of science based on empiricism, and the primary long-term goal is the complete explanation of behavior.

Traditionally, the rationalists have followed Aristotle's distinction between the "necessary" and the "accidental." For the rationalist, scientific explanation consists of formulating formal laws of the regular, the normal, the necessary. Accidental or contingent events merely facilitate, inhibit, or deflect; they are not the essentials of science. Thus competence, as the formal explanation of cognition and its development, has often been the sine qua non for the rationalists, and the manner in which this competence is affected by various task and situational factors has been treated as unimportant (for an important contemporary exception, see Pascual-Leone & Sparkman, 1980).

For the philosophical empiricists, scientific explanation has consisted of attempting to derive laws from contingent observable events. From the manifold possible events that might affect behavior, specific variables are selected for experimental scrutiny under the assumption that discoverable cause–effect relations will ultimately yield generalized causal laws. Thus from this perspective, explanation begins and ends with a primary focus on antecedent variables. Investigations explore the presumed causal relation between factors such as memory, attention, individual differences, task features, situational features, and the behavior to be explained. Formal explanations (i.e., competence models) are viewed, at best, as temporary expedients to be granted only until observation and experimentation reveal the "real" (causal) explanations. At worst, formal explanations are viewed as a flight of fantasy (mentalism) that obscure the search for "real" explanations and beg the "real" questions.

Research Strategies

A marriage of these two historical rivals requires a continuing recognition of the independent but significant contribution of each partner, of the differential roles each plays in the overall explanatory effort, and of the ways in which cooperation rather than competition may be enhanced throughout this process. This recognition yields three relatively independent but interrelated research strategies. First is the strategy of formulating the idealized abstract models that constitute competence at each level of cognitive development. This is the problem of representing the organization—not the content—of knowledge at each level. Here the ambiguities and vagueness of specific psychological processes and specific task and situational anomalies are subsumed to the interest of providing general representations of the necessary fundamental features of thought.

The second research strategy involves the validation of the proposed competence. Here the research task is that of conducting studies which control individual psychological processes, task variables, and situational variables in an effort to demonstrate that the proposed competence has empirical as well as rational meaning. Thus, for example, it might be asked whether Piaget's logical grouping "primary addition of classes" is an adequate representation of the 8- to 10-year-old's understanding of problems involving hierarchical classes. This question may be answered empirically, providing that such processes as memory demand, attention, and emotional state and such task factors as stimulus saliency, response bias, and task instructions are carefully controlled. It should also be noted that well-controlled studies that fail to validate a particular representation of competence may provide clues for more adequate representations; they do not, however, as noted earlier, falsify the view that competence is a necessary component of a theory of cognitive development.

These first two research strategies focus on competence and bracket or control performance variables, and these will be discussed briefly when we turn to an examination of specific issues and research. The third strategy accepts competence as represented and focuses on performance variables themselves in an effort to discover how these are related to and determine the application of competence in actual thought or behavior. Here it is explicitly recognized that predictions concerning thought or behavior based on competence are distorted by various performance variables. Within this strategy, several research questions may be raised. One question asks how performance variables might enhance or retard the rate of the acquisition of competence. A second question asks how performance variables might activate the application of recently acquired competence. A third question asks how performance variables might facilitate or hinder the utilization of competence given that competence has been fully consolidated and stabilized. While the first of these questions is beyond the scope of this chapter, we will examine some issues and research related to the latter two questions.

Criticisms

Before turning directly to an examination of various issues and investigations generated by these research strategies, several criticisms of the legitimacy of a competence–performance distinction should be addressed. A common criticism maintains that sufficient constancy has not been demonstrated to warrant the inference of competence. The problem here is the question of what constitutes "sufficient" constancy, and unfortunately the answer often lies in the eye of the beholder. That is, in a good deal of psychological work significant correlations between performances on various tasks in the order of .40 or above are accepted as meaningful in suggesting a reasonable degree of constancy across variable conditions. On the other hand, for the critic who starts from the assumption that variability is the rule, the counterargument can always be made that such correlations do not account for a sufficient amount of the variance to move away from a situationist perspective. Even as the correlations increase, the same counterargument can be made. For example, in reviewing the work of Inhelder, Sinclair, and Bovet (1974), Brown and Desforges (1979) note that reported correlations between various tests of operational thought were significant but "none accounts for more than 50 per cent of the variance and we believe that Piaget's descriptions of stages leads us to expect much more coherent structures than these results suggest" (p. 107). The point is that, except at the extremes, the issue is not resolvable by data. For the critic who comes to the enterprise with epistemological biases against the acceptance of competence, there will always be enough variance left over to mount an attack which proposes we ignore constancy and focus on variability. As we mentioned earlier, the competence–performance distinction is an attempt to break this vicious cycle of either-or explanation through the recognition that both constancy and variability are legitimate facts of psychological life.

A second criticism of the legitimacy of the competence–performance distinction also focuses on competence and is also generated from epistemological biases. This criticism maintains that since competence is necessarily inferred from behavior, all we have in reality is behavior or performance. As a consequence, the criticism continues, the competence–performance distinction diverts us from our primary focus which should be a careful analysis of what the child actually does in performing various tasks and how this is affected by cultural and stimulus determinants. It is beyond the scope of this chapter to elaborate the epistemological reasons for postulating some form of competence (Overton, 1975, 1976; Overton & Reese, 1973, 1981), but it should be clear that such a criticism is deeply rooted in the empiricist precept that all knowledge and hence all theoretical concepts (i.e., competence) must be reducible to observations. The competence–performance distinction, on the other hand, requires some reasonable degree of acceptance of the rationalists' precept that theory, although based on observations, can never be reduced to mere observation.

As a third criticism of the competence–performance distinction, it is noted that the distinction is frequently employed as a speculative post hoc explanation in situations where research failed to generate predicted results. Thus to take a simple example, an investigation may examine two age groups under the expectation that the older group will demonstrate formal operational competence on a variety of tasks. Both groups perform significantly more poorly than would be expected on any reasonable criteria of formal thought, and the researcher explains the lack of demonstrated competence of the older group by asserting that it was probably masked by some unknown performance variables. The use of the competence–performance distinction in this post hoc fashion is indeed open to legitimate criticism, but the criticism is appropriate only to this particular usage and not to others.

This third criticism is legitimate in cases where competence is not well articulated and perhaps not appropriate. For example, competence models have not been elaborated for psychometric intelligence, yet it is not uncommon to hear intelligence testers complain that the competence of various individuals or groups of individuals has not been adequately tapped because of interfering performance factors. What this means, in effect, is that the IQ scores did not attain the level the tester had expected based on some covert, implicit rationale. The competence–performance distinction is viable only to the extent that we have in hand a relatively well articulated and reasonably validated explicit model of competence. If competence is proposed in a fashion that is vague to the point of having no specifiable empirical implications, then it does not constitute formal explanation and plays no meaningful role in scientific explanation.

A fourth and final criticism is related to the third, but its resolution is different. Here research is explicitly designed to examine empirically a potentially meaningful competence–performance relationship. However, the performance variables tested do not demonstrate a relation to competence. The researcher again introduces a post hoc explanation proposing other possible effective performance variables. Stone and Day (1980) criticize this approach by suggesting, "It would always be possible to specify additional performance factors as potential explanations of a child's failure to reveal a particular competence" (p. 333), and as a consequence, the claim that the child possesses a particular competence is unfalsifiable. There are two problems with this criticism. First, it should be noted that the validity of competence is assessed by different research approaches than is the exploration of the relationship between competence and performance factors. Second, the exclusion of some potential performance variable can reasonably be viewed as a positive feature of the competence–activation/utilization approach. An example may best illustrate the point: Liben and Golbeck (1980), noting sex differences favoring males in the child's behavior on Piagetian spatial tasks, raised the question of whether these were the result of competence differences (e.g., a developmental lag in the emergence of competence for females) or were due to some activation/utilization variables. The researchers selected alternative forms of the task as a potentially effective activation/utilization variable. However, this variable failed to reduce the sex differences, and the authors proposed that "other performance factors may be operating differentially for males and females" (1980, p. 596). If a number of investigations fail to find activation/utilization factors that ameliorate the difference, the weight of evidence will move to the view that the difference is competence-based. However, it is a perfectly reasonable research program to search for alternative activation/utilization explanatory variables.

To summarize, competence models present abstract systems of formal features of behavior potentialities in the field of cognition. Competence models do not include the actual psychological processes or environmental factors that generate behavior in any specific situation. Activation/utilization (performance) models are designed to specify the psychological processes and environmental events that result in the application of competence or the distortion of behavioral predictions that would be made on the basis of competence models. Keeping the nature of the two models distinct facilitates the employment of several research strategies designed to ultimately integrate the roles played by each component and provide more complete explanations of cognitive development.

THE FORMULATION, VALIDATION, AND DIAGNOSIS OF COMPETENCE

Formulation

The task of formulating each qualitative level (i.e., stage) of competence is theoretical in nature and requires both a rational and empirical approach. In essence, the theorist in formulating a particular competence proposes (rational analysis) that certain principles, rules, or logicomathematical structures are adequate general representations or models of the actual organization of the mature cognitive behaviors (empirical observation) of the period under consideration (Feldman & Toulmin, 1976; Overton, 1975).

An issue of primary importance to the construction of competence is the level of generality at which competence is formulated. A theory such as Piaget's postulates competencies that model the general form of logical thought across a range of specific concepts. The advantage of this approach is that it provides breadth of (formal) explanatory scope and hence theoretical power. The disadvantage is that the very generality may lead to the empirical results that the actual understanding both within and between specific concepts is sensitive to a number of task, stimulus, and individual difference factors. Thus, for example, Piaget's grouping "primary addition of classes" models the general logical similarity in children's reasoning required for the solution to a number of tasks entailing class inclusion or the hierarchical ordering of classes. However, a number of studies have found that a variety of task variables affect the actual understanding of hierarchical classes (Trabasso, Isen, Dolecki, McLanahan, Riley, & Tucker, 1978; Winer, 1980). From the perspective of the competence–performance distinction, however, this apparent disadvantage becomes salutary as the task, situational, and individual difference factors become part of the activation/utilization model.

If competence is formulated at a less general level, then the relationship between competence and actual behavior forms a tighter unity. This approach has been termed "task analysis," and it entails the generation of principles or strategies that are presumed to underlie specific tasks or concepts (Siegler, 1980). Thus, for example, Gelman and Gallistel (1978) formulated five principles to account for numerical counting, and Siegler (in press) formulated separate rule sets for several tasks (balance scale, projection of shadows, and probability tasks) that derive from the Piagetian formal operational literature (see Wilkening & Anderson, 1980, for an alternative rule assessment methodology). Siegler and Richards (1979) have also produced a rule set for the concept of time. Two cautionary notes should be made regarding this type of approach. First, care should be taken to avoid focusing merely on superficial levels of performance (Keating, 1979; Siegler, 1980). Second, it should be recognized that as the number of task specific competence models increase, theoretical power decreases. To take the extreme case, imagine that for every specific concept, a distinct competence was formulated. All theoretical generality would be lost. Rules would function primarily as simple descriptive summaries of the specific behaviors, and again, the field would have moved away from a position which valued both formal and contingent explanation to a position claiming complete contingent determination. Thus do the distinct epistemologies subtly move back to a stance of either-or rivalry.

Validation and Diagnosis

Following formulation of the idealized competence at each qualitative level or stage, research is conducted to establish various facets of the validity of the competence models. This is a complex methodological enterprise since it entails the assessment of the sequential nature of the competence models (see Tomlinson-Keasey, in press; Wohlwill, 1973, Ch. 9) both within and across cognitive levels or stages as well as the construct validity of each competence model as an integrated system. With respect to Piagetian theory, investigations of stage sequences include a large portion of the total research that has been conducted. Recently, many such studies recognized and/or addressed the methodological problems involved in identifying invariant concurrences and sequences (Hooper, Toniolo, & Sipple, 1978; Hooper, Wanska, Peterson, & DeFrain, 1979; Niebuhr & Molfese, 1978; Tomlinson-Keasey, Eisert, Kahle, Hardy-Brown, & Keasey, 1979). With respect to the construct validity of the Piagetian competence models, studies by Tomlinson-Keasey et al. (1979), concerning the concrete operational model, and Neimark (1975a,b) and Shayer (1979), concerning the formal operational model, suggest that regardless of final decisions concerning specific logical features of the models, each represents a good approximation or a mapping of the coherence of thought that occurs during these periods. (For an opposing perspective, see Brown & Desforges, 1979.)

For purposes of the present discussion, the issue of interest is neither the specific methods appropriate to establishing the validity of the various competence models nor the ultimate outcome of such research (see Tomlinson-Keasey, in press, for a review). Rather, at issue is the question of how the competence–performance distinction facilitates the process of validation. The answer is that for purposes of validation, variables that form features of the activation/utilization model are controlled in such a way that we may cut through performance in an attempt to establish the psychological reality of the underlying competence. Flavell (1977) treats this as the issue of diagnosing the child's cognitive abilities in the context of controlling factors (performance variables) that might otherwise lead to an underestimation (i.e., a false negative error when the child has the ability but testing procedures lead to the inappropriate conclusion that he or she does not) or an overestimation (i.e., a false positive error when the child does not have the ability but testing procedures lead to the conclusion that he or she does) of the child's actual ability. Here, however, in order to differentiate the competence models from specific skills or abilities (i.e., avoid confusing competence and performance models) as suggested by Stone and Day (1980), and similarly to avoid the view that it is possible to assess pure knowledge (Miller, 1978), the issue is conceptualized as validating competence models through control of performance variables.

Potentially relevant differences in psychological processes, states, and styles that can lead to false negative evaluations of the validity of competence models include selective attention (Miller, 1978), attentional or mental capacity (Miller, 1978; Pascual-Leone, 1976a), memory (Stone & Day, 1980; Flavell, 1977), ability to comprehend verbal instructions and produce verbal explanations (Braine, 1968; Brainerd, 1973), motivation and emotional states (Flavell, 1977), and variation in information-processing approaches associated with cultural factors and cognitive style (Neimark, 1979, 1981; Overton & Meehan, 1981; Pascual-Leone, 1976b). These organismic variables in turn interact with situational and task factors such as complexity of task instructions (Danner & Day, 1977), figurative task features, stimulus salience, and complexity (Miller, 1978; Scardamalia, 1977; Winer, 1980), and task familiarity (Neimark, 1979).

The introduction of false positive evaluations arise primarily from experimental situations that permit such factors as guessing and response biases to be introduced or which employ tasks that permit solution strategies inappropriate to the competence being validated (Flavell, 1977). This latter factor deserves special mention because although it introduces a false positive into the specific study under examination, it also has the rather ironic effect of suggesting the nonvalidity of the sequential nature of the competence model being assessed. Thus, for example, although the several competence models proposed by Piaget are not tied to specific ages, given their assumed sequential nature, it would obviously prove embarrassing to the theory if behavioral implications of concrete operational competence were found in children at ages 3 to 4 years or if formal operational competence were discovered at age 6 or 7. The possibility of obtaining such effects spuriously occurs through the use of "simplified" tasks (Larsen, 1977) or tasks which permit solution strategies appropriate to lower levels of competence than that being assessed (e.g., figurative solutions in the assessment of logical competence). It is known, for example, that lower animals can discriminate and match to sample various numerical arrays. This task and the skill it demonstrates, however, are functionally distinct from the logical competence which represents the child's understanding of number concepts during the concrete operational period (see Piaget, 1968b, for criticism of number tasks on which very young children have been found to "conserve"). Similarly, demonstrations that some forms of transitivity (Flavell, 1977) or class inclusion (Winer, 1980) tasks can yield figurative solutions do not speak to the issue of the logical competence proposed by Piaget. The general point at issue here is that although competence models cannot be reduced to specific behavioral assessment procedures (Feldman & Toulmin, 1976), reasonably articulated models do present structural demand characteristics. If appropriate assessment of the competence models are to be made, then task selection must proceed by careful a priori analysis of these characteristics, and not on the basis of simple surface appearances.

It should be clear from the foregoing that the validation of competence models through the control of performance factors is not a simple matter that can be dealt with in a single study or series of studies. One methodological approach that can facilitate the process entails the use of training. Recently, a number of investigators have begun to present the view that under certain conditions training procedures are best interpreted as a method for uncovering or diagnosing competence rather than inducing the acquisition of new behaviors (Flavell, 1977; Gelman, 1978; Gelman & Gallistel, 1978; Holland & Palermo, 1975; Hornblum & Overton, 1976; Overton & Newman-Hornblum, 1980). This interpretation is conditional upon two factors. The first of these is that the training should not include specific instructions on the criterion task. The second factor involves the extent of training and extent of responsiveness to the training. That is, to the degree that the subject improves significantly on the competence criterion task following minimum exposure to training, it is reasonable to infer that training, in essence, acted as a vehicle for cutting through performance factors to uncover competence. Inversely, to the degree that extensive training leads to minimum improvement, it is fair to assume the absence of competence. Finally, in those cases when extensive training or specific instructions lead to improved performance, traditional but cautionary (see Wohlwill, 1973, pp. 317–321) interpretations concerning training would be appropriate. Specific examples of the use of training as a diagnostic tool will be presented in the following sections.

In this section, the emphasis has been on controlling performance factors in an effort to reach decisions concerning the validity of competence models. In the following sections, the competence models are accepted as represented, and the focus is on analysis of how various performance factors are related to and determine the application of competence in actual thought and behavior. As stated earlier, we will not consider the issue of how performance variables influence the rate of the behavioral acquisition represented by the competence models. This is a topic which covers a great deal of the developmental literature (see Beilin, 1971, 1976, 1978; Wohlwill, 1973, Ch. 11) and is far beyond the scope of this chapter. The following sections will examine how performance factors may function to activate recently acquired competence and how performance factors facilitate or hinder the utilization of consolidated competence.

THE ACTIVATION OF COMPETENCE

Perhaps one of the most interesting and general findings to emerge from the cognitive developmental literature over the past decade is that specific environmental agents or conditions tend to influence behavior only to the extent that the competence relevant to these behaviors is already present (see Beilin, 1971, 1976, 1978; Gelman, 1978; Gelman & Gallistel, 1978; Inhelder & Sinclair, 1969; Strauss, 1972). This is demonstrated, for example, in the Genevan findings that training effects depend upon the child's developmental level, that is, on the competence that the child has at its disposal (Inhelder, Sinclair, & Bovet, 1974). One interpretation of this general finding is that many of the specific experimental effects demonstrated with respect to cognitive skills may not be functionally related to the acquisition of competence at all but, rather, they may serve to activate recently acquired competence (Overton, 1976; Overton, Wagner, & Dolinsky, 1971). This interpretation suggests, as has been detailed elsewhere (Overton, 1978; Overton & Reese, 1973, 1981), that the acquisition of competence is subject to laws that are quite distinct from those which explain the actual demonstration of competence in overt behavior. An analogy to the development and actual functioning of anatomical structures is illustrative. The eye, for example, has a necessary developmental course that is determined by a series of interactional effects within the differentiating embryo, and this results in a fully formed structure. The eye does not function at this time; however, it is ready and awaits specific environmental events, which occur following birth, to be activated and make adaptive responses. Thus anatomical structures are subject to what Carmichael has termed the "law of anticipatory function" (1970, p. 448); that is, structures are complete and ready to function prior to their first adaptive responses, and these responses are determined by external agents distinct from those factors that explain the development itself.

It is assumed that cognitive structures follow this same general course. A basic competence is developed in accordance with necessary principles (Overton & Reese, 1973, 1981). Following this development, there ensues a phase during which specifiable environmental agents, that is, performance or activation variables, result in the activation of the structures and thus result in significantly improved task behavior (Overton, 1976). This proposal is compatible with, but distinct from, the early competence–performance distinction formulated by Flavell and Wohlwill (1969). In the present case, the focus is on how specific activation variables interact with a given competence. Flavell and Wohlwill, on the other hand, focused primarily on sequence issues relative to the acquisition of competence itself. To integrate the two approaches, we may say that our interest is on activation variables as they are related to Flavell and Wohlwill's phases 2 and 3 of any competence model (i.e., situations when the competence is present but is not always reflected in performance).

The strongest support for a competence–activation model comes from investigations that include an age-by-conditions design. The optimal design includes three age groups: one group that on theoretical or a priori grounds should not have the target competence, a second group that possesses nonactivated competence, and a third group that demonstrates competence in a variety of situations. The experimental prediction is that the presumed activation variable (condition) will significantly affect only the second group's competence-related behavior. It has also been suggested that pretested cognitive level be substituted for age groups in order to attain a more direct measure of competence status (i.e., precompetent, transitionally competent, and fully competent) and thus avoid issues involving an age by cognitive level confounding. Although this proposal often has merit, it can also generate theoretical and methodological problems. As an illustration, the issue emerges as to which specific feature of competence should be pretested. If the target competence is, for example, cross or multiple classification, then pretesting for conservation raises the question of the exact competence relations between cross classification and conservation. On the other hand, a direct pretesting of the target competence may minimize the impact of subsequent activation experience.

Evidence for activation may also be obtained from a conditions design in which the subject population consists of a single age group that is presumed to possess latent competence. For example, within Piaget's theory, concrete operational competence develops between approximately 5 to 8 years of age, and thus this represents the appropriate group where activation measures should be effective. It is obvious, however, that findings from such a single age group can be taken as no more than suggestive, since any obtained effects might also be applicable to younger or older groups and hence not imply activation of competence. In the following discussion, we will examine several factors that support or suggest the value of the competence–activation model.

A strong candidate as a class of activation variables is the figurative features or stimulus saliency of the task. If the young child's (5 years of age) competence is characterized primarily by a figurative approach to problems and the older child's competence is operational in nature, then a decrease in figurative task features or an increase in logical task features should facilitate the activation of operational competence (see Inhelder & Piaget, 1964). Reasoning from this position, Overton and Brodzinsky (1972) examined the effect of reduced figurative task features on the solution of cross-classification or multiple-classification problems. Three groups of children (4 to 5, 6 to 7, and 8 to 9 years of age) were presented with either of two forms of a matrix completion task. One form was the standard task which has strong figurative demands, and the second form reduced the figurative features of the task while maintaining the logical task requirements. In support of the activation hypothesis, only the transitional group (6- to 7-year-olds) showed any effect of task differences. This group's performance was significantly improved under the logical task structure.

In a recent paper, Odom (1978) has argued that since with an increase in stimulus saliency, 4-year-olds can solve matrix completion problems, figurative processes may provide an alternative conception to Piaget's position concerning cognitive change. However, it should be noted that an increase in stimulus saliency with respect to this task actually lowers the probability that the test is measuring Piaget's operational or logical competence model (see also, Overton & Jordan, 1971).

Reducing figurative task features as a means of activating operational competence has also been examined with respect to spatial skills. Brodzinsky, Jackson, and Overton (1972) tested children at 6, 8, and 10 years of age on modified versions of Piaget's three mountains task (single-object array and multiple-object arrays). Half the children at each age received the standard form of presentation, while half had the scene perceptually shielded from their view as they chose pictorial representations of several alternative spatial perspectives of the scene. Results supported the activation hypothesis in that performance was enhanced under the shielded condition for the older ages. This effect was primarily limited to the multiple arrays. Brodzinsky and Jackson (1973) later replicated the general finding and went on to examine the interaction between shielding and the internal complexity of the object arrays. Walker and Gollin (1977) conducted a similar study but found that shielding led to a reduction of egocentric-type errors in a 4-year-old group. Although this finding does not support the activation hypothesis, it should be noted that this study employed only a single-object array and therefore may implicate an earlier level of competence than operational competence (see Flavell, Everett, Croft, & Flavell, 1981; Masangkay, McCluskey, McIntyre, Sims-Knight, Vaughn, & Flavell, 1974).

Reducing figurative task features by the procedure of perceptual shielding has also been employed in studies of conservation. Here, however, the issue is complicated by the fact that degree of shielding of perceptual conflict features can in fact transform the problem to one that does not require operational competence (i.e., pseudoconservation, Piaget, 1968a). That is, if conservation entails a conceptual quantitative invariance in the face of perceptual conflict, then the absence of perceptual conflict effectively destroys the structural task demands. Beilin (1978) and Miller (1978) have reviewed investigations of conservation shielding, and each suggests interpretations compatible with the activation hypothesis. Of greatest relevance is a study by Strauss and Langer (1970) which demonstrated screening effects were most pronounced for transitional children.

In an early study of class inclusion, Wohlwill (1968) found that reducing distracting perceptual cues by presenting inclusion questions in verbal form significantly enhanced performance of transitional children. Although Winer (1980) has suggested that this activation effect may be primarily due to the additional verbal cues, the perceptual reduction effect has not been ruled out by appropriate designs focusing on this feature. In general, a number of class inclusion studies have found various figurative or stimulus saliency effects; however, these have not been adequate to test the activation hypothesis, because of either the absence of appropriate designs and controls or because the type of tasks employed tap figurative processes rather than the logical competence of class inclusion (see Winer, 1980, for a review).

With respect to conservation, Miller (1978) has reviewed a number of variables beyond figurative and stimulus saliency features which might serve an activation function. These include the interest value and familiarity of task material, the number of task components, and the type and degree of transformation needed for solution. Miller, in turn, relates these and other variables mentioned earlier to processes of selective attention and attentional capacity (i.e., activation/utilization model processes). Although the evidence reviewed is compatible with the activation hypothesis (i.e., the effect of the variables are generally related to cognitive level), it is only suggestive due to the lack of designs appropriate to generate specific experimental tests of activation.

With respect to the importance of familiar task content in the activation of competence, it should be noted that some researchers (e.g., Uzgiris, 1964; Zimmerman & Lanaro, 1974) explain the phenomenon of horizontal decalage (i.e., the fact that the various conservation notions are acquired at different times and in a certain order) as due to the differing degrees of familiarity the child has with the materials involved. For example, the child may deal with number-related activities more frequently than weight-related activities. Thus, according to Zimmerman and Lanaro (1974), it is not surprising that the acquisition of number conservation precedes the acquisition of weight conservation.

Another variable that may produce activation effects is task instructions. White and Glick (1978) tested kindergarten, first-, second-, and third-grade minority children on liquid conservation following instructions that varied according to whether they alerted the children to possible deception and whether the deception was embedded in the context of a story. Results indicated enhanced performance for the story-deception group, and this effect was limited to the transitional first and second graders.

As described earlier, training effects have increasingly been interpreted as affecting behavior only to the extent that the target competence is already present. Thus although training does not indicate specifically which activation variables are relevant, training may function to activate latent competence or to disclose that competence is not available. Beilin (1971, 1976, 1978) and Strauss (1972, pp. 337–340) review evidence relevant to this interpretation. Here we will examine some research not included in earlier reviews.

In an examination of cross or multiple classification, Overton and Kirschner (1972) trained children 4, 6, 8, and 10 years of age on 16 matrix completion problems. Training consisted of corrective feedback, and the experimental effects were limited to the 6-year-old group. Here, corrective feedback provided by the experimenter resulted in significant error reduction. Another study of cross classification (Parker, Sperr, & Rieff, 1972) is particularly interesting because it supports the activation model being described, despite the fact that neither earlier research nor competence considerations played any apparent role in the conceptualization of the investigation. In this study, children 5½, 6½, and 7½ years of age were presented with either a 12-step sequenced feedback training program, a single-step feedback training program, or a control condition. Performance on a transfer task demonstrated significant training effects only at the 6½- and 7½-year level. Halford (1980) has recently extended this work and suggests direct training effects occur at approximately the 5½-year level when mental age, rather than chronological age, is employed.

Several investigations which examined training effects with respect to class inclusion competence are also relevant to the activation hypothesis. In one study, Youniss (1971) extended an earlier class inclusion investigation conducted by Ahr and Youniss (1970). In the later study, children were pretested on a series of tasks considered to be precursors to class inclusion and then received a correction training procedure. The results supported a conclusion that the training was effective for those children who already had acquired the prerequisite structural competence. Langer (in press) attempted to induce recognition of a contradiction prior to class inclusion testing. He found that only children who were transitional benefited from the training. Similarly, Inhelder, Sinclair, and Bovet (1974) found that the effect of multiple training procedures on class inclusion was limited by the child's initial level of competence.

To this point the activation hypothesis has been discussed exclusively with respect to concrete operational competence. This has been dictated by the fact that virtually all the research conducted on this issue has been directed to this particular competence. This focus has been limited less by theoretical considerations than by the fact that traditionally it has been more difficult to determine a reasonable time interval during which formal operational competence is latently available to the child (e.g., see Sims-Knight, 1979). There are however several studies that are related to the issue of the activation of formal operational competence. In one study, Moshman (1977) tested tenth- and eleventh-grade children for their understanding of logical implication and disjunction. Arguing from Flavell and Wohlwill's (1969) four-phase model, Moshman predicted that the negation of these logical forms—considered as a performance variable—would be unrelated to early and late phases of competence but would suppress the normal activation of competence during phase three. The results supported this prediction. This study is unique both in its use of the phase model and in suggesting a performance variable that may hinder activation.

Two studies have been conducted which focus on the activation of logical conditional reasoning through the manipulation of task instructions. O'Brien and Overton (1980) presented third-grade, seventh-grade, and college students with a series of conditional inference problems. Following a procedure introduced by Wason (1964), half the subjects at each age level received evidence that contradicted possible faulty inferences. Significant improvement was found for the college group following the introduction of the contradictory evidence, and this effect generalized to new conditional-reasoning tasks. In an attempt to establish a more precise point at which this manipulation first becomes effective, and hence more clearly relevant to activation, O'Brien and Overton (in press) replicated the study with eighth and twelfth graders. In this study, significant improvement and generalization to new tasks was found for the twelfth-grade group. Research is currently in progress testing the intervening grades.

In conclusion, the research described in this section illustrates the heuristic value of an approach which considers contingent effects as determinants of competence activation rather than as determinants of competence itself. Obviously, further systematic empirical research employing appropriate designs and controls is needed to support and elaborate this perspective. Of particular value would be research focusing on the generality of activation effects across tasks and across time. In this section we have focused on how several task, instructional, and general training features affect recently acquired latent competence. In the following section we will examine how these features, as well as various individual difference factors, affect the expression of maturely functioning competence.

THE UTILIZATION OF COMPETENCE

Although there are divergent views on this issue (see Flavell & Wohlwill, 1969; Flavell, 1977, p. 115), the only necessary implication of any competence model of behavior is that the behaviors represented can be performed across a variety of content areas. That is, a competence model represents capacity, and issues of spontaneity of application or level of proficiency (Flavell, 1977) are features of the activation/utilization model. Thus given that competence has been activated, the question remains as to whether there are systematic factors which affect the conditions under which the competence is applied. This brings us to a discussion of utilization variables.

Conceptually, the distinction between activation and utilization, although at times somewhat arbitrary, is roughly analogous to the distinction drawn by the experimental ethologists and behavioral embryologists (Bateson, 1978; Gottlieb,

1976) between environmental factors that facilitate development and environmental factors that maintain mature behavioral systems. Obviously, a single variable or set of variables might serve neither, one, or both of these functions. At the level of research, if a particular factor affects only phases 2 and 3 of Flavell and Wohlwill's four-phase model, then it serves only an activation function. If, on the other hand, the factor serves primarily to enhance relatively adequate performance or is effective well after the time mature competence has been demonstrated for most individuals, then its function is that of utilization. Finally, to the extent that a factor is effective both at the time of latent competence and continuously effective across an extended time, it has both activation and utilization characteristics. As an example, in the O'Brien and Overton (1980, in press) studies described earlier, a purpose of the research was to explore the question of the activation of competence. However, if future research demonstrates that the contradictory instructional effects are also present at, for example, the tenth- and eleventh-grade level in addition to their impact at the twelfth-grade and college level, this factor would reasonably be considered an activation/utilization variable.

A theoretically systematic approach to the examination of the relation between competence and utilization has been proposed by Pascual-Leone (see Chapman, 1981, for a summary). Pascual-Leone's formulation is essentially a general theory of cognitive development. He begins from a basic assumption of Piaget's competence models, but the actual behavior of a subject is determined by the interaction of competence schemes and specific organismic factors termed "silent operators." These include the subject's previous learning experiences, perceptual field effects (i.e., stimulus effects), affective and personality factors, and mental effort (M operator). The silent operators, or what we would term utilization factors, are weighted with respect to the probability of producing a competence-related behavior. The relation between competence and utilization factors is then applied to specific issues such as (1) the effects of a task's information load on the performance of that task; (2) the role of perceptual field effects in inhibiting or facilitating performance of that task; (3) the problem of horizontal decalages (Pascual-Leone & Sparkman, 1980); and (4) the problem of style and other individual difference effects (Pascual-Leone, 1976b).

Several investigations of formal operational competence have suggested that reduction of task ambiguity through modified instructions, repeated presentation, or task modifications enhances utilization (Danner & Day, 1977; Kuhn, Ho, & Adams, 1979; Martorano, 1977; Stone & Day, 1978, 1980). Further, Martorano (1980) has shown that figurative factors (arrangement of stimuli and visual noise) suppress successful utilization, and Martorano and Zentall (1980) suggest that prior experience designed to focus attention on critical task dimensions may serve an activation/utilization function.

Of particular interest with respect to activation/utilization are the Stone and Day (1978, 1980) studies. These investigators assessed the formal operational competence of 9-, 11-, 12-, 13-, and 14-year-olds using Inhelder and Piaget's (1958) bending rods task. Initially, modified instructions and repeated task presentations were employed as activation variables to assess whether subjects (1) lacked competence, (2) possessed available but latent competence, or (3) spontaneously applied competence. Task modifications were also presented and this variable affected performance of both the latent and spontaneous groups; that is functioned for activation/utilization. Several additional measures, including tests of memory capacity, selective attention, and cognitive style (field independence), were employed. Although memory and attention did not differentiate the latent and spontaneous group, the spontaneous group performed significantly better on a test of field independence. This variable did not differentiate the latent and noncompetent groups. This finding strongly supports the view that cognitive style functions primarily as a utilization variable.

Studies which have further examined the relationship between cognitive style and formal operational competence are reviewed by Neimark (1979). These have assessed reflection–impulsivity (Riebman & Overton, 1977), field dependence/independence (Lawson, 1976; Linn, 1978; Saarni, 1973), or both (Neimark, 1975b). Neimark (1979) points out that the results from this work are consistent with the view of formal operational thinkers as being reflective, analytic, and field independent. Unfortunately, although they are suggestive, these studies were not specifically designed to yield direct support for the activation/utilization functions of cognitive style.

An investigation by Brodzinsky (1980) on reflection–impulsivity and concrete operational competence (spatial perspective-taking) is more specifically relevant to the issue of cognitive style and activation/utilization. On tests of 4-, 6-, and 10-year-olds, reflectivity was associated with performance on spatial perspective-taking only at the 6- and 8-year levels. This suggests that cognitive style may have served an activation function. However, for the 6-year-old group, cognitive style was related to specific error patterns (egocentric and adjacent errors), while for the 8-year-old group, style was related to overall accuracy. Although the effect for the younger group might reflect an earlier entrance into and faster development through the transition period for the reflective children, the effect reported for the 8-year-olds appears more consistent with a utilization function. For the 10-year-old group, a ceiling effect precluded style-related task differences. Liben's (1978) finding that field dependence is negatively related to performance on Piagetian spatial tasks at a time well past normal consolidation of concrete operational competence, that is, twelfth grade, provides additional support for interpreting cognitive style as a utilization variable.

In concluding this section, we examine two topic areas in which group performance differences are relevant to the issue of competence and its application. One topic of interest concerns the fate of competence in the elderly. Hornblum and Overton (1976), following Bearison (1974), argued that poor performance on concrete operational tasks (conservation) among the elderly was due to a failure of utilization rather than a lack of competence. This view was supported when, following brief exposure to corrective feedback, the performance of these elderly subjects was significantly enhanced on near and far transfer tasks as compared to subjects in a control group. Levin and Overton (1977) extended the general argument by demonstrating that following training for cognitive flexibility (set-breaking), the performance of elderly subjects on a spatial perspective-taking task improved significantly.

A second topic involves findings of individual differences for competence-related behaviors. Sex differences are illustrative of the general issue. Several investigations have found sex differences with respect to both concrete and formal operational competence (Chapman, 1975; Dale, 1970; Douglas & Wong, 1977; Elkind, 1961,

1962; Keating & Schaeffer, 1975; Liben, 1978; Liben & Golbeck, 1980; Martorano & Zentall, 1980; Riebman & Overton, 1977). When this finding occurs, females generally perform more poorly than males. This result is open to two interpretations: females may lag behind males developmentally in the acquisition of competence or the differences may be the product of activation/utilization variables which depress the expression of competence.

Liben (1978) and Liben and Golbeck (1980) examined the problem of sex differences on Piagetian spatial tasks from the perspective of the competence–performance distinction. As mentioned earlier, Liben and Goldbeck (1980) presented children in grades 3, 5, 7, 9, and 11 with modified forms (activation/utilization variables) of the spatial tasks. However, as in the original Liben (1978) study, sex differences again emerged. Although this finding is consistent with the developmental-lag hypothesis, it also provides the basis for moving in new directions with respect to other potential activation/utilization variables. Thus following Elkind's (1961, 1962) suggestion that sex difference in operational competence may reflect different social-role expectations for males and females, Overton and Meehan (1981) examined the formal operational competence of 13-year-olds in relation to sex roles and learned helplessness. This study found both variables to be significantly related to the expression of competence. Similar results for sex role are reported by Signorella and Jamison (1978) and Jamison and Signorella (1980) for spatial tasks. Along similar lines, Newman-Hornblum, Attig, and Kramer (1980) recently found that when the content of Piagetian concrete and formal operational tasks is modified to match the sex-role orientation of elderly subjects, task performance is affected.

CONCLUSION

The competence–activation/utilization approach described in this chapter offers two advantages over traditional approaches to the understanding of development. First, it provides a framework within which traditionally rival forms of explanation can be integrated. Competence and changes in competence are explained through formal explanation, activation and utilization are explained through contingent causal explanation. Second, the approach suggests several research strategies that can be applied in an effort to differentiate and evaluate the role of various necessary component processes in the functioning of the developing individual. The research reviewed suggests that this provides a more integrated and realistic perspective than one that focuses on a single component and treats other features as noise or error.

As a final note we would like to emphasize that although Piaget's competence theory has been focal in this chapter, the general approach described is applicable whenever a competence model is generated. In the field of cognitive development, as Siegler (1980) has noted, the use of task analyses to generate rule systems that describe the competence of the child is becoming increasingly popular. Given the assumption that such rule systems will prove to have reasonable scope, the issue of the conditions under which they are activated and utilized will be open to the approach and strategies discussed in this chapter.

REFERENCES

Ahr, P. R., & Youniss, J. Reasons for failure on the class inclusion problem. *Child Development,* 1970, **41**, 131–143.

Bandura, A. *Social learning theory.* Englewood Cliffs, N.J.: Prentice-Hall, 1977.

Bateson, P. P. G. How does behavior develop? In P. P. G. Bateson & P. H. Klopfer (Eds.), *Perspectives in ethology* (Vol. 3). New York: Plenum, 1978.

Bearison, D. The construct of regression: A Piagetian approach. *Merrill-Palmer Quarterly,* 1974, **20**, 21–30.

Beilin, H. The training and acquisition of logical operations. In M. Rosskopf, L. P. Stette, & S. Taback (Eds.), *Piagetian cognitive-developmental research and mathematical education.* Washington, D.C.: National Council of Teachers of Mathematics, 1971.

Beilin, H. Constructing cognitive operations linguistically. In H. W. Reese (Ed.), *Advances in child development and behavior* (Vol. 11). New York: Academic, 1976.

Beilin, H. Inducing conservation through training. In G. Steiner (Ed.), *Psychology of the 20th century (Vol. 7); Piaget and beyond.* Zurich: Kindler (German), 1978, pp. 260–289.

Bowers, K. S. Situationism in psychology: An analysis and a critique. *Psychological Review,* 1973, **80**, 307–336.

Braine, M. S. The ontogeny of certain logical operations: Piaget's formulation examined by nonverbal methods. In I. E. Sigel & F. H. Hooper (Eds.), *Logical thinking in children: Research based on Piaget's theory.* New York: Holt, Rinehart, and Winston, 1968, pp. 164–206.

Brainerd, C. J. Judgments and explanations as criteria for the presence of cognitive structures. *Psychological Bulletin,* 1973, **79**, 172–179.

Brodzinsky, D. M. Cognitive style differences in children's spatial perspective taking. *Developmental Psychology,* 1980, **16**, 151–152.

Brodzinsky, D. M., & Jackson, J. P. Effects of stimulus complexity and perceptual shielding in the development of spatial perspectives. Paper presented at the Biennial Meeting of the Society for Research in Child Development, Philadelphia, 1973.

Brodzinsky, D. M., Jackson, J. P., & Overton, W. F. Effects of perceptual shielding in the development of spatial perspectives. *Child Development,* 1972, **43**, 1041–1046.

Brown, G., & Desforges, C. *Piaget's theory: A psychological critique.* London: Routledge and Kegan Paul, 1979.

Carmichael, L. Onset and early development of behavior. In P. H. Mussen (Ed.), *Carmichael's manual of child psychology* (Vol. 1). New York: Wiley, 1970, pp. 447–563.

Chapman, M. Pascual-Leone's theory of constructive operators: An introduction. *Human Development,* 1981, **24**, 145–155.

Chapman, R. H. The development of children's understanding of proportions. *Child Development,* 1975, **46**, 141–148.

Chomsky, N. *Aspects of the theory of syntax.* Cambridge, Mass.: MIT Press, 1965.

Cronbach, L. J. Beyond the two disciplines of scientific psychology. *American Psychologist,* 1975, **30**, 116–127.

Dale, L. G. The growth of systematic thinking: Replication and analysis of Piaget's first chemical experiment. *Australian Journal of Psychology,* 1970, **22**, 277–286.

Danner, F. N., & Day, M. C. Eliciting formal operations. *Child Development,* 1977, **48**, 1600–1606.

Douglas, J. A., & Wong, A. C. Formal operations: Age and sex differences in Chinese and American children. *Child Development,* 1977, **48**, 689–692.

Elkind, D. Quantity conceptions in junior and senior high school students. *Child Development,* 1961, **32**, 551–560.

Elkind, D. Quantity conceptions in college students. *Journal of Social Psychology,* 1962, **57**, 459–465.

Feldman, C. F., & Toulmin, S. Logic and the theory of mind. In W. J. Arnold (Ed.), *Nebraska symposium on motivation—1975*. Lincoln: University of Nebraska Press, 1976, pp. 409–476.

Flavell, J. H. Developmental studies of mediated memory. In H. W. Reese & L. P. Lipsitt (Eds.), *Advances in child development and behavior* (Vol. 5). New York: Academic, 1970.

Flavell, J. H. *Cognitive development*. Englewood Cliffs, N.J.: Prentice-Hall, 1977.

Flavell, J. H., Everett, B. A., Croft, K., & Flavell, E. R. Young children's knowledge about visual perception: Further evidence for the Level 1–Level 2 distinction. *Developmental Psychology*, 1981, **17**, 99–103.

Flavell, J. H., & Wohlwill, J. Formal and functional aspects of cognitive development. In D. Elkind and J. Flavell (Eds.), *Studies in cognitive development: Essays in honor of Jean Piaget*. New York: Oxford University Press, 1969.

Gelman, R. Cognitive development. *Annual Review of Psychology*, 1978, **29**, 297–332.

Gelman, R., & Gallistel, C. R. *The child's understanding of number*. Cambridge, Mass.: Harvard University Press, 1978.

Gergen, K. J. The emerging crisis in life-span developmental theory. In P. B. Baltes & O. G. Brim (Eds.), *Life-span development and behavior* (Vol. 3). New York: Academic, in press.

Gottlieb, G. Conceptions of prenatal development: Behavioral embryology. *Psychological Review*, 1976, **83**, 215–234.

Halford, G. S. A learning set approach to multiple classification: Evidence for a theory of cognitive levels. *International Journal of Behavioral Development*, 1980, **3**, 409–422.

Holland, V. M., & Palermo, D. S. On learning "less": Language and cognitive development. *Child Development*, 1975, **46**, 437–443.

Hooper, F. H., Toniolo, T. A., & Sipple, T. S. A longitudinal analysis of logical reasoning relationships: Conservation and transitive inference. *Developmental Psychology*, 1978, **14**, 674–682.

Hooper, F. H., Wanska, S. K., Peterson, G. W., & DeFrain, J. The efficacy of small group instructional activities for training Piagetian concrete operations concepts. *The Journal of Genetic Psychology*, 1979, **134**, 281–297.

Hornblum, J. N., & Overton, W. F. Area and volume conservation among the elderly: Assessment and training. *Developmental Psychology*, 1976, **12**, 68–74.

Hull, C. L. *Principles of behavior*. New York: Appleton-Century, 1943.

Inhelder, B., & Piaget, J. *The growth of logical thinking from childhood to adolescence*. New York: Basic Books, 1958.

Inhelder, B., & Piaget, J. *The early growth of logic in the child*. New York: Norton, 1964.

Inhelder, B., & Sinclair, H. Learning cognitive structures. In P. Mussen, J. Langer, & M. Covington (Eds.), *Trends and issues in developmental psychology*. New York: Holt, Rinehart, and Winston, 1969.

Inhelder, B., Sinclair, H., & Bovert, M. *Learning and the development of cognition*. Cambridge, Mass.: Harvard University Press, 1974.

Jamison, W., & Signorella, M. L. Sex-typing and spatial ability: The association between masculinity and success on Piaget's water-level task. *Sex Roles*, 1980, **6**, 345–353.

Keating, D. P., Adolescent thinking. In J. P. Adelson (Ed.), *Handbook of adolescence*, New York: Wiley, 1979.

Keating, D. P., & Schaeffer, R. A. Ability and sex differences in the acquisition of formal operations. *Developmental Psychology*, 1975, **11**, 531–532.

Kuhn, D., Ho, V., & Adams, C. Formal reasoning among pre- and late adolescents. *Child Development*, 1979, **50**, 1128–1135.

Langer, J. Dialectics of development. In T. G. Bever (Ed.), *Regressions in development: Basic phenomena and theories*. Hillsdale, N.J.: Erlbaum, in press.

Larsen, G. Y. Methodology in developmental psychology: An examination of research on Piagetian theory. *Child Development,* 1977, **48,** 1160–1166.

Lawson, A. E. Formal operations and field independence in a heterogeneous sample. *Perceptual and Motor Skills,* 1976, **42,** 981–982.

Levin, B., & Overton, W. F. Perspective-taking and mental set among the aged: Assessment and training. Paper presented at the American Psychological Association Meeting, San Francisco, 1977.

Lewin, K. *Field theory in social science: Selected theoretical papers.* New York: Harper and Brothers, 1951.

Liben, L. S. Performance on Piagetian spatial tasks as a function of sex, field dependence, and training, *Merrill-Palmer Quarterly,* 1978, **24,** 97–110.

Liben, L. S., & Golbeck, S. L. Sex differences on Piagetian spatial tasks: Differences in competence or performance? *Child Development,* 1980, **51,** 594–597.

Linn, M. C. Influence of cognitive style and training on tasks requiring the separation of variables schema. *Child Development,* 1978, **49,** 874–877.

Martorano, S. A developmental analysis of performance on Piaget's formal operational tasks. *Developmental Psychology,* 1977, **13,** 666–672.

Martorano, S. Elicitation of a combinatorial problem solving strategy. Unpublished manuscript, 1980.

Martorano, S., & Zentall, T. R. Children's knowledge of the separation of variables concept. *Journal of Experimental Child Psychology,* 1980, **30,** 513–526.

Masangkay, Z. S., McCluskey, K. A., McIntyre, C. W., Sims-Knight, J., Vaughn, B. E., & Flavell, J. H. The early development of inferences about the visual percepts of others. *Child Development,* 1974, **45,** 357–366.

Miller, G. A. Some comments on competence and performance. In D. Aaronson & R. W. Rieber, *Developmental psycholinguistics and communication disorders.* New York: New York Academy of Sciences, 1975, pp. 201–204.

Miller, P. H. Stimulus variables in conservation: An alternative approach to assessment. *Merrill-Palmer Quarterly,* 1978, **24,** 141–160.

Mischel, W. Toward a cognitive social learning reconceptualization of personality. *Psychological Review,* 1973, **80,** 252–283.

Moshman, D. Consolidation and stage formation in the emergence of formal operations. *Developmental Psychology,* 1977, **13,** 95–100.

Neimark, E. D. Intellectual development during adolescence. In F. D. Horowitz (Ed.), *Review of child development research* (Vol. 4). Chicago: University of Chicago Press, 1975, pp. 541–594. (a)

Neimark, E. D. Longitudinal development of formal operations thought. *Genetic Psychology Monographs,* 1975, **91,** 171–225. (b)

Neimark, E. D. Current status of formal operations research. *Human Development,* 1979, **22,** 60–67.

Neimark, E. D. Toward the disembedding of formal operations from confounding with cognitive style. In I. Sigel, R. Golinkoff, & D. Brodzinsky (Eds.), *Piagetian theory and research: New directions and applications.* Hillsdale, N.J.: Erlbaum, 1981.

Newman-Hornblum, J., Attig, M., & Kramer, D. Sex-relevant Piagetian tasks and cognitive competence in the elderly. Paper presented at the 88th annual convention of the American Psychological Association, Montreal, 1980.

Niebuhr, V. N., & Molfese, V. J. Two operations in class inclusion: Quantification of inclusion and hierarchical classification. *Child Development,* 1978, **49,** 892–894.

O'Brien, D., & Overton, W. F. Conditional reasoning following contradictory evidence: A developmental analysis. *Journal of Experimental Child Psychology,* 1980, **30,** 44–61.

O'Brien, D. & Overton, W. F. Conditional reasoning and the competence-performance issue: A developmental analysis of a training task. *Journal of Experimental Child Psychology,* in press.

Odom, R. D. A perceptual-salience account of décalage relations and developmental change. In L. S. Siegel & C. J. Brainerd (Eds.), *Alternatives to Piaget's theory: Critical essays on the theory*. New York: Academic, 1978, pp. 111–130.

Overton, W. F. Piaget's theory of intellectual development and progressive education. In J. R. Squire (Ed.), *A new look at Progressive Education*. Washington Association for Supervision and Curriculum Development, 1972, pp. 88–115.

Overton, W. F. General systems, structure, and development. In K. Riegel and G. Rosenwald (Eds.), *Structure and transformation: Developmental aspects*. New York: Wiley Interscience, 1975, pp. 61–81.

Overton, W. F. Environmental ontogeny: A cognitive view. In K. Riegel & J. Meacham (Eds.), *The developing individual in a changing world* (Vol. 2). *Social and environmental issues*. The Hague: Mouton, 1976.

Overton, W. F. Klaus Riegel: Theoretical contribution to concepts of stability and change. *Human Development,* 1978, **21**, 360–363.

Overton, W. F., & Brodzinsky, D. Perceptual and logical factors in the development of multiplicative classification. *Developmental Psychology,* 1972, **6**, 104–109.

Overton, W. F., & Jordan, R. Stimulus preference and multiplicative classification in children. *Developmental Psychology,* 1971, **5**, 505–510.

Overton, W. F., & Kirschner, B. The role of activation in the development of multiplicative classification skills. Unpublished manuscript, SUNY, 1972.

Overton, W. F., & Meehan, A. Individual differences in formal operational thought: Sex-role and learned helplessness. Paper presented at the 11th annual symposium of the Jean Piaget Society, Philadelphia, 1981.

Overton, W. F., & Newman-Hornblum, J. Cognitive intervention research: A competence-performance model. Paper presented at the 33rd annual scientific meeting of the Gerontological Society of America, San Diego, 1980.

Overton, W. F., & Reese, H. W. Models of development: Methodological implications. In J. R. Nesselroade & H. W. Reese (Eds.), *Life-span developmental psychology: Methodological issues,* New York: Academic Press, 1973, pp. 65–86.

Overton, W. F., & Reese, H. W. Conceptual prerequisites for an understanding of stability–change and continuity–discontinuity. *International Journal of Behavioral Development,* 1981, **4**, pp. 99–123.

Overton, W. F., Wagner, J., & Dolinsky, H. Social class differences and task variables in the development of multiplicative classification. *Child Development,* 1971, **42**, 1951–1958.

Parker, R., Sperr, S., & Rieff, M. Multiple classification: A training approach. *Developmental Psychology,* 1972, **7**, 188–194.

Pascual-Leone, J. (1976a). A view of cognition from a formalist's perspective. In K. Riegel & J. Meacham (Eds.), *The developing individual in a changing world*. The Hague: Mouton, 1976. (a)

Pascual-Leone, J. On learning and development, Piagetian style, II: A critical historical analysis of Geneva's research programme. *Canadian Psychological Review,* 1976, **17**, 289–297. (b)

Pascual-Leone, J., & Sparkman, E. The dialectics of empiricism and rationalism: A last methodological reply to Trabasso. *Journal of Experimental Child Psychology,* 1980, **29**, 88–101.

Piaget, J. *On the development of memory and identity*. Barre, Mass.: Clark University Press and Barre Publishers, 1968. (a)

Piaget, J. Quantification, conservation, and nativism. *Science,* 1968, **162**, 976–979. (b)

Postman, L. Association and performance in the analysis of verbal learning. In T. R. Dixon & D. L. Horton (Eds.), *Verbal Behavior and General Behavior Theory*. Englewood Cliffs, N.J.: Prentice-Hall, 1968, pp. 550–571.

Reese, H. W. Verbal mediation as a function of age level. *Psychological Bulletin,* 1962, **59**, 502–509.

Riebman, B., & Overton, W. F. Reflection–impulsivity and the utilization of formal operational thought. Paper presented at the Biennial Meeting of the Society for Research in Child Development, New Orleans, 1977.

Saarni, C. I. Piagetian operations and field independence as factors in children's problem solving performance. *Child Development,* 1973, **44**, 338–345.

Scardamalia, M. Information processing capacity and the problem of horizontal decalage: A demonstration using combinatorial reasoning tasks. *Child Development,* 1977, **48**, 28–37.

Shayer, M. Has Piaget's construct of formal operational thinking any utility? *British Journal of Educational Psychology,* 1979, **49**, 265–276.

Siegler, R. S. Recent trends in the study of cognitive development: Variations on a task analytic theme. *Human Development,* 1980, **23**, 278–285.

Siegler, R. S. Developmental sequences within and between concepts. *Society for Research in Child Development Monographs,* in press.

Siegler, R. S., & Richards, D. D. Development of time, speed and distance concepts. *Developmental Psychology,* 1979, **15**, 288–298.

Signorella, M. L., & Jamison, W. Sex differences in the correlations among field dependence, spatial ability, sex role orientation, and performance on Piaget's water-level task. *Developmental Psychology,* 1978, **14**, 689–690.

Simms-Knight, J. E. Training second order operations in preadolescents and adolescents. Paper presented at the Meeting of the American Psychological Association, New York, 1979.

Stone, C. A., & Day, M. C. Levels of availability of a formal operational strategy. *Child Development,* 1978, **49**, 1054–1065.

Stone, C. A., & Day, M. C. Competence and performance models and the characterization of formal operational skills. *Human Development,* 1980, **23**, 323–353.

Strauss, S. Inducing cognitive development and learning: A review of short-term training experiments, I. The organismic developmental approach. *Cognition,* 1972, **1**, 329–357.

Strauss, S., & Langer, J. Operational thought inducement. *Child Development,* 1970, **41**, 163–175.

Tomlinson-Keasey, C. Structures, functions and stages: A trio of formal operational constructs. In S. Modgil (Ed.), *Jean Piaget: Consensus and controversy.* England: Praeger, in press.

Tomlinson-Keasey, C., Eisert, D. C., Kahle, L. R., Hardy-Brown, K., & Keasey, B. The structure of concrete operational thought. *Child Development,* 1979, **50**, 1153–1163.

Trabasso, T., Isen, A. M., Dolecki, P., McLanahan, A. G., Riley, C. A., & Tucker, T. How do children solve class-inclusion problems? In R. S. Siegler (Ed.), *Children's thinking: What develops?,* Hillsdale, N.J.: Erlbaum, 1978, pp. 151–180.

Tulving, E., & Pearlstone, Z. Availability versus accessibility of information in memory for words. *Journal of Verbal Learning and Verbal Behavior,* 1966, **5**, 381–391.

Uzgiris, I. Situational generality of conservation. *Child Development,* 1964, **35**, 831–841.

Walker, L. D., & Gollin, E. S. Perspective role-taking in young children. *Journal of Experimental Child Psychology,* 1977, **24**, 343–357.

Wason, P. C. The effect of self-contradiction on fallacious reasoning. *Quarterly Journal of Experimental Psychology,* 1964, **16**, 30–34.

Werner, H. Process and achievement: A basic problem of educational and developmental psychology. *Harvard Educ. Review,* 1937, **7**, 353–368.

White, D. E., & Glick, J. Competence and the context of performance. Paper presented at the eighth annual symposium of the Jean Piaget Society, Philadelphia, 1978.

Wilkening, F., & Anderson, N. H. Comparison of two rule assessment methodologies for studying cognitive development. (Center for Human Information Processing Tech. Rep. No. 94). University of California, San Diego, 1980.

Winer, G. A. Class-inclusion reasoning in children: A review of the empirical literature. *Child Development,* 1980, **51**, 309–328.

Wohlwill, J. F. Responses to class-inclusion questions for verbally and pictorially presented items. *Child Development,* 1968, **39**, 449–465.

Wohlwill, J. F. *The study of behavioral development.* New York: Academic, 1973.

Youniss, J. Classificatory schemes in relation to class inclusion before and after training. *Human Development,* 1971, **14**, 171–183.

Zimmerman, B. J., & Lanaro, P. Acquiring and retaining conservation of length through modeling and reversibility cues. *Merrill-Palmer Quarterly,* 1974, **20**, 145–161.

12

A BEHAVIORAL APPROACH TO CHILDREN'S LEARNING

James A. Sherman
Professor
Department of Human Development and Family Life
University of Kansas

Behavior of children becomes systematically more complex, more varied, and more appropriate as they grow older. The goals of developmental psychology are to describe these changes and to delineate the mechanisms which are responsible. Historically, two general approaches have been utilized to account for systematic changes in children's behavior. One approach has emphasized the role that environmental variables, events in the children's physical and social world, play in controlling or influencing growth of behavior. Another approach has focused on the development of mechanisms or states internal to the child, such as concepts, traits, stages, and competencies, which control or organize behavioral development. These two general approaches are not mutually exclusive. An emphasis on the development of internal states does not preclude the possible role of environmental events in affecting these states, or at least in providing the basic material by which they are developed. Similarly, an emphasis on the environment as the direct causal agent of behavioral development does not rule out the possible influence of events internal to the child. Nevertheless, the emphasis placed on the determinants of behavioral development, environmental or internal, and the resulting conception of development can differ considerably.

The first part of this chapter explores some of the assumptions, problems, and results of a behavioral learning view of the developmental process, a view that relies primarily upon the presumed influence of environmental events for that development. The development of behaviors involved in social interaction are emphasized. The second part of the chapter looks at some of the possible effects of adopting a behavioral learning approach on the developmental theory that results, what is accepted as the criterion of "understanding" development, and which behavioral phenomena are chosen for study and analysis.

Donald M. Baer, Aletha Huston, and Jan Sheldon-Wildgen provided both encouragement and thoughtful suggestions in preparing this chapter. I am very grateful.

A BEHAVIORAL LEARNING APPROACH TO DEVELOPMENT

Learning is a process in which behavior changes progressively as a result of interactions with environmental events. Let us consider the following example: suppose that we supervise a preschool classroom that attempts to foster the development of appropriate social behaviors of the children enrolled. One of the children in the classroom, Sam, appears to be a social isolate. He rarely talks to or approaches other children, never plays games with them, and does not respond to any of their invitations to participate in activities. Most of the time Sam sits in a corner by himself, simply watching the other children or engaging in solitary play or asking the preschool teachers for help. This is not an uncommon problem, and several studies have shown effective ways to develop greater amounts of social interaction with such children (Allen, Hart, Buell, Harris, & Wolf, 1964; Hart, Reynolds, Baer, Brawley, & Harris, 1968).

The most common technique for developing social interaction involves two components. First, the teachers and other children present numerous opportunities to engage in social interaction. The teachers might prompt the social interaction by arranging for one or more children to play an attractive activity or game close to Sam, and then lead him to it and suggest that he might want to play also. The other children (prompted earlier by the teachers) might ask Sam to join them or offer him materials with which to play the game. Second, the teachers would warmly approve of any attempts by Sam to play with the other children. These attempts initially might include only approaching the other children and standing nearby, then looking at the other children, accepting and holding materials offered by the other children, and later perhaps actually manipulating the material in conjunction with the other children in the context of the game. Thus the teacher would shape or reinforce closer and closer approximations to the desired behavior of playing with other children. (This assumes, of course, that the teachers' praise and attention was a reinforcing stimulus for Sam.)

These techniques are commonplace, commonsensical, and reasonably well documented methods for developing rudimentary social interactions among children who do not display them. Looked at carefully, however, the results may not be easy to analyze using a behavioral account of learning. Let us assume that Sam's developing social interactions are a result of the prompts by the teachers and the teachers' approval provided contingent upon Sam's behaviors which more and more closely approximated desirable social interactions. If so, the pertinent question for a simplistic behavioral analysis is, What specific behaviors were prompted and reinforced? Suppose that we find that walking up to other children, looking at them, accepting a toy car offered, and sitting among a group of children were behaviors that were directly prompted and/or reinforced. We note, however, that other behaviors not directly prompted began to appear, apparently for the first time, in Sam's repertoire. Sam began to smile at other children, talk to them, and offer them materials with which to play. When these occurred, they were, of course, praised by the teachers, but the question is, Where did they come from in the first place, to be praised by the teachers? It seems clear that behaviors that were not directly operated upon, at least until their initial appearance, nevertheless appeared.

Let us carry this example one step further by assuming that the teachers prompt and reinforce Sam to assemble puzzles with other children, to pull and be pulled by other children in wagons, to play jump rope, and to sit together and talk about pets. After these experiences, we observe that Sam begins to do many things with other children that he did not do before: he rides tricycles in a "wagon train," he helps push other children on the swings, and he plays hide and seek. Again, we observe the occurrence of behavior that Sam has never displayed previously and that has never been directly prompted, at least by the teachers.

Learning Mechanisms and Response Classes

How can we account for the occurrence of behaviors that apparently have not been taught directly? A number of possibilities exist. Some of these untaught behaviors may simply have had no opportunity to occur previously. Since Sam was rarely in the presence of other children, he may not have had an opportunity to smile at, talk to, and ride tricycles with other children. Thus simply increasing the amount of time that Sam was with other children may have increased the probability of other behavior. This account does not explain the existence of these other behaviors in Sam's repertoire but only their previous absence and their appearance later. It may be the case that the other children prompted and reinforced these new behaviors in Sam unbeknown to the teachers. This possibility is a variant on the notion of a "behavioral trap" proposed by Baer and colleagues (Baer & Wolf, 1970; Baer, Rowbury, & Goetz, 1976) in which development of a particular set of skills exposes a child to a set of environmental events hitherto not experienced, which serves to maintain behavior or promote new behavior in the absence of planned contingencies. It also may be the case that, together with an increased opportunity to engage in these other untaught behaviors, there exists a class of responses such that the strengthening of some members results in the strengthening of others as well. Thus increasing the likelihood that Sam approaches other children, looks at them, and accepts their offers of play materials also increases his probability of smiling at, talking to, and sharing materials with them. Similarly, directly strengthening cooperative play with wagons, on swings, and putting together puzzles may also result indirectly in increased cooperative play in other ways as well. Although this possibility describes the increased amounts of other untaught behaviors, it does not specify how these behaviors become part of such a hypothetical response class.

An example of the apparent development or formation of a response class is provided by several studies that taught imitation to children who previously did not imitate (Baer, Peterson, & Sherman, 1967; Garcia, 1976; Lovaas, Berberich, Perloff, & Schaeffer, 1966; Nordquist & Wahler, 1973). In one of these studies (Baer et al., 1967), retarded children were systematically taught to perform a response topographically similar to that demonstrated by a model; thus the model might raise a hand and the child would be taught to raise a hand immediately following. The initial teaching of the hand-raise response required both prompting (often manual guidance of the child's hand) and reinforcement of successively closer and closer approximations to the desired behavior. Eventually, when the model demonstrated a hand raise, the child did also. At this point, however, it was clear that nothing approaching a response class could be shown, for if the model demonstrated a response of clapping hands, the child either did nothing or raised a hand. Over time, however, the child was taught, through prompting and reinforcement, to match responses of clapping hands, putting a spoon in a cup, tapping the table with a wooden block, standing up and sitting down, and many others. In the course of this teaching, children began to display what might reasonably be called a developing response class: *they began to match new responses demonstrated by the model even though these new responses had never been taught before and even though these responses were topographically different than responses taught earlier.* For example, after being taught the series of responses listed above, a child might match a demonstration of putting the hands flat on the knees and continue to match this demonstration each time it was shown even though there was no direct prompting or reinforcement for doing so. Further, continued matching of these "novel" imitative responses seemed dependent upon reinforcement for other imitative responses, for when reinforcement was no longer provided or was provided but not contingent on some matching responses, all matching responses decreased in strength. Thus it seems fair to say that a response class of matching or imitation was the result of the training, in that strengthening or weakening some responses affected others as well. Put another way, the probability of occurrence of particular behaviors was not necessarily a result of what events impinged directly upon those behaviors but upon others.

It seems easy to understand how teaching a child to match some responses would result in the child learning to match other responses as well. After all, if we teach someone to "do as we do" in a number of instances, should we not expect that the person will do as we do in other instances as well? At the same time, it is difficult to point to the learning or behavioral mechanisms which characterize or underlie this process. One might argue that some sort of response induction or generalization occurred such that strengthening some responses, say involving the fingers, hands, and arms, thereby strengthened other responses similar in topography, thus accounting for the increased probability of "untaught" behavior. There is, in fact, some evidence to indicate that the topography of the imitative responses taught directly does influence the type of "untaught" imitative responses which result (Garcia, Baer, & Firestone, 1971). The response-induction argument is not entirely convincing, however, because some "untaught" imitative responses occur which appear to differ considerably in topography, if not in the parts of the body involved in their production. More important, the relatively precise correspondence between the model's behavior and the child's suggests something more than the presence of unstructured same-body-part response induction. It is as if the child is responding to be exactly similar to the model's behavior, at least with those body parts, as judged by an observer sitting off to the side.

A large number of behaviors that children display are similar to the example of imitation; following instructions, using appropriate verb tenses, being helpful to others, playing cooperatively, being aggressive, being creative, and others. Each of these presumed behavior classes is typically made up of a large number (perhaps an infinite number) of topographically different response members, yet these members covary in strength together rather than functioning independently.

How Response Classes Are Acquired

In teaching retarded children to imitate a model's behaviors, it is easy to see how the response class was formed: through the direct instruction, using prompting and reinforcement, by the model. If response classes are pervasive ways in which behavior is organized in children, how are they developed in the natural environment? At this point, we can only speculate, although several possibilities exist. One, as in the example of imitation, is through direct instruction by other people. In a large number of situations children are told how to behave. Parents, for example, often directly instruct their children in table manners and in appropriate ways of behaving in social situations, such as meeting other people and responding to invitations by others. Our educational system is based largely on the assumption that a wide range of academic skills can be taught through direct instruction that often includes verbal descriptions, demonstrations, prompting, practice, and feedback.

A second possible way that response classes might be developed is through observational learning. A great deal of evidence indicates that children can learn new behavior simply by watching others and by observing the actions of other people and consequences of these actions (Bandura, 1962; Thelen & Rennie, 1972). There is also evidence to suggest that children's behavior is affected by actions depicted on films and television (Stein & Friedrich, 1975). Considering that children may spend several hours each day watching television (Stein & Friedrich, 1975), this could constitute a powerful source of information about what response classes to use in various situations, the members of a response class, and the probable environmental consequences for these responses.

Third, it seems possible that some response classes might be formed directly by environmental consequences. For example, if a child wants to play with a toy currently in the possession of another child, there are a variety of ways to play with the toy. The child can ask the other for the toy, the child can bring another toy and offer to trade, the child can attempt to join in the game and thus have access to the desired toy, or the child can simply take the toy away from the other child. In particular situations some of these responses may be successful, resulting in access to the toy without any negative consequences from others in the environment, whereas others will not be successful. To the exent that there is consistency in the responses that are successful and unsuccessful, at least in that situation, one might expect a response class of these successful responses to be formed.

Undoubtedly, there are additional ways in which response classes might be formed. These class-forming mechanisms probably do not operate independently of each other; for example, a child may be told by his parents that the proper way to meet another is to look at the other person, say "Hello," and perhaps smile and shake hands. The child may also see other people engage in some or all of these behaviors and see that the other person smiled in return, shook hands, and acted in a friendly manner. Further, when the child engaged in appropriate meeting and greeting behaviors, the other person may respond in a positive manner and the parents may praise the child afterward. Thus there may be multiple opportunities and mechanisms that operate to form this particular response class.

Defining Properties of Response Classes

Up to this point the notion of a response class has been discussed by example, without any formal defining characteristics. The example of the development of imitation may be useful here. Once the retarded children had been taught to match a number of different responses demonstrated by the model, two effects became apparent. First, the children began to match new responses that had never been demonstrated before or directly taught (at least by the model). Thus the children began to perform new responses. Second, the children continued to match these new responses even though there was no reinforcement for doing so. Continued matching of these new responses, however, was not entirely independent of reinforcement, for when reinforcement was no longer provided for any matching responses, the children eventually stopped matching all responses. Thus reinforcement for some responses maintained other topographically different responses that never had direct consequences.

It is tempting to use these two characteristics, the occurrence of novel responses and the maintenance of some responses by operations performed on others, as defining characteristics of a response class. It seems likely, however, that these two qualities characterize only *one* type of response class: a generative response class or rule-governed response class, such as that exemplified by imitative behavior.

A number of "behaviors" seem good candidates as generative response classes: imitation, following instructions, much of language including forming plurals, use of verb tenses and basic sentence formation, completing long-division problems in arithmetic, and solving algebraic equations. In each of these examples, there are rules that characterize the responses that make up the class, such that teaching sufficient exemplars from the class will allow production of novel exemplars that have never been directly taught, as was the case when the children began to display new imitations which had not been directly taught.

There also may be responses that belong to the same class but cannot be characterized by abstract rule. For example, there are a number of ways a person can indicate agreement with or attention to what another person is saying in a conversation: by head nods, saying "umm humm," "yeah," or "I see." It does not seem likely that teaching a person a number of these exemplars alone would lead the person to use yet another topographically different form of indicating agreement or attention by, for example, smiling. This type of response class, then, would not display generative characteristics.

Thus the single common characteristic of both generative and nongenerative response classes is that all responses in the class can be strengthened or weakened by the same operation applied to only some of them. In the example of imitation, reinforcement provided contingent on some imitative responses served to maintain all imitative responses, even those that were not directly reinforced. To the extent that environmental consequences of behavior, such as positive reinforcement, control specific responses, it would seem that the ultimate control over the formation of response classes rests with these environmental consequences. There are, however, environmental events that precede the occurrence of behavior which can exert at least momentary or temporary control over responses and response classes. Their ability to continue to exert this control is determined, of course, by whether particular behaviors that occur in response to them have consistent consequences.

Again, the development of imitation with retarded children provides a good example. The children were taught to imitate demonstrations of a model in the context of a certain set of stimuli. The training was conducted in an isolated room; there was only one model; and the model looked up, made eye contact with the child, and said, "Do this," prior to demonstrating a response. It is very likely that some or some combination of these stimuli controlled the occurrence of imitative behavior. If a different model demonstrated a response to the child in a different room without first making eye contact and saying, "Do this," it is probably unlikely that the child would match the response. Thus the immediately preceding events could affect the probability of imitative behavior. Of course, whether or not the stimuli present in the training setting would continue to control the occurrence of imitative behavior and whether new stimuli could begin to control imitative behavior would depend upon what kinds of consequences there were for imitative behavior in those stimulus conditions. Momentarily at least, though, the stimulus conditions correlated with reinforcement for some responses of the class could set the occasion for the response class in general.

As in the example above, there can be a variety of stimuli that control the probability of occurrence of members of a response class or set the occasion for their occurrence. These stimuli can be components of the physical environment (the isolated room), the social environment (the model), or actions by others (the instruction "Do this"). Probably all these types of stimuli and others can operate to set the occasion for the occurrence of members of a response class in the normal environment. Further, there may be a number of different stimuli which set the occasion or increase the likelihood of occurrence of a particular class of responses. For example, the behaviors involved in meeting a new person, such as making eye contact, smiling, saying "hello," introducing yourself, and shaking hands, may be produced by a variety of stimuli, such as when you are in a room alone and an unknown person comes in, when someone says "you really should meet ———," when an unknown person standing next to you at a party smiles and makes eye contact, and when someone says, "I would like to meet ———." Thus the stimuli that set the occasion for a particular response class may vary in form but still affect the same responses.

Stimulus Classes

The previous description suggests that a behavioral learning account of development also requires an analysis of the formation of stimulus classes that set the occasion for various response classes. As was suggested for the formation of response classes, various past experiences of children may be responsible for the formation of stimulus classes. Direct instruction by others, observation of the situations in which others behave, and direct experience with various environmental consequences of behavior in different situations all may teach which responses should occur and where and when. Parents often directly teach their children how they should interact with others, and in doing so they specify in which situations certain behaviors are appropriate. For example, parents may specify that smiling at, talking to, and accompanying a guest that comes to the house are appropriate and desirable. In contrast, the parents may also specify to their children that engaging in these behaviors toward a stranger who approaches them on the street is not appropriate and possibly even dangerous. Children not only observe the actions of other people, live and on television, but also the situations in which the behavior occurred and often the consequences that the behaviors have had for the other people. Children respond in a variety of situations. In some situations, the behavior results in positive consequences, whereas in others the same behavior results in no noticeable consequences or negative consequences. In these ways, and perhaps others as well, the experiences of children may teach them the essential rules of social interaction, both the behaviors and the situations in which behaviors are appropriate.

"UNDERSTANDING" BEHAVIOR AND BEHAVIOR CHOSEN FOR STUDY

The preceding account has emphasized the role that environmental variables play in organizing and controlling behavioral development, as contrasted to the development of states or structures internal to the child which might similarly organize and control development. Both approaches attempt to account for the increasing skill and complexity of behavior as children grow older. Both are represented by constructs, based on observation of regularities in behavior, which go beyond the actual observed behavior to account for the occurrence of new behaviors in the repertoires of children. An environmental approach, in its purest form, suggests that ultimate control over behavioral development rests with the events that impinge upon children. A structural approach assumes that states internal to the child modify, organize, and ultimately control the overt behavior displayed by children.

While both environmental and structural approaches attempt to account for the same basic phenomenon, the increasingly complex skills of children, acceptance of one general approach over the other seems to produce fundamentally different outcomes. The most obvious difference is the focus on overt behavior or internal states as the defining content of developmental theory. An environmental account, emphasizing the direct interaction between environmental events and behavior, assumes that the overt behavior of children is the basic object of study. A structural approach, emphasizing the development of mechanisms or states internal to children, regards the overt behavior of children as an indirect reflection of these mechanisms or interactions among them. Thus in a structural account the overt behavior of children is the appropriate object of study only insofar as this behavior directly and accurately reflects the operation of internal mechanisms. There are, though, other, less obvious differences which result from an acceptance of an environmental or structural account of child development. The remainder of this chapter will discuss two additional differing outcomes among the many which exist: differences in defining what "understanding" behavior means and differences in the selection of behavioral phenomena for study.

Understanding Development

Understanding behavioral development can mean many different things. It can mean careful description of the form and sequence in which various behaviors occur in the normal development of children; it can mean the ability to predict accurately the occurrence of behavior at some particular time or in some particular situation; it can mean accounting for or explaining the behavior within a system of logically consistent constructs; or it can mean the ability to control and, therefore, to teach the behavior. Which of these meanings of "understanding" is accepted as a goal is undoubtedly determined by many factors. One factor surely is the set of assumptions which underlie the approach to understanding development. A structural approach assumes that internal states control development and that observable behavior is an indirect reflection of these states. The basic goal of a structural approach, then, is to describe these states and to analyze their functions and interactions. This results in the development of a set of logically consistent constructs that explain development and, in turn, leads to such systems being cited as the primary meaning of understanding behavior. An environmental approach assumes that behavior is directly influenced or produced by environmental events, which suggests direct manipulation of these events to observe their effects on behavior. To the extent that the manipulations have systematic effects, behavior can be produced or taught, which leads to an acceptance of control as the criterion for what understanding behavior means. The acceptance of control over behavior as the criterion for what understanding development means, in turn, seems to lead to the selection of particular types of behavior as the object of study.

If understanding the developmental process involves the ability to produce or control behavior, it might seem reasonable that any class of behavior in response to any class of stimuli might be chosen for study, perhaps based on availability, ease of measurement, and ease of manipulation. Certainly these factors can determine which behaviors and stimuli are selected for analysis. Other factors, however, also influence the behaviors and stimuli chosen for study.

Let us contrast the selection of behaviors for analysis from structural and environmental points of view. A structural account, concerned primarily with the interactions of mechanisms internal to the child, would lead to an analysis of those behaviors that most clearly or best exemplify the operations of those mechanisms. Thus if conservation of mass is of interest, it can be assessed using any material (e.g., clay, mud, bubble gum) about which a child indicates a choice (e.g., pointing, saying "that one," selecting objects from an assortment). Similarly, if one examines the effects of anxiety on memory, the behaviors exemplifying memory can be listing numbers previously heard, labeling the sequence of pictures seen earlier, reciting a story, or any other behavior that describes or reproduces some event experienced earlier in time.

In the examples cited above, the behaviors chosen for analysis were selected because of their theoretical relevance to the internal mechanisms being studied. In contrast, there are no such clear theoretical criteria for selecting the behaviors chosen for study within an environmental analysis. Instead, the decision needs to be made on other grounds. Certainly, practical considerations such as ease of observation and manipulation will be involved, but in the absence of clear theoretical standards to control the choice of behavior for study, other considerations become relevant. These include the pervasiveness of the behavior in the normal repertoire of children, the social importance of the behavior, and the extent to which the behaviors are judged to be basic to or prerequisite for the development of other socially important behaviors. It is these other considerations that lead to an analysis of behavior chosen on the basis of *social values rather than theoretical relevance.*

One of the leading environmental approaches to the understanding of behavioral development is the field known as the experimental analysis of behavior. In the late 1950s and early 1960s there was a major interest in extrapolating findings of basic research with laboratory animals to human behaviors of clinical and social importance. These initial extensions of basic research resulted in a large number of successful case studies, which in turn led to larger scale demonstrations of the use of environmental-behavioral procedures to analyze, modify, and develop socially important human behaviors such as language, academic performances of elementary- to college-age students, self-care skills, toileting skills, a variety of appropriate social behaviors, and many others. The success of these extrapolations (and modifications of techniques) is documented by the formation of at least 15 journals since 1963 devoted primarily, or in large part, to the analysis and application of environmental manipulations to problems of applied social interest (e.g., *Behaviour Research and Therapy, Journal of Applied Behavior Analysis, Behavior Modification, Journal of Behavioral Medicine*).

I suggest that the development of a strongly applied orientation within one of the major environmental approaches to the understanding of behavioral development was not an historical accident. It appears, instead, to be a direct result of assumptions about the direct causal role that environmental variables play in controlling behavior, the resulting development of a powerful technology to modify or produce behavior, and the absence of strong theoretical influences to study any convenient behavior as long as it was directly relevant to theory, thus allowing social considerations to affect that choice. Certainly there remains within the field of the experimental analysis of behavior a continuing emphasis on the analysis of basic behavioral phenomena. Nevertheless, the largest growth in the field has undoubtedly been in the direction of analyzing behavior and stimuli of applied importance. This applied emphasis, though not inevitable, was highly likely given the fundamental assumptions inherent in the environmental approach to understanding behavioral development.

A logical extension of the influence of social considerations in the selection of behavioral phenomena for analysis is the notion of social validity as described by Wolf (1978). He has presented examples and strong reasons for the desirability of systematically incorporating social influences in the choice of goals, or behaviors chosen for study, as well as in the choice of procedures used to influence the behaviors and in the analysis of the effects of these procedures.

SUMMARY

A behavioral approach to children's learning emphasizes the direct causal role that environmental events, such as models, prompts, and reinforcement, play in developing new behavior. Although a great deal of children's learning can be accounted for by the direct interaction between their behavior and their environment, there can be indirect effects as well. In many cases behaviors occur and are maintained in particular situations even though these behaviors were not directly taught in those situations. The concept of response classes provides one possible account of such indirect effects.

The defining characteristic of a response class is that events impinging directly on only some members of the class affect other members similarly. Thus stimuli which directly strengthen or weaken some members of a class in a particular situation, indirectly strengthen other members as well in that situation. Further, some response classes have generative characteristics in that teaching sufficient behavioral examples of the class can produce new and topographically different examples of the class which have never been directly taught.

Certain response classes have been experimentally established in laboratory settings by direct instruction. It seems possible that response classes might also be established in the natural environment through systematic instruction by others, as well as through observational learning and through contact with environmental events in the process of trial-and-error learning.

In a behavioral account of children's learning, the focus is on overt behavior of children as the defining content of developmental theory and on systematic attempts to manipulate the environment to observe the effects on behavior. To the extent that manipulations have systematic effects, behavior can be produced or taught, which leads to an acceptance of control as the criterion for what understanding behavioral development means. The ability to produce or control behavior, combined with the absence of strong theoretical criteria for selecting behaviors for study, leads, in turn, to an analysis of behaviors selected because of their social importance rather than their theoretical relevance. Thus the last 20 years of behavioral analysis has been characterized by the development of a significant applied technology and the corresponding formation of a number of new journals devoted to problems of applied social interest.

REFERENCES

Allen, K. E., Hart, B. M., Buell, J. S., Harris, F. R., & Wolf, M. M. Effects of social reinforcement on isolate behavior of a nursery school child. *Child Development,* 1964, **35,** 511–518.

Baer, D. M., Peterson, R. F., & Sherman, J. A. Development of imitation by reinforcing behavioral similarity to a model. *Journal of the Experimental Analysis of Behavior,* 1967, **10,** 405–416.

Baer, D. M., Rowbury, T. G., & Goetz, E. M. Behavioral traps in the preschool. In A. Pick (Ed.), *The sixth Minnesota symposium on child psychology.* Minneapolis: The University of Minnesota Press, 1976, pp. 3–27.

Baer, D. M., & Wolf, M. M. The entry into natural communities of reinforcement. In R. Ulrich, T. Stachnick, & J. Mabry (Eds.), *Control of human behavior* (Vol. II). New York: Scott, Foresman, 1970, pp. 319–324.

Bandura, A. Social learning through imitation. In M. R. Jones (Ed.), *Nebraska symposium on motivation.* Lincoln, Nebr.: University of Nebraska Press, 1962, pp. 211–269.

Hart, B. M., Reynolds, N. J., Baer, D. M., Brawley, E. R., & Harris, F. R. Effects of contingent and non-contingent social reinforcement on the cooperative play of a preschool child. *Journal of Applied Behavior Analysis,* 1968, **1,** 73–76.

Garcia, E. The development and generalization of delayed imitation. *Journal of Applied Behavior Analysis,* 1976, **9,** 499.

Garcia, E., Baer, D. M., & Firestone, I. The development of generalized imitation within topographically determined boundaries. *Journal of Applied Behavior Analysis,* 1971, **4,** 101–112.

Lovaas, O. I., Berberich, J. P., Perloff, B. F., & Schaeffer, B. Acquisition of imitative speech by schizophrenic children. *Science,* 1966, **151,** 705–707.

Nordquist, V. M., & Wahler, R. G. Naturalistic treatment of an autistic child. *Journal of Applied Behavior Analysis,* 1973, **6,** 79–87.

Stein, A. H., & Friedrich, L. K. Impact of television on children and youth. In E. M. Hetherington, J. W. Hagen, R. Kron, & E. H. Stein (Eds.), *Review of child development research,* (Vol. 5), Chicago: University of Chicago Press, 1975, pp. 183–256.

Thelen, M. H., & Rennie, D. L. The effect of vicarious reinforcement on imitation: A review of the literature. In B. A. Maher (Ed.), *Progress in experimental personality research* (Vol. 6). New York: Academic Press, 1972.

Wolf, M. M. Social validity: the case for subjective measurement or how applied behavior analysis is finding its heart. *Journal of Applied Behavior Analysis,* 1978, **11,** 203–214.

13

CHILD LANGUAGE
What Children Know and How

Mabel L. Rice
Department of Human Development and Family Life
University of Kansas

Between the ages of 12 and 24 months, regardless of the language spoken to them, children begin to talk, to converse with other people. Within the space of a few years they master an incredibly complex system of communication, in a spontaneous, effortless fashion. This occurs in cultures that encourage children to talk and dote on their every attempt, and also in cultures where children's talking is not highly valued. While this developmental phenomenon has been common knowledge for a long time, it remains a matter of considerable mystery. A growing body of evidence, collected during the intensive period of study of the past 20 years, reveals the complexity of the child's task and the remarkable facility with which language is acquired.

The modern study of child language can be traced to two happenings of the 1950s. (See Rieber & Vetter, 1980, for an historical overview of work preceding the modern period.) A new area of study, psycholinguistics, emerged from a meeting of three linguists and three psychologists interested in language who were brought together at a summer seminar in 1951 (Brown, 1958). This new field of inquiry focused on the intimate relationship between language and associated mental processes and how the study of one can increase our understanding of the other. One psycholinguist, Noam Chomsky, brought issues of *language acquisition* to the forefront with his theoretical work first published in 1957, followed by a modified version in 1965 (see Lyons, 1970, for an overview of Chomsky's work). He addressed the two major questions that have continued to dominate the literature: What is the nature of children's language skills (what do they know); and how is such knowledge acquired?

Chomsky's work emphasized formal linguistic structures, the syntax or grammar. He characterized human language as creative, "open ended," and rule governed. Therefore, children learn rules for combining words, not some rotely acquired set of sentences; they derive the structural regularities of their language from the utterances of others and then make use of those rules to create their own sentences. Another assertion was that the general principles that determine grammatical rules are evident in all human languages. There are certain universal properties of languages. A most controversial and provocative conclusion was that children are biologically programmed to acquire language. According to Chomsky, the environment plays a relatively minor role in accounting for how children learn to talk.

These four points of Chomsky's—that is, the theoretical primacy of formal linguistic structures (grammar), the rule-governed nature of language, the existence of linguistic universals, and biological factors as determinants of language acquisition—have each inspired a line of studies. Given the limits of this chapter, only the most controversial of the issues, the first and last ones, will be discussed here. In regard to the other two topics, suffice it to say that a number of studies conducted during the 1960s confirmed that children's grammatical knowledge demonstrates regularities indicative of underlying rules (e.g., Brown & Fraser, 1963; Brown, Cazden, & Bellugi, 1969), and there is evidence suggesting that certain aspects of grammatical acquisition are common to different languages (e.g., Slobin, 1970; for a current review, see Bowerman, 1981). Also in regard to the limits of this chapter, it will not include a description of the particulars of what linguistic skills children master by which stages of development. Several good sources of this information are available for the interested reader (Bowerman, 1978; Dale, 1976; Clark & Clark, 1977; Bloom & Lahey, 1978; deVilliers & deVilliers, 1978).

Opposition was provoked by Chomsky's assumption that description of the formal structure of language captures the essence of language and is the most, if not the only, illuminating way to approach the study of psycholinguistics. Alternate theoretical models have been proposed that have received recent support from child language researchers. These models provide a framework for describing the nature of a child's linguistic knowledge.

WHAT CHILDREN LEARN: COMMUNICATIVE COMPETENCE

Consistent with the then dominant theories of linguistics, Chomsky relegated meanings (semantics) to a secondary role in the study of language and did not address more general rules for language use, such as how language is modified to suit the social context. It has since become apparent that semantics and social rules represent important components of children's knowledge.

Semantics

Considerations of the meanings expressed in language dominated the child language literature during the 1970s. New linguistic theories appeared that linked syntax with semantic structures. Investigators such as Fillmore (1968) and McCawley (1971) argued that not only are syntactic and semantic structures interrelated but semantic components are also more fundamental. These new means of describing linguistic knowledge were readily applicable to child language, especially at the earliest stages of acquisition.

A dominant question was the nature of children's first syntactic knowledge, their first word combinations. The notions of "telegraphic speech" (Brown & Fraser, 1963) and "pivot grammar" (Braine, 1963) were the accepted characterizations of children's first sentences. Both were based on the forms children used, that is, the words alone, without regard for what the child intended the words to mean.

However, such descriptive systems were unable to account for many of the subtleties in children's knowledge; a famous example is the utterance of a 21-month-old girl: "mommy sock" (Bloom, 1970, pp. 5–6). Bloom pointed out that the same utterance was produced in different contexts, with apparently different meanings. In one context the comment referred to the mother's sock, but in the other context, the mother was putting on the girl's sock. Whereas grammatical categories based on surface forms, such as pivot/open, did not capture such differences, semantic categories, such as possessor/object and agent/object, did provide a means of distinguishing between the two utterances, with possessor/object describing the former context of "mommy sock," agent/object the latter. Careful analysis of the transcripts of the utterances of a number of young children led to the conclusion that descriptive systems based on semantic categories were most appropriate for revealing the regularities and continuities evident in early syntactic acquisition (see Brown, 1973, pp. 63–245, for a thorough review and discussion).

A consideration of meaning also extended to another kind of linguistic knowledge, that of lexical items. Clark's (1973) work on the acquisition of referential word meanings was the first in what has become a large literature exploring what children know about the words they use and understand. The study of children's word meanings has proved to be a particularly productive area of inquiry. The facility with which children learn words provides a very rich set of data. Carey (1978) reports that by the age of 6 the average child has learned over 14,000 words, which works out to an average of about nine new words a day, or almost one per waking hour. Out of this large corpus of data, certain families of words have received a great deal of study. The nature of young children's early representations of word meanings has illuminated issues of linguistic theory (e.g., Carey, 1978; Gentner, 1978) and has revealed the complex interrelationship between underlying conceptual understanding and linguistic meanings (e.g., Anglin, 1977; Nelson, 1977; Clark, 1978; Clark, 1980), an issue that will be discussed further in a later section of this chapter.

Sociolinguistics/Pragmatics

The study of what children know about linguistic meanings reflected one response to Chomsky's emphasis on the primacy of formal structures. Another group of language scholars objected to Chomsky's disregard of a speaker's knowledge of how to adjust language to suit the social context (Campbell & Wales, 1970; Hymes, 1971, 1972). They argued that an important part of a child's language learning involves knowing the social conventions for the use of language, such as how to make requests, to address someone, to tell a story or joke, to initiate and conduct a conversation, or to adjust one's comments to the listener's perspective. This kind of knowledge is referred to as sociolinguistic, or pragmatic. It has been added to grammar and meanings as accepted components of a child's linguistic knowledge. This entire set of skills has come to be referred to as "communicative competence" (Hymes, 1972).

The inclusion of sociolinguistic knowledge as a topic of child language study involves more than just the addition of another area of interest. It also entails a set of underlying premises and methodologies unlike those that have dominated traditional American linguistics (Ervin-Tripp & Mitchell-Kernan, 1977, pp. 1–3). Investigators of syntactic and semantic knowledge made the following assumptions: the linguistic context was sufficient to establish a speaker's knowledge; the sentence was the highest level of analysis; linguistic rules were regarded as relatively constant; and one function of language dominated the operating models, that of assertion and its alleged modifications, negation, passivization, and interrogation. In contrast, the study of sociolinguistics presumes that social context influences both the interpretation of language and rules for its use (e.g., the sentence "It is cold in here" can be interpreted as a statement of fact in some contexts or a request for action in others); the discourse unit, consisting of several comments (sentences or phrases), is a more appropriate level of analysis than the sentence for some studies (for example, initiation of a conversation often requires more than a single sentence); variability in linguistic rules is a normal property of language, with rules varying according to social features such as age, sex, and setting; language functions are diverse, including imperatives along with assertions, and vary culturally and developmentally. Thus the study of communicative competence brings not only new kinds of linguistic knowledge to the child language literature but also different questions, research methodologies, and conventions for interpreting data.

Beginning with the latter half of the 1970s (e.g., Bates, 1976; Ervin-Tripp & Mitchell-Kernan, 1977), a number of studies have appeared investigating the social dimensions of what children know about language. A growing body of evidence indicates that children as young as 3 and 4 years of age are remarkably adept at adjusting their language use and interpretations to the context of the social situation. They modify their communication patterns according to the perceived status, abilities, or perspectives of their conversational partner: they use a "baby talk" register when talking to infants or toddlers (Sachs, in press); they adjust their requests according to the rank or age of addressee, softening requests to the form of hints ("I sure do like ice cream") and polite forms ("please") for addressees of greater rank, status, or age (Ervin-Tripp, 1975); and they manipulate words like "him," "it," "will," "can," "this," and "that" to take into account the listener's perspective and information shared with the speaker (Shields, 1978). Preschool children can adjust their language to correspond to different roles, such as "mommy talk," "daddy talk," and "doctor talk," when acting out these roles (Andersen, 1977). They can carry on a conversation, following the rules for introducing, maintaining, and switching topics (e.g., Keenan, 1974; Dore, 1977). Young children keep track of new and old material as they converse, using "a" and "the" correctly to mark the distinction (Maratsos, 1975; 1979). They are able to describe sequences of everyday events (Nelson & Gruendel, 1979) and relate stories (Watson-Gegeo & Boggs, 1977). Also, children learn setting-specific communication codes, such as the rules that apply to the school setting (Dore, in press). Obviously, preschool children's knowledge of the social dimensions of language is not limited to a few isolated cases; instead, they demonstrate a comprehensive set of skills, with patterns of regularity indicative of underlying rules. It is clear that broadening the notion of what children know about language to include social context has opened up another rich area of inquiry.

It may seem that, with the inclusion of language structure, meanings, and rules for social adjustments, all aspects of a child's communicative skills have been addressed. However, there are possibilities for further broadening of our concepts of a child's communicative knowledge. The modern child lives in a world where verbal language is embedded in other systems of communicative symbols. There is increasing realization of the extent of children's involvement with a number of communicative media, many of which incorporate verbal language as part of a larger symbol system. For example, television communicates its messages by means of a complex set of representational codes, one of which is verbal language (Rice, Huston, & Wright, in press), and of course, written language is based on verbal language. Studies of media-presented language and the acquisition of verbal language are currently separate endeavors, although there is beginning to be an extension of developmental psycholinguistic perspectives and methodologies to studies of the media. Olson (1980) argues that there are close parallels between the properties of oral-ritualized language and the language of textbooks; he concludes that textbook language constitutes a "distinctive linguistic register." Bloome and Ripich (1979) analyzed the language in children's television commercials from a sociolinguistic perspective and report a systematic within-commercial progression from comments tied to plot or social context to comments tied to the product. They call for further investigation of the subtleties of language intended to sell products to children. I have begun to map the nature of the verbal language presented in children's television programs and how linguistic forms relate to the other representational codes of television (Rice, 1979; Rice, Huston, & Wright, in press). Preliminary evidence suggests that producers of programs intended for child audiences adjust their communicative codes in one of two patterns: they rely heavily on visual codes with minimal verbal encoding or, for those programs with many verbal messages, they simplify the complexity of the language and build in considerable redundancy. The linguistic and nonlinguistic features of children's television appear to interact in a systematic fashion designed to correspond to the mental abilities of the intended child viewers.

Just as features of social context can determine how verbal language is to be interpreted, so could conventions of media presentations influence the functions and meanings of language. It remains to be determined if some media, such as television, have unique conventions for the use of verbal language, if the language of television corresponds to a limited set of the features found in verbal language, or if the current models for describing verbal language also represent what happens in television programming.

Summary

The last decade has been a time of considerable change in the topics addressed by child language scholars. The trend has clearly been an expansion of the kinds of knowledge regarded as important, from the initial relatively narrow focus on the formal structures of language, to a consideration of the meanings expressed in word combinations and then the lexical items, to a concern with rules for social use of linguistic knowledge. Projecting the tendency to broaden the concepts of what children know about communication, one prediction is the extension of child language methods and models to media-presented language. As each of the components of a child's communicative competence has been studied, the findings reveal an often surprisingly sophisticated grasp of rules of communication by very young children. As more is known about what children know, the nature of their linguistic skills, the subtlety and richness of their apparently effortless achievements are made more apparent, and more fascinating. The question that immediately comes to mind is, How do children manage to learn these various dimensions of language?

HOW COMMUNICATIVE COMPETENCE IS ACQUIRED

The Role of Cognition

Competing Hypotheses. The questions of what children know linguistically and how they come to know it have been closely tied to each other. Chomsky's consideration of certain characteristics of how children acquire grammar led him to conclude that innate, species-specific language abilities accounted for children's rapid, spontaneous grammatical learning. However, other investigators objected to this conclusion, noting that the structural grammatical knowledge of young children was not an entirely unique kind of learning but, instead, demonstrated striking parallels to their sensorimotor knowledge of how to organize their world (e.g., Sinclair, 1971; Greenfield, Nelson, & Salzman, 1972), thereby linking the acquisition of grammar with more general cognitive learning.

As the "what" emphasis shifted to semantics and the meanings expressed in children's earliest utterances, support grew for the idea that underlying cognitive knowledge accounted for children's linguistic understandings. Many investigators (e.g., Brown, 1973) turned to Piaget's account of young children's cognitive knowledge as a model well suited to describe the parallels between nonlinguistic and grammatical knowledge, especially during the time of transition from nonlinguistic to linguistic means of communicating. Within a Piagetian framework, language acquisition is a matter of the child's figuring out which linguistic devices express what he already knows nonlinguistically. Meanings are allegedly used to decipher language (e.g., Macnamara, 1972), an interpretation that has become known as the "strong cognition hypothesis." It has inspired a considerable body of literature exploring the relationship between children's conceptual and linguistic understandings.

A number of findings support the idea that language builds directly on a cognitive base. Several investigators observed that children will use a new form to express a meaning that has previously been expressed with a simpler form; for example, when children first use verbs, they do not inflect them. Yet, given the communicative context, an observer can detect different meanings, such as reference to an action or state of temporary duration, immediate past, or a statement of the child's immediate wish or intention. Brown (1973) reports that these early meanings are the ones that are first indicated in verb markings, that is, progressive ("-ing"), past tense, and the catenative verbs, "gonna," "wanna," "hafta." Slobin (1973, pp. 184–85) formulated the principle "new forms, old functions" to describe this phenomenon and also noted that new functions (meanings) are first expressed by old forms. For example, when children learn to express rejection and then denial as categories of negation, they use the same forms, for example, "no," that had earlier been used only for nonexistence (Bloom, 1970). The principle "new forms, old functions; new functions, old forms," suggests that children are aware of certain concepts before they mark them in their language. Other evidence comes from the study of word meanings, which indicates that children use words to refer to objects and events in a manner unlike adults' use of the same forms (Clark,

1973; Nelson, 1974). Children sometimes use a given word to refer to a range of referents wider than an adult's (e.g., "daddy" for all men), to a more restricted range than adult usage (e.g., "up" only for their own activities), or even to make up their own novel words. Their own concepts seem to direct children's use of words, not how they have heard the words used by adults. The way in which children interpret language also supports the idea of conceptual control. Their early comprehension of various language forms reveals systematic errors that indicate strategies based upon the child's knowledge of the nature of the things in the real world (e.g., Clark, 1980).

However, a number of writers noted that the idea of underlying cognition accounting for language is an overly simplistic account of the complexities involved. One problem is that this account tends to gloss over the mapping problem. Roughly speaking, the mapping problem refers to the imperfect match between nonlinguistic and linguistic knowledge and the problems involved in getting from one kind of knowledge to the other. For example, some linguistic information seems to be unique to language, without ties to underlying cognition. Among the examples cited in the literature is the observation that children learn such formal linguistic rules as the grammatical acceptability of "put the hat on," "put it on," and "put on the hat," whereas "put on it" is not acceptable (McCawley, 1974). Slobin (1973) noted that when a child learns to express a particular idea linguistically depends upon the complexity of the available linguistic forms; if a child were learning two languages, he would first express a particular idea in the language with the simplest forms for that meaning. Cromer (1974) observed that the linguistic means used to express a particular concept (such as reference to the self) evolved in the direction of greater linguistic complexity even though the meanings stayed the same (children first use their own names to refer to themselves, then "me," then "I"). These discrepancies between linguistic and cognitive knowledge led Cromer (1974, 1976) to conclude that while cognition can account for underlying meanings, it does not account for the acquisition of formal linguistic devices, a position referred to as the "weak cognition hypothesis." More recently, Cromer (1981) has called for reconsideration of the possibility that certain uniquely human mental-processing abilities are implicated in children's acquisition of the grammatical aspects of language, a position consistent with Slobin's (1979, p. 76) beliefs. The current interest in "innate language-related abilities" is differentiated from Chomsky's earlier claims by an emphasis on the interaction among language-specific and general cognitive abilities and environmental factors.

Other scholars approached the mapping problem from another perspective, their concerns centering on the fact that since children know far more nonlinguistically than they do linguistically, they must figure out which distinctions are important for language. For example, while categories such as "agent," "patient," and "possession" are significant linguistically, Schlesinger (1977) has argued that these categories are not self-evident; they are not granted to the child in some environmentally ready-made package. An example provided by Schlesinger (1977, p. 156) is the child's problem in determining the boundaries of the agent concept. (The category "agent" is invoked in linguistic rules for word order, such as agent–action–object, in the case of English). If Mummy handing the bottle to the child is an instance of "agentness," then is Mummy holding the bottle also an instance, or is the bottle containing the milk also an instance of "agentness"? Children must infer how to organize their nonlinguistic knowledge into the categorical bundles that conform to the distinctions encoded in their particular native language. Schlesinger (1977) supported the conclusion expressed earlier by Blank (1974) and Bowerman (1976) that cognition alone does not account for the meanings expressed in language. Instead, there is an *interaction* between the language a child hears and his nonlinguistic understandings; language can serve to introduce a child to concepts he might not have come upon otherwise, even at the earliest stages of language acquisition.

Another modification to the strong cognition hypothesis arose in response to the premise that cognitive knowledge necessarily precedes the acquisition of related linguistic knowledge. Evidence appeared indicating that nonlinguistic concepts, such as a full mastery of object permanence, were not always attained prior to ostensibly related linguistic achievements, such as the ability to name or talk about objects (e.g., Corrigan, 1978). Based on a study correlating a number of linguistic and nonlinguistic measures obtained in longitudinal observations of 25 infants from 9 to 13 months of age, Bates and her colleagues (1977) concluded that the relationship between cognition and language is that of "local homologies." Certain cognitive and linguistic skills within particular content domains tend to be mastered by young children at approximately the same time, although the order of appearance may vary. The skills are presumed to be connected by means of a common, deeper underlying system of cognitive operations and structures that is biased toward neither.

All in all, the work exploring the relationship between cognition and language during the past 10 years has demonstrated the enormous complexity of language learning and the intricate nature of the interaction between nonlinguistic and linguistic knowledge. There is general agreement that while there are strong parallels between cognition and grammar, more is involved in learning formal linguistic structures than general conceptual growth alone. Some sort of human linguistic-processing mechanisms may account for children's grammatical acquisition; furthermore, the meanings expressed in language are not isomorphic with nonlinguistic knowledge. Instead, it is quite likely that young children must actively restructure nonlinguistic concepts to conform to the distinctions captured in language. Another conclusion is that the acquisition process does not follow an invariant sequence of first cognition, then language; sometimes certain linguistic skills appear before ostensibly related cognitive performances.

The Measurement Problem. The next decade promises to bring further clarifications. One of the limitations of much of the available work is the measurement of cognitive knowledge alleged to be related to language acquisition. Such general cognitive structures as object permanence have not proved to be particularly helpful, in part because they are usually defined in general terms (cf. Corrigan, 1978). Instead, it would be more productive to focus on the linkages between particular nonlinguistic knowledge and the linguistic forms used to express that meaning. In most of the work exploring such specific linkages, the underlying cognitive knowledge is inferred from the manner in which children use and understand words; there is no independent measure of nonlinguistic concepts (e.g. Clark, 1973, 1980; Nelson, 1974). However, separate measures of cognition and language were developed in recent studies involving the domains of color terms (Rice, 1978; 1980) and spatial relationships (Johnston, 1979). While both studies support the idea that underlying cognitive knowledge facilitates language acquisition, the extent of support differed. Johnston's cross-sectional data consistently supported the idea that nonlinguistic understandings of "in front of" and "in back of" preceded and directed the acquisition of associated locative terms. However, in the case of color term learning within a training context, nonlinguistic categorical knowledge, while generally helpful, was not necessary; some children who did not sort objects according to color prior to training of color terms could readily learn to refer correctly to the color of objects. The difference in findings between the two studies suggests that the relationship between cognition and language may differ somewhat according to the particular content domain and/or the experimental methods employed to infer the relationship.

Production and Comprehension Differences. An additional complexity is suggested by evidence that the two performance modes, production and comprehension, may tap into underlying cognition in a different manner. The clearest indication of this phenomenon is a finding in the Rice (1980) study of color terms. When children who demonstrated related conceptual knowledge (i.e., sorted objects according to color) were trained to produce correct color terms, they spontaneously acquired comprehension competence, whereas children without such nonlinguistic knowledge mastered only correct production while failing comprehension tasks. Mastery of comprehension required conceptual understandings, whereas children could learn to label correctly the color of objects when asked "What color is this?" without knowledge of associated nonlinguistic concepts. Other hints of a discrepancy between production and comprehension have been reported: the findings of Bates et al. (1977) that sensorimotor cognitive measures correlated with comprehension and production in slightly different ways and the observation that children's referential use of words in comprehension and production tasks is not the same (Nelson, Rescorla, & Gruendel, 1978). The different performance demands of linguistic production and comprehension and their ties to nonlinguistic concepts will certainly be an important research topic during the 1980s.

Social Cognition. So far the work investigating the role of nonlinguistic understandings in how children acquire language has centered around grammar and referential meanings. As attention shifts to the social dimensions of children's linguistic competence, new considerations arise. The kinds of cognitive knowledge that have been linked with grammar and referential meanings are object-based categories and certain abstract cognitive structures, such as object permanence. These concepts do not seem to capture what is involved when a child knows how to adjust language to her conversational partner's perspective, or when she uses a different set of linguistic responses for persons with higher status than with her peers or subordinates, or when she relates a story or engages in a conversation. Instead, other kinds of cognitive knowledge are implicated. For example, a child's knowledge about other persons (a general understanding of "peopleness," such as knowing that people have wants, likes, and dislikes) is intimately linked with her ability to adjust her communication to conform to what her listener can understand (Shields, 1978); children form social categories that influence their use of language (Ervin-Tripp, 1978); and children can organize their mental representations in a temporal manner that they can call upon when narrating events or conducting a lengthy communicative interchange (Nelson, n.d.). As these examples suggest, child language investigators are beginning to identify the cognitive correlates of children's sociolinguistic knowledge. The manner in which the two domains influence each other is presently a matter of conjecture; however, there is reason to believe that language may serve to introduce children to certain social distinctions (see Rice, in press, for a discussion of these issues).

To put the preceding discussion in a nutshell, diverse kinds of conceptual understandings are closely associated with the various dimensions of language in intricate patterns of acquisition that are presently only partially revealed. Granted this conclusion, one still wonders how a child acquires the particular kinds of knowledge. Obviously, neither cognition nor language spring forth in some autonomous fashion but, instead, unfold and develop by means of a child's interactions with the real world.

The Role of the Environment

The significance attributed to the environment depends upon what one believes about the contribution of the child to the language acquisition process, his general cognitive abilities, and possible innate language-specific abilities. Those who regard the child as an active participant in the language-learning process, one who draws upon all his real-world knowledge and possible certain processes that are finely-tuned for language learning, regard the child's interactions with the world as important sources of information, because such interactions contribute to both non-linguistic and linguistic knowledge *and* understanding of how to solve the many mapping problems. This view is dominant in current child language writings.

In earlier behavioral accounts of language acquisition, the environment was accorded a major role: models of language structure were supplied by the environment, along with events to ensure that the models were learned. According to this account, a relatively passive child would observe others use language and then learn it herself by means of imitation and reinforcement. This account has generally been rejected on the grounds that it does not account for the rule-governed nature of children's grammatical competence, and is also not supported by observations of children's patterns of imitation (e.g., some children seldom imitate others' speech) and reinforcement (parents do not provide reinforcement contingent on the grammatical correctness or incorrectness of their children's speech).

Chomsky minimized the significance of the environment in his emphasis on the role of innate language structures. The environment served only to provide the raw data, the forms of the particular language that a child would learn. A child actively processed these forms to discover grammatical rules, a process guided by innate principles of linguistic operation. As the child learned grammar, it was not necessary to draw upon more general real-world knowledge. Furthermore, Chomsky asserted that the child's data base, the speech she heard, was of a "degenerate quality" (1965, p. 58), with "fragments and deviant expressions of a variety of sorts" (1965, p. 201).

Speech to Young Children. This assertion was challenged by a number of researchers. By the mid-1970s a large body of evidence was available documenting that, contrary to Chomsky's claims, the environment does provide appropriate language models for young children. Adults (Broen, 1972; Snow, 1972) and even children as young as 4 years old (Shatz & Gelman, 1973) make many modifications in their speech to young children. These modifications are all in the direction of simplifying the language the young child hears: short, simple sentences; restricted selection of vocabulary; frequent repetitions; slow rate, with pauses to mark phrases and linguistic units; and models of dialogue interchanges. These special aspects of speech to children are all in the direction of correspondence with the particular psycholinguistic competencies of young children. Furthermore, speech is usually presented to young children in the context of nonlinguistic situational cues and social support which ensure that the meanings are obvious (Snow, in press).

While the phenomenon of "baby talk" has been extensively documented, it is not yet clear what to make of it. Several questions arise: Do such modifications account for language acquisition? Are they necessary for language learning, or is their contribution only facilitative? What features of the communicative situation contribute to a child's language learning?

In regard to the first question, several developmental psycholinguists have pointed out that even if the parental input were perfectly tailored to the child's mental capacities, that correspondence does not explain how a child infers underlying regularities not apparent in surface forms and then learns to match nonlinguistic knowledge with linguistic rules (Slobin, 1979, pp. 104–106; Newport, Gleitman, & Gleitman, 1977). Evidence consistent with this skepticism is the observation of considerable cultural variability regarding the amount of talk thought to be appropriate between adults and children (Schieffelin & Eisenburg, in press). The cultures from which most of the data have been collected, that is, Western, industrialized, middle- to upper-class cultures, seem to fall at the extreme end of encouraging children to talk just for the sake of talking, yet children in cultures who receive far less directed input manage to learn to talk.

Facilitative effects have been attributed to at least one feature of adult–child communicative interactions, that of semantically contingent speech. Roughly speaking, adult speech is semantically contingent with a child's utterance if it is somehow based on what the child said, for example, a repetition, expansion, clarification, comment, or affirmation. Several studies have found that the more semantically contingent speech a child hears, the greater the facilitation of her language acquisition (Snow, in press). The facilitative effects of maintaining child-initiated topics seems to be at least partly attributable to the congruence with the child's general interests and cognitive abilities as well as more language-specific feedback.

Indirect, Observational Learning. The current literature addressing environmental sources of information focuses exclusively on direct interchanges between a child and an adult, in which the adult, usually a parent, is directing her communication exclusively to the child. Yet such occasions are not the only source of language-relevant information for a child and may actually constitute only a small portion of the information available to a particular child. The emphasis on direct interactions has overlooked the many opportunities a child has for indirect learning as she observes adults talk to adults, the interactions of other children, and other children with adults. There is reason to believe that indirect, observational learning may play an important role in children's language acquisition. Evidence supporting this argument is the finding that children are able to learn the basic structures and meanings of language with only minimal access to direct interchanges (as in the case of hearing children of deaf parents, reported by Schiff, 1979, and children in cultures where adults do not talk with children). Furthermore, some kinds of linguistic learning (such as knowing how to use swear words) are most likely acquired in an observational context (see Rice, in press, for a more complete discussion).

Recognition of the utility of observational learning in the language acquisition process broadens the scope of real-world experiences regarded as relevant to language learning. One candidate for inclusion is television. To the extent that children can learn about language by observing how others use it, it is possible that they may learn something useful when viewing television, especially those programs that adjust the complexity of language to correspond to the competencies of a child audience.

However, a reminder of the potential usefulness of indirectly acquired information does not resolve the basic question of how children manage to acquire language. While it suggests a richer source of information, it does not account for what children do with it, how they manage to get beyond the surface characteristics of language (and its use in social context) to the underlying regularities, or how they solve the mapping problems.

Current Issues. The study of the environment has not provided any immediate solutions for the problem of how children acquire language. The language a child hears, even if simplified, does not provide a blueprint of the code she eventually masters; she has to infer relationships and rules not explicitly given. No particular environmental feature or event, or set of such, has been identified as critical to the child's ability to learn language, beyond a minimum of exposure to the native language. Yet the study of the child's world has been productive. The nature of the child's task is made apparent by a description of the information available to her. When one compares the task and environmental circumstances with what is learned, one can infer some of the intervening mental processes; for example, if certain linguistic knowledge can be acquired only in observational circumstances, one can infer acquisition via processes of indirect learning.

Current work suggests a number of topics in need of immediate exploration. Among them are the possibility of differential environmental effects for different kinds of linguistic knowledge, at different stages of development. It may well be that certain kinds of sociolinguistic knowledge, such as knowing how to speak politely, may be heavily influenced by language input and interpersonal events. It also seems likely that a child's responsiveness to environmental factors may vary as a function of age and social maturation. Undoubtedly, a child's experiences with the world influence language acquisition by means of a complex interaction with more general cognitive (including social cognition) processes, in conjunction with possible language-specific abilities that may be especially important at the very earliest stages of acquisition. A challenge for the study of development is to unravel the relative contributions of each of these factors.

CONCLUDING REMARKS

One of the purposes of this rather cursory overview of the developmental psycholinguistic literature was to capture the sense of excitement that presently pervades the study of child language. New models of what the child knows linguistically; a large body of empirical evidence demonstrating the diversity and complexity of the linguistic knowledge of even very young children; new explanations of how children come to acquire language; attempts to interrelate real-world experiences, cognitive learning, and linguistic acquisition; the collection of data across different language and cultural contexts—all these accomplishments of the past 15 years generate an optimistic enthusiasm for the topic. The optimism seems to be based on the intuition that the changes are in the right direction; the complex, broadly based current characterizations of child language seem to be more consistent with reality than earlier, more restricted ideas of what to study. The enthusiasm is that generated by the challenge of immediate research questions, the belief that the expansion of knowledge will continue, and an appreciation of how the study of child language is important not only insofar as it illuminates a significant phenomenon of child development but also as it contributes to our understanding of the nature of language and the functioning of the human mind.

REFERENCES

Anderson, E. Learning to speak with style: A study of the sociolinguistic skills of children. Unpublished Ph.D. dissertation, Stanford University, 1977.

Anglin, J. *Word, object and conceptual development.* New York: Norton, 1977.

Bates, E. *Language and context: The acquisition of pragmatics.* New York: Academic Press, 1976.

Bates, E., Benigni, L., Bretherton, I., Camaioni, L., & Volterra, V. Cognition and communication from 9–13 months: A correlational study. Program on Cognitive and Perceptual Factors in Human Development Report No. 12, Institute for the Study of Intellectual Behavior, University of Colorado, Boulder, Colo., 1977.

Blank, M. Cognitive functions of language in the preschool years. *Developmental Psychology,* 1974, **10**, 229–245.

Bloom, L. *Language development: Form and function in emerging grammars.* Cambridge, Mass.: MIT Press, 1970.

Bloom, L., & Lahey, M. *Language development and language disorders.* New York: Wiley, 1978.

Bloome, D., & Ripich, D. Language in children's television commercials: A sociolinguistic perspective. *Theory into Practice,* 1979, **18**(4), 220–225.

Bowerman, M. Semantic factors in the acquisition of rules for word use and sentence construction. In D. Morehead & A. Morehead (Eds.), *Directions in normal and deficient child language.* Baltimore: University Park Press, 1976.

Bowerman, M. Semantic and syntactic development: A review of what, when, and how in language acquisition. In R. L. Schiefelbusch (Ed.), *Bases for language intervention.* Baltimore: University Park Press, 1978, pp. 97–189.

Bowerman, M. Cross-cultural perspectives on language development. To appear in H. C. Triandis (Ed.), *Handbook of cross-cultural psychology.* Boston: Allyn & Bacon, 1981.

Braine, M. D. S. The ontogeny of English phrase structure: The first phase. *Language,* 1963, **39**, 1–14.

Broen, P. A. The verbal environment of the language-learning child. American Speech & Hearing Association Monographs, No. 17, 1972.

Brown, R. *Words and things.* New York: The Free Press, 1958.

Brown, R. *A first language: The early stages.* Cambridge, Mass.: Harvard Press, 1973.

Brown, R., Cazden, C., & Bellugi, U. The child's grammar from I to III. In J. P. Hill (Ed.),

Minnesota symposia on child psychology (Vol. 2). Minneapolis: University of Minnesota Press, 1969, pp. 28–73.

Brown, R., & Fraser, C. The acquisition of syntax. In C. N. Cofer & B. Musgrave (Eds.), *Verbal behavior and learning: Problems and processes.* New York: McGraw-Hill, 1963, pp. 158–201.

Campbell, R., & Wales, R. The study of Language Acquisition. In J. Lyons (Ed.), *New Horizons in linguistics.* England: Penguin Books, 1970.

Carey, S. The child as word learner. In M. Halle, J. Bresnan, & G. Miller (Eds.), *Linguistic theory and psychological reality.* Cambridge, Mass.: MIT Press, 1978.

Chomsky, N. *Syntactic structures.* The Hague: Mauton, 1957.

Chomsky, N. *Aspects of the theory of syntax.* Cambridge, Mass.: MIT Press, 1965.

Clark, E. V. What's in a word? On the child's acquisition of semantics in his first language. In T. E. Moore (Ed.), *Cognitive development and the acquisition of language,* New York: Academic Press, 1973, pp. 65–110.

Clark, E. V. Strategies for communicating. *Child Development,* 1978, **49**(4), 953–959.

Clark, E. V. Here's the *top:* Nonlinguistic strategies in the acquisition of orientational terms. *Child Development,* 1980, **51**, 329–338.

Clark, H., & Clark, E. *Psychology and language.* New York: Harcourt Brace Jovanovich, 1977.

Corrigan, R. Language development as related to stage 6 object permanence development. *Journal of Child Language,* 1978, **5**, 173–189.

Cromer, R. F. The development of language and cognition: The cognition hypothesis. In B. Foss (Ed.), *New perspectives in child development,* Baltimore: Penguin Books, 1974, pp. 184–252.

Cromer, R. F. The cognitive hypothesis of language acquisition and its implications for child language deficiency. In D. Morehead & A. Morehead (Eds.), *Directions in normal and deficient child language.* Baltimore: University Park Press, 1976.

Cromer, R. F. Reconceptualizing language acquisition and cognitive development. In R. L. Schiefelbusch & D. Bricker (Eds.), *Early language: Acquisition and intervention.* Baltimore: University Park Press, 1981.

Dale, P. *Language development: Structure and function* (2nd ed.). New York: Holt, Rinehart & Winston, 1976.

deVilliers, J. G., & deVilliers, P. A. *Language acquisition.* Cambridge: Harvard University Press, 1978.

Dore, J. Oh them sheriff: A pragmatic analysis of children's responses to questions. In S. Ervin-Tripp & C. Mitchell-Kernan (Eds.), *Child discourse.* New York: Academic Press, 1977.

Dore, J. The pragmatics of conversational competence: Two models, a method and a radical hypothesis. In R. L. Schiefelbusch & J. Pickar (Eds.), *Communicative competence: Acquisition and intervention.* Baltimore: University Park Press, in press.

Ervin-Tripp, S. Speech acts and social learning. In K. Basso & H. Shelby (Eds.), *Meaning in anthropology.* Albuquerque: University of New Mexico Press, 1975.

Ervin-Tripp, S. Whatever happened to communicative competence? In *Proceedings of the Linguistics Forum.* University of Illinois, summer 1978.

Ervin-Tripp, S., & Mitchell-Kernan, C. *Child discourse.* New York: Academic Press, 1977.

Fillmore, C. J. The case for case. In E. Bach & R. T. Harms (Eds.), *Universals in linguistic theory.* New York: Holt, Rinehart & Winston, 1968.

Gentner, D. On relational meaning: The acquisition of verb meaning. *Child Development,* 1978, **49**(4), 988–998.

Greenfield, P., Nelson, K., & Saltzman, E. The development of rule-bound strategies for manipulating seriated cups: A parallel between action and grammar. *Cognitive Psychology,* 1972, **3**, 291–310.

Hymes, D. Competence and performance in linguistic theory. In R. Huxley & E. Ingram

(Eds.), *Language acquisition: Models and methods.* New York: Academic Press, 1971, pp. 3–24.

Hymes, D. On communicative competence. In J. G. Pride & J. Holmes (Eds.), *Sociolinguistics.* Harmondsworth: Penguin Books, 1972.

Johnston, J. R. A study of spatial thought and expression: In back and in front. Unpublished Ph.D. dissertation, University of California, Berkeley, 1979.

Keenan, E. O. Conversational competence in children. *Journal of Child Language,* 1974, **1**, 163–183.

Lyons, J. *Chomsky.* Glasgow, Scotland: Collins, 1970.

McCawley, J. D. Prelexical syntax. In R. J. O'Brien (Ed.), *Monograph Series on Languages and Linguistics.* 22nd Annual Round Table. Washington, D.C.: Georgetown University Press, 1971.

McCawley, J. D. (Dialogue with) James McCawley. In H. Parret (Ed.), *Discussing language.* The Hague: Mouton, 1974.

Macnamara, J. Cognitive basis of language learning in infants. *Psychological Review,* 1972, **79**, 1013.

Maratsos, M. *The use of definite and indefinite reference in young children.* Cambridge, England: Cambridge University Press, 1976.

Maratsos, M. Learning how and when to use pronouns and determiners. In P. Fletcher & M. Garman (Eds.), *Language acquisition,* Cambridge: Cambridge University Press, 1979.

Nelson, K. Concept, word and sentence: Interrelations in acquisition and development. *Psychological Review,* 1974, **81**, 267–285.

Nelson, K. The conceptual basis for naming. In J. Macnamara (Ed.), *Language learning and thought,* 117–136. New York: Academic, 1977.

Nelson, K. Social cognition in a script framework. Unpublished paper, City University of New York, n.d.

Nelson, K., & Gruendel, J. M. From personal episode to social script: Two dimensions in the development of event knowledge. Paper presented at the biennial meetings of the Society for Research in Child Development, San Francisco, 1979.

Nelson, K., Rescorla, L., & Gruendel, J. Early lexicons: What do they mean? *Child Development,* 1978, **49**, 960–968.

Newport, E. L., Gleitman, H., & Gleitman, L. R. Mother, I'd rather do it myself: Some effects and non-effects of maternal speech style. In C. E. Snow & C. A. Ferguson (Eds.), *Talking to children: Language input and acquisition.* Cambridge: Cambridge University Press, 1977, pp. 109–150.

Olson, D. R. On the language and authority of textbooks. *Journal of Communication,* 1980, **30**, (1), 186–196.

Rice, M. L. The effect of children's prior nonverbal color concepts on the learning of color words. Unpublished Ph.D. dissertation, University of Kansas, 1978.

Rice, M. L., Television as a medium of verbal communication. Presented as a part of a symposium entitled "Children's Processing of Information from Television." Annual Convention of American Psychological Association, New York, 1979.

Rice, M. *Cognition to language: Categories, word meanings and training.* Baltimore: University Park Press, 1980.

Rice, M. L. Cognitive aspects of communicative development. In R. L. Schiefelbusch & J. Pickar (Eds.), *Communicative competence: Acquisition and intervention.* Baltimore: University Park Press, in press.

Rice, M. L., Huston, A. C., & Wright, J. C. The forms of television: Effects on children's attention, comprehension, and social behavior. In D. Pearl, L. Bouthilet & J. Lazar (Eds.), *Television and behavior: Ten years of scientific progress and implications for the 80's.* Washington, D.C.: U.S. Govt. Printing Office, in press.

Rieber, R. W., & Vetter, H. Theoretical and historical roots of psycholinguistic research. In R. W. Rieber (Ed.), *Psychology of Language and Thought.* New York: Plenum, 1980.

Sachs, J. Children's play and communicative development. In R. L. Schiefelbusch & J. Pickar (Eds.), *Communicative competence: Acquisition and intervention*. Baltimore: University Park Press, in press.

Schieffelin, B. B., & Eisenberg, A. R. Cultural variation in dialogue. In R. L. Schiefelbusch & J. Pickar (Eds.), *Communicative competence: Acquisition and intervention*. Baltimore: University Park Press, in press.

Schiff, N. B. The influence of deviant maternal input on the development of language during the preschool years. *JSHR* 1979, 22, 581–603.

Schlesinger, I. M. The role of cognitive development and linguistic input in language acquisition. *Journal of Child Language,* 1977, 4, 153–169.

Shatz, M., & Gelman, R. The development of communication skills: Modifications in the speech of young children as a function of listener. *Monographs of the Society for Research in Child Development,* Serial No. 152, 1973, 38, (5).

Shields, M. M. The child as psychologist: Construing the social world. In A. Lock (Ed.), *Action, gesture and symbol: The emergence of language*. London: Academic, 1978, pp. 529–581.

Sinclair, H. Sensorimotor action patterns as a condition for the acquisition of syntax. In R. Huxley & D. Ingram (Eds.), *Language acquisition: Models and methods*. New York: Academic, 1971.

Slobin, D. I. Universals of grammatical development in children. In W. Levelt & G. B. Flores d'Arcais (Eds.), *Advances in psycholinguistic research*. Amsterdam: North Holland, 1970.

Slobin, D. I. Cognitive prerequisites for the development of grammar. In C. A. Ferguson & D. I. Slobin (Eds.), *Studies of Child Language Development,* New York: Holt, Rinehart and Winston, 1973, pp. 175–208.

Slobin, D. I. *Psycholinguistics* (2nd ed.). Glenview, Ill.: Scott, Foresman, 1979.

Snow, C. E. Mothers' speech to children learning language. *Child Development,* 1972, 43, 549–565.

Snow, C. E. Parent–child interaction and the development of communicative ability. In R. L. Schiefelbusch & J. Pickar (Eds.), *Communicative competence: Acquisition and intervention*. Baltimore: University Park Press, in press.

Watson-Gegeo, K. A., & Boggs, S. T. From verbal play to talk story: The role of routines in speech events among Hawaiian children. In S. Ervin-Tripp & C. Mitchell-Kernan (Eds.), *Child Discourse*. New York: Academic, 1977, pp. 67–90.

14

THE DEVELOPMENTAL STUDY OF CHILDREN'S MEMORY

Lynn S. Liben
Department of Psychology
University of Pittsburgh

One common approach to children's memory has been the direct application of theories and methods from adult cognitive psychology. The prototypical case is one in which adults (or, more accurately, college students) are found to perform some task in some manner and then children are tested to determine their performance on these tasks. This approach has been valuable in providing descriptive data concerning the boundaries of children's memory, and thus investigations from this tradition are reviewed briefly in the section "Children's Memory in the Context of Adult Models of Memory."

These traditions are, however, purposely underrepresented here, in part because extensive reviews are available elsewhere and in part because these traditions do not provide an optimal fit with a volume devoted to human development across the life span. First, when adult cognition is used as the yardstick, children cannot help but look incompetent, a view finding increasing disfavor among developmentalists (e.g., Flavell, 1977; Gelman, 1978; Siegler, 1978). Second, the models imported from adult cognition typically assume that the information-processing system is static.[1] The developing child, however, is undergoing profound changes in a variety of domains. That these changes will interact profoundly with memory seems self-evident. Thus the research and theory reviewed in the section "Children's Memory in the Context of General Cognitive Development" has been chosen for its relevance for the question: What aspects of cognitive development, in general, are useful for predicting or understanding developmental changes in memory, in particular?

Just as it is inappropriate to limit the study of children's memory to models of adult cognition, it is equally inappropriate to link it to a single sociohistorical con-

---

[1] There are increasing numbers of exceptions to this and other generalizations concerning models of adult cognition. Most important, the state descriptions of cognitive processes that had dominated early work in cognitive psychology are increasingly being replaced by self-modifying cognitive systems. Excellent discussions of this trend and its implications for developmental psychology, as well as suggested readings, may be found in Klahr (1979).

text. These issues are discussed in the section "Contextual Effects on Children's Memory." Finally, the concluding section contains a brief discussion of conceptualizing memory as an independent variable and of the importance of studying children's memory within the context of individual and cultural development more generally.

CHILDREN'S MEMORY IN THE CONTEXT OF ADULT MODELS OF MEMORY

Much of the empirical work on children's memory during the past two decades may be mapped directly onto the model formulated by Atkinson and Shiffrin (1968) in which permanent, structural components of memory (the sensory register, short-term store, and long-term store) are distinguished from temporary control processes (e.g., rehearsal) used to transfer information across these structural components.[2] Developmental research within this tradition has not found evidence for substantial capacity changes with age: studies of developmental changes in the sensory register remain inconclusive (see Hoving, Spencer, Robb, & Schulte, 1978); analyses of short-term store have generally led to the conclusion that there is little developmental increase in capacity (e.g., Belmont & Butterfield, 1969; Chi, 1976), although the opposite conclusion has also been drawn (e.g., Friedrich, 1974). Investigations of long-term storage capacity have been avoided, because, with the possible exceptions of changes due to physical maturation or deterioration with age (see Poon, Fozard, Cermack, Arenberg, & Thompson, 1980), the capacity of long-term memory is thought to be functionally unlimited. (Its contents and organization, however, change.)

In contrast to the apparent lack of developmental differences in the structural components of memory, important developmental differences have been found in control processes. For example, developmental differences in the use and type of rehearsal strategies have been found by direct methods, such as observing lip movements during the study period (Flavell, Beach, & Chinsky, 1966) or recording the words spoken when children are asked to externalize their rehearsal process (Kellas, McCauley, & McFarland, 1975; Ornstein & Naus, 1977), and by indirect methods, such as inferring differential use of cumulative rehearsal from differential primacy effects in serial learning curves (e.g., Atkinson, Hansen, & Bernbach, 1964; Hagen & Kingsley, 1968; Kingsley & Hagen, 1969; although see Huttenlocher & Burke, 1976; Siegel, Allik, & Herman, 1976). Similarly, developmental differences in the use of semantic categorization have been observed directly by noting the extent to which stimulus materials are physically sorted into categories (e.g., Moely, Olson, Halwes, & Flavell, 1969) and indirectly by examining the extent to which free-recall output is clustered into semantic categories (e.g., Cole, Frankel, & Sharp, 1971; Moely, 1977).

Studies on the use of rehearsal, clustering, and other memory strategies support the generalization that older children are more likely than younger children to employ appropriate memory strategies. However, young children are able to use these

---

[2] The "depth of processing" model proposed by Craik and Lockhart (1972) has also inspired much research on children's memory; for example, see Hagen, Jongeward, and Kail (1975); Naus, Ornstein, and Hoving (1978); and Ornstein and Corsale (1979).

strategies when the experimenter provides appropriate hints, with recall improving concomitantly. Thus young children appear to suffer a "production" rather than a "mediation" deficiency (see Flavell, 1970). Recent work has been addressed to discovering the boundary conditions of these developmental differences, for example, by determining if they persist when the tasks are made more age appropriate and gamelike (DeLoache, 1981; Perlmutter, 1981); or when stimulus materials are determined in part by the children themselves (Tenney, 1975); or when feedback concerning the effectiveness of a particular strategy is provided (Ringel & Springer, 1980).

In summary, the research findings from this tradition support the generalization that while quantitative increments in capacity probably play a relatively small role in accounting for developmental improvements in memory, increasing facilities with control processes appear to be important. More extensive reviews of the work leading to these conclusions may be found in Hagen, Jongeward, and Kail (1975), Hagen and Stanovich (1977), Kail (1979), and Ornstein and Naus (1978).

CHILDREN'S MEMORY IN THE CONTEXT OF GENERAL COGNITIVE DEVELOPMENT

In the work reviewed above, developmental changes in memory were considered relatively independently of developmental changes in other areas of cognition. This approach is reminiscent of the treatment of memory as a "faculty" that operates in relative isolation from other cognitive and affective components of the individual. More recently, however, cognitive psychologists have begun to examine memory in the context of other cognitive phenomena such as the subject's knowledge base, schemata, and so on (e.g., see Neisser, 1976). In a parallel fashion, development of memory is examined here in relation to cognitive development more generally, including the development of logical operations, knowledge, representation, and metacognition.

Logical Operations

The program of research most directly addressed to the relationship between general cognitive development and memory is that initiated by Piaget and Inhelder (1973). As part of their general view that operations affect all contents (e.g., time, space, number) and processes (e.g., perception, imagery), Piaget and Inhelder (1973) hypothesized that memory, too, is influenced by underlying operations. Thus if children at different levels of operative development are asked to remember the objectively identical stimulus, their memories (reproductions, reconstructions, and recognitions) should be qualitatively different.

To demonstrate this phenomenon, Piaget and Inhelder asked children to reproduce an array of size-seriated sticks. Consistent with children's presumed operative levels, 3-year-olds' drawings showed no patterning (random arrangements); 4- and 5-year-olds' responses showed partial order (seriated subgroups; alternating large and small sticks), while 6- and 7-year-olds' reproductions were perfectly seriated. Without re-presenting the stimulus, Piaget and Inhelder asked children to reproduce the original stimulus again about 6 to 8 months after the initial recall session. Surprisingly, 74% of the children produced drawings that were more seriated at the later session. This "long-term memory improvement" was attributed to the underlying operative development presumed to have occurred during the intervening period.

Subsequent investigators have also found evidence for cross-sectional differences in reproductions and/or for longitudinal improvement using stimuli related to concepts of seriation (Altemeyer, Fulton, & Berney, 1969; Dahlem, 1968, 1969; Murray, 1981; Murray & Bausell, 1970), conservation (Murray, 1981), and space (Furth, Ross, & Youniss, 1974; Liben, 1974, 1975). Although several criticisms of the Genevan interpretations of these data have been raised, particularly with respect to the evidence for long-term memory improvement (see Liben, 1977a,b, 1980; Maurer, Siegel, Lewis, Kristofferson, Barnes, & Levy, 1979), the data do support the hypothesized influence of operations on memory within restricted time periods.

In addition to seeking to replicate the Genevan findings, investigators have also become interested in determining whether operative level, as assessed by performance on Piagetian tasks, is related to memory for stimuli developed outside the confines of Piagetian theory. This approach may be illustrated by work concerning children's memory for prose using a paradigm in which subjects are given separate, but related, sentences and later asked to make "old/new" judgments in a recognition task. Initial work with adults (Barclay, 1973; Bransford, Barclay, & Franks, 1972; Bransford & Franks, 1971) and subsequent work with second- and fourth-grade children (Paris & Carter, 1973) showed that subjects falsely (and confidently) "recognize" statements that are, in fact, new, but which can be inferred from the premises that were actually given. With respect to operative development, Liben and Posnansky (1977) suggested that the tendency to (falsely) recognize transitive inferences in prose memory should be restricted to children who understand transitive relations (if $A > B$, and $B > C$, then $A > C$), an understanding linked to concrete operations. To examine the proposed relationship, tests of transitive reasoning and inferential prose memory were given. The influence of the specific lexical form of the test items was also examined by including some recognition test items that were both semantically true and lexically similar to the original premises, and others that were semantically true but lexically different. For example, if premise sentences were "Mary is taller than Jane. Jane is taller than Susan," a true and lexically similar recognition test sentence would be "Mary is taller than Susan," whereas a true but lexically different test sentence would be "Susan is shorter than Mary."

Results showed that third graders were, indeed, basing their false recognition of true inferences on semantic meaning but that kindergarten and first-grade children were basing false recognitions on the superficial structure of the sentence; that is, the relational term employed in the premise sentences. Since it was these younger children who had shown variability in the ability to make transitive inferences, a relationship between logical skills and semantic inferencing was precluded in this study.

More recently, however, Johnson and Scholnick (1979) did demonstrate such a relationship, at least for one of the logical domains studied. Third- and fourth-grade children were given class inclusion and seriation tasks. Children who were successful on both, neither, or on only one of these tasks were given stories that allowed either inclusion or seriation inferences. Consistent with the hypothesized relationship between logical operations and sentence memory, true seriation inferences were falsely recognized more by children who had earlier been successful on the logical seriation tasks than by children who had failed the seriation tasks. The results from the inclusion inferences, however, did not provide evidence for the hypothesized link between operativity and memory: few false recognitions of true inclusion inferences occurred, and there was no evidence that their occurrence related to an ability to answer inclusion questions in the diagnostic phase. Johnson and Scholnick (1979) suggest that it may be necessary for a particular logical skill to be firmly established before it influences cognitive performance in another domain such as memory.

In addition to those studies specifically designed to investigate the relationship between operations and memory, research conducted for other purposes may also be examined for relevant data. For example, as part of his work on scientific reasoning, Siegler (1976; 1978) has tested children's ability to encode relevant information in the balance scale problem. Whereas 3-year-old children appear to encode (remember) neither weight nor distance accurately, 5-year-olds typically encode only weight, and 8-years-olds encode both. Interestingly, Siegler (1978) suggests that it may be deficiencies in memory that contribute to the young child's ability to encode. Yet in the context of the present discussion, one might turn that explanation around and suggest that the encoding problems *are* memory problems, themselves derived from immature logical operations. For example, the ability to perform multiple-classification tasks (an ability associated with concrete operations, see Inhelder & Piaget, 1964) might well relate to the child's ability to remember both weight and distance information in the balance scale task. Findings by Chap and Ross (1979) and by Liben (1981b) provide empirical support for the influence of operative level on encoding.

In summary, the two programs of research discussed above—Piagetian memory research and inferential prose memory—provide support for the hypothesis that developmental changes in memory reflect developmental changes in cognitive structure more generally. As noted elsewhere (Liben, 1980), developmental changes in a variety of other domains (e.g., large-scale spatial memory; scene memory) may also be profitably examined in relation to operative development.

Knowledge

Developmentalists of all theoretical persuasions recognize that as children get older, they acquire additional knowledge. The distinction between development and learning drawn by Piaget (1964), for example, is an acknowledgment that age-related changes include not only the development of logical operations (discussed above) but also the acquisition of knowledge about the physical and social world (see Liben, 1981a). An even stronger role is attributed to knowledge by learning theorists who argue that the age-related changes that have been attributed to structural changes in logical operations can be more parsimoniously attributed to increments in relevant information. Gagné (1968), for example, has analyzed performance on conservation tasks in this manner. Thus although theorists may disagree about the centrality of knowledge as an explanatory variable, they universally endorse the descriptive statement that older children know more than younger children. These knowledge differences may, in turn, account for age-related differences in memory, as discussed below. The first section below concerns knowledge about general organizing frameworks ("structural knowledge") and the second concerns knowledge about specific information ("content knowledge"). This division is made for organizational purposes only, and should not be taken to imply that the two kinds of knowledge are isolated from one another.

Structural Knowledge. The revelance of structural knowledge for children's memory may be illustrated by developmental work on story grammars. Story grammars are formal models used to describe the structure of stories and, in addition, may serve as a model of the knowledge that individuals apply in constructing, comprehending, and remembering stories. Developmental research has been concerned primarily with determining whether formal story grammars have psychological reality for children (see Johnson & Mandler, 1980; Stein, 1979; Stein & Glenn, 1979, for detailed reviews of this literature). In the initial studies (e.g., Stein & Glenn, 1975), children were told stories with an "ideal" structure; that is, stories containing episodes in the following order: presentation of a problem (initiating event), the character's reaction to that problem (internal response), a goal-directed action (attempt); a result of that action (consequence); and a response to that consequence by the protagonist (reaction). As predicted, recall across the various episode categories differed systematically. Interestingly, there were no developmental differences in the relative saliency of these episode categories, suggesting that the same structural system was underlying both younger and older children's recall. Other developmental differences were, however, evident. Older children recalled more statements overall, were more sensitive to internal responses of the story character, and included more new information (inferences) in their recall protocols.

In more recent studies, investigators have examined the consequences of using stories that do *not* conform to the ideal structure, accomplished, for example, by deleting particular episode categories or by rearranging the order of the episodes. Across the ages tested, children generally recall stories in a way that conforms more to the expected (ideal) story sequence than to the presented (deviated) sequence. Again, although the ideal structure seemed to influence recall at all ages tested, some developmental differences have been found. For example, both first- and fifth-grade children's recall was impaired when the initiating event had been deleted, but only first graders showed a recall decrement when the consequence was deleted (Stein & Glenn, 1977). Developmental differences between first- and fifth-grade children were also found with respect to their ability to remember stories in which there were sequence deviations that were marked rhetorically, as, for example, in positioning the consequence at the beginning rather than at the end of the story, but following it by the marker, "This happened because . . ." With marked inversions, fifth-grade children remembered the inversions as well or *better* than the standard stories, whereas first-grade children remembered the inversions as well or *worse* than the standard stories (Stein & Nezworski, 1978). Both Stein (1979) and Mandler and DeForest (1979) suggest that young children who have had less experience with atypical story structures (e.g., flashbacks) may be more dependent on expected sequences than are older children. Furthermore, it may be that as children increasingly come to understand causality, they become more and more able to understand causal links contained in stories, even if the relevant information is presented in jumbled order.

Stein and Glenn (1977) also scored recall protocols for the addition of new information when episodes had been deleted. Of particular interest was the finding that the majority of new information added was of the same category type as the information that had been deleted (for initiating event, attempt, and consequence). In general, fifth graders added more new information than first graders, with two exceptions, specifically, with stories in standard form (no deletions) and with stories with the reaction deleted. As in the findings from inverted stories discussed above, Stein and Glenn suggest that the developmental differences in the amount of accurate information recalled is probably due to the older children's better ability to find causal links and/or greater familiarity with disconnected stories.

In conclusion, story grammars do seem to have some psychological reality for children, as evidenced by different patterns of recall across categories; by disruptive effects on recall as a result of deleting or reordering categories; and by the tendencies for children to reorder or add information to recreate a story in standard form. Further, the same patterns of performance with respect to episode categories are evident in younger as well as older children. Interestingly, investigators in this area typically emphasize the lack of developmental differences. It seems likely, however, that the interesting changes have simply occurred prior to the first grade, since by this age, children (at least the middle-class children typically tested) have already had extensive exposure to formal stories. In addition, many first graders are likely to have the concrete operations needed to understand and recall the causal and temporal relations found in these stories. In addition to testing younger children, it would be useful if future studies examined the relationship between story recall and logical abilities with chronological age held constant.

A second kind of structural knowledge that has been studied developmentally is script knowledge. A script is "a structure that describes an appropriate sequence of events in a particular context. A script is made up of slots and requirements about what can fill those slots" (Schank & Abelson, 1977, p. 41). In their developmental work on scripts, Nelson and her colleagues (Nelson, 1978; Nelson & Gruendel, 1981) have studied a variety of scripts, such as going to a restaurant, getting dressed, and having a birthday party. Contrary to initial expectations that very young children would produce "disjointed, unordered, idiosyncratic accounts" (Nelson & Gruendel, 1981, p. 8), 3- and 4-year-old children were found to be able to produce scripts with reasonable order and detail (Nelson, 1978). Preschoolers can discriminate between *general* events ("What happens when you have dinner at home?" and *particular* events ("What happened when you had dinner yesterday?") (Hudson & Nelson, discussed in Nelson & Gruendel, 1981). The developmental differences that do appear in script knowledge seem to involve increases in complexity and detail rather than changing patterns.

Presumably, more detailed script knowledge could account for older children's tendency to recall more events in stories than younger children, because as script knowledge becomes more detailed, there is an increasing likelihood that a particular event named in a particular story will find a match in the child's existing script. Findings from a study by McCartney (1980) may be viewed in this light. Kindergarten and first-grade children were given stories about a typical evening in a young child's life that included a dinner script, a television event, and a going-to-bed script. Children were asked to make a tape of the story 2 minutes after hearing it. Their recalls were judged with respect to the number of events accurately recalled, as well as with respect to the intrusion of new events, modifications in phrasing, and similar dimensions. In general, findings suggested that anchor events and main events were remembered better than filler events and that children sequenced recall properly. Consistent with the hypothesized importance of knowledge, intrusions and modifications seemed to reflect the individual's knowledge base. For example, some children said the story character had gone "upstairs to bed," presumably reflecting the subject's own bedroom location since no location had been mentioned in the story. Consistent with the notion that developmentally increasing detailed scripts may support increasingly detailed memories, younger and older children were found to remember main events comparably, but older children were more likely to remember subsidiary events than were younger children.

In summary, even very young children have considerable structural knowledge, including knowledge of story structures and common events. Although the developmental data do not suggest radical qualitative changes in the nature of this knowledge as children get older (at least within the age range tested), data do suggest that older children's structural knowledge is more detailed and more complex than younger children's, changes that may support older children's better performance on a variety of memory tasks.

Content Knowledge. In addition to expanding and refining structural knowledge, individuals also acquire more and more specific information as they grow older. Having or not having particular information about a topic (or having it in weaker or stronger forms) may also effect developmental differences in memory performance. Virtually all investigators are sensitive to this issue insofar as they are careful to select materials that are familiar to children across the age range tested. For example, in selecting items for a free-recall task, investigators choose words that are within the vocabulary level of the subjects tested. However, while younger and older children might be equally knowledgeable about the referent of a particular word, the richness of this knowledge may differ significantly. For example, older and younger children might be equally facile in selecting the appropriate picture for the word "robin," but might be differentially able to list features of robins (e.g., beak color; type of song; color of eggs) and might fit robins into a different knowledge structure (e.g., simply one of many birds, versus a placement of "robin" in a hierarchical structure containing ground birds, sea birds, other animals, flying objects, and so on). In short, even when memory stimuli are familiar to children of different ages, the subtleties of this familiarity may vary considerably.

Only recently have investigators begun the potential impact of these knowledge differences on memory performance. Consistent with the hypothesized relationship between richness of knowledge and memory, for example, are findings of Richman, Nida, and Pittman (1976) who found that younger children's recall was inferior to older children's when meaningfulness of stimuli (CVC words) was permitted to vary across grades (as a result of using the same words at all ages), but that it was equivalent to older children's recall when the associative value of stimuli was equated across ages (by using different words at each age). Similarly, Ornstein and Naus (1979) found that the usual relationship between age and the use of active memory strategies is open to modification as a function of stimulus familiarity: young children were found to rehearse actively with highly familiar words, while older children showed relatively passive rehearsal with relatively unfamiliar (recently learned) words. Further evidence that the tendency to rehearse depends upon the degree to which stimuli are known comes from a study by Liben and Drury (1977) who found that preschoolers showed primacy effects in a serial probe task with well-known animals, but not with lesser-known letter stimuli. Similarly, deaf children's difficulty in using category cues at retrieval has also been attributed to the extent to which knowledge about category membership is readily available (Liben, 1979). If a rich context is available for encoding memory stimuli, a broader range of memory strategies may be relevant. Furthermore, as identification and categorization of individual items become increasingly automatic, more time is available for the application of the particular memory strategies chosen.

The relevance of the richness of semantic knowledge for memory is also demonstrated in research on inferential memory by Paris and Upton (1974; see review in Paris, 1975). Children from kindergarten through fifth grade heard a series of paragraphs and were then asked questions that tapped both verbatim and inferential information. Developmental increases were found in the tendency to infer information from particular words, for example, inferring the presence of a bird from the phrase "the flapping of wings" (i.e., semantic entailment) or inferring the use of a scissors from the phrase "she cut up the paper into little pieces" (i.e., implied instrument). Importantly, Paris and Upton found that recall of paragraphs was better in subjects who construct inferences from the events actually described than in subjects who do not draw such inferences.

Memory performance is affected not only by individuals' knowledge about language (discussed above), but also by knowledge about particular facts. This effect has been demonstrated in the adult memory literature by Sulin and Dooling (1974) who gave subjects biographical passages about characters identified as historical figures (Helen Keller; Adolf Hitler) or as unknown individuals (Carol Harris; Gerald Martin). Subjects in the former condition were more likely to "recognize" (falsely) new items that were congruent with what they already know about Keller or Hitler. A comparable effect with second- and fifth-grade children in passages about Abraham Lincoln or Howard Baker has been demonstrated by Landis (1978).

In a related study, Brown, Smiley, Day, Townsend, and Lawton (1977) read a passage about an escaping character to children in grades 3, 5, and 7. Some children were told that the escapee was a convict (George), while others were told it was the chimpanzee hero (Galen) of the television program "Planet of the Apes." On a subsequent recognition test, children were uniformly excellent at recognizing those sentences that had actually appeared in the story. They rejected, with moderate confidence, probe items that were incongruent with the context (Galen or George). Congruent probes, however, were accepted almost as confidently as were original sentences. Although no age differences were found in these judgments, only older children were able to discriminate between actual items and congruent foils on a later forced-choice task.

The general lack of developmental differences motivated Brown et al. (1977) to conduct a second study in which a recall task was substituted for the presumably less sensitive recognition task. Orientation to passages was accomplished by preliminary sessions in which children (grades 2, 4, and 6) were taught either about "Eskimo Targa," "Indian Targa," or, as a control, about the Spanish. One week later, children listened to a story called "Tor of the Targa" which they were then asked to recall in their own words. Analyses of the number of story units recalled showed increasing recall with grade; better recall with relevant orientation (Eskimo or Indian Targa) than with irrelevant orientation (Spanish); and systematic increases in recall as the importance of the story unit increased (as rated by independent judges). Also found was a significant interaction between grade and importance level: very important units were recalled equivalently across grades, but less important units were recalled significantly better by older children. The analysis of intrusions showed increasing numbers of theme-relevant intrusions with age.

The findings from these studies provide evidence that recall is enhanced when a relevant framework is provided. It may be, as Brown et al. (1977) point out, that one of the reasons that older children traditionally recall more than younger children is that older children are better able to identify relevant contexts for comprehension and recall. In support of this interpretation is the finding that older children's recall protocols contained more theme-relevant intrusions than younger children's. The results also help to identify the boundaries of developmental effects. Interestingly, Brown et al. emphasize the surprising lack of age-related differences in patterns of performance, which suggests that the kinds of constructive processes found in adults (e.g., Bransford & Johnson, 1972; Dooling & Lachman, 1971) are also evident in children, at least by second grade. (Again, as with story grammars, we do not have relevant data on appreciably younger children.) Despite these similarities, however, developmental differences were evident and should not be overlooked. For example, the finding that grade interacted with the importance level of story units is reminiscent of McCartney's (1980) finding, discussed earlier, that young children are as good as older children in remembering main script events but are less facile at remembering subsidiary information. Both sets of data suggest that with development, memory can rely upon increasingly differentiated knowledge. Finally, it should be recognized that more striking developmental differences in memory as a function of increments in knowledge might be found if the stimuli used were, in fact, selected to maximize age-related differences in the knowledge base. This was not the case for either of the two studies discussed here. The highly thematic foils concerning Lincoln used by Landis (1978), for example, were equally well known to both age groups; the information about the Targa provided by Brown et al. (1977) was apparently more or less equivalent in the two age groups. A more compelling indication of whether differences in the knowledge base of older and younger children effect differences in memory would be possible only in studies that employ materials that are empirically demonstrated (or experimentally manipulated) to be differentially known to older and younger children.

A promising source for experimental paradigms to investigate the effects of differential knowledge of materials comes from the adult literature on experts and novices. Of particular interest here is the finding that experts are better at recalling information related to their expertise than are novices, across domains as diverse as physics (e.g., Chi, Feltovich, & Glaser, 1980), baseball (e.g., Chiesi, Spilich, & Voss, 1979), and chess (e.g., DeGroot, 1965, 1966). Importantly, experts' superiority cannot be attributed to their generally superior memory. For example, experts and novices perform equivalently in reconstructing positions of chess pieces that have been placed randomly on the chess board (Chase & Simon, 1973).

In an extension of this research paradigm to children, Chi (1978) hypothesized that children would remember more than adults if the children were expert in the tested domain. As predicted, children recruited from a local chess tournament recalled significantly more positions of chess pieces in an immediate memory task, and took fewer trials to reconstruct the complete chess board, than did adults who had only minimal knowledge of chess. In related work investigating the effects of knowledge within subjects, Chi (1979) found that a child expert on the topic of dinosaurs performed better when memory tasks were related to dinosaurs that were better known than lesser known (where better- and lesser-known sets of dinosaurs were distinguished by the complexity and density of the child's link-node semantic network structures). Extensions of this kind of work should be useful in helping to determine the extent to which knowledge differences influence memory performance and in how these knowledge bases change systematically with age.

Representation

In addition to changes in logical processes and in knowledge, another profound developmental change in cognition that may have an important impact on memory, concerns the systems or modes used to represent information. Although it would be impossible to review the theory and research concerning the development of representation here (see Mandler, in press, for such a review), several developmental trends are noteworthy. As Piaget (1970) has emphasized, one central developmental progression in representation is the increasing separation of the representation from the referent. Even young infants (and lower-order species) can recognize a referent via an index that is connected directly to the referent (e.g., as in recognizing that smoke indicates fire). Only children beyond the sensorimotor period are able to re-present something in its absence, first by motivated symbols (such as drawings or gestures that are in some way isomorphic with the referent) and later by arbitrary symbols or signs (such as words). Piaget is less concerned with the mode of a representation (e.g., graphic, linguistic) than with its origins (motivated versus arbitrary). Werner and Kaplan (1963) similarly emphasize the developmentally increasing differentiation among self, object, and symbol.

Other theorists have stressed the general trend from motoric and visual forms of representation to linguistic forms of representation (as in the enactive, iconic, symbolic representational sequence suggested by Bruner, 1964). This developmental pattern is generally considered uncontroversial with respect to changes from infancy to toddlerhood, although the sequence is more controversial with respect to changes in later childhood. Kosslyn (1978a,b), for example, argues that children's representations become increasingly linguistic as a function of repeated encounters with their own images, but Rohwer (1970) stresses the initial superiority of verbal over imaginal representation due to the relatively greater systematicity and coherence of the former system.

Despite controversies with respect to order and importance of various representational modes (see, in particular, Kosslyn, 1978a,b; Kosslyn & Bower, 1974; Perlmutter & Myers, 1975; Reese, 1977), two major trends seem robust. The first is that there is an increasing availability of different representational modes. As a result, for any particular memory stimulus, older children are more likely to find a match between the form of the stimulus and their existing representational system, thus enhancing memory. An additional consequence of the increasing multiplicity of representational modes is that it permits encoding in more than one system simultaneously. This, in turn, adds redundancy to storage and hence also serves to enhance memory (as in the dual-encoding hypothesis formulated by Paivio, 1971; see also Pylyshyn, 1978).

The second trend is a special case of the first, specifically, that one of the new representational modes is linguistic. Linguistic representation may have unique qualities for memory storage. Unfortunately, developmentalists have traditionally avoided the potentially dramatic effects of the shift from preverbal to verbal modes by failing to integrate research questions and methods across the relevant chronological ages. That is, those who study infant memory tend to stop when infants begin to acquire language (e.g., the review by Fagan, Chapter 5, ends with 10-month-old infants), while researchers who study children's memory typically begin with preschoolers. One exception to this generalization is found in conceptualizations of infantile amnesia which stress the preverbal to verbal shift and which are now receiving attention from developmentalists (see Nelson & Ross, in press). In addition, some recent empirical work is beginning to address this gap, as in the work on naturalistic observations of children's memory (see Perlmutter, 1981) and in work on memory for spatial locations (e.g., Acredolo, 1981).

In short, quantitative and qualitative developmental changes in the modes available for the representation of information and in the connections among these modes offer increased flexibility and power for remembering. To understand these relationships, however, more systematic attention to the links between the development of representation and the development of memory is needed.

Metacognition

Another area of general cognitive development that may be expected to have important implications for children's memory concerns "metacognition." Most broadly, metacognition refers to conscious reflection on the processes and products of one's own thought. Reflectivity of this kind has been shown to undergo substantial development during childhood (e.g., Piaget, 1976). Children become increasingly knowledgeable about a variety of areas, including language (e.g., Gleitman, Gleitman, & Shipley, 1972); attention (Miller & Bigi, 1979); comprehension (Markman, 1973); and, of particular relevance here, memory (e.g., Kreutzer, Leonard, & Flavell, 1975).

In an excellent review of the issues and research on metamemory, Flavell and Wellman (1977) suggest that there are two major developments. First, children become increasingly able to prepare for future retrieval. For example, 11-year-olds, but not 4-year-olds, respond differently when asked to look at items carefully versus when they are asked to examine items for future recall (Appel, Cooper, McCarrell, Sims-Knight, Yussen, & Flavell, 1972).

Second, children become increasingly knowledgeable about the person, task, and strategy variables that affect memory performance. Within the domain of person variables, for example, older children are more aware that memory skills may vary across people of different ages and skills (e.g., Kreutzer, Leonard, & Flavell, 1975) and, in addition, are better able to monitor transient states, such as in recognizing that something is known but temporarily forgotten (Wellman, 1977). In the domain of task variables, children become increasingly able to discriminate among differentially memorable individual items, as in understanding that words with concrete referents will be easier to recall than words with abstract referents. Similarly, children become better at distinguishing among differentially memorable groups of items, as, for example, in knowing that a semantically categorizable list will be easier to recall than an unrelated list (e.g., Kreutzer et al., 1975). Finally, with respect to strategy variables, children become better in reporting strategies useful for future retrieval, as in writing memos to oneself (e.g., Kreutzer et al., 1975) and for facilitating current retrieval, as in searching for a lost item between the last point at which the item was known to be in possession and the point at which it was discovered to be missing (e.g., Drozdal & Flavell, 1975). In summary, there is considerable evidence for developmental advances in metamemory. (See Flavell & Wellman, 1977; Wellman, 1981; for reviews of additional empirical work, and Brown, 1978 for a discussion of this literature from the perspective of instruction.)

The assumption is generally made that advances in the ability to reflect upon memory are accompanied by increments in actual performance on memory tasks. Surprisingly, however, this assumption has received little empirical support. Wellman, Drozdal, Flavell, Salatas, and Ritter (1975), for example, found that while in some cases, children who reported more metamemorial knowledge were more likely to use related strategies on actual memory tasks than were children who did not report this metamemorial knowledge, in other cases, the link between metamemory judgments and memory performance was evident at older ages only. Flavell and Wellman (1977) interpret these results as indicating that even after metamnemonic knowledge is acquired, it must become coordinated with memory behavior. At still later stages of development, metamemorial knowledge may become even less accessible while memory behavior improves, because of the increasing automaticity in the application of memory strategies. Other empirical data have not shown clear developmental patterns, however. For example, neither the study by Kelly, Kofsky, Travers, and Johnson (1976), nor by Cavanaugh and Borkowski (1980) revealed consistent relationships between performance on metamemory tasks and actual recall at any age. Wellman (1981), however, has suggested that investigators may be using simplistic, strawman models of metamemory, and that with more sophisticated models and new empirical methods, more meaningful relationships between the two may well be found. What *is* certain is that the conclusion reached by Flavell and Wellman in 1977 is still true: "The relationship of knowledge about memory to actual mnemonic behavior, and developmental changes in this relationship, are complicated subjects, but subjects worthy of future research" (p. 29).

Conclusions

The evidence reviewed in this section suggests that developmental improvements in memory are tied to cognitive advances more generally. It is important to recognize that the four developmental areas discussed—logical operations, knowledge, representation, and metacognition—do not exhaust the domains that could have been considered. Furthermore, boundaries between these areas are fuzzy. For example, story grammars reflect causal and temporal relations (and hence draw upon logical operations); the ability to appreciate the memorability of categorizable lists rests upon the logical ability to classify; the possession of "factual" knowledge (such as the invariant horizontality of liquid) depends upon operational structures (such as a Cartesian coordinate system, see Piaget & Inhelder, 1956). In short, the effects of cognitive development on memory are far more complex than discussion of any single variable, taken in isolation, might imply.

CONTEXTUAL EFFECTS ON CHILDREN'S MEMORY

One of the important contributions of a life-span approach to development is that it emphasizes that individuals are always developing in a particular cultural-historical milieu, which is, itself, continually changing over time (e.g., Huston-Stein & Baltes, 1976; Lerner & Busch-Rossnagel, 1981). Other traditions, including dialectical psychology (e.g., Datan & Reese, 1977; Riegel, 1975), ecological psychology (e.g., Barker & Wright, 1955; Bronfenbrenner, 1977), and Soviet psychology (e.g., see Meacham, 1972, 1977) have also emphasized the importance of context in understanding developmental phenomena (see also Gibbs, 1979). Empirically, these influences have been studied most commonly in cross-cultural research (e.g., see article by Laboratory of Comparative Human Cognition, 1979).

Cross-cultural approaches to the study of memory development are important at the descriptive level for determining the universality of phenomena. Less obviously, but at least as important, cross-cultural work is valuable for suggesting necessary procedures for making cross-sectional age comparisons. For example, just as it is necessary to use materials that are functionally equivalent across cultures in making cross-cultural comparisons (Malpass, 1977), it is also important to make materials functionally equivalent across ages in making age comparisons. Young children, like the Kpelle (Scribner, 1974), may fail to use a clustering strategy on a free-recall task, not because they lack that strategy in the abstract but rather because the experimenter-provided basis for categorization does not match their own. Other methodological issues of cross-cultural work discussed by Malpass (1977) may (and should) be considered in developmental work as well.

Cross-cultural work on memory development is also valuable for testing the relative contribution of variables that are hopelessly confounded within this culture, such as chronological age and years of schooling. Wagner (1974, 1975, 1978), for example, has given various memory tasks to schooled and unschooled children and adults from urban and rural settings in Mexico and Morocco. The data have led Wagner to conclude that subjects' use of rehearsal strategies in laboratory memory tasks is not significantly related to age but, rather, to experiential factors such as formal schooling and, to a lesser extent, to living in an urban environment. Similarly, in studies of free recall, the use of semantic clustering appears to increase significantly with schooling rather than with chronological age (Sharp & Cole, 1975). Related findings and issues are discussed in two excellent reviews of the cross-cultural work on memory development by Cole and Scribner (1977) and by Wagner (1981) and in several recent monographs (Kagan, Klein, Finley, Rogoff, & Nolan, 1979; Sharp, Cole, & Lave, 1979; Stevenson, Parker, Wilkinson, Bonnevaux, & Gonzalez, 1978).

Finally, cross-cultural comparisons may illuminate the impact of historical changes that, because of their slow rates, preclude longitudinal investigation. For example, as societies develop written language, the demands on internal memory strategies may decline. Contemporary studies of cultures with strong oral traditions (e.g., Goody, 1977; Greenfield, 1972; Ross & Millsom, 1970; Wagner, 1981) may shed light on the impact of this change. (It is interesting to speculate how changes in our own technology may eventually alter memory demands for the majority of the population. Hand-held calculators, for example, may make memorization of multiplication tables obsolete; video- and audio-recording devices may limit need for internal and written storage; and so on.)

In summary, cross-cultural research on children's memory is important for establishing the universality of various memory phenomena, for suggesting appropriate methodologies that should be used to make comparisons across ages, and for helping to determine which variables are actually responsible for the age-linked changes in memory performance. Without work of this kind, both the description and explanation of memory development in children will be incomplete at best.

CONCLUSIONS

The data reviewed in this chapter provide evidence that children become increasingly facile at remembering as they get older and suggest that these improvements may be understood, in part, as the consequence of more general cognitive development. The approach taken here is, therefore, one in which memory development is treated as the dependent variable. Alternatively, one might treat children's memory as the independent variable, exploring ways in which changes in memory lead to change other cognitive domains; Trabasso and his colleagues (e.g., Bryant & Trabasso, 1971; Trabasso, 1977), for example, take this approach in arguing that young children's difficulties in making transitive inferences result from inadequate memory of the premises, rather than from inadequate logic. At an even broader level of analysis, it has been suggested that the child's operational progression (or the ability to employ "increasingly complex and powerful executive strategies," Case, 1978, p. 37) may be attributed to increases in the size of working memory (Case, 1978; Pascual-Leone, 1970). Of course, neither approach need exclude the other; one might ask both how cognitive change affects, and is affected by, memory development.

More generally, it must be noted that "explaining" developmental changes in memory by reference to developmental changes in cognitive development is only a limited form of explanation. We still require explanations of how and why the child's logical operations develop, how knowledge is acquired, how representational modes develop and interconnect, how the child becomes increasingly conscious of his or her own thought, and so on, "hows" that constitute the core issues for theories of cognitive development. Nevertheless, while we will never fully understand children's memory until we understand the individual and cultural contexts in which that memory is embedded, the approaches and research discussed here have at least begun to identify some of the reasons why older children usually remember more than their younger friends.

REFERENCES

Acredolo, L. P. Small- and large-scale spatial concepts in infancy and childhood. In L. S. Liben, A. H. Patterson, & N. Newcombe (Eds.), *Spatial representation and behavior across the life span.* New York: Academic, 1981.

Altemeyer, R., Fulton, D., & Berney, K. Long-term memory improvement: Confirmation of a finding by Piaget. *Child Development,* 1969, **40,** 845–857.

Appel, L., Cooper, R., McCarrell, N., Sims-Knight, J., Yussen, S., & Flavell, J. The development of the distinction between perceiving and memorizing. *Child Development,* 1972, **43,** 1365–1381.

Atkinson, R., Hansen, D., & Bernbach, H. Short-term memory with young children. *Psychonomic Science,* 1964, **1,** 255–256.

Atkinson, R., & Shiffrin, R. Human memory: A proposed system and its control processes. In K. Spence & J. Spence (Eds.), *The psychology of learning and motivation: Advances in research and theory* (Vol. 11). New York: Academic, 1968.

Barclay, J. The role of comprehension in remembering sentences. *Cognitive Psychology,* 1973, **4,** 229–254.

Barker, R. G., & Wright, H. F. *Midwest and its children.* New York: Harper & Row, 1955.

Belmont, J., & Butterfield, E. The relations of short-term memory to development and intelligence. In L. P. Lipsitt & H. W. Reese (Eds.), *Advances in Child Development and Behavior* (Vol. 4). New York: Academic, 1969.

Bransford, J., Barclay, J., & Franks, J. Sentence memory: A constructive versus interpretive approach. *Cognitive Psychology,* 1972, **3,** 193–209.

Bransford, J., & Franks, J. The abstraction of linguistic ideas. *Cognitive Psychology,* 1971, **2,** 331–350.

Bransford, J., & Johnson, M. Contextual prerequisites for understanding: Some investigations of comprehension and recall. *Journal of Verbal Learning and Verbal Behavior,* 1972, **11,** 717–726.

Bronfenbrenner, U. Toward an experimental ecology of human development. *American Psychologist,* 1977, **32,** 513–531.

Brown, A. L. Knowing when, where, and how to remember: A problem of metacognition. In R. Glaser (Ed.), *Advances in instructional psychology.* Hillsdale, N.J.: Erlbaum, 1978.

Brown, A., Smiley, S., Day, J., Townsend, M., & Lawton, S. Intrusion of a semantic idea in children's comprehension and retention of stories. *Child Development,* 1977, **48,** 1454–1466.

Bruner, J. The course of cognitive growth. *American Psychologist,* 1964, **19,** 1–15.

Bryant, P., & Trabasso, T. Transitive inferences and memory in young children. *Nature,* 1971, **232,** 456–458.

Case, R. Intellectual development from birth to adulthood: A neo-Piagetian interpretation. In R. S. Siegler (Ed.), *Children's thinking: What develops?* Hillsdale, N.J.: Erlbaum, 1978.

Cavanaugh, J. C., & Borkowski, J. G. Searching for metamemory-memory connections: A developmental study. *Developmental Psychology,* 1980, **16,** 441–453.

Chap, J., & Ross, B. M. Memorizing and copying visual patterns: A Piagetian interpretation. *Journal of Genetic Psychology,* 1979, **134,** 193–205.

Chase, W. G., & Simon, H. A. Perception in chess. *Cognitive Psychology,* 1973, **4,** 55–81.

Chi, M. T. H. Short-term memory limitations in children: Capacity or processing deficits? *Memory & Cognition,* 1976, **4,** 559–572.

Chi, M. T. H. Knowledge structures and memory development. In R. Siegler (Ed.), *Children's thinking: What develops?* Hillsdale, N.J.: Erlbaum, 1978.

Chi, M. T. H. Exploring a child's knowledge of dinosaurs: A case study. Paper presented at the Society for Research in Child Development, San Francisco, March 1979.

Chi, M. T. H., Feltovich, P. J., & Glaser, R. Representation of physics knowledge by experts and novices. Learning Research and Development Center, University of Pittsburgh, Technical Report, No. 2, March 1980.

Chiesi, H. L., Spilich, G. J., & Voss, J. F. Acquisition of domain-related information in relation to high and low domain knowledge. *Journal of Verbal Learning and Verbal Behavior,* 1979, **18,** 257–274.

Cole, M., Frankel, F., & Sharp, D. Development of free recall learning in children. *Developmental Psychology,* 1971, **4,** 109–123.

Cole, M., & Scribner, S. Cross-cultural studies of memory and cognition. In R. V. Kail, Jr., & J. W. Hagen (Eds.), *Perspectives on the development of memory and cognition.* Hillsdale, N.J.: Erlbaum, 1977.

Craik, R. I. M., & Lockhart, R. S. Levels of processing: A framework for memory research. *Journal of Verbal Learning and Verbal Behavior,* 1972, **11**, 671–684.

Dahlem, N. Reconstructive memory in kindergarten children. *Psychonomic Science,* 1968, **13**, 331–332.

Dahlem, N. Reconstructive memory in kindergarten children revisited. *Psychonomic Science,* 1969, **17**, 101–102.

Datan, N., & Reese, H. W. (Eds.), *Life-span developmental psychology: Dialectical perspectives on experimental psychology.* New York: Academic Press, 1977.

deGroot, A. *Thought and choice in chess.* The Hague: Mouton, 1965.

deGroot, A. Perception and memory versus thought: Some old ideas and recent findings. In B. Kleinmuntz (Ed.), *Problem solving.* New York: Wiley, 1966.

DeLoache, J. S. Naturalistic studies of memory and metamemory. In M. Perlmutter (Ed.), *Naturalistic studies of children's memory.* San Francisco: Jossey-Bass, 1981.

Dooling, D. J., & Lachman, R. Effects of comprehension on retention of prose. *Journal of Experimental Psychology,* 1971, **88**, 216–222.

Drozdal, J. G., Jr., & Flavell, J. H. A developmental study of logical search behavior. *Child Development,* 1975, **46**, 389–393.

Flavell, J. Developmental studies of mediated memory. In L. P. Lipsitt & H. W. Reese (Eds.), *Advances in child development and behavior* (Vol. 5). New York: Academic, 1970.

Flavell, J. *Cognitive development.* Englewood Cliffs, N.J.: Prentice-Hall, 1977.

Flavell, J., Beach, D. H., & Chinsky, J. M. Spontaneous verbal rehearsal in a memory task as a function of age. *Child Development,* 1966, **37**, 283–299.

Flavell, J., & Wellman, H. Metamemory. In R. V. Kail, Jr. & J. W. Hagen (Eds.), *Perspectives on the development of memory and cognition.* Hillsdale, N.J.: Erlbaum, 1977.

Friedrich, D. Developmental analysis of memory capacity and information-encoding strategy. *Developmental Psychology,* 1974, **10**, 559–563.

Furth, H. G., Ross, B. M., & Youniss, J. Operative understanding in children's immediate and long-term reproductions of drawings. *Child Development,* 1974, **45**, 63–70.

Gagné, R. M. Contributions of learning to human development. *Psychological Review,* 1968, **75**, 177–191.

Gelman, R. Counting in the preschooler: What does and does not develop. In R. S. Siegler (Ed.), *Children's thinking: What develops?* Hillsdale, N.J.: Erlbaum, 1978.

Gibbs, J. C. The meaning of ecologically oriented inquiry in contemporary psychology. *American Psychologist,* 1979, **34**, 127–140.

Gleitman, L. R., Gleitman, H., & Shipley, E. F. The emergence of the child as grammarian. *Cognition,* 1972, **1**, 137–164.

Goody, J. Mémoire et apprentisage dans les societes avec et sans écriture: La transmission du Bagre. *L'Homme,* 1977, **17**, 29–52.

Greenfield, P. M. Oral or written language: The consequences for cognitive development. *Language and Speech,* 1972, **15**, 169–178.

Hagen, J. W., Jongeward, R., & Kail, R. V., Jr. Cognitive perspectives on the development of memory. In H. W. Reese (Ed.), *Advances in child development and behavior* (Vol. 10). New York: Academic Press, 1975.

Hagen, J. W., & Kingsley, P. Labeling effects in short-term memory. *Child Development,* 1968, **39**, 113–121.

Hagen, J. W., & Stanovich, K. E. Memory: Strategies of acquisition. In R. V. Kail, Jr., & J. W. Hagen (Eds.), *Perspectives on the development of memory and cognition.* Hillsdale, N.J.: Erlbaum, 1977.

Hoving, K. L., Spencer, T., Robb, K., & Schulte, D. Developmental changes in visual information processing. In P. A. Ornstein (Ed.), *Memory development in children.* Hillsdale, N.J.: Erlbaum, 1978.

Huston-Stein, A., & Baltes, P. B. Theory and method in life-span developmental psychology: Implications for child development. In H. W. Reese & L. P. Lipsitt (Eds.), *Advances in child development and behavior* (Vol. 11). New York: Academic, 1976.

Huttenlocher, J., & Burke, D. Why does memory span increase with age? *Cognitive Psychology,* 1976, **8**, 1–31.

Inhelder, B., & Piaget, J. *The early growth of logic in the child.* New York: Norton, 1964.

Johnson, J. W., & Scholnick, E. K. Does cognitive development predict semantic integration? *Child Development,* 1979, **50**, 73–78.

Johnson, N. S., & Mandler, J. M. A tale of two structures: Underlying and surface forms in stories. CHIP Report 80, University of California, San Diego, 1980.

Kagan, J., Klein, R. E., Finley, G. E., Rogoff, B., & Nolan, E. A cross-cultural study of cognitive development. *Monographs of the Society for Research in Child Development,* 1979, **44**, No. 180.

Kail, R. V. *The development of memory in children.* San Francisco: Freeman, 1979.

Kellas, G., McCauley, C., & McFarland, C. E. Developmental aspects of storage and retrieval. *Journal of Experimental Child Psychology,* 1975, **19**, 51–62.

Kelly, M., Scholnick, E. K., Travers, S. H., & Johnson, J. W. Relations among memory, memory appraisal, and memory strategies. *Child Development,* 1976, **47**, 648–659.

Kingsley, P. R., & Hagen, J. W. Induced versus spontaneous rehearsal in short-term memory in nursery school children. *Developmental Psychology,* 1969, **1**, 40–46.

Klahr, D. Self modifying productions systems as models of cognitive development. Paper presented at the biennial meetings of the Society for Research in Child Development, San Francisco, 1979.

Kosslyn, S. M. Imagery and cognitive development: A teleological approach. In R. S. Siegler (Ed.), *Children's thinking: What develops?* Hillsdale, N.J.: Erlbaum, 1978. (a)

Kosslyn, S. M. The representational-development hypothesis. In P. A. Ornstein (Ed.), *Memory development in children.* Hillsdale, N.J.: Erlbaum, 1978. (b)

Kosslyn, S. M., & Bower, G. H. The role of imagery in sentence memory: A developmental study. *Child Development,* 1974, **45**, 30–38.

Kreutzer, M., Leonard, C., & Flavell, J. An interview study of children's knowledge about memory. *Monographs of the Society for Research in Child Development,* 1975, **40**, No. 159.

Laboratory of Comparative Human Cognition. *American Psychologist,* 1979, **34**, 827–833.

Landis, T. Y. Children's reconstructive memory for prose. Unpublished master's thesis, Pennsylvania State University, 1978.

Lerner, R. M., & Busch-Rossnagel, N. A. Individuals as producers of their development: Conceptual and empirical bases. In R. M. Lerner & N. A. Busch-Rossnagel (Eds.), *Individuals as producers of their development: A life-span perspective.* New York: Academic Press, 1981.

Liben, L. S. Operative understanding of horizontality and its relation to long-term memory. *Child Development,* 1974, **45**, 416–424.

Liben, L. S. Long-term memory for pictures related to seriation, horizontality, and verticality concepts. *Developmental Psychology,* 1975, **11**, 795–806.

Liben, L. S. Memory in the context of cognitive development: The Piagetian approach. In R. V. Kail, Jr., & J. W. Hagen (Eds.), *Perspectives on the development of memory and cognition.* Hillsdale, N.J.: Erlbaum, 1977. (a)

Liben, L. S. Memory from a cognitive-developmental perspective. In W. F. Overton & J. M. Gallagher (Eds.), *Knowledge and development,* (Vol. 1), *Advances in research and theory.* New York: Plenum, 1977. (b)

Liben, L. S. Free recall by deaf and hearing children: Semantic clustering and recall in trained and untrained groups. *Journal of Experimental Child Psychology,* 1979, **27**, 105–119.

Liben, L. S. The effect of operativity on memory. Paper presented in H. Beilin (Chair), *The operative basis of reading, memory, information processing and self instructions,* Symposium presented at the Southeastern Conference on Human Development, Alexandria, Virginia, April 1980.

Liben, L. S. Contributions of individuals to their development during childhood: A Piagetian perspective. In R. M. Lerner & N. A. Busch-Rossnagel (Eds.), *Individuals as producers of their development: A life-span perspective.* New York: Academic Press, 1981. (a)

Liben, L. S. Copying and remembering pictures in relation to subjects' operative levels. *Developmental Psychology,* 1981, **17**, 357–365. (b)

Liben, L. S., & Drury, A. Short-term memory in deaf and hearing children in relation to stimulus characteristics. *Journal of Experimental Child Psychology,* 1977, **24**, 60–73.

Liben, L. S., & Posnansky, C. J. Inferences on inference: The effects of age, transitive ability, memory load, and lexical factors. *Child Development,* 1977, **48**, 1490–1497.

McCartney, K. A. The influence of script-based knowledge upon children's story recall. Paper presented at the Southeastern Conference on Human Development. Alexandria, Virginia, April, 1980.

Malpass, R. S. Theory and method in cross-cultural psychology. *American Psychologist,* 1977, **32**, 1069–1079.

Mandler, J. M. Representation. In J. H. Flavell & E. M. Markman (Eds.), *Cognitive development.* New York: Wiley, in press.

Mandler, J. M., & DeForest, M. Developmental invariance in story recall. CHIP Report 78, University of California, San Diego, 1978.

Mandler, J. M., & DeForest, M. Is there more than one way to recall a story? *Child Development,* 1979, **50**, 886–889.

Markman, E. Factors affecting the young child's ability to monitor his memory. Unpublished doctoral dissertation, University of Pennsylvania, 1973.

Maurer, D., Siegel, L. S., Lewis, T. L., Kristofferson, M. W., Barnes, R. A., & Levy, B. A. Long-term memory improvement? *Child Development,* 1979, **50**, 106–118.

Meacham, J. The development of memory abilities in the individual and society. *Human Development,* 1972, **15**, 205–228.

Meacham, J. Soviet investigations of memory development. In R. V. Kail, Jr., & J. W. Hagen (Eds.), *Perspectives on the development of memory and cognition.* Hillsdale, N.J.: Erlbaum, 1977.

Miller, P. H., & Bigi, L. The development of children's understanding of attention. *Merrill-Palmer Quarterly,* 1979, **25**, 235–263.

Moely, B. Organizational factors in the development of memory. In R. V. Kail, Jr., & J. W. Hagen (Eds.), *Perspectives on the development of memory and cognition.* Hillsdale, N.J.: Erlbaum, 1977.

Moely, B., Olson, F., Halwes, T., & Flavell, J. Production deficiency in young children's clustered recall. *Developmental Psychology,* 1969, **1**, 26–34.

Murray, F. B. Long-term memory of operativity figures. *Journal of General Psychology,* 1981, **105**, 85–92.

Murray, F. B., & Bausell, R. Memory and conservation. *Psychonomic Science,* 1970, 21 (6), 334–335.

Naus, M. J., Ornstein, P. A., & Hoving, K. L. Developmental implications of multistore and depth-of-processing models of memory. In P. A. Ornstein (Ed.), *Memory development in children.* Hillsdale, N.J.: Erlbaum, 1978.

Neisser, U. *Cognition and reality.* San Francisco: Freeman, 1976.

Nelson, K. How children represent knowledge of their world in and out of language: A preliminary report. In R. S. Siegler (Ed.), *Children's thinking: What develops?* Hillsdale, N.J.: Erlbaum, 1978.

Nelson, K., & Gruendel, J. Generalized event representations: Basic building blocks of cognitive development. In A. Brown & M. Lamb (Eds.), *Advances in developmental psychology* (Vol. 1). Hillsdale, N.J.: Erlbaum, 1981.

Nelson, K., & Ross, G. The general and specifics of long-term memory in infants and young children. In M. Perlmutter (Ed.), *Naturalistic studies of children's memory*. San Francisco: Jossey-Bass, 1981.

Ornstein, P. A., & Corsale, J. Process and structure in children's memory. In G. J. Whitehurst & B. J. Zimmerman (Eds.), *The functions of language and cognition*. New York: Academic Press, 1979.

Ornstein, P. A., & Naus, M. J. Rehearsal processes in children's memory. In P. A. Ornstein (Ed.), *Memory development in children*. Hillsdale, N.J.: Erlbaum, 1978.

Ornstein, P. A., & Naus, M. J. Effects of the knowledge base on children's rehearsal and organizational strategies. Paper presented at the Biennial Meetings of the Society for Research in Child Development, San Francisco, 1979.

Paivio, A. *Imagery and verbal processes*. New York: Holt, Rinehart and Winston, 1971.

Paris, S. G. Integration and inference in children's comprehension and memory. In F. Restle, R. Shiffrin, J. Castellan, H. Lindman, and D. Pisoni (Eds.), *Cognitive theory* (Vol. 1). Hillsdale, N.J.: Erlbaum, 1975.

Paris, S. G., & Carter, A. Semantic and constructive aspects of sentence memory in children. *Developmental Psychology*, 1973, **9**, 109–113.

Paris, S. G., & Upton, L. R. The construction and retention of linguistic inferences by children. Paper presented at the Western Psychological Association meeting, San Francisco, April 1974.

Pascual-Leone, J. A mathematical model for the transition rule in Piaget's developmental stages. *Acta Psychologica*, 1970, **32**, 301–345.

Perlmutter, M. (Ed.) *Naturalistic studies of children's memory*. San Francisco: Jossey-Bass, 1981.

Perlmutter, M., & Myers, N. A. Young children's coding and storage of visual and verbal material. *Child Development*, 1975, **46**, 215–219.

Piaget, J. Piaget's theory. In P. Mussen (Ed.), *Carmichael's manual of child psychology*. New York: Wiley, 1970.

Piaget, J. *The grasp of consciousness*. Cambridge, Mass.: Harvard University Press, 1976.

Piaget, J., & Inhelder, B. *The child's conception of space*. New York: Norton, 1956.

Piaget, J., & Inhelder, B. *Memory and intelligence*. New York: Basic Books, 1973.

Poon, L. W., Fozard, J. L., Cermack, L. S., Arenberg, D., & Thompson, L. W. (Eds.). *Memory and aging*. Hillsdale, N.J.: Erlbaum, 1980.

Pylyshyn, Z. W. When is attribution of beliefs justified? *The Behavioral and Brain Sciences*, 1978, **1**, 592–593.

Reese, H. W. Imagery and associative memory. In R. V. Kail, Jr., & J. W. Hagen (Eds.), *Perspectives on the development of memory and cognition*. Hillsdale, N.J.: Erlbaum, 1977.

Richman, C. L., Nida, S., & Pittman, L. Effects of meaningfulness on child free-recall learning. *Developmental Psychology*, 1976, **12**, 460–465.

Riegel, K. F. Toward a dialectical theory of development. *Human Development*, 1975, **18**, 50–64.

Ringel, B. A., & Springer, C. J. On knowing how well one is remembering: The persistence of strategy use during transfer. *Journal of Experimental Child Psychology*, 1980, **29**, 322–333.

Rohwer, W. D. Images and pictures in children's learning: Research results and educational implications. *Psychological Bulletin*, 1970, **73**, 393–403.

Ross, B. M., & Millsom, C. Repeated memory of oral prose in Ghana and New York: *International Journal of Psychology*, 1970, **5**, 173–181.

Schank, R. C., & Abelson, R. P. *Scripts, plans, goals, and understanding*. Hillsdale, N.J.: Erlbaum, 1977.

Scribner, S. Developmental aspects of categorized recall in a West African society. *Cognitive Psychology*, 1974, **6**, 475–494.

Sharp, D. W., & Cole, M. The influence of educational experience on the development of cognitive skills as measured in formal tests and experiments. (Final report to Office of Education, Grant OEG 1965, 1975.)

Sharp, D. W., Cole, M., & Lave, C. Education and cognitive development: the evidence from experimental research. *Monographs of the Society for Research in Child Development,* 1979, **44**, No. 178.

Siegel, A., Allik, J., & Herman, J. The primacy effect in young children: verbal fact or spatial artifact? *Child Development,* 1976, **47**, 242–247.

Siegler, R. S. Three aspects of cognitive development. *Cognitive Psychology,* 1976, **4**, 481–520.

Siegler, R. S. The origins of scientific reasoning. In R. S. Siegler (Ed.), *Children's thinking: What develops?* Hillsdale, N.J.: Erlbaum, 1978.

Stein, N. L. How children understand stories: A developmental analysis. In L. Katz (Ed.), *Current topics in early childhood education* (Vol. 2). Hillsdale, N.J.: Ablex, 1979.

Stein, N. L., & Glenn, C. G. A developmental study of children's recall of story material. Paper presented at the Society for Research in Child Development, Denver, 1975.

Stein, N. L., & Glenn, C. G. The role of structural variation in children's recall of simple stories. Paper presented at the Society for Research in Child Development, New Orleans, 1977.

Stein, N. L., & Glenn, C. G. An analysis of story comprehension in elementary school children. In R. O. Freedle (Ed.), *New directions in discourse processing* (Vol. 2). Hillsdale, N.J.: Ablex, 1979.

Stein, N. L., & Nezworski, M. T. The effects of linguistic markers on children's recall of stories: A developmental study. (Tech. Rep. No. 72.) Urbana: University of Illinois, Center for the Study of Reading, March 1978.

Stevenson, H. W., Parker, T., Wilkinson, A., Bonnevaux, B., & Gonzales, M. Schooling, environment and cognitive development: A cross-cultural study. *Monographs of the Society for Research in Child Development,* 1978, **43**, No. 175.

Sulin, R. A., & Dooling, D. J. Intrusion of a thematic idea in retention of prose. *Journal of Experimental Psychology,* 1974, **103**, 255–262.

Tenney, Y. J. The child's conception of organization and recall. *Journal of Experimental Child Psychology,* 1975, **19**, 100–114.

Trabasso, T. The role of memory as a system in making transitive inferences. In R. V. Kail, Jr., & J. W. Hagen (Eds.), *Perspectives on the development of memory and cognition.* Hillsdale, N.J.: Erlbaum, 1977.

Wagner, D. A. The development of short-term and incidental memory: A cross-cultural study. *Child Development,* 1974, **45**, 389–396.

Wagner, D. A. The effects of verbal labeling on short-term and incidental memory: A cross-cultural and developmental study. *Memory and Cognition,* 1975, **3**, 595–596.

Wagner, D. A. Memories of Morocco: The influence of age, schooling, and environment on memory. *Cognitive Psychology,* 1978, **10**, 1–28.

Wagner, D. A. Culture and memory development. In H. C. Triandis & A. Heron (Eds.), *Handbook of cross-cultural psychology* (Vol. 4). Boston: Allyn & Bacon, 1981.

Wellman, H. M. Tip of the tongue and feeling of knowing experiences: A developmental study of memory monitoring. *Child Development,* 1977, **48**, 13–21.

Wellman, H. Metamemory revisited. Symposium presentation at the biennial meetings of the Society for Research in Child Development, Boston, April 1981.

Wellman, H. M., Drozdal, J. G., Jr., Flavell, J. H., Salatas, H., & Ritter, K. Metamemory development and its possible role in the selection of behavior. In G. A. Hale (Chair), *Development of selective processes in cognition.* Symposium presented at the biennial meeting of the Society for Research in Child Development, Denver, 1975.

Werner, J., & Kaplan, B. *Symbol formation: An organismic developmental approach to language and the expression of thought.* New York: Wiley, 1963.

15

SELF-CONTROL AND SELF-REGULATION IN CHILDHOOD

Charlotte J. Patterson
University of Virginia

The universe constantly and obediently answers to our conceptions.

Thoreau

Self-regulation, as I will employ the term here, involves the successful achievement, through intentional action, of personally selected aims. The aims themselves, and the actions through which they are realized, may be as varied as the individuals who conceive them, and as different as the people who carry them out. Whatever is done freely to fulfill an intention or intentions can, however, in this sense be regarded as self-regulated activity. In my usage, then, the realm of self-regulation is coterminous with the realm of effective, intentional action.

A great deal of self-regulated behavior must therefore be said to consist of nothing but ordinary, everyday activities. We go to bed at a chosen hour at night, and arise at a chosen hour in the morning. We dress ourselves, feed ourselves, and prepare ourselves for the anticipated events of the day. All these activities, and countless others like them, however conventional and habitual they may be or may have become, are done to realize personal intentions, and they are most often done effectively. What is first of all remarkable, then, is how successful attempts at self-regulation normally seem to be, and how little effort they generally seem to require. They are so ordinary for an adult that they often go almost unnoticed.

There are times, however, when something seems to go wrong. I find that I do not wish to retire at my appointed hour or that I am still sleepy when the alarm wakens me in the morning. The clothes I wish to wear are soiled or spotted; what I expect to eat for breakfast has already disappeared from the refrigerator. At points such as these, achievement of a goal is blocked, or the realization of one aim interferes with satisfaction of another; effective self-regulation is challenged. A problem emerges into awareness, and the need for "self-control" becomes evident. It is a breakdown, or the threat of a breakdown, in the normal stream of self-regulated behavior which makes what we call self-control seem necessary. Perhaps it is the very infrequency with which such problems arise that makes them seem so noticeable when they do.

Thus one might regard self-control as the "figure" and self-regulation as the "ground" of voluntary activity. The need for self-control emerges only against a background of voluntary action because it is only out of attempts to achieve personal goals that the possibility of failure to do so can arise. It is only in the context of the belief that I normally, or at least sometimes, *am* in control of my behavior that the possibility that I am *not* could seem real or threatening, or indeed have any meaning at all. It is only my normal ability to fit my behavior easily to self-articulated intentions or rules which makes it seem remarkable when special or extraordinary efforts become necessary. Hence the emergence of self-control as an issue—the very possibility of taking "extraordinary" measures to make behavior fit with preestablished rules—depends on the "ordinary" ability to do so almost without thinking. One might say that the problem of self-control rests on the assumption of self-regulation.

In this sense, one might think of self-control, when it is successful, as lying at the edges of self-regulation, defining its boundaries. A person in need of self-control is like a man standing on a precipice in a storm, not wanting to fall, but unsure of his footing, who wants nothing so much as firm ground under his feet. Thus one might say that the enduring problem of voluntary action is "to move away from the edge," or, more accurately, "to extend the edges outward"—to make what was dangerous seem secure, to make what was difficult seem easy, and then to begin the task again in a different spot. The adult's attempts at what we call "self-mastery," then, can be seen as efforts to answer the question: To what extent is self-regulation possible? That there can be no final answers to such a question is, I take it, one of the reasons why life can be such a challenge.

Into this very rough sketch, I now want to place the young child. As with the adult, what is first of all remarkable is how very much a child has already achieved. Even a young child has gained considerable mastery over his or her own behavior and over the behavior of the physical and social worlds in which he or she lives. Much of his or her ordinary activity may be seen as self-regulated in that it allows for the smooth and relatively untroubled realization of many personal intentions. By early childhood, then, we can say that a background of self-regulation has been established.

But there are also, and perhaps more often or more notably for the child than for us, breakdowns in the habitual exercise of self-regulation. The achievement of goals is sometimes blocked. Different aims or intentions come into conflict so that realization of one precludes satisfaction of another. Such breakdowns are not new in themselves for the child. From the moment that we have said an intention exists, the failure to realize it has become a real possibility. Frustration can seem very real even, or perhaps especially, on the faces of infants.

What is new in early childhood, of course, is language. More specifically, what is new here is the child's growing ability to represent intentions to himself or to herself through the use of language. As children, we become able to verbalize our goals for action, and to evaluate these actions against the goals we have voiced. What we acquire in this way is the ability to formulate, remember, and follow rules for our own behavior. Hence it is tempting, but would be wrong, to say that what we also acquire in childhood is the possibility of and the need for self-control.

The important thing about the growth of linguistic skill in this context, however, is not that it makes failures of self-regulation or the need for self-control *possible,* but that it makes them so much more *evident.* The enunciation, the *saying* of a rule serves to fix it, to make it more permanent, more memorable. In this sense, expressing an intention in words makes it easier to use it as a criterion against which to judge subsequent behavior. Thus the growth of linguistic skill in the young child serves, as it were, to bring problems of self-control into the foreground.

Not only does the growth of language make problems of self-control more obvious, and in this sense "create" them, but it can also serve as a tool for solving them. One of the most striking things about language in this regard is the flexibility of thinking which it affords. As George Kelly (e.g., 1955) used to insist, what can be construed can always be reconstrued and what can be described in one way can always be described in another. Moreover, our ideas about things and events and people—our constructions and descriptions of them—can have profound effects on our actions toward them. Then, too, as Luria (1961) and Vygotsky (1934/1962, 1978) before him emphasized, speech can be used more directly to plan and control our ongoing actions. The acquisition of linguistic skill, then, seems not only to make problems of self-control more evident, and in this way to bring them "up out of the background," but also to provide new ways to solve these problems, and thus to make them recede again. As we know from the results of research, however, it is first the problems, and only later their solutions, which appear in the course of development.

There are, of course, far more studies of children's self-control than can be reviewed, or even mentioned here (for other reviews, from a variety of standpoints, see Bandura, 1977; Karoly, 1977; Masters & Mokros, 1974; Meichenbaum, 1977; Mischel, 1974, 1979; Mischel & Patterson, 1978; O'Leary & Dubey, 1979; and Zivin, 1979). Rather than attempt a comprehensive overview, then, I hope instead to provide a brief introduction to research on the question of how and under what circumstances very young children may be able to use their developing linguistic and cognitive skills to further their efforts at self-control. Even within this more circumscribed area, I will be forced to leave out a great deal. Nevertheless, I hope that enough will be said to justify my claim that the success of even young children's attempts at self-control is profoundly affected by the symbolic activities which they undertake.

The research I will present below falls into three major areas: the effects of imaginal transformations, the effects of self-instructional plans, and children's knowledge about self-control processes. After glancing briefly at laboratory findings in each of these areas, I will consider their applicability to children's behavior in naturalistic settings and then attempt to outline their relevance to some of the ideas sketched above.

EFFECTS OF IMAGINAL TRANSFORMATIONS

Children have often been described as very "stimulus-bound" creatures. What the use of this term should suggest, I think, is both that young children often seem to be trapped by their interpretations of the world and that their interpretations often seem likewise to be trapped in the here and now of immediate circumstances. In what ways can language help the child to escape these traps?

To begin to answer this question, I want to consider some of the results from a program of research conducted by Walter Mischel. In order to study these matters, Mischel and his colleagues (e.g., Mischel & Ebbesen, 1970; Mischel, Ebbesen, & Zeiss, 1972) developed a paradigm with which to examine children's behavior in voluntary delay of reward. In this paradigm, preschool children, most of them about 4 years old, were first allowed to choose whether to receive a less preferred reward (e.g., a small pretzel) immediately or to wait for a more preferred but delayed reward (e.g., two or three marshmallows). If they chose the delayed reward, they were required to sit alone in a room until the waiting period (15 to 20 minutes in duration) was over. Having chosen this option, however, the children nevertheless remained free to reconsider at any time, terminate waiting, and receive the nonpreferred reward. Thus the child's position—freely taken—clearly called for self-control. The issue of principal interest was how long each child would be able to sustain voluntary waiting for the delayed reward.

Initial findings (Mischel & Ebbesen, 1970) appeared to show that attention to the actual rewards themselves inhibited children's ability to delay. Although children waited, on the average, over 11 minutes when the reward objects had been taken from the room, they averaged only about a minute if they waited with both the immediate and the delayed rewards in front of them. Attention to the desired but unavailable reward objects undermined children's efforts at self-control perhaps, these investigators suggested, because it added to the already frustrative character of the delay situation.

Subsequent studies confirmed that for these young children, focusing on the reward objects was likely to prove debilitating for efforts at self-control. It also appeared that attention to other, distracting subjects might enhance the ability to wait. For example, Mischel, Ebbesen, and Zeiss (1972) found that, whether or not the reward objects were actually present in front of the children, instructions to think about the rewards resulted in lower delay times than instructions to think about pleasant but unrelated activities (e.g., "swinging on a swing, with mommy pushing me"). These results were consistent with the idea that focusing on the rewards might add to the frustrating nature of the situation and hence inhibit delay. Thinking about pleasant, or "fun," things, on the other hand, might divert the child's attention from frustrating aspects of the situation and hence make waiting less aversive.

Subsequent findings, however, served to alter this picture in certain ways. Mischel and Moore (1973) found that although attention to the actual reward objects inhibited self-control performances, attention to symbolic presentations of the rewards (e.g., on slides) actually facilitated delay. To interpret this result, Mischel and Moore suggested a distinction between *motivational* and *cue* functions of rewards. They argued that abstract images of the rewards were more likely to serve an informational, or cue, function, reminding the child of the rewards for waiting but without generating undue frustration. The actual reward objects themselves, the investigators argued, were more likely to remind the child of his or her motivation to consume the rewards and hence to make delay more aversive and frustrating. Thus the effects of presentation mode on children's self-control were attributed to the ways of thinking which symbolic versus real presentations of the rewards suggested (Mischel & Moore, 1973; Mischel, 1974).

This interpretation has been confirmed in a number of subsequent studies (e.g., Mischel & Baker, 1973; Mischel & Moore, 1980; Moore, Mischel, & Zeiss, 1976).

In one study, for example (Mischel & Baker, 1973), children were instructed to think about the reward objects in either consummatory or nonconsummatory ways during the delay interval. In the consummatory condition, the experimenter asked the child to focus on the most delicious qualities of the rewards for which they were waiting (e.g., the crunchy, salty taste of a pretzel or the chewy, sweet taste of a marshmallow). In the nonconsummatory condition, the experimenter asked them instead to transform the rewards mentally into similar but inedible objects (e.g., by thinking of pretzel sticks as long, thin brown logs or of marshmallows as puffy, white clouds). Regardless of which set of instructions they were given, children were faced with the actual reward objects while they waited. Consistent with theoretical expectations, children who mentally transformed the rewards in nonconsummatory ways were able to wait over twice as long as those who focused instead on their consummatory qualities.

Similarly, Moore, Mischel, and Zeiss (1976) exposed children either to slide-presented images of the rewards or to the actual rewards themselves, and also varied the instructions about cognitive focus. Half the children in each group were asked to think of the rewards as real objects, and half were asked instead to think of them as symbolic representations (images). Consistent with the other findings, results showed that thinking of the reward objects as real inhibited, while thinking of them as symbolic representations facilitated children's ability to delay gratification. In other words, self-control was primarily affected not by the physical reality in front of the child but by the child's mental representation of it. Thus even these young children were able to use their cognitive capabilities in ways that facilitated their achievement of self-selected goals.

Results of other recent research (e.g., Miller & Karniol, 1976a,b; Patterson & Carter, 1979) demonstrate that the kinds of cognitive strategies that will be effective vary as a function of the kinds of self-control situations in which they are employed. For example, Bruce Carter and I noted that while some self-control situations require that one do nothing but wait for the passage of time, others call for the active completion of work. We reasoned that although attention to motivational properties of rewards might be frustrative and debilitating when one has only to wait, it might actually serve to energize and facilitate relevant activities when one must work.

In one study (Patterson & Carter, 1979), we compared attentional determinants of preschool children's self-control in these two kinds of situations. Children either waited or worked for delayed rewards, and they did so either in the presence or the absence of the reward objects. Consistent with earlier findings, presence of the reward objects inhibited self-control when it involved passive waiting. When self-control called for active work, however, presence of the rewards facilitated performance. A follow-up study (also reported in Patterson & Carter, 1979) demonstrated that presence of the rewards was equally frustrating for children who waited and for those who worked. Although there was no obvious outlet for such frustration for children who had to wait, however, it apparently served to energize the activities of those who had to work.

Overall, then, the findings indicate both that young children were able to use their cognitive abilities to transform difficult self-control situations into easier ones and that the nature of effective transformations depended strongly on the kind of situation in which they were to be employed. Since it was mainly cognitive activities classified as attentional and imaginal in nature that were of interest in these studies, it would be easy to overlook the mediating role of language. To do this, however, would be a mistake, because in every case it was the child's linguistic skill which made instructional manipulation of his or her cognitive activities possible. It was only through linguistic means that the experimenters were able to suggest the various strategies to the children, and it was only through linguistic means that the children were able to understand these suggestions. Even in these circumstances, then, the child's newly acquired linguistic skills played an important role.

EFFECTS OF SELF-INSTRUCTIONAL PLANS

Critical as language is for the kinds of results I have been describing, I want to turn now to consider a line of research which demonstrates even more directly the role of linguistic skills in facilitating children's self-control. Research in this area has examined children's ability to use language as a means by which to control or direct their own behavior (see, for example, Bem, 1967; Hartig & Kanfer, 1973; Kanfer & Zich, 1974; Meichenbaum, 1977; Sawin & Parke, 1979; Toner, Lewis, & Gribble, 1979; Toner & Smith, 1977; and Zivin, 1979). Since I cannot hope to present a representative sample of research here, I will focus instead on a series of studies which I conducted in collaboration with Walter Mischel.

To examine the effects of verbal self-instructions on children's self-control, we (Patterson & Mischel, 1975) first developed a resistance-to-temptation situation for use with preschool children. In this paradigm, children were offered attractive rewards for completing a long and repetitive task (e.g., copying letters of the alphabet onto a page) and also warned that an attractive "Clown Box" (a talking box, painted to resemble a clown) might try to distract them while they were working. While the children were working, an attractive but standardized distraction was introduced (i.e., the talking Clown Box). The measures of principal interest assessed how well each child resisted the Clown Box's distractions so as to continue working and to receive the promised rewards.

In our first studies using this paradigm (Patterson & Mischel, 1975), the experimenter suggested one or more cognitive or self-instructional plans to some of the children before they began to work. For example, one of these plans was, "When Mr. Clown Box says to look at him and play with him, then you can just look at your work, not at him, and say, 'No, I can't; I'm working.' " Results of two studies revealed that children did in fact use such plans (i.e., they did verbalize the plans while working), that children who did so were more effective in resisting the Clown Box's temptations than those who did not, and that children who had been instructed about only one such plan resisted the temptation as effectively as those who had been given a group of three. Clearly, children were able to employ these self-instructional techniques to good effect.

In thinking about these results, we focused on the plan which had been most effective in our initial studies (namely, the one quoted above) and tried to understand what factors might have been responsible for its success. It seemed, in a sense, as though it contained two plans in one. One aspect of it focused attention on the task to be accomplished, while not explicitly suggesting inhibition of attention to the temptation; we termed this the *task-facilitating* component of the plan. The other aspect seemed to forbid attention from being directed toward the temptation, without explicitly focusing on the task to be done; we called this the *temptation-inhibiting* aspect of the plan. Which of these two aspects of the plan was responsible for its success in facilitating the children's efforts at self-control?

In our next study (Patterson & Mischel, 1976) we attempted to answer this question by systematically varying the kinds of plans we suggested to the children. Using the paradigm we had developed, we offered some children a task-facilitating plan (e.g., "When Mr. Clown Box says to look at him . . . you can just look at [your work] and say, 'I'm going to look at my work' "). Others were offered a temptation-inhibiting plan (e.g., "When Mr. Clown Box says to look at him . . . you can just *not* look at him and say, 'I'm not going to look at Mr. Clown Box' "). Still other groups of children were offered either both plans together (i.e., our original, "combination" plan) or none at all. We then assessed the effects of these plans on children's ability to sustain goal-directed work in the face of the Clown Box's tempting distractions.

Our results showed, as before, that preschool children were more than willing to use such self-instructional plans but that whether or not a plan facilitated their self-control performances depended upon its specific content. The temptation-inhibiting plan was more effective than its task-facilitating counterpart. Children who employed the temptation-inhibiting plan spent less time attending to the Clown Box, worked longer, and completed more work than did children who had not been given a plan, but children using the task-facilitating plan did not differ from those who had not been given a plan. Moreover, those who received the combination of both plans did no better than those who used the temptation-inhibiting plan alone. Clearly, the crucial element was use of a plan which directed children's attention away from the source of temptation.

In our subsequent efforts to analyze the effects of verbal planning and self-instruction on children's self-control (Mischel & Patterson, 1976), we noted that a distinction can be made between the organization, or *structure,* of plans on the one hand and their content, or *substance,* on the other. One important structural characteristic of plans is the level of detail with which they are stated. Plans may range from unelaborated statements of intention (e.g., "I'll wash the dog today") to highly articulated sets of specific steps ("As soon as I've finished breakfast, I'll go to the store to buy the special shampoo, then . . .").

One general structural question of interest to us, then, was, To what extent does the success of plans in facilitating children's self-control depend upon their being elaborated in great detail before the occasion for their execution? Accordingly, in our next study (Mischel & Patterson, 1976), we offered some children plans that specified exactly what verbalizations they should employ; we called these *elaborated* plans. Other children were offered plans that specified the nature but not the specific contents of such verbalizations; we called these *unelaborated* plans.

We also wanted to continue our examination of ways in which substantive characteristics of plans affect their ability to facilitate self-control. To do this, we compared the effects of temptation-inhibiting, task-facilitating, and reward-oriented plans. A reward-oriented plan was one that focused on the rewarding consequences for successful self-control. We studied each of these three types of plans in both elaborated and unelaborated forms, so there were six groups of children in all. In addition, however, we tested some children who were given no plan, and some who were given a plan that was irrelevant to the self-control situation (e.g., reciting a nursery rhyme).

The results were exceptionally clear. Compared to performances of children who received an irrelevant plan or no plan at all, resistance to temptation was facilitated only among those who received the temptation-inhibiting or the reward-oriented plans, and only when these plans were presented in elaborated form. Thus the results not only confirmed our earlier findings about the importance of substantive elements of plans but also indicated the importance of their structural characteristics.

I was intrigued by the fact that although these preschool children used elaborated plans effectively, they seemed unable to employ the unelaborated variants of these plans to any advantage. On the one hand, it seemed reasonable to assume some advantage in stating plans at a relatively high level of abstraction, since a generalized plan will certainly be applicable to a wider variety of situations than any of its more detailed variants. On the other hand, it was clear that successful use of a generalized plan requires that its user be capable of filling in appropriate details at the moment of implementation. Furthermore, it was clear from our results that preschool children are not always able to do this effectively.

Hence efficient use of plans seems to require that a balance be achieved between the desire for generality and the need for elaboration and detail. The nature of this balance should depend not only on the pressures and demands of the specific self-control situation that is of interest but also on the age and cognitive sophistication of the person involved. One might expect, in other words, that as children grow older, they will become more capable of employing generalized plans to facilitate self-control. Results of a recent study (Carter, Patterson, & Quasebarth, 1979) confirm this expectation.

Overall, then, findings from these studies show that even young children are able to direct their own behavior successfully using self-instructional plans. What they said to themselves while engaged in efforts to achieve self-control had a tremendous influence on the degree of success they enjoyed. At the same time, however, the efficacy of any particular self-instructional plan depended on the substantive nature or content of the plan, the structural or organizational level at which it was presented, and the age of children who were to use it. Thus the results of these studies help to clarify the ways in which children's linguistic skills may be brought to bear to help them solve practical problems of conduct and to achieve self-control.

CHILDREN'S KNOWLEDGE ABOUT SELF-CONTROL PROCESSES

One question which I have thus far not considered is why children who appear quite able to use various effective strategies and plans for self-control nevertheless often fail to do this independently. Why, in other words, do the same young children, who appear so remarkably capable of using strategies when offered them, fail to come up with them on their own? The first answer to this question, of course, is that they do not always fail. In almost all the studies we have considered, one very interesting aspect of the results was that even some very young children spontaneously invented a variety of different strategies (see Mischel, 1974; Mischel & Patterson, 1978). Having said this, however, one must still acknowledge that many children did fail. Why?

Clearly, one possible reason might be children's lack of knowledge about self-control processes. They might not have known that strategies would be needed or that any one particular strategy would be more effective than another. While there is little evidence with regard to the first possibility, the second has received some research attention (Mischel, Mischel, & Hood, 1978a,b; Yates & Mischel, 1979), and the findings do in fact suggest notable limitations in children's knowledge about self-control.

In one study, for example, Mischel, Mischel, and Hood (1978a) asked children to choose the kinds of conditions that would facilitate self-control in a delay of gratification situation (cf. Mischel, 1974; Mischel & Ebbesen, 1970). They asked children whether they would find it easier to wait if the rewards were covered or if they were exposed during the delay interval, if they thought about consummatory or if they thought about nonconsummatory aspects of the rewards, and other questions of this nature. Interestingly, preschool children did not seem to recognize the advantages of covering the rewards or of thinking about nonconsummatory aspects of the rewards; their choices were essentially random. Older children, however, demonstrated far greater knowledge about these variables; third graders, and especially sixth graders, gave many more correct replies than would be expected by chance alone. These results suggest that the young child's failure to employ effective self-control strategies may be attributable to the fact that he or she knows so little about them—that is, to the fact that the child cannot distinguish effective from ineffective strategies even when directly asked to do so.

Perhaps even more here than elsewhere, one feels the need for further study. It is intriguing that at an age when children use relatively few self-control strategies, they also appear to know relatively little of the likely relevance of such strategies to achievement of their goals. Yet what exactly are we entitled to conclude from this? That knowledge about strategies is required before spontaneous use of them can emerge? Might it not just as easily be the other way around—that knowledge itself arises out of early, but unsystematic "use"?

To voice these questions is not to criticize this research; quite on the contrary, it is the best research which teaches us how to ask new questions. The tantalizing "new" question which this research places on the horizon concerns the relationships between knowledge about self-control variables and the success of actual self-control performances—the relationships, in other words, between a child's knowledge and his or her action. Expressing the question in this way reminds us at once, of course, that it is not really as new as it might initially have seemed, but also that it is at least as important as any that we have asked thus far. I, for one, will be anxiously awaiting the reports of further research on this question.

APPLICATION TO NATURALISTIC SETTINGS

Having discussed some of the results of laboratory research on children's use of self-control strategies, I now want to consider very briefly their applicability to problems of self-control in other settings. To what extent can the laboratory findings be applied to aid in the solution of children's practical problems of conduct in naturalistic settings? There are by now many studies attesting to the relevance of findings like those we have been considering (see Meichenbaum, 1977; O'Leary &

Dubey, 1979; Pressley, in press), but I will be able to focus on only one example here.

Sagotsky, Patterson, and Lepper (1978) studied elementary school children's self-control in the context of an individualized, self-paced mathematics program. In this study, we asked some children to use a simple self-control strategy, which we termed *self-monitoring*. We told these children that they should record instances of off-task behavior (e.g., talking, playing around) during the mathematics period and that they should regard each such recorded instance as a cue to return to work.

As described above, Walter Mischel and I (Patterson & Mischel, 1976) had documented the effectiveness of a similar strategy for facilitating young children's sustained on-task behavior in the face of varied distractions. In that study, use of the strategy did not affect the number of separate times that children were distracted but did enable them to return to work more quickly after each distraction. Hence in the classroom situation—where occasional distraction is inevitable—we (Sagotsky et al., 1978) expected that use of this strategy might make children more conscious of off-task behavior and more able to return to work quickly after having become momentarily distracted. If so, then children using this strategy should study more and progress more rapidly through the mathematics curriculum.

The results confirmed our expectations. Children using the self-monitoring strategy showed greater increases both in study time and in academic achievement than either children who used no strategy or children who were asked only to set appropriate study goals for themselves. These gains were maintained over a 5-week period, in the absence of external feedback or rewards contingent on study behavior. In short, children were quite successful in employing the self-monitoring strategy to facilitate desired behavior in the classroom setting. Especially in light of the relatively low costs of implementing self-control procedures of this kind, the prospects for use of such methods in naturalistic settings appear promising.

CONCLUSIONS

What conclusions are to be drawn from research in these areas? I think it is abundantly clear, even from the limited sample of findings considered here, that young children are able to use their linguistic skills both to direct their own activities and to transform their cognitive representations of self-control situations. Moreover, these self-produced directions and transformations can increase the likelihood that children will achieve their self-selected goals, even under difficult circumstances.

Such competence has often been called an ability to escape "stimulus control." Since this notion can be a very slippery one, however, it may be worth pointing out that what, if anything, the child has "escaped" in the greater number of successful efforts at self-control we have considered here is not objects or events in the world but rather his or her own initial interpretations of them. It is doubtless true that "self-control" would often seem easier if the world itself could be rearranged. What is most remarkable about children's successes in these self-control situations, however, is that they were achieved *not* by escaping or transforming the world in a physical sense (i.e., *not* by changing the "world itself") but by transforming it in *thought*. As long as this ability does not fail, the child will, I think, never face a truly implacable world.

As we know only too well, however, the ability does often fail. To what are we to attribute such failures? Carelessness, lack of interest, flagging motivation, and the like must always be considered. Beyond these, however, sheer ignorance—or should we say innocence?—looms large as an explanation. It is difficult to know just how rough a road may be until one has traveled on it, and almost all the roads to self-control are new ones for the child. How is a child to know that *this* task will require extraordinary efforts, planning, and the conscious use of strategies, whereas *that* task can be accomplished through ordinary means? And even if a child should know such things, on what grounds should he or she be able to select a strategy that would work? As the research clearly shows, there is no all-purpose candidate, and a choice among strategies is no mean task. Moreover, even if the child should hit upon an effective strategy, how would he or she keep it in mind, remember it long enough to carry it out? These kinds of knowledge, these skills, are hardly the sorts of things we would expect to be innate; rather, they are—we hope—the fruits of experience.

In this regard, the parallels with research findings in many areas of cognitive development are striking. For example, studies of memory development have revealed that young children are not inclined to produce many relevant memory strategies but that their memory performances are facilitated if they can be induced to do so (Brown, 1975; Flavell, 1977). Reasons for children's failures have been located in their disinclination to monitor memory stores and in their lack of knowledge about memory-relevant variables (Brown & DeLoache, 1978; Flavell & Wellman, 1977). Research on the development of communication skills has likewise demonstrated young children's deficiencies in strategy use and their improved performances when they do adopt the relevant strategies (Glucksberg, Krauss, & Higgins, 1975; Patterson & Kister, 1980). Explanations for these findings have again been found both in young children's failures to notice communication breakdowns and in their ignorance about communication-relevant variables (Flavell, 1980; Kotsonis & Patterson, 1980; Markman, 1980; Patterson & Kister, 1980).

As others (e.g., Brown & DeLoache, 1978; Flavell, 1976, 1977b; Mischel, 1979) have remarked, such parallels should not seem surprising. Although at one level it is true that the use of strategies seems to be in service of very different goals when we consider self-control as opposed to memory or communication, at another level we might say that there is "really" only one goal. Insofar as they are all methods for expanding the area over which effective, intentional control can be exercised, the use of a self-control strategy is essentially the same as the use of a memory strategy or a communication strategy. In this sense, then, there is only one overriding problem, and it is expansion of the realm of self-regulation. Both in our experiments and in our everyday lives, it is with this central problem that we so often see children struggling.

It is too easy, though, to focus exclusively on the child's difficulties. They are particularly evident to us, no doubt, because they are not *our* difficulties. But, of course, we have our own, and I think the shape if not the content of the problem is the same for us as it is for the child. The problem, as I have said, is to "extend the edges" of self-regulation—to go, as we might now say, from self-regulation to self-control, and back again, and again, and again.

In this connection, we might wish to recall that it was Thoreau (1854/1951, p. 24) who wrote, in *Walden*, "Man's capacities have never been measured, nor are we to judge what he can do by any precedents, so little has been tried." This sounds like a challenge, and it is; but it should also serve as a reminder. We are led to ask: If this is true of us, how much more true must it be of the child? Yet, it is so obvious "how little" of life a child has tried that we may overlook the many possibilities inherent in our own greater opportunities. Believing in the significance of these opportunities and of their possibilities, then, I should like also to ask of Thoreau's words: If they are true of the child, how much more true must they be of ourselves? It was also Thoreau (1854/1951, p. 92) who wrote, "I never knew, and never shall know, a worse man than myself."

REFERENCES

Bandura, A. *Social learning theory*. Englewood Cliffs, N.J.: Prentice-Hall, 1977.

Bem, S. L. Verbal self-control: The establishment of effective self-instruction. *Journal of Experimental Psychology*, 1967, **74**, 485–491.

Brown, A. The development of memory: Knowing, knowing about knowing and knowing how to know. In H. W. Reese (Ed.), *Advances in child development and behavior* (Vol. 10). New York: Academic, 1975.

Brown, A., & DeLoache, J. Skills, plans, and self-regulation. In R. Siegler (Ed.), *Children's thinking: What develops?* Hillsdale, N.J.: Erlbaum, 1978.

Carter, D. B., Patterson, C. J., & Quasebarth, S. J. Development of children's use of plans for self control. *Cognitive Therapy and Research*, 1979, **4**, 407–413.

Flavell, J. H. The development of metacommunication. Paper presented at the Symposium on Language and Cognition, Twenty-first International Congress of Psychology, Paris, July 1976.

Flavell, J. H. *Cognitive development*. Englewood Cliffs, N.J.: Prentice-Hall, 1977. (a)

Flavell, J. H. Metacognitive development. Paper presented at the NATO Advanced Study Institute on Structural/Process Theories of Complex Human Behavior, Banff, Alberta, Canada, 1977. (b)

Flavell, J. H. Cognitive monitoring. In W. P. Dickson (Ed.), *Children's oral communication skills*. New York: Academic, 1981.

Flavell, J. H., & Wellman, H. M. Metamemory. In R. V. Kail & J. W. Hagen (Eds.), *Perspectives on the development of memory and cognition*. Hillsdale, N.J.: Erlbaum, 1977.

Glucksberg, S., Krauss, R. M., & Higgins, E. T. The development of referential communication skills. In F. D. Horowitz (Ed.), *Review of child development research* (Vol. 4). Chicago: University of Chicago Press, 1975.

Hartig, M., & Kanfer, F. H. The role of verbal self-instructions in children's resistance to temptation. *Journal of Personality and Social Psychology*, 1973, **25**, 259–267.

Kanfer, F. H., & Zich, J. Self-control training: The effects of external control on children's resistance to temptation. *Developmental Psychology*, 1974, **10**, 108–115.

Karoly, P. Behavioral self-management in children: Concepts, methods, and directions. In M. Hersen, R. M. Eisler, & P. M. Miller (Eds.), *Progress in behavior modification* (Vol. 5). New York: Academic, 1977.

Kelly, G. *The psychology of personal constructs*. New York: Norton, 1955.

Kotsonis, M. E., & Patterson, C. J. Teaching a game to a friend: Normal and learning disabled children's knowledge about communication. Unpublished manuscript, University of Virginia, 1980.

Luria, A. *The role of speech in the regulation of normal and abnormal behavior*. New York: Liveright, 1961.

Markman, E. Comprehension monitoring. In W. P. Dickson (Ed.), *Children's oral communication skills*. New York: Academic, 1981.

Masters, J. C., & Mokros, J. Self-reinforcement processes in children. In H. W. Reese (Ed.), *Advances in child behavior and development* (Vol. 9). New York: Academic, 1974.

Meichenbaum, D. *Cognitive behavior modification.* New York: Plenum, 1977.

Miller, D. T., & Karniol, R. The role of rewards in externally and self-imposed delay of gratification. *Journal of Personality and Social Psychology,* 1976, **33**, 594–600. (a)

Miller, D. T., & Karniol, R. Coping strategies and attentional mechanisms in self-imposed and externally imposed delay situations. *Journal of Personality and Social Psychology,* 1976, **34**, 310–316. (b)

Mischel, H. N., Mischel, W., & Hood, S. Q. The development of knowledge about self-control. Unpublished manuscript, Stanford University, 1978. (b)

Mischel, H. N., Mischel, W., & Hood, S. Q. The development of knowledge of effective ideation to delay gratification. Unpublished manuscript, Stanford University, 1978. (a)

Mischel, W. Processes in delay of gratification. In L. Berkowitz (Ed.), *Advances in Social Psychology* (Vol. 7). New York: Academic, 1974.

Mischel, W. On the interface of cognition and personality. *American Psychologist,* 1979, **34**, 740–754.

Mischel, W., & Baker, N. Cognitive appraisals and transformations in delay behavior. *Journal of Personality and Social Psychology,* 1975, **31**, 254–261.

Mischel, W., & Ebbesen, E. Attention in delay of gratification, *Journal of Personality and Social Psychology,* 1970, **16**, 329–337.

Mischel, W., Ebbesen, E., & Zeiss, A. Cognitive and attentional mechanisms in delay of gratification. *Journal of Personality and Social Psychology,* 1972, **21**, 204–218.

Mischel, W., & Moore, B. Effects of attention to symbolically presented rewards upon self-control. *Journal of Personality and Social Psychology,* 1973, **28**, 172–179.

Mischel, W., & Moore, B. The role of ideation in voluntary delay for symbolically presented rewards. *Cognitive Therapy and Research,* 1980, **4**, 211–221.

Mischel, W., & Patterson, C. J. Substantive and structural elements of effective plans for self-control. *Journal of Personality and Social Psychology,* 1976, **34**, 942–950.

Mischel, W., & Patterson, C. J. Effective plans for self-control. In W. A. Collins (Ed.), *Minnesota Symposia on Child Psychology* (Vol. 11). Hillsdale, N.J.: Erlbaum, 1978.

Moore, B., Mischel, W., & Zeiss, A. Comparative effects of the reward stimulus and its cognitive representation in voluntary delay. *Journal of Personality and Social Psychology,* 1976, **34**, 419–424.

O'Leary, S. G., & Dubey, D. R. Applications of self-control procedures by children. *Journal of Applied Behavior Analysis,* 1979, **12**, 449–465.

Patterson, C. J., & Carter, D. B. Attentional determinants of children's self-control in waiting and working situations. *Child Development,* 1979, **50**, 272–275.

Patterson, C. J., & Kister, M. C. Development of listener skills for referential communication. In W. P. Dickson (Ed.), *Children's oral communication skills.* New York: Academic, 1981.

Patterson, C. J., & Mischel, W. Plans to resist distraction. *Developmental Psychology,* 1975, **11**, 369–378.

Patterson, C. J., & Mischel, W. Effects of temptation-inhibiting and task-facilitating plans on self-control. *Journal of Personality and Social Psychology,* 1976, **33**, 209–217.

Pressley, G. M. Increasing children's self-control through cognitive interventions. *Review of Educational Research,* in press.

Sagotsky, G., Patterson, C. J., & Lepper, M. R. Training children's self-control: A field experiment in self-monitoring and goal-setting in the classroom, *Journal of Experimental Child Psychology,* 1978, **25**, 242–253.

Sawin, D. B., & Parke, R. D. Development of self-verbalized control of resistance to deviation. *Developmental Psychology,* 1979, **15**, 120–127.

Thoreau, H. D. *Walden.* New York: Bramhall, 1854/1951.

Toner, I. J., Lewis, B. C., & Gribble, C. M. Evaluative verbalization and delay maintenance behavior in children. *Journal of Experimental Child Psychology,* 1979, **28**, 205–210.

Toner, I. J., & Smith, R. A. Age and overt verbalization in delay-maintenance behavior in children. *Journal of Experimental Child Psychology,* 1977, **24**, 123–128.

Vygotsky, L. S. *Thought and language.* Cambridge, Mass.: MIT Press 1934/1962.

Vygotsky, L. S. *Mind in society.* Cambridge, Mass.: Harvard University Press, 1978.

Yates, B. T., & Mischel, W. Young children's preferred attentional strategies for delaying gratification. *Journal of Personality and Social Psychology,* 1979, **37**, 286–300.

Zivin, G. *The development of self-regulation through private speech.* New York: Wiley, 1979.

16

ADULTS AS SOCIALIZING AGENTS

Beverly I. Fagot
University of Oregon

The role of the adult, particularly the parent, in socializing the child seems obvious. Parents are responsible for the very maintenance of the child's life, and until the child is required to enter school, parents can have almost total control over every aspect of the child's existence. Yet our studies of the relationships between child-rearing practices and child behavioral outcomes show contradictory results. Over different studies, specific child-rearing variables, such as feeding patterns and types of punishment, have not related systematically to specific child behaviors. However, several large-scale studies, such as Baumrind (1971) and Bayley and Schaefer (1964), have suggested two major dimensions of child rearing; hostility versus warmth and permissiveness versus restrictiveness. While there has been quite good agreement on the existence of the two dimensions, as Maccoby (1980) points out, there is still disagreement as to their effects, perhaps because each consists of several aspects which may not be highly correlated. For instance, parental control involves consistency of enforcement, level of expectation of the parent, restriction of the child's exploration, and arbitrary use of power by the parent. It is perhaps no wonder that it is difficult to draw any clear-cut conclusions concerning the effects of the two dimensions, because each research study has focused on a slightly different aspect of control or warmth. The same inconclusive flavor emerges from the literature on the effects of another set of adult socializers in the child's life—teachers. As a matter of fact, the two major dimensions of warmth and control, with slightly different names, have arisen in the teacher research (Flanders, 1970), and there is the same pattern of contradictory results.

SEX-DETERMINED DIFFERENCES IN ADULT TREATMENT

I would now like to use the area of sex-role differentiation as an example of the problems that arise when attempting to study how children are influenced by others in their environment. Theories of sex-role development have always assumed that boys and girls are treated differently by their parents and that sex-role development is influenced by the different conditions and treatments that children subsequently receive within the family situation. Yet after an extensive review of the sex differences literature, Maccoby and Jacklin (1974) came to the following conclusions concerning differential shaping of sex typed behaviors: "We have been able to find very little evidence to support it in relation to behaviors other than sex typing as very narrowly defined (e.g., toy preferences). The reinforcement contingencies for the two sexes appear to be remarkably similar" (Maccoby & Jacklin, 1974, p. 342). Maccoby and Jacklin did conclude that boys receive more intense socialization pressures than girls, but this conclusion has not been discussed as much as the lack of differential reinforcement. Block (1978) feels that such acceptance of the null hypothesis about differential rearing is premature for several reasons. First, the studies that Maccoby and Jacklin reviewed were primarily concerned with children 6 and under. Yet many sex-typical behaviors begin to have salience for parents only when the children enter school or even at adolescence. Second, most of the studies reviewed focused upon the mother, but we have evidence that the father, more than the mother, feels sex typing is important (Block, VonDerlippe, & Block, 1973). Third, concepts such as dependence have different meanings in different studies and sometimes in the same study. How could we hope to have consistent findings when the constructs themselves are defined differently by each researcher and the behaviors used to measure the construct show radical changes across studies and across ages of the children? Finally, all studies were considered equally by Maccoby and Jacklin, without regard for differences in sample size or the care with which appropriate statistical procedures were used. In fact, in some of these studies, the testing of sex differences appeared to be post hoc and unrelated to the purpose of the study. Block asserts that the existence of sex-determined socialization remains an open question.

How could highly respected researchers come to such different conclusions concerning the same set of studies? To understand, it is necessary to examine why certain variables were chosen for the study of sex-role socialization. Some variables were chosen specifically from psychoanalytic theoretical predictions. For instance, Freud predicted that girls should show weaker superego development than boys, due to the incomplete resolution of the Oedipal situation. Therefore, development of conscience was often studied as a function of sex of the child as well as parental values. Differences in aggressiveness and dependency were also predicted from theories of sex-role development; therefore, it was assumed that these behaviors would be treated differently for boys and girls. There were two problems using the broad constructs derived from Freudian theory. In some cases, the predicted sex differences did not exist, at least at the age children were studied. For instance, among young children at least, girls appear to develop a conscience earlier and to have more control over their behaviors than boys (Sears, Maccoby, & Levin, 1957). Second, sex differences may exist but not as a result of differential parent attitudes. Alternatively, it may be that attitudes of parents toward aggression have little to do with the development of aggression, but that the role models parents and others provide to the child have more influence. Finally, parents might be responding to innate sex differences that elicit treatment differences. For whatever reason, the studies derived from Freudian-based theory that examined parental effects on large systems of behaviors have been inconclusive.

There are really two different issues involved in the controversy: do adults treat boys and girls differently and if such differential socialization does take place, does it contribute substantially to sex-role differentiation? This chapter will look at the simpler question of whether adults do use different socialization techniques with boys and girls, and, if so, the immediate effects of these differences. The first step in understanding the process of sex-role differentiation should be the careful study of the relationship between attitudes and behaviors.

My study on toddlers and their parents (Fagot, 1978) illustrates the perils of relying exclusively on parent attitudes concerning sex-role differentiation or on parents' reports of how they treat children. In this study, I observed toddlers and their parents interacting in their homes, asked the parents to rate behaviors appropriate for boys and girls (sex stereotypes), and also interviewed parents on their values and socialization practices concerning sex roles. I observed three broad categories of child behavior often considered important in the development of sex roles: aggression, adult dependency, and large motor activities. It was possible from the observations to determine which child behaviors were actually more typical of one sex, so that those behaviors for which there were empirical differences were called *sex-preferred* (although I now use the term *sex typical*, as it does not have as great a value loading). Directly observed parent reactions to the behaviors were classified as positive, negative, or neutral. Parents did not differ in their reactions to sex-stereotyped behaviors (except in the case of doll play, which was also sex typical), but they did differ in their reactions to sex-typical behaviors. Parents reacted more positively to boys engaged in active motor behavior and to girls when they asked for help. Attitudes toward sex-role development as measured in the interview did not appear to influence the parents' behavioral responses to sex-typical behaviors. This is not to say that sex stereotypes were unrelated to sex-typical behaviors, for there was a significant correlation of .32 between sex-stereotyped behaviors and sex-typical behaviors in their children. However, this is not an impressive relationship and leaves a great deal of the variance to be explained by other factors.

The results of this study suggest that parents are not fully aware of their socialization techniques and that to rely only on their values and attitudes without examining their behavior as well gives a false picture of the actual socialization process. The study also suggests that for young children, the sex stereotypes held by parents are not yet congruent with their own reactions to their child's behaviors. The behaviors which elicit sex-determined parental reactions are perhaps precursors to later behaviors considered important for sex-role development.

Preschool teachers also have an impact on the development of young children, and there has been considerable research on the differential treatment of girls and boys by preschool teachers. Preschool teachers are almost all female, and there has been some concern that this would make it more difficult for young boys to develop appropriate sex-role behaviors. Fagot and Patterson (1969) did find that female preschool teachers tended to react positively to female-typical behaviors in both boys and girls and not male-typed behaviors. This finding has been interpreted to suggest that providing male models in the preschool might make it easier for boys to adopt male identities and to adjust to preschool environments. However, the situation is much more complex than that. Fagot (1981a) observed male and female preschool teachers' reactions to boys and girls in their classrooms. Half these teachers had at least 3 years teaching experience and half were new teachers. There were a few teacher sex differences, with males joining play more than females and giving more favorable comments, but there were many more differences between experienced and inexperienced teachers than between males and females. Experienced teachers, both males and females, showed the same pattern of responses as in the Fagot and Patterson study (1969). When children were engaged in sex-typed behaviors, the teachers reacted positively to female-typical behaviors and ignored male-typical behaviors for both sexes. Inexperienced teachers of both sexes tended to react to children in a more sex-typed way, responding more to boys in male-typed behaviors and girls in female-typed behaviors. The responses of experienced teachers appeared to be determined by their definition of appropriate pupil behavior rather than being determined by the sex of the child.

EFFECTIVENESS OF PEER AND ADULT REINFORCEMENT

The fact that boys over the year do not become less male sex typical opens up the whole question as to just how effective are teachers in changing a child's behavior. I have attempted to evaluate the effectiveness of teacher and peer feedback to preschool children who attend small informal play groups twice a week (Fagot, submitted). The children were observed in the natural play group setting. The target-child's behavior and any reactions to the behavior were recorded over 10-minute intervals. In one study of 40 children, play behaviors were coded as male typical, female typical, or neutral; and positive, negative, or "no" response from male and female peers and teachers were recorded. Comparing the effects of positive versus negative reactions is one way to analyze the effectiveness of a peer or teacher response. As might be expected, boys were given positive feedback from other boys for male behaviors but not female behaviors, while teachers rewarded their neutral and female behaviors, but not male. The important question, though, was, What effect did this differential response from peers versus teachers have on the probability of the target-boy continuing his behavior? Basically, for boys the only responder who made a difference was another boy. The probability of a boy's behavior (typical male, female, or neutral) continuing was significantly higher with a positive response from a male peer, while female peers and teachers, even male teachers, did not affect the boy's behavior. For girls, teachers' and other girls' responses had a significant effect with positive responses resulting in a greater likelihood of continuation than negative responses.

We also examined what happened when a child performed a behavior and got no response. A confidence interval was established around the no-reaction probability, and the probabilities of behavior continuing after positive or negative responses were compared to it. In this case, only one category showed a difference. When a boy was engaged in female-typical behaviors, a negative response from another boy was significantly more likely to terminate the behavior. For boys, the male peer group was most influential, while for girls, both adult teachers and female peers helped maintain behaviors.

The lack of immediate teacher effects on boys is supported in other types of studies. Feldbaum, Christenson, and O'Neal (1980) studied assimilation of new children into ongoing preschool groups. They found that teachers initiated equal numbers of supportive responses to new boys and girls, but only the girls used the teachers as a source of support, while the boys directed their attention to the male peer group.

What about long-term effects? Is there any sign that peer and teacher pressure changes a child's responses over a period of time? Fagot (1981b) studied a group of 67 children attending play groups over a period of 1 year. Factor scores from behavioral observations had been obtained on these children. The children were quite consistent over a period of a year on most factors. However, peer and teacher reactions did seem to have an effect in some cases. When children showed high levels of nontask behavior, teachers would often intervene to try to increase their participation in the group. Such children showed low correlations on the nontask factor over the period of a year, suggesting that the teachers' interventions were successful. Again, the teachers' interventions in children's sex-role behaviors were not particularly effective in changing behavior, but the boys' responses to other boys were successful. One starts to understand the difficulty of studying adult socialization effects when you realize that these children were 2-year-olds, supposedly an age when peer pressure is not yet effective and adult pressure has maximum effect.

BIDIRECTIONALITY AND THE DEVELOPMENT OF SOCIAL INTERCHANGES

Most of the research reviewed so far has concerned differences in adult treatment and related such differences to children's behaviors. However, as pointed out by numerous theorists (Bell, 1977; Cairns, 1979; Sears, 1951), it is more accurate to characterize socialization, not as a unidirectional effect from adult to child, but as a bidirectional system where both the child and the adult affect the behavior of the other and the system itself.

A pitfall needs to be noted concerning bidirectionality. Some recent articles have appeared to suggest that the child is in control of the parent, some to the point of sounding like a reworking of the old movie, *The Bad Seed*. Such child-oriented approaches would seem as doomed to failure to predict effects as the old adult-directed models, because they also fail to acknowledge the reciprocity of the relationship.

The concept of separate child effects and parent effects is not new, but rather than speaking of child effects or adult effects as if each were independent of the other and might somehow, if only we could untangle the consequences, be understood separately, now we think of the adult and child as members of a system which is interdependent. If one member of the system behaves in some way, the other member will respond to that behavior. Those behaviors and responses will change the system and make some slightly different set of effects possible during the next sequence of interactions.

Cairns (1979) suggests that interchanges between two individuals develop a synchrony so that the act of one individual supports and is coordinated with the activity of the other individual. One would expect that the ability to enter into synchrony in an interchange would depend upon the skill of both members of the relationship. Studies with older peers and with adult dyads suggest the positive interchanges are characterized by such synchrony of prosocial responses, while negative interchanges are also in synchrony but seemingly more controlled by responses to negative acts.

What is needed is a way to ask questions that gives structure to the study of the socialization effects. One important area where more information is needed is on the nature of interchanges between children and adults. Most studies of adult–child interchanges have used children with some problem, such as excessive aggression (Patterson, 1976) or "difficult" temperaments (Thomas, Chess, & Birch, 1968), and we do not know how the interchanges of these children differ from those of normal children. There have been some recent attempts to look at the development of social interchanges, both adult–child, and child–child. Holmberg (1980) studied the development of social interchanges, both positive and assertive, in children from 12 to 42 months of age with familiar teachers and familiar peers. She found that even 12-month-old children appeared socially competent (i.e., were involved in social initiations and subsequent elaborated interchanges) when interacting with an adult teacher. When interacting with peers, the nature of interaction was very different for the youngest and oldest children. By 42 months, peer interchanges resembled those of adult–child interchanges, with positive initiations and subsequent elaborated interchanges predominating. For the youngest children, however, initiations were as likely to be negative as positive and were much fewer in number. Also, there were rarely elaborated interchanges in the youngest dyads. For the youngest children, Holmberg suggests it was the adults who maintained the interchange, and that as the children learned to initiate, the adults would decrease control of the interchange. This finding makes sense, for one would expect that adults, with their greater social skills and more complex cognitive worlds, would play a controlling role in interchanges with very young children.

The term *control* should be used with care. There have been suggestions that infants "control interchanges" because adults must adjust to their limited repertoire. It is true that the limited behavioral repertoires of infants force adults into limited behavioral patterns, but this is a passive form of control. Infants certainly have some more active forms of control, such as gaze aversion, which terminate interchanges, but again, they are limited in comparison to an adult's repertoire. Adults have the flexibility to adapt to the infant, and in this sense, they actively control the interchange. In this sense, the adult has more control over the interchange.

While Holmberg's data give us information about timing of the development of social interchanges, we still do not know much about the process of such development. There have been several hypotheses as to how social interactions are learned. One hypothesis is that behaviors toward adults, particularly the mother, develop as part of attachment (Ainsworth, Blehar, Waters, & Wall, 1978), while relationships with peers develop in conjunction with the development of object skills (Rubenstein & Howes, 1979). Eckerman, Whatley, and Kutz (1975) suggest that interchanges develop first with caretakers and then are used in peer play. Another suggestion is that infants use similar social skills with both peers and adults, and differences in interactions are due to the adult's greater skill at maintaining the interaction. Vandell (1980) recently attempted to test these three sets of hypotheses. Children from 6 to 12 months were studied in pairs of infants and mothers, half the time with toys present and half with toys absent. The mothers were told to set the infants down and not initiate interchanges. While there were some differences in mother interactions and peer interactions in that infants did vocalize, look at and touch the mothers more than each other, there were many more similarities. The high degree of similarity between the adult and peer interaction may be a function of the experimental instructions. It is probably a rare adult–child interchange where the adult does not take the initiative and direct the interchange for a child below the age of a year. Both the Holmberg and Vandell et al. studies suggest that children learn one set of social responses which they attempt to use with increasing frequency during the second year of life. However, to me it also seems clear that something different than simple addition of responses must occur before the age of 3. The child appears to make a cognitive leap and no longer just responds appropriately but also initiates appropriately. It is as if the child now understands the nature of social interchanges and enters into them with the object of developing synchronous interchanges. At this time, the nature of adult–child socialization changes, with the adult decreasing control of the interchange and allowing the child more say in the direction of interchanges.

To sum up, by the age of 3, children can enter into truly bidirectional synchronous interchanges. Nevertheless, we must keep in mind that the adult–child relationship is still unequal. The child has a very rudimentary understanding of the world, and adults direct the content of most interchanges. The adult has sets of ideas and attitudes concerning what the child should do at particular times. The responses of the individual child undoubtedly modify the nature of interactions that take place and, in the long run, may modify expectations about the child. Still, in our rush to look for child effects, we must not forget that adults are working from complex cognitive structures which permit them to use many means of control with the child.

LIMITATIONS ON ADULT POWER

If adults are so powerful, then why are they not more successful at producing the effects they wish in their children? It seems to me that one clear reason is that the attitudes that adults hold and the behaviors they use when child rearing are not closely related, particularly when they are dealing with a young child. Parents may believe strongly in rational explanations and may think that they will always use this technique to teach their children. However, for the child under the age of 3, it is often impossible to depend upon reasoning. It is necessary to use physical means of restraint and restructuring of the environment or the child will not survive. Furthermore, parents learn ways of control which may not be related to their cognitive structures. So, though parents are more powerful than children, they also are likely to be inconsistent in their teaching and rearing techniques simply because they themselves are not aware of the different levels of interaction they use with young children.

Another reason why there is such low predictability from child-rearing techniques and values used in the early years is that the parents are also changing and developing. Parents' values do not remain static. Undoubtedly, what a parent wants and expects of a child is modified by the child's behavior. In addition, there are changing cultural trends in raising children, as well as pressures from families and friends, which may change a parent's expectations. Also, as the parents mature, their values may change. Adult lives are not static, and their later values and behavior may influence children as much or more than the methods they used in early childhood.

Finally, there is some indication that there is a long delay between the occurrence of specific parent practices and their effects on children's gender-role behaviors. There is some evidence that the young child learns a set of highly visible sex stereotypes and adopts corresponding behaviors. It is only as the child develops a more complex cognitive structure that cultural stereotypes are distinguished from the specific attitudes and behaviors endorsed by parents. This may mean that the effect of the parent's individual values and behaviors on sex-role development is not very noticeable during the early years. For instance, Meyer (1980) found that 10- to 12-year-old girls' sex-role attitudes and aspirations correlated significantly with their mothers' attitudes but that attitudes of 6- to 8-year-old girls did not correlate with their mothers' attiudes. Instead, the younger girls had significantly more stereotyped sex-role attitudes and aspirations. However, even if their gender-role behaviors are not immediately affected, children who are encouraged only in very sex-stereotyped behavior patterns should differ in other respects from those who are encouraged to try out a broad repertoire of behaviors. Their sex-typing scores might be similar, but other measures of cognitive and social behaviors might show differences which, in the long run, would add up to significant parental socialization effects.

If one puts the complex peer and adult influences together with the fact that one needs to conceptualize the effect of the child on the adult, the fact that the child is changing rapidly as a result of his or her own cognitive development, and that the adult's attitudes and behaviors are not always well matched, then it becomes understandable why some individuals turn from socialization research in despair! However, what is needed in the area is not despair or a retreat to simpler models and problems but projects that are designed to take account of the complexities of the process. What this means is that we need measures of ongoing process and of the cognitive structures of both the adults and children in the socialization dyad. From social learning and ethological research, we have developed a sophisticated interactional methodology and from research in cognitive development, reasonable methods for studying the individual's mind. Unfortunately, most projects use one or another of these approaches, and consequently, we understand only pieces of the socialization process.

To illustrate how this approach applies to the study of sex-role development, I would like to use my current research project as an example of how problems in the area can be attacked. I am interested in the development of sex role in children from 1 to 3, the age range when the child is coming to understand his or her own sex and is starting to develop behaviors and attitudes which result in the differences between the sexes. It is quite clear that sex-role development is a cognitive process in which the child acquires an understanding of what it means to be a male or female and a set of learned behaviors and attitudes which can be considered by the child or others as appropriate for a member of that sex. The process has two major outcomes, gender identity and gender role (Money & Ehrhardt, 1972). Gender identity, the perception of oneself as male or female, is very difficult to change. Gender role, the attitudes and behaviors by which individuals indicate their sex to others, on the other hand, is always open to modification and changes constantly throughout the life span. Both processes are taking place simultaneously and we do not know much about their relationship. We know very little about how environmental variables influence gender identity, for most socialization work has concentrated on the shaping of gender-role behaviors by others in the child's environment. The adoption of certain gender-role behaviors is hypothesized to have consequences for the child's learning of a whole range of social and emotional behaviors (Block, 1979), so that whatever model is used when looking for differences in cognitive and emotional skills must be sensitive not only to the sex of the child but also the gender role behaviors adopted by the child.

OLD FINDINGS, NEW DIRECTIONS

Originally, I started out with the rather traditional viewpoint that if I could understand how parents treat their children, then I would understand sex-role development. In watching parents interact with their children in natural settings, I certainly understand a good deal more about sex-role development than I did as a graduate student embarking upon my dissertation research. Parents react more favorably when children engage in same-sex-typical behaviors. That is, boys' parents encouraged gross motor and manipulative play as well as sex-stereotyped boy play, and they discouraged female-typical toy play and dependency behaviors. Girls' parents encouraged dependent behaviors, nurturing play behaviors, but discouraged active motor play (Fagot, 1978). However, from observing in preschools, I came to see that peers and teachers also had a strong influence on sex-role development. For boys, the male peer group provides an extremely conservative influence toward sex stereotyping, and the male peer group effectively stops exploration of certain kinds of behaviors (Fagot, 1977). Teachers in preschools try to encourage boys to try out school-type activities, but at least with very young children, they are not too successful (Fagot, submitted).

However, the external pressures provided by the environment cannot give a complete picture of sex-role development. For instance, I find that children must have at least a rudimentary understanding that there are two sexes before they react to peer pressures (Fagot, in preparation). Both behavioral and cognitive variables are important for all participants in the process: the child, the parents, peers, and teachers.

One real difficulty with such a project concerns how one handles this large amount of data. The traditional socialization approach has been to correlate scores and provide for the reader a summary of relationships. However, in this project, the first step to be taken is to look at different methods of socialization used by parents of boys and girls as well as differences in attitudes. Then the consequences of different patterns of parental behavior and attitudes can be examined using a conditional probability paradigm. For example, when comparing parents of girls who are nonstereotyped in their attitudes to parents of girls who are stereotyped, one can analyze parents' behaviors in the two groups to determine whether the parents are affected by their attitudes or the sex of their child, or some combination. Child behaviors can be examined as a function of variations in the parent variables. Is a child whose parents have nonstereotyped attitudes and behave in a nonstereotyped fashion more likely to try out a broad range of behaviors than a child from professed nonstereotyped parents who behave in a stereotyped fashion? At the same time, specific adult–child interactions can be studied from the observation data. Finally, we can try to determine just how specific child variables influence adults, who, in turn, react to specific children in different ways. In other words, rather than assume that parent treatment is responsible for child differences, we can identify interactions between child behavior and parent treatment.

To be more speculative, what effect might parents play at various stages of the child's sex-role development? In the age period from 1 to 3, I would expect that the child's attitudes and concepts are not necessarily related to the parent's. During the early years, the parent's reactions to specific behaviors should be related to the child's willingness to try out different kinds of behaviors. The child's gender role probably proceeds more or less unrelated to parent attitudes and values during this time, because the child has recognized only a broad outline of a complicated process and picks out as important only very obvious sex stereotypes. However, the parents' reactions to specific child behaviors will have long-term consequences. Parents who encourage the child to learn varied activities will have children who do not show the kind of gender-related insufficiencies mentioned by Baumrind (1979) —lack of control and prosocial behaviors in males, and overcontrol and lack of assertiveness and self-confidence in girls. In the late middle-childhood years, when children can conceptualize sex typing in a more complex way, they should be influenced by their parents' sex-role attitudes and behaviors. This hypothesis suggests that children will come to resemble their parents at 10 years of age rather than 4. Parents are influential in the sex-role socialization process at two different periods; for young children they either encourage or discourage a broad behavior repertoire which is either limited by the sex of the child or not. As children gain the cognitive ability to understand sex role in an individual way in middle childhood, then parental attitudes should influence the child.

Therefore, in order to predict what place adults have in the socialization of sex role, we must specify the age and cognitive level of the child at whom we are looking and the part of the sex-role process in which we are interested. While it is undoubtedly true that parental socialization plays some role in the gender-identity process and that something goes awry in some individuals who end up confused as to gender identity or identified with the wrong sex, we really do not know how to predict such confusion or to deal with its consequences. We know somewhat more about gender-role behaviors, but even here, we must always take into account children's peer group experiences as well as their experiences with adults. Finally, we need to specify just how the child's own characteristics modify the behavior and values of the adults in his or her environment.

REFERENCES

Ainsworth, M. D. S., Blehar, M. C., Waters, E., & Wall, S. *Patterns of attachment.* Hillsdale, N.J.: Erlbaum, 1978.

Baumrind, D. Current patterns of parental authority. *Developmental Psychology Monograph,* 1971, **4**, (1, pt 2).

Baumrind, D. Sex related socialization effects. Paper presented at the meetings of the Society for Research in Child Development, San Francisco, 1979.

Bayley, N., & Schaefer, E. S. Correlations of maternal child behaviors with the development of mental abilities: Data from the Berkeley Growth Study. *Monographs of the Society for Research in Child Development,* 1964, vol. **29**.

Bell, R. Q. Socialization findings re-examined. In R. Q. Bell & L. V. Harper (Eds.), *Child effects on adults.* Hillsdale, N.J.: Erlbaum, 1977.

Block, J., VonDerLippe, A., & Block, J. H. Sex-role and socialization patterns: Some personality concomitants and environmental antecedents. *Journal of Consulting and Clinical Psychology,* 1973, **41**, 321–341.

Block, J. H. Another look at sex differentiation in the socialization behaviors of mothers and fathers. In J. Sherman & F. Denmark (Eds.), *Psychology of women: Future directions of research.* New York: Psychological Dimensions, 1978.

Block, J. H. Socialization influences on personality development in males and females. American Psychological Association, New York, September 1–5, 1979.

Block, J. H., & Block, J. The role of ego-control and ego resiliency in the organization of behavior. In W. A. Collins (Ed.), *Minnesota symposium on child psychology* (Vol. 13). Hillsdale, N.J.: Erlbaum, 1979.

Cairns, R. B. *Social development: The origins and plasticity of interchanges.* San Francisco: Freeman, 1979.

Eckerman, C., Whatley, J. C., & Kutz, S. Growth of social play with peers during the second year of life. *Developmental Psychology,* 1975, **11**, 42–49.

Fagot, B. I. Consequences of moderate cross-gender behavior in preschool children. *Child Development,* 1977, **48**, 902–907.

Fagot, B. I. The influence of sex of child on parental reactions to toddler children. *Child Development,* 1978, **49**, 459–465.

Fagot, B. I. Male and female teachers: Do they treat boys and girls differently? *Sex Roles,* 1981, **7**, 203–271. (a)

Fagot, B. I. Continuity and change in play styles as a function of sex of the child. *International Journal of Behavioral Development,* 1981, **4**, 37–43. (b)

Fagot, B. I. Just how effective are peers and teachers in shaping sex role behavior in early childhood? Submitted.

Fagot, B. I. The relationship among sex-typical behaviors, gender labels, and gender identity in toddler children (paper in preparation).

Fagot, B. I., & Patterson, G. R. An in vivo analysis of reinforcing contingencies for sex-role behaviors in the preschool child. *Developmental Psychology,* 1969, **1**, 563–568.

Feldbaum, C. L., Christenson, T. E., & O'Neal, E. C. An observational study of the assimilation of the newcomer to the preschool. *Child Development,* 1980, **51**, 497–507.

Flanders, N. A. *Analyzing teacher behavior.* Reading, Mass.: Addison-Wesley, 1970.

Holmberg, M. C. The development of social interchange patterns from 12 to 42 months. *Child Development,* 1980, **51**, 448–456.

Maccoby, E. E. *Social development: Psychological growth and the parent-child relationship.* New York: Harcourt Brace Jovanovich, 1980.

Maccoby, E. E., & Jacklin, C. N. *The psychology of sex differences.* Stanford: Stanford University Press, 1974.

Meyer, B. The development of girls' sex-role attitudes. *Child Development,* 1980, **51**, 508–514.

Money, J., & Ehrhardt, A. A. *Man & woman, boy & girl.* Baltimore: The Johns Hopkins University Press, 1972.

Patterson, G. R. The aggressive child: Victim and architect of a coercive system. In L. A. Hammerlynck, L. C. Handy, & E. J. Mash (Eds.), *Behavior modification and families: Theory and research.* New York: Bruner Mazell, 1976.

Rubenstein, J. C., & Howes, C. Caregiving and infant behavior in day-care and homes. *Developmental Psychology,* 1979, **15**, 1–24.

Sears, R. R. A theoretical framework for personality and social psychologists. *American Psychologist,* 1951, **6**, 476–482.

Sears, R. R., Maccoby, E. E., & Levin, H. *Patterns of childrearing.* Evanston, Ill.: Row, Petersen, 1957.

Thomas, A., Chess, S., & Birch, H. *Temperament and behavior disorders in children.* New York: New York University Press, 1968.

Vandell, D. L. Sociability with peer and mother during the first year. *Developmental Psychology,* 1980, **16**, 355–361.

17

EMPATHY AND PROSOCIAL BEHAVIOR IN CHILDREN

Mark A. Barnett
*Associate Professor
Department of Psychology
Kansas State University*

CONCEPTUALIZATIONS OF EMPATHY

Empathy has frequently been suggested as an important factor in the development and expression of prosocial behavior in children. However, the concept of empathy has been defined differently by various researchers interested in altruism and related interpersonal behaviors. The central issue on which the conceptual debate focuses concerns the extent to which empathy encompasses an affective, as well as a cognitive, component. Some writers (Borke, 1971, 1973; Buckley, Siegel, & Ness, 1979; Chandler & Greenspan, 1972; Dymond, 1949; Greenspan, Barenboim, & Chandler, 1976) define empathy as the cognitive ability to recognize and understand the thoughts, feelings, or intentions of another individual. Within this framework, various terms—including cognitive perspective taking, person perception, role-taking, and social cognition—have been proposed as either closely aligned or synonymous with empathy. Alternatively, other writers (Feshbach, 1978b; Hoffman, 1975a; Iannotti, 1979; Leiman, 1978; Mussen & Eisenberg-Berg, 1977; Sawin, 1979; Staub, 1978) view empathy as the vicarious experiencing of the emotional state of another individual. From this perspective, while particular cognitive abilities are seen as necessary for an empathic response to occur, the "sharing" of another's affect, particularly a needy other's sadness, is considered the critical component of empathy and the one which serves to motivate subsequent altruistic behaviors.

According to "affective empathy" theorists, the empathically aroused individual may aid the target of his or her arousal so as to alleviate the mutually experienced distress (Aronfreed, 1970; Hoffman, 1975a, 1977a). Similarly, an empathically distressed observer may be motivated to help an unfortunate other by anticipating vicarious pleasure following the helpful act. Hoffman (1977a) claims that empathic distress, even if experienced primarily as aversive to the self, differs from usual self-serving motives in that (1) it is aroused by another person's misfortune, rather than one's own, (2) a major goal of the subsequent behavior is to help the other, not merely oneself, and (3) the potential for gratification in the observer is contingent on acting to alleviate the other's distress. Therefore, whether helping is motivated by the anticipated cessation of negative affect, the anticipated experience of positive affect, or some combination of both, it is the initial vicarious experiencing of the other's distress, rather than the mere cognitive awareness of the other's misfortune or affective state, which is believed to give empathy its unique prosocial quality.

ANTECEDENTS OF EMPATHY

While theories abound concerning the factors associated with the development of empathy in children, little empirical research on the topic is presently available. Consistent with the notion that empathy involves both cognitive and affective components, most of the attention concerning the antecedents of empathy has focused on the child's developing capacity or inclination to experience, interpret, and respond to the feelings of another individual.

Hoffman (1977b) suggests that during the first week of life, infants demonstrate an affective orientation to other infants which may represent a "constitutionally based, early precursor of empathy" (p. 299). One- and two-day-old infants have been found to show distress and cry to the sound of another infant's cry; interestingly, no such response was made to a computer-simulated cry of equal intensity (Sagi & Hoffman, 1976; Simner, 1971). While the infant's "primitive empathic response" to another's apparent distress appears quite spontaneous and intense, it cannot be considered to even approach a mature empathic response because the infant lacks a cognitive awareness of the self or other.

Another suspected early antecedent of empathy concerns the intense affective relationship between the caretaker and the infant (Hoffman, 1977a; Sullivan, 1953). This classical-conditioning view posits that the caretaker's affective state may be "transmitted" to the infant through physical handling, as well as other verbal and nonverbal modes of communication. For example, if the mother is uncomfortable while feeding, her stiffened arms or unsure grip may elicit distress in the infant. In future interactions, the mother's verbal and nonverbal cues, which were initially associated with her discomfort, may function as conditioned stimuli that evoke the young child's distress. Moreover, stimulus generalization is believed to occur such that the inclination to experience empathic responses to other individuals' distress takes root at this time. As with the notion of "primitive empathy," however, early classical conditioning can provide, at best, only a partial explanation of the development of empathy since the self–other distinction in the very young child is blurred.

Models of the child's increasing capacity to experience empathy have emphasized the development of the requisite cognitive skills for experiencing and appropriately interpreting vicarious affective arousal. According to Hoffman (1975a, 1977a), the child's emerging ability to comprehend the distinction between self and other and the awareness that other individuals have internal states and feelings separate from one's own set the stage for higher levels of empathic responding. Similarly, Feshbach's (1978b) three-component model states that the truly empathic response requires (1) the ability to discriminate and identify the affective states of other individuals, (2) the ability to assume the perspective or role of the other, and (3) the evocation of a shared emotional response. From these perspectives, the child's developing cognitive capacities, as well as experiences and training which promote role-taking abilities, are believed to enhance the child's awareness of and responsiveness to another's distress. Some support for the latter contention comes from studies of role-taking training in young children (Feshbach, 1978a; Iannotti, 1978; Staub, 1971). In one investigation, Staub (1971) encouraged a group of kindergartners to understand and express the feelings of victims and helpers by having them act out both roles in various contrived situations; in the other conditions, the children either discussed helpful acts without role-playing or enacted scenes unrelated to helping (control group). Observations made following training revealed that girls in the victim–helper role-playing group responded more frequently to the (recorded) cries of distress of another child in an adjoining room than did girls in the control group. In addition, boys in the former group demonstrated greater generosity to a needy other than did same-sex peers in the control group, despite a lack of special training with respect to sharing. In contrast to the facilitating effects of victim–helper role-play training, which were found to persist over a 1 week interval, the discussion of helpful acts without role-playing was found to have little influence on helping or sharing.

In addition to the accumulation of perspective-taking skills and experiences, Feshbach (1975a) suggests that "those aspects of the socialization process that relate to the experience, expression and restraint of feelings are highly relevant to the development and manifestation of empathy" (p. 26). In a similar vein, Hoffman (1975b) contends that parental use of the induction disciplinary technique, which emphasizes the negative consequences of the transgressor's behavior for the feelings of others, is likely to "enlist the child's proclivities for empathy" (p. 234). While "socialization of empathy" research is scarce, some correlational studies (Barnett, Howard, King, & Dino, 1980; Barnett, King, Howard, & Dino, 1980; Eisenberg-Berg & Mussen, 1978; Feshbach, 1975b) have explored the relationship between indices of (1) the parents' characteristics or socialization styles and (2) the offspring's empathic tendency.

In a retrospective investigation of the antecedents of empathy (Barnett, Howard, King, & Dino, 1980), 72 college undergraduates scoring at the high and low extremes on the Mehrabian and Epstein (1972) empathy scale completed a questionnaire which requested information about their early socialization experiences. The questionnaire required individuals to rate each parent separately on the extent to which mother (father) had (1) expressed affection to him or her, (2) discussed his or her, that is, the parent's, feelings, (3) discussed the feelings of other individuals, and (4) encouraged the subject to discuss his or her own feelings. (The patterns of response to the latter three items were found to be very similar; therefore, responses to these items were combined into a single "emphasis on feelings" scale.) Finally, each subject was presented with a brief definition of empathy ("the tendency to be sensitive to and responsive to the feelings of other individuals") and was asked to rate his or her mother (father) on a scale from "not at all empathic" to "very empathic." Highly empathic individuals reported that their parents had been more affectionate with them and had discussed their own and other individuals' feelings to a greater extent than did less empathic individuals; moreover, highly empathic individuals rated their parents as more empathic than did their less empathic counterparts. In general, the factors believed to enhance the development and expression of empathy were reported to have been more prevalent in the mothers' prior interaction with the subjects than the fathers'. In addition, the females reported that (1) their mothers had discussed feelings with them to a greater extent than did males and (2) their parents had generally been more affectionate with them than did males. This pattern of sex differences may have contributed to the finding that, although inclusion in the study was restricted to individuals scoring in the high and low empathy extremes for each sex, females' scores on the empathy measure were still significantly higher than those of males.

Similar findings appeared in a study of high school students who rated their parents' child-rearing practices and filled out an affective empathy questionnaire (Eisenberg-Berg & Mussen, 1978). Mothers of highly empathic males were perceived as affectionate, nonpunitive, nonrestrictive, egalitarian, receptive to discussions of problems, and as setting high standards. Females' empathy was not correlated with reported child-rearing practices, perhaps because their empathy scores were generally higher than those of males.[1]

Although the pattern of findings in these studies indicates that the early socialization experiences of high and low empathic individuals may differ on a variety of dimensions relevant to the awareness and expression of affect, some caution must be exercised in interpreting these retrospective data. For example, individuals differing in empathic disposition may tend to differentially recall or distort their parents' characteristics or socialization practices. Similarly, some of the reported sex differences may have reflected cultural stereotypes concerning empathy and the expression of affect. Although a more comprehensive assessment of the antecedents of empathy might incorporate a parental "socialization diary" (see Zahn-Waxler, Radke-Yarrow, & King, 1979) and naturalistic observations of the child's affect-laden interactions with various socialization agents (including parents, siblings, and peers), it should be noted that such approaches are vulnerable to their own brand of distortion and bias. Moreover, in studying the antecedents of social-emotional development, the individual's own perception of his or her parents' characteristics and various socialization experiences is important information which cannot be replaced by the perceptions of others.

RESEARCH ON EMPATHY AND PROSOCIAL BEHAVIOR

Correlational Studies

Studies of the relationship between the young child's empathic tendency and prosocial behavior have typically utilized the Feshbach Affective Situations Test (Feshbach & Roe, 1968), or some modified version of it, as the measure of empathy. This measure consists of four pairs of slide sequences in which young children are shown in situations designed to elicit four different emotions: happiness, sadness, anger, and fear. A brief narration describing the events in the slides accompanies each sequence. After viewing a complete slide sequence, the child is asked, "How do you feel?" or "Tell me how you feel," and the responses are recorded for later analysis. Each response is rated on the following 3-point scale for the degree to which it matches the affect of the child featured in a slide sequence: 0 = incorrect response; 1 = generally correct emotion on the positive–negative affect dimension; 2 = correct, specific emotion. The total empathy score for each child is thus the sum for all four emotions (range = 0–16).

[1] See Hoffman (1977c) for a discussion of sex differences in empathy and related interpersonal behaviors.

Although some positive findings have been reported (Feshbach, 1973; Marcus, Telleen, & Roke, 1979), a majority of studies have found scores on the Feshbach and Roe measure to be unrelated to various prosocial indices (e.g., Eisenberg-Berg & Lennon, 1980; Fay, 1970; Levine & Hoffman, 1975; Sawin, 1979). One factor which may have served to attenuate the dispositional empathy–prosocial behavior relation in some studies is the selection of a prosocial index, such as cooperation on a game (Levine & Hoffman, 1975), which may be relatively devoid of empathy-arousing cues. In addition, the measure of empathy may be suspect. Concerning the latter point, several attacks (Eisenberg-Berg & Lennon, 1980; Hoffman, 1977a; Leiman, 1978; Sawin, 1979) have recently been leveled against the Affective Situations Test, including criticism of the measure's (1) emphasis on the correctness, rather than on the intensity, of the empathic response, (2) verbal bias, which may yield scores confounded with the child's cognitive ability, age, or sex, (3) assumption that a child's "emotional bond" to a story character develops rapidly and is easily manipulated, and (4) vulnerability to demand characteristics and social desirability.

While little empirical research is available on other dispositional measures of affective empathy in children, the use of nonverbal indices of empathic arousal may avoid some of the difficulties encountered in self-report procedures. Observational studies (Buck, 1975; Hamilton, 1973; Leiman, 1978) indicate that many young children will spontaneously match their own facial expressions to those of sad characters depicted in affect-laden slide or film presentations. Other researchers (Eisenberg-Berg & Neal, 1979; Murphy, 1937; Strayer, 1980; Zahn-Waxler & Radke-Yarrow, 1979; Zahn-Waxler, Radke-Yarrow, & King, 1979) have noted that toddlers and preschoolers frequently respond to another child's misfortune by staring with an anxious expression or by displaying other nonverbal signs of distress. Furthermore, Murphy (1937) observed that when a young child helped, his or her affective response typically diminished; when a child failed to help, however, the nonverbal cues of distress were generally prolonged. Finally, Sawin (1979) reports that an empathy index based on ratings of first graders' facial expressions and tone of voice in response to the Feshbach and Roe (1968) slides was a better predictor of altruism (sharing prize marbles with a less fortunate child) than was the conventional empathy score based on self-reported affect.

In addition to facial expression and tone of voice, physiological indices of empathic arousal, which have been found to be associated with helping in adults (Krebs, 1975), might also be used effectively in studies with young children. However, when some children are confronted with the distress of another individual, particularly in a strange laboratory setting, changes in galvanic skin response or heart rate may reflect a startle response or fear rather than vicarious affect arousal. The use of multiple measures of empathy in future studies, in addition to providing insight into individual differences in modes of arousal, may enable researchers to predict the child's prosocial behavior better than any single index.

Experimental Studies

Experimental studies of the empathy–altruism relation have typically investigated the effects of directing the child's attention to an unfortunate other's distress on subsequent generosity. One difficulty with experimental research has been that empathy has generally been assumed to be aroused rather than directly measured.

The results of recent investigations (Eisenberg-Berg & Geisheker, 1979; Howard & Barnett, 1981) indicate that children's charitability may be enhanced by donation requests which make the distress of the needy other salient. In one investigation (Howard & Barnett, 1981), 161 four- to eight-year-old children played a simple game and received 30 plastic tokens which were described as redeemable for prizes upon completion of the study. The children were informed that there were some peers who would not be able to receive any prizes, but that they might share their tokens with them if they wished. Utilizing a procedure adapted from Stotland (1969), children were randomly assigned to one of two conditions in which they were either (1) encouraged to focus on and share the feelings of the less fortunate others or (2) encouraged to think about the less fortunate children but with no mention of feelings being made. Each child was then given an opportunity to donate some prize tokens, in private, to the other children. Following the 1 minute donation period, a brief postexperimental questionnaire was administered to assess the child's affective response to the plight of the needy others. The results indicated that children in the empathy-arousal condition reported feeling sadder and demonstrated a higher incidence of donating than did children in the instructional set condition devoid of affective cues.[2] These findings are consistent with studies indicating that the development of generosity and consideration for others is enhanced by inductive socialization techniques which direct the child's attention to the feelings of others (Dlugokinski & Firestone, 1974; Hoffman & Saltzstein, 1967).

In an experimental study in which a direct "in-task" assessment of empathy was attempted, Leiman (1978) had 85 five- and six-year-old children watch a brief videotaped presentation of a same-sex actor who initially talked to, played games with, and directed questions to the viewer. During the final segment of the presentation, the actor discovered that a favored marble collection had been stolen and a prolonged close-up of the actor's face reflected extreme sadness. The child's own facial expression while observing the actor's distress was videotaped and subsequently rated for the occurrence of affect matching. The helping measure was the extent to which the child worked on a "marble donation machine" to produce marbles for the actor during a 4-minute free-play period. Leiman found that empathic children, who had been rated as matching the actor's sad facial expression, produced more marbles than did a group whose facial expression had been rated as neutral and nonempathic.

Leiman's (1978) results demonstrate that young children who empathize with a needy other may show a heightened tendency to assist that individual. Other recent findings (Barnett, Howard, King, & Dino, 1981; Barnett, King, & Howard, 1979) suggest that eliciting empathy may promote a generalized inclination to aid others, including individuals who may not have served as the original source of concern. In one investigation (Barnett et al., 1979), 85 seven- to twelve-year-old children were awarded 30 prize tokens and then asked to discuss happy, sad, or affectively neutral incidents that had been experienced either by themselves or another child. Following the inducement of affect, the children were given the opportunity to share their

---

[2] In a related study, Dlugokinski and Firestone (1974) found that an inductive appeal to donate to charity, which emphasized "empathy and consideration of others' needs" (p. 24), was particularly effective with children who perceived their mothers as utilizing an inductive socialization style.

experimental earnings, in private, with some less fortunate children. The results indicated that children who had discussed another child's misfortune, which presumably served to heighten empathic arousal, donated more of their experimental earnings to others than did children who had been saddened by discussing personal misfortune. The mean number of prize chips shared by children in the remaining conditions (1) fell midway between the "sad other" and "sad self" groups and (2) did not differ significantly from one another, apparently because they did not differentially elicit empathic concerns relevant to helping. Focusing children's attention on another's misfortune and sadness, rather than on their own, may thus serve to make the feelings and needs of others generally more salient and thereby enhance their tendency to behave altruistically. However, the results of a recent study (Barnett, Howard, Melton, & Dino, in press), utilizing a dispositional empathy index based on teacher and peer ratings, indicate that inducing sadness about another individual may heighten the subsequent charitability of high, but not low, empathic children.

QUALIFIERS OF EMPATHY–ALTRUISM RELATION

The present chapter has indicated that the vicarious experiencing of an unfortunate other's sadness is associated with heightened prosocial behavior in children. However, several limitations and qualifications of the empathy–altruism relation should be noted. The experience of empathic arousal does not ensure that a helpful behavior will be enacted. Children who experience distress and concern when confronted with a needy other may fail to offer assistance if they perceive little personal responsibility to help or feel incompetent to intervene effectively (Aronfreed, 1968). In addition, the young child who has limited helping experience may not have learned that effective helping can serve to alleviate the other's, as well as one's own, distress. On some occasions, the child may become overaroused by empathic distress, such that egoistic concerns predominate over a concern for the other; under such conditions, the child may come to avoid or even derogate the victim rather than assist. While a moderate degree of empathic arousal may prove to optimally motivate helping, children need to be given the encouragement and interpersonal skills necessary to translate even heightened empathic arousal into appropriate prosocial behavior.

When a helpful behavior is enacted, this does not ensure that it was motivated by the experience of empathic arousal. Although research on preschoolers' reasoning about their own spontaneous prosocial behavior (Eisenberg-Berg & Neal, 1979) indicates that helping is frequently mediated by "an empathic concern" (p. 229), internalized helping norms, values, and beliefs of a less affective nature undoubtedly also serve to elicit the child's altruistic acts. In addition, innumerable personality and situational variables influence the child's generosity and concern for others (see review by Mussen & Eisenberg-Berg, 1977). For example, exposure to a model who helps or shares increases children's prosocial behavior even when no effort has been made to arouse the observer's affective or cognitive reactions (Bryan, 1975). Therefore, while empathy may be an important motivator of helping, it is perhaps best viewed as only one of many complexly intertwined factors which influence the development and expression of prosocial behavior.

CONCLUSION: THE NEED FOR A BROADER VIEW OF EMPATHY

To a large extent, research on empathy and prosocial behavior in children has focused on affective and behavioral responses to the distress or sadness of others. As Zahn-Waxler and Radke-Yarrow (1979) have emphasized, however, empathy is a process that involves the entire range of emotions and is an essential element in the child's affective bond with other individuals. By expanding the range of empathic responses explored and encompassing empathy within the broader framework of affective and interpersonal development, we stand to gain a greater understanding of the manner in which children come to experience, interpret, and respond to various emotions in themselves and others.

Although some preliminary steps have been taken to view empathy more broadly, much remains to be discovered concerning its role in the child's evolving interpersonal behavior. In a naturalistic study of preschoolers' affective behavior (Strayer, 1980), empathy in response to the emotional displays of peers was found to occur at markedly different rates across the affect categories of happy, sad, angry, and hurt. Indeed, the children were found to emit more empathic responses to happy displays than to the other affect categories combined. In addition, particular "empathic behavioral responses" (p. 818) were found to be consistently associated with particular affective states in others. Sharing, for example, was a prevalent response to sad displays, whereas information gathering and offering reassurance were the predominant responses to displays of hurt. Future research should be extended to incorporate other age groups and emotions in order to delineate regularities in the development and expression of particular empathic responses. Such studies will undoubtedly also uncover individual differences in children's tendencies or willingness to respond to particular emotions in others. In this regard, Zahn-Waxler and Radke-Yarrow (1979) have reported stable and patterned individual differences in empathic responses among 1- and 2-year-olds as well as individual continuity to age 7 in the child's intensity and mode of response to others' emotions. Assessing the antecedents of such individual differences will serve to expand our understanding of empathy and its function in the affective and social development of the child.

Finally, little is known of the influence of empathy on specific interpersonal behaviors other than helping and aggression (e.g., Feshbach, 1978b). Again, the need for a broader view of the role of empathy in the child's interactions with others seems clear. Future research on the development and expression of empathy in children could be incorporated within such diverse areas of study as leadership, friendship, and interpersonal problem solving. While other emotions and social skills undoubtedly play an important part in these and others areas, the extent of the child's sensitivity and responsiveness to the diverse feelings of others will likely be found to assume a central role in the ability to establish and maintain satisfying interpersonal relationships.

REFERENCES

Aronfreed, J. *Conduct and conscience: The socialization of internalized control over behavior.* New York: Academic, 1968.

Aronfreed, J. The socialization of altruistic and sympathetic behavior: Some theoretical and experimental analyses. In J. Macauley & L. Berkowitz (Eds.), *Altruism and helping behavior*. New York: Academic, 1970, pp. 103–126.

Barnett, M. A., Howard, J. A., King, L. M., & Dino, G. A. Antecedents of empathy: Retrospective accounts of early socialization. *Personality and Social Psychology Bulletin*, 1980, **6**, 361–365.

Barnett, M. A., Howard, J. A., King, L. M. & Dino, G. A. Helping behavior and the transfer of empathy. *Journal of Social Psychology*, 1981, **115**, 125–132.

Barnett, M. A., Howard, J. A., Melton, E. M., & Dino, G. A. Effect of inducing sadness about self or other on helping behavior in high and low empathic children. *Child Development*, in press.

Barnett, M. A., King, L. M., & Howard, J. A. Inducing affect about self or other: Effects on generosity in children. *Developmental Psychology*, 1979, **15**, 164–167.

Barnett, M. A., King, L. M., Howard, J. A., & Dino, G. A. Empathy in young children: Relation to parents' empathy, affection, and emphasis on the feelings of others. *Developmental Psychology*, 1980, **16**, 243–244.

Borke, H. Interpersonal perception of young children: Egocentrism or empathy. *Developmental Psychology*, 1971, **5**, 263–269.

Borke, H. The development of empathy in Chinese and American children between three and six years of age: A cross-cultural study. *Developmental Psychology*, 1973, **9**, 102–108.

Bryan, J. H. Children's cooperation and helping behaviors. In E. M. Hetherington (Ed.), *Review of child development research*. Vol. 5. 127–182. Chicago: Univ. of Chicago Press, 1975.

Buck, R. W. Nonverbal communication of affect in children. *Journal of Personality and Social Psychology*, 1975, **31**, 644–653.

Buckley, N., Siegel, L. S., & Ness, S. Egocentrism, empathy, and altruistic behavior in young children. *Developmental Psychology*, 1979, **15**, 329–330.

Chandler, M. J., & Greenspan, S. Ersatz egocentrism: A reply to H. Borke. *Developmental Psychology*, 1972, **7**, 104–106.

Dlugokinski, E. L., & Firestone, I. J. Other centeredness and susceptibility to charitable appeals: Effects of perceived discipline. *Developmental Psychology*, 1974, **10**, 21–28.

Dymond, R. F. A scale for measurement of empathic ability. *Journal of Consulting Psychology*, 1949, **14**, 127–133.

Eisenberg-Berg, N., & Geisheker, E. Content of preachings and power of the model/preacher: The effect on children's generosity. *Developmental Psychology*, 1979, **15**, 168–175.

Eisenberg-Berg, N., & Lennon, R. Altruism and the assessment of empathy in the preschool years. *Child Development*, 1980, **51**, 552–557.

Eisenberg-Berg, N., & Mussen, P. Empathy and moral development in adolescence. *Developmental Psychology*, 1978, **14**, 185–186.

Eisenberg-Berg, N., & Neal, C. Children's moral reasoning about their own spontaneous prosocial behavior. *Developmental Psychology*, 1979, **15**, 228–229.

Fay, B. The relationships of cognitive moral judgment, generosity, and empathic behavior in six and eight year old children. Unpublished doctoral dissertation, University of California, Los Angeles, 1970.

Feshbach, N. D. Empathy: An interpersonal process. Paper presented at the meeting of the American Psychological Association, Montreal, 1973.

Feshbach, N. D. Empathy in children: Some theoretical and empirical considerations. *The Counseling Psychologist*, 1975, **5**, 25–30. (a)

Feshbach, N. D. The relationship of child-rearing factors to children's aggression, empathy and related positive and negative social behaviors. In J. DeWit & W. W. Hartup (Eds.), *Determinants and origins of aggressive behavior*. The Hague, Netherlands: Mouton, 1975, pp. 427–436. (b)

Feshbach, N. D. Empathy training: A field study in affective education. Paper presented at the meeting of the American Educational Research Association, Toronto, 1978. (a)

Feshbach, N. D. Studies of empathic behavior in children. In B. A. Maher (Ed.), *Progress in experimental personality research* (Vol. 8). New York: Academic, 1978, 1–47. (b)

Feshbach, N. D., & Roe, K. Empathy in six- and seven-year-olds. *Child Development*, 1968, **39**, 133–145.

Greenspan, S., Barenboim, C., & Chandler, M. J. Empathy and pseudo-empathy: The affective judgments of first- and third-graders. *Journal of Genetic Psychology*, 1976, **129**, 77–88.

Hamilton, M. L. Imitative behavior and expressive ability in facial expression of emotion. *Developmental Psychology*, 1973, **8**, 138.

Hoffman, M. L. Developmental synthesis of affect and cognition and its implications for altruistic motivation. *Developmental Psychology*, 1975, **11**, 607–622. (a)

Hoffman, M. L. Moral internalization, parental power, and the nature of parent-child interaction. *Developmental Psychology*, 1975, **11**, 228–239. (b)

Hoffman, M. L. Empathy, its development and prosocial implications. In C. B. Keasey (Ed.), *Nebraska symposium on motivation* (Vol. 25). Lincoln: University of Nebraska Press, 1977, pp. 169–217. (a)

Hoffman, M. L. Personality and social development. *Annual Review of Psychology*, 1977, **28**, 295–321. (b)

Hoffman, M. L. Sex differences in empathy and related behaviors. *Psychological Bulletin*, 1977, **84**, 712–722. (c)

Hoffman, M. L., & Saltzstein, H. D. Parent discipline and the child's moral development. *Journal of Personality and Social Psychology*, 1967, **5**, 45–57.

Howard, J. A., & Barnett, M. A. Arousal of empathy and subsequent generosity in young children. *Journal of Genetic Psychology*, 1981, **138**, 307–308.

Iannotti, R. J. Effect of role-taking experiences on role taking, empathy, altruism, and aggression. *Developmental Psychology*, 1978, **14**, 119–124.

Iannotti, R. J. The elements of empathy. Paper presented at the meeting of the Society for Research in Child Development, San Francisco, 1979.

Krebs, D. L. Empathy and altruism. *Journal of Personality and Social Psychology*, 1975, **32**, 1134–1146.

Leiman, B. Affective empathy and subsequent altruism in kindergarteners and first graders. Paper presented at the meeting of the American Psychological Association, Toronto, 1978.

Levine, L. E., & Hoffman, M. L. Empathy and cooperation in four-year-olds. *Developmental Psychology*, 1975, **11**, 533–534.

Marcus, R. F., Telleen, S., & Roke, E. J. Relation between cooperation and empathy in young children. *Developmental Psychology*, 1979, **15**, 346–347.

Mehrabian, A., & Epstein, N. A measure of emotional empathy. *Journal of Personality*, 1972, **40**, 525–543.

Murphy, L. B. *Social behavior and child personality*. New York: Columbia University Press, 1937.

Mussen, P., & Eisenberg-Berg, N. *Roots of caring, sharing, and helping: The development of prosocial behavior in children*. San Francisco: Freeman, 1977.

Sagi, A., & Hoffman, M. L. Empathic distress in newborns. *Developmental Psychology*, 1976, **12**, 175–176.

Sawin, D. Assessing empathy in children: A search for an elusive construct. Paper presented at the meeting of the Society for Research in Child Development, San Francisco, 1979.

Simner, M. L. Newborn's response to the cry of another infant. *Developmental Psychology*, 1971, **5**, 136–150.

Staub, E. The use of role-playing and induction in children's learning of helping and sharing behavior. *Child Development*, 1971, **42**, 805–816.

Staub, E. *Positive social behavior and morality: Social and personal influences* (Vol. 1). New York: Academic, 1978.

Stotland, E. Exploratory investigations of empathy. In L. Berkowitz (Ed.), *Advances in experimental social psychology* (Vol. 4). New York: Academic, 1969, pp. 271–314.

Strayer, J. A naturalistic study of empathic behaviors and their relation to affective states and perspective-taking skills in preschool children. *Child Development,* 1980, **15**, 815–822.

Sullivan, H. S. *The interpersonal theory of psychiatry.* New York: Norton, 1953.

Zahn-Waxler, C., & Radke-Yarrow, M. A developmental analysis of children's responses to emotions in others. Paper presented at the meeting of the Society for Research in Child Development, San Francisco, 1979.

Zahn-Waxler, C., Radke-Yarrow, M., & King, R. A. Child rearing and children's prosocial initiations towards victims of distress. *Child Development,* 1979, **50**, 319–330.

18

CHILDREN'S FRIENDSHIPS

Wyndol Furman
Department of Psychology
University of Denver

A friend is a person who knows how you really feel. Friends keep you company and talk to you when you're not feeling good. You can't always talk with your Mom and Dad so you need someone who is no relation to talk to. If you didn't have friends, you'd probably be down in the dumps and you'd just be watching TV all day.

Pat, age 12.

You need to have friends because you can't just shut yourself off in a closet and say, I'll just work and do nothing socially. You can have lots of fun with friends. You can go to movies with them. You can get mad with them too and yell and all. A friend is someone who will listen and you can trust and you enjoy being with. So, you need to have friends so you're not just a plain person; you can have fun and be mad.

Bill, age 13.

Friends are very important to Pat and Bill. They're more than playmates; they're special. They listen and understand what you mean; they stick up for you; and they care for you. Like Pat and Bill, developmental psychologists have come to recognize the significance of these relationships in children's development. For children of any age, having friends is considered a significant social achievement, an index of social competence, and a sign of mental health (Hartup, 1978). Currently, studies are being conducted on numerous aspects of this topic, such as developmental changes in friendship, children's conceptions of friendship, and children's behavior with friends. The present chapter provides an overview of this research on children's friendships.

THE NATURE OF FRIENDSHIP

When asked what a friend is, even preschool children recognize that it is "someone you like." This feeling of affection is central to friendship; children are willing to

---

Preparation of this chapter was supported by Grant No. BSN-8014668 from the National Science Foundation.

do special favors for their friends *because* they like them. Affection is the "glue" which binds the relationship together.

Friendships have several additional defining characteristics which differentiate them from other relationships. Friendships are exclusively dyadic. Affectionate feelings can occur at a group level, but they are qualitatively different from the dyadic bond. Additionally, although young children's friendships are sometimes one-sided, mature ones are reciprocal. Unlike relationships with parents and siblings, friendships are voluntary. One chooses with whom to be friends and may terminate any such relationship at any time. Consequently, friendships are more fragile.

Friendships also entail a set of social rules and norms which dictate appropriate behaviors and obligations for the participants. As one second grader put it, "You don't tell someone you're their best friend and then go beat them up." Although some overlap exists, the rules for friendships are distinct from those for other relationships, such as acquaintanceships or parent–child relationships (Cabral, Volpe, Youniss, & Gellert, 1977; Furman & Bierman, 1981a).

Friendships should be differentiated from friendly relations or acquaintanceships (Kurth, 1970). In friendships the affective tie is stronger and the obligations greater; friendships involve more extensive and more intimate interactions; they are more personal (vs. role) relationships, and thus the sense of uniqueness is greater. Friends are also more likely to be similar in sex, age, and status. All friends were once acquaintances, but most acquaintances do not become friends.

Friendship and popularity are also not identical. Popularity refers to the general degree of liking by the peer group, while a friendship is a specific relation between two people. One does not need to be popular to have friends. In fact, the determinants of popularity and individual friendship selections are by no means isomorphic. For example, popular children tend to be sociable and outgoing, but one's friends are not necessarily so (Masters & Furman, 1981).

Finally, the study of friendship should not be equated with the study of the behavior of friends. Instead, it is the study of a *relationship* (cf. Furman, 1981a). This relationship serves to organize the behaviors, motives, and attributions of the children. Relationships have continuity over time and the participants develop expectations of each other. Even identical behaviors have different meanings in different relationships. For instance, Cabral, Volpe, Youniss and Geller (1977) asked children what effect a breach of confidence would have if it were done by a parent, a friend, or an acquaintance. The effects of the violation varied among the three. Parents were assumed to have broken the confidence for the sake of the child and the effect on the relationship was minimal. On the other hand, a breach of confidence was seen as seriously harming a friendship. Finally, acquaintances were permitted to break confidences, but that act would preclude the relationship from developing into a friendship. Thus the meaning and impact of a behavior depend upon the relationship in which it occurs.

THE DEVELOPMENTAL COURSE OF FRIENDSHIPS

Toddlers. Although it may not be appropriate to say that toddlers have friendships, many of the first steps toward friendships emerge here. Pairs of "friends" participate in sustained reciprocal and complimentary interactions more frequently than do unfamiliar toddlers (Howes & Mueller, in press). Even 12-month-old infants engage in more positive affect and proximity seeking with "friends" than with strangers (Lewis, Young, Brooks, & Michalson, 1975). Toddlers also differentially initiate interactions with their various peers (Vandell, 1978). Mutual or stable preferences for a playmate, however, do not occur until 2 years of age and, even then, are atypical. Many of the preceeding findings may be a reflection of the toddler's differentiation between familiar and unfamiliar peers rather than one between friends and others. At the same time, such a discrimination is a prerequisite for the establishment of friendships.

Preschool Children. Unlike toddlers, preschool children can be said to have friends. These friends are typically neighbors or classmates whom they like and associate with frequently. The friendship centers around the shared play activities. Children have been found to engage in twice as many reinforcing and neutral interactions with their stated friends than with other peers at nursery school (Masters & Furman, 1981). Children this age, however, do not have a real concept of a relationship which supercedes these enjoyable interchanges (Furman & Bierman, 1981b). Friendships can even be exchanged for material goods; for example, "I'll be your friend if you gimme a cookie." Consequently, friendships can be unstable and changing from day to day. Children both break up and make up with their friends quite readily. Although most preschool friendship are short-lived, some do last for a long time. Regardless of their length, these early relationships are important to children's social development.

School-Aged Children. During the school years, more mature forms of friendship emerge. According to Sullivan (1953), young children's friendships are egocentric and are established to enhance status or meet one's own needs. Around 9, however, children begin to develop deeper friendships, or "chumships." These relationships are characterized by genuine affection and love for each other; children develop a real sensitivity to what matters to their chums. Sullivan proposed that a chumship is a major milestone in development. It gives children a means of expressing interpersonal intimacy, serves to validate their sense of self-worth, and teaches them a sense of humanity.

Available data support Sullivan's theory. Boys with chums have higher self-esteem and are more altruistic than those without one. Similarly, as they approach junior high, children begin to emphasize the importance of intimacy, loyalty, and character admiration to friendships (Bigelow & La Gaipa, 1980; Furman & Bierman, 1981a).

Children's friendships are also characterized by sex differences. Girls tend to develop a deep intimate relationship with one or two other girls, while boys establish friendships with many other boys (Eder & Hallinan, 1978; Maccoby & Jacklin, 1976). Often boys' friendships are with various members of their gang. For girls, the few "intensive" friendships reflect the importance of peer relations to them, while for boys, the "extensive" ones do (Waldrop & Halverson, 1975).

Adolescents. Children continue to have same-sexed chumships in adolescence, but the relationships become even more intimate in nature. Self-disclosure, trust, commitment, respect, and similar value structures are all central in these highly personal relationships.

Same-sexed friendships also facilitate the transition to heterosexual relationships (Dunphy, 1963). At the beginning of adolescence, children cluster together in unisexual cliques. Subsequently, these cliques begin interacting with each other, forming heterosexual crowds. These crowds serve as the organizing force behind dances, parties and other gatherings. More important, the crowd's activities provide means for establishing heterosexual relationships. Young adolescents can meet and interact with the other sex in a relatively protected setting. They begin dating and going places together as couples. Early dating relationships are usually very superficial and ritualized, leaving same-sexed friendships as the primary source of peer support and guidance (Douvan & Adelson, 1966). As they approach adulthood, however, adolescents begin to develop intimate romantic relationships that can be characterized as a special kind of friendship.

CONCEPTIONS OF FRIENDSHIP

Not only do children's friendship change developmentally, but their conceptions of the nature of friendship change as well. Children of different ages have qualitatively different perceptions of the rules, obligations, and functions of friendship. To illustrate this research, a recent study by the author will be described (Furman & Bierman, 1981a). This investigation focused on five commonly reported domains of friendship conceptions: (a) support, (b) association, (c) intimacy, (d) similarity, and (e) affection. Two kinds of descriptions were distinguished within each domain. One kind referred to behaviors or overt characteristics, while the other referred to dispositional characteristics, such as intentions or traits. For example, in the domain of support, references to acts of helping or sharing would be considered to be behavioral, while statements about being considerate or dependable would be dispositional. A list of the behavioral and dispositional features in each domain is presented in Table 18.1. Previous research suggested that the dispositional features would emerge at a later developmental point than the behavioral ones (Livesley & Bromley, 1973).

To test these hypotheses, second-, fourth-, and sixth-grade children were administered three measures: (a) an open-ended interview in which they were asked questions such as "What is a friend?" or "What should you do to be a friend?" (b) a recognition task in which they heard a series of 10 vignettes about friends and, after each, were asked why the children were friends, and (c) a questionnaire in which they rated the importance of each of the 10 features.

A summary of the results is shown in Table 18.1. Each of the five behavioral features were mentioned by the majority of children at all three grade levels; consequently, few developmental changes were observed. In contrast, developmental increases were consistently observed in three of the dispositional features. Some changes were also observed on the feature of dispositional association (faithful/loyal). This general pattern of results is similar to that obtained by other investigators (Berndt, in press; Bigelow, 1977; Bigelow & La Gaipa, 1975).

The fact that the five behavioral features were well-established by the second grade suggests that the initial formation of friendship conceptions begins at an earlier age. Preschool children do, in fact, have a rudimentary concept of friendship

TABLE 18.1. Developmental Changes in Friendship Conceptions

Features	Task		
	Open-Ended Interview	Story Recognition	Questionnaire (Friends)
Behavioral			
Support (share/help)	n.s.	n.s.	$2 < 4 + 6$
Association (be with/companionship)	$2 < 6$	n.s.	n.s.
Intimacy (self-disclosure/talk)	$2 + 4 < 6$	n.s.	n.s.
Similarity (common activities/mutual play)	$2 + 4 > 6$	n.s.	n.s.
Affection (overt liking)	$2 < 6$	n.s.	n.s.
Dispositional			
Support (dependable/considerate)	$2 < 4 + 6$	$2 + 4 < 6$	$4 < 6$
Association (faithful/loyal)	$2 + 4 < 6$	n.s.	n.s.
Intimacy (trust/understand)	$2 < 4 + 6$	$2 < 4 + 6$	$2 < 6$
Similarity (interests/beliefs)	n.s.	n.s.	n.s.
Affection (accept/admire)	$4 < 6$	$2 < 4 + 6$	$2 < 6$
Others			
Miscellaneous	n.s.	n.s.	

NOTE: Numbers represent grade levels. Numbers which are separated by $>$ or $<$ are significantly different at $p < .05$; n.s. = not significant. From Furman and Bierman (1981a).

(Hayes, 1978). When asked why they liked their best friend, many of them mention simple overt features such as common activities, general play, propinquity, and physical possessions. Subsequent research indicates that the concrete features of common activities and propinquity are already apparent by 4 or 5 years of age, while relational features, such as affection (liking) and support (helping and sharing), increase in saliency around 6 or 7 (Furman & Bierman, 1981b).

Thus as children grow older, their conceptions of friendship undergo significant transformations. Some of these changes can be attributed to cognitive development. For example, the emergence of dispositional features may reflect the child's increasing ability to abstract and draw inferences from behavior. Social perspective-taking has also been hypothesized to be an important determinant (Selman & Jacquette, 1977); that is, as children become less egocentric, they should be more able to recognize the reciprocal, or bilateral, nature of friendship. Although promising, these theories about the cognitive factors involved have not been extensively tested yet.

Almost no theoretical or empirical consideration has been given to the role of social experiences in the development of friendship conceptions. In fact, only one study has examined the relation between friendship expectations and social behavior. There, unpopular children were found to have immature conceptions (Wood, 1976). Thus it appears that social concepts, cognitive ability, and social behavior are interrelated, but the specific nature of these relations remains to be determined.

THE FRIENDSHIP PROCESS

The developmental course of a friendship can be divided into five phases: (a) selecting potential friends, (b) becoming acquainted, (c) making friends with an acquaintance, (d) deepening or maintaining an existing friendship, and (e) growing apart and ending a relationship. Each of these phases is described subsequently, but before proceeding, it should be noted that some phases do overlap and clear demarcation points between phases often do not exist.

Selecting Potential Friends

Most research is focused on the role of similarity in the selection process. Substantial data indicate that children tend to choose friends from those with similar overt characteristics. For instance, friends are usually the same age (Hartup, 1970). Apparently, the give-and-take essential to a successful friendship is easier to achieve if the two are equal in status and age.

Typically, children develop friendships with peers of the same sex (cf. Asher, Oden, & Gottman, 1977). The vast majority of preschooler's friendship choices are same-sexed (Challman, 1932; Masters & Furman, 1981), and if anything, the sex cleavage increases during elementary school. Asher (1973) found that 95% of school children's choices were same-sexed.

Children are also more likely to choose friends from among their own race (Asher et al., 1977; Carter, DeTine-Carter, & Benson, 1980). Even in 1939, though, the sex cleavage was found to be stronger than the race cleavage (Criswell, 1939). The most recent data indicate that the racial cleavage remains, but the difference in preference for same- and cross-race peers is relatively small (Singleton & Asher, 1979).

Aside from the research on age, sex, and race, the evidence for the similarity of friends is not very strong. For instance, some studies indicate that friends have similar intellectual abilities, while other research does not (Hartup, 1970). Data on the similarity of sociometric status are also mixed (Hartup, 1970). Izard (1960) reported that the personality profiles of friends were more similar than those of randomly selected pairs, but the correspondence was only significant on 3 of the inventory's 15 scales. Kandel (1978) found some promising evidence of similarity in the behavior of adolescent friends, especially with regard to drug usage. Even here, though, the correlations were modest (r's = .30 to .40).

Perhaps clearer indications of similarity would be obtained if some alternative approaches to the issue were tried. The kind of similarity important to a friendship may vary developmentally or from relationship to relationship. Friends may need to be similar in characteristics significant to the two, but not necessarily in other ways. For example, children who enjoy sports may want their friends to play as well; children who have little interest in sports may not be concerned about whether their friends do or don't play.

The perception of similarity may also be as important as the actual degree of similarity. Davitz (1955) found that children at camp perceived their activity preferences to be closer to their most liked peer than to their least liked peer, but in actuality they were not more similar. Perhaps similarity is important in early phases of friendship formation, but not at later points. Initially, shared interests could give the relationship a foundation on which to develop, while at a later point some differences would be more acceptable. Alternatively, similarity may continue to be important throughout the course of the relationship, but the kind of similarity which is significant may change as the relationship does. Similarity on superficial characteristics and interests may be sufficient at first, but as the relationship deepens, similarity on more personal characteristics may be needed (Duck, 1973).

Except for the research on similarity, few studies have focused on the determinants of friendship selections. Research on popularity may, however, be germane (cf. Hartup, 1970). Sociability, cooperativeness, socioempathy, good adjustment, and other positive personality characteristics have all been found to be correlated with sociometric status and may well be important determinants of friendship selections. It is important, though, that the determinants of popularity and friendship selections not be assumed to be isomorphic. Masters and Furman (1981) found that overall rates of reinforcing and neutral interactions were correlated with popularity but did not predict which persons a specific child named as friends. Instead, children's selections were more influenced by the specific interactions between them and the other. Thus perhaps popularity correlates are only predictive of specific friendship selections if they apply to that particular relationship. For example, if cooperative with another, a child may be more likely to be described as that person's friend. If cooperative with most others, but not that person, the child isn't likely to be chosen by him or her.

The selection process has been described here as a distinct phase in friendship formation. To some degree, it is. Children are selective about with whom they want to initiate a relationship. Typically, they are interested in peers of the same age, sex, and race. Others who do not fulfill these characteristics are usually precluded from becoming a friend. In this sense the selection phase is the first one in friendship formation (Berscheid & Graziano, 1979).

On the other hand, the selection process continues beyond the initial screening. A child learns about another's social ability, cooperativeness, and other behaviors toward him or her by experiencing these acts as the two become acquainted. Consequently, the child's interest in another may fluctuate or diminish as a relationship emerges. In this sense the selection process is not a discrete phase.

Becoming Acquainted

The acquaintanceship process has been studied by comparing dyads of unfamiliar and familiar children during play sessions. Compared to acquainted ones, unacquainted pairs of preschool children play together less frequently and in less sophisticated manners (Doyle, Connolly, & Rivest, 1980; Jormakka, 1976). Unacquainted pairs also make fewer directive overtures and are less successful when they do. On the other hand, sharing information about oneself and inquiries about their partner's interests and background are more commonly observed in unfamiliar dyads than in familiar dyads (Jormakka, 1976).

These comparisons between acquainted and unacquainted pairs provide some indication of the general changes resulting from becoming acquainted. Jormakka (1976) has also outlined some of the very first steps in the process by examining changes from the beginning to the end of a first encounter. She found that as the session proceeded, unacquainted children began to talk more and laugh more.

Their descriptions of the play also increased, while "hovering" behaviors, such as standing immobile and gazing, decreased.

Similarly, Gottman and Parkhurst (1980) found sequential changes in the interactions of unacquainted preschool children. In this series of studies 3- to 6-year-old children were paired with either a best friend or a stranger. Each pair interacted in one child's home. Their conversations were taped and subsequently coded into a set of three classification schemes. On the basis of their results, Gottman and Parkhurst hypothesized an eight-stage sequence of the acquaintanceship process. They proposed that dyads must successfully go through early stages, such as show and tell or social comparison, before proceeding to the later stages of stereotyped and extended fantasy play. Interestingly, they found marked developmental changes between the younger (3- to 4½-year-old(s)) and older preschoolers (4½- to 6-year-old(s)). The younger children appeared to make every effort to be close friends with their new partner—a strategy which was either very successful or disastrous. In contrast, the older children took the cautious route of "getting along" rather than trying to become intimate immediately. Apparently, the older children had begun to acquire adults' social routines for interacting with strangers.

These findings are consonant with the author's observations of the acquaintanceship process in dyads of third-grade children (Furman, 1981b). Unfamiliar pairs started very tentatively. First, they silently observed each other and then began sharing superficial information about themselves. Subsequently, they started talking more and began playing beside each other. In some cases, the pairs were playing together by the end of the session. In contrast, familiar pairs engaged in high rates of mutual play from the start. Approval, disapproval, and directive overtures were all more common in the familiar pairs.

The findings of these studies can be integrated within a theory of acquaintanceship postulating four central processes: (a) disclosure and discovery of coorientation, (b) establishment of a mutual enterprise, (c) individuation of the relationship, and (d) development of an affective bond. The first three processes are expected to occur in some general hierarchical order. In the first phase the disclosure of personal information provides the children with expectations about each other. These expectations maximize the likelihood of reinforcing interchanges and lead to the discovery of similarities. Such similarities serve as the basis for establishing a mutual activity (phase 2). Here, the participation of the two in a common activity serves as the substance or content of the relation. The specific details of the common enterprise are not as important as that the two have found a means of interacting hospitably. Once such a means has been firmly established, the relationship can become individuated (phase 3). The two can develop the specific kind of relationship they desire. Requests, expressions of positive and negative feelings, and other efforts to direct the course of the relationship all become more common. The final process is the establishment of an affective bond. This bond is expected to develop throughout but becomes increasingly stronger as the relationship becomes individualized.

Making Friends with an Acquaintance

Almost no research has examined the process of acquaintanceship over a longer time span. Recently, Furman and Childs (1981) studied the development of relationships among 143 young adolescents at a summer camp. On the second, fifth, and seventh day of camp, the children completed a questionnaire which assessed six facets of their relationships with each of their cabinmates. The features which were assessed were (a) support, (b) intimacy, (c) companionship, (d) similarity, (e) affection, and (f) quarreling. On the seventh day, children were also asked how much of a friend each peer was. Although a week is a short time in the course of most relationships, the data suggest that the day-long encounters in the camp expedited the acquaintanceship process. Even by the second day, children had begun to differentiate among those receiving high, medium, and low scores on the friendship question. As the days passed, the differentiation increased. For "high" friends, the ratings increased from day 2 to day 5 on affection and to some degree on companionship, intimacy, and quarreling. The ratings on all six features were greatest on day 7. Simultaneously, the children's ratings of "low" friendships decreased over the three administrations.

Data were also collected on children who were acquainted with each other prior to the camp. As expected, the "high" friendship group ratings were relatively stable across the three administrations. These relationships had a prior history and, thus, were not likely to change significantly within a 7-day period. Interestingly, the ratings of the "low" friendship group of acquainted children did decrease over time. This temporal trend may be an indirect reflection of the emergence of new relationships. As children become acquainted with new peers, they seem to lose interest in old acquaintances to whom they weren't close.

Aside from this study, very little longitudinal research has been conducted on either the acquaintanceship process or the friendship formation process. Our knowledge is limited to comparisons between "endpoints" (i.e., friends vs. strangers or acquaintances). While important, such comparisons do not directly assess the processes involved in the development of a relationship. Longitudinal research is needed to determine when different features or processes "clock in" and what the transitional stages of a relationship are.

Maintaining and Deepening a Friendship

Longitudinal studies on this phase of a relationship do not exist, but some interesting research has been conducted on the behavior of friends. Friendship seems to facilitate social responsiveness. When watching a funny movie together, friends laugh, smile, talk, and look at each other more often than strangers do; furthermore, there is a greater concordance in the affective expression of friends (Foot, Chapman, & Smith, 1977). Several investigators have compared the behavior of friends and other children on structured tasks, such as block building (Newcomb, Brady, & Hartup, 1979; Philp, 1940). Typically, no differences are observed in task performance, but friends talk more and express more affect than acquaintances or strangers do. Thus friendship is manifested in the style, as well as the content, of the interaction.

Growing Apart and Ending a Relationship

Two studies have examined the explanations children give for ending a relationship. Austin and Thompson (1948) found that a lack of recent contact, a quarrel, and "incompatibility" were the most common reasons provided by sixth graders for no longer describing someone as a friend. More recently, Bigelow and La Gaipa (1980) reported some interesting developmental trends in the reasons provided for breakups. Nine-year-old children tended to attribute a breakup to conflicts or quarrels, while 13-year-olds gave reasons of disloyalty or disrespect for the other. These trends seem parallel to the developmental changes in friendship conceptions reported previously. These investigations provide important information about the explanations for a relationship's decay, but as yet, nobody has studied the actual behavior of friends during the conflict and dissolution phase.

A FINAL WORD

This chapter has covered research on numerous aspects of friendship. Material has ranged for observational studies of toddler's friendships to investigations of the similarity of adolescent friends' personalities. We have reviewed research on both the process of meeting someone and the reasons for ending a relationship. One issue remains to be addressed—the significance of friendships.

With the exception of Sullivan (1953), traditional personality theorists principally focused on the importance of the parent–child relationship in development; little was said about peers or friends. Even today, many textbooks in developmental psychology include relatively little discussion of the role of peers.

Peer interactions and friendships, however, are not secondary aspects of a child's life. They have a fundamental role in social and cognitive development. Play and other peer interactions teach children social and interpersonal skills. These interchanges also give them practice in adult roles and contribute to their problem-solving and cognitive abilities (Bruner, Jolly, & Sylva, 1976). Peers also provide diverse models to the child; a child uses friends to develop and try out an identity, values, and beliefs. In general, peer interactions are a proving ground for social behaviors.

Peers and friends also contribute to many "parental" functions. Peers help in caretaking, provide affection, promote cognitive growth and learning, and contribute to the general socialization of the child. Peers have been found to contribute to sex-role development, moral development, the acquisition of norms and values, and the formation of a self-concept (Furman & Buhrmester, in press). Although at times conflicts occur, peer and parental values are usually concordant (Douvan & Adelson, 1966; Hartup, 1970; Langworthy, 1959).

At the same time, peers are not simply surrogate parents. Peers and friends provide unique contributions to the social development of the child. For example, in his theory of moral development, Piaget (1932) proposed that interactions with equals lead children to recognize that rules are the product of group decisions and are not sacred or immutable. Peers are also greatly involved in the development of the control of aggression and the acquisition of sexuality (Hartup, 1976). Unlike parent–child interactions, peer interactions are egalitarian in nature; consequently, their contribution is unique.

Many of these important peer interactions occur within the context of friendships. Moreover, friendships themselves make several distinct contributions. They facilitate the transition from family relations to autonomy (Campbell, 1969).

Friendships give children the critical experience of being involved in an intimate relationship with an equal. Only here can they learn numerous social competences, such as appropriate self-disclosure or satisfactory means of resolving conflicts with an intimate. Most important, children are provided a sense of belonging and affection in their friendships. Here is a person who likes, cares, and even loves you even though under no obligation to do so. To be fully appreciated, loyalty, acceptance, and authenticity need to be experienced.

Friendships are also significant social achievements; consequently, the establishment of new friendships is likely to enhance one's self-esteem. Conversely, the absence of friendships has serious effects on mental health. Numerous studies have documented that unpopular children or children with few or no friends are not well-adjusted (cf. Hartup, 1970). Furthermore, the impact of inadequate childhood peer relations can remain throughout life. Poor peer adjustment has been found to be associated with juvenile delinquency, psychoses, and other psychiatric problems (Cowen, Pederson, Babijian, Izzo, & Trost, 1973; Roff, 1963; Roff, Sells, & Golden, 1972). Friendships can also have a rehabilitating effect. Emotionally disturbed children who have a best friend are at a much better prognosis than those without one; this difference is even apparent 10 to 15 years later (Sundy & Kreyberg, 1968).

Although the positive impact of friendships is well-documented, it is important to recognize that children encounter negative experiences with their friends as well. Pain, jealousy, and rejection are sometimes felt as well as affection, trust, and support. In fact, children will probably experience the gamut of positive and negative emotions in any such intense relationship. Similarly, because friends are important sources of social comparison, it is possible that competition, as well as comparison, may be intensified between friends. Children may even compete with each other for friends. As yet, few data exist about these "negative" experiences or how the status of the relationship may affect such experiences.

Thus contemporary psychological descriptions of friendship are somewhat idealized. In part, this unbalanced perspective can be attributed to the frequent reliance on self-report techniques. We believe that when children are asked to describe a friendship, they will be much more likely to report the positive aspects of the relationship rather than the negative ones. More frequent use of direct observational techniques should help provide a more balanced picture of all facets of friendships. Particularly needed are studies comparing children's self-reports and their actual behavior with friends.

Although an idealized picture of friendships is not accurate, it is clear that friendships provide an arena for a wide range of essential experiences for the growing child. Even though the study of children's friendships has only a short history, the significance of such relationships is already apparent. Friendships change markedly through development, and yet at all ages they are a central facet of a child's life. Clearly then, friendships are more than just child's play.

REFERENCES

Asher, S. T. The influence of race and sex on children's sociometric choices across the school year. Unpublished manuscript, University of Illinois, 1973.

Asher, S. T., Oden, S. L., & Gottman, J. M. Children's friendships in school settings. In

L. G. Katz (Ed.) *Current topics in early childhood education* (Vol. I). Hillsdale, N.J.: Ablex, 1977.

Berndt, T. J. Relations between social cognition, nonsocial cognition and social behavior: The case of friendship. In J. H. Flavell & L. D. Ross (Eds.), *New directions in the study of social-cognitive development,* in press.

Berscheid, E., & Graziano, W. The initiation of social relationships and interpersonal attraction. In R. L. Burgess & T. L. Huston (Eds.), *Social exchange in developing relationships.* New York: Academic Press, 1979.

Bigelow, B. J. Children's friendship expectations: A cognitive-developmental study. *Child Development,* 1977, **48**, 246–253.

Bigelow, B. J., & La Gaipa, J. J. Children's written descriptions of friendship: A multidimensional analysis. *Developmental Psychology,* 1975, **11**, 857–858.

Bigelow, B. J., & La Gaipa, J. J. The development of friendship values and choice. In H. C. Foot, A. J. Chapman, & J. R. Smith (Eds.), *Friendship and social relations in children.* New York: Wiley, 1980.

Bruner, J. S., Jolly, A., & Sylva, K. (Eds.). *Play: Its role in development and evolution.* Harmondsworth, Middlesex: Penguin, 1976.

Cabral, G., Volpe, J., Youniss, J., & Gellert, B. Resolving a problem in friendship and other relationships. Unpublished manuscript, Catholic University of America, 1977.

Campbell, E. Q. Adolescent socialization. In D. A. Goslin (Ed.), *Handbook of socialization theory and research.* Chicago: Rand McNally, 1969.

Carter, D. E., DeTine-Carter, S. L., & Benson, F. W. Interracial acceptance in the classroom. In H. C. Foot, A. J. Chapman, & J. R. Smith (Eds.), *Friendship and social relations in children.* Chichester, England: Wiley, 1980.

Cowen, E. L., Pederson, A., Babijian, H., Izzo, L. D., & Trost, M. A. Long-term follow-up of early detected vulnerable children. *Journal of Consulting and Clinical Psychology,* 1973, **41**, 438–446.

Criswell, J. H. A sociometric study of race cleavage in the classroom. *Archives of Psychology,* 1939, **235**, 1–82.

Davitz, J. R. Social perception and sociometric choice in children. *Journal of Abnormal and Social Psychology,* 1955, **50**, 173–176.

Douvan, E., & Adelson, J. *The adolescent experience.* New York: Wiley, 1966.

Doyle, A., Connolly, J., & Rivest, L. The effect of playmate familiarity on the social interactions of young children. *Child Development,* 1980, **51**, 217–223.

Duck, S. W. Similarity and perceived similarity of personal constructs as influences on friendship choice. *British Journal of Social and Clinical Psychology,* 1973, **12**, 1–6.

Dunphy, D. C. The social structure of urban adolescent peer groups. *Sociometry,* 1963, **26**, 230–246.

Eder, D., & Hallinan, M. T. Sex differences in children's friendships. *American Sociological Review,* 1978, **43**, 237–250.

Foot, H. C., Chapman, A. J., & Smith, J. T. Friendship and social responsiveness in boys and girls. *Journal of Personality and Social Psychology,* 1977, **35**, 401–411.

Furman, W. Some observations on the study of personal relationships. Paper presented at a conference sponsored by the National Science Foundation on Boundary Areas in Psychology, Vanderbilt University, 1981. (a)

Furman, W. The process of becoming acquainted in middle childhood. Unpublished manuscript, University of Denver, 1981. (b)

Furman, W., & Bierman, K. L. Children's conceptions of friendship: A multimethod study of developmental changes. Unpublished manuscript, University of Denver, 1981. (a)

Furman, W., & Bierman, K. L. Developmental changes in young children's conceptions of friendship. Unpublished manuscript, University of Denver, 1981. (b)

Furman, W., & Buhrmester, D. The contribution of peers and siblings to the parenting process. In M. Kostelnik (Ed.), *Patterns of supplementary parenting* (Vol. 2). *Child nurturance* New York: Plenum, in press.

Furman, W. & Childs, M. K. A temporal perspective on children's friendship. Paper presented at meetings of Society for Research in Child Development, Boston, 1981.

Gottman, J., & Parkhurst, J. A developmental theory of friendship and acquaintanceship. In A. Collins (Ed.), *Minnesota symposia on child psychology* (Vol. 13). Hillsdale, N.J.: Erlbaum, 1980.

Hartup, W. W. Peer interaction and social organization. In P. H. Mussen (Ed.), *Carmichael's manual of child psychology* (3rd ed., Vol. 2). New York: Wiley, 1970.

Hartup, W. W. Peer interaction and the behavioral development of the individual child. In E. Schopler & R. J. Reichler (Eds.), *Psychopathology and child development*. New York: Plenum, 1976.

Hartup, W. W. Children and their friends. In H. McGurk (Ed.), *Childhood social development*. London: Metheun, 1978.

Hayes, D. S. Cognitive bases for liking and disliking among preschool children. *Child Development*, 1978, **49**, 906–909.

Howes, C., & Mueller, E. Early peer friendships: Their significance for development. In W. Spiel (Ed.), *The psychology of the twentieth century*. Zurich: Kindler, in press.

Izard, C. E. Personality similarity and friendship. *Journal of Abnormal and Social Psychology*, 1960, **61**, 47–51.

Jormakka, L. The behavior of children during a first encounter. *Scandanavian Journal of Psychology*, 1976, **17**, 15–22.

Kurth, S. B. Friendships and friendly relations. In J. McCall (Ed.), *Social relationships*. Chicago: Aldine, 1970.

Langworthy, R. L. Community status and influence in a high school. *American Sociological Review*, 1959, **24**, 537–539.

Lewis, M., Young, G., Brooks, J., & Michalson, L. The beginning of friendship. In M. Lewis & L. A. Rosenblum (Eds.), *Friendship and peer relations*. New York: Wiley, 1975.

Livesley, W. J., & Bromley, D. B. *Person perception in childhood and adolescence*. London: Wiley, 1973.

Maccoby, E. E., & Jacklin, C. N. *The psychology of sex differences*. Stanford, California: Stanford University, 1974.

Masters, J. C., & Furman, W. Popularity, individual friendship selection, and specific peer interactions among children. *Developmental Psychology*, 1981, **17**, 344–350.

Newcomb, A. F., Brady, J. E., & Hartup, W. W. Friendship and incentive condition as determinants of children's task-oriented social behavior. *Child Development*, 1979, **50**, 878–881.

Philp, A. J. Strangers and friends as competitors and cooperators. *Journal of Genetic Psychology*, 1940, **57**, 249–258.

Piaget, J. *The moral judgment of the child*. Glencoe, Ill.: The Free Press, 1932.

Roff, J. Childhood social interaction and young adult psychosis. *Journal of Clinical Psychology*, 1963, **19**, 152–157.

Roff, M., Sells, S. B., & Golden, M. M. *Social adjustment and personality development in children*. Minneapolis: University of Minnesota Press, 1972.

Selman, R. L., & Jacquette, D. Stability and oscillation in interpersonal awareness: A clinical-development analysis. In C. B. Keasy (Ed.), *The Nebraska symposium on motivation* (Vol. 25). Lincoln: University of Nebraska Press, 1977.

Singleton, L. C., & Asher, S. R. Racial integration and children's peer preferences: An investigation of developmental and cohort differences. *Child Development*, 1979, **50**, 936–941.

Sullivan, H. S. *The interpersonal theory of psychiatry*. New York: Norton, 1953.

Sundby, M. S., & Kreyberg, P. C. *Prognosis in child psychiatry*. Baltimore: Williams & Wilkins, 1968.

Waldrop, M. F., & Halverson, C. F. Intensive and extensive peer behavior: Longitudinal and cross-sectional analyses. *Child Development*, 1975, **46**, 19–26.

ADOLESCENCE

19

FOUR ASPECTS OF ADOLESCENT COGNITIVE DEVELOPMENT

David Moshman
University of Nebraska-Lincoln

Edith Neimark
Douglass College, Rutgers

To the extent that adolescence represents the transition between childhood and adulthood, our own adult conceptualization of this transition tells us something not only about children and adolescents but also about our conceptions of ourselves. Thus in the area of cognition, our theories of what and how children must develop to think as we do inevitably reflect what we see as most fundamental about our own thinking. Theoretical disputes, then, are partially a function of whether we view the mature knower as, say, logician, scientist, philosopher, or artist.

For present purposes, we will assume that the developing adolescent, like the developing child and the developing adult, is all of these and more.

THE DEVELOPING LOGICIAN

Empirical study of the development of deductive reasoning has focused on age changes in a number of different aspects of logic, including (a) conclusions, the particular deductive results individuals produce or accept given various sets of premises; (b) processes, the actual step-by-step mental processing of premise information leading to observed patterns of deduction; (c) structures, how processes and knowledge are assumed, on the basis of some theory, to be organized; and (d) metalogic, the individual's conceptions regarding the nature of logical reasoning.

Conclusions

Not surprisingly, with increasing age, people are increasingly likely to reach conclusions consistent with formal logic. Nevertheless, it is clear not only that preschool children *can* reach correct conclusions (Ennis, 1976; O'Brien & Shapiro, 1968) but that even older adolescents and college students may *fail* to do so depending on a variety of form, content, and other variables related to the task they are given (e.g., Roberge & Paulus, 1971; Wason & Johnson-Laird, 1972). Thus although one can argue for a gradual increase in logicality during adolescence, the data do not show a discrete qualitative shift during adolescence from alogical to logical conclusions. Moreover, there is increasing awareness of the inadequacies of focus upon conclusions for clarification of the underlying reasoning processes (Braine, 1978).

Processes

Suppose you were given the premises "If the object is red, then it is small" and "If the object is small, then it is in the blue box." Research on logical processes is less interested in *whether* the conclusion reached is dictated by formal logic (i.e., "If the object is red, then it is in the blue box") than in *how* it is reached, regardless of conclusion. It is widely, though not universally, assumed that conclusions are reached by combining the premises according to some system of rules. Many theorists have attempted a description of what those rules might be (e.g., Braine, 1978; Falmagne, 1975; Osherson, 1975, 1976). If a developmental change in the nature of logical processes took place during the adolescent years, then one would expect to find differences between the rule system accounting for deductive processes in children and the rule system describing the processes in late adolescence. While there is general evidence for broadening and enrichment of the rule system (Siegler, 1976; Taplin, Staudenmeyer, & Taddonio, 1974), many question whether this enlargement constitutes qualitative change. Even the careful and thoughtful work of Osherson (1975, 1976) on this matter yields rather equivocal results. Osherson himself cautiously interprets his data as providing little evidence for qualitative change at the process level but raises the possibility that qualitative differences do obtain at a deeper structural level or at the level of metalogic (cf. Osherson & Markman, 1975).

Structures

At a more abstract structural level, results seem less equivocal. Piaget's theory (Inhelder & Piaget, 1958) suggests that one should expect a qualitative shift during adolescence from comprehension of logical propositions in terms of concrete classification operations to comprehension based on formal second-order classifications of classifications and on a more flexible structure for organizing component operations. Consider, for example, the propositions (a) "The book is paperbound and fiction," (b) "The book is paperbound or fiction or both," and (c) "If the book is paperbound, then it is fiction." One might expect a concrete thinker to interpret all these statements as though the logical connective were "and," since this requires a simple multiplication of the class of paperbacks and the class of fiction to yield the class of paperback fiction. By contrast, a formal thinker, capable of second-order operations, would be able to show reasoning based on interpretations of b and c in terms of classes of classes: For b, paperbound fiction + paperbound nonfiction + hardbound fiction; for c, paperbound fiction + hardbound fiction + hardbound nonfiction. Of course, we cannot determine how an individual construes a proposition simply by direct inquiry or by seeing what conclusion is reached from a particular set of premises. A number of investigators, however, have presented subjects with sets of inference tasks carefully constructed to enable the investigator to determine the interpretations applied by the subject from the subject's pattern of performance across tasks. The results of such studies have shown that the sort of differentiations among logical propositions requiring second-order (formal) operations is absent in childhood; that—as a consequence—very systematic and predictable errors are observed; and that there is substantial development during adolescence in the ability to make such differentiations (e.g., Moshman, 1979a; Staudenmayer & Bourne, 1977; Sternberg, 1979a). As noted above, however, even older adolescents and college students frequently show the sorts of interpretations described as typical of children. In addition, they, too, are affected by a variety of form, content, and task variables. Thus although the data do suggest a qualitative reorganization of the logical concepts of middle childhood on the part of a substantial proportion of adolescents, this does not mean that all older adolescents and adults consistently apply a mode of reasoning different from that of children. The existence of a structure is no guarantee that it is exclusively or invariably reflected in performance.

Metalogic

The final aspect of adolescent logic involves not simply the use of logic to reach conclusions but rather one's conceptions about the nature of logic. Thus, for example, most grade school children, given the premises "Elephants are bigger than dogs" and "Dogs are bigger than mice," would have little difficulty reaching the conclusion "Elephants are bigger than mice." They would, however, be likely to reject the argument "Mice are bigger than dogs, and dogs are bigger than elephants; therefore, mice are bigger than elephants" on the ground that both premises and conclusion are obviously contrary to fact. As is implied by the terms *concrete* and *formal* operations, Piaget's theory emphasizes the adolescent's increasing ability to focus upon the form of the argument regardless of the content, as well as increased awareness of the appropriateness of such a focus, in evaluating validity of inference. Thus in the case of the example cited, the adolescent is able to see that the second argument is logically identical to the first and equally valid, whereas the child is incapable of doing so. Ability to differentiate form from content and to focus upon the former, without regard to the latter, enables the adolescent to operate upon a plane of abstraction in which the implications of purely hypothetical possibilities may be exhaustively explored. This opens up a whole new universe of systematic reflection and system building, a major feature of adult thought which is lacking in child thought.

THE DEVELOPING SCIENTIST

Abilities at system building and testing are central to the development of scientific reasoning, a process which Inhelder and Piaget (1958) selected as the paradigm for their initial investigation of adolescent thought. Research investigating the developing adolescent as a scientist includes work on (a) knowledge of specific content domains, (b) mastery of general organizing principles or schemes, and (c) knowledge about the nature of theories and theory testing.

Specific Content Domains

Understanding of empirical phenomena presumably involves both general development of broad cognitive competencies and more focused learning about specific scientific domains. Learning in specific domains has recently become the focus of a good deal of educational research motivated primarily by concern with improving instruction. Most of the work has dealt with mastery of physical sciences, but there has been some exploration of natural and social sciences as well (cf. Peel, 1971; Keats, Collis, & Halford, 1978). In most of this work emphasis is placed upon structural learning (i.e., learning to reason like a chemist, mathematician, biologist, etc.) rather than accumulation of particular content material.

Development of Schemes

Inhelder and Piaget (1958) identify eight formal operations concepts, or schemes, which cannot be directly inferred from experience but which underlie much of scientific reasoning. These include combinations, proportions, coordinations of frames of reference, mechanical equilibrium, probability, correlation, multiplicative compensations, and conservations (e.g., of energy) which transcend empirical verification. To date the schemes most directly related to probability and statistics have been the more commonly investigated. Available evidence shows a general pattern of substantial development during adolescence but, nonetheless, imperfect performance by adults on tasks assessing (a) combinatorial competencies, such as systematic generation of all possible combinations or permutations from a set of elements (e.g., Martorano, 1977; Scardamalia, 1977); (b) comprehension and use of proportions (e.g., Seigler & Vago, 1978; Fischbein, 1975); (c) conceptualization of probabilities as relations between some actual, desired, or potential reality and a broader set of hypothetical possibilities (e.g., Chapman, 1975); and (d) understanding of correlations as probabilistic relations between variables (e.g., Martorano, 1977; Neimark, 1975b).

Metatheoretical Knowledge

Ability to think about theories and the testing of them will be considered in terms of five components (Moshman, 1979b). The first and most frequently investigated is the ability to control variables. Inhelder and Piaget place considerable emphasis upon the understanding that investigation of the effect of a variable requires that it be manipulated while all other variables are held constant. A large body of evidence attests to improvement during adolescence in the ability to design and execute unconfounded experiments (cf. reviews by Kuhn & Phelps, 1979; Levine & Linn, 1977), and more recent evidence shows a similar trend in the ability to reach appropriate conclusions from existing multivariable information (Kuhn & Brannock, 1977). In a more naturalistic situation where concern shifts from obtaining information to controlling results, Tschirgi (1980) has shown a pragmatic tendency among grade school and college students to adjust the strategy for manipulation of variables (either holding all but one constant or varying all but one) in relation to the intended outcome (repeating success or eliminating failure). Tschirgi's results are consistent with Moshman's (1979b) suggestion that even adults have difficulty in distinguishing the practical use of theories from the formal testing of them, a second basic metatheoretical competency. It seems likely, however, that a differentiation of the requirements of application from those of testing does develop during adolescence.

A third metatheoretical competency is the understanding that a theory is properly tested by attempts at its disproof rather than by accumulation of supporting evidence: a falsification strategy (Popper, 1959). There is extensive evidence that even adults commonly fail to adopt the falsification strategy (e.g., Kuhn & Phelps, 1979; Mynatt, Doherty, & Tweney, 1977; Wason & Johnson-Laird, 1972), but the ability to use it does seem to develop during adolescence (Moshman, 1979a). This development may be related to the ability to distinguish between theory and data; that is, to realize that the former differs from the latter not only in being more general but also in being hypothetical, rather than actual, and in being more readily falsifiable than verifiable. Unfortunately, very little of the considerable research on a falsification strategy has looked at it developmentally.

Strauss and Kroy (1977) suggest two additional metatheoretical competencies that seem to be of importance but concerning which there is little evidence, developmental or otherwise. One is an orientation toward parsimony, that is, a preference for simple, elegant, comprehensive explanations, over those that have been protected against frequent falsification through the addition of post hoc assumptions and corrections. The other is idealization, the ability to recognize and neutralize distorting aspects of reality in order to construct an idealized theory (e.g., conservation of motion or the ideal gas law) that does not correspond directly to anything observable but nevertheless helps explain such observables. One might speculate, in anticipation of relevant research, that development of such metatheoretical knowledge would fit the common pattern of a qualitatively new understanding appearing in many adolescents but remaining far from universally applied for the population as a whole.

THE DEVELOPING PHILOSOPHER

The preceding discussions of the adolescent's metalogical and metatheoretical knowledge illustrate development in a direction which would, if pursued and further reflected upon, turn the adolescent into a philosopher of science (as opposed to a scientist). This section expands on the conception of the developing philosopher by viewing the adolescent as (a) a philosopher of ethics and (b) an epistemologist, metaphysician, and philosopher of mind.

Ethical Philosophy

The predominant theory of moral development over the past 2 decades has been Kohlberg's (1969) cognitive-developmental formulation. Kohlberg's theory and methodology both reflect his view of the developing individual as a budding ethical philosopher who reflects with increasing sophistication on increasingly abstract considerations of moral dilemmas and coordinates these reflections into a coherent and systematic ethical theory. Three general levels of moral judgment are differentiated: (I) a preconventional, or premoral, level in which judgments are based not on moral considerations but on punishment and social exchange; (II) a conventional level involving an internalized morality based on peer relations or societal institutions; and (III) a postconventional, or principled, level of autonomous and transcendent moral principles. Although the theory has elicited strong support (e.g., Rest et al., 1978), it has also attracted a great deal of serious criticism (Kurtines & Greif, 1974). Discussion here will focus on two major sets of criticisms which seem to be constructive in intent and to get at important issues in the adolescent's ethical philosophy.

The first issue involves the charge of sex bias in Kohlberg's stages and particularly with respect to moral judgments at the conventional level (which usually emerges during adolescence). It is commonly found that the modal stage of reasoning achieved by adult women is the first stage within this level, that is, Kohlberg's stage 3, an orientation emphasizing helping others, earning social approval, maintaining reciprocal social relationships, and in general being a nice person. The modal stage for men, on the other hand, is the upper stage of the conventional level, Kohlberg's stage 4, an orientation emphasizing more abstract obligations, respect for authority and law, and maintenance of the social order. Kohlberg's theory can explain the apparently more advanced moral development of males in terms of differential sex-role socialization and experiences, leading to the expectation that more advanced moral development in women may be expected to the extent that changes in society provide them with more abstract morally relevant experiences in broader societal domains.

Longitudinal evidence from Holstein (1976), however, raises problems for this explanation. Her data show a tendency for males to skip from preconventional stage 2 directly to stage 4, whereas females who move beyond stage 3 are likely to skip directly to stage 5. Although, as Holstein herself notes, methodological considerations mandate caution in interpreting these data, they nonetheless suggest the possibility that stages 3 and 4 are not hierarchically ordered levels of understanding through which all developing moralists must pass (to the extent that they advance at all), but rather alternative realizations of a broader conventional level morality.

Interviews with a variety of women, reported by Gilligan (1977), support this interpretation by indicating the relatively greater emphasis by women, compared to men, on considerations of responsibility, care, and interpersonal context over abstract universal ethical truths at all levels of development. Gilligan highlights this difference by contrasting the biblical Abraham, who was willing to sacrifice his child's life to affirm the primacy and totality of his faith, with the mother who verifies her motherhood to Solomon by a readiness to sacrifice truth, as well as her personal claim, for the sake of her child's life. Recent revisions of Kohlberg's theory by Kohlberg and his associates have attempted to take these considerations into account (Colby, 1978). Sex differences in development of moral understanding have been reported by other writers as well: Hoffman (1979) interprets them as indicating the moral superiority of females; Eron (1980) recommends feminine socialization for boys and girls as a means to combat aggressiveness.

A second major issue is the observation of regression in late adolescence from stage 4 to stage 2. As with the evidence on sex differences, Kohlberg himself initially accepted this as a genuine phenomenon, arguing that it was a temporary regression facilitating the later construction of postconventional stage 5 judgments (Kohlberg

& Kramer, 1969). Recently, however, the regression has been viewed as a spurious decline because what superficially resembles stage 2 reasoning is found upon closer analysis to be quite different from the premoral reasoning of young children. It is now treated as a transitional stage, 4½, in which the conventional stage 4 morality of adolescence is rejected but the principled stage 5 thinking that will eventually replace it has yet to be constructed (Kohlberg, 1973; Turiel, 1974). This reformulation has been incorporated into the most recent "official" revision of Kohlberg's theory, which, in attempting to differentiate more rigorously the structure of moral reasoning from factors of content and function, drops stage 6 and adds a number of new substages and transitional stages (Colby, 1978).

Nonetheless, criticisms remain. Gibbs (1979) suggests that Kohlberg has overdifferentiated moral stages from cognitive-structural changes in other areas and suggests, instead, a view of adolescent moral judgment focusing on (a) detaching oneself from ongoing exchange and immediate interests in that exchange in order to take an impartial, spectator, perspective on the moral aspects of the interaction and (b) expansion of this detached perspective to apply not only to face-to-face relationships but also to more abstract social systems and institutions. Other writers (e.g., Perry, 1970) tend to focus on the relativism of postconventional morality. Murphy and Gilligan (1980) propose two alternative postconventional avenues for incorporation of relativity: (a) a formal approach which derives moral solutions from a system of formal principles and concepts (e.g., social contract) and (b) a contextual approach deriving from the responsibility for and consequences of moral choice. Post hoc analysis of longitudinal data suggests development of contextual relativism to be precipitated by an intense personal dilemma.

Epistemology, Metaphysics, and Philosophy of Mind

These are abstruse matters, of course, but not beyond the scope of adolescent concerns. A notable attempt to investigate the natural epistemologies of college students was undertaken by Perry (1970), who reported a trend from (a) a view of knowledge as absolute, established truths; to (b) recognition of multiple perspectives and doubt regarding the legitimacy of choosing among them; to (c) recognition that even given the legitimacy of diverse perspectives, ideas must be justified and may be evaluated relative to each other; to (d) finding a basis for commitment even in a relativistic world. Broughton (1978) has pursued the study of natural philosophies to include a broader range of ages (preschool through college graduates) and philosophical issues (self–world, mental–material, physical–social, reality–appearance, knower–known). Though the responses of adolescents are typically a far cry from the sophistication of professional philosophers, their recognition of the basic issues generally places them closer to the philosopher than to the philosophical innocence of the young child. Consider, for example, the following interchange in which an interviewer attempts to raise basic epistemological and metaphysical issues with a 5-year-old:

> "How do you know that chest is really over there?"
> Answer: "They put it there."
> "How do you know that house is over there?"
> Answer: "That's where they built it." (Broughton, 1978, p. 79)

As for the mind, one 10-year-old notes that it is "A lot smarter than the other parts of your body" (Broughton, 1978, p. 84).

Broughton indicates that adolescents, by contrast, make much sharper distinctions than do younger children between the mental and the physical, between reality and appearance, and between subjectivity and objectivity. Moreover, they increasingly tend to see the problematical nature of the concept of truth. However, Broughton's interviews suggest that development may proceed to still higher levels and in some cases involve sophisticated rejections of these adolescent dualisms. For one 22-year-old, "Mind is matter that knows how to do tricky things." Self, moreover, is merely "a philosophical invention . . . somebody made up to sell deodorant" (Broughton, 1978, p. 94). Thus in the life span story of the developing philosopher, adolescence seems to fall somewhere around the midpoint of a complex process of differentiations and reintegrations, of skepticism and reaffirmations, of relativism and commitment. Chandler (1978) discusses how adolescent cliquishness, conformity, stereotyping, ideological conversions, and overabstraction may be, at least in part, the understandable affective reactions of budding philosophers to their disconcerting new insights into subjectivity, relativism, and epistemic isolation. There is obviously much still to be learned about adolescent philosophies and their affective consequences.

THE DEVELOPING ARTIST

Although there is much less information concerning this aspect of adolescent development than the three just considered, preliminary data suggest that it will merit more extended study. Studies of the interpretation of art (Seefeldt, 1979; Kenney & Nodine, 1979) show clear developmental trends away from literal, content-based evaluation to focus upon form, artist's intent, and features of design with increasing age. All these trends are what one would expect to find if art interpretation were a cognitively influenced activity reflecting the same features of cognitive development manifest in more obviously logical areas. On the other hand, Gardner (1979) in his work on the developing artist focuses not upon the role of logic but upon specific characteristics of various symbol systems and media. The figurative aspects of cognitive development, as contrasted to the operative aspects, have tended to be neglected, and art appreciation or creation constitute natural areas for their investigation.

Recently, cognitive psychologists and students of language have become interested in the subject of metaphor which, because of relations to the logical process of analogy, would be expected to reflect changes found in other areas of cognitive development. Although there has been criticism (Ortony, Reynolds, & Arter, 1978) of the conceptualization of developmental research on metaphor, existing developmental evidence on interpretation of metaphor (e.g., Cometa & Eson, 1978; Gallagher, 1978) and of poetry (Hardy-Brown, 1979) reflect the same developmental shift in focus from content to form which was found in studies of art appreciation. This research, along with accumulating evidence on the development of analogical reasoning (Sternberg & Rifkin, 1979) promises a further broadening of our understanding of adolescent cognition.

CONCLUSIONS: THE STATE OF THE ART

This examination of adolescent thought in four disparate areas has revealed a number of common developmental trends suggesting a general pattern of intellectual advance during adolescence. First there is a finer differentiation of components concomitant with their integration into a more flexibly organized whole. This trend is clearly exemplified in the development of logic and its application to other areas, as, for instance, testing a scientific hypothesis by seeking disconfirmatory evidence. The second trend, increasing detachment of thought from both context and personal viewpoint, results in both greater generality and abstraction. Decontextualization is reflected, for example, in the shift of focus from content to form in evaluation of art, logic, morality, or science. This fundamental shift in the determinants of judgment underlies the appearance of such varied attainments as metaphor and ethical judgment. Depersonalization is, similarly, broadly manifest in, for example, the ability to take a variety of viewpoints in ethical judgment as well as in the ability to view self objectively in the many forms of metacognitive evaluation. Current theories address themselves to description and explanation of these developmental trends.

Beyond question, Piaget's theory of formal operations (Inhelder & Piaget, 1958) remains the preeminent account of adolescent cognition. There are now so many good reviews of the theory and of relevant research that we have not attempted to cover that territory again (e.g., Brainerd, 1978a; Cowan, 1978; Flavell, 1977; Keating, 1979; Kuhn, 1979; Neimark, 1975a, 1979). There is, however, increasing criticism of the theory from a number of directions. One result of these criticisms has been a growing awareness of differences in metatheoretical perspective (Brainerd, 1978b) and theoretical intent. Piaget's theory is, in terms of one emerging distinction, a competence theory: it focuses upon the general structure of adolescent abilities as an idealization of what sorts of cognitive capacities should under optimal conditions be inferable from behavior patterns in a variety of tasks. This goal is very different from that of performance theories, which attempt an accurate description of observed behavior in specific situations. Many current theoretical attempts are directed to development of a performance theory of adolescent cognition.

The attempts at elaboration of a performance theory are proceeding in one of two directions. The first, guided largely by information-processing and computer simulation approaches toward the study of adult cognition, consists in analyzing task behavior into basic component skills. One example of this approach is the work of Sternberg (1977, 1979b), whose componential analysis identifies elementary information processes operating upon symbols (e.g., encoding, recoding, specific strategies for organizing information, etc.) which are quite general across classes of tasks or may be specific to a given subtask. Selection and ordering of components is governed by metacomponent or control processes. Developmental changes tend to be reflected in control processes. A related approach is reflected in the work of Siegler (1976, 1980), who analyzes errors in task performance into hierarchies of rule usage. Higher-level rules correspond to analysis into more alternative variables and better elaborated schemes for their integration.

A second approach to construction of performance theory is reflected in the neo-Piagetian approach of Pascual-Leone (1980), who introduces classes of variables moderating performance. Among the variables included are consequences of prior experience and training as reflected in content knowledge and organized skills (structural learning), stimulus salience features of specific task materials and context, and an individual difference factor related to the cognitive style of field dependence–independence. At present, Pascual-Leone and his students are unique in addressing the issue of individual differences, but there is mounting interest in the subject (e.g., Lunzer, 1978), and relevant work is beginning to appear (see Neimark, 1979, 1981, for reviews).

Although performance theories tend to deemphasize qualitative changes associated with cognitive development, and to emphasize continuities over time, it is not now clear that they are incompatible with Piagetian approaches, to the extent that a "crucial experiment" for which the two approaches would yield different outcome predictions is attainable. The two approaches do, however, lead to different research emphases and procedures. While the current state of "controversy" may lead to seeming confusion, it is likely that the broadening scope and variety of research will ultimately lead to a more detailed picture of the nature of cognitive development during adolescence.

REFERENCES

Braine, M. D. S. On the relation between the natural logic of reasoning and standard logic. *Psychological Review,* 1978, **85**, 1–21.

Brainerd, C. J. *Piaget's theory of intelligence.* Englewood Cliffs, N.J.: Prentice-Hall, 1978. (a)

Brainerd, C. J. The stage question in cognitive-developmental theory. *Behavioral and Brain Sciences,* 1978, **2**, 173–213. (b)

Broughton, J. Development of concepts of self, mind, reality, and knowledge. In W. Damon (Ed.), *Social cognition.* San Francisco: Jossey-Bass, 1978.

Chandler, M. J. Adolescence, egocentrism, and epistemological loneliness. In B. Z. Presseisen, D. Goldstein, & M. H. Appel (Eds.), *Topics in cognitive development: Language and operational thought* (Vol. 2). New York: Plenum, 1978.

Chapman, R. H. The development of children's understanding of proportions. *Child Development,* 1975, **46**, 141–148.

Colby, A. Evolution of a moral-developmental theory. In W. Damon (Ed.), *Moral development.* San Francisco: Jossey-Bass, 1978.

Cometa, N. S., & Eson, M. E. Logical operations and metaphor interpretation: A Piagetian model. *Developmental Psychology,* 1978, **49**, 649–659.

Cowan, P. A. *Piaget with feeling: Cognitive, social, and emotional dimensions.* New York: Holt, Rinehart & Winston, 1978.

Ennis, R. H. An alternative to Piaget's conceptualization of logical competence. *Child Development,* 1976, **47**, 903–919.

Eron, L. D. Prescription for reduction of aggression. *American Psychologist,* 1980, **35**, 244–252.

Falmagne, R. J. *Reasoning: Representation and process.* Hillsdale, N.J.: Erlbaum, 1975.

Fischbein, E. *The intuitive sources of probabilistic thinking in children.* Synthese Library, Vol. **85**, Dordrecht: Reidel, 1975.

Flavell, J. H. *Cognitive development.* Englewood Cliffs, N.J.: Prentice-Hall, 1977.

Gallagher, J. M. The future of formal thought research: The study of analogy and metaphor. In B. Z. Presseisen, D. Goldstein, & M. H. Appel (Eds.), *Topics in cognitive development: Language and operational thought* (Vol. 2). New York: Plenum, 1978.

Gardner, H. Developmental psychology after Piaget: An approach in terms of symbolization. *Human Development,* 1979, **22**, 73–88.

Gibbs, J. C. Kohlberg's moral stage theory: A Piagetian revision. *Human Development,* 1979, **22**, 89–112.

Gilligan, C. In a different voice: Women's conceptions of self and of morality. *Harvard Educational Review,* 1977, **47**, 481–517.

Hardy-Brown, K. Formal operations and the issue of generality: The analysis of poetry by college students. *Human Development,* 1979, **22**, 127–136.

Hoffman, M. L. Development of moral thought, feeling, and behavior. *American Psychologist,* 1979, **34**, 958–966.

Holstein, C. Development of moral judgment: A longitudinal study of males and females. *Child Development,* 1976, **47**, 51–61.

Inhelder, B., & Piaget, J. *The growth of logical thinking: From childhood to adolescence.* New York: Basic Books, 1958.

Keating, D. P. Adolescent thinking. In J. P. Adelson (Ed.), *Handbook of adolescence.* New York: Wiley, 1979.

Keats, J. A., Collis, K. F., & Halford, G. S. *Cognitive development.* New York: Wiley, 1978.

Kenney, J. L., & Nodine, C. F. Developmental changes in sensitivity to the content, formal, and affective dimensions of painting. *Bulletin of the Psychonomic Society,* 1979, **14**, 463–466.

Kohlberg, L. Stage and sequence: The cognitive-developmental approach to socialization. In D. A. Goslin (Ed.), *Handbook of socialization theory and research.* New York: Rand McNally, 1969.

Kohlberg, L. Continuities in childhood and adult moral development revisited. In P. B. Baltes & L. R. Goulet (Eds.), *Lifespan developmental psychology* (2nd ed.). New York: Academic, 1973.

Kohlberg, L., & Kramer, R. Continuities and discontinuities in childhood and adult moral development. *Human Development,* 1969, **12**, 93–120.

Kuhn, D., The significance of Piaget's formal operations stage in education. *Journal of Education,* 1979, **161**, 34–50.

Kuhn, D., & Brannock, J. Development of the isolation of variables scheme in experimental and "natural experiment" contexts. *Developmental Psychology,* 1977, **13**, 9–14.

Kuhn, D., & Phelps, E. A methodology for observing development of a formal reasoning strategy. In D. Kuhn (Ed.), *Intellectual development beyond childhood.* San Francisco: Jossey-Bass, 1979.

Kurtines, W., & Greif, E. B. The development of moral thought: Review and evaluation of Kohlberg's approach. *Psychological Bulletin,* 1974, **81**, 453–470.

Levine, D. I., & Linn, M. C. Scientific reasoning ability in adolescence: Theoretical viewpoints and educational implications. *Journal of Research in Science Teaching,* 1977, **14**, 371–384.

Lunzer, E. A. Formal reasoning: A reappraisal. In B. Z. Presseisen, D. Goldstein, & M. H. Appel (Eds.), *Topics in cognitive development* (Vol. 2). New York: Plenum, 1978, pp. 47–76.

Martorano, S. C. A developmental analysis of performance on Piaget's formal operations tasks. *Developmental Psychology,* 1977, **13**, 666–672.

Moshman, D. Development of formal hypothesis-testing ability. *Developmental Psychology,* 1979, **15**, 104–112. (a)

Moshman, D. To *really* get ahead, get a metatheory. In D. Kuhn (Ed.), *Intellectual development beyond childhood.* San Francisco: Jossey-Bass, 1979. (b)

Murphy, J. M., & Gilligan, C. Moral development in late adolescence and adulthood: A critique and reconstruction of Kohlberg's theory. *Human Development,* 1980, **23**, 77–104.

Mynatt, C. R., Doherty, M. E., & Tweney, R. D. Confirmation bias in a simulated research environment. *Quarterly Journal of Experimental Psychology,* 1977, **29**, 85–95.

Neimark, E. D. Intellectual development during adolescence. In F. D. Horowitz (Ed.), *Review of child development research* (Vol. 4). Chicago: University of Chicago Press, 1975. (a)

Neimark, E. D. Longitudinal development of formal operations thought. *Genetic Psychology Monographs*, 1975, **91**, 171–225. (b)

Neimark, E. D. Current status of formal operations research. *Human Development*, 1979, **22**, 60–67.

Neimark, E. D. Toward the disembedding of formal operations from confounding with cognitive style. In I. E. Sigel, D. M. Brodzinsky, & R. M. Golinkoff (Eds.), *New directions in Piagetian theory and practice*. Hillsdale, N.J.: Erlbaum, 1981.

O'Brien, P. C., & Shapiro, B. J. The development of logical thinking in children. *American Educational Research Journal*, 1968, **5**, 531–542.

Ortony, A., Reynolds, R. E., & Arter, J. A. Metaphor: Theoretical and empirical research. *Psychological Bulletin*, 1978, **85**, 919–943.

Osherson, D. N. *Logical abilities in children: Reasoning in adolescence: Deductive inference* (Vol. 3). Hillsdale, N.J.: Erlbaum, 1975.

Osherson, D. N. *Logical abilities in children: Reasoning and concepts* (Vol. 4). Hillsdale, N.J.: Erlbaum, 1976.

Osherson, D. N., & Markman, E. Language and the ability to evaluate contradictions and tautologies. *Cognition*, 1975, **3**, 213–226.

Pascual-Leone, J. On learning and development, Piagetian style: II. A critical historical analysis of Geneva's research programme. *Canadian Psychological Review*, 1976, **17**, 289–297.

Pascual-Leone, L. Constructive problems for constructive theories: The current relevance of Piaget's work and a critique of information-processing simulation psychology. In H. Kluwe & R. Spada (Eds.), *Developmental models of thinking*. New York: Academic, 1980.

Peel, E. A. *The nature of adolescent judgment*. New York: Wiley, 1971.

Perry, W. L. *Forms of intellectual and ethical development in the college years: A scheme*. New York: Holt, Rinehart and Winston, 1970.

Popper, K. R. *The logic of scientific discovery*. London: Hutchinson, 1959.

Rest, J. R., Davison, M. I., & Robbins, S. Age trends in judging moral issues: A review of cross-sectional, longitudinal, and sequential studies of the defining issues test. *Child Development*, 1978, **49**, 263–279.

Roberge, J. J., & Paulus, D. H. Children's class and conditional reasoning abilities. *Developmental Psychology*, 1971, **4**, 193–200.

Scardamalia, M. Information processing capacity and the problem of horizontal decalage: A demonstration using combinatorial reasoning tasks. *Child Development*, 1977, **48**, 28–37.

Seefeldt, F. M. Formal operations and adolescent painting. *The Genetic Epistemologist*, 1979, **8**(3), 5–6.

Siegler, R. S. Three aspects of cognitive development. *Cognitive Psychology*, 1976, **8**, 481–510.

Siegler, R. S. Recent trends in the study of cognitive development: Variations on a task-analytic theme. *Human Development*, 1980, **23**, 278–285.

Siegler, R. S., & Vago, S. The development of a proportionality concept: Judging relative fullness. *Journal of Experimental Child Psychology*, 1978, **25**, 371–395.

Staudenmayer, H., & Bourne, L. E. Learning to interpret conditional sentences: A developmental study. *Developmental Psychology*, 1977, **13**, 616–623.

Sternberg, R. J. *Intelligence, information processing, and analogical reasoning: The componential analysis of human abilities*. Hillsdale, N.J.: Erlbaum, 1977.

Sternberg, R. J. Developmental patterns in the encoding and combination of logical connectives. *Journal of Experimental Child Psychology*, 1979, **28**, 469–498. (a)

Sternberg, R. J. The nature of mental abilities. *American Psychologist*, 1979, **34**, 214–230. (b)

Sternberg, R. J., & Rifkin, B. The development of analogical reasoning processes. *Journal of Experimental Child Psychology*, 1979, **27**, 195–232.

Strauss, S., & Kroy, M. The child as logician or methodologist? A critique of formal operations. *Human Development*, 1977, **20**, 102–117.

Taplin, J. E., Staudenmayer, H., & Taddonio, J. L. Developmental changes in conditional reasoning: Linguistic or logical? *Journal of Experimental Child Psychology,* 1974, **17**, 360–373.

Tschirgi, J. E. Sensible reasoning: A hypothesis about hypotheses. *Child Development,* 1980, **51**, 1–10.

Turiel, E. Conflict and transition in adolescent moral development. *Child Development,* 1974, **45**, 14–29.

Wason, P. C., & Johnson-Laird, P. N. *Psychology of reasoning: Structure and content.* Cambridge, Mass.: Harvard University Press, 1972.

20
ADOLESCENT DRUG AND ALCOHOL BEHAVIORS

Patricia B. Sutker
Department of Psychiatry and Behavioral Sciences
Medical University of South Carolina

Increasing focus has been given to conceptual models describing life-cycle stages and the usefulness of developmental theories for understanding specific sets of behaviors, emotions, and cognitions. Only recently has attention turned to the crises associated with mid-life and the elderly years, while adolescence has long been recognized as a time of multifaceted change, heightened stress, and vulnerability for emergence of nonconformity, including drug abuse. In describing adolescence, social scientists inevitably refer to alterations in attitudes and values, family and peer interactions, psychosocial expectations, and status from dependence to independence. Not the least of adolescent age-related crises are biological-gender changes which affect attitudes about the self-system and the interpersonal environment.

Explanations of adolescent drug use and abuse are best cast in a theoretical framework which emphasizes the multidetermined nature of behavior, societal roles in defining the legitimacy of behaviors, and objective specification of behaviors to be studied over time. Looking at definitions, use and abuse descriptions are derived quantitatively from incidence, prevalence, and frequency data or from qualitative judgments of use extent and consequences among subgroups labeled socially deviant. Descriptive methods and explanatory positions vary greatly depending upon researcher discipline, that is, psychologist, sociologist, anthropologist. However, for psychologists, the range of behaviors, cognitions, and emotions antecedent to, associated with, and subsequent to discrete drug phenomena is of critical interest.

DRUG USE PATTERNS

Spurred by illicit drug use of epidemic proportions in the 1960s, researchers now closely monitor national drug preferences and use patterns. Surveys of high school,

---

This work was supported in part by grant AA 04042-02 from the National Institute on Alcohol Abuse and Alcoholism, ADAMHA. Special appreciation is expressed to Anne T. Patsiokas, Albert N. Allain, and Pamela J. Thompson for their invaluable assistance. The author is now Chief, Psychology Service at the Veterans Administration Medical Center in New Orleans, Louisiana.

household, and community respondents serve as primary vehicles for data collection, and health service and criminal justice records also provide sources for estimating drug behavior severity. Study methodologies vary, and drug patterns change over temporal and situational parameters. Yet results are similar in suggesting widespread use of licit and illicit chemicals among adolescent boys and girls. Interpretations of survey and other self-report data must be tempered with caution, and Beschner and Treasure (1979) pointed to the confusion which can arise from imprecise definitions of such terms as prevalence, incidence, current use, use frequency, recent use, and total use times. Other researchers have also detailed methodological issues to be considered in evaluating drug use and drinking practice descriptions (Lavenhar, 1979; Marden, Zylman, Fillmore, & Bacon, 1976).

Prevalence Estimates

Adolescent use prevalence (drugs ever used) was described by Fishburne, Abelson, and Cisin (1980) who surveyed households nationwide and by Johnston, Bachman, and O'Malley (1979) who sampled American high school seniors. Though estimates depend on study and sample characteristics, there is concensus that alcohol, cigarettes, and marijuana are the drugs most frequently tried. Alcohol prevalence rates for high school seniors are cited as high as 93% by Johnston et al. (1979) and 90% by Oetting and Goldstein (1979) for native Americans. Cigarette exposure rates range from 45–75% for girls and 46–76% for boys (Beschner & Treasure, 1979), and marijuana rates are estimated at 31% for 12- to 17-year-olds (Fishburne et al., 1980), 69% for college students (Jessor & Jessor, 1977), and 44% for 14- to 18-year-old nonstudents (Smart & Blair, 1980). Moderately high rates of stimulant and barbiturate experimentation and lower figures for heroin and cocaine were provided by Beschner and Treasure (1979), but Fishburne et al. (1980) reported that 5.4% of 12- to 17-year-olds tried cocaine, 0.5% sampled heroin, and 3.2% used opiates other than heroin.

Frequency Descriptions

Increasing drug involvement follows a steplike, but not necessarily inevitable, progression as described recently by Kandel, Kessler, and Margulies (1978) and Rittenhouse.[1] These and other researchers emphasized the importance of cigarettes and other licit drugs as precursors to later illicit drug use. Beschner and Treasure (1979) indicated that at least one-fourth to one-third of youths use cigarettes and marijuana frequently, and Beschner and Friedman (1979) stated that approximately 15 million youths are regular marijuana users or smoke at least once a week. However, adolescents most prefer beer and wine and consume alcohol more regularly than any other drug. In the Johnston et al. (1979) survey, 72% of seniors admitted alcohol use in the previous month and 7% declared daily use. Further, Donovan and Jessor (1978) reported that roughly 5% of seventh-grade boys and girls admitted problem drinking by study definitions, and nearly 40% and 21% of twelfth-grade boys and girls, respectively, were so classified. Greatest

---
[1] Rittenhouse, J. D. Learning drug use: From "legal" substance to marijuana and beyond. Paper presented at the meeting of the American Psychological Association, Montreal, September, 1980.

drinking increases occurred between grades 7 and 8 for boys and 8 and 9 for girls.

Experimentation with amphetamines, hallucinogens, barbiturates, cocaine, and opiates is relatively widespread, but regular use of illicit drugs is more limited (Green, 1979). Johnston et al. (1979) reported that less than 1% of seniors admitted daily illicit drug use other than marijuana. Stimulant use rates are second to marijuana, and surveys indicate increased experimental and current use of opiates and cocaine. Johnston et al. (1979) found a rise in cocaine use from 9% in 1975 to 15% in 1979, but McCoy, McBride, Ruse, Page, and Clayton (1979) described a trend for the numbers of heroin addicts in the United States to stabilize from 1974 to 1977 at approximately 550,000. However, opiates (excluding heroin) have become popular additions to the repertoire of the multiple-drug user, and obvious factors mask identification of both cocaine and opiate regular users, that is, nonrepresentation in school data, reluctance to self-report.

Age, Gender, Ethnicity

Prevalence and regular use rates vary depending upon user age, gender, and ethnicity. Problems, such as nonrepresentative time of data gathering, inaccurate measurements of drug or alcohol amount and frequency, difficulties in estimating intoxication, gender differences in self-disclosing, and lack of attention to body weight and time for drug consumption, render generalizations from even the best survey studies tenuous. Yet most researchers find drug abusers older than nonabusers, increased use among older groups, and relationships between drug type and ages at first and regular use. Initial drug contacts, except marijuana, alcohol, and cigarettes, typically occur during the last 3 years of high school, whereas the initial use rate for alcohol is 56% prior to high school and 30% for marijuana (Johnston et al., 1979). Prevalence estimates of marijuana use were cited by Fishburne et al. (1980) as 8% of 12- to 13-year-olds, 32% of 14- to 15-year-olds, 51% of 16- to 17-year-olds, and 68% of 18- to 25-year-olds, and Donovan and Jessor (1978) found the percentages of students admitting problem drinking increased with grade progression.

Early 1970s high school surveys often indicated greater drug experimentation and involvement among boys than girls (Hager, Vener, & Stewart, 1971; Matchett, 1971; Smart & Fejer, 1972), but more recent results suggest attenuation of gender differences (Beschner & Treasure, 1979; Smart & Blair, 1980; Wechsler & McFadden, 1976). Findings are not entirely consistent, and differences are related to age and extent or type of drug use. For example, Bowker (1974) found 21% more male than female youths were described as highly involved in drug use, and Johnston et al. (1979) reported that 9.6% of senior boys admitted daily drinking compared to 4% of girls. Gender differences in alcohol consumption patterns have been confirmed by Rosenbluth, Nathan, and Lawson (1978) who observed beer drinking in a bar setting and found that among 18- to 22-year-old college students, men consumed more alcohol and drank at faster rates than women, whether in groups or dyads.

Gender differences in drug use patterns are not always apparent, depending on the subjects sampled. Johnston et al. (1979) found overall marijuana use higher among male seniors, and more boys (12.7%) admitted daily use than girls (7.3%). Annual prevalence estimates for inhalants, cocaine, hallucinogens, and heroin were also found to be about 2 times as high for male as female students, but rates were about equal for stimulants and tranquilizers. Among chronic illicit drug users in residential treatment, women reported similar ages at first drug used, frequency of drug use, and drug preferences as did their male peers (Sutker, Archer, & Allain, 1978), and Klinge, Vaziri, and Lennox (1976) found no significant differences in use preference, duration, frequency, or patterns among male and female adolescent psychiatric inpatients. Gender differences in drug use sources, activities to acquire drugs, and other drug-related phenomena are reviewed elsewhere (Sutker, in press).

Among adults, ethnicity has been associated with specific use patterns, with chronic drug-abusing blacks reporting use of fewer drug categories, showing preference for depressants over stimulants, and engaging in drug use at older ages than whites similar in education, intelligence, and background (Sutker et al., 1978). However, Green (1979) indicated that traditional black–white differences in drug use may be diminishing. Dembo, Burgos, Des Jarlais, and Schmeidler (1979) explored use of seven drug classes among black, white, and Puerto Rican junior high schoolers in New York city. With controls imposed for socioeconomic status, they found no drug involvement differences which could be explained simply by ethnicity. O'Donnell, Voss, Clayton, Slatin, and Room (1976) found similar marijuana prevalence rates among blacks and whites born in 1953–1954, with whites reporting higher rates for heroin, stimulants, psychedelics, and sedatives and becoming more similar to blacks in current use rates for cocaine and opiates.

Few researchers have explored native American adolescent-drinking practices, and generalizations across Indian tribes are problematic. Waddell and Everett (1980) described drinking behaviors among Southwestern Indians, and Oetting and Goldstein (1979) reviewed person, peer, family, and cultural factors associated with adolescent native American drinking. Topper (1980) studied drinking patterns among adolescent Navajos living on the reservation and found that drinker age and gender and the drinking context were important determinants of behavior associated with drinking and conceptualizations about drinking. Heavy consumption was seen as normal for young Navajo men, and alcoholism was defined as drinking alone. Other researchers compared junior and high school Indian and white adolescent drug patterns and found Indian youth more likely to experiment with alcohol, marijuana, and some illicit drugs but perhaps no more inclined to continue use (Cockerham, 1977; Oetting & Goldstein, 1979).

FACTORS ASSOCIATED WITH DRUG BEHAVIORS

Studies of drug phenomena are plagued by impreciseness, derived from imperfect samples, and often characterized by singular perspectives. Methodological flaws are striking in most cross-sectional and longitudinal work, and cross-sectional designs are particularly limited for testing cause–effect relationships. More often than not, social scientists have focused predominantly on variables inherent in person or environment systems, prompting critical reviewers to elaborate on the weaknesses of univariate models (Braucht, Kirby, & Berry, 1978; Gorsuch & Butler, 1976; Nathan & Lansky, 1978). Suffice to say, the literature supports sophisticated explanations of initial and sustained drug use. Complex person systems, described in terms of biological (including genetic) and psychological characteristics, are suggested to interact with the social and physical environment systems in which behavior unfolds. Explanatory positions also take into account behavior systems, or relationships between drug-related and nonrelated activities. Although causative links have not been established firmly between predisposing variables and initial or continued use, the literature is rich in offering hypotheses for testing (see Kandel, 1978; Sutker & Archer, in press). Having briefly summarized epidemiological work, this chapter reviews factors thought to be etiologically significant with an attempt to maintain an interactionist, multidisciplinary perspective.

Factors Associated with Person Systems

Much of the research exploring person characteristics among drug abuser samples is built on person deficit assumptions derived from personological trait theories. While Fiske (1979) cautioned that the study of persons may never attain scientific status, Mischel (1979) expressed optimism regarding the potential usefulness of increased integration of cognitive and personological constructs in the study of persons. Despite discrepant viewpoints regarding the worth of person variables and their investigation, we will review some of the factors related to person systems which have been implicated theoretically to explain drug behavior etiology and progression. Our purpose is to discuss both cognitive and more traditional personality concepts, such as motives and depression, and to suggest ways in which they may contribute to explanatory perspectives of drug behaviors.

Attitudes Toward Drug Behaviors. Adolescent drug behaviors are influenced by personal beliefs and values regarding drug experimentation and use, although attitudes and behaviors are not necessarily convergent. Attitudes toward use are affected by use experience and expectations, and a complex interaction has been found among other person traits, social support systems, drug availability, other environment factors, perceived use functions, and attitudes toward use per se (Sadava & Forsyth, 1977a,b). For example, Kohn and Annis (1978) interpreted their path-analysis of high school data to support a model which assumes that attitudes affect use directly but that use is affected by perceived use functions and other factors. Kandel et al. (1978) suggested that adolescents who hold favorable attitudes toward marijuana are more likely to initiate its use, and adolescents who start using marijuana are more often characterized by anticonformist ideology and minor delinquency. Peer influences and beliefs favorable to marijuana were the strongest predictors of marijuana initiation, whereas poor relations with parents, personal deviance, and exposure to drug-using parents and peers were more important for other illicit drug initiation.

Little is known about relationships between attitudes toward drugs other than marijuana and their use. However, Kandel (1978) suggested that drug use initiation was preceded by beliefs that use of the specific drug was not harmful. Johnston et al. (1979) questioned seniors about the harmfulness and illegality of drug use and found that substantial numbers perceived regular illicit use as harmful; that is, 88% for heroin, 82% for LSD, and 70% for amphetamines, barbiturates, and cocaine. Only 42% believed that regular marijuana use represented risk, and significant numbers judged experimental or occasional use of other illicit drugs as nonrisky. Though regular illicit use was disapproved by most seniors, most seniors indicated that private illicit drug use should be decriminalized. Preble and Casey (1969) suggested that attitudes and values regarding highly illicit drug use are closely tied to the drug subculture for young male heroin addicts.

Drug Use Motives. Motives for drug use are a complex function of such factors as pharmacological or perceived effects, drug-taking set characteristics, person states and traits, and drug use experiences. Initial use reasons differ from those for sustained use, and use motives cannot be assumed to be consistent across situations for specific drug types or within individuals. Literature is accumulating which describes drug effects within the laboratory situation with the aim of understanding use motives, expected consequences, and subsequent emotional and cognitive changes. For example, it has been hypothesized that alcohol is consumed for its tension-reducing properties, but studies show that response to alcohol among young adults is a complex function of expectancy, dose, set, gender, modeling, and other factors (Abrams & Wilson, 1979; Dericco & Garlington, 1977; Higgins & Marlatt, 1973, 1975; Lied & Marlatt, 1979; Sutker, Allain, Brantley, & Randall, 1981; Wilson & Abrams, 1977).

Naditch (1975) studied relationships among psychopathology, acute adverse marijuana and LSD reactions, and use motives in predominantly white, male college students and examined three primary initial use reasons—curiosity or pleasure, reluctant use from peer pressure, and self-medication. Pleasure and self-medication motives were associated with marijuana use, but motives were reversed in order for LSD. Motives were also related to drug type by Bowker (1977) who found that among seventh to twelfth graders, college students, and adults physical motives for satisfaction and relaxation prompted use of such drugs as aspirin, narcotics, and tranquilizers, whereas intellectual and adventure/curiosity motives were linked to amphetamine and hallucinogen use. Bowker's results were qualified by age differences, and youths admitted greater use to satisfy adventure/curiosity motives than adults. Ethnicity and gender may also be related to specific use motives. For example, men may be more likely to use alcohol for "coping" and women to associate coping motives with psychotherapeutics (Parry, Cisin, Balter, Mellinger, & Manheimer, 1974). Reasons for initiating opiate use are generally similar for men and women (Chein, Gerald, Lee, & Rosenfeld, 1964; Ellinwood, Smith, & Vaillant, 1966), and Sutker et al. (1978) found neither gender nor ethnicity was associated with reason for first drug, alcohol, or opiate use among chronic illicit drug users.

Motives may be related to adverse outcomes, use progression, and other person characteristics. Researchers have found an association between motives for self-perception alteration or therapeutic intent and unfavorable use outcomes (Carman, 1979; Naditch, 1975). Summarizing data on marijuana use and nonuse motives, Weinstein (1978) suggested associations between "drug-related" meanings for initial use or nonuse and "person-oriented" meanings for continued use. He concluded that initial use was justified by curiosity and nonuse motivated by avoidance of legal problems, physical injury, addiction possibility, and moral dilemmas. Habitual use was most often prompted by needs for self-fulfillment, pleasure, emotional relief, and social acceptance. Kandel et al. (1978) found that reasons for marijuana use which suggested person effects, that is, increase self-understanding, reduce depression, were related to subsequent use of other illicit drugs, and Cutter and Fisher (1980) pointed to relationships between personal drinking motives, parental attitudes toward drinking, and maternal intoxication and self-esteem. Though intriguing, the study of use motives across situations and drug types is yet in the early stages.

State and Trait Characteristics. Whether cross-situation person characteristics exist or can be measured has been argued (Bem & Allen, 1974; Bem & Funder, 1978; Mischel, 1977). This issue aside, several person constructs, other than the cognitive ones mentioned above, have been related to drug behaviors. Studies of person states and traits applicable to adolescent drug use have typically employed cross-sectional or longitudinal measurement of characteristics reflecting general personality functioning. Although assumptions regarding personality deviance and psychopathology are often contrived from such work, there are precious little data which may be used to predict drug initiation and frequent use from specific measures of psychopathology. Researchers have concentrated on concepts related to self-esteem, social nonconformity, sensation seeking, and alienation using indices of personality functioning. Making a transition to characteristics which may fall within ranges defined as psychopathology, investigators have hypothesized that depression, interpersonal maladjustment, sociopathy, and other disorders are related to drug behaviors.

Whether or not low levels of self-esteem lead to multiple interpersonal problems and drug behaviors specifically has been a focus for theory and research. Bromley (1977) hypothesized that self-system dispositional features, that is, beliefs, aspirations, values, are relatively stable and give rise to consistent, regular adjustment strategies. Others suggested that interpersonal events mold self-esteem or that the causal direction is from interactions to self-esteem. Looking at relationships between self-esteem and interpersonal problems among boys in the sophomore, junior, and senior high school years, Kahle, Kulka, and Klingel (1980) found support for the position that low self-esteem precedes interpersonal problems. Although path-analysis and panel studies lend themselves readily to testing self-esteem hypotheses, most studies attempting to identify self-system predictors or correlates of adolescent drug use have been less than adequate and yielded conflicting results.

Haagen (cited in Kandel, 1978) found that marijuana use rates among college juniors were related to scores on measures of personality traits administered at freshman orientation. Low self-esteem characterized students who reported frequent use in their junior year, but use extent at the time of personality testing may be questioned. Ahlgren and Norem-Hebeisen (1979) compared drug abuse groups with dysfunctional institutionalized and control high school aged youths and concluded that results were consistent with the view that low self-esteem preceded drug use. Others have not confirmed such findings (McKenna-Hartung, Hartung, & Baxter, 1971; O'Malley, 1975; Stokes, 1974). Associating socioemotional reasons for drinking with perceptions of parent–parent distance within a small sample of undergraduates, Cutter and Fisher (1980) suggested that family conditions leading to poor self-esteem necessarily but not sufficiently preceded problem drinking.

Gorsuch and Butler (1976) concluded that drug abusers are more behaviorally deviant than their nondrug-using peers; however, the extent to which cognitive or behavioral nonconformity acts as precursor to drug experimentation, chronic use, or use by drug type is unclear. Huba, Wingard, and Bentler (1979) studied interaction patterns among seventh- to ninth-grade drug users, their peers, and significant adults and relationships with self-reported drug use. Social deviance indicators were not related to use or social interaction dimensions, and Huba et al. (1979) suggested that among youths who initiated drug use, specific interpersonal nets were not intrinsically linked to maladaptive, antiachievement patterns. Yet Sadava and Forsyth (1977b) found that greater tolerance for deviance, combined with social support and reduced negative use functions, predicted change toward stable marijuana use among undergraduates. Most studies show nondrug users as more conventional than users of alcohol (Moos, Moos, & Kulik, 1977; Potvin & Lee, 1980), marijuana (Brook, Lukoff, & Whiteman, 1977; Crain, Ertel, & Gorman, 1975; Hochman & Brill, 1973; Hogan, Mankin, Conway, & Fox, 1970), and LSD (Khavari, Mabry, & Humes, 1977). Work is required to determine the relative and interactive contributions of nonconforming cognitive strategies and behavioral deviance in influencing drug behaviors, time of use initiation, and use motives and extent.

Sensation seeking is a person trait frequently cited as related to drug behaviors, although investigators have questioned whether sensation seeking, at least as measured by Zuckerman's revised (1978) Sensation-Seeking Scale (SSS), reliably reflects an enduring disposition (Ridgeway & Russell, 1980). Zuckerman (1972) showed that SSS scores were correlated with higher alcohol consumption among college students—a finding replicated by Schwarz, Burkhart, and Green (1978). Other studies linked SSS scores with public drunkenness arrest chronicity (Malatesta, Sutker, & Treiber, 1981) and drug use in college students (Segal, 1975; Zuckerman, Bone, Neary, Mangelsdorff, & Brustman, 1972), adult workers (Khavari et al., 1977), and hospitalized veterans (Kilpatrick, Sutker, Roitzsch, & Miller, 1976). Carrol and Zuckerman (1977) found that SSS scores were related to drug choice and use patterns among young adult drug abuse patients. Similarly, Sutker et al. (1978) showed that sensation-seeking levels were related to time of use initiation, variation in use, and use motives. For example, low sensation seekers remembered peer influence motives for first alcohol use as opposed to high sensation seekers who cited pleasure and curiosity (Sutker et al., 1978).

Using the California Psychological Inventory, Hogan et al. (1970) found college marijuana users described as socially poised, open to experience, and concerned with others' feelings as well as impulsive, pleasure seeking, and rebellious. Nonusers were characterized as responsible, rule abiding, inflexible, conventional, and narrow in interests—descriptions similar to those reported by Green and Haymes (1973). Looking at pleasure-hedonistic motives among undergraduates, Stokes (1974) found the sensual-hedonistic factor related to use of tobacco, alcohol, marijuana, hallucinogens, prescription drugs, and nonprescription tranquilizers and stimulants but not to coffee, tea, or nonprescription sedatives and analgesics. Hochman and Brill (1973) reported that undergraduate chronic marijuana users showed acculturation into a relativistic, gratification-oriented, stimulus-seeking value system but no evidence of deteriorated functioning or adaptive style. Khavari et al. (1977) also found that psychedelic use, depending on type, was associated with needs to seek out new and unconventional experiences but that such variables were not equally suited to predict use across drug types. Since the bulk of findings are descriptive of drug use–sensation-seeking relationships among young adults of 18 to 25 years, their applicability to early adolescence must be tested.

Researchers have hypothesized that social alienation, or a sense of isolation from major social systems, predicts marijuana and other illicit drug use. Jessor, Young,

Young, and Tesi (1970) showed that alienation and personality attributes reflecting frustration, dissatisfaction, and powerlessness were related to amount of alcohol intake, frequency of drunkenness, and person effect functions among American, but not Italian, male youths. Jessor and Jessor (1978) also reported that prospective marijuana users scored high on personal alienation, and Horman (1973) showed that alienation was related to nonmedical drug use among liberal arts students. Oetting and Goldstein (1979) found youthful native American drug users alienated from both Indian and white cultures but not from their peers. In contrast, Sadava and Forsyth (1977b) indicated that alienation contributed modestly, if at all, to predicting transitions in marijuana use. Other investigators have reported no relationships between drug use and alienation or anomie (O'Malley, 1975; Potvin & Lee, 1980).

For the most part, studies among adolescents have included measures of general personality functioning, and results may not be extended legitimately to derive assumptions about psychopathology. Among high school students, Kandel et al. (1978) found that more deviant personal characteristics predicted illicit drug use other than marijuana, and Paton and Kandel (1978) indicated that depressive mood was positively related to illicit multiple drug use among girls and whites and negatively associated among black and Puerto Rican boys. Oetting and Goldstein (1979) described drug use among Indian youths as associated with greater social–personal problems, and Crain et al. (1975) found college barbiturate abusers more emotionally disturbed on the Omnibus Personality Inventory than marijuana users or nonusers. In a unique study, Halikas, Goodwin, and Guze (1972) applied the psychiatric interview to ferret out relationships between psychopathology and marijuana use among samples of college-aged nondrug users and their marijuana using friends. Both groups showed a high incidence of psychopathology, though with different patterns. Marijuana users were more often diagnosed as sociopathic compared to their friends who showed a preponderance of affective disorders.

The Minnesota Multiphasic Personality Inventory (MMPI) has been used to characterize psychopathology among adult drug abusers (Sutker & Archer, 1979). With the exception of the Halikas et al. (1972) study, the work of Naditch (1975), and the MMPI investigations reviewed below, a paucity of research has been devoted to understanding patterns of psychopathology and relationships with adolescent drug behaviors. Naditch (1975) found that psychopathology was associated with particular kinds of drug use and use outcomes, or specifically, that psychological maladjustment and regression were predictive of LSD use even after the effects of use motives were partialled out. Unfortunately, Naditch's work and the bulk of related studies focused on late-adolescent and adult samples. For example, Smart and Jones (1970) compared the MMPI profiles of young adult groups of LSD users and matched controls and found significantly higher elevations among LSD users on scales 8, 9, 4, and 1, with 96% classified as abnormal compared to 46% of nonusers and a preponderance of character disorders (23%) as opposed to controls (2%).

Amphetamine abusers were found to produce significantly higher scores on scales F, 8, 7, 4, 2, and 9 than high school controls, suggesting to Brook, Szandorowska, and Whitehead (1976) that they were likely to experience impoverished interpersonal relationships, inadequacy feelings, and distrust of authority. However, comparisons between hospitalized amphetamine abusers ranging in age from

15 to 25 years and high school students seem unlikely to yield meaningful results. Two studies have compared adolescent, drug-abusing, psychiatric inpatients with similarly aged nondrug-abusing inpatients. Burke and Eichberg (1972) showed significant differences between male abusers and nonabusers on almost every MMPI scale but no differences for female groups, and Klinge, Lachar, Grisell, and Berman (1978) found highly similar MMPI profile elevations between gender-mixed groups of extensive users and nonusers from similar socioeconomic backgrounds. Comparing youthful inhalant users with polydrug noninhalant users and nondrug users admitted to a psychiatric emergency room, Korman, Trimboli, and Semler (1980) found evidence of more aggressive behaviors, greater social disruption, and a wider range of cognitive deficits among inhalant users. Obviously, there remains a need to explore the relationships among adolescent drug behaviors, early manifestations of psychopathology, and subsequent psychological adjustment.

Factors Associated with Environment Systems

Interpersonal interaction patterns influence adolescent drug behaviors, but the relative contributions of peer and family systems versus other aspects of the environment have not been carefully delineated. One task is, of course, to specify particular factors in the social environment and general cultural milieu which influence drug behaviors. To their credit, some investigators exploring these questions have shunned a simplistic approach and attempted to organize person, environment, and behavior systems with respect to their relative, interactive, and/or independent contributions (Huba et al., 1979; Jessor & Jessor, 1978; Kandel et al., 1978; Sadava & Forsyth, 1977a,b).

Peer Interactions and Influences. Peer group and friendship circles strongly influence adolescent drug behaviors, depending on drug type, user age, and degree of involvement. Work by Huba et al. (1979) showed that young drug-using adolescents associate with peers and adults who use similar drugs. Other cross-sectional data suggest that teenagers who drink are more likely to have drinking friends (Lassey & Carlson, 1980; Wechsler & Thum, 1973) and that there is a strong association between personal and perceived friends' use (Adler & Lotecka, 1973; Bowker, 1974; Oetting & Goldstein, 1979). Kandel and her colleagues (Kandel, 1978; Kandel et al., 1978) found that perceived peer drug use, self-reported peer use, and perceived peer tolerance for use were important for predicting use by drug type (particularly marijuana initiation) and that extensive drug discussions with friends preceded all types of illicit drug initiation. Although peer influences may attenuate during late adolescence or adulthood, Bowker (1974) found college drug users preoccupied with similar peer concerns as the younger samples studied by Kandel. Kandel (1978) also indicated that peers influence rates of increased marijuana use, and Green's (1979) review suggests that extensive drug use may be more influenced by personal and family factors than peer behaviors.

Within the past 5 years, laboratory experiments have demonstrated drug use modeling affects. Caudill and Marlatt (1975) found that heavy social drinkers exposed to heavy-drinking models drank significantly more alcohol than those exposed to low-consumption or no-model conditions. Peer model drinking rates also have been shown to influence subject drinking rates, regardless of instruction sets (Dericco

& Garlington, 1977; Garlington & Dericco, 1977). Reid (1978) found that heavy- or light-drinking male models significantly affected drinking rates of male subjects observed in a natural setting, and Lied and Marlatt (1979) reported that subjects exposed to heavy-drinking models consumed more than those exposed to light drinkers, with this effect more pronounced among heavy as opposed to light social drinkers (particularly men).

Family Interactions and Influences. Family influences cited as important to adolescent drug use are parent modeling of prescription and nonprescription use, parent values toward social conformity and use, and quality of parent–parent and parent–child relationships. Although many studies constitute simple user–nonuser comparisons, complex interrelationships among familial and peer variables have been assessed. Work can be cited in support of the notion that parent use modeling, whether cigarette, alcohol, prescription or other drug, encourages adolescent use (Adler & Lotecka, 1973; Lassey & Carlson, 1980; Smart & Fejer, 1972). There tends to be a positive association between parent use of psychoactives and alcohol (especially for both parents) and adolescent use, particularly for illicit drugs other than marijuana and alcohol (Kandel et al., 1978). Parent attitudes conveying approval for or ambivalence toward drinking and drug use are also related to higher adolescent use rates (Cahalan, Cisin, & Crossley, 1969; Kandel et al., 1978; Prendergast, 1974).

The likelihood of drinking or drug use may be less if adolescents perceive themselves as close to family members and loved by both parents (Lassey & Carlson, 1980; Streit, Halsted, & Pascale, 1974; Wechsler & Thum, 1973). Prendergast and Schafer (1974) suggested that, rather than parent attitudes toward drinking or drinking modeling, parent attitudes and behavior toward the adolescent as a person are related to drinking. Prendergast (1974) also characterized the family typology conducive to marijuana use as incorporating a well-educated, prescription drug-using father and a mother who was neither strictly controlling nor strongly disapproving of marijuana use. Drug abusers may also perceive their parents as more hostile, granting of autonomy, unloving, uncommunicative, and distant than nonusers' parents (Lassey & Carlson, 1980; Streit et al., 1974). Although it has been contended that drug abusers come from less intact families than nonusers, evidence here is less convincing (Kirk, 1979), particularly in light of increased rates of divorce and single parenting.

Chein and his colleagues (1964) described lack of emotional closeness and familial affection, hostility between parents, and a higher incidence of broken homes among heroin addict families. They described fathers of male addicts as immoral models in terms of criminal, unfaithful, and alcoholic behaviors and mothers as the most important parental figure for male addicts. Rathus, Fichner-Rathus, and Siegel (1977) found that young male drug abusers viewed their mothers as less honest, fair, and valuable and their fathers as less nice, honest, strong, and kind than did nonusers. Whether such descriptions show addict home life to be different from other groups with similar socioeconomic backgrounds is yet to be determined. However, evidence is growing that children who become chronic drug users often report an early history of physical and sexual assault, especially women (Densen-Gerber & Benward, 1976; Ellinwood et al., 1966), and at least two studies (Graeven & Schaef, 1978; Klinge et al., 1976) describe sex-specific differences in the impact of negative home factors on illicit drug behaviors. Hence it appears that modeling influences, qualified by sex-type considerations, are important for initial and less extensive drug use, whereas increased drug involvement may suggest greater family disturbance and varying impact, depending on user gender.

Situation and Context Factors. Some investigators speculate that neither peer nor parent influences constitute the best predictors of drug behaviors and point to aspects of the context, that is, drug availability, drinking place, life stress. If a drug is unavailable, problems regarding its use do not exist, and drug availability strongly influences consumption (Gorsuch & Butler, 1976; Huba et al., 1979; Sadava & Forsyth, 1977b). In many respects, the iatrogenic user has responded to drug availability in a context which necessitated at least temporary use. Emphasizing situation factors, Smart and Gray (1979) and Smart, Gray, and Bennett (1978) pointed to drinking away from home and in cars as highly related to adolescent drinking. Reporting on drinking practices among high schoolers, Smart et al. (1978) found that most students drink, but infrequently. Of those who are frequently drunk, they drink away from home and in cars, and their parents are unaware of their drinking patterns. Although these researchers suggested that parent modeling may be important at drinking outset, the place and extent of drinking were more important in predicting heavy drinking.

Animal self-administration studies demonstrate the intrinsically rewarding or euphoric effects of such drugs as opiates and cocaine even in the absence of negative stimulation (Woods, 1978). However, drinking and drug use have been hypothesized to represent responses to life stress and resulting intrapsychic anxiety. Gorsuch and Butler (1976) speculated that physicians self-administer drugs to reduce anxiety associated with stress (Vaillant, Sobowale, & McArthur, 1972), and female alcoholics have suggested that alcohol consumption was used to reduce discomfort coinciding with the premenstruum and menstruum (Belfer, Shader, Carroll, & Harmatz, 1971; Podolsky, 1963). Both examples illustrate application of drive reduction theory, or specifically, the tension-reduction hypothesis which incorporates two notions: individuals drink or use drugs to relieve tension and drugs provide that relief. Using measures of work and life-change stress, perceived alcohol use functions, and estimates of quantity/frequency and drunkenness, Sadava, Thistle, and Forsyth (1978) found that increased alcohol use was associated with more positive and less negative use functions with little evidence of stress-related drinking. Conversely, in a study with several methodological limitations, Duncan (1977) found that drug-dependent adolescents remembered more life stress events during the year prior to initial illicit drug use than those generated by junior and high school normative groups. De Leon and Wexler (1973) also presented a rationale explaining opiate use as a means of coping with anxiety associated with sexual expression.

Adolescent drug-using subcultures and the societal whole represent sources of influence for drug behaviors and vice versa. In a classic application of life history interviews to study the lives and activities of New York City heroin addicts, Preble and Casey (1969) refuted the notion that adolescents use illicit drugs to escape stress. They presented heroin users as daily engaged in meaningful activities and relationships which are reinforced and maintained by the addict subculture. The meaning of heroin use then is provided by interactions within the subculture itself and lies in gratification of accomplishing a never-ending series of exciting tasks. This rationale is also presented in the heroin addict toast "Honky-Tonk Bud" interpreted by Agar (1971). To appreciate the influence of the larger sociocultural milieu, one need only reflect upon the sweeping counterculture adolescent drug epidemic of the late 1960s. During this time, even the children of affluent parents found themselves invested in illicit drugs—a phenomenon which has led to greater tolerance for marijuana use and increased discussion of its decriminalization.

Behavior Systems and Drug Use

Fiske (1979) reminded psychologists that we are about the study of behavior and relationships between well-defined, discrete behaviors at that. He encouraged behavioral scientists to study phenomena within controlled laboratory settings or naturally in the field after the ethologists' example. Although the complications involved in field study of alcohol and drug behaviors may not be insurmountable (Rosenbluth et al., 1978; Sutker, 1974), social scientists have relied predominantly on subject reports of drug and related behaviors (sometimes with verification from other sources) and laboratory work to specify behavior–behavior relationships. Albeit there are problems with self-report data and artificial experimental situations, the emphasis on a range of discrete behaviors is important for drug research, in that use and abuse occur within an overall context which incorporates classes of activities with little apparent but perhaps relevant relation to drug use.

Hundleby made a strong case for studying behavior–behavior relationships and drug abuse phenomena among high school students.[2] Using samples of approximately 1000 boys and girls, he intercorrelated and factored 43 behavior items within gender groups and derived seven areas: social, outdoor, religious, home related, athletic, sexual, and delinquent. Delinquent and sexual behaviors were positively correlated with drug use, but scholastic and religious behaviors were negatively related. A strong association was found between social behaviors and drug use with few gender differences. Jessor and Jessor (1977) also reported positive relationships between political activism, marijuana use, sexual behavior, alcohol consumption, and general deviance and negative associations between these behavior sets and religious and school performance activities. The several behavior systems reviewed below have been explored in relation to drug use, with focus directed toward school attendance and performance, religious involvement, deviant or delinquent acts, and various kinds of drug use.

School Attendance and Performance. Cohen and Santo (1979) offered a synergistic explanation that drug abuse may lead to educational failure, education system deficiencies and academic failure may predispose to drug use and abuse, and the two in concert may be worse than either alone. Poor school attendance and performance are easily confounded, yet considerable evidence suggests that low values and expectations for achievement predict marijuana use (Jessor & Jessor, 1978). Kandel (1975) found that multiple users of illicit drugs or single drug users performed more poorly in school than marijuana-only users or nonusers, and Johnston

---

[2] Hundleby, J. D. Using other behaviors in the study of adolescent drug usage. In R. P. Schlegel (Chair), Social psychological aspects of nonmedical drug use. Symposium presented at the meeting of the American Psychological Association, August 31, 1978.

et al. (1979) reported that college-bound seniors admitted less drug use than non-college-bound youths. Although what has been described as the amotivational syndrome may impact here, research has not identified the point at which extent of marijuana use interferes with high school or college achievements and grade point averages (Brill & Christie, 1974; Hochman & Brill, 1973). Studying relationships between academic performance and drug use patterns in college men, Mellinger, Somers, Brazell, and Manheimer (1978) found no effects among moderate users but adverse outcomes among small groups of multiple drug users which could not be explained by prior characteristics. Students using illicit drugs, such as heroin and cocaine, with regularity may fail to complete high school and are infrequently included in survey studies (McCoy et al., 1979).

Religious Involvement. Survey data suggest that adolescent alcohol and drug users are less religious than their nonusing counterparts, and church attendance and religion meaningfulness predict illicit drug nonuse in cross-sectional and panel studies (Gorsuch & Butler, 1976; Moos et al., 1977). Some researchers have found that denomination preference was unrelated to use (Green, 1979), and others reported conflicting results. In Rohrbaugh and Jessor's (1975) study, religious fundamentalism did not account for the hypothesized controlling effect of religiosity on deviant behaviors, but Schlegel and Sanborn (1979) showed that fundamentalist Protestant high school students were less likely to drink than liberal Protestants or Roman Catholics. Oetting and Goldstein (1979) found less drug involvement among native American adolescents who attended religious ceremonies and perceived themselves as religious. Thus regardless of affiliation, church-attending adolescents seem less likely to use alcohol and drugs.

Delinquent and Deviant Behaviors. Illicit drug use is positively and strongly related to other forms of illegal behavior, and delinquency typically precedes drug use (Johnston, O'Malley, & Eveland, 1978). Adolescent delinquency also predicts adult antisocial involvement and drug dependence (Robins, 1966; Vaillant, 1966). Looking to preadolescence, Nylander (1979) followed up children requiring mental health treatment over 10 to 20 years and found that antisocial behaviors predictive of later drug use were noted as early as 9 years of age. Egger, Webb, and Reynolds (1978) also showed that a general delinquency factor and family background factors accounted for most of the variance in data derived from a retrospective comparison of 200 opiate addicts in treatment and 200 matched controls. Groups did not differ in reports of early unhappiness, feelings of inadequacy, or enjoyment of sports activities, and delinquent and drug behaviors were evident as predictors as early as 12 years. Similarly, Oetting and Goldstein (1979) found that Indian drug-using students were more likely to have talked back to teachers, cheated on tests, or purposefully damaged property than nonusers who espoused more conservative values.

One type of drug behavior increases the probability of another, and use is a necessary antecedent to abuse. Kandel and Faust (1975) pointed to a sequence of at least four drug involvement stages: beer or wine; cigarettes or hard liquor; marijuana; and other illicit drugs. The two stages of legal use were considered necessary intermediates between nonuse and marijuana, and few youths were seen to progress to other illicit use without marijuana experience. Use intensity at prior stage may also be related to use of a higher-ranked drug or development of drug complications. For example, Fillmore (1974) reported that quantity estimates of adolescent drinking were predictive of later problem drinking among men but not necessarily women. Kandel et al. (1978) studied the antecedents of entry in three sequential drug use stages, that is, hard liquor, marijuana, and other drugs. Previous minor delinquency and cigarette, beer, and wine use were important for later liquor use, while personal beliefs and values favorable to marijuana and association with marijuana-using peers were strongest marijuana use predictors. Other variables aside, one type of deviance seems to lead to another with a complex interplay of person–environment and behavior factors. The extent to which cognitive nonconformity or tolerance for deviance exists apart from delinquency or drug behaviors is a matter of interest and may spark inquisitive minds to research.

Other Behaviors. Several researchers have specified some of the nondeviant behaviors associated with drug use. The work of Hundleby (Note 2) is an example, and Egger et al. (1978) showed that involvement in sports activities and social interactions was positively associated with drug use. These findings illustrate the necessity of avoiding a negative set in evaluating the person antecedents of drug use. Indeed, most of the work predicting later drug use from early signs derives from retrospective behavioral evaluations among children or adolescents referred for therapeutic intervention (Nylander, 1979; Robins, 1966). In a unique study, Kellam, Ensminger, and Simon (1980) identified characteristics of first-grade black students which were related to alcohol and drug abuse 10 years later. Higher scores on intelligence measures were associated with alcohol or drug use, and antecedents were clearer for boys than girls. Shy students were found to use drugs least often in later years, while students described as aggressive in the first grade used drugs most often later.

Multivariate Descriptions and Explanations

The findings of investigators applying a multivariate approach have been integrated in this chapter, but the work of two groups requires special mention. Jessor and his colleagues extended application of a sociopsychological theory of youthful problem behavior and deviance to longitudinal investigation of drug use among high school and college students. Three important systems were conceptualized for study, or that of personality, perceived environment, and behavior. Incorporating descriptive, predictive, and associative evidence, these investigators suggested that problem-behavior theory as an explanatory framework for predicting variation in marijuana use received considerable support through finding replication and convergence. In this context, they described adolescents likely to engage in marijuana use as more concerned with personal autonomy, less interested in conventional institutional goals, jaundiced in their views of society, and more tolerant of trangressions. Such youths perceived less parent support, less compatibility between parent and friend expectations, greater influence of friends relative to parents, and greater support of and models for drug use.

Extending this approach, Sadava and Forsyth (1977a) proposed a field-theoretical model of person–social–environment systems for predicting marijuana use. Distinctions were made between proximal and distal variables in person and environment systems and among four drug use criteria, for example, frequency, polydrug, time as user, and adverse drug outcomes. In this scheme, marijuana users showed a use-prone pattern different from nonusers with contributions from the proximal environment, that is, high support for use, drug availability, parent and peer use models, and distal environment, that is, older age, lower grade point averages. Such proximal person characteristics as high positive functions for use and distal person characteristics as high values for independence, peer conformity, and low delay of gratification were among the variables found to be predictive. From personological considerations, these multivariate approaches could be made more useful by expanded focus on person traits and states, as well as health variables.

SUMMARY STATEMENT AND IMPLICATIONS

Recent findings show that drug use is common on junior high, high school, and college campuses among male and female students. The average age of drug initiation is young, particularly for alcohol, cigarettes, and marijuana, and differences in drug behaviors related to gender, ethnicity, and geography may well be attenuating. Hunt, Farley, and Hunt (1979) described periods of new drug use as sporadic and sudden for entire school populations and suggested that these may be followed by minimal new use. However, there is no indication that adolescent drug use will decline, and in fact, evidence points to the contrary. For example, alcohol consumption and cigarette smoking are licit substances. The sociocultural context of American society has also become more permissive toward certain types of illicit drug use evidenced by the movement for decriminalization of marijuana use. Further, most youths do not view drug use as problematic and may endorse illicit drug use if privacy is maintained. Although drug use may always be with us in increasing proportions, the critical question is whether abuse phenomena will continue to escalate.

Several research groups have contributed to our knowledge of the developmental stages of youth drug use and their correlates and antecedents. Less is known about the psychological and biological consequences of drug use initiated in adolescence or drug-behavior sequencing through the life stages of men and women. Contributions have emerged from laboratory investigations of drinking phenomena, naturalistic studies of drinking practices or stimuli associated with illicit drug use, and cross-sectional descriptions of person or environment characteristics. Yet there is a need for expanded, multivariate, multimethod research executed with sound theoretical underpinnings in naturalistic and laboratory settings. Although innumerable studies may be conceived, the deficiency in prospective work among children to identify early precursors to drug behaviors is striking. Without prospective investigation of children across the variety of home situations represented in American society, there will be incomplete understanding of drug use origins. To cut across populations of high school and college students is but a beginning. Longitudinal work is required which combines household sampling and multimethod, multitrait strategies with multidisciplinary exploration of individual, family, and cultural variables.

Users of illicit drugs other than marijuana represent a target for treatment and judicial systems, but few investigators have applied prospective approaches to uncover the antecedents of illicit drug involvement or attempted to span the gap between experimental or frequent use and the maladaptive patterns of multiple or chronic illicit drug users. For example, there is little overlap between the populations represented by the work of Kandel et al. (1978) and the heroin addicts described by Preble and Casey (1969). Illicit drug users may be found in the households of highly populated American cities or remote outposts, but they are often absent or expelled from school—hence underrepresented in most of the panel work described. Conversely, high school or college students are infrequently sampled among treatment patients or juvenile offenders. Little is known about those individuals who fall intermediate to the groups most often studied or about the interim stages between infrequent or manageable use and socially or physically handicapping drug involvement.

Drug behaviors and abuse phenomena may be seen as highly complex, learned patterns which are acquired and maintained by a complex interaction of physiological, psychological, and social factors. Aside from impacting on possible genetic predispositions to drug behaviors or biochemical aberrations, prevention and treatment strategies must target on person factors which precede or mediate drug use, distal or proximal environmental antecedents, drug or other behaviors important in use progression, and drug use outcomes and consequences. Turning to prevention, the most widely used methods have been education programs emphasizing factual information on drug effects and consequences, "rap" sessions, and encounter/sensitivity groups. Dembo and Miran (1976) identified three prevention thrusts, including focus on moral considerations, scientific evidence, and alternative ways to gain satisfaction, and they distinguished education or information programs from those designed to encourage development of interpersonal skills and values for healthy living. Prevention programs have also been discussed by Braucht, Follingstad, Brakarsh, and Berry (1973).

From a behavioral standpoint, prevention is conceptualized as modification of the person–social–environment antecedents to drug use to lessen or obviate their influence. With emphasis on personal responsibility, rather than irreversible disease processes, this view is compatible with efforts to impact on any of the demonstrated antecedents to drug behaviors; that is, parent modeling, depressive cognitions, peer pressure, drug availability, media advertisement, delinquency. For example, youths are experimenting with drugs at earlier ages; hence primary prevention efforts must be undertaken to influence childhood cognitions, emotions, and behavior. Work to postpone drug experimentation and use will be at least partially effective, in that one drug behavior leads to another in progression and later age of start predicts better chances for remission. From current findings, we know it is possible to modify age of first cigarette or alcohol use by affecting parent attitudes, discipline practices, and modeling. Education and training among young parents may be as important as requiring family participation in school immunization programs.

A good example of prevention programming for young adults is that of Miller (1979) who emphasized a general push toward healthy living. Based on the notion that young adults represent an ideal target for teaching responsible drinking, Miller's program employs performance feedback, behavior rehearsal, and self-management to ensure moderate, appropriate drinking. Developed for problem drinkers, the Foy, Miller, Eisler, and O'Toole (1976) drink refusal training may also have extended application to adolescents and their parents. Personal involvement prevention strategies seem to be viewed more favorably by adolescents, and Dembo and Miran (1976) found that with the exception of "rap" sessions and ex-addict talks (which may exert paradoxical effects), youths were unimpressed with school drug abuse prevention programs.

Insufficient attention has been given to identification of children or adolescents with social–personal problems and to tailoring treatment efforts for youthful patients. Expulsions and suspensions are ineffective mechanisms for managing unacceptable behaviors, and once drug users are excluded from the supervision and involvement of the school environment, they may be lost until treatment is required. Few ready-made treatment alternatives are available for youth, and most programs focus on adult alcoholics and drug abusers. More often than not, adolescents identified as chronic drug abusers are apprehended by law enforcement officials and remanded to training institutions or adolescent prisons—an unhappy outcome facilitative of future criminal and drug behaviors (Winslow, 1976). Such institutions often constitute a setting which combines drug availability, peer pressure, lack of stimulation, and other incentives to expanded drug experimentation, that is, opiates.

Smith, Levy, and Striar (1979) recently pointed to the limited suitability of available treatment services and environments for adolescent drug users. They discussed a range of treatment services which varied from traditional individual and group counseling, specialized chemotherapy, acupuncture, and meditation options, alternative approaches such as free clinics, residential programs including therapeutic communities and halfway houses, and school projects. One example is the treatment offered by Odyssey House, an international drug abuse organization, in which adolescents and adults may be integrated within therapeutic communities with special provisions for adolescent programming or adolescents are housed in communities designed for youths alone (Rohrs, Murphy, Goldsmith, & Densen-Gerber, 1972).

This summary only scratches the surface in describing drug phenomena origins, expression, and consequences. Additional work is needed to understand, predict, and control drug behaviors, and prevention and treatment resources must be tailored to youthful needs. Whether such programs are implemented successfully depends upon societal and governmental response, but even the most promising efforts are not applied in a vacuum. Children and adolescents are exposed to parent, peer, and media use models as well as drug sanctions based largely on appeal to authority regarding consumption and use consequences. Research has not substantiated notions that legal drugs are less harmful physiologically than those defined as illicit, and the push for decriminalization of chemicals in vogue with more affluent youths reflects insensitivity to use patterns and consequences among the disadvantaged and poor. In our society, relatively mild penalties are associated with driving under intoxicant influence compared to possession of controlled substances, and opiate self-administration is judged criminal but drunkenness pictured with levity. Surely American youths are not oblivious to such arbitrary values and discrepant behaviors. Therefore, we must use what we know about drug phenomena to evaluate and modify definitions of use legitimacy, related enforcement efforts, and prevention and treatment strategies so that we can impact more meaningfully on individual and collective drug behaviors. In the final analysis, progress in impacting on drug behaviors can never transcend and, in fact, only reflects the thoughts and activities of most people.

REFERENCES

Abrams, D. B., & Wilson, G. T. Effects of alcohol and social anxiety in women: Cognitive versus physiological processes. *Journal of Abnormal Psychology,* 1979, **88,** 161–173.

Adler, P. T., & Lotecka, L. Drug use among high school students: Patterns and correlates. *International Journal of the Addictions,* 1973, **8,** 537–548.

Agar, M. H. Folklore of the heroin addict: Two examples. *Journal of American Folklore,* 1971, **84,** 175–185.

Ahlgren, A., & Norem-Hebeisen, A. A. Self-esteem patterns distinctive of groups of drug abusing and other dysfunctional adolescents. *International Journal of the Addictions,* 1979, **14,** 759–777.

Belfer, M. L., Shader, R. I., Carroll, M., & Harmatz, J. S. Alcoholism in women. *Archives of General Psychiatry,* 1971, **25,** 540–544.

Bem, D. J., & Allen, A. On predicting some of the people some of the time: The search for cross-situational consistencies in behavior. *Psychological Review,* 1974, **81,** 506–520.

Bem, D. J., & Funder, D. C. Predicting more of the people more of the time: Assessing the personality of situations. *Psychological Review,* 1978, **85,** 485–501.

Beschner, G. M., & Friedman, A. S. *Youth drug abuse: Problems, issues, and treatment.* Lexington, Mass.: Heath, 1979.

Beschner, G. M., & Treasure, K. G. Female adolescent drug use. In G. M. Beschner & A. S. Friedman (Eds.), *Youth drug abuse: Problems, issues, and treatment.* Lexington, Mass.: Heath, 1979.

Bowker, L. H. Student drug use and the perceived peer drug environment. *International Journal of the Addictions,* 1974, **9,** 851–861.

Bowker, L. H. Motives for drug use: An application of Cohen's typology. *International Journal of the Addictions,* 1977, **12,** 983–991.

Braucht, G. N., Follingstad, D., Brakarsh, D., & Berry, K. L. Drug education: A review of goals, approaches, and effectiveness, and a paradigm for evaluation. *Quarterly Journal of Studies on Alcohol,* 1973, **34,** 1279–1292.

Braucht, G. N., Kirby, M. W., & Berry, G. J. Psychosocial correlates of empirical types of multiple drug abusers. *Journal of Consulting and Clinical Psychology,* 1978, **46,** 1463–1475.

Brill, N. Q., & Christie, R. L. Marihuana use and psychosocial adaptation: Follow-up study of a collegiate population. *Archives of General Psychiatry,* 1974, **31,** 713–719.

Bromley, D. B. Natural language and the development of the self. In H. E. Howe & C. B. Keasey (Eds.), *Nebraska Symposium on Motivation* (Vol. 25). Lincoln: University of Nebraska Press, 1977.

Brook, J. S., Lukoff, I. F., & Whiteman, M. Peer, family and personality domains as related to adolescents' drug behavior. *Psychological Reports,* 1977, **41,** 1095–1102.

Brook, R., Szandorowska, B., & Whitehead, P. C. Psychosocial dysfunctions as precursors to amphetamine abuse among adolescents. *Addictive Diseases: An International Journal,* 1976, **2,** 465–478.

Burke, E. L., & Eichberg, R. H. Personality characteristics of adolescent users of dangerous drugs as indicated by the MMPI. *The Journal of Nervous and Mental Disease,* 1972, **154,** 291–298.

Cahalan, D., Cisin, I., & Crossley, H. M. *American drinking practices* (Monograph No. 6). New Brunswick, New Jersey: Rutgers Center of Alcohol Studies, 1969.

Carman, R. S. Motivations for drug use and problematic outcomes among rural junior high school students. *Addictive Behaviors,* 1979, **4,** 91–93.

Carrol, E. M., & Zuckerman, M. Psychopathology and sensation seeking in "downers," "speeders," and "trippers": A study of the relationship between personality and drug choice. *International Journal of the Addictions,* 1977, **12,** 591–601.

Caudill, B. D., & Marlatt, G. A. Modeling influences in social drinking: An experimental analogue. *Journal of Consulting and Clinical Psychology,* 1975, **43,** 405–415.

Chein, I., Gerald, D. L., Lee, R. S., & Rosenfeld, E. *The road to H: Narcotics, delinquency, and social policy.* New York: Basic Books, 1964.

Cockerham, W. C. Patterns of alcohol and multiple drug use among rural white and American Indian adolescents. *International Journal of the Addictions,* 1977, **12,** 271–285.

Cohen, A. Y., & Santo, Y. Youth drug abuse and education: Empirical and theoretical considerations. In G. M. Beschner & A. S. Friedman (Eds.), *Youth drug abuse: Problems, issues, and treatment.* Lexington, Mass.: Heath, 1979.

Crain, W. C., Ertel, D., & Gorman, B. S. Personality correlates of drug preference among college undergraduates. *International Journal of the Addictions,* 1975, **10,** 849–856.

Cutter, H. S. G., & Fisher, J. C. Family experience and the motives for drinking. *International Journal of the Addictions,* 1980, **15,** 339–358.

De Leon, G., & Wexler, H. K. Heroin addiction: Its relation to sexual behavior and sexual experience. *Journal of Abnormal Psychology,* 1973, **81,** 36–38.

Dembo, R., Burgos, W., Des Jarlais, D. D., & Schmeidler, J. Ethnicity and drug use among urban junior high school youths. *International Journal of the Addictions,* 1979, **14,** 557–568.

Dembo, R., & Miran, M. Evaluation of drug prevention programs by youths in a middle-class community. *International Journal of the Addictions,* 1976, **11,** 881–903.

Densen-Gerber, J., & Benward, J. *Incest as a causative factor in anti-social behavior: An explanatory study.* New York: Odyssey Institute Studios, 1976.

Dericco, D. A., & Garlington, W. K. The effect of modeling and disclosure of experimenter's intent on drinking rate of college students. *Addictive Behaviors,* 1977, **2,** 135–139.

Donovan, J. E., & Jessor, R. Adolescent problem drinking: Psychosocial correlates in a national sample study. *Journal of Studies on Alcohol,* 1978, **39,** 1506–1524.

Duncan, D. F. Life stress as a precursor to adolescent drug dependence. *International Journal of the Addictions,* 1977, **12,** 1047–1056.

Egger, G. J., Webb, R. A. J., & Reynolds, I. Early adolescent antecedents of narcotic abuse. *International Journal of the Addictions,* 1978, **13,** 773–781.

Ellinwood, E. H., Jr., Smith, W. G., & Vaillant, G. E. Narcotic addiction in males and females: A comparison. *International Journal of the Addictions,* 1966, **1,** 33–45.

Fillmore, K. M. Drinking and problem drinking in early adulthood and middle age. An exploratory 20-year follow-up study. *Quarterly Journal of Studies on Alcohol,* 1974, **35,** 819–840.

Fishburne, P. M., Abelson, H. I., & Cisin, I. *National survey on drug abuse; main findings: 1979.* Rockville, Md.: National Institute on Drug Abuse, 1980.

Fiske, D. W. Two worlds of psychological phenomena. *American Psychologist,* 1979, **34,** 733–739.

Foy, D. W., Miller, P. M., Eisler, R. M., & O'Toole, D. H. Social-skills training to teach alcoholics to refuse drinks effectively. *Journal of Studies on Alcohol,* 1976, **37,** 1340–1345.

Garlington, W. K., & Dericco, D. A. The effect of modeling on drinking rate. *Journal of Applied Behavior Analysis,* 1977, **10,** 207–211.

Gorsuch, R. L., & Butler, M. C. Initial drug abuse: A review of predisposing social psychological factors. *Psychological Bulletin,* 1976, **83,** 120–137.

Graeven, D. B., & Schaef, R. D. Family life and levels of involvement in an adolescent heroin epidemic. *International Journal of the Addictions,* 1978, **13,** 747–771.

Green, J. Overview of adolescent drug use. In G. M. Beschner & A. S. Friedman (Eds.), *Youth drug abuse: Problems, issues, and treatment.* Lexington, Mass.: Heath, 1979.

Green, L. L., & Haymes, M. Value orientation and psychosocial adjustment at various levels of marijuana use. *Journal of Youth and Adolescence,* 1973, **2,** 213–231.

Hager, D. L., Vener, A. M., & Stewart, C. S. Patterns of adolescent drug use in middle America. *Journal of Counseling Psychology,* 1971, **18,** 292–297.

Halikas, J. A., Goodwin, D. W., & Guze, S. B. Marihuana use and psychiatric illness. *Archives of General Psychiatry,* 1972, **27,** 162–165.

Higgins, R. L., & Marlatt, G. A. Effects of anxiety arousal on the consumption of alcohol by alcoholics and social drinkers. *Journal of Consulting and Clinical Psychology,* 1973, **41,** 426–433.

Higgins, R. L., & Marlatt, G. A. Fear of interpersonal evaluation as a determinant of alcohol consumption in male social drinkers. *Journal of Abnormal Psychology,* 1975, **84,** 644–651.

Hochman, J. S., & Brill, N. Q. Chronic marijuana use and psychological adaptation. *American Journal of Psychiatry,* 1973, **130,** 132–139.

Hogan, R., Mankin, D., Conway, J., & Fox, S. Personality correlates of undergraduate marijuana use. *Journal of Consulting and Clinical Psychology,* 1970, **35,** 58–63.

Horman, R. E. Alienation and student drug use. *International Journal of the Addictions,* 1973, **8,** 325–331.

Huba, G. J., Wingard, J. A., & Bentler, P. M. Beginning adolescent drug use and peer and adult international patterns. *Journal of Consulting and Clinical Psychology,* 1979, **47,** 265–276.

Hunt, L. G., Farley, E. C., & Hunt, R. G. Spread of drug use in populations of youths. In G. M. Beschner & A. S. Friedman (Eds.), *Youth drug abuse: Problems, issues, and treatment.* Lexington, Mass.: Heath, 1979.

Jessor, R., & Jessor, S. L. *Problem behavior and psychosocial development.* New York: Academic, 1977.

Jessor, R., & Jessor, S. L. Theory testing in longitudinal research on marijuana use. In D. B. Kandel (Ed.), *Longitudinal research on drug use.* Washington, D.C.: Hemisphere, 1978.

Jessor, R., Young, H. B., Young, E. B., & Tesi, G. Perceived opportunity, alienation and drinking behavior among Italian and American youth. *Journal of Personality and Social Psychology,* 1970, **15,** 215–222.

Johnston, L. D., Bachman, J. G., & O'Malley, P. M. *1979 highlights: Drugs and the nation's high school students: Five year national trends.* Rockville, Md.: National Institute on Drug Abuse, 1979.

Johnston, L. D., O'Malley, P. M., & Eveland, L. K. Drugs and delinquency: A search for causal connections. In D. B. Kandel (Ed.), *Longitudinal research on drug use.* Washington, D.C.: Hemisphere, 1978.

Kahle, L. R., Kulka, R. A., & Klingel, D. M. Low adolescent self-esteem leads to multiple interpersonal problems: A test of social-adaptation theory. *Journal of Personality and Social Psychology,* 1980, **39,** 496–502.

Kandel, D. Some comments on the relationship of selected criteria variables to adolescent illicit drug use. In D. J. Lettieri (Ed.), *Predicting adolescent drug abuse: A review of issues, methods and correlates.* Rockville, Md.: National Institute on Drug Abuse, 1975.

Kandel, D. B. Convergences in prospective longitudinal surveys of drug use in normal populations. In D. B. Kandel (Ed.), *Longitudinal research on drug use.* Washington, D.C.: Hemisphere, 1978.

Kandel, D., & Faust, R. Sequence and stages in patterns of adolescent drug use. *Archives of General Psychiatry,* 1975, **32,** 923–932.

Kandel, D. B., Kessler, R. C., & Margulies, R. Z. Antecedents of adolescent initiation into stages of drug use: A developmental analysis. *Journal of Youth and Adolescence,* 1978, **7,** 13–40.

Kellam, S. G., Ensminger, M. E., & Simon, M. B. Mental health in first grade and teenage drug, alcohol, and cigarette use. *Drug and Alcohol Dependence,* 1980, **5,** 273–304.

Khavari, K. A., Mabry, E., & Humes, M. Personality correlates of hallucinogen use. *Journal of Abnormal Psychology,* 1977, **86,** 172–178.

Kilpatrick, D. G., Sutker, P. B., Roitzsch, J. C., & Miller, W. C. Personality correlates of polydrug use. *Psychological Reports,* 1976, **38,** 311–317.

Kirk, R. S. Drug use among rural youth. In G. M. Beschner & A. S. Friedman (Eds.), *Youth drug abuse: Problems, issues, and treatment.* Lexington, Mass.: Heath, 1979.

Klinge, V., Lachar, D., Grisell, J., & Berman, W. Adolescence: Effects of scoring norms on adolescent psychiatric drug users' and nonusers' MMPI profiles. *Adolescence,* 1978, **13**, 1–10.

Klinge, V., Vaziri, H., & Lennox, K. Comparison of psychiatric inpatient male and female adolescent drug abusers. *International Journal of the Addictions,* 1976, **11**, 309–323.

Kohn, P. M., & Annis, H. M. Personality and social factors in adolescent marijuana use: A path-analytic study. *Journal of Consulting and Clinical Psychology,* 1978, **46**, 366–367.

Korman, M., Trimboli, F., & Semler, I. A comparative evaluation of 162 inhalant users. *Addictive Behaviors,* 1980, **5**, 143–152.

Lassey, M. L., & Carlson, J. E. Drinking among rural youth: The dynamics of parental and peer influence. *International Journal of the Addictions,* 1980, **15**, 61–75.

Lavenhar, M. A. Methodology in youth drug abuse research. In G. M. Beschner & A. S. Friedman (Eds.), *Youth drug abuse: Problems, issues, and treatment.* Lexington, Mass.: Heath, 1979.

Lied, E. R., & Marlatt, G. A. Modeling as a determinant of alcohol consumption: Effect of subject sex and prior drinking history. *Addictive Behaviors,* 1979, **4**, 47–54.

McCoy, C., McBride, D. C., Ruse, B. R., Page, J. B., & Clayton, R. R. Youth opiate use. In G. M. Beschner & A. S. Friedman (Eds.), *Youth drug abuse: Problems, issues, and treatment.* Lexington, Mass.: Heath, 1979.

McKenna-Hartung, S., Hartung, J., & Baxter, J. Self and ideal self-concept in a drug-using subculture. *Journal of Personality Assessment,* 1971, **35**, 461–471.

Malatesta, V. J., Sutker, P. B., & Treiber, F. A. Sensation seeking and chronic public drunkenness. *Journal of Consulting and Clinical Psychology,* 1981, **49**, 292–294.

Marden, P., Zylman, R., Fillmore, K. M., & Bacon, S. D. Comments on "A national study of adolescent drinking behavior, attitudes, and correlates." *Journal of Studies on Alcohol,* 1976, **37**, 1346–1358.

Matchett, W. F. Who uses drugs? A study in a suburban public high school. *Journal of School Health,* 1971, **41**, 90–93.

Mellinger, G. D., Somers, R. H., Bazell, S., & Manheimer, D. L. Drug use, academic performance, and career indecision: Longitudinal data in search of a model. In D. B. Kandel (Ed.), *Longitudinal research on drug use.* Washington, D.C.: Hemisphere, 1978.

Miller, P. M. Behavioral strategies for reducing drinking among young adults. In H. T. Blane & M. E. Chafetz (Eds.), *Youth, alcohol, and social policy.* New York: Plenum, 1979.

Mischel, W. On the future of personality assessment. *American Psychologist,* 1977, **32**, 246–252.

Mischel, W. On the interface of cognition and personality beyond the person–situation debate. *American Psychologist,* 1979, **34**, 740–754.

Moos, R. H., Moos, B. S., & Kulik, J. A. Behavioral and self-concept antecedents and correlates of college-student drinking patterns. *International Journal of the Addictions,* 1977, **12**, 603–615.

Naditch, M. P. Relation of motives for drug use and psychopathology in the development of acute adverse reactions to psychoactive drugs. *Journal of Abnormal Psychology,* 1975, **84**, 374–385.

Nathan, P. E., & Lansky, D. Common methodological problems in research on the addictions. *Journal of Consulting and Clinical Psychology,* 1978, **46**, 713–726.

Nylander, I. A 20-year prospective follow-up study of 2164 cases at the child guidance clinics in Stockholm. *Acta Paediatrica Scandinavica,* 1979, **68**, Supplement 276, 1–45.

O'Donnell, J. A., Voss, H. L., Clayton, R. R., Slatin, G. T., & Room, R. G. W. *Young men and drugs—a nationwide survey.* Rockville, Md.: National Institute on Drug Abuse, 1976.

Oetting, E. R., & Goldstein, G. S. Drug use among native American adolescents. In G. M. Beschner & A. S. Friedman (Eds.), *Youth drug abuse: Problems, issues, and treatment.* Lexington, Mass.: Heath, 1979.

O'Malley, P. M. Correlates and consequences of illicit drug use (Doctoral dissertation, University of Michigan, 1975). *Dissertation Abstracts International,* 1975, **36**, 3011B. (University Microfilm No. 75-29, 302)

Parry, H. J., Cisin, I. H., Balter, M. B., Mellinger, G. D., & Manheimer, D. L. Increased alcohol intake as a coping mechanism for psychic distress. In R. Cooperstock (Ed.), *Social aspects of the medical use of psychotropic drugs.* Toronto: Addiction Research Foundation, 1974.

Paton, S. M., & Kandel, D. B. Psychological factors and adolescent illicit drug use: Ethnicity and sex differences. *Adolescence,* 1978, **13**, 187–200.

Podolsky, E. The woman alcoholic and premenstrual tension. *Journal of the American Medical Women's Association,* 1963, **18**, 816–818.

Potvin, R. H., & Lee, C-F. Multistage path models of adolescent alcohol and drug use: Age variations. *Journal of Studies on Alcohol,* 1980, **41**, 531–542.

Preble, E., & Casey, J. J. Taking care of business: The heroin user's life on the street. *International Journal of the Addictions,* 1969, **4**, 1–24.

Prendergast, T. J. Family characteristics associated with marijuana use among adolescents. *International Journal of the Addictions,* 1974, **9**, 827–839.

Prendergast, T. J., & Schaefer, E. S. Correlates of drinking and drunkenness among high-school students. *Quarterly Journal of Studies on Alcohol,* 1974, **35**, 232–242.

Rathus, S. A., Fichner-Rathus, L., & Siegel, L. J. Behavioral and familial correlates of episodic heroin abuse among suburban adolescents. *International Journal of the Addictions,* 1977, **12**, 625–632.

Reid, J. B. Study of drinking in natural settings. In G. A. Marlatt & P. E. Nathan (Eds.), *Behavioral approaches to alcoholism.* New Brunswick, N.J.: 1978, Rutgers Center of Alcohol Studies.

Ridgeway, D., & Russell, J. A. Reliability and validity of the Sensation-Seeking Scale: Psychometric problems in Form V. *Journal of Consulting and Clinical Psychology,* 1980, **48**, 662–664.

Robins, L. N. *Deviant children grown up.* Baltimore: Williams & Wilkins, 1966.

Rohrbaugh, J., & Jessor, R. Religiosity in youth: A personal control against deviant behavior. *Journal of Personality,* 1975, **43**, 136–155.

Rohrs, C. C., Murphy, J. P., Goldsmith, B., & Densen-Gerber, J. The phenomenon of adolescent addiction. *Journal of Forensic Sciences,* 1972, **17**, 522–524.

Rosenbluth, J., Nathan, P. E., & Lawson, D. M. Environmental influences on drinking by college students in a college pub: Behavioral observation in the natural environment. *Addictive Behaviors,* 1978, **3**, 117–121.

Sadava, S. W., & Forsyth, R. Person-environment interaction and college student drug use: A multivariate longitudinal study. *Genetic Psychology Monographs,* 1977, **96**, 211–245. (a)

Sadava, S. W., & Forsyth, R. Turning on, turning off, and relapse: Social psychological determinants of status change in cannabis use. *International Journal of the Addictions,* 1977, **12**, 509–528. (b)

Sadava, S. W., Thistle, R., & Forsyth, R. Stress, escapism and patterns of alcohol and drug use. *Journal of Studies on Alcohol,* 1978, **39**, 725–736.

Schlegel, R. P., & Sanborn, M. D. Religious affiliation and adolescent drinking. *Journal of Studies on Alcohol,* 1979, **40**, 693–703.

Schwarz, R. M., Burkhart, B. R., & Green, S. B. Turning on or turning off: Sensation seeking or tension reduction as motivational determinants of alcohol use. *Journal of Consulting and Clinical Psychology,* 1978, **46**, 1144–1145.

Segal, B. Personality factors related to drug and alcohol use. In D. J. Lettieri (Ed.), *Predicting adolescent drug abuse: A review of issues, methods, and correlates.* Rockville, Md.: National Institute on Drug Abuse, 1975.

Smart, R. G., & Blair, N. L. Drug use and drug problems among teenagers in a household sample. *Drug and Alcohol Dependence,* 1980, **5**, 171–179.

Smart, R. G., & Fejer, D. Drug use among adolescents and their parents: Closing the generation gap in mood modification. *Journal of Abnormal Psychology,* 1972, **79**, 153–160.

Smart, R. G., & Gray, G. Parental and peer influences as correlates of problem drinking among high school students. *International Journal of the Addictions,* 1979, **14**, 905–917.

Smart, R. G., Gray, G., & Bennett, C. Predictors of drinking and signs of heavy drinking among high school students. *International Journal of the Addictions,* 1978, **13**, 1079–1094.

Smart, R. G., & Jones, D. Illicit LSD users: Their personality characteristics and psychopathology. *Journal of Abnormal Psychology,* 1970, **75**, 286–292.

Smith, D., Levy, S. J., & Striar, D. E. Treatment services for youthful drug users. In G. M. Beschner & A. S. Friedman (Eds.), *Youth drug abuse: Problems, issues, and treatment.* Lexington, Mass.: Heath, 1979.

Stokes, J. P. Personality traits and attitudes and their relationship to student drug using behavior. *International Journal of the Addictions,* 1974, **9**, 267–287.

Streit, F., Halsted, D. L., & Pascale, P. J. Differences among youthful users and nonusers of drugs based on their perception of parental behavior. *International Journal of the Addictions,* 1974, **9**, 749–755.

Sutker, P. B. Field observations of a heroin addict: A case study. *American Journal of Community Psychology,* 1974, **2**, 35–42.

Sutker, P. B. Drug-dependent women: An overview of the literature. In G. Beschner, B. G. Reed, & J. Mondanaro (Eds.), *Treatment services for drug-dependent women* (Vol. 1). Rockville, Md.: National Institute on Drug Abuse, 1981.

Sutker, P. B., Allain, A. N., Brantley, P. J., & Randall, C. L. Acute alcohol intoxication, negative affect, and autonomic arousal in women and men. *Addictive Behaviors,* 1981, **6**, 403–410.

Sutker, P. B. & Archer, R. P. Opiate abuse and dependence disorders. In H. E. Adams & P. B. Sutker (Eds.), *Comprehensive handbook of psychopathology.* New York: Plenum, in press.

Sutker, P. B., & Archer, R. P. MMPI characteristics of opiate addicts, alcoholics, and other drug abusers. In C. S. Newmark (Ed.), *MMPI: Current clinical and research trends.* New York: Praeger, 1979.

Sutker, P. B., Archer, R. P., & Allain, A. N. Drug abuse patterns, personality characteristics, and relationships with sex, race, and sensation seeking. *Journal of Consulting and Clinical Psychology,* 1978, **46**, 1374–1378.

Topper, M. D. Drinking as an expression of status: Navajo male adolescents. In J. O. Waddell & M. W. Everett (Eds.), *Drinking behaviors among Southwestern Indians.* Tucson: University of Arizona Press, 1980.

Vaillant, G. E. A 12-year follow-up of New York narcotic addicts. *Archives of General Psychiatry,* 1966, **15**, 599–609.

Vaillant, G. E., Sobowale, N. C., & McArthur, C. Some psychologic vulnerabilities of physicians. *New England Journal of Medicine,* 1972, **287**, 372–375.

Waddell, J. O., & Everett, M. W. (Eds.) *Drinking behavior among Southwestern Indians.* Tucson: University of Arizona Press, 1980.

Wechsler, H., & McFadden, M. Sex differences in adolescent alcohol and drug use: A disappearing phenomenon. *Journal of Studies on Alcohol,* 1976, **37**, 1291–1301.

Wechsler, H., & Thum, D. Teenage drinking, drug use and social correlates. *Quarterly Journal of Studies on Alcohol,* 1973, **34**, 1220–1227.

Weinstein, R. M. The avowal of motives for marijuana behavior. *International Journal of the Addictions,* 1978, **13**, 887–910.

Wilson, G. T., & Abrams, D. B. Effects of alcohol on social anxiety and physiological arousal: Cognitive versus pharmacological processes. *Cognitive Therapy and Research,* 1977, **1**, 195–210.

Winslow, R. W. (Ed.) *Juvenile delinquency in a free society* (3rd ed.). Encino, Calif.: Dickenson, 1976.

Woods, J. H. Behavioral pharmacology of drug self-administration. In M. A. Lipton, A. DiMascio, & K. F. Killam (Eds.), *Pharmacology: A generation of progress.* New York: Raven, 1978.

Zuckerman, M. Drug usage as one manifestation of a "sensation seeking" trait. In W. Keup (Ed.), *Drug abuse: Current concepts and research.* Springfield, Ill.: Thomas, 1972.

Zuckerman, M. Sensation-seeking. In H. London & J. Exner (Eds.), *Dimensions of personality.* New York: Wiley, 1978.

Zuckerman, M., Bone, R. N., Neary, R., Mangelsdorff, D., & Brustman, B. What is the sensation-seeker? Personality trait and experience correlates of the Sensation Seeking Scales. *Journal of Consulting and Clinical Psychology,* 1972, **39**, 308–321.

21

ADOLESCENT AGGRESSION

Herbert C. Quay
University of Miami

Since the 1950s, concern on the part of both behavioral scientists and the public at large about aggression in adolescents has been largely motivated by the "delinquency rate," especially in urban areas. There is no doubt that the extent to which youths are involved in criminal or delinquent activities is a serious social problem. In 1979 there were reported to the FBI (Federal Bureau of Investigation, 1979) 2,143,369 arrests[1] of persons under 18, this number accounting for 22.5% of all reported arrests. These arrests are further broken down by age in Table 21.1.

The adolescent years from 15 through 17 are responsible for almost 17% of the total 1979 arrests. For violent crime (defined by the FBI as murder, forcible rape, robbery, and aggravated assault) arrests of those under 18 constituted 20.1% of the total of all arrests for this category of crime. Ages 15 through 17 accounted for 15% of the total. However, while there were 87,375 arrests of persons under 18 for violent crimes, there were almost 9 times as many arrests in this age group (751,421) for property crimes.

Of all arrests in 1979, 84.3% were males; of those under 18 arrested, about 74% were males. Thus while crime is a predominantly male enterprise, there are proportionately more females arrested who are under 18 than above that age.[2] It is of interest to note that, of those arrested for violent crimes in 1979 who were under 18, 35% were females; this compares to 32% in 1970.

The preoccupation of researchers and theorists with male deviance seems to reflect the 1970 and earlier statistics where females accounted for only between 10 and 15% of arrests. Thus while all the research and theory to be discussed later in this chapter pertain to males, it is clear at the outset that attention should be given to aggression in adolescent females as well.

Perhaps because of the publicity frequently given to violent offenses perpetrated by juveniles, there has also been a strong tendency on the part of both the lay public and behavioral scientists to equate delinquency and aggressive behavior. However, the majority of delinquents are not, in fact, physically aggressive, assaultive, or violent. As shown above, arrests for property crimes are about 9 times more

---

[1] Since an individual may be arrested more than once, these figures pertain to arrests, not persons.
[2] With respect to total arrests, the actual number of females increased 24% from 1970 to 1979 while arrests of males increased only 3.4%.

TABLE 21.1.

	10 and under	11–12	13–14	15	16	17	18
Arrests	74,652	136,754	450,637	407,152	515,979	558,195	595,798
% total arrests	0.8	1.4	4.7	4.3	5.4	5.9	6.3

frequent than arrests for violent offenses. Knowing that a youth is delinquent tells us nothing about his behavior and personality except that a label has been given to him by a court. It is critical that the student of behavior recognize that "juvenile delinquency" is a concept which has its basis in law rather than in psychology; it is not a label which implies that those who carry it are homogeneous in behavioral and/or psychological characteristics.

The behavioral study of delinquency is further complicated by the fact that the label is applied only to those who are apprehended and subsequently processed by the juvenile justice system. The extent of undetected or unrecorded, but nevertheless legally delinquent, acts has been estimated to be vastly greater in number than those acts for which arrests have been made. Thus studying adjudicated delinquents means studying only a small portion of those who violate the law.

Finally, the justice system itself does not always react to all those who commit law violations in the same way; age, ethnicity, social class, and family structure may determine which one of two adolescents who have committed the same offense becomes an officially recorded "delinquent." Empey (1978) has provided a recent review of both unrecorded delinquency and the reaction of social control agencies to delinquent behavior.

All these problems make it unprofitable for the psychologist to study delinquency, per se. In order to contribute to an understanding of lawbreaking in youth, the psychologist can study the genesis and development of behavioral and psychological *characteristics* related to, but certainly not exclusively associated with, lawbreaking, such as aggression, moral and cognitive development (see Chapter 19), and substance abuse (see Chapter 20).

Alternatively, attention can be given to *individuals* who manifest dimensions or patterns of deviant behavior frequently found among legally defined delinquents. These patterns will be outlined below and two of them are the major focus of this chapter.

DIMENSIONS OF DEVIANT BEHAVIOR IN ADOLESCENCE

Over the past 40 years, evidence has accumulated from multivariate statistical (predominantly factor analytic) studies pointing to the presence of four major dimensions of deviant behavior among delinquent as well as otherwise deviant adolescents. Throughout the years, these dimensions have been given different names by different researchers; the use of labels given by the present author in a recent detailed review (Quay, 1979) will be continued here.

The first dimension, called *conduct* disorder, involves such behavior as fighting, assault, defiance, disobedience, destructiveness, negativism, uncooperativeness, irritability, profanity, dishonesty, and irresponsiblity. This pattern clearly involves acting against persons and the environment and has been referred to by Achenbach as an "externalizing" disorder (Achenbach & Edelbrock, 1978).

A second dimension of externalizing behavior, called *socialized aggressive* disorder, involves group stealing, gang membership, truancy from home and school, and close ties to a peer group that has often been labeled delinquent. While this chapter will focus on these two patterns of aggression, two others are dimensions also frequently found.

The dimension labeled *anxiety withdrawal* subsumes anxiety, tension, timidity, social isolation, hypersensitivity, sadness, and feelings of inferiority. This pattern involves retreat, rather than attack, and Achenbach (Achenbach & Edelbrock, 1978) has suggested the term *internalizing*. A fourth dimension, not clearly either externalizing or internalizing, labeled *immaturity* involves problems in attention, excess daydreaming, clumsiness, passivity, sluggishness, and lack of perseverance. In younger children, the pattern is probably that now called *attention deficit* disorder (American Psychiatric Association, 1980).

We should note at this point that while the need to consider these dimensions of disorder as separate and distinct is compelling, most early investigators utilized the broader labels of "delinquent" or "emotionally disturbed" in selecting cases for study. The result of this lumping of cases together as if they were all homogeneous obscured differences among these groups and often resulted in the failure to find any differences between the deviant and nondeviant samples (see earlier reviews by Schuessler & Cressey, 1950; Quay, 1955).

It is also important to recognize that those patterns of adolescent deviance that may be found among those legally delinquent or those who have committed a clearly aggressive act may also be found among adolescents whose deviance is defined in terms of mental health or educational problems, rather than criminality. The two externalizing patterns, which we will discuss in detail below, may be found in clinics, hospitals, and special schools as well. Furthermore, one must also be cognizant of the fact that the deviant is differentiated from the nondeviant more in terms of quantity of deviance rather than in terms of quality of deviance.

CONDUCT DISORDER

The multivariate statistical studies in which this pattern has invariably emerged have provided us with an oft-replicated set of intercorrelated behavioral characteristics. However, these studies do not tell us how many adolescents there are who manifest this pattern, nor do they tell us how the pattern is related to such important psychological variables such as intelligence, cognition, learning, and motivation. Neither do they provide data as to causes, cures, or outcome. These multivariate studies have only provided a basis for identifying homogeneous dimensions upon which individuals can be placed for further study; we are dependent upon subsequent research for a psychological understanding of these patterns in terms of their relations to psychologically meaningful variables, such as cognitive and interpersonal behavior and patterns of family interaction. To extend the meaning of these dimensions of disorder, we must turn to studies which have examined the relation of these dimensions to other variables.

Cognitive Development

While academic achievement among those with conduct disorder is often well below expectation, the best evidence suggests that general intelligence, especially when measured by nonverbal tests, is not meaningfully below that for nondeviant controls (see Quay, 1979). However, when investigators have looked at more specific aspects of intellectual functioning, especially those related to the ability to think abstractly and to reason about moral issues, deficiencies in these aspects of cognition for those with conduct disorder have been found in comparison to both normals and those with other patterns of disorder (Bear, 1979; Fodor, 1972; Jurkovic & Prentice, 1974; Richman & Lindgren, 1981).

Response to Social Reward

The finding that the behavior of adolescents with conduct disorder is not greatly influenced by social approval would be important for both cause and treatment, since much learning in childhood and adolescence take place in a social context and is reinforced by the social approval of relevant adults and peers. Furthermore, many methods of behavior change depend upon verbal intervention between therapist and client. Thus the extent to which there may be some decrement in responsiveness to social rewards in adolescents who are seemingly alienated from both peers and adults has been a matter of some interest. While the results in terms of the reinforcing power of social approval have not always been as straightforward as predicted, there is some evidence for lack of response to social rewards among adolescents with conduct disorder (Stewart, 1972; see also Dietrich, 1976).

Stimulation Seeking

Extrapolating from a theory that psychopathic behavior in adults reflects an inordinate need for sensory stimulation (Quay, 1975), two studies have found evidence for boredom susceptibility and stimulation seeking in conduct-disordered adolescents (Orris, 1969; Skrzypek, 1969). This line of research has more recently been extended to younger children with basically similar findings (De Myer-Gapin & Scott, 1977; Whitehill, De Myer-Gapin & Scott, 1976). It is important to note here that one cannot equate conduct disorder in adolescence to psychopathy in adults. As we have pointed out above, conduct disorder is a dimensional concept so that it can be present in varying degrees of severity. Thus extrapolation of theories about adult psychopathy to cases of extreme conduct disorder in adolescents is warranted only if these notions are subsequently tested empirically.

The finding that such conduct-disordered adolescents are apparent stimulation seekers may have relevance to a variety of factors, such as substance abuse (see Chapter 20) and the need for treatment approaches that combine control with stimulus enhancement (e.g., Ingram, Gerard, Quay, & Levinson, 1970).

Physiological Factors

One aspect of the stimulation-seeking hypothesis suggests that the basis for psychopathic "stimulus hunger" may lie in some form of physiological underarousal (see Quay, 1965; 1977a). Concomitant with prior research demonstrating reduced autonomic nervous system arousal in adult psychopaths, Borkovec (1970) found a lowered galvanic skin response (GSR) for conduct-disordered adolescents when they were compared to anxious-withdrawn and socialized-aggressive boys—all of whom were institutionalized delinquents. Similar results with regard to GSR reactivity were later reported by Siddle, Nicol, and Foggitt (1973).

With regard to adult psychopathy, deficits in the learning of passive avoidance or behavioral inhibition have been suggested by Trasler (1978) and by Fowles (1980). Most recently, dysfunction in the septal area of the brain has been postulated as the underlying neurophysiological mechanism (Gorenstein & Newman, 1980). These theories have been tested only indirectly with adult psychopaths and not at all with conduct-disordered adolescents.

With those cases of extreme and persistent aggressive behavior, seemingly refractory to the consequences of their actions, either some drive to increase arousal at whatever cost and/or some dysfunction in the mechanism whereby one learns to inhibit punished acts are appealing concepts. At this time, there is enough evidence on the favorable side of both hypotheses that necessitates each being taken seriously and further tested.

Other Variables

There have been a number of studies which have contrasted the behavior of institutionalized delinquents classified as conduct disordered with those classified as belonging to other subgroups. These studies have generally found the conduct disordered more likely to be repeaters, to present more disciplinary problems in the institution, to be less likely to succeed on probation and in diversion and work-release programs, and to show more malingering and manipulative behavior (see Quay, 1979, for a review). In short, this group performs less well than do others in programs designed to reduce their deviant behavior.

Lueger (1980) looked at the relationship between cheating (in a laboratory setting) and dimensions of disorder. Those subjects who did cheat had higher scores than those who did not on both conduct disorder and socialized aggression. Additional analyses compared high scorers on conduct disorder with high scorers on anxiety withdrawal; significantly more of the former than the latter cheated.

Differing ways in which deviant children and adolescents perceive others in their environments have been addressed in two recent studies. Nasby, Hayden, and De Paulo (1980) found a significant positive correlation between perceptions of social stimuli as displays of hostility and scores on the conduct-disorder dimension in two separate experiments. They interpreted their results as supporting the hypothesis that a generalized and marked attribution to infer hostility became greater as aggression scores increased.

Spencer (1980) studied the responses of four delinquent subgroups to the Role Construct Repertory Test (Kelly, 1955). The usual variables on which this test was scored did not differ significantly among the groups. However, when the data were analyzed in terms of the characteristics of persons in the environment which had been used by subjects to describe the stimulus figures on the test, some differences among the groups did appear. Both the conduct-disorder and socialized-aggressive boys saw the youth counselor as an authority figure whereas for the anxious-withdrawn group, the counselor was in a cluster of persons clearly having an advisor role. Only in the case of the socialized aggressives did the peer group include "a friend who gets in trouble often"; this finding was consonant with the picture of the socialized aggressive as involved with "delinquent" peers. Both these studies suggest a tendency on the part of conduct-disordered youth in particular to perceive others as hostile and/or authoritarian.

SOCIALIZED AGGRESSION

There has been little empirical study of the behavior and personality characteristics of gang delinquents despite their high visibility to the public and the extent to which they have been the target of sociological theorizing (see Empey, 1978, for a review). The dimension of socialized aggression has evolved mainly, but not exclusively, out of studies of adolescents who were legally delinquent and involves group activities of a "delinquent" nature and allegiance to peers rather than adults. However, to assume that all who manifest this pattern are members of delinquent gangs or that all members of delinquent gangs engage solely in socialized-aggressive forms of deviance would be in error.

In one of the rare examples of empirical research, Short, Tennyson, and Howard (1963) factor analyzed a checklist of activities upon which members of delinquent gangs had been rated. They found that delinquent activities could be resolved into five independent dimensions which they labeled "conflict" (fighting, assault, concealed weapons), "stable corner activities" (sports, social activities, gambling), "stable sex" (intercourse, petting, hanging on corner), "retreatest" (homosexuality, narcotics, marijuana), and "authority protest" (auto theft, driving without a license, public nuisance, runaway). They noted that actual criminal behaviors were spread over all five factors; no set of clearly criminal activities emerged as a single factor. Short et al. also demonstrated that not all the behavior of the gangs was aggressive; only the conflict and authority protest patterns would seem to qualify as aggressive. There were also differences beween gangs as to the predominant pattern of activities.

While one could speculate that boys heavily involved in conflict behavior would be high on the psychological dimension of conduct disorder, while boys involved in authority protest would be high on socialized aggression, Cartwright, Howard, and Reuterman (1980) failed to find any relationships between different forms of gang behavior and personality dimensions of gang members as measured by Cattell's (1955) Objective–Analytic Personality Factor Battery. However, this set of measures is not an ideal assessment battery to test for relationships between dimensions of deviance and dimensions of gang behavior. When gang members were compared with members of socially approved boys' clubs, the gang boys were found to be less assertive, exuberant, realistic, and self-reliant than were the club members. In passing, it should be noted that a series of studies by Cartwright and his colleagues (see Cartwright et al., 1980) did not find relationships, between empirically derived dimensions of gang activities and either the coherence of the gang as a social group or socioeconomic variables, that were as straightforward as had been expected.

Correlates of Socialized Aggression

In the very few studies which have looked at socialized-aggressive boys in contrast to cases manifesting other patterns of deviance, the interest of the investigator has most often centered on the other group(s), particularly the conduct disordered. By and large, those studies already reviewed above have suggested that socialized aggressives are much closer to nondeviant adolescents in cognitive style than are those with conduct disorders. However, as already noted, this dimension has been positively related to cheating behavior (Lueger, 1980). What does seem to be consistent is a greater susceptibility of these adolescents to influence by peers rather than adults (Akamatsu & Farudi, 1978).

THE DYNAMICS OF AGGRESSIVE DISORDERS IN FAMILY INTERACTION

The assessment of parent–child interaction and the effects of child-rearing practices (or lack thereof) on child behavior is an area that is fraught with methodological difficulties (see Hetherington & Martin, 1979). Certainly parents affect the behavior of their adolescents, but the reverse is true as well, so that there are interactions between the parties and concern must be given to the "direction of effect" (see Bell, 1981) with respect to interpreting research findings.

An extensive survey of the literature led Hetherington and Martin (1979) to conclude that "deviant parents have deviant children" (p. 257). Parents of aggressive children are more likely to be maladjusted, inconsistent, arbitrary, incompetent, and to provide models for aggressive, even criminal, behavior. Parental negativism and the use of physical punishment and permissiveness for aggression continue to be found to be instigators of aggressive behavior in children (Olweus, 1980).

While the majority of studies have used the broad classificatory labels of delinquent or disturbed, some few studies have examined the differential correlates of conduct disorder versus socialized aggression. The most complex study (Hetherington, Stouwie, & Ridberg, 1971) looked at actual patterns of family interaction and found that, in two-parent families, the parents of socialized-aggressive boys were seen as more permissive than those whose sons had conduct disorders. In addition the father of socialized-aggressive delinquent boys were dominant, while the mothers were sometimes passively resistant. In families with conduct-disordered sons, the father was also dominant, but the mother was more active. With regard to the boys themselves, there was greater participation in the family interaction among the socialized aggressives who were often unable to compromise with their father's decisions. The conduct-disordered boys participated to a lesser extent but did disagree with and aggress against the mother.

It is of interest that in a study in which adolescent and young adult offenders were asked about their attitudes toward their parents, the socialized-aggressive group was negative about their fathers but not their mothers, while the conduct-disorder group was negative about both (Megargee & Golden, 1973).

FREQUENCY OF AGGRESSIVE DISORDER

Establishing the prevalence of behavior disorder is a difficult undertaking due to problems associated with both defining a disorder and deciding who has one. Nevertheless, a number of studies using differing approaches converge fairly well on a rate of about 15% for all types of disorders in adolescents in the general population. When disorders are subdivided into two groups it appears that anxiety-withdrawal and aggressive disorders each account for about 40% of total disorders; thus conduct disorder and socialized aggression together have a frequency of about 6% in representative general population (see Graham & Rutter, 1973; Lavik, 1977; Leslie, 1974; for summaries of studies of the epidemiology of behavior disorder in both childhood and adolescence, see Graham, 1979; Rutter, 1980).

There are no studies which provide separate prevalence estimates for conduct disorder versus socialized-aggressive disorder, as epidemiologic investigations have tended to lump these two together. Furthermore, none of the general population studies have been conducted in the United States.

THE PERSISTENCE OF AGGRESSIVE DISORDER

Because a disorder is found to be present in an adolescent does not mean that it necessarily arose during the adolescent period. In fact, with regard to conduct disorder in particular, there is considerable evidence for continuity from ages around 8 to 10 on into adolescence (West & Farrington, 1977; Graham & Rutter, 1973; Lefkowitz, Eron, Walder, & Huesmann, 1977; Olweus, 1979). Some cases of aggressive disorder apparently do arise de novo during adolescence (Rutter, Graham, Chadwick, & Yule, 1976). As to the correlates of early and late-arising disorders, Rutter (1980) has suggested that family difficulties discussed above are more closely associated with the earlier onset. Factors which may be specifically related to onset in adolescence are largely unknown.

There are a number of methodological problems related to the determination of the persistence of disorders from adolescence to adulthood (see Robins, 1979). However, the most generally accepted conclusions are that (1) most aggressive children do not become antisocial adults, (2) most antisocial behavior in adults was preceded by similar behavior in children, (3) the severity and frequency of child and adolescent aggression is a better predictor of poor adult functioning than any other form of childhood behavior, and (4) the actual behavior of the child is more predictive of adult functioning than is family background or social class (see Mitchell & Rosa, 1981; Robins, 1978, 1979; West & Farrington, 1973, 1977). In the only study directly comparing the later behavior of conduct-disorder compared to socialized-aggressive delinquents, Henn, Bardwell, and Jenkins (1980) found that the socialized-aggressive group was less likely to have been convicted of a crime or imprisoned in later life. This finding, which clearly requires replication, would suggest that socialized-aggressive disorder is not as good a predictor of adult outcome as is conduct disorder.

While there is predictability, the fact that most aggressive children do not become adult criminals deserves reiterating. In her early study, Robins (1966) found that of those cases actually referred to a child guidance clinic for antisocial behavior, only 44% had a subsequent court appearance for a major crime—compared however to only 3% of controls and 12% of clinic cases with internalizing disorders. Only 30% of Mitchell and Rosa's (1981) parent-identified aggressive cases later became criminals; this compared to 9% who became criminals who were not rated earlier by parents as aggressive. Furthermore, among those (in a large sample) simply identified as delinquent, without reference to degree of aggressiveness, Wolfgang, Figlio, and Sellin (1973) found that only 54% offended more than once.

TREATMENT

While treatment approaches are usually categorized either according to the modality (e.g., individual psychotherapy, transactional analysis) or setting (e.g., institution, school), these distinctions often obscure more than they illuminate. Individual psychotherapy may vary from psychoanalysis requiring years to complete to very short-term counseling. What goes on under the rubric of institutional treatment may also range from a total-institution token economy to a weekly "town meeting." Furthermore, as we have shown elsewhere (Quay, 1977b) the treatment purported to be implemented may actually have been delivered in such weak fashion as to be practically nonexistent.

An additional problem in evaluating treatment is that the exact nature of the disorders suffered by the clients may not be specified in terms other than the general label of delinquent or emotionally disturbed. If, in fact, the cases treated are heterogeneous, then the results of the evaluation may be negative overall but the effect on some subgroup of clients may have been quite positive (see Palmer, 1975).

As a result of these problems we can frequently only guess as to the extent to which a treatment has positively affected cases of conduct disorder and/or socialized aggression. Furthermore, it is impossible to determine whether different treatments might have had different effects in the two dimensions of disorder.

Psychotherapy

Tramontana (1980) critically reviewed studies of individual, group, and family therapy with adolescents which had been published during 1967–1977 and found that the overall rate of positive outcome was 75% with psychotherapy versus 39% without psychotherapy. Since over half the 15 experimental studies reviewed focused on delinquents, some of these studies no doubt included aggressive adolescents. However, it is still impossible to tell how effective psychotherapy was with the aggressive disorders since some or many of the subjects likely had other forms of disorder. Earlier studies, however, have suggested that persisting aggressive behavior is a poor prognostic sign (e.g., Pichel, 1974). As Tramontana (1980) noted, future psychotherapy research should more fully describe the criteria by which subjects were selected.

Institutional Treatment

Studies of institutional treatment which have provided anything approximating a rigorous evaluation are very limited in number. An excellent study comparing institutional programs utilizing transactional analysis and behavior modification with delinquent boys classified according to a system based on interpersonal maturity (see Warren, 1977) has been reported by Jesness (1975). Positive changes in cases seemingly analogous to conduct disorder and socialized aggression were reported both on within-institution measures and subsequent recidivism.

Within a group-home setting ("achievement place") the utilization of behavior-modification techniques has been reported to have had positive effects on children and adolescents who had had repeated contacts with junvenile authorities (see Braukmann, Kirigin & Wolf, 1976), although information as to the nature and degree of aggressive behavior is not known.

Community Treatment

Interventions categorized under this rubric generally involve more than individual or group psychotherapy but do not involve residential placement. Various programs have utilized combinations of counseling, academic remediation, vocational training, and job placement. While characteristics of clients are rarely specifically described, some recent reports indicate successful intervention, often in terms of reducing subsequent offenses (e.g., Lee & Haynes, 1980; Palmer, 1974; Quay & Love, 1977; Seidman, Rappaport, & Davidson, 1980; see also the volume by Ross & Gendreau (1980) for a collection of intervention studies with both children and adolescents).

PREVENTION

While various forms of intervention offer some promise, the ideal would be to prevent the development of excess aggression while not stifling normal assertiveness and initiative. Based on the factors related to the genesis of both forms of aggression, true prevention would involve changes in family structure, organization, and child-rearing practices in lower-class families. Additionally, large-scale social intervention designed to reduce poverty, social disorganization, and excessively large families would be needed.

Short of true prevention, programs utilizing interventions in the preschool and early school years appear to have promise (e.g., Cowen, Gesten, Orgel, & Wilson, 1979; Rickel & Dyhdalo, in press), as they may forestall the development of persisting patterns of aggressive behavior.

CONCLUSIONS

The evidence is quite clear in providing a picture of conduct disorder in adolescents as a pattern of behavior that is seriously pathological. Those youths who are extreme on this dimension behave in verbally and physically aggressive ways and are problems to whatever agents of society with which they have contact. Their social relations are poor, and they tend to see others as hostile and authoritarian. Their ability to reason abstractly and their moral judgment are deficient even when compared to others who are also legally delinquent. They do not tolerate routine and boredom and seek out excitement. Physiological factors, as well as rejection, conflict, and deviant role models within the family, have been implicated in the genesis of this disorder. While certain interventions appear to hold some promise, the current picture is one of considerable persistence of conduct disorder from late childhood into adolescence and on to adult life.

While we know less about socialized aggression, the picture as it now stands is not so bleak. The behavior is less seriously pathological in that interpersonal relations and interests are maintained, cognitive functioning and moral judgment do not seem to be impaired, and physiological factors have not been implicated. What little evidence there is suggests a prognosis that is more favorable than that for conduct disorder. All this is not to say that adolescents manifesting socialized aggression are not serious social problems who must be dealt with by appropriate agents of society. What is clear is that studies need a great deal more research into the nature and genesis of this pattern of aggression.

REFERENCES

Achenbach, T. M., & Edelbrock, C. G. The classification of child psychopathology: A review and analysis of empirical efforts. *Psychological Bulletin,* 1978, **85**, 1275–1301.

Akamatsu, T. J., & Farudi, P. A. Effects of model status and juvenile offender type on the imitation of self-reward criteria. *Journal of Consulting and Clinical Psychology,* 1978, **46**, 187–188.

American Psychiatric Association. *Diagnostic and statistical manual of mental disorders* (3rd ed.). Washington, D.C.: American Psychiatric Association, 1980.

Bear, G. G., II. The relationship of moral reasoning to conduct problems and intelligence. Unpublished doctoral dissertation, University of Virginia, 1979.

Bell, R. Q. Symposium; Parent, child and reciprocal effects: New experimental approaches. *Journal of Abnormal Child Psychology,* 1981, **9**, 299–301.

Borkovec, T. D. Autonomic reactivity to sensory stimulation in psychopathic, neurotic and normal delinquents. *Journal of Consulting and Clinical Psychology,* 1970, **35**, 217–222.

Braukmann, C. J., Kirigin, K. A., & Wolf, M. M. Achievement Place: The researcher's prospective. Paper presented at the meeting of the American Psychological Association, Washington, D.C.: 1976.

Cartwright, D. S., Howard, K. I., & Reuterman, N. A. Multivariate analysis of gang delinquency: IV. Personality factors in gangs and clubs. *Multivariate Behavioral Research,* 1980, **15**, 3–22.

Cattell, R. B. *The objective–analytic personality factor battery.* Champaign, Ill.: Institute for Personality and Ability Testing, 1955.

Cowen, E. L., Orgel, A. R., Gesten, E. L., & Wilson, A. B. The evaluation of an intervention program for young school children with acting-out problems. *Journal of Abnormal Child Psychology,* 1979, **7**, 381–396.

DeMyer-Gapin, S., & Scott, T. J. Effects of stimulus novelty on stimulation seeking in antisocial and neurotic children. *Journal of Abnormal Psychology,* 1977, **86**, 96–98.

Dietrich, C. Differential effects of task and reinforcement variables on the performance of three groups of behavior problem children. *Journal of Abnormal Child Psychology,* 1976, **4**, 155–171.

Empey, L. T. *American delinquency: Its meaning and construction.* Homewood, Ill.: Dorsey, 1978.

Federal Bureau of Investigation. *Uniform crime reports for the United States.* Washington, D.C.: USGPO, 1979.

Fodor, E. M. Delinquency and susceptibility to social influence among adolescents as a function of level of moral development. *Journal of Social Psychology,* 1972, **86**, 257–260.

Fowles, D. C. The three arousal model: Implications of Gray's two-factor learning theory for heart rate, electrodermal activity, and psychopathy. *Psychophysiology,* 1980, **17**, 87–104.

Gorenstein, E. E., & Newman, J. P. Disinhibitory psychopathology: A new perspective and a model for research. *Psychological Review,* 1980, **87**, 301–315.

Graham, P. Epidemiological studies. In H. C. Quay & J. S. Werry (Eds.), *Psychopathological disorders of childhood* (2nd ed.). New York: Wiley, 1979.

Graham, J., & Rutter, M. Psychiatric disorder in the young adolescent: A follow-up study. *Proceedings of the Royal Society of Medicine,* 1973, **66,** 1226–1229.

Henn, F. A., Bardwell, R., & Jenkins, R. L. Juvenile delinquents revisited: Adult criminal activity. *Archives of General Psychiatry,* 1980, **37,** 1160–1163.

Hetherington, E. M., & Martin, B. Family interaction. In H. C. Quay & J. S. Werry (Eds.), *Psychopathological disorders of childhood* (2nd ed.). New York: Wiley, 1979.

Hetherington, E. M., Stouwie, R., & Ridberg, E. H. Patterns of family interaction and child rearing attitudes related to three dimensions of juvenile delinquency. *Journal of Abnormal Psychology,* 1971, **77,** 160–176.

Ingram, G. L., Gerard, R. E., Quay, H. C., & Levinson, R. B. An experimental program for the psychopathic delinquent. *Journal of Research in Crime and Delinquency,* Jan. 1970, pp. 24–30.

Jesness, C. F. Comparative effectiveness of behavior modification and transactional analysis programs for delinquents. *Journal of Consulting and Clinical Psychology,* 1975, **43,** 758–779.

Jurkovic, G. J., & Prentice, N. M. Relation of moral and cognitive development to dimensions of juvenile delinquency. *Journal of Abnormal Psychology,* 1977, **86,** 414–420.

Kelly, G. *The psychology of personal constructs.* New York: Norton, 1955.

Lavik, N. J. Urban–rural differences in rates of disorder. A comparative psychiatric population study of Norwegian adolescents. In P. J. Graham (Ed.), *Epidemiological approaches in child psychiatry.* London: Academic Press, 1977.

Lee, R., & Haynes, N. M. Project CREST and the dual-treatment approach to delinquency: Methods and research summarized. In R. R. Ross & P. Gendreau (Eds.), *Effective correctional treatment.* Toronto: Butterworths, 1980.

Lefkowitz, M. M., Eron, L. D., Walder, L. O., & Huesmann, L. R. *Growing up to be violent: A longitudinal study of the development of aggression.* New York: Pergamon, 1977.

Leslie, S. A. Psychiatric disorder in the young adolescents of an industrial town. *British Journal of Psychiatry,* 1974, **125,** 113–124.

Lueger, R. J. Person and situation factors influencing transgression in behavior-problem adolescents. *Journal of Abnormal Psychology,* 1980, **89,** 453–458.

Megargee, E. I., & Golden, R. E. Parental attitudes of psychopathic and subcultural delinquents. *Criminology,* Feb. 1973, pp. 427–439.

Mitchell, S., & Rosa, P. Boyhood behavior problems as precursors of criminality; A follow-up study. *Journal of Child Psychology and Psychiatry,* 1981, **22,** 19–33.

Nasby, W., Hayden, B., & DePaulo, B. M. Attributional bias among aggressive boys to interpret unambiguous social stimuli as displays of hostility. *Journal of Abnormal Psychology,* 1980, **89,** 459–468.

Olweus, D. Stability of aggressive reaction patterns in males: A review. *Psychological Bulletin,* 1979, **86,** 852–875.

Olweus, D. Familial and temperamental determinants of aggressive behavior in adolescent boys: A causal analysis. *Developmental Psychology,* 1980, **16,** 644–660.

Orris, J. B. Visual monitoring performance in three subgroups of male delinquents. *Journal of Abnormal Psychology,* 1969, **74,** 227–229.

Palmer, T. The Youth Authority's Community Treatment Project. *Federal Probation,* 1974, **38,** 3–14.

Palmer, T. Martinson revisited. *Journal of Research in Crime and Delinquency,* 1975, **12,** 133–152.

Pichel, J. A long-term follow-up of 60 adolescent psychiatric out patients. *American Journal of Psychiatry,* 1974, **131,** 140–144.

Quay, H. C. (Ed.) *Juvenile delinquency: Theory and research.* Princeton, N.J.: Van Nostrand, 1955.

Quay, H. C. Psychopathic personality as pathological stimulation-seeking. *American Journal of Psychiatry*, 1965, **122**, 180–183.

Quay, H. C. Psychopathic behavior: reflections on its nature, origins and treatment. In F. Weizmann & I. Uzgiris (Eds.), *The structuring of experience*. New York: Plenum, 1977. (a)

Quay, H. C. The three faces of evaluation: What can be expected to work. *Criminal Justice and Behavior*, 1977, **4**, 341–354. (b)

Quay, H. C. Classification. In H. C. Quay & J. S. Werry (Eds.), *Psychopathological disorders of childhood* (2nd ed.). New York: Wiley, 1979.

Quay, H. C., & Love, C. T. The effects of a juvenile diversion program on rearrests. *Criminal Justice and Behavior*, 1977, **4**, 377–396.

Richman, L. C., & Lindgren, S. D. Verbal mediation deficits: Relation to behavior and achievement in children. *Journal of Abnormal Psychology*, 1981, **90**, 99–104.

Rickel, A. U., & Dyhdalo, L. A two year follow-up study of a preventive mental health program for preschoolers. *Journal of Abnormal Child Psychology*, 1982, in press.

Robins, L. N. *Deviant children grown up*. Baltimore: Williams & Wilkins, 1966.

Robins, L. N. Sturdy childhood predictors of adult antisocial behavior: Replications from longitudinal studies. *Psychological Medicine*, 1978, **8**, 611–622.

Robins, L. N. Follow-up studies. In H. C. Quay & J. S. Werry (Eds.), *Psychopathological disorders of childhood* (2nd ed.). New York: Wiley, 1979.

Ross, R. R., & Gendreau, P. *Effective correctional treatment*. Toronto: Butterworths, 1980.

Rutter, M. *Changing youth in a changing society*. Cambridge, Mass.: Harvard University Press, 1980.

Rutter, M., Graham, P., Chadwick, O. F. D., & Yule, W. Adolescent turmoil: Fact or fiction. *Journal of Child Psychiatry and Psychology*, 1976, **17**, 35–56.

Sandberg, S. T., Rutter, M., & Taylor, E. Hyperkinetic disorder in psychiatric clinic attenders. *Developmental Medicine and Child Neurology*, 1978, **20**, 279–299.

Schuessler, K. R., & Cressey, D. R. Personality characteristics of criminals. *American Journal of Sociology*, 1950, **55**, 476–484.

Seidman, E., Rappaport, J., & Davidson, W. S., III. Adolescents in legal jeopardy: Initial success and replication of an alternative to the criminal justice system. In R. R. Ross & P. Gendreau (Eds.), *Effective correctional treatment*. Toronto: Butterworths, 1980.

Short, J. F., Jr., Tennyson, R. A., & Howard, K. I. Behavior dimensions of gang delinquency. *American Sociological Review*, 1963, **28**, 411–428.

Siddle, D. A. T., Nicol, A. R., & Foggitt, R. H. Habituation and over-extinction of the GSR component of the accenting response in anti-social adolescents. *British Journal of Social and Clinical Psychology*, 1973, **12**, 303–308.

Skrzypek, G. J. Effect of perceptual isolation and arousal on anxiety, complexity preference, and novelty preference in psychopathic and neurotic delinquents. *Journal of Abnormal Psychology*, 1969, **74**, 321–329.

Spencer, F. W. Cognitive characteristics of four subgroups of delinquent males. *Criminal Justice and Behavior*, 1980, **7**, 387–399.

Stewart, D. J. Effects of social reinforcement on dependency and aggressive responses of psychopathic, neurotic and subcultural delinquents. *Journal of Abnormal Psychology*, 1972, **79**, 76–83.

Tramontana, M. G. Critical review of research on psychotherapy outcome with adolescents: 1967–77. *Psychological Bulletin*, 1980, **88**, 429–450.

Trasler, G. Relations between psychopathy and persistent criminality—Methodological and theoretical issues. In R. D. Hare & D. Schalling (Eds.), *Psychopathic behaviour: Approaches to research*. New York: Wiley, 1978.

Warren, M. Q. Correctional treatment and coercion: the differential effectiveness perspective. *Criminal Justice and Behavior*, 1977, **4**, 355–376.

West, D. J., & Farrington, D. P. *Who becomes delinquent?* London: Heinemann, 1973.

West, D. J., & Farrington, D. P. *The delinquent way of life.* London: Heinemann, 1977.

Whitehill, M., DeMyer-Gapin, S., & Scott, T. J. Stimulation-seeking in antisocial preadolescent children. *Journal of Abnormal Psychology,* 1976, **85**, 101–104.

Wolfgang, M. E., Figlio, R. M., & Sellin, T. *Delinquency in a birth cohort.* Chicago: University of Chicago Press, 1973.

22

ADOLESCENT SUBCULTURE IN THE SCHOOLS

Kent A. McClelland
University of Miami

Just about 20 years ago sociologist James S. Coleman (1961) published the results of one of the most ambitious studies ever undertaken of American high schools. On the basis of surveys conducted among students, parents, and teachers from 10 Midwestern high schools, Coleman concluded that adolescents in high school tend to be cut off from contact with adults and that this isolation has spawned "a set of small teen-age societies, which focus teen-age interests and attitudes on things far removed from adult responsibilities, and which may develop standards that lead away from those goals established by the larger society" (1961, p. 9). These adolescent societies share an adolescent subculture, Coleman said, that emphasizes athletic success and popularity, rather than academic work. Coleman viewed the situation with some alarm, and he recommended a number of ways that parents and schools could diminish the power of the adolescent subculture and reassert the priority of academics (1961, Ch. 11). This chapter traces the controversies arising from Coleman's depiction of an "adolescent society" and tries to assess the extent to which Coleman's view can still be applied to American high schools in the 1970s and 80s.

OVERVIEW OF RESEARCH ON HIGH SCHOOL STUDENT SUBCULTURE

Coleman was by no means the first to discern an adolescent subculture that repudiated adult interests. Coleman's work, in fact, stands squarely in the tradition of sociological functionalism, as articulated by Talcott Parsons (1954) who in 1942 had described an American "youth culture" based on athletics, having a good time, and participation in school-sponsored social activities. Parsons saw the youth culture as "a product of tension in the relationships of younger people and adults," but he also observed that "the youth culture has important positive functions in easing

---

The author would like to thank Charles F. Longino, Aaron Lipman, and Craig McEwen for their help and comments in the preparation of this chapter.

the transition from the security of childhood in the family of orientation to that of full adult in marriage and occupational status" (1954, pp. 93, 101). From the functionalist point of view a society undergoing rapid social change cannot afford to have children following too closely in the footsteps of their parents. Thus an adolescent's participation in youth culture, in spite of some negative aspects, was seen as a natural part of growing up in modern society. C. Wayne Gordon (1957), in a monograph entitled *The Social System of the High School,* provided the classic study of a high school from this perspective.

The term *subculture,* which Coleman (1961) used in preference to *youth culture,* had come into currency among sociologists in the late forties. Milton M. Gordon had in 1947 provided what has since come to be regarded as the orthodox definition of subculture as

> ... a concept used here to refer to a sub-division of a national culture, composed of a combination of factorable social situations such as class status, ethnic background, regional and rural or urban residence, and religious affiliation but forming in *their combination a functioning unity which has an integrated impact on the participating individual.* (Gordon, 1970, p. 32, emphasis in original)

Gordon went on to give as examples of persons participating in subcultures the lower-class black Protestant from the rural South or the middle-class Jew from the urban Northeast. He noted that the middle-class, white, Midwestern Protestant could not be said with any certainty to participate in a subculture, because it was not clear that the impact of this combination of social identities on the individual was in any way different from participation in the national culture as a whole (1970, p. 34).

Sociologists quickly adopted the concept of subculture and just as quickly began to apply it to groups that were not homogeneous ethnically and religiously, not confined to a particular region of the country, nor composed of people of both sexes and all ages, as Gordon's definition had specified. As early as 1946, researchers studying delinquency were describing the "delinquent subcultures" of teenaged gangs that engaged in deviant and aggressive behaviors (see Chapter 21), and the use of the idea of subculture for describing and explaining delinquency is still popular today, with Albert Cohen's (1955) theory representing the landmark study in this area. The concept of subculture was also applied to certain occupations, especially deviant ones (see Sutherland, 1970, and Hollingshead, 1970). By the early sixties, even elderly people were being described as belonging to a subculture of their own (Rose, 1962; Longino, McClelland, & Peterson, 1980). The common thread in these various applications of the term *subculture* was that the group in question was seen as having a distinctive pattern of norms, values, dialect or jargon, ways of using material objects, and so forth, in other words, a culture that was different from the national culture in many respects, though similar in others. How different the culture or group had to be, in order to qualify as a subculture, was never clearly stated. Thus by the time of Coleman's (1961) study of high schools, the idea of subculture had proved itself a highly flexible, if perhaps imprecise, weapon in the sociologist's arsenal of analytical concepts.

Coleman's (1961) study differed from earlier treatments of youth culture, or adolescent subculture, mainly in its pessimistic tone. Working from the functionalist premise that social systems developing in relative isolation will evolve distinctive norms and values (see Hollingshead, 1970). Coleman concluded that, in high schools, the process had gone too far. The emergent adolescent subculture tended to alienate students from their parents. Moreover, the values of the subculture subverted the major function of the school: academic achievement. Current historians of education might find Coleman's implicit assumption that the sole aim of education was academics a bit amusing in light of the previous 5 or 6 decades of progressive education, which had stressed social and moral, rather than purely academic, education of children (see Cremin, 1964). Coleman's thinking was evidently influenced, however, by the calls for better science training and a return to the basics that had followed the Soviet launching of Sputnik in 1957. The educational solutions that Coleman offered, all couched in the language of a dedicated social engineer, also seem to have been influenced by the spirit of earnest reformism that characterized the early sixties (cf. Conant's, 1959, plans for expanding the size of high schools.)

While Coleman's work generated immediate controversy (e.g., Berger, 1963a; Matza, 1964) and has continued to be widely cited, the arrival of the hippies in 1965–67 (see Irwin, 1977) and the ensuing youth movement of the late sixties gave a new focus for the concerns of those social scientists interested in adolescents. Quite a voluminous literature was written in response to the "counter culture" (Roszak, 1969) that swept across college campuses during this period. Much of this literature deals with the psychological backgrounds of campus radicals (Keniston, 1968, 1971), campus political demonstrations (Lipset, 1972; Becker, 1970), or the "generation gap" (Mead, 1970; Bengston, 1970). The movement on college campuses also had an impact on the high schools (cf. Liberale & Seligson, 1970; Gross & Osterman, 1971), but most of the relevant literature pertains to college-age youth, and this chapter will not treat it in any detail.

With the decline of the youth movement in the early seventies, interest in adolescence on the part of social scientists seemed to wane as well. A report of a presidential science advisory committee chaired by Coleman (Panel on Youth, 1974) and two yearbooks of the National Society for the Study of Education (Gordon, 1974; Havighurst & Dreyer, 1975) marked the effective end to the body of American literature devoted to the sixties' youth movement.

At the same time as American scholars have apparently been losing interest in issues related to adolescent subculture, a good deal of lively and provocative work on the subject has been produced by a group of British Marxist sociologists. The theoretical background of their position can be traced to Cohen's (1955) seminal work on the formation of delinquent subcultures. Cohen argued that groups of boys, faced with the common psychological problem of inability to obtain status in the eyes of the wider community, would solve this problem by evolving sets of subcultural norms and values that support delinquent behavior as an alternative source of status. This basic theory was refined and given a Marxist twist by P. Cohen (1972) in a paper examining the various subcultures of working-class youths (mods, skinheads, etc.) that grew up in London's East End in the fifties and sixties. The paper described how urban redevelopment partially destroyed what had been a stable working-class neighborhood and left many youths unemployed or with poor economic prospects. He then argued that the youths reacted to the contradictions and strains imposed upon their "parent" working-class culture by seeking magical solutions in the area of subcultural style.

Other British sociologists seized on the idea of subculture styles as representing imaginary or ideological solutions to unresolvable material problems. Hebdige (1979) extended the analysis to the punk and reggae subcultures of the seventies. Willis (1978) focused on motorbike boys and hippies, while Brake (1980) has provided an extensive analysis and review of this literature. Collections of articles in this vein are contained in Hall and Jefferson (1976) and in Mungham and Pearson (1976). Tanner (1978) reviews and offers a critique of some of these works.

American readers may have some difficulties with the British Marxist group's insistence that most youth subcultures (except the hippies—see Clarke, Hall, Jefferson, & Roberts, 1976) are working class in origin. Social class has tended to be less clearly defined in the United States than in England. The material on British subcultures does have relevance to American high schools, however, to the extent that subcultural styles regularly cross the Atlantic in the form of media coverage and pop music. The new wave rock music, for instance, which is currently popular in the United States, got its start with British punk rock in the mid-seventies.

A few recent American studies of high school students deserve special notice. Cusick (1973a,b) has produced an insightful participant-observation study of a high school in the seventies. Larkin (1979), in another participant-observation study, has arrived independently at much the same sort of analysis as recent British sociologists offer. Irwin (1977) has chronicled the rise and fall of the California-surfing subculture of the fifties and the hippie subculture of the sixties. While Irwin's work deals mainly with school dropouts, he does describe how the surfing style spread through Los Angeles high schools and how the hippie culture then recruited from northern California high schools (1977, pp. 131–135). Boocock (1980, pp. 212–241) has recently provided a comprehensive review of the literature on adolescent subculture.

SUMMARY OF THE ARGUMENT TO BE PRESENTED

Our purpose in the remainder of this chapter will be to review the theoretical and empirical objections that have been raised to Coleman's (1961) treatment of adolescent subculture in high schools and to examine these objections in light of the evidence from more recent studies of high schools and high school aged adolescents. Coleman's main points can be summarized in the form of five propositions:

1. Adolescent subculture has become increasingly different from adult culture, as teenagers have more and more formed a separate "small society."
2. Adolescent subculture is a unified and pervasive phenomenon in which almost every high school student from nearly every high school in the country participates.
3. Adolescent subculture alienates high school students from their parents and orients them toward their peers.
4. Adolescent subculture promotes behavior that is antagonistic to the most important goals (i.e., academic) of the high school.
5. Consequently, adolescent subculture stands in opposition to the culture of the wider society. As such, adolescent subculture constitutes a social problem.

Coleman, of course, did not phrase his conclusions in quite the unqualified language of these propositions, but they do represent the essence of his position. Very possibly, an important reason for the great volume of attention given to Coleman's book over the last 2 decades has been that he framed his argument in this simple and straightforward manner (while his data were impressively complex). Nevertheless, most subsequent authors have been critical of Coleman's thesis; moreover, most have argued that Coleman's viewpoint was, in fact, oversimplified.

Each of the five propositions that summarize Coleman's point of view has been subsequently attacked. Our objective here will be to detail and to evaluate these attacks. We will treat propositions one and two together. These propositions deal with the unity of adolescent subculture and its increasing separation from adult culture. Most authors who have objected to Coleman's vision of an adolescent subculture that is pervasive and unified have seen, instead, two or more competing subcultures with differing degrees of separation from the wider culture. Thus objections to propositions one and two have ordinarily gone hand in hand.

Propositions three, four, and five also naturally fall together. Proposition three deals with opposition between adolescent subculture and parents; four, with opposition to the school; and five, with opposition to the wider culture (an implicit consequence of propositions three and four). Many authors have accepted, in principle, points one and two of Coleman's thesis but have argued that Coleman exaggerated the degree of opposition between high school student and parents or between student and the school. Other authors, usually those seeing multiple adolescent subcultures, have noted that parents themselves come from a variety of subcultures and that the school organization possesses a multiplicity of goals, and, thus, that Coleman's simplified view of events failed to do justice to the complexity of the actual situation. In the same vein, critics of proposition five have argued that the national culture is a complex, many-stranded, often internally self-contradictory phenomenon, so that apparent opposition between one adolescent subculture or another and the wider culture may mask a deeper unity. We will take up these arguments one by one in subsequent sections. Because a good deal of the pertinent literature deals only with parents and peers or else with opposition between adolescent subculture and the schools, we will treat propositions four and five in separate, but parallel, sections.

The final section of this chapter argues that the controversy over adolescent subculture has laid bare some basic weaknesses in sociological subculture theory. We end with a discussion of the implications of these conclusions for future investigations of the social world of the high school.

COLEMAN'S CRITICS

The Separation and Unity of Adolescent Subculture

Coleman and the functionalist observers of high school who preceded him (e.g., Gordon, 1957) saw high school classes as having a complicated social structure, usually with a large number of subgroups or cliques (see Coleman, 1961, pp. 175–182). This structural complexity, however, formed a unified social system, dominated in each high school by a "leading crowd" of the most athletic boys and most popular girls (1961, pp. 34–43, 97–172). Students, whether members of the leading crowd or not, shared common cultural assumptions and values. The adolescent subculture was seen as being basically similar from high school to high school as well, although some value differences were noted for different schools (Coleman, 1961, pp. 58–96, 279–310). For instance, in some schools the status system was apparently based solely on the popularity that accompanied athletic prowess, while in others, the "well-rounded" student who combined good grades with athletics had an advantage. None of the schools studied, however, had a status system that rewarded academic achievement alone.

Growing unity of the adolescent subculture, in Coleman's view, was occurring in conjunction with a growing separation between the culture of the wider society and the cultural world of the high school. Popular music, movies, and the teenage market were more and more catering to the special interests and values of the high school student (1961, p. 4), and although he admitted he had no proof, Coleman believed that the gulf between adolescents and adults was widening. As he put it, "These young people speak a different language. What is more relevant to the present point, the language they speak is becoming more and more different" (1961, p. 3).

The issue of the unity of adolescent subculture was one on which several alternatives to Coleman's view were already available. Hollingshead (1949) in his pioneering study of the "Elmtown," Indiana, high school (one of the schools Coleman also chose to study) had found a sharp discontinuity between groups of high school students based on social class. Working-class students had been seen as rebellious, likely to drop out of school, and likely to engage in delinquent activities, while middle-class students had fallen into the athletics-and-popularity mold of adolescent subculture described by Coleman. Research on delinquency in the fifties (e.g., Cohen, 1955) concentrated on delinquent subculture, which was seen as very different from the subculture of the average nondelinquent teenager. (See Chapter 21)

The degree of separation perceived between adult culture and adolescent subculture depended on which adolescent subculture one was talking about. Delinquent subcultures, of course, were thought of as antithetical to the values of the wider culture. Yinger, for instance, coined the term *contracultures* to be applied to subcultures of delinquency (1960). The high school subcultures of more conforming adolescents, on the other hand, were seen as much less separate from the wider culture. In response to Coleman, Berger (1963a) argued that the average high school student's interest in athletics and lack of interest in academic studies was a trait shared with most of the adults in the community. Thus the allegedly distinctive adolescent subculture only reflected community norms. Matza (1964b) agreed that Coleman had exaggerated the degree of separation between adult culture and the subculture of the majority of conventional adolescents.

By the time that Coleman's work was appearing, other observers had taken the dichotomy between the subculture of conforming youth and the delinquent subculture of rebellious youth and turned it into one of several typologies of multiple adolescent subcultures by subdividing either the conformity or rebellion category. Matza (1961) suggested three forms of youthful rebellion: delinquency; radicalism, or the advocacy of political revolution; and bohemianism, the romantic rejection of conventional moral and aesthetic values. In a later work Matza (1964b) continued by subdividing the conventional adolescent subculture into several "traditional" styles: scrupulosity, meticulous adherence to religious values (especially among Catholic youth); studiousness; and concentration on sports. Burton Clark (1962) expanded the dichotomy between conventionality and delinquency to a three-way categorization of high school subcultures: the fun subculture, similar to Coleman's description; the academic subculture, corresponding to Matza's studious style; and the delinquent subculture (pp. 244–270). Later, Jere Cohen (1979) reanalyzed some of Coleman's data and found student behavior traits clustering into three factors which Cohen interpreted as corresponding to the fun, academic, and delinquent subcultures of Clark's suggested typology.

Other typologies of adolescent subcultures have continued to appear. Young (1971) offered the typology of conformist, delinquent, and bohemian youth cultures. Brake (1980) denied that conformist youth could even be considered a subcultural category, since there is relatively little conflict between their values and those of the wider society (p. 23). Among rebellious youth, however, Brake distinguished four subcultural types: delinquent subcultures, cultural rebellion (or bohemianism), reformist political movements, and political militancy (or radicalism; 1980, pp. 4–5). Matza, Young, and Brake would, no doubt, have agreed that bohemian and politically oriented subcultures draw more adherents from among older youth than high school students. Nevertheless, all of them held that Coleman's vision of a unified adolescent subculture was oversimplified even when applied only to high school aged adolescents.

The trend in the last decade has been toward viewing adolescent subcultures as so many and so fragmented that no simple typology is possible. This view seems to have arisen in response to the youth movement of the late sixties that began to disintegrate in the early seventies. The arrival of the long-haired hippies on the scene in the sixties made it clear to all observers that the stereotype of the leather-jacketed "greaser," which had been so popular in the late fifties and early sixties, could no longer serve even to typify even rebellious youth. When the hippie movement lost momentum and splintered into isolated groups of Jesus freaks, underground radicals, drug takers, black militants, and individuals interested only in doing their own thing, the image of a relatively unified adolescent subculture was irretrievably shattered.

For example, Gottlieb and Heinsohn (1973) saw "a wide variety of cultures and subcultures" in American public high schools and complained that American social science had given too much attention to white, middle-class students, so that little was known about the cultures of minorities and working-class adolescents. Participant-observation studies in high schools during the seventies seemed to underscore the proposition that student subcultures were fragmented. Palonsky, for instance, portrayed the high school social structure he studied as composed of a number of small, unrelated groups with vividly contrasting cultural styles (1975). Larkin (1979) painted much the same picture of the school he visited. Cusick summed up his participant observation in a high school as follows:

> Personally, I thought that having worked in public schools for eight years, I knew something about the "adolescent society." What I saw in Gates, however, surprised me. I had previously seen adolescents in any particular school as socially homogenous. While I knew there were more or less prestigious students, achievers, and deviants, I had thought that they had more in common with one another, and had some common interests about which they communicated. That wasn't true of

students in Gates. The groups there really were discrete social units, narrowly bounded not only by age, sex, and neighborhood, but chiefly by interests. . . . Where is this "adolescent society" of which we speak? The data suggest that rather than a society or even a subculture, there is a fragmented series of interest groups revolving around specific items and past patterns of interaction. These groups may be the important social referent, not some mythical subculture or "adolescent society." (1973b, p. 161)

In another study, Todd (1979) gave a sketchy report of a study of a high school that revealed, among a large number of only weakly differentiated student subgroups, two main subcultures: a "citizens" group, who participated extensively in officially sponsored school activities and a "tribe" group, who rejected many school values and activities but who nevertheless were not actively deviant in behavior.

If high schools are, in fact, less well integrated socially and culturally than in the past, the reasons may have to do with changes in the structure of the high school, as well as with the growth and decline of youth movements as discussed earlier. Following Conant's (1959) recommendation that high schools be consolidated, the average size of high schools has increased to several thousand students instead of several hundred. Moreover, busing to achieve racial desegregation, which began in the late sixties, has brought together, in many cases, groups of students that formerly attended separate high schools. In short, the same trends that have led to greater fragmentation of social life and impersonality in social relations for the wider society may well have had their effects on high schools.

It is possible, however, that the image of fragmentation in adolescent subcultures has been overdrawn by recent observers. Rigsby and McDill (1975) point out, with some justice, that the researcher's perception of number of student subcultures may depend in part on the level of analysis undertaken. Coleman was concerned mainly with comparing schools and found much similarity between them in the broad outlines of their cultures. Subsequent researchers have rarely had data on as many different schools and, thus, have usually concentrated on describing what differentiates students within a single school.

Nevertheless, the evidence presented here shows that Coleman's version of a unified, separate adolescent subculture has been rejected by most later observers, who opted either for some typology of contrasting adolescent subcultures or, more recently, for the position that adolescent subcultures are too many and varied to catalog. Murdock and McCron (1976, p. 14) have even claimed that Coleman's unified model of subculture failed to fit 4 of the 10 high schools in his original study.

Since Coleman's time, we conclude, the processes of growth, consolidation of districts, and desegregation are combining with the unsettling influence of the youth movement in the late sixties to make high schools ever more culturally diverse. Whether the retrenchment, nostalgia, and apathy of the late seventies have led, in some high schools, to the return of a simpler cultural outlook and social structure reminiscent of the fifties is a question not yet dealt with in the literature on adolescents.

Cultural Opposition between Adolescents and Adults

Coleman buttressed his position on the antagonism between adolescent subculture and adult culture by producing data from questionnaires administered to about

7500 high school students in his study. One of the questions he asked the students in the 10 high schools he studied was whether it would be harder to take your parents' disapproval, your teachers' disapproval, or breaking with your friend. Slightly over half the students responded that their parents' disapproval would be hardest to take, while between 40 and 45% of the sample felt that breaking with a friend would be more difficult. Only a tiny percentage regarded their teacher's disapproval as most serious. Members of the leading crowd, as Coleman defined it in each school, were a bit more likely than others to see breaking with a friend hardest to take, and Coleman interpreted these results as indicating that the peer group subculture tended to pull students away from their parents and, to an even greater degree, from their teachers (1961, pp. 5–6).

In another section of the questionnaire, Coleman asked students how they would like to be remembered at their schools. From the alternatives presented, about 45% of the boys chose "athletic star." Slightly fewer than a third of the boys preferred to be remembered as "brilliant students," and the remainder as "most popular." The girls got to choose from among "brilliant student," "leader in activities," and "most popular." Each alternative was selected by about a third of the sample, although "brilliant student" came in last. By contrast, a parallel question was asked to a small sample of parents of these students, and a majority of parents responded that they would prefer their child to be remembered as a brilliant student (considerably more often for boys than girls). Coleman again interpreted these results as demonstrating the strong differences between adolescent subculture and adult culture (1961, pp. 28–34).

Coleman made little distinction between the cultural values of parents, the school, and the general society. His implicit assumption was that all shared in viewing academic achievement as the one goal of paramount value. Other literature in the area has usually treated opposition between peers and parents and adolescent antagonism toward the school as separate issues, however, and we will deal with these issues separately as well. The final subsection in this part of the chapter looks beyond parents and the school to the culture of the general society and the degree to which that national culture and adolescent subcultures can be said to be opposed.

Opposition of Parents and Peers. Coleman's suggestion that high school students were becoming increasingly oriented to peers and alienated from parents has been attacked more often than defended in the last 2 decades. The first skirmish in the attack came actually several years before the appearance of Coleman's volume, when Elkin and Westley (1955, 1957) reported that the small sample of middle-class Canadian youth they studied were very much more oriented to their parents than their peers and that an "oppositional" peer group did not exist among them. Therefore, they concluded, adolescent subculture was not universal. The notion that relations between parents and teenagers were normally stormy or stressful had not originated with Coleman, of course; the idea goes back at least to G. Stanley Hall, who in 1904 produced a voluminous work on adolescence (Hall, 1969). More recently, Davis (1940) had expounded, from a functionalist perspective, "the sociology of parent–youth conflict."

The Elkin and Westley line of argument was applied directly to Coleman's findings by Epperson (1964), who contended that the way Coleman had asked his questions had stacked the cards in favor of his conclusion. Specifically, Epperson pointed out that Coleman had asked students to weigh the seriousness of breaking off relations with a friend against merely incurring the disapproval of parents. Epperson thought that balancing friend's disapproval against parents' disapproval or else leaving home against breaking with the friend might be fairer tests. Even so, Epperson argued, Coleman's actual findings (as opposed to his conclusions) indicated that parents were more important than peers to high school students.

Coleman's findings concerning the relative preference of high school students for academics versus athletics came into question as well. Hillman (1969) tried to replicate Coleman's question with a small sample from a middle-class high school in the mid-sixties. She found that being remembered as a brilliant student ranked first among this sample, with student leaders and children of parents with white-collar occupations showing the greatest preference for the academic alternative. Snyder (1969) reported another attempt at replicating Coleman's findings, this time in a study with a longitudinal design. Although a small sample of high school seniors had, like Coleman's sample, preferred to be remembered for their athletic, rather than academic accomplishments, when about half this sample were followed up 5 years later, they expressed views much more favorable to academics. Moreover, a preference for athletics, activities, or popularity in high school had in no way precluded subsequent academic achievement in college. Snyder concluded that the effects of any adolescent subculture were only temporary.

Kandel and Lesser (1972) continued the attack on Coleman's conclusions about opposition between parents and peers. Reporting on survey results from a sample of about a thousand students from several American high schools and a similar number of Danish students, Kandel and Lesser argued that relationships between adolescents and their parents were typically "close and harmonious" and that, contrary to what Coleman had implied, an orientation to peers and an orientation to parents were not mutually exclusive; students were often closely aligned with both parents and peers or, in a few cases, alienated from both (1972, pp. 7–8, 86). Rutter, Graham, Chadwick, and Yule (1976) summarized the literature attacking Coleman on this issue and then added some evidence from clinical studies of British adolescents. Normal adolescents reported very little in the way of serious trouble with their parents, and Rutter et al. concluded that the stereotype of stormy relations between parents and teenagers was largely a myth.

Not all the literature has been unequivocally critical of Coleman's stand on parent–peer relations, however; several authors have concluded that both parents and peers exert considerable influence on the average teenager. Brittain (1963) suggested that the relative influence of parents and peers would depend on the situation, with peers having more influence on current activities and parents, on future plans. Kandel and Lesser, in the study described in the paragraph above, found more or less this pattern (1972, pp. 120, 135). Larson (1972a,b), in a study of about 1500 Oregon high school students, found that most students tended to be oriented to the wishes of their parents. In hypothetical decision-making situations, however, students tried to balance parent and best-friend demands. For the most serious situations, Larson interpreted his results as showing that the adolescents rejected both parent and peer influence and consulted their own consciences. By contrast, Floyd and South (1972) in a similar type of study reported that peer pressures generally outweighed parental influence. In sum, no easy generalization that either parents or peers have paramount influence over adolescents seems to be supported by the available evidence.

Findings from another body of literature relating to parent and peer influences are not any more clear-cut. Sociologists have been particularly interested in how high school students make decisions about going to college, because educational plans and aspirations are seen as an important intervening variable in the transmission of socioeconomic status from one generation to another (Sewell & Hauser, 1975). Haller and Butterworth (1960) found that, even controlling for social status, the educational aspirations of adolescent best-friend pairs were correlated with each other; they concluded that peers had an influence on the decision to go to college. Alexander and Campbell (1964) reported similar results. McDill and Coleman (1965) reported on a follow-up survey with part of the original Coleman sample; they found that membership in the school's elite peer group was more predictive of college plans for seniors than it had been when they were underclassmen, although it did not overbalance the influence of parents' expressed wishes even in the senior year. McDill and Coleman interpreted these results as indicating that the adolescent subculture exerted increasing influences on students the longer they were in high school. Kandel and Lesser (1969, 1972) found the high school students in their sample to have educational plans more similar to those articulated for them by their mothers than to those of their best friends. Williams (1972) surveyed a sample of 5000 Canadian students in grade 10 and followed them up in grade 12. He found that, when other factors were controlled, a measure of peer group aspirations in grade 10 had no statistical effect on the student's own educational aspirations as a senior. Alexander and Eckland (1975) presented a statistical model that showed peers and parents with roughly equal influence on educational expectations. Picou and Carter (1976) distinguished influences in the form of directly expressed advice or expectations from the more subtle influences of modeled behavior. They reported that, for their sample of over 3000 Louisiana high school students, direct expressions of parental preferences about going to college had greater impact on aspirations than friends' direct expressions of encouragement or lack of it. Nevertheless, modeled behavior, in the form of friends' own college plans, had the strongest statistical influence on the student's aspirations (Picou & Carter, 1976). This collection of contradictory results defies any simple summary. Apparently, under various sets of conditions, either parental pressures, or peer influences, or both may be important in determining a high school student's plans for further education.

Review of the literature on opposition between parent and peer influences has left us in something of a muddle: Coleman's extreme position that adolescent subculture has alienated students from their parents must clearly be rejected, but the proposition that high school students are immune to peer influence has hardly been established in its place. Some adolescents have been reported to have been oriented mainly to their parents, some to peers, some to both, and some to neither. There seems to be nothing sensible to be said of the typical high school student on this subject.

The muddle lasts only as long as one accepts, with Coleman, the implicit assumption that we are talking about a single adolescent subculture that is more or less uniform across the nation and around the world. If, as was argued in the preceding section, there are a number of increasingly fragmented adolescent subcultures with different degrees of separation from and opposition to parent culture, then the contradictory results are exactly what one would expect. It becomes nonsensical to worry about the *typical* high school student, because one does not exist. Instead, the behavior of any given high school student must be understood in terms of the particular subculture in which he participates. Future examinations of the problem of parent versus peer influence will need to be conducted in these somewhat more sophisticated terms.

Opposition to the School. Just as Coleman saw the adolescent subculture setting teenagers in conflict with their parents, he saw the values of the adolescent subculture as corrosive of the central value of the school—academic achievement. Teachers' opinions, Coleman reported, had little influence on high school students in comparison with the influence of parents and peers (1961, p. 5). Students would rather have been remembered as star athletes or leaders in activities than as brilliant students, and athletics and activities overshadowed good grades as criteria of popularity among other students (1961, pp. 28–30, 43–50). Nevertheless, Coleman's own data gave only weak support to the proposition that the adolescent subculture and the school were at odds. Popular students claimed to spend the same amount of time doing homework as the less popular students did (1961, pp. 43–50); in nearly every high school, students from what Coleman designated as the leading crowd earned better-than-average grades and were more likely than other students to plan on going to college (1961, pp. 82, 114–115). Moreover, in 3 of the 10 schools studied, academic achievement was seen as having high importance in determining who belonged to the leading crowd (1961, pp. 86–88).

Other evidence accumulated to reinforce the idea that there was no great degree of inherent opposition between high school student subculture and academics. For instance, Rehberg and Schafer (1968) found a positive relationship between athletic participation and plans to go to college. Spady (1970), reporting on a survey of about 300 West Coast high school students in their senior year and then 4 years later, found that participation in high school extracurricular activities was associated with academic achievement in college. Students who participated in *both* activities and athletics did even better in college (controlling other factors), but participation in athletics alone led to lower-than-expected educational achievement (Spady, 1970). Snyder (1969) found a similar connection between activities and later educational achievement. Polk and Hafferty (1972) performed a factor analysis on attitude data from 1800 students in an Oregon high school and found that items having to do with involvement with peers loaded on a separate factor from items related to academic achievement and commitment to school. This they interpreted as indicating that student subculture had little effect, either positive or negative, on the academic side of school. By the early seventies, it seemed clear that Coleman had overstated his case (Campbell, 1969, p. 833). Even if high school athletics were still extremely popular (Eitzen, 1975), participation in high school athletics and activities were not necessarily impediments to educational achievement.

A more convincing version of subcultural conflict between students and the school was offered by several British researchers in the sixties (Hargreaves, 1967; Sugarman, 1967, 1968; Lacey, 1970). These authors did not try to make any generalizations about a single adolescent subculture embracing all students but instead suggested a bipolar division between conventional, conforming students who did well in school and rebellious students who were academic failures. Lacey gave a particularly vivid account of how, even in the highly selective "grammar" schools, students who were banished to the lower tracks, or "streams," developed an "antigroup" culture and adopted as their values the inverted values of the school (1970, pp. 57–58). Sugarman (1967) insisted that the students who belonged to the antischool subculture not only misbehaved in class but were also more likely to adopt the styles of teenage rebellion, such as smoking, dating, and a taste for rock music. More recently, British studies in this vein have focused on working-class adolescents and on the battle between the teacher and students over the maintenance of order in the classroom (Reynolds, 1976; Willis, 1977; Marsh, Rossner, & Harre, 1978).

Polk and his colleagues (Polk & Pink, 1971; Polk & Schafer, 1972) brought the British perspective, as articulated by Sugarman (1967, 1968), into the analysis of data from students in an American high school from the Pacific Northwest. Like Sugarman, they found that students most alienated from the high school in terms of grades and attitudes were also the most likely to participate in teenage forms of deviance such as smoking and cruising in cars (Polk & Pink, 1971). Frease (1973), using data from the same Oregon study, reported that juvenile delinquency was more closely correlated with school grades than with social class for this sample. Noblit (1976), again on the basis of Polk's data, argued that the identities conferred by the school, by its branding some students successes and others failures, were an extremely important factor in promoting delinquency. Thus delinquent behavior and other less serious, but still annoying, teenage habits were seen in this perspective as an expression of rebellion against the school.

Murdock and Phelps (1972), in a review of British research on antischool subcultures, cautioned against a too simple equation of youth culture with opposition to school. Any given student could be highly oriented both to the demands of the school and the subculture of his peers, they argued, or else simultaneously disengaged from both. Moreover, a distinction must be made between the "street culture" of football games and fights and the "pop culture media" of records, radio, and television. Either cultural style could be a vehicle for expressing the student's dissatisfaction with school, but different groups would typically make use of the different styles; the disaffected working-class male would choose street culture, while the middle-class female would express her rebellion by means of pop culture. Murdock and Phelps found, for example, that commitment to school was negatively correlated with use of pop media culture in a predominantly middle-class school but positively correlated with adherence to pop media culture in a mostly working-class secondary school (1972).

The insistence of Murdock and Phelps (1972) that several different youth cultures must be considered, in assessing the effects of youth culture on the schools, fits in well with the conclusion of earlier sections of this chapter that adolescent subcultures have become increasingly fragmented and that they differ greatly in their opposition to adult norms. Some other recent studies reinforce this position, with respect to subcultures and schools. Damico (1975, 1976) studied clique formation in the ninth-grade class of an American high school and found that clique membership was a better predictor of grades than was a standardized aptitude test. Nevertheless, cliques were as likely to pull the grades of their members up as down, and clique membership had little effect on attitudes toward the school. Rutter, Maugham, Mortimore, and Ouston (1979) reported that British secondary schools with large intakes of low-ability students were more likely, other factors held constant, to experience problems with student delinquency and "contraschool peer groups" (1979, pp. 199–203). Hindelang (1976) made the case that many delinquent adolescents were not a part of any subcultural group at all; furthermore, he argued, subcultural theorists have continued to overemphasize group delinquency, because official reports tend to underestimate the number of lone delinquents. Noblit (1976) concluded that delinquency was promoted more by labels of failure conferred by the school than by association with other delinquents. None of these results is compatible with Coleman's assumption that adolescent subculture is a monolithic entity with uniform effects on the student.

The overall conclusions to be drawn from this summary of the literature on opposition between adolescent subcultures and schools depend on whether we are discussing conforming or nonconforming subcultures. The conforming subcultures based on athletics and school activities appear to be largely irrelevant to academic performance or mildly supportive of it. After all, such activities are officially sponsored by the school organization and could not long continue if they presented a serious obstacle to academics. Nonconforming or delinquent behavior, on the other hand, does appear to have a close relationship to the experience of being labeled by the school as academically incompetent. Most of the authors cited here, however, see the rebellion as the effect of the school's label, rather than as the source of the academic failure. Nevertheless, once the pattern is established, delinquent behavior and academic failure must be mutually reinforcing.

Coleman, in overstating the impact of adolescent subculture on the student, may have understated the power of schools. As Campbell (1969, p. 837) has noted, in situations where the family has lost power over the adolescent, both the peer group *and* the school have an opportunity to extend their influence. Kamens (1977) argued that any student subculture could only be organized around identities other than those officially supported by the school. Schools as institutions, we would conclude, are relentlessly engaged in the process of sorting students, on the basis of their behavior and academic ability, into categories of good and bad, success and failure. Such officially conferred identities will encourage interaction between certain types of students (especially in schools that track good and bad students into different classrooms), but there is no reason to expect that academic prowess will play much of a role in the content of the subcultural identities students construct for each other. The point of the subcultural identities will be, as Cohen (1955) argued, to compensate for the indignities implicit in the official school labels. Because the school is engaged in sorting students into different categories with different image problems and different resources (bright students in the fast track and dumb students in the slow track, for instance), the school itself will be contributing to the proliferation and fragmentation of student subcultures. Once formed, the student subcultures may do more to contribute to the stability of the school than to threaten it, since subcultures provide a symbolic negation of the definitions of inferiority that the school inevitably imposes on most students.

Opposition with the Wider Culture. Coleman (1961), as we have noted earlier, made little distinction between the values of parents, the school, or the whole society; he seems to have proceeded on the assumption that academic achievement was of paramount importance for everybody involved, except members of the adolescent society in the school. Thus adolescent opposition to academics, if a problem for parents and the school, was implicitly a problem for the entire society, as well. Several subsequent authors have attacked Coleman's assumption by pointing to value conflicts between parents and the schools or between schools and the wider society. Other authors have stressed the existence of internal contradictions within the national culture, so that instances that appear to demonstrate the opposition of adolescents to dominant cultural values are shown, on closer examination, to unite adolescents with other groups in society. Observers who see adolescent subcultures as multiple and fragmented sometimes make the point that the culture of the whole society is at least as complex and fragmentary. To these issues we now turn.

To assume that the cultural orientations of parents and of the school system are identical is to ignore the social-class differences that separate parents and teachers. While teachers are, by definition, middle-class in outlook (usually lower-middle class in background), a large proportion of parents are working-class. Miller (1970) pointed to the continuity between the subculture of the working class as a whole and the behavior of working-class adolescents. Miller argued that the delinquency of working-class boys was a reaction to certain "focal concerns" shared by working-class men of all ages. These focal concerns—"areas or issues which command wide-spread and persistent attention and a high degree of emotional involvement"—included trouble, toughness, smartness (or ability to "con" someone else), excitement, fate (or luck), and autonomy (Miller, 1970, pp. 56, 57). Miller maintained that these concerns were much less salient to the middle class. Several British authors (e.g., Cohen, 1972; Murdock, 1975) have elaborated on the theme that British adolescent subcultures have, for the most part, developed in response to economic pressures and cultural contradictions that have affected the whole working class. Willis (1977) makes a convincing argument that the misbehavior of working-class boys in the classroom mirrors the horseplay of working-class men on the shop floors of factories, and, thus, that the school experience is preparing these boys to accept the most undesirable jobs. In the view of these authors, disorder in the schoolroom is simply a reflection of the tension between workers and bosses in the wider society. Although no one has, as yet, developed the theme, it seems plausible to suggest that some misbehavior by upper-middle-class adolescents reflects the (narrower) cultural gulf between their parents and the lower-middle-class teacher.

Coleman's premise that the wider community agreed with the schools' prime valuation on academics has also come under attack. In this case, adolescent indifference to academic pursuits was seen as shared by the wider community. Berger (1963a), for example, claimed that most parents and other community members were no more interested in academics over sports than high schoolers were. Thus Coleman's adolescent subculture was not really a deviant body of norms. A more complicated argument of the same type was advanced by Matza and Sykes (1961) who held that rebellious adolescents were only appropriating "subterranean values" of the wider society. These values included a "search for kicks . . . disdain of work . . . desire for the big score . . . [and] acceptance of aggressive toughness as the proof of masculinity" (1961, p. 715). The values were considered subterranean, because many in the middle class showed by their behavior that they shared them, even if it was bad form to admit it. By this reckoning, adolescents could be opposed to the values of the school but in agreement with the values of the whole community.

Some authors have taken this line of argument a step further by suggesting that the drawing of subcultural distinctions based on age is a mistake. They argue, in other words, that adolescent subcultures are not confined to adolescents. Berger (1963b), for instance, maintained that labeling a subculture adolescent or youthful led to confusion, since many older people adhere to roughly the same hedonistic norms (e.g., working-class laborers, bohemian businessmen, show business personalities, etc.). David M. Smith argued on much the same lines:

> Whilst not wishing to denigrate subcultural theories which purport to explain behaviour relating to some youths, the use of the term youth culture is particularly unfortunate, I would claim, because it implies that youth [age] is a crucial variable in the explanations they offer. This is not necessarily the case. (1976, p. 373)

Unless almost all adolescents and virtually no one else took part in what was called subcultural behavior, Smith saw no use for the term. Keyes (1976), in his entertaining book for the popular press, suggested a variant of the arguments above. High school, he claimed, is the one experience shared by almost every American adult and, thus, has had a decided impact on adult institutions. Other authors (e.g., Parsons, 1954; Young, 1971) have pointed out that much of adolescent rebellion consists of appropriating behavior patterns, such as sexuality, smoking, and drinking, that adults take for granted. The conclusion of this line of argument is that presumed subcultural boundaries dividing adolescents from adults are not matters of true cultural difference but, at most, matters of cultural emphasis.

Several authors with a Marxist point of view have suggested that the ways in which adolescents try to express their opposition to adult society only demonstrate their subjugation to the basic economic values of the wider culture. Young (1971), for instance, argued that advanced capitalist economies have substituted norms of hedonism and consumption for the former Puritan norms of hard work and thriftiness and that, thus, adolescent irresponsibility is a reflection of the dominant values of the economy. Several years previous, Henry (1963) had expounded much the same thesis in a study of high school students. A teenage hangout, Henry argued, was best viewed as an institution contributing to the growth of the gross national product. Larkin (1979) has described the high school years as a period of "forced consumption" and has contended that much of the behavior of high school students should be interpreted as a reaction to the economic pressures put on them to learn how to be insatiable consumers.

We come to the conclusion that so-called adolescent subcultures cannot be clearly distinguished from the complex and often internally self-contradictory adult culture in which they are embedded. Discussions of the opposition between youth and adult culture lose their point in that context. A question remains, however; if this line of argument is true, how did the idea of opposition between adult and adolescent culture ever arise?

The simple answer is that, though their cultures are not necessarily in conflict, individual adults and adolescents often are. Conflict and cultural difference are not the same thing. Structured conflicts can be part of the glue that binds group to group and, thus, holds society together. True cultural differences, on the other hand, are far more likely to produce complete bafflement than anger in a social encounter. The participants are literally speaking different languages, and interaction becomes virtually impossible. Adolescents, by contrast, are in nearly full command of the verbal and nonverbal languages of adults. As Campbell has remarked, rebellious adolescents tend to select styles of behavior that express in exaggerated fashion their opposition to adults (1969, p. 841). The angry reactions of adults to each teenage excess are proof that communication has been achieved. But they are proof, at the same time, of the shared culture that underlies the communication.

A second answer to the question of why adolescent and adult cultures have been thought to be oppositional is that some adolescent subcultures worthy of the name have, in fact, arisen, both in the United States and in England. Irwin (1977) described the "grand scenes" of surfing and the hippie movement as full-fledged youth subcultures. The British Teddy boys, mods, skinheads, and punks are other examples (see Brake, 1980; Hebdige, 1979; Hall & Jefferson, 1976; Mungham & Pearson, 1976). The religious cults of the early seventies may also qualify (Foss & Larkin, 1979). It is worth noting that all these developed outside of the schools, although they recruited from among school attenders. According to Hebdige (1979), the cultural creativity of these movements lay in their ability to take everyday items of clothing or transportation and, by reassembling them into an original style, make use of the products of society to express their antagonism against it. A vivid example of this process was the motorcycle gang's ability to turn leather jackets and motorbikes into symbols of doom (Willis, 1977). The original styles of music, speech, and dress were quickly picked up, however, by the popular media and commercial advertisers who retailed them in the form of packaged imitations to the adolescent mass markets, a process which simultaneously destroyed the internal vitality of the subculture (Hebdige, 1979; Brake, 1980; Irwin, 1977).

A number of observers have suggested that adolescent participants in these genuine subcultures, as well as many other adolescents, have often sought to enlarge their vocabulary of oppositional symbols by imitating black subculture (Hebdige, 1979; Brake, 1980; French, 1978). It is interesting to note, however, that black teenagers have been almost entirely absent from the literature on adolescent subcultures. A few recent participant observation studies (Cusick, 1973b; Palonsky, 1975; Larkin, 1979) have reported the presence of black student groups in integrated high schools, but none has explored the possibility of subcultural division between black parents and youths.

In sum, then, this survey of the literature on opposition between adolescent subcultures and adult culture has suggested that adult culture is so many-stranded that there is little about the culture of adolescents that is unique. Nevertheless, opposition between adults and youths has been present; much of the opposition appears to be founded, however, precisely on the cultural similarity between adults and youths, rather than on their differences (cf. Slater, 1976). A few full-fledged adolescent subcultures have arisen in the recent past but have quickly succumbed to absorption by the wider culture, although they have enriched the symbolic vocabulary of opposition between adolescent and adult.

CONCLUSIONS

In this chapter we have reviewed the critical reaction to James S. Coleman's landmark study on high schools—*The Adolescent Society* (1961). We have argued that most observers of high schools since Coleman have rejected his portrayal of a relatively unified and distinctive adolescent subculture in high schools. Later observers have tended to view high schools as the meeting place of a wide variety of subcultural strains, some of which oppose parents and the school, others of which do not, and all of which reflect one element or another of the complex adult culture. Whether or not Coleman's depiction was accurate for schools of the fifties, the last 2 decades of growth in school size and development of new styles of deviance (such as drug use in the high schools) have made generalizations about high school students seem more and more tenuous.

In spite of the critical debate, however, we suspect that many people will continue to think in terms of a single adolescent subculture. There is a great temptation to do so, partly because the word *culture* has come into widespread use in everyday language, as well as in the technical language of social science. In ordinary speech, culture is as often used a label for a group of people as for a way of life. Consequently, the easy assumption has sometimes been made by social scientists, as well as others, that every identifiable group has a culture or at least a subculture. Thus the concepts of group and culture have become confused.

Clark (1974) has pointed out that social structural analysis, which is concerned with groups, and cultural analysis are, in fact, two different ways of describing the same human behaviors; in social structural analysis the focus is on patterns of contact between people, in cultural analysis, on the meanings of actions and communications. Even though the same behaviors serve as raw material for both types of analysis, the constructs derived from the two analyses must be kept conceptually distinct. In particular, it is a mistake to assume that the boundaries of a group and a culture necessarily coincide. Cultures are ways of life and systems of meaning that develop over long periods of time. Groups tend to form and dissolve more rapidly, so that cultures occasionally outlive the groups that supported them (as, for example, the culture of the ancient Romans).

When a group of students forms in a high school, for instance, the individuals involved already share a culture, or interaction would be virtually impossible. Becker and Geer (1960) have described the cultural equipment that individuals bring to a newly formed group as "latent culture," but for the group in the beginning stages, latent culture is the only culture. After the students have been together for a period of time, they will typically develop some ways of saying and doing things that are peculiar to the group itself which arise from their shared history of experiences. This unique group-based culture, however, will represent only a tiny percentage of the cultural forms and material actually used by the group in day-to-day interactions, unless the group stays together for quite a long time. Nelson and Rosenbaum (1972), for example, asked students in a Wisconsin high school to list the slang terms they used for things like money, popularity, cars, clothes, and so forth. Seniors knew more slang than freshmen, and the total number of slang terms reported numbered in the hundreds. This is quite impressive, until one remembers that these students (presumably) had working vocabularies that numbered in the tens of thousands of words. Moreover, many of the slang terms listed would be recognized by the average adult.

Because the term subculture has never been very precisely defined, almost any degree of distinctiveness in cultural forms might be described by someone as subcultural. That high school students' behavior is distinctive in some ways from that of the rest of us, nobody would doubt. On the other hand, that such patterns of difference are so consistent or so thoroughgoing to merit the weighty term subculture—

which connotes, at the very least, a separate way of life—is very much open to question. Perhaps it would make more sense to adopt Irwin's (1977) term *action scene* to apply to an arena of social action and meaning that envelopes the individual only temporarily, as high schools do. In any case, adolescence is too brief an episode in a life, and high schools are too deeply embedded in the dominant culture, for the application of the term subculture to life in high schools to bring anything but confusion. Future observers will need to take a conceptually more sophisticated approach to the typical patterns and problems of adolescence.

REFERENCES

Alexander, C. N., & Campbell, E. Q. Peer influences on adolescent educational aspirations and attainments. *American Sociological Review,* 1964, **29**(4), 568–575.

Alexander, K., & Eckland, B. K. Contextual effects in high-school attainment process. *American Sociological Review,* 1975, **40**(3), 402–416.

Becker, H. S. (Ed.). *Campus power struggle.* Chicago: Aldine, 1970.

Becker, H. S., & Geer, B. Latent culture: A note on the theory of latent social roles. *Administrative Science Quarterly,* 1960, **5**, 304–313.

Bengston, V. L. The generation gap: A review and typology of perspectives. *Youth and Society,* 1970, **2**(1), 7–31.

Berger, B. M. Adolescence and beyond. *Social Problems,* 1963, **10** (spring), 394–408. (a)

Berger, B. M. On the youthfulness of youth cultures. *Social Research,* 1963, **30**, 319–342. (b)

Boocock, S. S. *Sociology of education: An introduction* (2nd ed.). Boston: Houghton-Mifflin, 1980.

Brake, M. *The sociology of youth culture and youth subcultures.* London: Routledge and Kegan Paul, 1980.

Brittain, C. V. Adolescent choices and parent–peer cross-pressures. *American Sociological Review,* 1963, **28** (June), 385–391.

Campbell, E. Q. Adolescent socialization. In D. A. Goslin (Ed.), *Handbook of socialization theory and research.* New York: Russell Sage Foundation, 1969, pp. 821–859.

Clark, B. R. *Educating the expert society.* San Francisco: Chandler, 1962.

Clark, M. On the concept of "sub-culture." *British Journal of Sociology,* 1974, **25**(4), 428–441.

Clarke, J., Hall, S., Jefferson, T., & Roberts, B. Subcultures, cultures and class: A theoretical overview. In S. Hall & T. Jefferson (Eds.), *Resistance through rituals.* London: Hutchinson, 1976, pp. 9–74.

Cohen, A. K. *Delinquent boys.* Glencoe, Ill.: Free Press, 1955.

Cohen, J. High-school subculture and the adult world. *Adolescence,* 1979, **14**(55), 491–502.

Cohen, P. Subcultural conflict and the working class community. *Working Papers in Cultural Studies,* 1972, **2** (spring), 5–52.

Coleman, J. S., with the assistance of Johnston, J. W. C., & Jonassohn, K. *The adolescent society: The social life of the teenager and its impact on education.* New York: Free Press of Glencoe, 1961.

Conant, J. B. *The American high school today.* New York: McGraw-Hill, 1959.

Cremin, L. A. *The transformation of the school: Progressivism in American education 1876–1957.* New York: Vintage Books, 1964.

Cusick, P. A. Adolescent groups and the school organization. *School Review,* 1973, **82**, 116–126. (a)

Cusick, P. A. *Inside high school: The student's world.* New York: Holt, Rinehart & Winston, 1973. (b)

Damico, S. B. Effects of clique membership upon academic achievement. *Adolescence,* 1975, **10**(37), 93–100.

Damico, S. B. Clique membership and its relationship to academic achievement and attitude toward school. *Journal of Research and Development in Education* 1976, **9**(4), 29–35.

Davis, K. Sociology of parent–youth conflict. *American Sociological Review,* 1940, **5** (August), 523–535.

Eitzen, D. S. Athletics in status system of male adolescents: Replication of Coleman's adolescent society. *Adolescence,* 1975, **10**(38), 267–276.

Elkin, F., & Westley, W. A. The myth of adolescent culture. *American Sociological Review,* 1955, **20**, 680–684.

Elkin, F., & Westley, W. A. The protective environment and adolescent socialization. *Social Forces,* 1957, **35**, 245–249.

Epperson, D. C. A reassessment of indices of parental influence in the adolescent society. *American Sociological Review,* 1964, **29**, 93–96.

Floyd, H. H., Jr., & South, D. R. Dilemma of youth: The choice of parents or peers or a frame of reference for behavior. *Journal of Marriage and the Family,* 172, **34**(4), 627–634.

Foss, D., & Larkin, R. Roar of the lemming: Youth, post-movement groups and the life construction crisis. In H. M. Johnson (Ed.), Religious change. (Special issue of *Sociological Inquiry, 49*). San Francisco: Jossey-Bass, 1979.

Frease, D. E. Delinquency, social class, and schools. *Sociology and Social Research,* 1973, **57**(4), 443–459.

French, R. L. Nonverbal patterns in youth culture. *Educational Leadership,* 1978, **35**(7), 541–546.

Gordon, C. W. *The social system of the high school.* Glencoe, Ill.: Free Press, 1957.

Gordon, C. W. (Ed.). *Uses of the sociology of education* (73rd yearbook of the NSSE Part II). Chicago: National Society for the Study of Education, 1974.

Gordon, M. M. The concept of the sub-culture and its application. In D. O. Arnold (Ed.), *Subcultures.* Berkeley, Calif.: Glendessary, 1970, pp. 31–36.

Gottlieb, D., & Heinsohn, A. L. Sociology and youth. *Sociological Quarterly,* 1973, **14**(2), 249–270.

Gross, R., & Osterman, P. (Eds.). *High school.* New York: Simon & Schuster, 1971.

Hall, G. S. *Adolescence.* New York: Arno, 1969.

Hall, S., & Jefferson, T. *Resistance through rituals: Youth subcultures in post-war Britain.* London: Hutchinson, 1976.

Haller, A. O., & Butterworth, C. E. Peer influence on levels of occupational and educational aspiration. *Social Forces,* 1960, **38**, 289–295.

Hargreaves, D. H. *Social relations in a secondary school.* London: Routledge & Kegan Paul, 1967.

Havighurst, R. J., & Dreyer, P. H. (Eds.). *Youth* (74th Yearbook of the NSSE, Part I). Chicago: National Society for the Study of Education, 1975.

Hebdige, D. *Subculture: The meaning of style.* London: Methuen, 1979.

Henry, J. *Culture against man.* New York: Vintage Books, 1963.

Hillman, K. G. Student valuation of academic achievement. *Sociological Quarterly,* 1969, **10**, 384–391.

Hindelang, M. J. With a little help from their friends: Group participation in reported delinquent behavior. *British Journal of Criminology,* 1976, **16**, 109–125.

Hollingshead, A. B. *Elmtown's Youth.* New York: Wiley, 1949.

Hollingshead, A. B. Behavior systems as a field for research. In D. O. Arnold (Ed.), *Subcultures.* Berkeley, Calif.: Glendessary Press, 1970, pp. 21–30.

Irwin, J. *Scenes.* Beverly Hills, Calif.: Sage, 1977.

Kamens, D. H. Institutional definitions and collective action: The concept of student as a source of school authority and student culture. *Youth and Society*, 1977, **9**(1), 55–78.

Kandel, D. B., & Lesser, G. S. Parental and peer influence on educational plans of adolescents. *American Sociological Review*, 1969, **34**, 212–222.

Kandel, D. B., & Lesser, G. S. *Youth in two worlds*. San Francisco: Jossey-Bass, 1972.

Keniston, K. *Young radicals: Notes on committed youth*. New York: Harcourt Brace and World, 1968.

Keniston, K. *Youth and dissent*. New York: Harcourt Brace Jovanovich, 1971.

Keyes, R. *Is there life after high school?* Boston: Little, Brown, 1976.

Lacey, C. *Hightown grammar: The school as a social system*. Manchester, England: University Press, 1970.

Larkin, R. W. *Suburban youth in cultural crisis*. New York: Oxford University Press, 1979.

Larson, L. E. The influence of parents and peers during adolescence: The situation hypothesis revisited. *Journal of Marriage and the Family*, 1972, **34**, 67–74. (a)

Larson, L. E. The relative influence of parent-adolescent affect in predicting the salience hierarchy among youth. *Pacific Sociological Review*, 1973, **15**(1), 83–102. (b)

Liberale, M., & Seligson, T. (Eds.). *The high school revolutionaries*. New York: Vintage Books, 1970.

Lipset, S. M. *Rebellion in the universities: A history of student activism in America*. London: Routledge & Kegan Paul, 1972.

Longino, C. F., McClelland, K. A., & Peterson, W. A. The aged subculture hypothesis: Social integration, gerontophilia and self-conception. *Journal of Gerontology*, 1980, **35**, 758–767.

McDill, E., & Coleman, J. S. Family and peer influences in college plans of high school students. *Sociology of Education*, 1965, **38**(2), 112–126.

Marsh, P., Rossner, E., & Harre, R. *The rules of disorder*. London: Routledge & Kegan Paul, 1978.

Matza, D. Subterranean traditions of youth. *The Annals of the American Academy of Political and Social Sciences*, 1961, **338** (November), 102–118.

Matza, D. Position and behavior patterns of youth. In Robert E. L. Faris (Ed.), *Handbook of modern sociology*. Chicago: Rand McNally, 1964.

Matza, D., & Sykes, G. M. Juvenile delinquency and subterranean values. *American Sociological Review*, 1961, **26**, 712–719.

Mead, M. *Culture and commitment: A study of the generation gap*. Garden City, N.Y.: Natural History Press, 1970.

Miller, W. B. Lower class culture as a generating milieu of gang delinquency. In D. O. Arnold (Ed.), *Subcultures*. Berkeley, Calif.: Glendessary, 1970, pp. 54–63.

Mungham, G., & Pearson, G. (Eds.) *Working class youth culture*. London: Routledge & Kegan Paul, 1976.

Murdock, G. Education, culture and myth of classlessness. In J. T. Haworth & M. A. Smith (Eds.), *Work and leisure*. London: Lepus, 1975, pp. 119–132.

Murdock, G., & McCron, R. Youth and class: The career of a confusion. In G. Mungham & G. Pearson (Eds.), *Working class youth culture*. London: Routledge & Kegan Paul, 1976, pp. 10–26.

Murdock, G., & Phelps, G. Youth culture and the school revisited. *British Journal of Sociology*, 1972, **23**, 478–482.

Nelson, E. A., & Rosenbaum, E. Language patterns within the youth subculture: Development of slang vocabularies. *Merrill-Palmer Quarterly of Behavior and Development*, 1972, **18**, 273–285.

Noblit, G. W. The adolescent experience and delinquency: School versus subcultural effects. *Youth and Society*, 1976, **8**(1), 27–44.

Palonsky, S. B. Hempies and squeaks, truckers and cruisers: Participant observer study in a city high school. *Educational Administrative Quarterly*, 1975, **11**(2), 86–103.

Panel on Youth of the President's Science Advisory Committee (J. Coleman, Chair). *Youth: Transition to adulthood.* Chicago: University of Chicago Press, 1974.

Parsons, T. Age and sex in the social structure of the United States. In *Essays in sociological theory* (rev. ed.). New York: Free Press, 1954, pp. 89–103.

Picou, J. S., & Carter, T. M. Significant-other influence and aspirations. *Sociology of Education,* 1976, **49**(1), 12–22.

Polk, K., & Hafferty, D. School cultures, adolescent commitments and delinquency. In K. Polk & W. E. Schafer (Eds.), *Schools and delinquency.* Englewood Cliffs, N.J.: Prentice-Hall, 1972, pp. 71–90.

Polk, K., & Pink, W. Youth culture and the school: Re-visited. *British Journal of Sociology,* 1971, **22**, 160–171.

Polk, K., & Schafer, W. E. (Eds.). *Schools and delinquency.* Englewood Cliffs, N.J.: Prentice-Hall, 1972.

Rehberg, R. A., & Schafer, W. E. Participation in interscholastic athletics and college expectations. *American Journal of Sociology,* 1968, **73** (May), 732–740.

Reynolds, D. When pupils and teachers refuse a truce: The secondary school and the creation of delinquency. In G. Mungham & G. Pearson, *Working class youth culture.* London: Routledge and Kegan Paul, 1976, pp. 125–138.

Rigsby, L. C., & McDill, E. L. Value orientation of high school students. In H. R. Stub (Ed.), *The sociology of education: A sourcebook* (3rd ed.). Homewood, Ill.: Dorsey Press, 1975, pp. 53–75.

Rose, A. M. The subculture of the aging: A topic for sociological research. *Gerontologist,* 1962, **2**, 123–127.

Roszak, T. *The making of a counter culture: Reflections on the technocratic society and its youth opposition.* New York: Doubleday, 1969.

Rutter, M., Graham, P., Chadwick, O. F. D., & Yule, W. Adolescent turmoil: Fact or fiction? *Journal of Child Psychology and Psychiatry and Allied Disciplines,* 1976, **17**(1), 35–56.

Rutter, M., Maugham, B., Mortimore, P., & Ouston, J., with Smith, A. *Fifteen thousand hours: Secondary schools and their effects on children.* Cambridge, Mass.: Harvard University Press, 1979.

Sewell, W. H., & Hauser, R. M. *Education, occupation and earnings: Achievement in the early career.* New York: Academic, 1975.

Slater, P. *The pursuit of loneliness.* Boston: Beacon Press, 1976.

Smith, D. M. The concept of youth culture: Re-evaluation. *Youth and Society,* 1967, **7**(4), 347–366.

Snyder, E. E. A longitudinal analysis of the relationship between high school student values, social participation, and educational-occupational achievement. *Sociology of Education,* 1969, **42**(3), 261–270.

Spady, W. G. Lament for the letterman: Effects of peer status and extracurricular activities on goals and achievement. *American Journal of Sociology,* 1970, **75**, 680–702.

Sugarman, B. Involvement in youth culture, academic achievement and conformity in school. *British Journal of Sociology,* 1967, **18**, 151–164.

Sugarman, B. Social norms in teenage boys' peer groups: A study of their implications for achievement and conduct in four London schools. *Human relations,* 1968, **21**, 41–58.

Sutherland, E. H. Behavior systems in crime. In D. O. Arnold (Ed.), *Subcultures.* Berkeley, Calif.: Glendessary, 1970, pp. 9–20.

Tanner, J. New directions for subcultural theory: An analysis of British working-class youth culture. *Youth and Society,* 1978, **9**(4), 343–372.

Todd, D. M. Contrasting adaptations to the social environment of a high school: Implications of a case study of helping behavior in two adolescent subcultures. In J. G. Kelly (Ed.), *Adolescent boys in high school.* Hillsdale, N.J.: Erlbaum, 1979, pp. 177–186.

Williams, T. H. Educational aspirations: longitudinal evidence on their development in Canadian youth. *Sociology of Education,* 1972, **45**(2), 107–133.

Willis, P. E. *Learning to labour: How working class kids get working class jobs.* Westmead, Eng.: Saxon House, 1977.

Willis, P. E. *Profane culture.* London: Routledge & Kegan Paul, 1978.

Yinger, M. Contraculture and subculture. *American Sociological Review,* 1960, **25**, 625–635.

Young, J. *The drugtakers.* London: MacGibbon & Kee, 1971.

23

ADOLESCENT CHILDBEARING IN THE UNITED STATES
Apparent Causes and Consequences

Catherine S. Chilman
School of Social Welfare
University of Wisconsin-Milwaukee

PREVALENCE

Although there is a general impression that the nation has recently experienced an "epidemic in adolescent childbearing," this is hardly the case. The birthrate for young women between the ages of 15 and 19 reached a peak in 1958 and has been declining fairly steadily since that time, though this decline was somewhat less before 1975 for teenagers than for older groups. Birthrates for older women (between the ages of 20 and 34) have been higher than those for teenagers over the past 25 years.

What *has* changed markedly for adolescents is an increase in the rate of childbearing outside marriage. Rates of births to unmarried adolescents rose more or less steadily between 1940 and 1978 with a particularly sharp increase for white teenagers between 1967 and 1971. During 1978, about 1.4% of white adolescents and 9% of black adolescents gave birth outside marriage. More strikingly, over one-fourth of the babies born to white teenagers and more than 80% born to black teenagers were born to unmarried women (National Center for Health Statistics, 1980). Another marked change since the late 1960s is the tendency for young mothers to keep their out-of-wedlock children rather than giving them up for adoption. Although over 80% of white adolescents relinquished their illegitimate children for adoption in 1965, this was true for less than 10% of this group in 1980.

CAUSES

The basic causes of adolescent parenthood are the following conditions and behaviors among teenagers: participation in intercourse, adequate fertility and timing of intercourse, contraceptive failures, lack of recourse to abortion, and the occurrence of a live birth. The causes of nonmarital adolescent parenthood are nonmarital coitus, the above conditions and behaviors, and the failure to place the baby for adoption or to marry before the child's birth.

Although the foregoing delineation of causes may seem obvious, it has not been obvious to a host of researchers in the past as well as in the present. Scores of studies have compared teenagers who become pregnant to those who don't, teenage mothers to teenagers without children. These comparisons have been made without looking at the sexual and contraceptive behaviors of the young women involved (Chilman, 1978; Wilson, 1980). However, a number of studies *have* looked at the pertinent behaviors summarized above. These are presented briefly in the following pages.

Intercourse

Adolescent participation in intercourse, especially nonmarital intercourse, increased dramatically among young white women and somewhat less so among young white men toward the end of the 1960s, and this trend continued into the 1970s.[1] It appears that this increase was caused primarily by the pervasive and profound changes that shook American society during this period—a period in which all established social institutions and values, including those pertaining to the family and sexual morality as well as sex roles, were called into question, especially by young people.

Studies of sexual behavior carried out between the 1920s and about 1966 generally showed that about 10 to 15% of young white women and about 25% of young white men had experienced coitus by the time they were seniors in high school. By 1968 or so, this proportion had increased to about 40% of 17- to 18-year-old young women and 50% of young men of that age.

This trend continued into the 1970s. The 1976 Zelnik and Kantner survey showed that, compared to 1971, a considerable increase had occurred in the prevalence of nonmarital intercourse among young women between the ages of 15 and 19. According to the 1976 survey, 63% of black teenagers and 31% of white ones between the ages of 15 and 19 were sexually experienced. Young black women were particularly apt to have their first intercourse when they were very young. For example, almost 40% of them were nonvirgins by age 15, compared to 14% of white youngsters. By age 19, 84% of black women had experienced intercourse compared to 49% of the white respondents (Zelnik & Kantner, 1977). These differences between the races tend to become smaller when the factor of differences in socioeconomic status is taken into account. For example, the prevalence of nonmarital intercourse among young black women whose fathers are college graduates is similar to the prevalence for young white women whose fathers have the same educational level (Kantner & Zelnik, 1972).

According to the Zelnik and Kantner survey done in 1976, nonmarital intercourse was most likely to occur at the home of the young woman's partner or in her home. The young women tended to have intercourse infrequently—several times a

---

[1] Data concerning the sexual behaviors of black teenagers before 1971 are exceedingly sparse. Large national surveys of the sexual and contraceptive behaviors of young white and young black women between the ages of 15 and 19 were carried out by Kantner and Zelnik in both 1971 and 1976. A somewhat similar survey for young men was conducted in 1979–80, but results are not available as of this writing.

TABLE 23.1. Summary of Major Factors Apparently Associated with Nonmarital Intercourse among Adolescents

Factors	Males	Females
Social situations		
Father having less than a college education	unknown	yes, for blacks
Low level of religiousness	yes	yes
Norms favoring equality between the sexes	probably	yes
Permissive sexual norms of the larger society	yes	yes
Racism and poverty	yes	yes
Migration from rural to urban areas	unknown	yes
Peer group pressure	yes	not clear
Lower social class	yes (probably)	yes (probably)
Sexually permissive friends	unknown	unknown
Single-parent (probably low-income) family	unknown	yes
Psychological		
Use of drugs and alcohol	yes	no
Low self-esteem	no[a]	yes[a]
Desire for affection	no[a]	yes[a]
Low educational goals and poor educational achievement	yes	yes
Alienation	no[a]	yes[a]
Deviant attitudes	yes	yes
High social criticism	no[a]	yes[a]
Permissive attitudes of parents	yes	yes
Strained parent–child relationships and little parent–child communication	yes	yes
Going steady; being in love	yes	yes
Risk-taking attitudes	yes[a]	yes[a]
Passivity and dependence	no[a]	yes[a]
Aggression; high levels of activity	yes[a]	no[a]
High degree of interpersonal skills with opposite sex	yes[a]	no[a]
Lack of self-assessment of psychological readiness	no[a]	yes[a]
Biological		
Older than 16	yes	yes
Early puberty	yes	yes (probably for blacks)

[a] Findings supported by only one or two small studies. Other variables are supported by a number of investigations. The major studies on which this table is based are Furstenberg (1976); Jessor and Jessor (1975); Sorenson (1973); Kantner and Zelnik (1972); Udry, Bauman and Morris (1975); Simon, Berger, and Gagnon (1972); Zelnik and Kantner (1977); Fox (1979); Cvetkovich and Grote (1977); Presser (1976).

month. Compared to 1971, there was an increased tendency for adolescent women to have had more than one sex partner and for the partner to be somewhat older (Zelnik & Kantner, 1977).

Table 23.1 shows factors that have been identified by research with high school subjects to be significantly associated with participation in nonmarital intercourse.

The factors summarized in Table 23.1 deserve careful consideration. Among other things, they reveal the importance of many variables—social, situational, psychological, and biological—as underlying causes of adolescent sexual behavior. The significance of the young persons' educational achievement level and his or her educational occupational goals was evident in a number of the studies. It seems likely that, for many young people, school failures plus discouragement about future achievement in school and work combine with other factors to stimulate a search for alternate and present satisfactions such as those often found in a sexual relationship. Adult arguments that such a relationship should be deferred until later when further education and goods jobs have been obtained are apt to carry little weight when future "success" seems dubious and the general cultural climate is both sexually stimulating and largely permissive.

Contraceptive Behaviors

According to a number of studies between 1968 and 1974, only about 50% of sexually active high school-age and college-age women used contraceptives at their first intercourse. When Zelnik and Kantner (1977) repeated their 1971 national survey of young women between ages 15 and 19, they found that contraceptive use had partly improved for both races. Forty percent of single, sexually active women in their study reported that they "always" used contraception compared to only 18% in 1971. However, in 1976, 20% said they had *never* used contraceptives (a decline in use between 1971 and 1976). It appears that those who never used contraceptives were particularly apt to be very young teenagers.

There was a marked increase in the use of the pill and a reduction in the use of the condom and withdrawal between 1971 and 1976. Also, the proportion of black women who used effective contraceptives increased considerably during these years. The improvement in this group was particularly marked.

Table 23.2 shows the variables associated with the failure of adolescent women to use effective contraceptives, as revealed in a number of studies during the 1970s (Zelnik & Kantner, 1979; Shah, Zelnik, & Kantner, 1975; Cvetkovich & Grote, 1977; Ladner, 1972; Miller, 1976; Luker, 1975; Lindeman, 1974; Rosen, Martindale, & Griselda, 1976; Hornick, Doran, & Crawford, 1979; Russ-Eft, Springer, & Beever, 1979; Zelnik & Kantner, 1979; Jorgenson, King, & Torrey, 1980; Presser, 1977). It is noteworthy that none of these studies were directed toward male subjects. In general, the entire field of contraception, abortion, pregnancy, childbirth, and childrearing tends to be focused on women, with scarcely a thought to the roles of men. This seems to be especially true of the topic of adolescent childbearing.

The part played by demographic variables, including socioeconomic status, race, and ethnicity are not shown with sufficient clarity in Table 23.2. However, research reveals that poverty status often breeds attitudes of fatalism, powerlessness, alienation, a sense of personal incompetence, and hopelessness in respect to striving for high educational and occupational goals. This is especially apt to be true when

TABLE 23.2. Reasons Revealed by Research for Failure of Adolescent Women to Use Contraceptives or to Use Effective Ones Consistently

Demographic variables

    Age less than 18
    Single status
    Lower-socioeconomic status
    Minority group member
    Not going to college
    Being a fundamentalist Protestant

Situational variables

    Not being in a steady, committed relationship
    Not having experienced a pregnancy
    Having intercourse sporadically and without prior planning
    Being in a high-stress situation
    Not having ready access to a free, confidential family-planning service that does not require parental consent
    Lack of communication with parents regarding all aspects of life, including sexual behavior and contraceptive use

Psychological variables

    Desiring a pregnancy; high fertility values
    Ignorance of pregnancy risks, of family-planning services
    Attitudes of fatalism, powerlessness, alienation, incompetence, trusting to luck
    Passive, dependent, traditional female role attitudes
    High levels of anxiety; low ego strength
    Lack of acceptance of the reality of one's own sex behavior; thinking coitus won't occur
    Risk-taking, pleasure-oriented attitudes
    Fear of contraceptive side effects and possible infertility
    Wrong assumptions about the "safe times" of the menstrual cycle

---

racism combines with poverty to reduce one's life chances (Ladner, 1971; Rainwater, 1970; Stack, 1973; Staples, 1973; Billingsley, 1970). As shown in Table 23.2, the above attitudes are values associated with failure to use effective contraceptives. They are also associated with early participation in nonmarital intercourse, as we discussed briefly in an earlier section of this chapter.

Much is often made of the inadequate knowledge many teenagers have about contraceptives and their own reproductive systems (especially the timing of ovulation). Sex education is the frequently recommended remedy. However, as shown in Table 23.2, knowledge deficits have not been found to be salient to contraceptive use. Then, too, research to date indicates that sex education, including contraceptive education, appears to have little impact on the contraceptive behavior of adolescents (Miller, 1976; Vener & Stewart, 1974; Goldsmith, Gabrielson, & Gabrielson, 1972; Cvetkovich & Grote, 1977). Recent experimental projects suggest that specially designed educational programs which deal with sexuality in the context of interpersonal relationships and include training in couple communication skills are more apt to be effective than the more usual approaches (Cvetkovich & Grote, 1977; Schinke & Gilchrist, 1977).

Problems in obtaining effective contraceptives (such as the pill, diaphram and jelly, or IUD for girls, condoms for boys) are not prominently mentioned by teenagers as birth control problems for them. However, it was not until the late 1960s that federal funds were available to help defray the costs of contraceptive services for teenagers. Many private physicians hesitate to provide contraceptives for adolescents without the consent of their parents, and free or low-cost clinics that will provide these services are not universally available, although their numbers have increased markedly since 1970 or so. (Further discussion of programmatic approaches to prevent or "treat" adolescent childbearing is largely beyond the scope of this chapter.)

Many further comments could be made about the findings presented in Table 23.2. Of particular note, however, is the centrality of a high level of self-esteem, easy parent–youth communication, and a warm, supportive couple relationship to the effective use of reliable contraceptives.

Abortion

Legalized abortion nationwide has been available in the United States since a Supreme Court decision in 1973. Between 1973 and 1979 federal aid was available to defray the cost of legal abortions to poor women including teenagers. However, Supreme Court action in 1980 confirmed that such funds might be restricted by the government and aid for abortion services was abolished or drastically reduced in most states. As of this writing (1980), it is too early to know how these recent developments have affected the incidence of abortion in this country. Between 1972 and 1976 the abortion rate among teenagers more than doubled. For young women under age 15, there were more abortions than live births in 1976. Although in earlier years abortion seemed to be more acceptable to white than to black women, national data for 1976 showed that the abortion rates were higher for blacks than for whites (Forrest, Tietze, & Sullivan, 1978).

Only a few recent studies are available regarding the social and psychological characteristics of teenage abortion users. Research by Evans, Selstad and Welcher (1976), provides information on a California sample of Anglo and Chicano single adolescents known to pregnancy clinics. Teenagers who elected abortion were likely to have been doing well in school before pregnancy and to have come from intact, non-Catholic families not dependent on welfare.

A few studies give clues that relief is the most usual immediate reaction following an abortion. Later adverse emotional reactions seem to occur more frequently among women with a previous history of depression and anxiety (Osofsky & Osofsky, 1972). Evans, Selstad, and Welcher (1976) found that abortees who later regretted their decision were more likely to be Chicano, Catholic, of low-socioeconomic origins, to be poor students in school, and to feel abortion was forced on them by their parents.

Nonmarital Births

As we have seen, there has been a marked increase in nonmarital births among adolescents over the past 40 years or so. Although this trend may seem alarming at first glance the problems it presents for teenage mothers and their children may be fewer than many people assume. This is largely owing to a host of social and economic changes that have occurred in recent years. Traditionally, unmarried mothers and their children have been often viewed as social pariahs, with a future of failure and rejection before them. Such views have been caused by a sexist, punitive society. The hypocrisy of this society can be noted in its usual ready acceptance of shotgun marriages for premaritally pregnant brides. However, as sketched earlier, the sexual and sex-role revolutions of the 1960s brought with them long overdue changes in respect to many aspects of women's sexual and sex-role behaviors. Double standard attitudes toward sexual morality were sharply reduced as well as earlier beliefs that teenagers (especially girls) were immoral if they participated in intercourse outside of marriage. Although nonmarital births were scarcely welcomed in most families, they tended to become less of a family disgrace during the late 1960s and 1970s than had generally been the case earlier, especially in white families and in middle- and upper-class families of both whites and blacks.

Then, too, the feminist movement of 1967 and the following years took the lead in stimulating many people, both men and women, to see that marriage need not be the central goal of a woman's life. In fact, the suspicion grew (and it was confirmed by research findings) that young mothers were frequently better off if they did not marry. They were more likely to go further in school (as of the early 1970s, schools receiving federal funds were prohibited from excluding or discriminating against pregnant girls and young mothers who were students in the school), to get child care and other assistance from their families, and to have fewer children while still being very young (see for example, Furstenberg, 1976; Card, 1978). Then, too, increasing employment opportunities for women, including mothers, whether married or not, meant that women did not have to seek economic security either through deference to the wishes of their parents (staying virginal or at least nonpregnant) or through finding and keeping a husband. Moreover, an expansion of the legal rights of women in recent years has meant that, among other things, mothers of young children could not be denied public assistance because they were single.

When the high rates of youth unemployment are considered, especially the rates of over 50% of inner-city minority group youths under age 21, it becomes easier to understand that many young men are not regarded as desirable husbands, in the economic sense, by the young women with whom they associate. Even if these youths do manage to find employment, their wages are apt to be so low that the young mothers can find better financial support for their children through public assistance (Ross & Sawhill, 1975; Moore & Caldwell, 1977). In 22 states, families lose assistance benefits when the father of the children is present in the home. Thus having children outside of marriage can be financially wise. Marriage seems even less desirable when unemployed and underemployed young people can readily observe the high rates of separation and divorce in homes undermined by the abrasions of poverty, unemployment, and underemployment. Even though few single, teenage women *plan* to become parents, the availability of various kinds of economic, social, familial, and educational support programs make it far more possible for today's adolescents to care adequately for themselves and their children and to avoid unwanted abortions or adoptive placements as well as marriages that have little rationale other than legitimizing a child's birth (Moore & Caldwell, 1977).

None of the above should be interpreted as an argument in favor of adolescent parenthood, either marital or nonmarital. Rather it is meant to challenge assumptions that belong to an earlier day and to present some facts about a changed society and the changes it has wrought in the situations and prospects of young unmarried mothers and their children.

CONSEQUENCES OF EARLY CHILDBEARING

Table 23.3 lists the apparent consequences of adolescent childbearing for parents as revealed by research.

Many of these studies reanalyzed previously obtained data from other large surveys. The reanalyses consisted of comparing various groups of males and females in respect to their age at which they first became parents. The careful use of sophisticated statistical methods in a number of these reanalyses resulted in necessary and important controls for the adverse effects of poverty and race on the educational, occupational, and family lives of young men and women irrespective of their ages at first childbearing. These methods of data analysis helped correct for such common beliefs as those that attribute later unemployment to adolescent parenthood. In actuality, young black men and women experience extremely high rates of unemployment (over 50% in the inner city, as noted above) whether or not they are (or were) adolescent parents.

A further consideration of findings presented in Table 23.3 indicates that, in general, the direct social and psychological effects of early childbearing, per se, appear to be fairly minimal for young people in many aspects of their later lives. Moreover, Baldwin and Cain (1980) report, in a review of pertinent research, that adverse effects of early childbearing on the health of the mother tend to disappear when free or low-cost, high-quality medical care is made readily available. Earlier observed negative health "consequences" seem to have been caused by the poverty of many of the young women and their resulting inability to get adequate obstetrical care, rather than by their age.

Consequences for Children of Adolescent Parents

We now turn to an allied, but different, consequence question: How are developmental outcomes of children correlated with maternal age at the time of the child's birth? At least six recent studies have sought answers to this question (Maracek, 1979; Furstenberg, 1976; Kellam, 1979; Sandler et al., 1979; Card, 1978; Dryfoos & Belmont, 1978).

Findings from these studies are summarized in Table 23.4.

A review of results shown in Table 23.4 points to some salient findings. These include the apparent importance of having more than one adult in the home to help care for young children, the seeming greater vulnerability of sons to deficits in parental care (and, perhaps, lack of a father figure in the home), the negative impact of family disruption and substitute parenting on a child's cognitive development, educational achievement and social behavior, and the tendency of children to repeat the life patterns of their parents. Similar findings have been obtained by other large bodies of research which seek to assess the effects of parenting behaviors on children.

It should be noted that the age of mothers at the child's birth is shown to have a

TABLE 23.3. Later Correlates of Early Childbearing for Parents, as Developed by Research: 1970–1980

Educational

    Adolescent pregnancy and/or parenthood is *one* factor in dropping out of high school.

    An early marriage is closely associated with dropping out of school; the critical factor is early marriage rather than pregnancy or parenthood.

    Adolescent mothers tend to have done poorly in school, had low school interests and goals *before* pregnancy occurred.

    Adverse educational effects for whites, especially if they became adolescent mothers when they were less than age 17.

    No direct educational effects for blacks.

    Availability of child care, especially by family members, is an important source of school return for blacks.

    Half or more of early school dropouts return to school in later life, often in early middle age.

Later family size

    Larger family size is likely, especially if first birth occurs before mother is age 15 or 17, particularly if the mother is black and especially if she marries.

    Slight tendency for adolescent parents to have aspirations for a larger family than is true for older parents.

Marriage and marriage disruption

    The majority of unmarried adolescent parents marry within a few years of the first child's birth.

    Early first marriages with or without adolescent pregnancy are more likely than later first marriages to dissolve in later life; *early marriage* rather than the timing of the birth appears to be the key variable in later marital disruption.

    Perhaps very early childbearing is associated with later marital disruption.

    Adolescent mothers are more likely to return to school if they fail to marry.

Labor force participation

    Later occupational status, hours worked, wages earned not directly associated with maternal age at first birth.

    Timing of first pregnancy appears to have no effect on occupational status of male partners.

    Young married men and women express somewhat more job satisfaction than singles.

    Young married women with children slightly less satisfied with their career prospects than young married women without children. No differences found between single parents and single nonparents.

    Adolescent parents, by age 30, somewhat more dissatisfied with jobs (30% dissatisfied) than young people who postponed parenthood.

Effects on welfare assistance

    Adolescent single mothers slightly more apt than adolescent marrieds or nonmothers of similar backgrounds to be dependent on AFDC when children very young, but this higher dependency rate is short-lived.

TABLE 23.3. (Continued)

In later years, women who had been adolescent mothers are especially apt to be welfare-dependent, if they are compared to the general population.

Availability of welfare assistance makes it possible for a sizable proportion of adolescent mothers to return to school.

Black adolescent mothers more apt than those who were older at first birth to have grown up in a family that received public assistance. Dependence on public assistance after child's birth most apt to occur for very young adolescent mothers.

Maternal behavior and attitudes with first-born children

Majority appear to be as competent and caring as older but otherwise comparable mothers (by race and socioeconomic status).

Majority of mother–infant interactive behaviors rated as appropriate but significantly more of mothers with first births at less than age 18 were rated as overly protective or too inattentive at child's 4-month-old testing; rating improved somewhat at 8-month observations.

No significant differences found between adolescent and postadolescent mothers in respect to accidents occurring to their children.

No significant differences in measured child-rearing attitudes or maternal perceptions of the temperament of their infants.

Adolescent mothers observed to be high on maternal warmth and physical interaction; lower than older mothers on verbal interaction.

SOURCES: Moore and Hofferth, 1978; McCarthy and Menken, 1979; Trussel and Menken, 1978; Morrison, 1978; Furstenberg, 1976; Furstenberg and Crawford, 1978; Kellam, 1979; Presser, 1976, 1980; Maracek, 1979; Sandler, McLaughlin, Sherrod, and Vietze, 1979; Russ-Eft et al., 1979.

slight, but fairly consistent, effect on the development of their children, especially their sons. This effect seems to be transmitted largely through the greater tendency of adolescent mothers to maintain one-parent households, to experience marital disruptions if they marry when they are very young, and to place their children with substitute parents.

SUMMARY

Although the rate of *nonmarital* births to teenagers has risen in recent years, the *overall* adolescent birthrate has fallen markedly since 1958. The causes of the rise in "illegitimate" births are a rise in nonmarital intercourse, nonuse or poor use of contraceptives, limited recourse to abortion, and a reduction in rates of teenage marriage and of release of babies for adoption. Research shows that a number of physical, social, psychological, and economic factors interact to cause the above behaviors.

Analyses of longitudinal data from a number of studies which statistically control for the adverse effects of poverty and racism indicate that parental age, as a factor by itself, is of little significance in the life outcomes for teenage parents and their children. For example, despite overly simplistic claims to the contrary, teen-

TABLE 23.4. Correlates of Early Childbearing for the Children of Adolescent Parents as Shown by Research: 1970–1980

Physical health

    No adverse effects owing to maternal age if high quality prenatal and later health care are available.

    Health of child tends to be better if two adults, rather than one, are in the home.

Cognitive development

    Significantly lower scores for children of adolescent mothers tend to become minimal when appropriate controls are instituted for adverse effects of poverty, racism, and family headship.

    Sons of adolescent parents more adversely effected than daughters in respect to cognitive development and educational achievement.

    At age 4 years, intelligence test scores slightly lower for sons of very young adolescent mothers.

    Presence of another adult in household reduces adverse cognitive development effects seemingly associated with adolescent single-parent families.

Socioemotional adjustment and personality characteristics

    No significant differences found during infancy, early childhood, or by age 12 or so if appropriate controls employed for effects of poverty, racism, and family headship.

    A small but relatively large proportion of children of adolescent mothers rated as having behavior problems (especially sons).

Overall behavioral development

    A somewhat larger proportion (about 9%) of children of adolescent mothers, compared to about 4% of other children, rated slow in overall development, boys seem to be particularly affected.

    Infant sons of adolescent black mothers have higher ratings of slow responses to mother.

Educational achievement of children

    No significant differences found in some studies; in others, slight differences largely explained by somewhat lower levels of cognitive development.

    At age 7, children of adolescent mothers more apt to repeat a grade; sons have somewhat lower reading achievement scores; a larger proportion of sons of very young adolescent mothers are rated as having a learning disturbance.

    Effects of adolescent parenthood seem greater for educational achievement than measured intelligence.

    Greater tendency for school adjustment problems in first grade tends to increase over the years, with severe behavioral and school problems more likely to be present by adolescence.

Family composition and structure

    Greater tendency for marital disruption, single-parent households, remarriage, so that child has a stepfather, or is cared for by neither parent and is often cared for by a grandparent or other relative.

    Slight tendency for children, as teenagers, to repeat the parental pattern of early childbearing and/or marriage as well as large families. Tendency remains even after effects of racism, poverty, and family headship are statistically controlled.

age parenting, in and of itself, is a minimal cause of dropping out of high school, youth unemployment, and welfare dependency. Public policies that provide adequate income and employment for children, youths, and families and that promote racial equality and a high quality of social, educational, and health (including family-planning and abortion) services are indicated by the research concerning the causes and consequences of adolescent childbearing. Teenage parenting is not a problem separate from other basic dysfunctions of contemporary society; it is one more piece in a total jigsaw puzzle that cannot be solved by single-issue, money-saving approaches (Chilman, 1979).

REFERENCES

Baldwin, W., & Cain, V. The children of teenage parents. *Family Planning Perspectives*, 1980, **12**, 34–43.

Billingsley, A. Illegitimacy and the black community. In *Illegitimacy: Changing services for changing times*. New York: National Council on Illegitimacy, 1970, pp. 70–85.

Card, J. *Long-term consequences for children born to adolescent parents*. Palo Alto, Calif.: American Institutes of Research, 1978.

Chilman, C. *Adolescent sexuality in a changing American society: Social and psychological perspectives*. Department of Health, Education and Welfare, U.S. Government Printing Office, Washington, D.C. Pub #(NIH) 79-1426, 1978.

Chilman, C. Teenage pregnancy: A research review. *Social Work*. 1979, **24**, 492–498.

Cvetkovich, G., & Grote, B. Current research on adolescents and program implications. Paper presented at Conference on Family Planning, Olympia, Wash., 1977.

Dryfoos, J., & Belmont, L. The intellectual and behavioral status of children born to adolescent mothers. Final report to Center for Population Research. National Institute of Child Health and Human Development. Bethesda, Md., 1978.

Evans, J., Selstad, G., & Welcher, W. Teenagers' fertility control behavior and attitudes before and after abortion, childbearing or negative pregnancy test. *Family Planning Perspectives*, 1976, **8**, 192–200.

Forrest, J., Tietze, C., & Sullivan, E. Abortion in the United States, 1966–1977. *Family Planning Perspectives*, 1978, **10**, 271–280.

Fox, G. The family role in adolescent sexual behavior. Paper given at Family Impact Seminar, Washington, D.C., 1979.

Furstenberg, F. *Unplanned parenthood: The social consequences of teenage childbearing*. New York: Free Press, 1976.

Furstenberg, F., & Crawford, A. Family support: Helping teenagers cope. *Family Planning Perspectives*, 1978, **10**, 323–333.

Goldsmith, S., Gabrielson, M., & Gabrielson, I. Teenagers, sex and contraception. *Family Planning Perspectives*, 1972, **4**, 51–59.

Hornick, J., Doran, L., & Crawford, S. Premarital contraceptive usage among male and female adolescents. *Family Coordinator*, 1979, **28**, 181–190.

Jessor, S., & Jessor, R. Transition from virginity to non-virginity among youth. *Developmental Psychology*, 1975, **11**, 473–484.

Jorgenson, S., King, S., & Torrey, B. Dyadic and social network influences on adolescent exposure to pregnancy risk. *Journal of Marriage and the Family*, 1980, **42**, 141–155.

Kanter, J., & Zelnik, M. Sexual Experiences of young unmarried women in the U.S. *Family Planning Perspectives*, 1972, **4**, 9–17.

Kellam, S. Consequences of teenage motherhood for mother, child and family in a black urban community. Progress reports to Center for Population Research. National Institute of Child Health and Human Development. Bethesda, Md., 1979.

Ladner, J. *Tomorrow's tomorrow: The black women.* Garden City, N.J., 1971.

Lindeman, C. *Birth control and unmarried young women.* New York: Springer, 1974.

Luker, K. *Taking chances: Abortion and the decision not to contracept.* Berkeley, Calif.: University of California Press, 1975.

McCarthy, J., & Menken, J. Marriage, remarriage, marital disruption and age at first birth. *Family Planning Perspectives,* 1979, **11,** 21–30.

Maracek, J. Economic, social and psychological consequences of adolescent childbearing. Report to the Center for Population Research, National Institute of Child Health and Human Development. Bethesda, Md., 1979.

Miller, W. Some psychological factors in undergraduate contraceptive use. Paper given at the 84th Convention, America Psychological Association, Washington, D.C., 1976.

Moore, K., & Caldwell, S. B. *Out of wedlock childbearing.* Washington, D.C.: The Urban Institute, 1977.

Moore, K., & Hofferth, S. *Consequences of age at first childbirth: Final research summary.* Washington, D.C.: Urban Institute, 1978.

Morrison, P. Consequences of late adolescent childbearing. Preliminary report to Center for Population Research. National Institute of Child Health and Human Development. Bethesda, Md., 1978.

National Center for Health Statistics. *Monthly Vital Statistics Report,* **29,** 1, Supplement, Hyattsville, Md., April 1980.

Osofsky, J., & Osofsky, H. The psychological reaction of patients to legalized abortions. *American Journal of Orthopsychiatry,* 1972, **42,** 48–60.

Presser, H. Social factors affecting the timing of the first child. Paper presented at the Conference on the First Child and Family Formation, Pacific Grove, Calif., March 1976.

Presser, H. Guessing and misinformation about pregnancy risk among urban mothers. *Family Planning Perspectives,* 1977, **9,** 234–236.

Presser, H. The social and demographic consequences of teenage childbearing for urban women. NICHD, Center for Population Research, Bethesda, Md., 1980.

Rainwater, L. *Behind ghetto walls.* Chicago: Aldine Publishing Co., 1970.

Rosen, R. A., Martindale, L., & Grisdela, M. Pregnancy study report. Wayne State University, Detroit, Mich., March 1976.

Ross, H., & Sawhill, I. *Time of transition: The growth of families headed by women.* Washington, D.C.: The Urban Institute, 1975.

Russ-Eft, D., Springer, M., & Beever, A. Antecedents of adolescent parenthood and consequences at age 30. *Family Coordinator,* 1979, **28,** 173–178.

Sandler, H., McLaughlin, J., Sherrod, K. F., & Vietze, P. Social-psychological characteristics of adolescent mothers and behavioral characteristics of their first-born infants. Report to the Center for Population Research, National Institute of Child Health and Human Development, Bethesda, Md., 1979.

Schinke, S., & Gilchrist, L. Adolescent pregnancy: An interpersonal skill training approach to prevention. *Social Work in Health Care.* 1977, **3,** 158–167.

Shah, F., Zelnik, M., & Kantner, J. Unprotected intercourse among unwed teenagers. *Family Planning Perspectives,* 1975, **7,** 39.

Simon, W., Berger, A., & Gagnon, J. Beyond anxiety and fantasy: The coital experiences of college youths. *Journal of Youth and Adolescence,* 1972, **1,** 203–222.

Sorenson, R. *Adolescent sexuality in contemporary America.* New York: World, 1973.

Stack, C. *All our kin: Strategies for survival in a black community.* New York: Harper & Row, 1973.

Staples, R. *The black woman in America: Sex, marriage and the family.* Chicago: Nelson-Hall, 1973.

Trussel, J., & Menken, J. Early childbearing and subsequent fertility. *Family Planning Perspectives,* 1978, **10,** 209–218.

Udry, J., Bauman, K., & Morris, N. Changes in premarital coital experience at recent decade of birth cohorts of urban America. *Journal of Marriage and the Family,* 1975, **37**, 783–787.

Vener, A., & Stewart, C. Adolescent sexual behavior in middle America revisited: 1970–1973. *Journal of Marriage and the Family,* 1974, **36**, 728–735.

Wilson, F. Antecedents of adolescent pregnancy. *J. Biosoc. Sci.,* 1980, **12**, 141–152.

Zelnik, M., & Kantner, J. Sexual and contraceptive experience of young unmarried women in the United States. *Family Planning Perspectives,* 1977, **9**, 55–73.

Zelnik, M., & Kantner, J. Reasons for non-use of contraceptives by sexually active women, ages 15–19. *Family Planning Perspectives,* 1979, **11**, 289–294.

ADULTHOOD

24

COGNITIVE DEVELOPMENT IN ADULTHOOD
Using What You've Got

Edith D. Neimark
Douglass College
Rutgers
The State University

The title "Cognitive Development in Adulthood" will strike many readers as a contradiction in terms. Isn't adulthood the long asymptote on the curve of cognitive development from infancy through adolescence? So far as psychological theory is concerned, the answer to that question is an unequivocal "yes." In the most widely accepted theory of cognitive development, that of Piaget, it is hypothesized that the highest level of schemes, as well as the highest level of organization of operative skills, are attained by the end of adolescence; that is, the development of basic competence is completed during this period. Social convention in most societies, on the other hand, generally makes a sharp distinction between adolescents and adults in assigning major responsibilities for conduct of the affairs of society. This universal practice implies a belief that development continues during the adult years.

What is it that is assumed to develop during this period? Common sense provides a ready answer: what develops is experience, knowledge, and—it is hoped—wisdom; in other words, the application and enhancement of basic competence in performance of the affairs of life.

Psychological theories commonly differentiate competence, a repertoire of basic cognitive capacities, from performance, their deployment in practice. Piagetian theory (Inhelder & Piaget, 1958) provides the best detailed description of the development of competence but provides little elucidation of its translation into performance. Description of the additional factors involved in determining performance has only recently become a focus of concern for cognitive theorists (see, e.g., Overton & Newman in the Childhood section of this book); perhaps the most advanced solution available to date is Pascual-Leone's (1980) theory of constructive operators. The remainder of this section will give a brief overview of the idealized picture of adult intellectual competence followed by an equally brief overview of some variables mediating performance. Some of the demonstrated determinants of adult thought will be considered in greater detail in the second part of this chapter.

The Nature of Adult Intellectual Competence at Its Best

The process of cognitive growth can be characterized globally as a course of increasing detachment, of both differentiation and integration of component skills, and of increasing self-knowledge as to the nature of one's intellectual capacities.

Let us clarify the course of detachment. There is, first of all, progressive detachment of knowledge from its context of experience: advancement from knowledge as action during the sensorimotor period to the almost exclusively symbolic plane of reflective abstraction. There is also detachment from the specificity of a unique experience to generalization about classes of events characterized in terms of their properties and relations; for example, from a pet dog to the class of all dogs, all canids, and so forth. Detachment also refers to increasing abstraction from the concrete here-and-now instance to a formal idealization of the class in question. Finally, detachment refers to a broadened frame of reference in which events are placed in a context not of personal reactions but of objective impersonal understanding: not simply how one feels about an event but what it means with respect to a body of organized knowledge.

The differentiation and integration of component skills has been characterized by Piaget in terms of (a) a new category of symbolic transformations, formal operations, operating upon (b) abstract units, propositions, organized into (c) a simple, flexible structure having the properties of a mathematical system, the INRC group. The observable consequence of this restructuring of thought is the appearance of hypothetico–deductive reasoning; that is, starting from any initial assumption one could imagine and explore its logical implications as a prelude to action. An additional characterization of formal operational thought invokes (d) eight formal operational schemes: general conceptual frameworks for dealing with such relations as proportion, probability, correlations, equilibrium, and coordination of different frames of reference. Arlin (1980) differentiates the formal operational thought of adolescence from that of adulthood in terms of consolidation and application of these schemes. For the adolescent mastery of the schemes is an intellectual goal; for adults the schemes become means toward further advancement of understanding. Consider, for example, the coordination of frames of reference scheme which underlies appreciation of relativity. Relativity is a complex concept; a revolutionary advance in theories of physics, anthropology, and ethics resulted from its incorporation. This is equally true of concepts deriving from other such schemes as dynamic equilibrium or probabilistic (rather than strict) determinism, both of which influence theoretical advances in physical, biological, and social domains.

It should be emphasized that this description of adult thought deals with the optimal capacity which may be attained; much early research and subsequent criticism of Piagetian theory focused upon mounting evidence (cf. Neimark, 1975) that many individuals do not operate on this hyperion plane. Partly as a result of this evidence, current research has turned to identification of some of the additional factors moderating observed intellectual performance. Before turning to a description of recent evidence, I shall review one theoretical attempt at classifying the determinants of performance.

Pascual-Leone's Theory of Constructive Operators

Pascual-Leone (1976, 1980) and his students (Case, 1974, 1978) describe performance in terms of the momentary state of a system of habitual schemes, H. The schemes in this repertoire serve three functions: figurative, transformational, and executive. The first two functions closely parallel the Piagetian distinction of figurative and operative aspects of thought (i.e., the form in which knowledge is represented or encoded as contrasted with the programs of mental transformation). The third, executive function, refers to the control (i.e., selection and sequencing) of operations: what Miller, Galanter, and Pribram (1960) call "plans." It is elaboration of executive schemes which is probably most characteristic of adult thought. The availability of schemes in the repertoire is moderated by a number of silent operators: M, C, L, F, A, B, I, each of which will be described below.

The M operator refers to attentional capacity: the maximum number of schemes that can be activated in support of an executive scheme. It is a capacity factor, somewhat analogous to Spearman's (1927) g, assumed to increase linearly with age from one activated operation at 3 to 4 years of age to a maximum of seven by age 16.

The C operator has to do with specific content knowledge (i.e., information concerning a specific knowledge domain). It is affected by individual differences in experience and prior learning.

A structural learning operator, L, refers to schemes for the organization and augmentation of content knowledge; for instance, knowing how to generate geometric proofs, the proper form for legal argument, medical diagnosis, scientific reasoning, and so forth.

Specific features of the immediate stimulus situation are reflected in the F (field) factor dealing with features salient for the individual that may facilitate or impede optimal performance.

There are two individual difference factors associated with the personality and style of a given individual: A refers to affective and motivational factors; B refers to biases and beliefs. Chief among the B factors are individual differences in field dependence or independence which, in turn, affect susceptibility to field factors and readiness for utilization of optimal M power.

The last, and least, elaborated operator I is associated with inhibitory factors.

The potential contribution of each of the operators should be sufficiently evident to obviate detailed explanation. For example, given two individuals of equal ability (whatever that may mean in practice) such that one has some familiarity with the material and one has not, the first will be expected to do better on the assigned task. This could account for the widely obtained finding that persons with little formal schooling, or women, tend to do poorly on the mechanical and chemical experimentation tasks devised by Inhelder for the study of formal operations. In this particular example it is likely that not only C, content, but also $L, A, B,$ and F factors may be involved. Specifically, the untutored individual may be unfamiliar not only with physics and chemistry experiments but also the general procedures of experimentation and may be intimidated by the task demands (which are rarely detailed explicitly in the instructions). In such a threatening context he or she may seize upon familiar aspects of the situation and assimilate them to readily available schemes—in other words, transform the experimenter's task into his or her own version of it. The resulting translation may be a very different task, the solution of which will probably be interpreted by the experimenter as evidence for lack of competence. The next section will review many probable examples of this sort of subject transformation of the experimenter's intent. The list of silent operators will be used as an organizing framework for this review despite occasional problems of arbitrariness or inappropriateness. Two additional difficulties should be noted at the outset. First, as suggested by the illustration, rarely is it the case that only a single factor is involved. Second, although Pascual-Leone's theory is used here as an organizing framework for presenting evidence, it is not the theory which motivated the original collection of the evidence used to illustrate it.

THE VICISSITUDES OF ADULT THOUGHT

Expert versus Novice

All societies impose upon normal adult members the requirement that they earn a livelihood. Inevitably each individual must specialize in certain tasks and roles at the expense of others. Despite the pervasiveness and universality of this societal expectation, there is relatively little research on the process of becoming an expert or on comparison of the skills displayed by experts versus amateurs (but see Simon & Simon, 1978). The early work of Bryan and Harter (1899) on telegraphers and of Book (1908) on typists suggested that acquisition of skill consists not in performing the same component skills with greater speed and efficiency but, rather, in formation of more compact units; for example, words or phrases in place of individual letters. In current parlance we would say that experts form bigger chunks of information. This conclusion is supported by research on chess masters by Binet (Simmel & Barron, 1966) and by Chase and Simon (1973). The former found that at increasing levels of skill the blindfolded player's representation of the board became more and more abstract (as contrasted with a literal visual image of the board). Chase and Simon found chess masters to be superior in recall of board positions only when pieces were arrayed in meaningful positions; with random placement of the pieces there was no difference between masters and poorer players. Piaget (1972) has speculated that accessibility of formal operations might well be affected by conditions of professional specialization, a prediction generally borne out by the evidence (e.g., Sabatini & Labouvie-Vief, 1979).

Recent research on solution of physics problems (Chi et al., in press) and on learning lists of computer programming terms (McKeithen et al., 1981) as a function of level of expertise suggests expertise to be characterized by use of a consistent organizing framework (of physical principles or programming procedures) for dealing with particular problems. This brief review indicates that the difference between expert and novice is not simply a matter of content knowledge, but also of how the knowledge is coded and organized.

The Role of Cultural Conventions

Vygotsky's (1962) theory of intellectual development assigns a large causal role to social factors. Luria (1976) took advantage of a unique opportunity to study the impact of such factors which arose in the early 1930s with the introduction of universal education throughout the Soviet Union. Various groups of individuals in Uzbekistan (an isolated province in central Asia) differing in degree of exposure to modern Western culture were tested on a variety of reasoning problems. The most isolated group, women observing strict Moslem tradition, failed on all the tests presented. A great deal of subsequent cross-cultural evidence based upon both logical-reasoning problems (e.g., Cole et al., 1971) and Piagetian tasks (e.g., Dasen, 1977) has provided consistent support for two generalizations first evident in Luria's work: (a) members of traditional cultures perform more poorly on reasoning tasks than do members of Western cultures and (b) within a given culture, quality of performance is an increasing function of number of years of formal schooling. Moreover, all cross-cultural evidence demonstrates a characteristic quality in the responses of nonliterate peoples which is well illustrated by the typical answer Luria obtained to his syllogism: "In the far north all bears are white; Novaya Zemlya is in the far north. What color are the bears there?" Uzbek women typically refused to answer, suggesting that Luria ask someone who had been there and had seen the bears for himself. Scribner (1977) calls this refusal to go beyond direct personal experience "empirical bias," and convincingly demonstrates that it influences not only response to reasoning problems but also the initial understanding of the problem itself. In discussion of everyday events, however, she found the same individuals to be fully capable of logical reasoning about not only concrete examples of familiar experience but also hypothetical variants of them. Comparable findings are reported by a number of cross-cultural investigators (cf. Glick, 1975, for a review). This repeated finding suggests that members of non-Western cultures may be lacking not in the capacity for adult reasoning but only in the structural learning required for expressing their reasoning in accepted Western form (or for differentiating empirical from logical questions).

Specific conventions for reasoning, such as formal logic or scientific method, are human inventions dating back to earlier writers; their further elaboration continues to occupy contemporary thinkers. During the course of schooling, literate individuals are deliberately instructed in the rules of formal reasoning (for algebra, geometry, sciences, etc.). It should not be surprising, therefore, that skill in formal reasoning is a function of amount of formal schooling or that even college students still have a good deal of difficulty with it. That structural learning of reasoning conventions is a necessary but not sufficient condition for manifestation of correct formal reasoning is demonstrated by a wealth of evidence to be sampled in the next section.

Biases in Adult Judgment, Reasoning, and Decision Making

The term *bias* refers to a systematic (as contrasted with random or unsystematic) source of error. *Error* can refer to disagreement with common observation or to deviation from a criterion. In the case of reasoning, there are criteria, or normative rules, which prescribe how the process should proceed. These normative rules provide a convenient starting point for laboratory study of adult reasoning. As will be shown below, however, available evidence demonstrates that the normative model does not provide an accurate description of observed performance. Since the observed deviations have been found to be consistent, lawful, and characterized by a logic of their own, they are properly classified as biases. Some commonly observed biases are described to illustrate that although adult reasoning is more powerful, and closer to prescribed convention, than child reasoning, it is nevertheless still subject to predictable error.

1. Errors in Formal Logic. Rules for deducing necessary consequences from one or more propositions, each of which is either true or false, are prescribed by formal logic. *Necessary* refers to logical (rather than empirical) necessity; *formal* denotes that truth or falsity of the conclusion derives from the form of the argument rather than the content of the premises. As might be expected, however, it is difficult even for sophisticated adults to ignore content totally.

Early research on syllogistic reasoning identified a number of systematic errors, for example: (1) atmosphere effect—a bias toward particular and negative conclusions given one particular or negative premise (Woodworth & Sells, 1935); (2) illicit conversion—treating the subject and predicate of a premise, for example, all A's are B's, as interchangeable, to all B's are A's (Chapman & Chapman, 1959; Revlis, 1975); (3) probabilistic inference—interpreting a premise which is generally true as universally true (Chapman & Chapman, 1959). Recent research focuses upon misinterpretation of quantifiers (Ceraso & Provitera, 1971) and going beyond presented information (Henle, 1962; Frase, 1972; Kintsch, 1974).

Syllogistic reasoning is of limited utility in daily life, but other varieties of formal reasoning, such as hypothetico–deductive reasoning and, especially, conditional reasoning, are widely applicable to reaching conclusions based on empirical evidence. They are the focus of much recent research (cf. Falmagne, 1975; Wason & Johnson-Laird, 1972). Here, too, a number of systematic sources of error have been identified, such as: (1) confirmation bias (Wason & Johnson-Laird, 1972)—a tendency to seek proof in verification rather than falsification of the complement (Moshman, 1979; Tweney et al., 1980); (2) one-to-one matching of terms (Wason & Evans, 1974–75); (3) restricted interpretation of quantifiers, such as "all" and "some" (Neimark & Chapman, 1975), or connectives, such as "if–then" (Taplin, Staudenmeyer, & Taddonio, 1974). Although, as was also true in the discussion of syllogistic reasoning, the systematic errors observed are here described as biases, it remains to be demonstrated that they cannot be explained in terms of failures of structural learning, salient field effects, or some combination of the two. The usual procedure for demonstrating that an error is, in fact, a bias is to demonstrate that it is impervious to explicit training or to variation of instructions and content. Some evidence (e.g., Tweney et al., 1980; Brehmer, 1980) is beginning to appear; as will be seen in the next section, relevant evidence that errors truly reflect bias is available for biases of judgment.

2. Judgment and Decision Making. The major issues of everyday life are more likely to concern questions of fact or probability and the attainment of practical goals rather than determination of logical necessity. The factors to be considered are better described in terms of a distribution along some quantitative continuum rather than by classes and complements (e.g., varying degrees of acceptability of a solution rather than acceptable or not acceptable). Moreover, the values assigned to costs, values, likelihoods of occurrence, and so forth, in a given problem are generally unique to a specific time, place, and set of conditions. Formal logic is of little value in the solution of such problems. The publication of *Theory of Games and*

Economic Behavior by von Neumann and Morganstern (1944) provided the framework for the first normative theories of decision making. Once again, the normative model provided inaccurate description of observed behavior. Attempts at a descriptive theory, such as Simon's (1957) theory of bounded rationality, assume that limitations on human information–processing capacity require the construction of simplified models for decisions. For recent discussions of decision models and research evidence relevant to them see Einhorn & Hogarth (1981) and the special issue of *Acta Psychologica* (1980).

One major component of all decision theories is assessment of the likelihood of uncertain events. Tversky and Kahneman (1974), in a classic paper, proposed the operation of three heuristic biases in judgment: representativeness, availability, and anchoring effects. Each heuristic, because it serves to simplify the process of judgment, is of practical value although it may also lead to error. The representativeness heuristic is called into play in situations requiring generalization to a broader universe or formulation of a principle on the basis of a limited set of observations. This is the sort of situation for which the discipline of inferential statistics provides normative rules. People, even trained statisticians, tend to be poor statisticians in practical situations. Their judgments appear to be governed by what might be called a "law of small numbers": an assumption that even a small and limited sample is representative of the population. Some other biases associated with the representativeness heuristic are the gambler's fallacy, unwarranted confidence in judgment based on small samples, and rejection of regression effects.

The availability heuristic deals with the effect upon judgment of the amenability of the phenomenon in question to representation by imagery, verbal encoding, and so forth, that is, equating event likelihood with ease of imagining an instance. Judgment of the likelihood of such events as floods, accidents, fatal illnesses, and so forth, have been shown to be better predicted by their frequency of press coverage than by their objective frequency of occurrence (Slovic, Fischoff & Lichenstein, 1980). Once again, "experts" seem to be as subject as lay persons to biases associated with availability.

Anchoring refers to the effect of an initial estimate (however arrived at) upon subsequent adjusted values of that estimate. For example, when asked to set confidence intervals for some statistic, such as the population of the United States, the number of foreign cars imported annually, and so forth, the interval in which most persons estimate the true value to fall is not only much too narrow but, also, frequently fails to include the true mean at all. Even for much simpler tasks, like the judgment of weights, distances, or temporal durations, anchoring effects have been shown (Helson, 1964).

Given all these sources of error in assigning values to factors to be incorporated in a final decision, it is not surprising that the final decision is rarely optimal. Some reported sources of error concern failure to utilize important relevant evidence, inconsistency in the pattern of judgment, and overconfidence in the probable likelihood of the correctness of the judgment or the course of action chosen (Slovic, Fischoff, & Lichenstein, 1977). In the face of accumulating evidence concerning the systematic nature of errors in judgment and decision making, it is natural to speculate that training in the nature of error bias and the means of its elimination could be useful and effective. Although Nisbett and Ross (1980) offer suggestions as to what a training program might include, there is not yet any evidence concerning the modification of bias through training.

Individual Differences in Cognitive Style as a Source of Bias

Psychologists seem to be in general agreement that individuals differ in their general approach to cognitive problems. There is less agreement upon identification of the dimensions of difference or on a classification scheme to incorporate them. The more popular categorizations of this individual difference, generally called *cognitive style,* are described by Goldstein and Blackman (1978). The present discussion will be confined to only one: Witkin's (Witkin & Goodenough, 1977a; Witkin, Goodenough, & Oltman, 1977) field dependence–independence component of psychological differentiation, largely because it is the one for which most relevant evidence on adults is available.

The terms *field dependence* and *field independence* refer to opposite poles on a continuum of degree of autonomy from external referents in both cognitive and social domains. As the names imply, field independence refers to autonomy from external referents, field dependence to dependence upon them. Field-independent individuals do well in tasks requiring cognitive restructuring, especially spatial restructuring (e.g., identifying hidden figures or maintaining the physical upright in a field of nonexistent or misleading perceptual cues); field-dependent individuals are more interpersonally oriented and sensitive to environmental cues. If these putative styles were to operate broadly over a variety of experimental contexts, then one would expect to find (a) that field-dependent individuals perform more poorly on the type of reasoning tasks used to assess formal operations competence and (b) that their performance is far more subject than that of field independents to variation of field factors such as task content, instructions, and so forth. The first prediction has been supported by a variety of experimental evidence. Although the second has been less frequently subjected to experimental tests, confirming evidence is beginning to appear (see Neimark, 1981, for a review of evidence relevant to both predictions). Whether style is subject to experimental modifications, as in the case of other biases, is an unanswered question.

Pascual-Leone (deRibaupierre & Pascual-Leone, 1979) assigns a motivational component to the dimension of relative autonomy from field factors (in addition to the relationship to field factors inherent in the definition of field dependence–independence). Field-independent individuals are assumed to seek opportunities to use their intellectual skills and to possess a well-integrated repertoire of executive schemes for deploying those skills. As a result, they tend to approach most problems in a thoughtful, logical manner. Field dependents, on the other hand, are more drawn to and skilled at social interactions. Although there is little direct evidence relevant to this assumption, it is compatible with the evidence of Witkin and his associates (Witkin & Goodenough, 1977b) on the role of cognitive style in occupational choice: in general, field independents chose occupations emphasizing individual attainment (e.g., the arts and sciences), whereas field dependents tend to prefer service occupations (e.g., nursing, social work).

The Role of Field Factors

In the preceding discussion it was noted that the effect of each performance factor was influenced by field effects such as the content and context of the experimental task. It may be concluded from this uniform finding that field effects play an important role in cognitive performance. But for each performance factor the nature of the interaction was that field effects played a less important role for individuals at the favorable end of the continuum than for those at the unfavorable end (e.g., individuals lacking in requisite content or structural learning, having a field dependent style, etc., perform well only when relevant stimulus features are salient). The same generalization may be applied to adult cognition in general (as compared with earlier developmental stages). The adult thinker is much less constrained by particular features of task content, context, or personal feelings with respect to them. This ability to transcend the particular and to approach each situation with detachment is popularly regarded as a defining characteristic of adult thought. Unfortunately, I know of no research specifically addressed to the assumption nor of any evidence directly relevant to it.

FROM THE LABORATORY TO THE REAL WORLD

Summary

To summarize the foregoing discussion, it was first noted that inherent, optimal, potential capacity of the adult thinker is a theoretical abstraction not to be directly observed but only inferred from performance. Performance may be facilitated or, more commonly, impeded by the operation of a number of additional factors. Among the factors described are content knowledge, structural learning, field factors, and individual difference factors of motivation and bias.

Current Status of Theory and Knowledge

No review of adult cognitive development could overlook the current status of the field: it barely exists as a recognized discipline. One reason for this woeful neglect is the current state of relevant theory. Most views of intelligence, as well as the most influential theory of cognitive development, Piaget's, tacitly assume that cognitive development ceases at the end of adolescence and that whatever changes may take place thereafter are best described as stability followed by decline of ability with advancing age. Recently, suggestions of a uniquely adult stage of development have begun to appear (Arlin, 1975, 1980; Broughton, 1977; Gruber & Voneche, 1976; Labouvie-Vief, 1980; Riegel, 1973). None of the proposals is sufficiently elaborated to constitute a directing theory, nor is there much comparability across theories. To date, none of the pretheoretical suggestions has generated either experimental test or theoretical elaboration. A vacuum remains to be filled.

A second major void is the lack of systematic evidence on cognitive development in adulthood. It is, alas, unlikely that this void will be filled in the near future. The lack of a powerful directing theory has already been noted. Brute empirical data gathering is also unlikely for at least two reasons. First, to the extent that adults are busy with matters of practical importance, they are rarely available as research subjects. Second, there is reason to question whether the sorts of activity most fully representative of adult thought can be reproduced under laboratory conditions. The laboratory provides a controlled, generally simplified, and artificial context for the observation of behavior. In real life, on the other hand, one is rarely confronted with a cleanly delimited situation where, for a fixed period of time, one is exclusively concerned with performance of a task whose requisites for solution are fully available. The tasks of life tend to be part of a jumble of concomitant, and sometimes conflicting, demands. At present, psychological theory has little to offer by way of clarification of a host of basic questions concerning how adult cognition is brought to bear on the tasks of life. How are demands identified and ordered? How does one codify experience, assess probabilities, assign priorities, recognize the need for additional information, or combine these components in planning, problem solving, and decision making?

While the elements of adult thought and the general structure of their organization have been suggested by systematic laboratory investigation, the generality of these findings remains to be tested. To cite but one of many obvious instances, simply by virtue of having lived longer than a child or adolescent, the adult has a wealth of experience and a resulting richer, more detached perspective. Society values that perspective, but psychologists have yet to describe its nature or impact. That assignment is typically left to novelists, philosophers, theologians, historians, and newspaper columnists.

REFERENCES

Arlin, P. K. Cognitive development in adulthood: A fifth stage? *Developmental Psychology,* 1975, **11**, 602–606.

Arlin, P. K. Adolescent and adult thought: A search for structures. Paper presented at meetings of the Piaget Society, Philadelphia, June 1980.

Book, W. F. *The psychology of skill.* Missoula, Mont.: University of Montana Press, 1908.

Brehmer, B. In one word: Not from experience. *Acta Psychologica,* 1980, **45**, 222–241.

Broughton, J. "Beyond formal operations": Theoretical thought in adolescence. *Teachers College Record,* 1977, **79**, 87–97.

Bryan, W. I., & Harter, N. Studies on the telegraphic language: The acquisition of a hierarchy of habits. *Psychological Review,* 1899, **6**, 345–365.

Case, R. Structure & Strictures: Some functional limitations on the course of cognitive growth. *Cognitive Psychology,* 1974, **6**, 544–573.

Case, R. Intellectual development from birth to adulthood: A neo-Piagetian interpretation. In R. S. Siegler (Ed.), *Children's thinking: What develops?* Hillsdale, N.J.: Erlbaum, 1978, pp. 37–72.

Ceraso, J., & Provitera, A. Sources of error in syllogistic reasoning. *Cognitive Psychology,* 1971, **2**, 400–410.

Chapman, L. J., & Chapman, J. P. Atmosphere effect re-examined. *Journal of Experimental Psychology,* 1951, **58**, 220–226.

Chase, W. G., & Simon, H. A. Perception in chess. *Cognitive Psychology,* 1973, **4**, 55–81.

Chi, M. T. H., Glaser, R., & Rees, E. Expertise in problem solving. In R. Sternberg (Ed.), *Advances in the Psychology of Human Intelligence,* Vol. 1. Hillsdale, N.J.: Erlbaum, in press.

Cole, M., Gay, J., Glick, J., & Sharp, D. *The cultural context of learning and thinking.* New York: Basic Books, 1971.

Dasen, P. R. Are cognitive processes universal? A contribution to cross-cultural Piagetian psychology. In N. Warren (Ed.), *Studies in cross-cultural psychology,* Vol. 3. New York: Academic, 1977, pp. 155–201.

deRibaupierre, A., & Pascual-Leone, J. Formal operations and M power: A neo-Piagetian investigation. In D. Kuhn (Ed.), Intellectual development beyond childhood. *New Directions for Child Development,* 1979, **5,** 1–43.

Einhorn, H. J., & Hogarth, R. M. Behavioral decision theory: Processes of judgment and choice. *Annual Review of Psychology,* 1981, **32,** 53–88.

Falmagne, R. J. *Reasoning: Representation and process.* Hillsdale, N.J.: Erlbaum, 1975.

Frase, L. T. Maintenance and control in the acquisition of knowledge from written materials. In R. O. Freedle & J. B. Carroll (Eds.), *Language comprehension and the acquisition of knowledge.* Washington, D.C.: Winston, 1972.

Glick, J. Cognitive development in cross-cultural perspective in F. D. Horowitz (Ed.), *Review of child development research* (Vol. 4). Chicago: University of Chicago Press, 1975, pp. 595–654.

Goldstein, K., & Blackman, S. *Cognitive style.* New York: Wiley, 1978.

Gruber, H., & Vonèche, J. Reflexions sur les operations formelles de la pensee. *Archives deo Psychologie,* 1976, **44,** 45–55.

Helson, H. *Adaptation-level theory: An experimental and systematic approach to behavior.* New York: Harper, 1964.

Henle, M. On the relation between logic and thinking. *Psychological Review,* 1962, **69,** 366–378.

Inhelder, B., & Piaget, J. *The growth of logical thinking.* New York: Basic Books, 1958.

Kintsch, W. *The representation of meaning in memory.* Hillsdale, N.J.: Erlbaum, 1974.

Labouvie-Vief, G. Beyond formal operations: Uses and limits of pure logic in lifespan development. *Human Development,* 1980, **23,** 141–161.

Luria, A. R. *Cognitive development: Its cultural and social foundations.* Cambridge, Mass.: Harvard University Press, 1976.

McKeithen, K. B., Reitman, J. S., Rueter, H. H., & Hirtle, S. C. Knowledge organization and skill differences in computer programmers. *Cognitive Psychology,* 1981, **13,** 307–325.

Miller, G. A., Gallanter, E., & Pribram, K. *Plans and the structure of behavior.* New York: Holt, Rinehart & Winston, 1960.

Moshman, D. Development of formal hypothesis-testing ability. *Developmental Psychology,* 1979, **15,** 104–112.

Neimark, E. D. Intellectual development during adolescence. In F. D. Horowitz (Ed.), *Review of child development research* (Vol. 4). Chicago: University of Chicago Press, 1975, pp. 541–594.

Neimark, E. D. Toward the disembedding of formal operations from confounding with cognitive style. In I. Sigel, D. Brodzinsky, & R. Golinkoff (Eds.), *Piagetian theory and research: New directions and applications.* Hillsdale, N.J.: Erlbaum, 1981, pp. 177–190.

Neimark, E. D., & Chapman, R. H. Development of the comprehension of logical quantifiers. In R. J. Falmagne (Ed.), *Reasoning: Representation and process.* Hillsdale, N.J.: Erlbaum, 1975, pp. 135–152.

Nisbett, R., & Ross, L. *Human inference: Strategies and shortcomings of social judgment.* Englewood Cliffs, N.J.: Prentice-Hall, 1980.

Pascual-Leone, J. Metasubjective problems of construction: Forms of knowing and their psychological mechanism. *Canadian Psychological Review,* 1976, **17,** 110–125.

Pascual-Leone, J. Constructive problems for constructive theories. In H. Spada & R. Kluwe (Eds.), *Developmental models of thinking.* New York: Academic, 1980.

Piaget, J. Intellectual evolution from adolescence to adulthood. *Human Development,* 1972, **15,** 1–12.

Revlis, R. Syllogistic reasoning: logical decisions from a complex data base. In R. J. Falmagne (Ed.), *Reasoning: Representation and process.* Hillsdale, N.J.: Erlbaum, 1975, pp. 93–134.

Riegel, K. F. Dialectic operations: The final period of cognitive development, *Human Development,* 1973, **16,** 346–370.

Sabatini, P. P., & Labouvie-Vief, G. Age and professional specialization in formal reasoning. Paper presented at the meetings of the Gerontological Society, Washington, D.C., 1979.

Scribner, S. Modes of thinking and ways of speaking: Culture and logic reconsidered. In P. N. Johnson-Laird & P. C. Wason (Eds.), *Thinking. Readings in cognitive science.* Cambridge: Cambridge University Press, 1977, pp. 483-511.

Simmel, M., & Barron, S. B. (Trans.) Binet, A. Mnemonic virtuosity: A study of chess players. *Genetic Psychology Monographs,* 1966, **74**, 127-162.

Simon, D. P., & Simon, H. A. Individual differences in solving physics problems. In R. S. Siegler (Ed.) *Children's thinking: What develops?* Hillsdale, N.J.: Erlbaum, 1978, pp. 325-348.

Simon, H. A. *Models of man: Social and rational.* New York: Wiley, 1954.

Slovic, P., Fischoff, B., & Lichenstein, S. Behavioral decision theory. *Annual Review of Psychology,* 1977, **28**, 1-39.

Slovic, P., Fischoff, B., & Lichenstein, S. Perceived risk. In R. C. Schwing & W. S. Albers, Jr. (Eds.), *Societal risk assessment: How safe is safe enough?* New York: Plenum, 1980.

Spearman, C. *The abilities of man.* New York: Macmillan, 1927.

Taplin, J. E., Staudenmayer, H., & Taddonio, J. L. Developmental changes in conditional reasoning: Linguistic or logical? *Journal of Experimental Child Psychology,* 1974, **17**, 360-373.

Tversky, A., & Kahneman, D. Judgment under uncertainty: Heuristics and biases, *Science,* 1974, **185**, 1124-1131.

Tweney, R. D., Doherty, M. E., Worner, W. J., Pliske, D. B., Mynatt, C. R., Gross, K. A., & Arkhelin, D. L. Strategies of rule discovery in an inference task. *Quarterly Journal of Experimental Psychology,* 1980, **32**, 109-123.

vonNeumann, J., & Morganstern, O. *Theory of games and economic behavior* (3rd ed.). Princeton, N.J.: Princeton University Press, 1953.

Vygotsky, L. S. *Thought and language.* Cambridge, Mass.: M.I.T. Press, 1962.

Wason, P. C., & Evans, J. StB. T. Dual processes in reasoning? *Cognition,* 1974/5, **3**, 141-154.

Wason, P. C., & Johnson-Laird, P. N. *Psychology of reasoning.* Cambridge, Mass.: Harvard University Press, 1972.

Witkin, H. A., & Goodenough, D. R. Field dependence revisited. Princeton, N.J.: Educational Testing Service, ETS Research Bulletin, RB-77-16, 1977. (a)

Witkin, H. A., & Goodenough, D. R. Field dependence and interpersonal behavior, *Psychological Bulletin,* 1977, **84**, 661-689. (b)

Witkin, H. A., Goodenough, D. R., & Oltman, P. K. Psychological differentiation: Current status. Princeton, N.J.: Educational Testing Service, ETS Research Bulletin, RB-77-17, 1977.

Woodworth, R. S., & Sells, S. B. An atmosphere effect in formal syllogistic reasoning. *Journal of Experimental Psychology,* 1935, **18**, 451-460.

25

DISCONTINUITIES IN DEVELOPMENT FROM CHILDHOOD TO ADULTHOOD
A Cognitive-Developmental View

Gisela Labouvie-Vief
Wayne State University

Both continuity and discontinuity need to be considered in defining developmental events and processes. Phenomena of development in adulthood are, however, often interpreted within a continuity framework, because theories of development are still addressed to the earlier part of the life span. Adult "maturity," as a consequence, is usually seen as a state in which the organism achieves an equilibrium some time between the early rush of childhood changes and terminal deteriorative biological processes.

Theories of child development—from which notions of adult maturity are derived—inherently stress notions of change, transformation, and discontinuity. Theories of adulthood, in contrast, have stressed the cessation of development. Several dramatic and unprecedented changes in our society are forcing developmental psychologists, however, to look differently at life-span development. The still prevalent, somewhat static, view of adult development may have been fairly appropriate in past, stable cultures, in which development could be equated with the individual's socialization into a predictable set of tasks and a static set of roles (Mead, 1970). In such societies, changes in adult roles appear to initiate with changing biological capacities of the organism (Neugarten & Hagestad, 1976). Recent advances in science and technology, however, in lengthening the life span and "graying" the population have made change an intrinsic aspect of adult development, as well.

Preparation of this paper was supported by funds of the Wayne State–University of Michigan Institute of Gerontology and by Research Career Development Award NIA 5 K04 AG00018 to the author.

Recent attempts to formulate theoretical models that speak directly to the activities and adaptations required of middle-aged and older adults have focused on two major issues. First, there is an increased concern with the fact that standards of maturity in adulthood can no longer be solely borrowed from theories aimed at child development. Second, there is an emerging realization that development cannot be defined any longer in a cultural, historical vacuum; it is inevitably embedded in a flow of historical events. If theories of child development, therefore, stress uniformity of growth, theories of adulthood also must address variability of adult adaptation.

In order to demarcate, therefore, the view outlined in this chapter, it may be most useful to contrast it with currently predominant notions of child development. Specifically, we will use Piaget's theory as a frame of reference and point of departure. Piaget's theory is not only the most widely known account of human development, but it also incorporates a number of assumptions that appear to guide most other developmental theories, such as those of Freud, Erikson, and Werner.

First and foremost, Piaget asserts that behaviors or processes can never be studied in isolation but must be looked at from the perspective of their systemic interdependence. Behavior is structured, and it is the resulting system of regulations among behaviors or acts that constitutes the object of analysis. Childlike language in an adult, for example, takes on profoundly different meanings depending on the spatial and temporal context of other behaviors: it could either indicate a regressive breakdown or—in the context of an adult talking to a child—an adaptive flexibility in which the complexity of the communicative act is tailored to the needs and capacities of the receiver.

The resulting emphasis on structure, organization, and form—that is, the coordinated regulation of systems of behavior—is a feature, we believe, which may help to clarify progressive change in adulthood. Two further features of Piaget's theory, however, present greater problems for interpreting adulthood. One of these is the connotation that psychological growth parallels and is constrained by biological growth. The other is the correlated connotation that developmental processes are equilibrated fairly early in the adult life span, when the major rush of biological growth is completed. Both these assumptions have been restrictive to adult development research as they suggest a view of adulthood that is essentially deficit-oriented. There are important reasons for modifying such a view.

STRUCTURES IN CONTEXT

Because Piaget sees psychological growth as an outgrowth of biological growth, development is assumed to be fairly invariant across cultural and subcultural groups. This assumption, in turn, has led to the belief that the tasks developed by Piaget for Swiss children have much broader validity. Indeed, they have come to serve as standards for judging developmental maturity in a more universal sense.

By these standards, adulthood displays intellectual stagnation and regression. This negative view of adulthood, however, overlooks certain methodological pitfalls inherent in the standards derived from Piaget's framework. Performance on Piaget's tasks is profoundly affected by education (e.g., Neimark in Chapter 24) and cultural heritage (see Greenfield, 1976; Scribner, 1979) and to that extent it is justified to question whether such deviations need to imply that those raised in other contexts than modern, Western educational systems are indeed deficient. The question, in other words, arises if and how social contexts can be incorporated into such universalistic models. On the one hand, then, Piaget's biological assumptions easily lead to a distortion of our view of adult development (see Labouvie-Vief, 1980a). On the other hand, however, we can highlight other aspects of his model as building blocks for a more comprehensive view of adult development. Specifically, as a structuralist and biologist, Piaget was interested in broad formal ways or operations by which individuals organize experience. He was less interested in the tasks by which these operations could be researched. Indeed, he was aware that those two aspects—form and content, or the specific tasks from which formal changes are inferred—cannot be truly disassociated but are interdependent, or "correlated."

While it is perfectly justifiable to talk of such operations as classification, permutation, and so forth, in an abstract manner, without referring to the specific contents or symbols that are being classified or permutated, these operations are always acquired in the context of specific material objects and cultural symbols. The cultural environment thus constrains those specific tasks and materials that can be formally organized or processed. Further, it may even limit the specific forms—or modes of operation—possible. Thus in judging the competencies of a particular culture or age group, one must attend carefully to the ecological demands of particular environments—otherwise, statements about deficit may arise as artifacts.

One such potential artifact derives from the fact that current tests of formal reasoning are heavily biased toward content which reflects a scientifically oriented academic education. It is possible that there may be other formal structures for people with experience in different content areas. Operational competence in children is influenced by the particular contents of their ecological settings (see Greenfield, 1976). Formal operational competence in adults is similarly affected by formal education (e.g., Blasi & Hoeffel, 1974).

A second potential artifact arises from the fact that formal operations, even allowing for content fluctuations, are highly specialized thought structures; there may well be "new and special structures that still remain to be discovered and studied" (Piaget, 1972, p. 11). Wason and Johnson-Laird (1972), for example, have submitted that the formal model underlying analyses of reasoning processes may not adequately represent the "practical" reasoning governing ordinary (e.g., semantic) discourse. The former is borrowed from propositional calculus; it is bivalent and truth-functional, and it may represent but a limiting case of practical inference. Thus even if the individual's inferential machinery is undisturbed by such performance variables as memory and load or abstract content, systematic deviations do not necessarily indicate flaws in inferential ability; rather, they suggest that models dealing solely with propositional thinking do not cover the full range of inductive and deductive competence. This would be particularly true when cultural or educational background is alien to principles of formal logic (see also Fillenbaum, 1977; Neisser, 1976; Olson, 1976).

Much of the variance in the psychological behavior of adults, as a consequence, may reflect diverse specializations related to history or culture (Schaie, 1979; Schaie & Labouvie-Vief, 1974). Sabatini and Labouvie-Vief (1979) have presented data to show that formal operational competence is affected more by occupation than by age. In their study, old scientists outperformed young scientists on a measure of formal reasoning; only a group of *non*scientists showed the familiar age-decline pattern.

Such evidence stands in opposition to a strictly biological model, showing it to be a limited metaphor for psychological development, which by its very nature is diversified and multihierarchic, channeling individuals into diverse ecological niches.

It is in this sense of failing to address the powerful differentiating effects of ecological contexts that Riegel (1976) criticized Piaget's model for relying on a "passive" conception of the environment. Where the environment is assumed to be relatively fixed and biologically determined, the individual's activity will not introduce variability into the system. Hence the major dimensions of development are assumed to be the same for all. If the environment is seen as active, in contrast, individuals, during the course of development or of cultural evolution, could transform it. They would thereby create new artifacts and symbols and make up altogether new contexts to which, in turn, they would react (see Leontiev, 1977; Vygotsky, 1978). While psychological development originates from biological development, the contexts upon which it feeds and which give it direction are human creations. In Vygotsky's words, this fact implies a change from an explanatory model that is primarily "zoological" to one that is inherently social—a change which becomes more and more important with increasing maturity.

DISEQUILIBRIUM, GROWTH, AND REGRESSION

From the previous argument, many apparently regressive changes in adult cognition reflect, in fact, the result of diverse specialization. This routing of adults into different sociocultural niches through different developmental pathways must be seen as intrinsic and necessary parts of theories of adulthood change. It is not sufficient, however, to deal with change in adulthood per se: the allowance for diversity does not itself challenge the assumption that adulthood is the asymptote of processes concerned primarily with childhood.

Most notions of adult maturity indeed assume that development reaches stability relatively early in the life span. This emphasis on stability derives from the developmental-structural assumption that psychological organization is to be studied from the aspect of equilibration or completion. Higher forms of development are characterized by an increased level of self-regulation in which the action of subsystems is constrained by the equilibrium-maintaining mechanism of the system as a whole. Overall, therefore, the system becomes less readily disequilibrated—that is, it is less affected by variations in context.

The ability of the organism to maintain stability is, no doubt, an essential component of its ability to survive. But so, paradoxically, is its ability to tolerate and create instability and to remain open to new information. Elsewhere (Labouvie-Vief, 1980, 1981) I have discussed this need for openness from a logical, biological, and psychological perspective and shown that to argue for a perfect closure of psychological organization at any point of development is, in effect, depriving the course of development of its major dynamism.

Within most theories of development, the achievement of stability has been seen to be a major goal. Hence the absence of conflict, the equilibration of thinking, the existence of a smooth fit between the organism and its context, and the appearance of normality have all connoted good developmental adaptation, or "maturity." By contrast, the existence of conflict or a degree of maladjustment have connoted a regressive breakdown of psychological organization.

Not all indications of regression and structural disorganization are the result, however, of a structural breakdown of the organism in a wider sense. Sometimes the loosening of organization may be indicative of transitory disintegrative processes which foreshadow, or are a concomitant of, accelerated development and new levels of integration.

Dabrowski (1970; Piechowski, 1975) has offered an interesting treatment of this issue in his theory of "positive disintegration." States of disintegration, in Dabrowski's framework, often signal the transition from lower to higher states of organization. Thus disequilibrium, in this theory, becomes a positive and necessary force in psychological development.

Several components of this emphasis on disequilibrium help to put the usual concern with stability in adult development into a somewhat altered perspective. First, it asserts that high levels of development or psychological maturity are not necessarily characterized by equilibrium, but may include high states of tension and symptoms—such as anxiety, depression, self-doubt, and ambivalence—that are not signs of maladaptation but rather of structural complexity (see also Brent, 1978). Second, one must clearly differentiate between transitional states of maladaptation which my be a precursor of accelerated development and those that are a sign of retarded or involutional development. The very concept of "stability" is thereby put into a more dynamic perspective which encompasses the individual's ability to change. A good example of this point is offered by Maas and Kuypers' (1974) longitudinal study of adulthood. In this study, one subsample of women was highly maladjusted in their thirties. Yet in their old age, these women had resolved their problems of self-worth and marital dissatisfaction, had created major changes in their lives, and enjoyed active and complex lives.

Third, therefore, it is necessary to distinguish those forms of regression which essentially are a part of growth from those which are part of developmental arrest or involution. The former kind of development, Dabrowski argues, is often earmarked by the coexistence of different subsystems—some more infantile and some more highly developed, some more intellectual and some more affective—which are not yet fully integrated and therefore coexist in a state of conflict and tension.

Finally, in this framework the eventual dynamic reorganization of those subsystems into a novel integration is not necessarily growth; it could also be decline (see Labouvie-Vief, 1981). As a new form of organization is being negotiated, earlier systems are reshaped and transformed, some being affirmed and enhanced, others denied and relegated to atrophy.

THE SOCIAL INTEGRATION OF COGNITION

It is important, therefore, to point out that not all instances of apparent adult decrement result from failure to account for diverse developmental pathways. Some may signify the emergence of new, fairly robust modes of organization of thought. Returning now to our earlier discussion of logical development—and even accepting Piaget's assertion that logic of some form is a milestone accomplishment of adolescence—we are led to question whether logic indeed is the apogee of mature thought: is the only point of adult development the perfection of logic?

It is interesting here to note that Piaget himself asserts that structures of formal thought are a specialized accomplishment of youth and become partially replaced and subordinated by new structures which arise from the constraints of adult life:

> With the advent of formal intelligence, thinking takes wings, and it is not surprising that at first this unexpected power is both used and abused . . . each new mental ability starts off by incorporating the world in a process of egocentric assimilation. Only later does it attain equilibrium through a compensating accommodation to reality. . . . Adolescent egocentricity is manifested by a belief in the omnipotence of reflection as though the world should submit itself to idealistic schemes rather than to systems of reality. (Piaget, 1967, pp. 63–64)

> True adaptation to society comes automatically when the adolescent reformer attempts to put his ideas to work. Just as experience reconciles formal thought with the reality of things, so does effective and enduring work, undertaken in concrete and well-defined situations, cure dreams. (Piaget, 1967, pp. 68–69)

What does this curing of dreams entail? Piaget (1970) has discussed it in general terms by pointing out that formalized systems must remain limited in validity. That is, a formalized logic, even though it is consistent in an internal sense (i.e., by providing rules of inference), must always draw on inferences which are not part of its system. Logic, in other words, relies on a context of validation. Or as Perry (1978) has put it:

> Reason reveals relations within any given context; it can also compare one context with another on the basis of metacontexts established for this purpose. But there is a limit. In the end, reason itself remains reflexively relativistic, a property which turns reason back upon reason's own findings. In even its farthest reaches then reason will leave the thinker with several legitimate contexts and no way of choosing among them—no way at least that he can justify through reason alone. (pp. 135–136)

The adolescent's relative facility with the manipulation of formal, symbolic logic per se, then, does not offer a way out of this dilemma. Indeed, it creates a kind of existential "vertigo" or "epistemological loneliness" (Chandler, 1978) from which youth attempts to retreat by the creation of apparent islands of emotional safety. Sensing a loss of logical certainty, they may construe certainty through adherence to authority and dogma, the support of peer groups, and so forth. The resulting slow erosion of "pure" logic as these preliminary strategies are given up is not a regressive phenomenon, therefore. Rather, it indicates the emerging search for a context within which logic is to be embedded, and this context is highly personal and interpersonal, rooted as it is in social reality and emotional pressures.

Perry (1978) has charted this change in college students moving from the relative insulation of their home environment to the pluralism of college life. These young people, with considerable conflict, slowly realized that logic *per se* did not provide any answers to such pressing problems as choice of career or partner. They found that emotional security was achieved only by realizing that their positions must be affirmed from within, that reason alone could not provide assurance.

Similarly, Gilligan (Gilligan & Murphy, 1980) demonstrated in the domain of moral reasoning that this rising postadolescent relativism should not be construed—as it is in Kohlberg's scheme of morality—as a regressive phenomenon. In her research, this relativism sprang from a realization that former "principled" universalistic reasoning was, after all, divorced from action and from an understanding of social reality. As one subject reflected, after about 7 years, on his earlier postconventional response to the moral dilemma of whether Heinz, a husband, should steal a drug to save his dying wife:

> This is a very crisp little dilemma and you can latch onto that principle pretty fast and in that situation you can say that life is more important than money. But then, when you reflect back on how you really act in your own life, you don't use that principle, or I haven't yet used that principle to operate on. And none of the people who answer that dilemma that way use that principle to operate on because they were blowing $7,000 a year for their education at Harvard instead of giving it to the Children's Fund to give porridge to the kids in Botswana and to that extent answering the dilemma with that principle is not hypocritical, it's just that you don't recognize it. I hadn't recognized it at the time, and I am sure they didn't recognize it either. (Gilligan & Murphy, 1980, p. 24)

We are arguing, therefore, that the major change from youth to mature adulthood is not merely one of perfecting logic. Instead, it consists in relating logic to other systems—action, affect, interpersonal relations—which up till now have developed in a largely parallel fashion (see also Riegel, 1973). This movement is similar to the one Vygotsky (1978) has charted in the younger child for the language system. Language, Vygotsky argues, first develops parallel to other subsystems (action, perception, etc.), but eventually these subsystems become interrelated, causing the function of language to change from mere significance to planning and guiding pragmatic action. Language becomes integrated as a tool into pragmatic systems.

We are proposing that logic, in the mature adult, becomes similarly integrated into pragmatics. On the one hand, this integration gives a higher power of internal coherence (logic) and wider temporal and spatial extension to thinking and behavior (see Chandler, Paget, & Koch, 1978; Erikson, 1968). On the other hand, logic thereby also comes to be constrained—in pursuing new goals, logic must become interrelated with pragmatic knowledge.

IMPLICATIONS

When adulthood is examined through the highly abstracted, logicalized schemes appropriate to youth and typical of most current research paradigms, the appearance of deficit and concrete modes of organization is overriding. Yet such a negative view of adulthood is necessary only if we continue to adhere to the notion that the unconstrained exercise of logical powers, per se, constitutes the highest standard of adaptability or maturity. If we are willing to admit that mature action and cognition need to incorporate elements of social responsibility and to include knowledge of affective and interpersonal elements of functioning (Schaie, 1977), then it is hardly surprising that the thinking of the adult may contain a flavor of concreteness and pragmatic knowledge.

If this changing theme indeed provides a new structural organization of thought, it will be necessary to reexamine the importance of purported deficits in adult cognition (see Labouvie-Vief, 1980, 1981). What in one context signals the decline in logical competence, in another context may well signify that logic has become subordinated under new goals: concern with the managing of personal and social resources, the moral value of modes of thinking and courses of action; in short, the social context of thought.

REFERENCES

Blasi, A., & Hoeffel, E. C. Adolescence and formal operations. *Human Development,* 1974, **17,** 344–363.

Brent, S. B. Individual specialization, collective adaptation and rate of environmental change. *Human Development,* 1978, **21,** 21–33.

Chandler, M. J. Relativism and the problem of epistemological loneliness. *Human Development,* 1975, **18,** 171–180.

Chandler, M. J., Paget, K. F., & Koch, D. The child's demystification of psychological defense mechanisms: A structural and developmental analysis. *Developmental Psychology,* 1978, **14,** 197–205.

Dabrowski, K. *Mental growth through positive disintegration.* London: Gryf, 1970.

Erikson, E. H. *Adulthood.* New York: Norton, 1978.

Fillenbaum, S. Mind your p's and q's: The role of content and context in some uses of and, or, and if. In G. H. Bower (Ed.), *The psychology of learning and motivation* (Vol. II). New York: Academic, 1977.

Gilligan, C., & Murphy, J. M. Development from adolescence to adulthood: The philosopher and the dilemma of the fact. In D. Kuhn (Ed.), *Intellectual development beyond childhood.* New York: Jossey-Bass, 1980.

Greenfield, P. M. Cross-cultural research and Piagetian theory: Paradox and progress. In K. F. Riegel & J. Meacham (Eds.), *The developing individual in a changing world. Historical and cultural issues* (Vol. 1). Chicago: Aldine, 1976.

Labouvie-Vief, G. Beyond formal operations: Uses and limits of pure logic in life-span development. *Human Development,* 1980, **23,** 141–161.

Labouvie-Vief, G. Proactive and reactive aspects of constructivism: Growth and aging in life-span perspective. In R. Lerner & N. Busch-Rossnagel (Eds.), *Individuals as producers of their own development.* New York: Academic, 1981.

Leontiev, A. N. *Probleme der Entwicklung des Psychischen;* 2. Aufl. Stuttgart: Fischer, 1977.

Maas, H. S., & Kuypers, J. A. *From thirty to seventy.* San Francisco: Jossey-Bass, 1974.

Mead, M. *Culture and commitment.* New York: Doubleday, 1970.

Neimark, E. D. Intellectual development during adolescence. In F. D. Horowitz (Ed.), *Review of child development research.* Chicago: University of Chicago Press, 1975.

Neisser, U. *Cognition and reality.* San Francisco: Freeman, 1976.

Neugarten, B. L., & Hagestad, G. O. Age and the life course. In R. H. Binstock & E. Shanas (Eds.), *Handbook of aging and the social sciences.* New York: Van Nostrand-Reinhold, 1966.

Olson, D. R. Culture, technology, and intellect. In L. Resnick (Ed.), *The nature of intelligence.* New York: Wiley, 1976.

Perry, W. I. *Forms of intellectual and ethical development in the college years.* New York: Holt, Rinehart and Winston, 1970.

Piaget, J. *Six psychological studies.* New York: Random House, 1967.

Piaget, J. *Structuralism.* New York: Basic Books, 1970.

Piaget, J. Intellectual evolution from adolescence to adulthood. *Human Development,* 1972, **16,** 1–12.

Piechowski, M. M. A theoretical and empirical approach to the study of development. *Genetic Psychology Monographs,* 1975, **92**, 231–297.

Riegel, K. F. Dialectical operations: The final period of cognitive development. *Human Development,* 1973, **16**, 346–370.

Riegel, K. F. The dialectics of human development. *American Psychologist,* 1976, **31**, 669–70.

Sabatini, P., & Labouvie-Vief, G. Age and professional specialization formal reasoning. Paper presented at 1979 Annual Meeting of the Gerontological Society, Washington, D.C., 1979.

Schaie, K. W. Toward a stage theory of adult development. *International Journal of Aging and Human Development,* 1977, **8**, 129–138.

Schaie, K. W. The primary mental abilities in adulthood: An exploration in the development of psychometric intelligence. In P. B. Baltes & O. G. Brims (Eds.), *Life-span development and behavior* (Vol. 2). New York: Academic, 1979.

Schaie, K. W., & Labouvie-Vief, G. Generational vs. ontogenetic components of change in adult cognitive behavior. *Developmental Psychology,* 1974, **10**, 305–320.

Scribner, S. Modes of thinking and ways of speaking: Culture and logic reconsidered. In R. O. Freedle (Ed.), *New directions in discourse processing* (Vol. 2). Norwood, N.J.: Ablex, 1979.

Vygotsky, L. S. Mind in Society. In M. Cole, V. John-Steiner, S. Scribner, & E. Souberman (Eds.), *Mind in society: The development of higher psychological processes.* Cambridge, Mass.: Harvard University Press, 1978.

Wason, P. C., & Johnson-Laird, P. B., *Psychology of reasoning: Structure and content.* Cambridge, Mass.: Harvard University Press, 1972.

26

MENTAL HEALTH IN THE ADULT YEARS

Barbara F. Turner
Castellano B. Turner
University of Massachusetts

The purpose of this chapter is to review prevailing concepts and recent empirical studies of mental health in early and middle adulthood. Our concern is with the implications of developmental changes in biological, psychological and social resources over this portion of the life course, and of historical changes that may differentially affect the mental health of successive birth cohorts. Age, gender, social class, and race differences, and their interactions with these developmental and contextual variables, will also be addressed.

Most definitions of mental health focus on the absence of mental illness or psychiatric symptomatology (Offer & Sabshin, 1974). Criteria of positive mental health—as distinct from absence of mental illness—include mastery of the environment, integration of the personality, perception of reality, subjective well-being, positive attitudes toward the self, self-actualization, and autonomy (cf. Birren & Renner, 1980; Jahoda, 1958). Definitions of mental health that emphasize ideal or optimal functioning—such as Maslow's (1954) self-actualization or Erikson's (1963) ego integrity—have received relatively little empirical study, however. They represent clinical ideals that are not expected to be fulfilled in reality, at least for most individuals.

Sociologists have traditionally focused on the demographic correlates of mental health, such as social class (Faris & Dunham, 1939; Hollingshead & Redlich, 1958; Langner & Michael, 1963), gender (Weissman & Klerman, 1979), and race (Dohrenwend & Dohrenwend, 1970). It has long been known that socially disadvantaged individuals are especially likely to exhibit maladaptive symptoms, but there is vast disagreement about why this is so.

Many life-span researchers are increasingly disenchanted with the predictive value for adult mental health of events in early childhood (e.g., Neugarten, 1978; Vaillant, 1977). The 1970s has been a decade of research on stress, coping, and adaptation. Life events have been viewed as stressors that require change in behavior and thus adaptive effort (e.g., Holmes & Rahe, 1967; Lowenthal & Chiriboga, 1973). Hundreds of studies using life-event scales have demonstrated that stress is indeed linked to mental health. It is less clear, however, that the cumulative readjustments or behavior changes required by life events cause changes in mental health (Chiriboga, 1980), and these questions have led to more sophisticated models. A variety of predisposing, antecedent, and mediating variables have been proposed—for example, biological, psychological, and contextual resources and deficits; distal and current stressor experiences; threat appraisal and coping strategies; and an array of outcome variables. At the same time, increasing dissatisfaction with existing life-event scales (cf. Chiriboga & Cutler, 1980) has led to their reappraisal.

It is not the purpose of this chapter to review the literature generated by the life-events approach. Recent reviews of this literature from the perspective of adult development and aging have been published by Chiriboga and Cutler (1980) and Hultsch and Plemons (1979). The theoretical frameworks underlying this research are useful, however, in examining mental health among adults.

ANTECEDENT AND MEDIATING VARIABLES

In any stress situation, some individuals wither, while others thrive. In this section, we are concerned with predisposing and mediating conditions which contribute to such variable results: (1) age-related changes in biological, psychological, and social resources and deficits; (2) early life history; and (3) cohort-related events.

Biological Resources and Deficits. These have been found to be related to subjective well-being. Before the age of 40, however, there is little variation in physical health and the biological–psychological connection is tenuous (Campbell, Converse, & Rodgers, 1976). After 40, biological deficits are more common. They are common enough to lower some people's feelings of well-being, but they are not ordinarily so incapacitating as to affect adaptation to stress.

It should be noted that biological resources and deficits are usually assessed by self-reports and that their relation to psychological distress is a reciprocal one (cf. Tessler & Mechanic, 1978). Even self-reports of relatively specific physical symptoms are, in part, dependent upon psychological state (Mechanic, 1980), suggesting caution in interpretation.

Psychological Resources and Deficits. Chiriboga and Lowenthal (1975) assessed the balance between psychological resources and deficits in a lower-middle-class sample of four groups of women and men approaching life transitions: high school seniors, newlyweds, middle-aged parents (average age of 48) whose child was about to leave home, and preretirees (average age of 58). Psychological resources included interpersonal accommodation—whether one dominates or accommodates to others; growth orientation—openness to change, akin to self-actualization; and capacity for mutuality, hope, and intelligence. Psychological deficits included self-rated emotional problems, behavioral symptoms, self-criticism, and Thematic Apperception Test (TAT) measures of anxiety and hostility.

It is notable that "growth orientation" was unusual in this sample; only 15% were rated high on this variable. Furthermore, the two middle-aged groups were significantly less growth-oriented than the two younger ones. The newlyweds were the most open to new experiences. This sample as a whole was primarily family-

and job-oriented. Lowenthal (1975) suggested that concepts like self-actualization might be more applicable to the talented, powerful, and privileged upper-middle- and upper-class elites.

When both psychological resources and deficits were considered, the happiest adolescents in this sample scored high on both. Among the newlyweds, however, the happiest were those with many resources but few deficits. The happiest middle-aged had few deficits and intermediate resources, while the happiest preretirees were low in both deficits and resources. In fact, the most *unhappy* in this oldest group were the psychically "complex"—those high in both resources and deficits, like the *happiest* high school seniors. Psychic complexity may thus be maladaptive in the absence of possibilities for self-expression. Older middle-aged adults in our society find few such possibilities (Chiriboga & Lowenthal, 1975). Opportunities for self-expression are more plentiful for upper-middle-class individuals and for youths.

At all life stages, the psychically complex are described as having active and complex life-styles; they cope "well with the diversity of stress which in-depth encounters with work and with people involve" (Fiske, 1980, p. 341). In fact, they may seek out and thrive upon stress. The psychically simple—those low in both resources and deficits—may avoid stressful events, and might react to environmental change with acute distress. The predictive value of this resource-and-deficit balance model for adaptation to stress has yet to be adequately tested, however. Conceptual and methodological problems involved are formidable.

Social Variables

The variables that enter into the resource–deficit balance include social supports and interpersonal relationships (e.g., family, friends, mutual self-help groups); demographic variables, such as marital status; socioeconomic variables (e.g., occupational status, education, income); and knowledge, motivation, and ability to solve problems. Interpersonal relationships are believed to reduce the impact of stress through provision of emotional and instrumental support (McCubbin, Joy, Cauble, Comeau, Patterson, & Needle, 1980). Both quantity and quality of interpersonal relationships are reciprocally linked to mental health. In fact, Cohler (1980), among others, has recently called attention to circumstances in which the availability of social networks or interpersonal relationships may be deleterious to mental health. One can have too much as well as too little in the way of interpersonal relations.

Age may play a part, also. Interpersonal resources may generally cushion the impact of stress in early through middle adulthood (Lowenthal et al., 1975) but not help older middle-aged women. A number of writers (Gutmann, 1964, 1975, 1977; Livson, 1976a; Lowenthal et al., 1975; Neugarten & Gutmann, 1964) suggest that, in midlife, women become more assertive, while men become more affiliative—interested in interpersonal relationships. Both genders increase in interiority, or attention to the satisfaction of intrapersonal needs (Neugarten, 1973). At the present point in history, middle-aged women with assertive needs have increasingly been able to express them in the world outside the home. Traditionally, however, women have been responsible for the maintenance of interpersonal relationships and primary-group social networks at all ages. This takes work and time. As a consequence, men may reap the benefits of extensive social networks that their wives provide. These networks can operate as buffers against stress for men, who are less responsible for their maintenance, but increase frustration and psychological distress for middle-aged women.

Marital status reflects the availability of social support. It has consistently been related to mental health. In virtually all studies, married people are less disposed to psychological distress than the unmarried (cf. Pearlin & Johnson, 1977). Most interpretations of these findings focus on the traumatic effects of the process of uncoupling (Bachrach, 1975) or suggest that psychological maladaptation is the cause, rather than the consequence, of singlehood: that the maladapted are less able to get married and to stay married. In a study of 2300 Chicago-area residents, however, Pearlin and Johnson (1977) concluded that the unmarried show more psychological distress (indexed by depression) than the married, primarily because the former are both more exposed to *and* more vulnerable to the strains of economic hardship, social isolation, and parental responsibilities. These relationships held when age, gender, and race were controlled. Depression was most common among nonmarried parents of children under 6 years of age, and adults under 40 are more likely to have preschool children than are adults over 40. Women are far more likely than men to have custody of children, so that in practice, unmarried mothers under 40 suffer more than men from the strain of parental responsibilities. Pearlin and Johnson (1977) conclude that "the links between marital status and depression are, in part, shaped by conditions rooted in broad social and economic structural arrangements and are not simply reflections of individual adjustments made on the basis of personality" (p. 714).

Early Life History

This is thought to influence vulnerability to stress. Many theories of personality development focus on the implications for adult adaptation of such early life experiences as parental loss (by death or divorce) or "rejection by mother." Parental loss early in life was strikingly characteristic of young men and of young and middle-aged women who reported considerable recent stress that concerned or overwhelmed them (Lowenthal & Chiriboga, 1975). Curiously, middle-aged men with the same loss seemed challenged rather than overwhelmed by their recent stress experiences. It is possible, therefore, that the negative effects of early parental loss can be compensated by long-term exposure to ego-enhancing experiences outside the home (Lowenthal & Chiriboga, 1975). Vaillant (1974) drew a similar conclusion from his longitudinal study of 95 graduates of an elite men's college whom he followed from their college years to the age of 47. Loss of a parent or rejection by a mother in childhood failed to discriminate between the men who showed the best and the worst psychological adaptation at 47. On the other hand, *overall* ratings of the childhood environment of these privileged men did predict psychological maladaptation in midlife. It was the number, rather than the kind, of childhood stresses that was associated with later poor mental health.

Over this century, in our society, early loss of a parent by death has grown less common, loss by divorce more common. Loss of a parent (typically, the father) changes a child's environment in many ways not considered by personality theorists. A typical effect of divorce for an ex-wife and children, for example, is a precipitate drop into poverty.

How are more recent stresses related to mental health between the ages of 21 and 65? As noted earlier, the past decade of research on life events has raised more questions than provided answers. Compared to younger adults, middle-aged and older adults report fewer life events altogether, and thus less stress. They also rate events as less disruptive (cf. Chiriboga & Cutler, 1980).

Such results appear counterintuitive, to say the least. To take just one example, the so-called male midlife crisis has been described as resulting from a pileup of negative stressors (Brim, 1976). Do the findings of life-event studies indicate, then, that midlife crises are the stuff of myth? Not necessarily. Many items in life-event scales are more appropriate to younger adults than to middle-aged and older persons (Lowenthal & Chiriboga, 1973). Life-event scales, in short, do not adequately sample events relevant to the lives of older adults.

One new direction of research in this area is the dimensionalization of life events. Distinctions are drawn between positive and negative events; between events that individuals have responsibility for bringing about and those over which they have no control; between subjective and objective events; and between gain and loss events. A second direction of current research deals with stressful experiences that are not captured by the life-event construct. Examples include "off-time" events (Neugarten & Datan, 1972), chronic "hassles" (Lazarus, 1980), and the non-occurrence of expected events (Beeson & Lowenthal, 1975).

Cultural Events

Such events as wars and economic depressions affect all individuals in a society but have differential effects on persons of different ages and statuses. Cultural events may affect mental health by altering directly the interpersonal context within which the individual develops or by affecting the structural forces that more indirectly influence mental health. Especially striking, perhaps, are the effects of cultural events upon occupational career contexts and upon cohort size and composition, which in turn have far-reaching ramifications (cf. Neugarten & Hagestad, 1976).

The Great Depression, for example, has affected the adaptive contexts and the adaptation of several cohorts (Elder, 1974, 1977). For example, the Berkeley Guidance Study members were born in 1928–29. When economic conditions were grimmest, they were very young children and totally dependent, which rendered them vulnerable to the sometimes chaotic family contexts created by economic misfortune. In contrast, Oakland Growth Study men, born in 1920–21, were less dependent and thus less vulnerable to family misfortune. At the same time, the Oakland sample was still too young during the 1930s, the height of the Depression, to enter the adult work force at the point when opportunities were most limited. Instead, their adolescent work experience increased their freedom from family and crystallized their vocational goals at an early age. Their ultimate occupational attainments were not depressed, therefore, despite the lowered educational attainments made necessary by the economic deprivation of their parents (Elder, 1974). Of importance for conceptions of positive mental health is Elder's suggestion that adversity, if encountered early—but not too early—may enhance competence and self-esteem which, in turn, makes adult mastery more likely.

Intervening Variables

Adults experiencing stress select and apply coping strategies. At this point, research on coping strategies is still in its infancy (Cohen, 1980; Horowitz & Wilner, 1980;

Pearlin & Schooler, 1978). From the viewpoint of psychoanalytic theory, coping and defense mechanisms operate outside of awareness. If true, self-reports would be invalid, assessments by clinicians difficult. Sociologists, on the other hand, view coping strategies as consciously deployed institutionalized solutions to normative life stress (Mechanic, 1974) subject to empirical investigation. The efficacy of coping strategies in avoiding stress or reducing harm depends, in part, on the degree of individual control that is realistically possible in a given situation. Many conditions of occupational life and treatment of life-threatening illnesses are examples of situations less amenable to individual control. In Pearlin and Johnson's (1977) study of coping strategies used by urban adults, young and old were equally likely to employ effective strategies. There was no evidence that older people selected either less effective or more passive coping mechanisms than did younger adults.

The following sections address age, gender, social class, and race differences in mental health during adulthood.

AGE

Patterns of change in adulthood occur within a complex and dynamic matrix of biological, social, contextual, and historical events, and age-related change must be disentangled from cohort and historical effects. One must keep in mind that most studies are cross-sectional, indicating age differences (if any) or cohort differences, rather than age changes. Nor are longitudinal studies free of confounding effects. The changes that appear in longitudinal studies may reflect time-of-measurement effects, including the effects of cultural change within the span of years that a longitudinal sample is studied, rather than the effects of maturation. (The more adequate cross-sequential and time-sequential designs are rare indeed.) Rapid social change, further, makes the generalization of longitudinal findings to later cohorts problematic.

Analyses of men's responses to objective personality tests repeated over periods of 6, 10, and 12 years reveal remarkable stability of mean level and rank ordering on various subscales at all ages (Costa & McCrae, 1977; Costa & McCrae, 1978; Costa, McCrae, & Arenberg, 1980). Included in these analyses are the Guilford-Zimmerman Temperament Survey adjustment-neuroticism scales (emotional stability, objectivity, friendliness, and personal relations). The ages at the initial testing ranged all the way from 17 to 85.

Such objective personality tests are based upon trait theories of personality, which generally assume stability in traits over time. Clinical approaches may be more sensitive to developmental changes, if they exist. Vaillant (1974), for example, reported that ego mechanisms of defense matured over time in his sample of highly privileged men. During their college years, these men were twice as likely to use *immature* defenses as mature ones; and between 36 and 50, they were 4 times as likely to use mature ones.

Investigators at the Institute of Human Development in Berkeley (e.g., Block in collaboration with Haan, 1971; Livson, 1975, 1976a, b) have traced the development of two samples over many years: the Berkeley Guidance Study (born 1927–28) and the Oakland Growth Study (born 1921). The 90-item California Q-sort was used to rate interview material, the primary data source. The personality profiles drawn of these two groups when they were in their 30s showed considerably enhanced coping capacity over those of the same men and women as adolescents. As adults, they also showed less narcissistic impulsivity.

A cluster analysis of data from the Berkeley sample when they were in their 40s showed that more than 70% of both the women and the men were psychologically healthy, with a comfortable, well-socialized style of functioning (Haan, 1976). In early adolescence, 48% of these men showed this personality organization; in late adolescence, 43%; and in the 30s, 37%. The corresponding figures for the women were 33%, 26%, and 48%. Thus the overall age trend reflects movement toward enhanced personal comfort and competence. The low point for the women was late adolescence, for the men, their 30s.

There is some suggestion from longitudinal studies, then, that undesirable characteristics decline from adolescence to midlife while competence increases. Cohort and contextual effects may account for these findings, however; all these respondents were born during the 1920s, and almost all enjoy upper-middle-class status as adults.

GENDER DIFFERENCES IN MENTAL HEALTH[1]

Sex status is so powerful an influence on the life chances, social roles, and lifestyles of individuals that patterns of adult development tend to be very different for women than for men.

The first issue addressed in this section is a gender comparison on criteria of positive mental health. When Maslow (1954) looked for self-actualized individuals, he had difficulty finding even one woman who met his criteria. In the lower-middle-class sample studied by Lowenthal et al. (1975) about 20% of the men, but only half as many women, were characterized by a growth-motivated approach to life. If "autonomy" is the desideratum, then women are still disadvantaged relative to men, for autonomy is a masculine goal, and the relatedness that women maintain may be seen as an indication of "inadequate maturity" (Notman, 1980).

Routes to psychological health differ for women and men. Using personality ratings done when the Oakland Growth study respondents were 50, Livson (1975, 1976b) divided the women and men at the mean on Block's (1961) Q-sort of psychological health. When Livson looked back at the records of those who were above the mean, seven women and seven men had also been high in psychological health at 40. Seventeen women and fourteen men had improved markedly between 40 and 50. The stable women and men evinced conventional gender-related personality characteristics. Their personalities appeared well suited to traditional female and male roles as the roles themselves evolved from adolescence to midlife. In contrast, the women and men whose psychological health improved in midlife had, as adolescents, shown personality characteristics inconsistent with traditional gender roles. In adolescence, the women were intellectual and the men emotionally expressive. At 40, these women were depressed, irritable, and prone to daydream, while the men evinced a form of hostile, exploitative hypermasculinity. Livson suggested that these men had suppressed their adolescent emotionality in order to fulfill

---

[1] Portions of this section are drawn from B. F. Turner, Sex-related differences in aging. In B. B. Wolman & G. Stricker (Eds.), *Handbook of developmental psychology*. New York: Prentice-Hall, in press.

sex-role expectations for high achievement. For both sexes, then, the price of the suppression of genotypic personality characteristics was a diminution of psychological health at the age of 40. By age 50, however, the men had integrated their emotional expressiveness with their sense of masculinity and were nurturant and sensual rather than anxiously hypermasculine. The women at 50 were autonomous, trusting, and goal-oriented.

The mechanism for this shift in gender-related personality characteristics that apparently led to better mental health is unclear. Changes in social roles (launching of children for women, coming to terms with occupational goals for men) are obvious possibilities, but change in time perspective or other age-related changes may also be involved (Livson, in press).

Turning to indices of everyday maladaptation, the preponderance of evidence in a variety of surveys indicates that women are more likely than men to suffer from psychological impairment. Women report more symptomatology and more emotional problems—in particular, depression—than do men (cf. Pearlin, 1975). At least one important longitudinal study, however, points to cohort differences in incidence of mental impairment among women. In 1974, Srole and Fischer (1978) reinterviewed 695 of the 1660 white adults, aged 20 to 59, first interviewed in 1954 for the Midtown Manhattan Study. Compared to women aged 40 to 59 in 1954, the mental health ratings of women aged 40 to 59 were significantly higher in 1974. In 1954, for example, 26% of the women 50 to 59 had been impaired versus only 15% of the men. Twenty years later, only 11% of the women who were now in their 50s were impaired, compared with 9% of the men. The male superiority in mental health of 1954 had all but disappeared. This finding does not necessarily herald similarly high levels of mental health for younger cohorts of women when they reach middle age. Women aged 35 to 39 sampled in a National Health Survey in the early 1970s, for example, were more likely to report symptoms of nervous breakdown than women at that age 12 years earlier.

A voluminous literature indicates that regardless of age, females in our society have lower self-esteem than do males (cf. Turner, 1979). Self-criticism is closely tied to self-esteem, and women are also more self-critical than men, a finding which does *not* reflect response bias (Gove & Geerken, 1977).

Gender differences in self-criticism, however, appear to diminish by late middle age (Turner, 1979). In a large sample survey (Gurin, Veroff, & Feld, 1960), women over 55 were less self-critical, relative to men of the same age, than were women under 55, although the older women were still more self-critical than their male age-mates. Similar findings emerged in Lowenthal et al.'s (1975) study of lower-middle-class women and men in four adult-life transitional stages. High school senior women circled many more attributes on an adjective-rating list that they disliked about themselves than did high school men. Gender differences in self-criticism were minimal in the late middle-aged group. What accounts for this diminution in self-esteem? The findings of these studies may, of course, indicate cohort differences rather than age changes.

Because so many stereotypically feminine traits are rated as socially undersirable by both women and men (Broverman, Broverman, Clarkson, Rosenkrantz, & Vogel, 1970; Lowenthal et al., 1975), it follows that if "femininity" declines and "masculinity" increases as women grow older, self-acceptance would increase. Indeed, masculinity in self-concept facilitates self-esteem among both genders (cf.

Spence & Helmreich, 1978). In the Lowenthal et al. (1975) study, those preretired women who had highly feminine self-concepts were considerably more self-critical than those who had less feminine self-concepts.

A number of factors appear to account for gender differences in nonclinical depression. Women's employment status is *not* among these; employed women and full-time housewives do not differ in depression (Pearlin, 1975). But women who are disenchanted with homemaking experience more depression, and disenchantment is greatest when there is a house full of very young children, especially when the mother is a single parent. Further, women are more likely than men to use passive coping strategies that are, in certain situations, less effective than the more active strategies used by men (Pearlin & Schooler, 1978).

SOCIAL CLASS

There are several "classic" research reports demonstrating the relationship of social class to mental illness (Faris & Dunham, 1939; Hollingshead & Redlich, 1958; Srole et al., 1962). Moreover, in a summary of 2 decades of research, Dohrenwend and Dohrenwend (1969) concluded that the overwhelming weight of evidence indicated that psychological disorders were more common among poor people than among middle- or upper-class people. The most recent evidence continues to support this conclusion (Kessler & Cleary, 1980). This does not mean that social class per se is the cause of mental disorder. The voluminous literature suggests that the relationship is a very complex one (Bullough & Bullough, 1972).

Researchers in this area acknowledge the contribution of constitutional factors (Dohrenwend & Dohrenwend, 1969) but emphasize the evidence for a link between social class, stress, and psychological disorders. Myers, Lindenthal, and Pepper (1975) found an uneven distribution of certain life events by social class, lower-class individuals experiencing more events that were both unpleasant and needed more readjustment. Kessler (1979), who seems to believe that variations in exposure to stressful experiences explain differential rates of emotional problems, has looked for factors which could modify the impact of such exposure. He cites (1) the interpretation of an event as stressful; (2) the availability and efficacy of options for responding; and (3) the presence of social support systems. One might speculate that early socialization provides different social classes with different repertoires of responses to threat. Members of the lower class learn less adequate ways to save themselves from emotional distress. This conclusion is reached by Pearlin and Schooler (1977) in their investigations.

Several other explanations offered for class differences in mental illness should be mentioned. First, Carr and Krause (1978) attempted to demonstrate that a number of response biases on scales and inventories of psychological well-being might explain the unequal reports of distress by class. Lower-class people, they believe, have a tendency to acquiesce to any descriptive statement. Class continued to be related to symptomatology, however, even when response bias was controlled. Second, when Myers and Bean (1968) did a follow-up on *Social Class and Mental Illness* (Hollingshead & Redlich, 1958), they found that lower-class subjects had received less effective treatment. Thus part of the social-class differential in mental illness rates could be differential rehabilitation. Since long-term confinement in state mental hospitals is the common treatment of the lower-class psychiatric patient and since such hospitals have long been regarded as iatrogenic, the social-class differential continues to grow.

Such an explanation does not, of course, apply to evidence about differential psychological distress in community samples. Moreover, even with regard to serious psychiatric problems, the argument is weakened if the social-class comparison is based on first admission rates rather than the general psychiatric population. Third, Honigman (1969) has made a persuasive argument for the notion that middle-class values and (by extension) styles of coping are used as the basis for judging the mental health of the lower classes and other disadvantaged groups. Such a process would naturally lead to labeling non-middle-class groups as deviants. Dill et al. (1980) notes that the coping strategies of low-income women are customarily labeled as illegitimate by professionals and agency personnel.

Maslow's (1954) theory, for example, specifies that human beings must satisfy more basic needs before being able to move on to the ultimate ideal of self-actualization. Following his definition, it is clear that the poor would have difficulty in being judged healthy. Much of life for the disadvantaged in a society must be spent in the attainment of basic needs. Self-actualization is thus rarely open to the poor.

RACE AND MENTAL HEALTH

Race differences in indices of mental health and psychological disturbance have been well-documented. Bullough and Bullough (1972) point to several diagnostic categories in which blacks are overrepresented, including schizophrenia, alcoholism, and drug addiction. Hare (1979) and Clark (1965) have reported that blacks have higher rates of mental hospital admissions as well as outpatient psychiatric contact. Dohrenwend and Dohrenwend's (1970) data from community surveys indicate greater emotional problems among blacks.

Explanations of these differences have followed a pattern similar to that for social class. For example, Prudhomme and Musto (1973) provide a historical review of attempts to explain both lower and higher incidence of mental illness as indications of the inherited constitutional inferiority of blacks. However, the bulk of evidence shows that race is not a primary etiological factor (Warheit, Holzer, & Arey, 1975). When social class is controlled, differences tend to disappear (Bullough & Bullough, 1972; Kessler, 1979).

In spite of the consistent demonstration that race differences in mental health are based on social-class differences, theoreticians and researchers have continued to seek explanations for the assumed race differences. The major focus of such formulations has been racism (Kardiner & Ovesey, 1951; Clark, 1965; Wilcox, 1973; Thomas & Lindenthal, 1979; Hare, 1979). The best case for such an interpretation can be made for essential hypertension, which is thought to reflect suppressed anger toward white society (Grier & Cobbs, 1968, 1971). Although blacks face discrimination, and this is stressful, it seems more parsimonious to account for the empirical data by social class.

There has been much interest in the factor of *self-esteem,* assuming that black Americans have low self-esteem. This assumption has persisted despite considerable contradictory evidence (Turner & Turner, 1974; Porter & Washington, 1979). Again, it is clear that the family disorganization and instability associated with inducing poor self-concepts in children is not limited to black families and is certainly more closely associated with social class (Coopersmith, 1967).

Parker and Kleiner (1966) have presented support for their hypothesis that mental health problems in urban, black communities is related to a combination of "goal-striving" stress, discrepancy from reference group achievement, and low self-esteem. Unfortunately, they included no comparison with whites in this study. One might reasonably expect that the same variables would be related to mental illness in white samples, too.

Geismar and Gerhart (1968) found that family functioning was highly influenced by social-class level. Differences in such functioning between black and white families largely disappear when social class is controlled. This suggests that the origins of variation in adult mental health may well be found in early family socialization. Psychological resources gained in early life experiences can be the key to understanding responsiveness to stress events, as well as many other behaviors.

CONCLUSION

In sum, it appears, first, that conceptions of positive mental health reveal a marked social-class, and also a gender, bias. Writers have reached no consensus or have not seriously addressed the issue of what self-actualization, for example, in a lower-class, working-class or even lower-middle-class context might look like.

Second, cohort differences in mental health seem most apparent in gender comparisons and least apparent for social-class and race comparisons. The latter is unsurprising, since there has been relatively little secular change, overall, in indices of class and race equality, such as relative income. As Pearlin and Schooler (1978), among others, have pointed out, many life problems faced by individuals are rooted in social and economic organization and are impervious to individual efforts to cope with or change. In some cases, therefore, mental health impairments reflect the failure of social systems rather than the shortcomings of individuals.

REFERENCES

Bachrach, L. L. Marital status and mental disorder: An analytical review. CHEW Publication No. (ADM) 75-217. Washington, D.C.: U.S. Government Printing Office, 1975.

Beeson, D. L., & Lowenthal, M. J. Perceived stress across life course. In M. F. Lowenthal, M. Thurnher, D. Chiriboga, & associates, *Four stages of life*. San Francisco: Jossey-Bass, 1975.

Bird, C. The best years of a woman's life. *Psychology Today,* June 1979, 20–22; 26.

Birren, J. E., & Renner, V. J. Concepts and issues of mental health and aging. In J. E. Birren & R. B. Sloane (Eds.), *Handbook of mental health and aging.* Englewood Cliffs, N.J.: Prentice-Hall, 1980.

Block, J. *The Q-sort method in personality assessment and psychiatric research.* Springfield, Ill.: Thomas, 1961.

Block, J., in collaboration with Haan, N. *Lives through time.* Berkeley, Calif.: Bancroft, 1971.

Broverman, I., Broverman, D. M., Clarkson, F. E., Rosenkrantz, P. S., & Vogel, S. R. Sex-role stereotypes and clinical judgments of mental health. *Journal of Consulting and Clinical Psychology,* 1970, **43**, 1–7.

Bullough, B., & Bullough, V. *Poverty, ethnic identity, and health care.* New York: Appleton-Century-Crofts, 1972.

Campbell, A., Converse, P. E., & Rodgers, W. L. *The quality of American life.* New York: Russell Sage, 1976.

Carr, L. G., & Krauss, N. Social status, psychiatric symptomatology, and response bias. *Journal of Health and Social Behavior,* 1978, **19**, 86–91.

Chiriboga, D. A. Introduction: Stress and coping. In L. W. Poon (Ed.), *Aging in the 1980's: Psychological issues.* Washington, D.C.: American Psychological Association, 1980.

Chiriboga, D. A., & Lowenthal, M. F. Complexities of adaptation. In M. F. Lowenthal, M. Thurnher, D. Chiriboga, & associates, *Four stages of life.* San Francisco: Jossey-Bass, 1975.

Clark, K. B. *Dark ghetto: Dilemmas of social power.* New York: Harper & Row, 1965.

Cohen, F. Coping with surgery: Information, psychological preparation, and recovery. In L. W. Poon (Ed.), *Aging in the 1980's: Psychological issues.* Washington, D.C.: American Psychological Association, 1980.

Cohler, B. Autonomy and interdependence in the family of adulthood: A psychological perspective. In H. Orbach & I. Hulicka (Chairs), *Aging, families and family relations: Behavioral and social science perspectives on our knowledge.* Symposium presented at the 33rd Annual Scientific Meeting of the Gerontological Society of America, San Diego, November 1980.

Coopersmith, S., *The antecedents of self-esteem.* San Francisco: W. Freeman, 1967.

Costa, P. T., Jr., & McCrae, R. R. Age differences in personality structure revisited: Studies in validity, stability and change. *International Journal of Aging and Human Development,* 1977, **8**, 261–275.

Costa, P. T., Jr., & McCrae, R. R. Objective personality assessment. In M. Storandt, I. C. Siegler, & M. F. Elias (Eds.), *The clinical psychology of aging.* New York: Plenum, 1978.

Costa, P. T., Jr., McCrae, R. R., & Arenberg, D. Enduring dispositions in adult males. *Journal of Personality and Social Psychology,* 1980, **38**, 793–800.

Dill, D., Feld, E., Martin, J., Beukema, P., & Belle, D. The impact of the environment on the coping efforts of low-income mothers. *Family Relations,* 1980, **29**, 503–509.

Dohrenwend, B., & Dohrenwend, B. P. Class and race as status-related sources of stress. In S. Levine & N. Scotch (Eds.), *Social stress.* Chicago: Archive, 1970.

Dohrenwend, B. P., & Dohrenwend, B. S. *Social status and psychological disorder: A causal inquiry.* New York: Wiley, 1969.

Elder, G. H., Jr. *Children of the Great Depression.* Chicago: University of Chicago Press, 1974.

Elder, G. H., Jr. Historical change in life patterns and personality. In P. B. Baltes & O. G. Brim, Jr. (Eds.), *Life-span development and behavior* (Vol. 2). New York: Academic, 1979.

Erikson, E. *Childhood and society.* New York: Norton, 1963.

Faris, R., & Dunham, H. W. *Mental disorders in urban areas.* New York: Hafner, 1939.

Fiske, M. Tasks and crises of the second half of life: The interrelationships of commitment, coping, and adaptation. In J. E. Birren & R. B. Sloane (Eds.), *Handbook of mental health and aging.* Englewood Cliffs, N.J.: Prentice-Hall, 1980.

Geismar, L. L., & Gerhart, V. C. Social class, ethnicity, and family functioning: exploring social issues raised by the Moynihan Report. *Journal of Marriage and the Family,* 1968, **30**, 480–482.

Gove, W. R., & Geerken, M. R. Response bias in surveys of mental health: An empirical investigation. *American Journal of Sociology,* 1977, **82**, 1289–1317.

Grier, W. H., & Cobbs, P. M. *The Jesus bag.* New York: McGraw-Hill, 1971.

Gurin, G., Veroff, J., & Feld, S. *Americans view their mental health.* New York: Basic Books, 1960.

Gutmann, D. An exploration of ego configurations in middle and later life. In B. L. Neugarten (Ed.), *Personality in middle and later life.* New York: Atherton, 1964.

Gutmann, D. L. Parenthood: A key to the comparative psychology of the life cycle. In N. Datan & L. Ginsberg (Eds.), *Life span developmental psychology: Normative life crises.* New York: Academic, 1975.

Gutmann, D. L. The cross-cultural perspective: Notes toward a comparative psychology of aging. In J. Birren & K. W. Schaie (Eds.), *Handbook of the psychology of aging.* New York: Van Nostrand, 1977.

Haan, N. Personality organizations of well-functioning younger people and older adults. *International Journal of Aging and Human Development,* 1976, **7**, 117–127.

Hare, N. The relative psycho-socio-economic suppression of the black male. In W. D. Smith, K. H. Burlew, M. H. Mosley, & W. M. Whitney (Eds.), *Reflections on black psychology.* Washington, D.C.: University Press of America, 1979.

Hollingshead, A. B., & Redlich, F. C. *Social class and mental illness.* New York: Wiley, 1958.

Holmes, T. H., & Rahe, R. H. The social readjustment rating scale. *Journal of Psychosomatic Research,* 1967, **11**, 213–218.

Honigmann, J. J. Middle class values and cross-cultural understanding. In J. C. Finney (Ed.), *Culture, change, mental health, and poverty.* New York: Simon & Schuster, 1969.

Horowitz, M. J., & Wilner, N. Life events, stress, and coping. In L. W. Poon (Ed.), *Aging in the 1980's: Psychological issues.* Washington, D.C.: American Psychological Association, 1980.

Hultsch, D. F., & Plemons, J. K. Life events and life-span development. In P. B. Baltes & O. G. Brim (Eds.), *Life-span development and behavior* (Vol. 2). New York: Academic, 1979.

Jahoda, M. *Current concepts of positive mental health.* New York: Basic Books, 1958.

Kardiner, A., & Ovesey, L. *The mark of oppression: A psychosocial study of the American Negro.* New York: Norton, 1950.

Kessler, R. C. Stress, social status, and psychological disorders. *Journal of Health and Social Behavior,* 1979, **20**, 259–272.

Kessler, R. C., & Cleary, P. D. Social class and psychological distress. *American Sociological Review,* 1980, **45**, 463–478.

Langer, T. S., & Michael, S. T. *Life stress and mental health.* New York: Free Press, 1963.

Lazarus, R. S. The stress and coping paradigm. In C. Eisdorfer, D. Cohen, A. Kleinman, & P. Maxim (Eds.), *Theoretical bases in psychopathology.* New York: Spectrum, 1980.

Lewis, A. Between guesswork and certainty in psychiatry. *Lancet,* 1958, **1**, 170–175.

Lieberman, M. A. Adaptive processes in late life. In N. Datan & L. H. Ginsberg (Eds.), *Life-span developmental psychology: Normative life crises.* New York: Academic, 1975.

Livson, F. B. Sex differences in personality development in the middle years: A longitudinal study. Paper presented at the 28th Annual Scientific Meeting of the Gerontological Society, Louisville, Ky., October 1975.

Livson, F. B. Coming together in the middle years: A longitudinal study of sex role convergence. In B. F. Turner (Chair), *The double standard of aging: A question of sex differences.* Symposium presented at the 29th Annual Scientific Meeting of the Gerontological Society, New York, 1976. (a)

Livson, F. B. Patterns of personality development in middle-aged women: A longitudinal study. *International Journal of Aging and Human Development,* 1976, **7**, 107–115. (b)

Livson, F. B. Paths to psychological health in the middle years: Sex differences. In D. H. Eichorn, J. A. Clausen, N. Haan, M. P. Honzik, & P. Mussen (Eds.), *Present and past in middle life.* New York: Academic, in press.

Lowenthal, M. F. Summary and implications. In M. F. Lowenthal, M. Thurnher, D. Chiriboga, & associates, *Four stages of life.* San Francisco: Jossey-Bass, 1975.

Lowenthal, M. F., & Chiriboga, D. Social stress and adaptation: Toward a life-course perspective. In C. Eisdorfer & M. P. Lawton (Eds.), *The psychology of adult development and aging*. Washington, D.C.: American Psychological Association, 1973.

Lowenthal, M. F., Thurnher, M., Chiriboga, D., & associates. *Four stages of life*. San Francisco: Jossey-Bass, 1975.

McCubbin, H. I., Joy, C. B., Cauble, A. E., Comeau, J. K., Patterson, J. M., & Needle R. H. Family stress and coping: A decade review. *Journal of Marriage and the Family*, 1980, **42**, 855–872.

Maslow, A. H. *Motivation and personality*. New York: Harper & Row, 1954.

Mechanic, D. Social structure and personal adaptation: Some neglected dimensions. In C. B. Coehlo, D. A. Hamburg, & J. E. Adams (Eds.), *Coping and adaptation*. New York: Basic Books, 1974.

Mechanic, D. The experience and reporting of common physical complaints. *Journal of Health and Social Behavior*, 1980, **21**, 146–155.

Myers, J. K., & Bean, L. L. *A decade later: A follow-up of social class and mental illness*. New York: Wiley, 1968.

Myers, J. K., Lindenthal, J. J., & Pepper, M. P. Social class, life events, and psychiatric symptoms. *Journal of Health and Social Behavior*, 1975, **16**, 421–427.

Neugarten, B. L. Personality changes in adulthood. Paper presented in the Master Lectures on the Psychology of Aging, 86th Annual Convention of the American Psychological Association. Toronto, Canada, August 1978.

Neugarten, B. L., & Datan, N. Sociological perspectives on the life cycle. In P. B. Baltes & K. W. Schaie (Eds.), *Life-span developmental psychology: Personality and socialization*. New York: Academic, 1973.

Neugarten, B. L., & Gutmann, D. L. Age-sex roles and personality in middle age: A thematic apperception study. In B. L. Neugarten & associates, *Personality in middle and late life*. New York: Atherton, 1964.

Notman, M. T. Changing roles for women at mid-life. In W. H. Norman & T. J. Scaramella (Eds.), *Midlife: Developmental and clinical issues*. New York: Brunner/Mazel, 1980.

Offer, D., & Sabshin, M. *Normality: Theoretical and clinical concepts of mental health* (rev. ed.). New York: Basic Books, 1974.

Parker, S., & Kleiner, R. J. *Mental illness in the urban Negro community*. New York: Free Press, 1966.

Pearlin, L. I. Sex roles and depression. In N. Datan & L. H. Ginsberg (Eds.), *Life-span developmental psychology: Normative life crises*. New York: Academic, 1975.

Pearlin, L. I., & Johnson, J. S. Mental status, life strains and depression. *American Sociological Review*, 1977, **42**, 704–715.

Pearlin, L. I., & Schooler, C. The structure of coping. *Journal of Health and Social Behavior*, 1978, **19**, 2–21.

Peck, R. F., & Berkowitz, H. Personality and adjustment in middle age. In B. L. Neugarten & associates (Eds.), *Personality in middle and late life*. New York: Atherton, 1964.

Pettigrew, T. F. *A profile of the Negro American*. Princeton: Van Nostrand, 1964.

Porter, J. R., & Washington, R. E. Black identity and self-esteem: A review of studies of black self-concept, 1968–1978. *Annual Review of Sociology*, 1979, 53–74.

Prudhomme, C., & Musto, D. F. Historical perspectives on mental health and racism in the United States. In C. V. Willie, B. Kramer, & B. S. Brown (Eds.), *Racism and mental health*. Pittsburgh: University of Pittsburgh Press, 1973.

Spence, J. T., & Helmreich, R. L. *Masculinity and femininity: Their psychological dimensions, correlates and antecedents*. Austin: University of Texas Press, 1978.

Srole, L., & Fischer, A. K. The Midtown Manhattan Study: Longitudinal focus on aging. In R. E. Weber (Organizer), *Lives in transition: Advancements in theory and research*. Symposium presented at the 31st Annual Meeting of the Gerontological Society, Dallas, Tex., 1978.

Tessler, R., & Mechanic, D. Psychological distress and perceived health status. *Journal of Health and Social Behavior,* 1978, **19**, 254–262.

Thomas, C. S., & Lindenthal, J. J. The depression of the oppressed. In R. C. Allen (Ed.), *Mental health in America: The years of crisis.* Chicago: Marquis Academic Media, 1979.

Turner, B. F. Psychological predictors of adaptation to the stress of institutionalization in the aged. Unpublished doctoral dissertation, The University of Chicago, 1969.

Turner, B. F. The self-concepts of older women. *Research on Aging,* 1979, **1**, 464–480.

Turner, B. F. Sex-related differences in aging. In B. B. Wolman & G. Stricker (Eds.), *Handbook of developmental psychology.* New York: Prentice-Hall, in press.

Turner, B. F., & Turner, C. B. Evaluations of women and men among black and white college students. *Sociological Quarterly,* 1974, **15**, 442–456.

Vaillant, G. E. *Adaptation to life.* Boston: Little, Brown, 1977.

Warheit, G. J., Holzer, C. E., & Arey, S. A. Race and mental illness: An epidemiologic update. *Journal of Health and Social Behavior,* 1975, **16**, 243–256.

Weissman, M. M., & Klerman, G. L. Sex differences and the epidemiology of depression. In E. S. Gomberg & V. Franks (Eds.), *Gender and disordered behavior: Sex differences in psychopathology.* New York: Brunner/Mazel, 1979.

Wilcox, P. Positive mental health in the black community: The black liberation movement. In C. V. Willie, B. Kramer, & B. Brown (Eds.), *Racism and mental health.* Pittsburgh: University of Pittsburgh Press, 1973.

27

FROM GREGARIOUSNESS TO INTIMACY
Marriage and Friendship over the Adult Years

Margaret Hellie Huyck
Illinois Institute of Technology

THE THEME: SOCIAL CONNECTEDNESS

Humans are uniquely and intensely social creatures. Other animals are also sociable; they gather together, form dominance hierarchies, mate, rear their young, cooperate in protecting insiders from strangers, and communicate with each other more than with outsiders (Wilson, 1980). However, humans have additional capacities of intellect, memory, and language which allow elaborated, self-reflective patterns of interpersonal relations to evolve.

In this chapter we consider some of the varieties of human interpersonal relationships. We do not discuss parent–child relationships but concern ourselves here with how adults in contemporary western society relate to each other affectively.

Affective relationships between adults have typically been analyzed in two ways: (1) as reflections of intrapersonal sentiments, needs, dispositions, or personality characteristics and/or (2) in terms of the institutional or quasi-institutional social forms recognized to meet individual and societal needs.

THE INDIVIDUAL PERSPECTIVE: LIKING AND LOVING

Most psychologists postulate a universal drive, motive, or need to relate emotionally with other humans. Individuals are often assumed or shown to be variable in the strength of various manifestations of this general characteristic of sociability. Thus Murray's theory of personality needs includes the nouns *affiliation* (the need to be in friendly association with others, to love, to cooperate) and *succorance* (the need to receive aid, protection, love, consolation, and guidance from another); other needs include power, achievement, nurturance, abasement, and so forth (Murray,

1938). The relative strength of such needs is presumed to motivate an individual toward or away from various patterns of action. An individual with strong needs for affiliation, for example, will seek out occasions to be in close collaboration with others; this may involve putting more energy into a love relationship than into education or preferring to work on a team rather than alone.

There is considerable evidence that the relative strength of affiliative needs is linked to sex and to age. Women, especially younger women, are, on the average, more compelled by their needs to form, maintain, and protect intimate relationships than are men. Young men are characterized by the quest for power and dominance in social relationships; in the second half of life, men in many cultures turn more toward love and expend less energy repudiating their own tenderness or their own needs for affiliation and nurturance (Gutmann, 1977; Jung, 1954; Levinson, 1978). Such an intrapsychic shift has been observed cross-culturally (Gutmann, 1977) and has been postulated to be linked to shifts in gender-specific demands associated with parenting dependent children (Gutmann, 1975).

Perhaps more important, several varieties of sociability are commonly distinguished in popular and scientific use. We often cannot be sure what a particular term means to the researchers or informant using it; this makes it more difficult to interpret research in this area. Generally, two broad categories are differentiated: liking and loving (Rubin, 1973).

One meaning of *liking* is the experience of emotional warmth and closeness to another person; it is this dimension which we will explore here. (The dimension of respect, also used with regard to liking, will not be considered here.)

Two kinds of *love* are frequently described, both reflecting inclinations within one person's mind or "heart." One variety of emotional experience is marked by the sense of intense need for the beloved, the desire to exclusivity, and the idealization of the beloved. Traditionally, this form of love is described as romantic, passionate, erotic, attachment; and, more disparagingly, as immature, deficiency love, transference love, or neurotic love. More recently, Dorothy Tennov (1979) described "limerent" love, characterized by its involuntarily compelling, intrusive quality, suddenness of onset, drive for exclusivity and reciprocation, and intensification through adversity. Tennov argues that limerence is a form of madness which otherwise very sane and rational people may develop, regardless of age, sex, or marital status. "While illogical, it is also normal" (Tennov, 1979, p. 180). Some individuals are nonlimerent and cannot comprehend the intensity of feelings and strivings expressed by limerent people.

Another form of love is described as more giving, autonomous, "caring," planful, and mutual. This is often regarded as "real" love, "mature" love, and the appropriate basis for long-term commitments. Tennov found that limerence sometimes evolved into love, but that many individuals experienced "companionate love" without ever feeling limerent. Unfortunately, much research does not distinguish these two varieties, especially among older adults.

In addition, *intimacy* is a third dimension of love which describes the relationship, rather than the sentiments expressed by either individual. The definitions and uses of what constitutes intimacy are variable, though most reflect the degree of close and confidential communication between people. However, no specific behavior or interaction has the inherent quality of intimacy (Macionis, 1978). This reminder is important when we consider the institutionalized ways of meeting intimacy needs; because one is married, or has sexual relations, does not necessarily mean one experiences intimacy in that relationship.

THE SOCIAL PERSPECTIVE: FRIENDS AND SPOUSES

Much of the research on interpersonal relations has focused on the standard cultural arrangements for meeting individual and social needs. Thus the research evidence is organized around the common social forms.

It is a matter of continued controversy whether individual feelings and social institutions are basically mutual and harmonious (as expressed in the sentiment, "Love and marriage go together like a horse and carriage"). An alternative view is that sentiment seldom survives institutionalization.

Socially recognized patterns contain social roles for the participants. The most formal, institutionalized form—marriage—has prescriptions for the role behavior of husband and wife written into religious and secular laws. Less institutionalized forms, such as friendship, divorce, and nonmarital love relationships, are characterized by emerging and/or more variable behavioral norms.

Patterns of individual and social experience vary along familiar dimensions. Most of the research reports sex differences. We are particularly interested here in differences related to age; such age-linked variability may reflect cohort experiences or developmental changes over the course of adulthood. In addition, variations associated with beliefs and practices termed *cultural* are often evident, reported in terms of social class, ethnicity, or religion.

Marriage

Participation. Most adults marry, at least once. Marriage is a recognized social institution in every society; the forms vary among cultures and over time within the same culture, but the core involves a publically recognized bond between a man and a woman. Their relationship is governed by legal and informal norms about the rights and obligations of being a husband or wife.

Marriage is best understood as a process, from initial attraction, through courtship, marriage, and, often, dissolution and remarriage. The phases of the "family life cycle" are somewhat linked to age; most individuals court and marry for the first time as young adults. However, it is clear that, with increasing healthy longevity and greater economic independence of women, the marital cycle may be repeated over the course of adulthood. As of now, we know relatively little about how the process of courtship is different for 20-year-olds and 60-year-olds, though intuitive sense, common observation, and limited research indicate that it is different (Jacobs & Vinick, 1979).

The evidence is fairly clear that potential partners sort themselves out on *homogamy;* similarity in terms of age, propinquity, education, social class, and intelligence seems to be the norm. There is some evidence for similarity in sex drive and less evidence about personality inadequacy and physical attractiveness (Murstein, 1980). Premarital pregnancy tends to suppress the effects of assortative mating on other variables; this is, obviously, an issue with younger rather than older couples.

On the other hand, older couples may mate more on the basis of availability and known companionability, since many remarriages occur within a friendship network. Overall, the evidence suggests that individuals tend to match up with someone whose perceived capacity to reward them is approximately equal to their own perceived rewardability, a pattern supporting the equity theory of mate selection (Murstein, 1980).

Dating does not necessarily lead to marriage. One study of young adults found that nearly half of 231 dating couples had broken up within 2 years of the initial study (Hill, Rubin, & Peplau, 1976). Most indicated that boredom, different interests, or the desire to be independent were major reasons for the breakup. Women emphasized interpersonal problems as contributing to the breakup; men listed geographical distance. Murstein (1980) speculated that men may be more interested in love, romance, and sex because they have "nothing to lose"; women, on the other hand, want marriage and are more likely to break up an unpromising relationship. Interestingly, a study of remarried older adults (all over 60) found the women more reluctant to reenter marriage, mostly because they were ambivalent about taking on another caretaking relationship (Jacobs & Vinick, 1979). It is not known, of course, how many women really resist remarriage because of that; the emphasis is more typically placed on the unavailability of men for older women.

Another study of young adults found that couples who made "good progress" during courtship (e.g., were still together months later) (1) were more accurate in predicting the partner's response to personality items; (2) tended to confirm the partner in the way the partner saw him or herself; and (3) were more compatible in roles (Murstein, 1972).

The primary shift in courtship patterns over the past decade has been in the rapid rise of cohabitation (Murstein, 1980). There has also been an increase in the overall prevalence of varied premarital sexual behaviors, particularly coitus (Clayton & Bohemeier, 1980). Most notable has been the decreased difference between men and women in reported premarital sexual behavior. For example, King, Balswick, and Robinson (1977) studied nonrandom samples of students at a large state university in the South in 1965, 1970, and 1975. Sixty-five percent of the men reported intercourse in 1965 and 1970; 74% by 1975. Among the women, 29% in 1965, 37% in 1970, and 57% in 1975 reported having intercourse. Another change noted in several research reports is a decrease in the average age at onset of coitus and an increase in the number of sexual partners among those who are experienced (Clayton & Bohemeier, 1980).

The changes in courtship patterns reflect delayed entrance into first marriage. The median age at first marriage among males was 22.5 years in 1963 and 23.0 in 1977; among women the median age increased from 20.3 to 21.1 years (Glick, 1980). The delayed entrance into first marriage is particularly notable among some groups of young women who apparently plan to marry later and have smaller families in order to accommodate the combination of occupational and domestic roles (Moore & O'Connell, 1978).

Another group of women with reduced rates of marriage are blacks. In 1970, there were around 1 million more black women than black men, with the sex ratio disparity greatest during the childbearing years (Jackson, 1971). As the black sex ratio has worsened, the proportion of black female–headed households has increased. Some scholars have suggested that polygyny be legitimized for such situations; others suggest that informal, consensual polygyny is already practiced even though regarded as an inferior alternative by black women (Scott, 1980).

More marriages are now ended by divorce than death, and more remarriages at each broad age range occur to divorced rather than widowed persons (Glick, 1980). The crude divorce rate in the United States increased from 2.5 to 5.3 per 1000 population between 1965 and 1979 (Carter & Glick, 1976). Current estimates are that nearly one-half of the couples now married will end up divorced. The likelihood of divorce is greatest during the first 7 years of marriage. In spite of an overall increase in the percentage of marriages ending in divorce, and in the increased percentage among the middle aged, the timing has not changed substantially (Troll, Miller, & Atchley, 1979).

The probability of divorce is related to age at marriage, with the lowest divorce rates for those marrying in their 20s and the highest among teenage marriages, particularly those where the bride was pregnant. According to Glick, the data on divorce suggest that "persons who have the personal and social attributes that help them to succeed in attaining the goal of high school or college graduation are also likely to succeed in entering marriage with a person who will be a satisfying marriage partner" (1980, p. 467).

Most divorced adults remarry—about three-fourths of the women and five-sixths of the men; others may make fairly long-term commitments which do not show up in remarriage statistics. Over half of those divorced remarry within 3 years. Data indicate that remarriage is faster for men than for women, for younger than older, for those who sought the divorce, for those with more substantial incomes, and for whites than blacks; among women, remarriage was more likely for poorly educated (less than high school) than better educated (1 year or more of college; National Center for Health Statistics, 1980). Whether or not a woman has children did not affect chances of remarriage for white women if they were aged 25 to 34 at the time of divorce; women divorced before age 25 were more likely to remarry if they were childless, and mothers divorced after age 35 were more likely to remarry than were nonmothers (Koo & Suchindra, 1980). Women over age 40 are especially disadvantaged in remarriage, since women survive longer than men, and men tend to remarry women who are several years younger than they are.

Marriage Styles. There are many strategies for characterizing marriage and family styles; because few have been applied systematically in families over the range of adulthood, it is difficult to identify cohort and/or developmental effects.

Overall, the research indicates that the marriage relationship remains central throughout adulthood for adults remaining married. Marital interaction styles have often been characterized on the dimension of power: the relative influence of the husband or wife in the relationship. At least three alternative types have been identified: husband as head and wife as complement; husband as senior partner and wife as junior partner; and husband and wife as equal partners (Scanzoni & Scanzoni, 1976). Among young adults, classification of the marriage as wife equal, junior partner, or complementary was associated with predictable differences on variables such as household task performance, fertility control, and sex-role preference (Scanzoni, 1980). Variations of such patterns were evident in the problem-solving behavior of more than 1000 couples studied for the first few years of marriage (Miller & Olson, 1978). The style of the marriage was usually set within the first year of marriage, and remained unchanged unless the couple sought therapy. Although 80% of the couples interviewed claimed to have "shared leadership cooperation," the researchers rated only 12% as having such an equalitarian relationship. The most common patterns were labeled husband-led disengaged and husband-led cooperative; six other variations were observed among 10% or less of the sample.

Age differences in attitudes about appropriate marriage and family relationships are evident. A national opinion survey found that older respondents expressed more traditional (e.g., husband as head) view of marriage, were more approving of sex-typed child rearing, and were more negative about feminist concerns; people over 65 were more extreme in these views than those 50 to 64, who were more conservative than younger adults (McGee & Evers, 1980). Presumably such differences largely reflect historical changes. It is not clear, however, that such public opinion statements correlate highly with actual patterns of interaction within a relationship.

In fact, more qualitative or clinical studies of marital interaction find evidence of midlife shifts in the relationship which reflect personality transformations described earlier. The move toward greater androgeny in later life is not reflected in reallocation of household tasks or more public marital-role behavior (Turner, 1981). Rather, it is revealed in such behavior as the satisfaction expressed by retired husbands about the increased companionship and nurturance, in complaints from wives about overly dependent husbands, in fantasies about the "perfect relationship," and in the kinds of problems presented in therapy or in the divorce courts (Gutmann, 1980; Mann, 1980; Turner, 1980).

Another dimension of marital style assesses the *boundaries* of the relationship, or who is included as "family" and what kinds of intimacy are endorsed within and outside of the defined boundaries.

On the one hand, multinational research over several decades has made it clear that adults are not isolated from kin living outside the household. Adults, married and unmarried, retain ties to adult children, aging parents, and siblings; the relationships are marked by both obligation and affection, are typically reciprocal, and are maintained in spite of geographical distance (Troll, Atchley, & Miller, 1979). Women maintain more contact with kin than do men (Lee, 1980). Involvement with kin also varies with stage in the life cycle. Young singles are more involved with kin than with their older parent(s); contacts with parents are increased when the young-adult child becomes parental; and older adults become more involved with kin as they are less involved in work and child rearing (Schulman, 1975).

On the other hand, the boundaries defining the marriage relationship as a distinct intimate tie have been reaffirmed. Social movements challenging traditional assumptions included invitations to try various alternatives to "traditional marriages." The most alluring of the alternatives seemed to be "open marriage" (O'Neill & O'Neill, 1972), where each partner is encouraged (or allowed) to develop independent interests and intimate relationships. After a decade of experience and research with this option in "flexible boundaries," several themes are apparent. Open marriages are more acceptable to younger, better-educated, less religious couples with few or no children (Wachowski & Bragg, 1980). Extramarital intimate friendships, sometimes including sex, are experienced as exciting, pleasurable, and growth enhancing; they are also tied to feelings of guilt and difficulties in maintaining multiple intimate relationships (Bunk, 1980; Ramey, 1975, 1976). Couples who maintain more openly flexible boundaries usually evolve special rules to minimize the threat to the marriage relationship; common strategies are to keep the marriage primary by (1) limiting the intensity of other relationships and (2) being open about the involvement (Bunk, 1980; Macklin, 1980). Thus research indicates that relatively few couples will opt for a marital style acknowledging sexual relationships outside the marriage. It is still not clear how nonsexual cross-sex friendships will be handled.

Marriage Satisfaction. Probably the most commonly studied dimension of marriage is expressed satisfaction with the marriage. It is generally, but not always, conceptualized as congruence between expectations and actual marital rewards (Spanier & Lewis, 1980). Often, several dimensions are assessed, such as adjustment, communication, marital happiness, integration, and satisfaction with the relationship. In the past decade, researchers have used larger samples; have developed better measures of marital quality; have included more husbands or couples; and have sometimes studied cohabiting couples as a variant of the dyadic relationship (Spanier & Lewis, 1980).

Most information on age-related changes in marital satisfaction are based on cross-sectional data. Many such studies report a U-shaped pattern, with marital satisfaction highest among newlywed nonparents, lowest among couples with teens living at home, and higher among older couples (Spanier, Lewis, & Cole, 1975). Data from husbands are rarer and less consistent than those derived from wives (Turner, 1981). Dissatisfaction is stronger and more consistent for wives than for husbands.

Attempts to explain this curvilinear pattern focus on methodological and theoretical points. The data are clearly flawed methodologically: different measures of marital satisfaction have been used in different studies; data from unrelated men and women are used to generalize about dyadic couples; post–child rearing and childless couples are often not distinguished; and cohort effects may mask or distort real changes (or stability) over time (Schram, 1979; Spanier & Lewis, 1980).

However, if one accepts the evidence of a general curvilinear pattern in satisfaction, several theories may account for it. Role strain may account for the lower marital satisfaction among parents, especially mothers, of dependent children, even though parenting may bring its own rewards (Spanier & Lewis, 1980). The very positive marital evaluations among elderly couples may reflect survival bias if less satisfied couples have already divorced; response bias tendencies (Campbell, Converse, & Rogers, 1976); or striving for cognitive consistency if those who have invested the most time, energy, and other resources value the relationship most (Spanier, Lewis, & Cole, 1975).

Perhaps the more interesting questions involve factors related to marital satisfaction and strain at different phases of adulthood. For example, a consistent theme of more qualitative studies of marriage is that women at all ages want more companionship and conversation than they receive (Turner, 1981). Self-disclosing conversational sharing is most common during courtship and the newlywed stage, and personal compatibility is an important basis for evaluating the quality of the relationship during the early years (Lowenthal, Thurnher, & Chiriboga, 1975). Young husbands are apt to complain about the dependency of their wives. Positive evaluations of spouses during the middle years seems more linked to marital-role performance—breadwinning and homemaking—than to personality (Lowenthal et al.,

1975). Strains are frequently evident during the middle years, as both partners negotiate a transition to greater androgeny (Gutmann, 1980; Livson, 1977; Mann, 1980). Older wives are apt to complain about the physical and emotional dependency of the husband. In very old age, couples may form a "symbiotic" relationship (Troll, 1971). As Turner observes (1981), the fact that people who have been married many years tend to die within a short time of each other suggests that they have been holding each other up as in the form of an arch, which collapses when either side falls.

Remarriage. In 1971, 24% of all marriages were remarriages; by 1977, it had risen to 32% (Price-Bohem & Balswick, 1980). Remarriage is more common following divorce than widowhood. Although remarriage occurs at all age groups, it is more likely among younger adults and for whites. Most couples remarrying after divorce regard their second marriages as distinctively different from their first; they report different expectations of marriage, more flexibility and sharing of householding, and less willingness to remain unhappily married (Furstenberg, 1980). Overall, remarried couples report about as much marital satisfaction, personal happiness, and worry as first marrieds; compared to first-marriage couples, remarried women are more likely to be very high in marital happiness and men are likely to be either very high or low in marital happiness (Weingarten, 1980).

The impact and quality of remarriage is obviously affected by the ages and life stage of the participants. Many remarriages involve dependent children. Such reconstituted families have special adjustment problems, some of which reflect the lack of norms regulating contacts between various current and former extended family members (Price-Bohem & Balswick, 1980).

Over the past decade, the number of people remarrying in later life (over age 60) has increased substantially, although the proportion of all remarriages which are later-life remarriages has not changed much (Treas & van Hilst, 1976). Older widowed women are less likely than widowed men to remarry. A recent intensive study of 24 remarried older couples by Jacobs and Vinick (1979) noted that the men were more unhappy and lonely as widowers and thus more likely to remarry. Women in this study were more likely than men to express negative feelings about remarriage, especially if they had nursed their previous mate through a terminal illness. Once remarried, however, they did not necessarily continue feeling negative, although serious problems in the marriage were more common for wives than for husbands. Men listed companionship and being cared for as prime reasons for marriage; women emphasized companionship and "love"; no woman listed being cared for as a reason for remarriage, though several mentioned having someone to care for.

Ethnic Variations. While most research reviewed above studied white, middle-class populations, somewhat better research is beginning to identify different patterns of meeting needs for intimacy, companionship, and caring. The review of this new research by Staples and Mirandé (1980) is summarized here.

Research on *black Americans* reflects a shift in perspective away from defining black families as deviant and pathological to seeing them as variant forms (e.g., extended-family units and "fictive kin") that may be culturally equivalent to the modal white forms. Several facts are notable. The majority of adult blacks are unmarried. The black divorce rate has increased by 130% in the past decade. There has been a sharp increase in marriages of black men and white women, along with a high dissolution rate of such marriages. The majority of black children are illegitimate. These trends reflect, in part, the scarcity of black men, and their difficulty in maintaining economic stability. There seems to be considerable variability in family forms with little evidence that a matriarchy exists in intact couple relationships.

Much research on *Chicano* family relations has stressed the importance of "machismo," either as a negative or positive influence in family life. This image is being revised however. Marital roles in the Chicano family seem to be predominantly equalitarian, across educational levels, urban–rural residency, and region. Chicano families are characterized by high fertility, and most children under 18 live with both parents in intact families. About 60% of the population is married; relatively more are single and fewer are divorced or widowed than in other ethnic groups.

Native Americans show great diversity in language and customs. They have generally been studied by white anthropologists, and much of the research still emphasizes Native Americans as cultural deviants posing problems for the larger society. Generally, they are like the black Americans in having high fertility, many out-of-wedlock births, many female-headed households, high unemployment, and extended families as the basic units.

Asian Americans constitute less than 1% of the American population. Cohort differences are often marked, since the oldest were often immigrants and younger groups again include recent immigrants. Overall, they have more conservative sexual values, lower fertility, fewer illegitimate births, and a more obligatory kinship system. It has proven difficult to maintain the traditional "honorable" status of the elders, and this is reflected in intergenerational tensions.

Friendship

There is little agreement about how to define or measure friendship during the adult years. Expectations about friend relationships are not legally defined, and careful researchers have found considerable variability in the kinds of relationships defined as friendships. A common assumption is that the number and quality of friends available is an important index of sociability and potential support. However, designating someone as a friend may not be a good predictor of actual interactions. For example, 156 older Manhattan hotel dwellers were asked to indicate who their friends were and, also, to describe in detail the kinds of daily contacts they had with various people (Cohen, Cook, & Raykowski, 1980). It was clear that many of the contacts who were not identified as friends were, nevertheless, involved in important interactions. Most interesting, a significant minority of nonfriend contacts were considered both "intimate" and "important" to the respondent, and many persons designated as friends were not considered important or intimate. Similar complex patterns were found in a study of urban widows (Lapata, 1977).

Same-Sex Friendships. Most friendships are same-sex; and researchers consistently identify sex differences in friendship styles. Much of the current research on friendship betrays a considerable bias toward using feminine styles of friendship as the "ideal" standard. Research generally shows that female friendships are likely to involve self-disclosure: shared feelings, thoughts, and vulnerabilities. The patterns of adult feminine friendships seem to have antecedents in observed childhood patterns, with girls playing in small clusters or pairs and spending a good deal of time discussing personal feelings. A study of friendship among women aged 14 to 80 found that all close friends were used for intimacy; that is, providing emotional support, validation, and an opportunity to "let off steam." Close friends were also used for help, status, and power (Candy, 1977). The complexity of female friendships is often blurred in emphasizing the "superior" intimacy and friendship skills of women.

By comparison, men are usually judged to be "deficient" in friendship skills. As the decade review of male friendship research stated bluntly, "This paper traces some of the antecedent sources of impoverished male friendship which often reach crisis levels in late life" (Tognoli, 1980, p. 273). Boys' friendship groups are larger than those of girls and are focused more on group activities. Adult men report more same-sex friendships than do women, but they are not close or intimate; conversation focuses rather on impersonal topics (such as sports, politics, and cars); (Tognoli, 1980). Adult male friendships seem to be based more on doing things together, particularly activities which require mutual cooperation (such as team sports, combat, or work productivity). Even in a same-sex discussion group, college men were less intimate and open with each other and talked less about themselves, their feelings, and the significance of their relationships than did same-sex groups of women (Aries, 1976). Men are likely to keep feelings of vulnerability to themselves, or to direct intimate self-disclosure toward a woman—girlfriend, wife, or lover. These patterns have been reported in men of all age groups, from adolescence to old age.

Using the feminine style of friendship as a model, several writers have tried to explain the factors which limit male friendships. As Tognoli (1980) summarized them, the most powerful deterrent is homophobia. The fear that close companionship among men may arouse sexual desires results in a system where only the "supermasculine" men (such as policemen and soldiers) are allowed to express close, comradely intimacy. Second, the competitiveness required by adherence to the work ethic virtually mandates concealing personal problems and fears, rather than sharing them with potential (or actual) rivals. (Women entering "masculine" or highly competitive work settings often report similar pressures to control feminine-style friendships and not to disclose personal information to anyone who is or who may become a status superior or inferior; such norms are quite inconsistent with typical feminine friendship patterns and can be a source of great stress.) Third, Tognoli speculates that the pattern of a man unloading all his personal problems onto his woman, and not reciprocating, may be another expression of power. Finally, male patterns of friendship may reflect the detachment, even alienation, from domestic space that is felt by many men. Lacking a sense of comfort in private dwelling space, men interact in more public meeting places, and the style of interaction reflects the setting. Tognoli and other writers advocating the "liberation" of men from male patterns of behavior argue for changing sex roles, and specifically working to counteract homophobia, competitiveness, and inexpressiveness among males.

Homosexual Relationships. Some men and women respond affectively and/or erotically to members of the same sex. The extent to which sexual fantasies and activities are homosexual varies greatly, as does the extent to which a homosexual preference is reflected in patterns of friendship or intimacy. Same-sex partnerships must be recognized as a variant family form, even though it is not legally acknowledged as such (Macklin, 1980).

Homosexual individuals strive to meet sociability and intimacy needs within a social context that generally degrades and rejects their partner preferences. It is, then, unclear how much of observed behavior reflects inner-psychological structures and how much is a response to the social status of a deviant. Several themes emerge from the research. First, homosexuals are diverse: some are social isolates, some maintain long-term marriagelike relationships, some maintain heterosexual marriages and are parents, and some are frankly promiscuous. Second, male homosexuals tend to have more partners than do lesbian women. Third, many homosexuals are not interested in changing their life-style, although they would prefer less social rejection because of their sexual-partner preference.

Cross-Sex Friendships. Cross-sex friendships seem to be governed by current patterns of norms and by opportunity. In America, the norms have allowed considerable, but not complete, access of men and women to each other, at all ages. There is often an implicit or explicit assumption that adult cross-sex friendships may or will become sexualized. Because of this assumption, cross-sex friendships are encouraged among unmarried people but are discouraged or regarded with suspicion among married adults.

One of the opportunity factors is that there is still much resistance to full sexual integration of the work place. Since many adults meet friends through work, this limits the likelihood of making cross-sex friendships. As work becomes sex-integrated, such friendships are more common, and new norms are emerging to meet the changing circumstances. These norms are not yet clear, however (Bell, 1979).

Discussion

The research reviewed here indicates that there are age differences in patterns of meeting intimacy needs; it is not yet clear which differences reflect socialization or historical effects and which reflect developmental transformations. In addition, fairly clear sex differences emerge in the research on marriage and friendship.

Such sex differences are usually explained, if at all, as the rather unfortunate consequence of "inadequate socialization" of males for intimacy. Within marriage, these sex differences may help account for the attractiveness of marriage for men and the dissatisfaction with marital communication among women. Recent social movements have stressed the desirability of men and women developing a wide range of potential interaction styles; in the case of close relationships, however, the "liberal" social scientists have emphasized how men should become more like women. Such a move would—though presumably beneficial for domestic relations—be potentially incompatible with work roles; and thus, ultimately, with domestic roles as well; these outcomes have not been systematically explored.

There has been relatively little consideration of the possibility that the patterns observed may be rooted in precultural, cross-species sexual differentiation (Wilson, 1980) and that the styles may be enjoyable and functional overall. Men may really enjoy conversing about sports, cars, and sex, just as women enjoy discussing interpersonal relationships. Men may derive self-esteem from their ability to keep their own counsel and deny feelings of vulnerability; these may, in fact, contribute to their ability to cope effectively with nearly inevitable competitive situations. The stress on task-oriented, mutual validation of masculinity may be an important source of support. The different styles of interaction may reflect one way dependencies between the sexes have been balanced: women have depended on men for material support, occupational achievement, and status for the family unit, and for defense of the household and nation when necessary; these are often in the service of much-valued motherhood. Men have depended on women for mothering their children, for comfort, and for emotional expressivity.

The challenge now is to explore the more qualitative meanings of social connectedness, as these are expressed by men and women over time, in varying cultural contexts. A further challenge is to reclaim pluralism, with different "standards of excellence" allowed as valid.

REFERENCES

Aries, E. Male–female interpersonal styles in all male, all female, and mixed groups. In A. G. Sargent (Ed.), *Beyond sex roles*. St. Paul: West, 1976.

Bell, R. *Marriage and family interaction* (5th ed.). Homewood, Ill.: Dorsey, 1979.

Bunk, B. Sexually-open marriage: Ground rules for countering potential threats to marriage. *Alternative Lifestyles,* 1980, **3** (August), 312–328.

Campbell, A., Converse, P. E., & Rodgers, W. L. *The quality of american life.* New York: Russell Sage, 1976.

Candy, S. What do women use friends for? In L. Troll, J. Israel, & K. Israel (Eds.), *Looking ahead: A woman's guide to the problems and joys of growing older*. Englewood Cliffs, N.J.: Prentice-Hall, 1977.

Carter, H., & Glick, P. C. *Marriage and divorce: A social and economic study* (Rev. ed.). Cambridge, Mass.: Harvard University Press, 1976.

Clayton, R. R., & J. L. Bohemeier. Premarital sex in the seventies. *Journal of Marriage and the Family,* 1980, **42**, 759–775.

Cohen, C. I., Cook, D., & Rajkowski, H. What's in a friend? Paper presented at the 33rd Annual Scientific Meetings of the Gerontological Society of America, San Diego, Cal., November 21–25, 1980. Abstract in *The Gerontologist,* 1980, **20**(5), 84.

Furstenberg, F. Reflections on remarriage. *Journal of Family Issues,* 1980, **1**, 443–453.

Glick, P. C. Remarriage: Some recent changes and variations. *Journal of Family Issues,* 1980, **1**, 455–478.

Gutmann, D. L. Parenthood: Key to the comparative psychology of the life cycle? In N. Datan & L. Ginsberg (Eds.), *Life span developmental psychology: Normative life crises*. New York: Academic Press, 1975, pp. 167–184.

Gutmann, D. L. The cross-cultural perspective. In J. Birren & K. W. Schaie (Eds.), *Handbook of the psychology of aging*. New York: Van Nostrand Reinhold, 1977, pp. 302–306.

Gutmann, D. L. The post-parental years: Clinical problems and developmental possibilities. In W. H. Norman & T. J. Scaramella (Eds.), *Mid-life: Developmental and clinical issues*. New York: Brunner/Mazel, 1980.

Hill, C. T., Rubin, Z., & Peplau, L. A. Breakups before marriage: The end of 103 affairs. *Journal of Social Issues,* 1976, **32**, 147–168.

Jackson, J. But where are the men? *Black Scholar,* 1971, **3**, 30–41.

Jacobs, R. & Vinick, B. *Re-engagement in later life: Re-employment and remarriage*. Stamford, Conn.: Greylock, 1979.

Jung, C. *The development of personality.* New York: Pantheon Books, 1954.

King, K., Balswick, J., & Robinson, I. E. The continuing premarital sexual revolution among college females. *Journal of Marriage and the Family,* 1977, **39**, 455–459.

Koo, H., & Suchindran, C. M. Effects of children on women's remarriage prospects. *Journal of Family Issues,* 1980, **1** (December), 497–515.

Lee, G. Kinship in the seventies: A decade review of research and theory. *Journal of Marriage and the Family,* 1980, **42**, 923–984.

Levinson, D. J., with Darrow, C. N., Klein, E. B., Levinson, M. H., & McKee, B. *The Seasons of a man's life.* New York: Knopf, 1978.

Livson, F. Coming out of the closet: Marriage and other crises of middle age. In L. Troll, J. Israel, & K. Israel (Eds.), *Looking ahead: A woman's guide to the problems and joys of growing older.* Englewood Cliffs, N.J.: Prentice-Hall, 1977.

Lopata, H. The meaning of friendship in widowhood. In L. Troll, J. Israel, & K. Israel (Eds.), *Looking Ahead: A woman's guide to the problems and joys of growing older.* Englewood Cliffs, N.J.: Prentice-Hall, 1977.

Lowenthal, M., Thurnher, M., & Chiriboga, D. *Four stages of life.* San Francisco: Jossey-Bass, 1975.

McGee, J., & Evers, M. Marriage, family, and feminism: Older people's views. *The Gerontologist,* 1980, Part II, **20**(5/November), 158.

Macionis, J. Intimacy: Structure and process in interpersonal relationships. *Alternative Lifestyles,* 1978, **1** (February), 113–130.

Macklin, E. Nontraditional family forms: A decade of research. *Journal of Marriage and the Family,* 1980, **42**, 905–922.

Mann, C. Mid-life and the family: Strains, challenges, and options of the middle years. In W. H. Norman & T. J. Scaramella (Eds.), *Mid-life: Developmental and clinical issues.* New York: Brunner/Mazel: 1980.

Miller, B. C. & Olsen, D. Typology of marital interaction and contextual characteristics: Cluster analysis of the I.M.C. Unpublished paper available from D. Olsen, Minnesota Family Study Center, University of Minnesota, Minneapolis, 1978.

Moore, M. J., & O'Connell, M. Perspectives on American fertility. Current Population Reports, Series P-23, No. 70. Washington, D.C.: U.S. Government Printing Office, 1978.

Murray, H. *Explorations in personality.* New York: Oxford University Press, 1938.

Murstein, B. I. Person perception and courtship progress among pre-marital couples. *Journal of Marriage and the Family,* 1972, **34**, 621–627.

Murstein, B. I. Mate selection in the 1970s. *Journal of Marriage and the Family,* 1980, **42**, 777–792.

National Center for Health Statistics. Births, marriages, divorces, and deaths for 1979. Monthly Vital Statistics Report, V. 28, No. 12, Provisional Statistics. Hyattsville, Md., U.S. Department of Health, Education, and Welfare, 1980.

O'Neill, N., & O'Neill, G. *Open marriage: A new life style for couples.* New York: M. Evans & Co., 1972; Avon Books, 1973.

Price-Bonham, S., & Balswick, J. O. The noninstitutions: Divorce, desertion, and remarriage. *Journal of Marriage and the Family,* 1980, **42**, 958–972.

Ramey, J. W. Intimate groups and networks: Frequent consequences of sexually open marriage. *The Family Coordinator,* 1975, **24**, 515–530.

Ramey, J. W. *Intimate Friendships.* Englewood Cliffs, N.J.: Prentice-Hall, 1976.

Rubin, Z. *Liking and loving: An invitation to social psychology.* New York: Rinehart and Winston, 1973.

Scanzoni, J. Contemporary marriage types: A research note. *Journal of Family Issues,* 1980, 125–140.

Scanzoni, L., & Scanzoni, J. *Men, women and change: A sociology of marriage and family.* New York: McGraw-Hill, 1976.

Schramm, R. W. Marital satisfaction over the family life cycle. *Journal of Marriage and the Family,* 1979, **41**, 7–12.

Scott, J. W. Black polygamous family formation. *Alternative Lifestyles,* 1980, **3** (February), 41–64.

Shulman, N. Life cycle variations in patterns of close relationships. *Journal of Marriage and the Family,* 1975, **37**, 813–821.

Spanier, G. B., & Lewis, R. A. Marital quality: A review of the seventies. *Journal of Marriage and the Family,* 1980, **42**, 825–839.

Spanier, G. B., Lewis, R. A., & Coles, L. C. Marital adjustment over the family life cycle: The issue of curvilinearity. *Journal of Marriage and the Family,* 1975, **37**, 262–275.

Staples, R., & Mirandé, A. Racial and cultural variations among American families: A decennial review of the literature on minority families. *Journal of Marriage and the Family,* 1980, 42, 887–903.

Tennov, D. *Love and limerence: The experience of being in love.* Briarcliff Manor, N.Y.: Stein and Day, 1979.

Tognoli, J. Male friendship and intimacy across the life span. *Family Relations,* 1980, **29**, 273–279.

Treas, J. & Van Hilst, A. Marriage and remarriage rates among older Americans. *The Gerontologist,* 1976, **16**, 132–136.

Troll, L. The family of later life: A decade of review. In C. Broderick (Ed.), *A decade of family research and action.* Minneapolis: National Council on Family Relations, 1971.

Troll, L., Atchley, R., & Miller, S. *Families in Later Life.* Belmont, Calif.: Wadsworth, 1979.

Turner, B. F. Sex-related differences in aging. In B. B. Wolman & G. Stricker (Eds.), *Handbook of developmental psychology.* Englewood Cliffs, N.J.: Prentice-Hall, 1981.

Turner, N. Divorce in mid-life: Clinical implications and applications. In W. H. Norman & T. J. Scaramella (Eds.), *Mid-life: Developmental and clinical issues.* New York: Brunner/Mazel, 1980.

Wachowski, D., & Bragg, H. Open marriage and marital adjustment. *Journal of Marriage and the Family,* 1980, **42**, 57–62.

Weingarten, H. Remarriage and well-being: National survey evidence of social and psychological effects. *Journal of Family Issues,* 1980, **1**, 533–559.

Wilson, E. *Sociobiology: The abridged version.* Cambridge, Mass.: Belknap/Harvard University Press, 1980.

28

PARENT AND CHILD
Generations in the Family

Gunhild O. Hagestad
College of Human Development
The Pennsylvania State University

LATE TWENTIETH CENTURY PARENT–CHILD RELATIONSHIPS

This chapter suggests that everyday language and imagery lag behind current demographic realities when it comes to parent–child relationships. Furthermore, social-behavioral research and theory addressing such relationships in no way have kept up with historical change in family composition. I argue that it is essential to recognize that within families, parent–child relationships evolve in a continuous chain, a set of interlocking careers. Such relationships are shaped by three kinds of development and change: individual lifetime, family time, and historical time (Aldous, 1978; Hareven, 1977). In a process which is quite similar to the flow of birth cohorts through a system of "life stations" (Neugarten & Hagestad, 1976; Riley et al., 1968), parent–child relationships evolve in a chain which is continuously renewed through the introduction of new generations. I furthermore argue that existing work on parents and children has tended to take a fragmented, discontinuous view, which in no way captures the duration, stability, and interrelated nature of parent–child interactions. Social scientists are challenged to develop theoretical and empirical approaches which are not limited to one stage of individual lifetime, one phase of family development, or one generation of parent–child pairs.

Past research has been limited to parent–child relationships at two extremes of the human life span. One tradition, by far the strongest one, is focused on young children and their parents. A more recent, still exploratory, effort is concentrated on aged parents and their children. Between these two bodies of work is a great void. We have taken an alpha and omega view of parent–child relationships. It is high time that we attempt to bridge the two.

Conceptual Ambiguity

In an undergraduate course, I discussed how children may serve as confidants for their parents. A puzzled student raised his hand and asked, "How can a *child* be a confidant?" His confusion, of course, stemmed from a basic ambiguity in our language. The term *child* has two very different referents: a person in the first stage of life and a person in a family role, reciprocal to the role of parent. In the latter case, one is a child as long as one has a parent, independent of age and life stage.

The term *parent* does not have the same kind of duality, but ambiguity still surrounds it. In social science publications we see discussions of "the loss of the parent role," "retirement from parenthood," and the "postparental phase." At a recent national meeting of family researchers, a paper reporting on an interesting and well-done piece of research referred to "former mothers." A member of the audience (most likely a mother!) asked for clarification, expressing surprise that so many middle-aged women had lost all their children! Although most of us would agree that parents are parents as long as they have living offspring, our ways of thinking about parents and children lag behind current demographic conditions.

The Impact of Demographic Change

In popular imagery, the phrase "parents and children" is likely to evoke a picture of adults between the ages of 20 and 40 with offspring aged 1 to 16. Very seldom, if ever, would it bring forth an image of individuals who all have grey hair and all are over the age of 60.

Our concepts and our language in no way match present family reality. Demographic change and altered rhythms in the family cycle have made it common for parent–child relationships to last 60 years or more. Only for one-quarter of that time will the "child" fit the chronological referent of the term. Furthermore, during the majority of shared life years between parents and children, the "child" is also likely to be a parent.

Increased general life expectancy, combined with smaller, more closely spaced families (Glick, 1977; Neugarten & Moore, 1968), have produced greater "life overlaps" between parents and children. These factors have also given us multigenerational families, in which three or four generations of parents and children coexist within a given lineage at the same time (Troll, 1971; Troll et al., 1979). Thus a family lineage will have several tiers of parent–child relationships, or what we from a life course perspective may call "generational stations."

This situation is historically novel. Not too long ago in American society, a sizable proportion of individuals lost one or both parents before they themselves reached adulthood (Glick, 1977; Uhlenberg, 1980). Today, it is not uncommon for parents and children to share experiences of adulthood, such as being a student or facing the challenge of a new job.

Not only has the population pyramid of families changed dramatically, but altered mortality and fertility patterns have created new psychological milieux within families, among generations. Such historical changes have led several authors to suggest that under conditions of high mortality, people were hesitant to form attachments, knowing that intergenerational, dyadic ties could not be counted on to endure (Aries, 1962; Blauner, 1966; Uhlenberg, 1980). Instead, they invested in the security of the family as a corporate group. It is argued that not only have we come to take long-term intergenerational bonds for granted and invest in them, but because of reduced fertility, we also have fewer individuals within each generation to invest in. Therefore, intergenerational relationships are not only more *extensive* now, they may also have become more *intensive* (e.g., Skolnick, 1978; Uhlenberg, 1980).

PAST WORK ON PARENT–CHILD RELATIONS: THE ALPHA AND OMEGA VIEW

A recent review of 1970s research on parent–child relationships (Walters & Walters, 1980), concludes that even in most recent work, the main focus is overwhelmingly on young children and their parents. The analytic focus is on the very beginning of a long relationship career, disregarding many life stages and generational stations. The focus is on the *alpha* of bonds. A secondary focus is the relationship between older people and their middle-aged children. This *omega* focus is typically not included in discussions of parent–child relations, but will be found under such headings as "older families," "families in later life," and "kin relations." In the two *Journal of Marriage and the Family's* decade review issues, which summarized and synthesized research and theorizing in the 1960s and the 1970s, the alpha and omega of parent–child relationships were addressed in separate articles. Yet, only the former was discussed under the heading "Parent–Child Relationships."

Four general conclusions can be drawn about the alpha and omega of parent–child relations and the study of them. First, there is a great data gap between the alpha and omega. We know very little about young adults and their parents; literally nothing about middle adults, aged 30 to 45, and their parents (Troll, 1971; Streib & Beck, 1980). Little or no effort has been made to link alpha and omega, either through cross-sectional or longitudinal research. Second, the alpha phase of the research is tilted toward the child; the omega research is focused on the parent. Third, most major theoretical building has been done within the alpha tradition. Work on older people and their children has, for the most part, been descriptive, "fact gathering." Troll noted this trend in her 1971 decade review; Streib and Beck (1980), who reviewed work in the following decade, see little change in this respect. Fourth, and most fundamentally, attempts to integrate generalizations advanced in the two traditions are literally nonexistent.

In part, the state of affairs outlined above may reflect disciplinary boundaries. The bulk of work on young children and their parents has been by scholars in developmental psychology; that on the second half of life by family sociologists and gerontologists. Furthermore, the two bodies of work often appear to have been generated by quite different sets of concerns. Much of the writing on young children and their parents have sought to build theory of early development. A good deal of the work on middle-aged adults and their parents has been fueled by practical or "political' concerns. Even the flurry of publications which came in response to Parsons' (1949, 1955) statements regarding "the isolated nuclear family" had a somewhat political flavor. It became a mission, a calling, to demonstrate that old people in this society are not isolated from their family—not abandoned by their children.

Shanas (1979) sees such myths as a multiheaded Hydra, a challenge to social research warriors. After nearly 2 decades of research we should be prepared to declare the beast slain. A series of studies in Western industrialized societies consistently found high rates of contact between old parents and at least one child (e.g., Adams, 1970; Firth et al., 1969; Hill et al., 1970; Knipscheer, 1980; Lopata, 1973; National Council on Aging, 1975; Olsen et al., 1976; Rosenmayr, 1973; Shanas et al., 1968; Shanas, 1979; Teeland, 1978; Townsend, 1957). Most of this research also found considerable exchange of goods and services between parents and adult offspring. We now need to move on in our work and ask about the *quality* of intergenerational bonds; the *meaning* of parent–child ties across adulthood. There are a number of new and pressing issues to address.

Within the recent population explosion among the very old (e.g., Siegel, 1979), many of whom require care and attention, it has become essential for social planners and policy makers to know how American families deal with their older members. It is also important to know the wider familial contexts of aged parents and their children. Clearly, a focus on omega dyads in isolation has severe limitations, as do cohort data on life course patterns, fertility and mortality trends (e.g., Gelfand et al., 1978; Treas, 1977). We need to recognize that families vary enormously in their generational constellations and demographic composition (Hagestad, 1981; Troll, 1970).

Scholars who take a life course perspective and examine human lives and relationships across their entire duration are also faced with another mission: to build bridges between the alpha and omega view of parents and children. Such bridges would benefit theoretical efforts in the fields of individual and family development as well as social planning. We need to consider the entire span of parent–child bonds and their constant transformation through individual development and generational turnover. As a modest start towards such integration, I present the sketch for a model of parent–child relationships across generations. Then I briefly examine some substantive and conceptual issues which can be "fed into" the model.

TOWARD A DYNAMIC VIEW OF PARENT–CHILD RELATIONSHIPS

Most existing work on parent–child relations has implicitly or explicitly addressed dyadic relationships, typically between mother and child. Schematically, then, the analytic focus can be represented as in Figure 28.1.

Above, I argued that in the alpha tradition, C has received more attention than P. Recently, however, there have been numerous calls for taking a reciprocal view of parent–child dyads, for example in the study of socialization, influence and attachment. In the omega focus, P has been given more attention than C, and few attempts have been made to take a dyadic, reciprocal view of such parent–child relations. The issues of reciprocity will be discussed in more detail below.

As a first step in bridging the alpha and omega parent–child relations, we need to put them in a model which allows us to discuss time and change on an individual, a family, and a societal level. (See Figure 28.2.) To begin with, we need to recog-

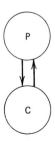

Figure 28.1

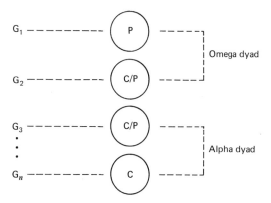

Figure 28.2

nize that a particular parent–child linkage is typically part of a multilink, vertical chain, in which several individuals simultaneously occupy the roles of both parent and child. For the sake of simplicity, let us examine direct linkages within a hypothetical lineage. The oldest living generation in a lineage, G_1, will be the only one who occupies the role of parent only. At the other end of the generational spectrum, G_n, which may be anywhere from one to six links removed, are lineage members who occupy only the role of the child. Individuals in the connecting links are both children and parents.

From a structural, static perspective, we can see each circle as representing one *generational station,* not dissimilar to age grades in society. We can also consider the number of stations which exist in a lineage at a given time.

From a dynamic, process perspective, we can trace the careers of individuals or parent–child pairs as they move across stations. The number of stations, however, is not fixed. When G_1 dies, the lineage will lose one link; when a new generation is born, a link is added. With the death of an omega parent, other members move one link "up." Thus we can study family lineage development by examining what Ryder (1965) called "demographic metabolism"—the entry of new members—the departure of old. Finally, we can look at the structure of generational stations over historical time, examining the number of stations commonly found, the relative number of people in the various stations, the age distance between stations, and the cultural significance of being in one particular station across different historical periods. In the early part of this chapter, I argued that the general historical trend has been toward increasing numbers of stations and increasing age of omega parents. Furthermore, because of recent fertility patterns, the average number of calendar years separating each station has decreased.

As I have argued elsewhere (Hagestad, 1981) it is extremely important to note, however, that families with identical numbers of generational stations can have dramatically different age/cohort structures and be in quite different phases of family development (Rodgers, 1962, 1973; Troll, 1971). A four-generation family may have an omega parent who is age 50; an alpha child age 2. Another four-generation lineage may have a 96-year-old omega mother and a 20-year-old alpha child. Two mothers in their 40s may have the same number of children, both with the oldest in the early 20s. Their positions in a system of generational stations, however, may be drastically different. One may be an omega parent of alpha children, with both her parents deceased and with no grandchildren. Another may be a grandmother as well as a granddaughter, with two generations of parents above her.

POSSIBLE RESEARCH STRATEGIES

Cross-Sectional Approaches

Using the concept of generational stations, we can examine parent–child relationships on three levels: individual parent or child, parent–child dyad, and family lineage.

On the first level we can compare behavior patterns, subjective perceptions, and expectations by parents and children who at a given time are in a particular generational station. We can compare individuals who are similar in age and life stage but different in their parent station. If our unit of analysis is the dyad, we can compare patterns of contact, exchanges, and affective closeness in pairs from different stations. Again, we can contrast pairs who at first glance appear to have similar age differentials between parent and child but who are embedded in contrasting lineage constellations of parent stations. For example, two parent–child pairs with a 50-year-old parent and a 20-year-old child may in one case represent the omega and alpha in their lineage, or they may, in another case, be surrounded by two other generations: two older and one younger. It would seem reasonable to expect that the two would operate differently, for example, in the allocation of time and resources or in the sense of personal vulnerability.

On a family level, we can describe lineage patterns on a set of variables—demographic, behavioral, or attitudinal. For example, we can describe families according to the number of parent stations the lineage has, the age distribution, and the number of individuals in each set of parent–child linkages. We can develop "family dependency ratios" and variables expressing the degree of lineage homogeneity on such dimensions as ethnicity, social class, and religious affiliation. Such family-level variables can subsequently be used as independent variables in research on the individual or dyad level.

Longitudinal Approaches

Most longitudinal work on parents and children has been over relatively short periods of time (e.g., 5 years) and within the same parent station (mostly alpha children and their parents). Very little work has traced individual parents and children or parent–child dyads across the life course. Even fewer attempts have been made to look not only at "developmental phases" but at the generational context of the phenomena under study.

Examining Contrasts and Change

There is an astonishing lack of research on parents and children over time. For example, nobody, to my knowledge, has examined "parenting styles" in the alpha station and related it to styles in later stations. Does the "authoritative" parent, in Baumrind's (1971) conceptual framework, have an easy time developing an adult–adult relationship with a child who is about to become a parent? Are there early parenting styles which make "the role reversal of later life"—a situation in which the parent becomes increasingly dependent on the child—easier to accept? Do parent–child pairs display consistency in interaction patterns, affective closeness, and attitude differences across the duration of their relationship? Where do we have the greater consistency: across the same units over time? across units in similar developmental phases? across units embedded in similar generational contexts? across individuals who share similar historical–cultural definitions of parent and child roles?

EXAMINING PARENT–CHILD PAIRS FROM A DYNAMIC, MULTIGENERATIONAL PERSPECTIVE

The location of parent–child pairs in a structure of generational stations has generally been neglected in past research. However, there is work which strongly suggests the usefulness—even necessity—of considering parent–child relationships within multiple links on a vertical lineage axis. I will choose a few examples from studies addressing relationships among adults.

Several authors have spoken of life cycle and generational "squeezes." In an often-quoted paper, Oppenheimer (1974) spoke of an imbalance between father's earning power and the needs of growing children. Recently, we have started to discuss another type of squeeze: that experienced by members of the middle generation in three-generation families. These individuals may simultaneously find themselves engaged in "parent caring" (Lieberman, 1978) and facing young adult offspring who need support in establishing themselves as adults (Hill et al., 1970; Silverstone & Hyman, 1976; Streib, 1972). Neugarten (1978) has added that an increasing number of families may actually have *two* middle generations: one middle aged, the other young–old. Thus there may be two generations of aging parents, which presents new strains on family resources (e.g., Brody, 1978).

An exploratory study by Smith (1981) found that when middle-aged women experienced a sense of overload in caring for aged parents, they tended to have young adult children who were "off track" in *their* life course development. For example, they had a son in his late 20s who was still without a job, or a daughter whose marriage had failed and had moved back into her old room. This study illustrates the importance of considering support patterns, as well as expectations, across more than one parent–child link.

Patterns of Reciprocity

Often, when middle-aged persons feel a generational squeeze, it is a result of the combined stresses, vulnerabilities, and strengths of three generations: the need for support by children who are in early adulthood; midlife changes in the parent; and increased need for sustenance by grandparents, sometimes also great-grandparents.

Thus if we want to examine patterns of exchange and reciprocity in parent–child pairs, it is essential to not only take a developmental view but also to consider the intergenerational context of a given pair. Hill et al. (1970) discuss patterns of support in three-generational families and suggest that balancing not only occurs *across spheres* of exchange, it also takes place across more than one intergenerational link.

Rosenmayr (1978) makes a strong statement on the importance of viewing exchanges across more than two generations—as a continuous process across family history:

> The chain of intergenerational bonds reaches a balance—if at all—only if viewed over more than two elements in the chain. The exchange that occurs within the family is never reciprocal with identical individuals, but functions as a transfer process. Parents who have yielded benefits to their children are not "adequately" rewarded by these children, the latter instead yield benefits to their offspring. . . . The exchange between generations is chronically one-sided, only transfer as an infinite process creates a balance.

While work on omega dyads has typically used the concept of reciprocity in discussions of intergenerational support, work on alpha relationships has emphasized reciprocity in socialization (e.g., Ahammer, 1973; Hartup, 1978; Hartup & Lempers, 1973). However, for both bodies of literature, Ahammer's (1973) conclusion still holds: "Even though in recent years increasingly more emphasis has been placed on the reciprocal nature of parent–child relationships, it has been surprisingly little researched" (p. 275).

THE CONTINUOUS PRIMARY TIE

When C. H. Cooley introduced the term *primary group,* his main example was the family. Cooley saw the family as primary in two senses of that word. Chronologically, it is our first human group. It is also primary, in the sense of chief, principal, prime. For most of us, family membership is primary throughout life and constitutes the key social anchor at the end of our lives, as it did at the beginning.

Most parents and children give primary relationships to each other for the duration of their shared lives. Our parents and our children represent what Sullivan (1947) called "significant others." Plath (1980) refers to them as *consociates:* "those around us whose lives run close, and parallel, to ours . . . intimates whom I engage and re-engage across long chains of interaction" (Plath, 1980, p. 289).

Influence between parent and child, therefore, is a life-long process. Indeed, it may *transcend* individual life spans. In a three-generational study carried out in the greater-Chicago metropolitan area (Hagestad et al., in progress; Halprin, 1979), we asked, "Who has the most over-all influence on you these days, would you say?" Respondents from all three generations were likely to say a parent. Among respondents in the grandparent generation, many named parents who had been dead for 30 years or more. Troll (1972) reports similar findings from her three-generational research. Plath (1980) reminds us of Victor Hugo's statement: "If you want to reform a man, start with his grandmother."

In a field where concepts such as internalization and identification are theoretical building blocks, we should be quite ready to recognize influence which is not limited to ongoing interaction. Indeed, we may need to remind ourselves of the common dictionary definition of *influence:* "capacity or power to produce effects on others by intangible or indirect means" (*Random House Dictionary of the English Language,* 1969). I would like to briefly examine influence between parents and children under two conceptual headings: *socialization* and *developmental reciprocity* (Klein et al., 1979).

Socialization

Most of our work on influence patterns in the family has been done under the conceptual umbrella of *socialization.* The last 2 decades have seen a series of publications which argue that socialization is an interactional, reciprocal process. Social scientists have been urged to abandon what Hartup (1978) calls "the social mold" view, in which one party is seen as the "socializer"; another as the "socializee." Within the family context, the old view stressed parents' socialization of their children, preparing them for adult roles (Goslin, 1969).

Recent work has called for research on the *mutual* influence between parent and child, with new attention to the children's effects on parents (e.g., Bell, 1968). This thrust is reflected in the title of Rheingold's (1969) pioneering paper: "The Social and the Socializing Infant." By now, a sizable body of research has examined reciprocal influence between parent and child. (For overviews, see Lerner & Spanier, 1978; Lewis & Rosenblum, 1974.) However, even with the theoretical impetus to reexamine family patterns of socialization, existing work has some distinct limitations.

First, it has tended to concentrate on very young children—mostly infants. Second, influence from child to parent has mostly been studied with regard to parenting behavior. Brim's (1968) conclusion in his review of research on adult socialization still seems to hold, more than a decade later: "Apparently no one has even asked in a systematic way of a sample of parents whether they perceive their children to have had any influence on their own values, life plans, personal desires, work orientation, or marital relationship" (pp. 213–214).

Very little work exists on influence both up and down generational lines, once children have made the transition into adulthood. However, a recent study of 148 Chicago-area families (Hagestad et al., in progress) found an impressive array of reported influence between adults and their parents. The study examined two kinds of parent–child dyads from each family: aged parent–middle-aged child and young adult child–middle-aged parent. In areas ranging from health practices and diet to political attitudes and views on parenthood, we found numerous attempts to shape the other, both up and down generational lines. In the past, when influence between young adults and their parents was discussed, it was often under the heading of intergenerational *transmission.* As that term suggests, we are back into a unidirectional "social mold" view of influence. Such research, often done in the area of political socialization, has tended to *infer* influence when similarities between parents and children are found in attitudes or political behavior. Seldom, if ever, is there consideration given to the possibility that such similarity may, in part, reflect influence "up," from the child to the parent (Troll & Bengtson, 1979).

This type of research is also peculiarly outcome-oriented. In work on adults and their parents, very little has been done to observe the *process* and *content* of influence, the way observational research has explored interactions between parents and young children. When both sides of the parent–child dyad are adults, socialization often becomes a process of mutual *negotiation* (Bengtson & Black, 1973). As Riley et al. (1972) have pointed out:

> This potential for reciprocal socialization increases as the child gains ability consciously to manipulate his environment. And the greater this potential, the more uncertain the outcome, as the possibility increases that the individual being socialized will succeed in legitimizing his own expectations—both for himself and the socializer—with the result that the goals and role performances for each are redefined. (p. 542)

As can be seen from their statement, these authors suggest that the more resources and rewards a person controls, the more "power" he has in a socialization process. They further suggest that the older children are, the more they have the choice of "opting out," which further represents power behind their expectations.

Most of us would agree that in early parent–child interaction, the dependency of the child and the superior resources of the parent create an imbalanced basis for negotiation. However, several authors, like Riley et al., argue that over the course of the parent–child relationship, we see a shift in "bargaining positions." They suggest that we see a changing asymmetry, from the parent being more powerful, to the child "in control." Brim (1968) states:

> The final outcome of the ever-growing influence of the child on the parent is the *gradual inversion of the relationship between the two,* as it shifts from the initial position in which the parent has complete responsibility and authority to the reverse, at a later age period, when the child has come to assume these same responsibilities of the parental role toward his aging and less able parents. (p. 214, emphasis added)

Bengtson and Kuypers (1971) apply Waller's principle of least interest to parent–child relations and speak of parents' *developmental stake*. These authors argue that after years of investing in their offspring, parents have more at stake in maintaining the relationship than do maturing children. They use the concept of developmental stake to address a fairly consistent finding from research on adults and their parents: that the older member of the pair will seek to minimize conflict and friction and report a more positive affective quality in the relationship than will the younger one (e.g., Bengtson & Black, 1973; Horowitz, 1979; Teeland, 1978). We may add that in the case of an omega parent, there clearly is not the *option* of investing "up," available to younger generations. At the other extreme, the alpha child cannot invest "down." This raises the question of when we might observe a shift in asymmetry patterns—when the child no longer is an *alpha?* In other words, do we tend to invest down in intergenerational links; so that when we enter generational stations where we are both children and parents, the parent role will take precedence? Such questions move us beyond considerations of socialization to much broader issues regarding influence among family members and the general significance of the parent–child tie to both partners in the relationship.

As was seen above, work on the meaning of parent–child bonds shows contrasts between the alpha and omega views. In the former, the focus tends to be on the significance of the relationship for the child; in the latter we discuss the importance of the bond to the parent. Gruda Skard (1965) pointed out quite some time ago that the term *maternal deprivation* should not be limited to descriptions of the child but also be extended to mothers. Her words have often not been heeded: witness recent research on divorce.

A review of such research (Hagestad & Smyer, 1980) found a limited and somewhat outdated view of parents and children. There is recognition that they are important to one another, but the assumption seems to be that the bond is more important for the child. Thus much of the literature on children of single-parent households focuses on what we might call "paternal deprivation." What is *not* considered is the possibility that the *father* may also be deprived, having lost day-to-day involvement in parenting. It appears that unidirectional views of developmental influences between parent and child still dominate research surrounding divorce (Levitin, 1979). No research has looked at *parents of divorce*—middle-aged and old people whose children have gone through marital disruption. In sum, we have not carried the assumption that parents and children are important to one another to its logical extension: that the parent–child bond is important to *both* parties and across the duration of the relationship. Consequently, we have not systematically examined how a divorce on either side of the bond may cause major disruption of expectations and available resources for the other. This brings us to the topic of complex, often unrecognized, ways in which parents and children "produce effects on each other through intangible or indirect means," to use the dictionary definition of influence.

Developmental Reciprocities

As parents and children, we find that our lives are intimately interconnected over a long period of time, that our "growth and self-realization will be shaped by theirs, and development is . . . "the outcome of mutual human cultivation, of the 'shadows' we cast upon each other's lives" (Plath, 1980, p. 288). Such interdependencies, or *developmental reciprocities* (Klein et al., 1979), have been the topic of countless literary works but have generally been neglected by social scientists. In the section on patterns of reciprocity, I briefly discussed the importance of taking a multigenerational view of available family resources—tangible and intangible. Let me briefly mention other ways in which our own life course progress hinges on the progress of others in *their* life-stage tasks.

In any culture, there are shared expectations about "the normal, expectable life cycle" (Neugarten, 1969). These expectations include the nature and timing of role transitions, rights and obligations, and available resources. In addition, families develop their own sets of such expectations (Hagestad, 1981; Sears, 1951; Turner, 1970). Life events which may warrant the label "crises," like divorce, upset these shared expectations. The specific expectancies which are left unmet depend upon life phase or the stage of family development at the time of the event.

Individuals' life course expectations involve not only their own development but also that of significant others whose lives are closely interconnected with their own. This is most clearly seen between parents and children.

Children count on a period of strength—almost invulnerability—in their parents, up to old age. Illness and deaths which upset such expectations represent more severe crises than when they come as expected, on time. Children have stake in their parents' presence and functioning. Recently, a colleague whose parents both died before he was 40 exclaimed: "I'm orphaned!" A listener responded that he was *too old* to be an orphan. The answer came quickly: "But I'm also too young to be next in line!" Translated into the present framework, his statement was "I am not ready to be an omega!" As his comment reflects, our subjective sense of time and aging is closely linked to the lives of family consociates. The bittersweet awareness of such family reflectors is seen in a common statement by parents who are trying to soften it through humor: "How can my children be so grown, when I have not gotten any older?"

Parents also build strong developmental expectations regarding children. Couples plan on "slowing down" when the children have finished school and found work. Often, their personal sense of accomplishment and security is built on the knowledge that the children have mastered *their* life tasks successfully. Erikson's concept of generativity in late adulthood is linked to the "reflected glory" obtained from the younger generation's successful mastery of the tasks of early adulthood (Plath, 1980). When children *do not* leave on time and do not "turn out right," parents are left with a sense of strain and personal failure (Nydegger & Mitteness, 1979; Wilen, 1979). Only when children have been successfully launched into adulthood do parents have the freedom to attend to some of *their* developmental concerns, such as signs of their own aging and old parents' increasing dependency. A major part of what Blenkner (1965) calls filial maturity—the readiness to accept a new dependence from parents—builds on a successful transformation of the relationship with the third generation—the children. Thus in order to better understand one life course, we need to consider at least three sets of parents and children.

Beyond work on early childhood, we know very little about how parents and children mutually shape each other's developmental milieux. We have little knowledge of what adults and their parents expect of one another, especially what they take for granted. A recent study of divorce in middle age (Hagestad & Smyer, 1980) suggests that research on stressful family events, especially when they are unexpected, may help us identify and examine subtle, unspoken expectations which parents and children hold regarding each other. Such expectancies have potentially powerful influence, both in their subjects and their objects.

We have a long way to go before we understand how and when parent–child ties form the warp in the fabric of human lives—the threads connecting alpha and omega: the beginning and the end.

REFERENCES

Adams, B. N. Isolation, function, and beyond. American kinship in the 1960's. *Journal of Marriage and the Family,* 1970, **32**, 575–597.

Ahammer, I. M. Social-learning theory as a framework for the study of adult personality development. In P. B. Baltes & K. W. Schaie (Eds.), *Life-span developmental psychology: Personality and socialization.* New York: Academic, 1973.

Aldous, J. *Family careers: Developmental change in families.* New York: Wiley, 1978.

Aries, P. *Centuries of childhood: A social history of the family.* New York: Vintage Books, 1962.

Baumrind, D. Current patterns of prental authority. *Developmental Psychology Monographs,* 1971, **4** (1), 1–102.

Bell, R. Z. A reinterpretation of the direction of effects in studies of socialization. *Psychological Review,* 1968, **75**, 81–95.

Bengtson, V. L. Generation and family effects in value socialization. *American Sociological Review,* 1975, **40**, 358–371.

Bengtson, V. L., & Black, O. Intergenerational relations and continuities in socialization. In P. Baltes & W. Schaie (Eds.), *Life span developmental psychology.* New York: Academic, 1973, pp. 208–235.

Bengston, V. L., & Kuypers, J. A. Generational difference and the "developmental stake." *Aging and Human Development,* 1977, **2**, 249–260.

Blauner, R. Death and social structure. *Psychiatry,* 1966, **29**, 378–394.

Blenker, M. Social work and family relationships in later life with some thoughts on filial maturity. In E. Shanas & G. F. Streib (Eds.), *Social structure and the family.* Englewood Cliffs, N.J.: Prentice-Hall, 1965.

Bortner, R. W., Bohn, C. J., & Hultsch, D. F. A cross-cultural study of the effects of children in parental assessment of past, present and future. *Journal of Marriage and The Family,* 1974, 370–378.

Brim, O. G. Adult socialization. In J. A. Clausen (Ed.), *Socialization and society.* Boston: Little, Brown, Inc., 1968.

Brody, E. M. The aging of the family. *Annals of the American Academy of Political and Social Sciences,* July 1978, pp. 438.

Firth, R., Hubert, J., & Forge, A. *Families and their relatives.* London: Routledge and Kegan Paul, 1969.

Gelfand, D. E., Olsen, J. K., & Block, M. R. Two generations of elderly in the changing American family: Implications for service. *The Family Coordinator,* 1978, **27**, 395–403.

Glick, P. C. Updating the family life cycle. *Journal of Marriage and the Family,* 1977, **39**, 5–13.

Glick, P. C. The future of the American family. *Current Population Reports,* Series P-23, 1979, No. 78.

Goslin, D. Introduction. In Goslin, D. (Ed.), *Handbook of socialization theory and research.* Chicago: Rand McNally, 1969.

Hagestad, G. O. Problems and promises in the social psychology of intergenerational relations. In R. Fogel, E. Hatfield, S. Kiesler, & J. March (Eds.), *Stability and change in the family.* New York: Academic, 1981.

Hagestad, G. O., Cohler, B., & Neugarten, B. L. *The vertical bond: influence and closeness in three generations of urban families,* in progress.

Hagestad, G. O., & Smyer, M. A. Midlife Divorce: Implications for Parent-Caring. Paper presented at the annual meeting of the American Orthopsychiatric Association, Toronto, Canada, 1980.

Halprin, F. *Patterns of influence between middle-aged children and their parents.* Paper presented at the annual meeting of the Gerontological Society, Washington, D.C., 1979.

Hareven, T. K. Family time and historical time. *Daedalus,* 1977, **106**, 57–70.

Hartup, W. W. Perspectives on child and family interaction: Past, present, and future. In R. M. Lerner & G. B. Spanier (Eds.), *Child influences on marital and family interactions: A life-span perspective.* New York: Academic, 1978.

Hartup, W. W., & Lempers, J. A problem in life-span development: The interactional analysis of family attachments. In P. B. Baltes & K. W. Schaie (Eds.), *Life-span developmental psychology: Personality and socialization.* New York: Academic, 1973.

Hill, R., Foote, N., Aldous, J., Carlson, R., & MacDonald, R. *Family development in three generations.* Cambridge, Mass.: Schenkman, 1970.

Horowitz, S. Middle-aged parents and their young adult offspring: Self perceptions of their affective bond. Paper presented at the annual meetings of the Gerontological Society, Washington, D.C., 1979.

Klein, D. M., Jargensen, S. R., & Miller, B. Research methods and developmental reciprocity in families. In R. M. Lerner & G. B. Spanier (Eds.), *Child influences on marital and family interaction: A life-span perspective.* New York: Academic, 1978.

Knipscheer, C. P. M. *Oude mensen en hun sociale omgeving.* Gravenhage: Vuga-Baekerij's, 1980.

Lerner, R. M., & Spanier, G. B. (Eds.) *Child influences on marital and family interaction: A life-span perspective.* New York: Academic, 1979, pp. 1–22.

Levitin, T. E. Children and divorce. *Journal of Social Issues,* 1979, **35**, 1–25.

Lewis, M., & Rosenblum, L. A. *The effect of the infant on its caregiver.* New York: Wiley, 1974.

Lieberman, G. L. Children of the elderly as natural helpers: Some demographic differences. *American Journal of Community Psychology,* 1978, **6**, 489–498.

Lopata, H. Z. *Widowhood in an American city.* Cambridge, Mass.: Schenkman, 1973.

National Council on Aging. *The myth and reality of aging in America.* New York: Harris.

Neugarten, B. L. Continuities and discontinuities of psychological issues into adult life. *Human Development,* 1969, **12**, 121–130.

Neugarten, B. L. The middle generations. In Ragan, P. K. (Ed.), *Aging parents.* Los Angeles: University of Southern California Press, 1979.

Neugarten, B. L., & Hagestad, G. O. Age and the life course. In R. Binstock & E. Shanas (Eds.), *Handbook of aging and the social sciences.* New York: Van Nostrand Reinhold, 1976, pp. 35–55.

Neugarten, B. L., & Moore, J. W. The changing age status systems. In B. L. Neugarten (Ed.), *Middle age and aging.* Chicago: University of Chicago Press, 1968, pp. 5–21.

Nydegger, C. N., & Mitteness, L. Role development: The case of fatherhood. Paper presented at the annual meeting of the Gerontological Society, Washington, D.C., 1979.

Olsen, H., Trampe, J. P., & Hansen, G. *Familieontakter i den Tidlige* Alderdom, Copenhagen: Socialforskningsinstitutet, 1976, Nr. K4.

Oppenheimer, V. K. The life cycle squeeze: The interaction of men's occupational and family life cycles. *Demography,* 1974, **11**, 227–245.

Parsons, T. The social structure of the family. In R. Anshen (Ed.), *The family: Its function and destiny.* New York: Harper, 1949, pp. 173–201.

Plath, D. W. Contours of consociation: Lessons from a Japanese narrative. In P. Baltes & O. Brim, Jr. (Eds.), *Life-span development and behavior* (Vol. 3). New York: Academic, 1980.

Random House Dictionary of the English Language. New York: Random House, 1969.

Rheingold, H. L. The social and socializing infant. In D. A. Goslin (Ed.), *Handbook of socialization theory and research.* Chicago: Rand McNally, 1969.

Riley, M. W., Foner, A., Moore, M. E., Hess, B., & Roth, B. K. *Aging and society: An inventory of research findings* (Vol. 1). New York: Russell Sage Foundation, 1968.

Riley, M. W., Johnson, M. E., & Foner, A. (Eds.). *Aging and society: A sociology of age stratification* (Vol. 3), New York: Russell Sage Foundation, 1972.

Robertson, J. F. Interaction in three generation families, parents as mediators: Toward a theoretical perspective. *International Journal of Aging and Human Development,* 1975, **6**, 103–108.

Rodgers, R. H. "Improvements in the construction and analysis of family life cycle categories." Western Michigan University, 1962.

Rodgers, R. H. *Family interaction and transaction: The developmental approach.* Englewood Cliffs, N.J.: Prentice-Hall, 1973.

Rosenmayr, L., Family relations of the elderly. *Zeitschrift für Gerontologie,* 1973, Band 6, Heft 4.

Rosenmayr, L. A view of multigenerational relations in the family. Paper presented at the 9th World Congress of Sociology, Uppsala, Sweden, 1978.

Ryder, N. The cohort as a concept in the study of social change. *American Sociological Review,* 1965, **30**, 843–861.

Sears, R. R. A. A theoretical framework for personality and social behavior. *American Psychologist,* 1951, **6**, 476–483.

Shanas, E. Social myth as hypothesis: The case of the family relations of old people. *Gerontologist,* 1979, **19**, 3–9.

Shanas, E., Townsend, P., Wedderburn, D., Friis, H., Milhoj, P., & Stehouwer, J. *Old people in three industrial societies.* New York and London: Atherton and Routledge Kegan Paul, 1968.

Siegel, J. S. Prospective trends in the size and structure of the elderly population, impact of mortality trends, and some implications. U.S. Census, *Current Population Reports,* Special Studies Series, 1979, 23.

Silverstone, B., & Hyman, H. K. *You and your aging parents.* New York: Pantheon, 1976.

Skard, A. G. Maternal deprivation: The research and implications. *The Journal of Marriage and Family,* 1965, **27**, 3.

Skolnick, A. *The intimate environment* (2nd ed.). Boston: Little, Brown, 1978.

Smith, L. Meeting filial responsibility demands in middle age. Unpublished M.A. Thesis, The Pennsylvania State University, 1981.

Streib, G. F. Older families and their troubles: Familial and social responses. *Family Coordinator,* 1972, **21**, 5–19.

Streib, G. F., & Beck, R. W. Older families: A decade review. *Journal of Marriage and the Family,* 1980, **42**, 937–956.

Sullivan, H. S. *Conception of modern psychiatry.* Washington, D.C.: W. H. White Psychiatric Foundation, 1949, pp. 18–21.

Teeland, L. *Keeping in touch.* Gothenburg, Sweden: University of Gothenburg Monograph, 1978.

Townsend, P. *The family life of old people.* New York: Free Press, 1957.

Treas, J. Family support systems for the aged. *The Gerontologist,* 1977, **17**, 486–491.

Troll, L. E. Issues in the study of generations. *Aging and Human Development,* 1970, **1**, 199–218.

Troll, L. E. The family of later life: A decade review. *Journal of Marriage and the Family,* 1971, **33**, 263–290.

Troll, L. E. Is parent–child conflict what we mean by the generation gap? *The Family Coordinator,* 1972, **21**, 347–349.

Troll, L. E. Grandparenting. Paper presented at the annual meeting of APA, New York, 1979.

Troll, L. E., Miller, S. J., & Atchley, R. C. *Families in later life.* Belmont, Calif.: Wadsworth, 1979.

Troll, L. E., & Smith, J. Attachment through the life-span: Some questions about dyadic bonds among adults. *Human Development,* 1976, **19**, 156–170.

Turner, R. H. *Family interaction.* New York: Wiley, 1970.

Uhlenberg, P. Death and the family. *Journal of Family History,* 1980, **5**, 313–320.

Walters, J., & Walters, L. H. Parent-child relationships: A review, 1970–1979. *Journal of Marriage and the Family,* 1980, **42**, 807–822.

Wilen, J. B. Changing relationships among grandparents, parents, and their young adult children. Paper presented at the annual meeting of the Gerontological Society, Washington, D.C., 1979.

BY THE SWEAT OF YOUR BROW

Renee Garfinkel
Clinic for Intellectual Function in the Aging
The Graduate Hospital
Philadelphia, Pennsylvania

Work and rest, the alternating balance, is a rhythm as primal as breathing in and breathing out. Remember the Bible story of Adam and Eve who disobeyed God and forfeited Paradise. Adam was punished for the transgression by the curse of work, Eve condemned to bear and rear children. The punishments pronounced in the book of Genesis to the first couple carried within them the seeds of their redemption: both Eve and Adam were sentenced to labor that is potentially joyful and satisfying, as well as difficult.

But the Old Testament chose to emphasize the curse and the burden of work and man's enslavement to it. Later writers and thinkers took varying perspectives; Karl Marx, for example, chose to think of work as the fountainhead of human dignity. Christian thought looked at work as resistance to the devil's temptation. Hannah Arendt thought of work as merely a stage in human development. No matter how one evaluates work, its undeniable importance in our civilization remains. We know that status, wealth, power, self-esteem, and even health and longevity are intimately related to one's work. One might speculate that work is the single most dominant influence in the life of an adult, touching on everything from speech patterns to friendship networks.

What Is Work? Freud is said to have described a mentally and psychologically healthy person as one who can love and work. It is an elegant criterion of mental health, deceptive in its apparent simplicity. Loving and working are two highly complex concepts that are understood on many different psychological levels simultaneously. The idea of work, like the idea of love, is so formidable that psychologists have generally avoided dealing with the concepts as wholes. We study patterns of attraction, or we examine particular aspects of work and behavior. But we have yet to develop a global understanding of what work is.

Tom Sawyer had an intuitive grasp of the meaning of work in the fence-painting scene when, by altering the context and the values of the situation, he transformed his chore into his friend's pleasure. Work is an activity that is compelled or required, and for which one is paid. For most workers, part of work's definition is that it is *not enjoyable* (see Kahn, 1974). Moving a piano is work; lifting 250-pound barbells is not. Interestingly, when workers were asked to define "occupation," a conceptual shift occurred in the workers sampled. Main defining characteristics of "occupation," along with pay, were enjoyment and the exercise of skills.

Our society offers no meaningful alternative to work:

1. Work defines our position in society. When we meet someone new, the first thing they inquire about (after our name) is, What do you do?
2. Work is the context in which we act out a major part of the human drama in such areas as:
   a. Competition (sales prizes)
   b. Territoriality ("A name on the door rates a Bigelow on the floor.")
   c. Bonding (being a team player)
   d. Nurturing (the mentor relationship)
3. Work is the opportunity for doing, creating, and achieving. A craftsperson and a corporate executive have in common the instrumental tasks of creating whether it be a chair or an empire.

Creating. The element of creating, or producing, is worth analyzing a little more closely, as it plays a pivotal role in the value of work for human adult development. When work lacks this element, it can be a monstrously destructive human effort. We learn this in the most horrifying sense from the myth of Sisyphus. Sisyphus, you will remember, is the pathetic mythical creature who is doomed to eternally roll a boulder up a mountain only to have it inevitably slide back down again. If he had only been able to build a pile of stones on the top for a little while . . . that would have made the difference between despair and satisfaction.

Beyond Creation. Work does provide the opportunities listed above, and more. If we probe beneath the surface of the observable in a person's life, we can find an underlying integrity. One might even look at the entire range of human activity as merely a backdrop for the acting out of unconscious agenda. In this sense, work is a major arena for wrestling with such questions as, Am I worthwhile? Am I alone? What is my purpose?

THE ROLE OF WORK IN ADULT DEVELOPMENT

Until very recently, the issue of work has been ignored by developmental psychological theorists and left to be contemplated by philosophers, theologians, or human factors engineers.

The neglect of the role of work in human development may be understood in the context of a discipline that has traditionally been biased toward childhood and adolescence. Developmental psychology has, until fairly recently, emphasized the early part of life. Observations, research, and theory focused on the early years of life with an implicit assumption that the human being was formed and fixed by about age 20. Work was not an important area of study for developmental psychologists who believed all postadolescent change to be relatively insignificant.

Even the seminal development theorist, Erik Erikson, omitted work completely from his conceptualization of the developmental tasks of adulthood (Erikson,

1950). That omission has recently been corrected by Vaillant (1977) and by Levinson et al. (1978), in their longitudinal studies of healthy adult males. Vaillant suggests a stage he calls *career consolidation* be added to Erikson's developmental sequence. Vaillant's career consolidation is congruent with the concept of developmental tasks and challenges in each stage of life. In middle adulthood the instrumental concerns of doing one's job—skillfully, successfully, appropriately—are pervasive. One of Vaillant's subjects described his 20s as the years in which he learned to get along with his wife and his 30s as the years he learned to do his job. Thus Vaillant's modification of Erikson's schema would suggest that after the adolescent has made some progress establishing an *identity,* the young adult can begin to deal with *intimacy* and establish significant relationships with his peers. With the capacity for relationships established, the middle adult attends to the task of work in the stage of career consolidation. Later on in adulthood, according to Vaillant, he would move on to the task of generativity. In this stage one becomes involved in the growth of the next generation. There is a shifting of concern from one's own career progress and achievement to that of younger people, and a growing involvement with the transmission of traditions and wisdom. In the stage of generativity one distills the value of earlier experience and uses it to help the generation on its way up.

A developmental perspective of work would be congruent with the concept of career. The orderliness of normative life events, including but not limited to work activities, has a strong impact on the quality of an individual life. Hogan (1978) found that men who deviated from the normative order of first completing school, getting a job, and then marrying were more likely to become separated or divorced than men who conformed to the normative sequence. If the developmental theorists are correct and each life stage contains its own agenda of tasks to be accomplished and issues to be resolved, then work must adapt over time to match the changing worker. The idea of career contains the movement necessary to convey the dynamic quality of the influence of work on the life of a person. The structure of a career—its timing, sequence, and progress—as well as its content, contributes to the evolution of the adult. Sociology refers to "occupational socialization" as a process occurring over time in which work experience fosters certain kinds of personal development. Lorence and Mortimer (1980) found that work autonomy (i.e., work that permits self-direction) fosters work involvement. Even mental flexibility, a component of intelligence, can be stimulated or stifled by work (see Kohn, 1973, 1976; Kohn & Schooler, 1978). The personal development engendered by work in turn becomes a factor in promoting the direction, order, and stability of career progress. By generalization and interaction the career trajectory influences the other facets of an individual's life, such as economics and family.

Erikson postulated that developmental process in each stage depended upon the successful resolution of the preceding stage; that is, in order to be able to establish intimate relations one must first have a sense of identity. The person who is confused about who he is could not possibly develop a viable intimate relationship and will remain fundamentally isolated. Similarly, a person who has not moved beyond the stage of intimacy and has never resolved the issues of work, or career consolidation, will be unable in later adulthood to deal with the task of generativity. Developmental psychologists would predict that such a person would stagnate in later adulthood because she would be psychologically unprepared for that stage in life.

Using a stage theory of adult development, let us try to examine some of the dynamics of the role of work in adult development. We find that the tension and flow in this area of human activity, not unlike that in many areas of life, is between the biological realm and the environment in which human beings live. We will see how psychological needs interact with the environment to fashion the growing human person.

Sandy Kaan was a minister's wife. At 21 she was a shy, dependent, reserved beauty who preferred to remain in the background. By age 45, Sandy was an assertive, confident leader of a national nonprofit organization.

What is the relationship between biological limitations and environmental possibilities in adult development?

One universal observation in adult development has been that of "role reversal" in middle adulthood. Men and women, it is said, become more like one another as they age. Women discover their intelligence, imagination, drive, and leadership qualities, and men discover their sense of aesthetics, their gentleness, and their ability to nurture (see Neugarten, 1973). Support for women's increasing instrumentality and men's increasing expressiveness with age was found in primitive cultures, where postmenopausal women were permitted to sit in council with the elders. (Neugarten & Guttman, 1958). We seem to have cross-cultural backing for a phenomenon that is frequently discussed in our culture: woman completes her tasks as homemaker and nurturer of the young and becomes a high-powered executive. It would appear to be a "natural" progression of psychological development, and indeed for many women, like Sandy Kaan (above), circumstances *do* follow that sequence.

On the other hand, a study of women physicians (Cartwright, 1977) followed 10 to 12 years after medical school found that they became more effective in their work and also became more aware of their loving, nurturing feelings. These women spent their early career training in competition and academic achievement. The phenomenon of expressiveness in this sample of women later on in adulthood may perhaps be related to delayed childbearing or to the effects of hands-on experience in the practice of medicine compared to their earlier theoretical training. Whatever the reason, the differences in pace and emphasis in the psychological development of this sample of physicians compared to other samples of adult females indicate the interweaving of environmental and biological influences in normal adult development. But no research known to the author has found a well-functioning adult who completely skipped the stage of career consolidation.

In the not so distant past, a very common psychiatric diagnosis would be given to women over 40 experiencing their first psychiatric disorder. That diagnosis was "involutional melancholia," a profound depression almost exclusive to postmenopausal women. This depressive disorder is characterized by the absence of precipitating environmental events, as well as the absence of an apparent biological basis. Perhaps involutional depression is the expression of the stagnation experienced by aging women who never had the opportunity to achieve or to create in some kind of work. If that is the case, then we would expect the incidence of midlife depression to decline as our culture permits more women to develop, achieve, and grow through all of life's stages.

Retirement: The Challenge of Generativity

Ben Marks had built a successful career in a large, multinational corporation. From clerk to engineer he served the company and it rewarded him well. Now an executive in his mid-fifties, Ben looks ahead toward retirement and he feels terrified.

We have been looking at adult development as a series of psychological choices and positions taken within the context of a given set of circumstances. Sometimes the set of circumstances and their constraints are primarily biological, as with the tendency for young adult women to develop their nurturing skills because of the biological imperative of bearing and caring for young children. When we come to the end of the stage we have been referring to as career consolidation, we find constraints that are largely sociological rather than biological. It is society that determines a time for the adult to withdraw to a less central role in the work world.

Retirement as an institutionalized life transition is a fairly recent phenomenon. In times past people worked at their trade until their health and their strength failed them. This is no longer the case. Studies on the successful retiree and on the different ways in which people shape their generative and retirement years are still relatively few. We know that in later adulthood, as in earlier periods in life, one has an advantage if one is healthy, economically secure, and well cared for. Beyond that, it appears that people who can maintain some involvement in their primary interest and with their primary social groups are people who retire well. Academicians who can still write, research, and study the field of their choice flourish in retirement. McPherson and Guppy (1979) found that people who had involvement in expressive types of organizations (an example of such a group might be a barbershop quartet) and who had a leisure orientation were likely to do well in retirement. This is not surprising when one considers that the active involvement in a leisure pursuit or in a voluntary organization provides a setting in which free time can be purposefully enjoyed. We have not yet explored how retirement is used by people for whom, for example, a preretirement minor interest grows into a major involvement, or people who choose other, less conventional, retirement lifestyles. The role of contemplation or inner-directed activities in late life has yet to be studied.

Just as in an earlier stage of development one may respond to the challenge of "intimacy versus isolation" by forming a great variety of interpersonal relationships, so, too, there are many different ways in which to be generative. The elder statesman is generative, and so is the retiree planting her garden and the foster grandparent, as well. Retirement is the next stage which needs to be explored by observing creative and vibrant adults who continue to fashion themselves out of the increased freedom and challenge that later adulthood offers.

Work as Environment

The work environment can be conceptualized as a stimulus to which one is exposed over a long period of time. Because of the prolonged exposure, the work environment has a pervasive and probably cumulative effect on adult development. Work exerts on enormous multifaceted influence on us simply as context. We review and judge one another within a particular setting, and to a surprising extent, the context determines what we see and consequently who we become.

This is illustrated in a study by Shinar (1978) which found that men and women are judged differently on such qualities as leadership and interpersonal adjustment depending upon the sex appropriateness or inappropriateness of their occupations. We see that work is the setting in which the worker is perceived in a certain way quite apart from her actual behavior in that setting: where the worker is, not only what she does, determines who she is judged to be by others. Feedback from other people is a vital source of information from which we learn our value. Growth and learning can take place when feedback is related to behavior, but when peers, supervisors, or subordinates respond to prejudiced attributes instead of to a worker's actual performance, the result is confusion and anger. When one is responded to in ways that bear little relationship to the content or the quality of one's work, a feeling of helplessness and despair can result. The sense of helplessness, if allowed to continue unabated, can ultimately contribute to depression. Thus the influence of work pervades our self-concept, even below the level of consciousness.

Work and Health

Joan Paulson is a divorced mother of two school-aged boys. A 35-year-old high school graduate, Joan works as a secretary in a brokerage house. Although she has no known medical condition and considers herself to be healthy, Joan takes painkillers on the average of 3 times a week.

From black lung prevention in miners to the installation of jogging tracks in executive suites, the work world has been increasingly accepting its responsibility for the health and longevity of workers. Sometimes the relationship between work and health appears obvious, as in the dangers of being a firefighter or a member of the bomb squad. But most of the time the relationship between work and health is neither clear nor immediate. (It took us 20 years to learn the dangers of working with asbestos.) Work affects our physical and mental health, our family lives, and our personalities in many subtle ways. We have only recently discovered the role of work stress on health. Not surprisingly, a good deal of the research in the area of work stress and health has been directed to the stresses experienced by highly visible, high-status workers such as male executives who are at risk for coronary heart disease. Although scientists and managers may share comparable affluence, it is the manager who is more likely to have a heart attack. Stress in the work place is the main source of the difference. Air traffic controllers are another well-known and well-paid group of people whose work stress has a profoundly destructive effect on their health.

Occupational stress and its health effects are not limited to well-paid professions—low-paid, monotonous, pressured work can be dangerous to one's health, as well. The famous Framingham heart studies found recently that it is not the high-achieving, hard-driving woman who is at risk for heart attacks, but rather her secretary sister with the hard-driving boss (Haynes & Feinlieb, 1980). One incisive study in the area of occupational stress and health was conducted in Denmark by sociologist Tage Kristensen and was called *Women's Work and Health* (1978). This was a follow-up on an earlier (1973) survey of the entire Danish national labor organization which found that women workers have more health disorders, more stress, and do not thrive as well as men. In her intensive study of working women's health problems, Kristensen found women workers concentrated in sedentary, monotonous, and fast-paced jobs (for example, seamstresses, and cashiers in supermarkets). Men in those same industries were far more likely than women to have varied work, such as supervising, inspecting, or repairing. The women's monotonous, rapid, and sedentary work was found to have a deeply insidious effect on health and resulted in chronic conditions, such as headaches, back pain, swollen ankles, and feeling stressed and nervous. One-third to one-fourth of the sample studied were frequent and regular users of pain killers! These chronic medical conditions are not recognized as occupational diseases as yet, so when the workers become too debilitated to work, they are not compensated with disability or workers' insurance.

The deadening effects of fast-paced, repetitive, sedentary, and mindless work have long been associated with alienation (see Geyer, 1972) and other assorted social evils. We now know that such work is an actual health hazard. Joan Paulson (the case above) is a victim of an occupational disorder. Like her brother, the shipbuilder who worked with asbestos, she is unwittingly exposing her system to stress it cannot accommodate. Over time, with continued exposure, Joan's pains will increase and her general health will deteriorate. If the work conditions don't change, Joan will probably have to retire early from work and will not be eligible for any sort of financial compensation for her losses.

Physically, mentally, emotionally, even intrapersonally and interpersonally, work pervades our lives. In many ways work can be compared to the family in its developmental significance: even by its *absence,* work affects us. The meaning, impact, and stress of unemployment extracts a significant toll in our culture. We are only beginning to untangle the complex dynamics of work in our lives. Indeed, the relative neglect of the developmental importance of work may in itself demonstrate that importance; perhaps we have been so intimately bound to our work lives that we have been unable to differentiate and perceive its effects. Work is the medium in which we spend most of our adult lives; we are formed by our work.

REFERENCES

Arbejdsmiljøgruppens rapport nr. 2. Arbejdsmiljøfondet, Copenhagen, Denmark, 1973.

Cartwright, L. K. Personality changes in a sample of women physicians. *Journal of Medical Education,* 1977, **52**, 467–474.

Erikson, E. H. *Childhood and society.* New York: Norton, 1950.

Geyer, R. F. *Bibliography alienation* (2nd ed.). Amsterdam: Netherlands Universities' Joint Social Research Centre, 1972.

Haynes, S. G., & Feinleib, M. Women, work and coronary heart disease: Prospective findings from the Framingham heart study. *American Journal of Public Health,* 1980, **70**, 133–141.

Hogan, D. The variable order of events in the life course. *American Sociological Review,* 1978, **43**, 573–586.

Kahn, R. On the meaning of work. Paper presented at the Cornell Conference on Mental Health, 1974.

Kanter, R. M. The job makes the person. *Psychology Today,* 89–90, 1976.

Kohn, M. L. Occupational experience & psychological functioning: An assessment of reciprocal effects. *American Sociological Review,* 1973, **38**, 97–118.

Kohn, M. L. Occupational structure and alienation. *American Journal of Sociology,* 1976, **82**, 111–130.

Kohn, M. L., & Schooler, C. The reciprocal effects of the substantive complexity of work and the intellectual flexibility: A longitudinal assessment. *American Journal of Sociology,* 1978, **84**, 24–52.

Kristensen, T. S. *Women's work and health.* Publication 9, Institute for Social Medicine, Copenhagen University, Denmark, 1978.

Levinson, D. J., with Darrow, C. N., Klein, E. B., Levinson, M. H., & McGee, B. *The seasons of a man's life.* New York: Knopf, 1978.

Lorence, J., & Mortimer, J. T. Work experience and work involvement. *The sociology of work and occupations,* in press.

McPherson, B., & Guppy, N. Pre-retirement lifestyle and the degree of planning for retirement. *Journal of Gerontology,* 1979, **34**, 254–263.

Neugarten, B. L. Personality change in late life: A developmental perspective. In C. Eisdorfer, & M. P. Lawton, (Eds.), *The psychology of adult development and aging.* Washington, D.C.: American Psychological Association, 1973.

Neugarten, B. L., & Guttman, D. Age, sex roles and personality in middle age: A thematic apperception study. *Psychological Monographs,* 1958, **72**, 1–33.

Shinar, E. H. Person perception as a function of occupation and sex. *Sex Roles,* 1978, **4**, 679–693.

Vaillant, G. E. *Adaptation to life.* Boston: Little, Brown, 1977.

AGING

30

MODERNIZATION AND AGING

Gordon E. Finley
Professor of Psychology
Florida International University

The purpose of this chapter is to examine modernization theories of aging in light of the available anthropological, sociological, and psychological evidence. Such theories and data are of great relevance not only to the developing world today but also to our immediate future. It must be noted at the outset, however, that the available data bases are uneven. In particular, cross-cultural psychological studies of aging are striking by their scarcity (Delgado & Finley, 1978; Finley, 1981; Gutmann, 1977, 1980; Heron & Kroeger, 1981; LeVine, 1978). By contrast, comparative anthropological and sociological aging studies began to appear in considerable number during the 1970s (Fry, 1980; Keith, 1980; Maddox, 1979; Missinne & Seem, 1979).

Thus this chapter will draw extensively upon the anthropological and sociological literatures. It will begin with the population profiles of the more and less developed regions of the world. This will be followed by a presentation of two theories of modernization and a consideration of these theories in light of three sets of evidence: ethnographic, demographic, and cross-cultural surveys of attitudes toward the aged. It is hoped that this chapter will spark interest in cross-cultural studies of aging among psychologists.

POPULATION PROFILES: MORE AND LESS DEVELOPED REGIONS OF THE WORLD

It frequently is helpful to begin with an image of the concrete realities under consideration. The population profiles presented in Figure 30.1 provide one such image. An extended discussion of worldwide demographic changes and their impact on the aged may be found in Hauser (1976).

The lower portion of Figure 30.1 represents what we traditionally think of as a

---

I would like to thank the Florida International University Foundation for a grant to undertake the literature review and Marta Pizarro for skillfully gathering the materials. I also am indebted to Anthony Glascock and Erdman Palmore for their comments on an earlier draft.

512 Modernization and Aging

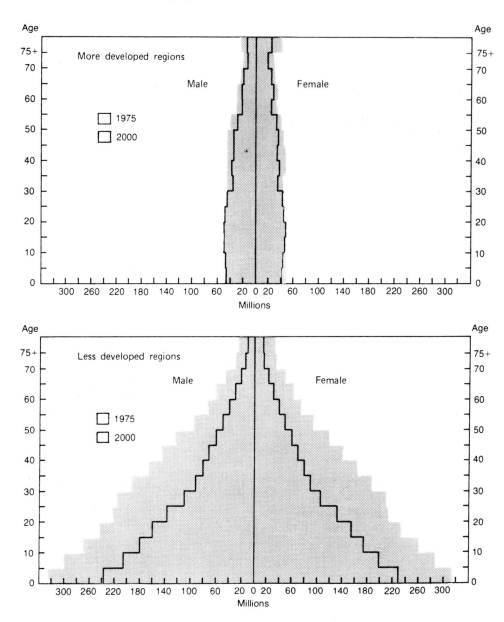

Figure 30.1. Age-Sex Composition of More Developed and Less Developed Regions, 1975 and 2000: Medium Series (United States Bureau of the Census. *Illustrative Projections of World Populations to the 21st Century,* 1979, Figure E.).

population pyramid. With males recorded on the left half of the figure and females on the right half, with actual data for 1975 indicated in black and with projections for the year 2000 indicated in gray shading, we find a nearly perfect pyramid with large numbers of children at the bottom and monotonically decreasing numbers of persons at each succeeding cohort. The lower portion of the figure refers to the less developed regions of the world.

The upper portion of Figure 30.1 represents the same data for the more developed regions of the world. Here, however, we do not find a population pyramid. Rather, we find a population rectangle with strikingly similar numbers of people at all stages of the life cycle. The population rectangle represents nations which today are at the most advanced stages of modernization, such as England, Sweden, and the United States. The population pyramid represents nations which today are at intermediate stages of modernization, such as India, Greece, and Bangladesh. Profiles for more technologically primitive societies are not shown since there currently is some dispute regarding the proportion of the population over 65 for both contemporary and historical groups.

The following historical overview helps to understand these profiles in light of the relationships among technological change, demographic change, and the implications of these changes for the present and future aged.

> Another revolution occurred with mechanization and industrialization within the last 750 years, and again the population grew, this time past the billion mark. More recently, it is generally recognized that sanitation, pesticides, and other public health measures have dramatically increased population over threefold during the last 100 years and thus have produced a fourth demographic revolution, by lowering infant mortality and introducing various life-prolonging practices. . . . another demographic revolution, based upon the shared knowledge that world population growth must be stabilized, is beginning to take shape. If it gains momentum, and there is a dramatic decline in fertility of many populations throughout the world, then problems of the aged will grow proportionately as fertility declines, and, of course, the aged population will grow even faster as longevity is extended. (Katz, 1978, p. 6)

Does the future hold an inverted population pyramid?

IMPACT OF MODERNIZATION ON THE AGED

Modernizing Nations

". . . the theory holds that with increasing modernization, the status of older people declines" (Cowgill, 1974, p. 124). Cowgill defines modernization as:

> . . . the transformation of a total society from a relatively rural way of life based on animate power, limited technology, relatively undifferentiated institutions, parochial and traditional outlook and values, toward a predominantly urban way of life based on inanimate sources of power, highly developed scientific technology, highly differentiated institutions matched by segmented individual roles, and a cosmopolitan outlook which emphasizes efficiency and progress. (1974, p. 127)

While, in a few nations, the process was gradual and self-generating, today it has become largely a process of abrupt imitation. Cowgill (1974) argues that modernization consists of the following four factors which cause a decline in the status of the aged.

Health technology first results in a younger population through the control of neonatal, infant, and child deaths, but ultimately it leads to a reduction of births and an aging of the population. Increased longevity leads to generational competition for jobs and scarce resources, retirement, and a lowered status for the elderly, in part, because of the work ethic.

Economic technology first affects the urban centers and the young who have migrated there to take up the new, more prestigious, and better-paying jobs. This creates job obsolescence for the aged, retirement, lower income, and a lower status for the elderly because of the work ethic and the new values of efficiency and progress.

Urbanization separates work from the home and youthful urban migrants from rural, traditional parents. This reduces the frequency and intimacy of contact, the amount of daily interdependence, and places the aged parents in a peripheral position relative to the nuclear families of their adult children. Upwardly mobile children surpass their parents, creating an inversion of status.

Education and the elimination of illiteracy explicitly favor the young. The young now have the valued knowledge of the new society. This again creates a status inversion between youth and their parents and a widening generation gap.

Agrarian Societies

A second version of modernization theory was formulated by Press and McKool (1972). It focuses on the positive status and treatment of the aged found in agrarian, peasant societies prior to advanced levels of modernization. They argue that the aged enjoy high status when the following conditions exist: a socioeconomically homogeneous community; a system wherein sequential roles entail progressively higher responsibility and authority; a life-cycle sequence of roles characterized by continuity; a system wherein the control of important family or community resources and processes are in the hands of the older members; a system wherein the old may engage in valued, useful, and important activities; and a system wherein the extended family is a viable residential or economic unit.

They also argue that there are four factors which generate prestige for the aged: an advisory factor which is reflected in the degree to which the advice of the elder actually is heeded; a contributory factor which is reflected in the extent to which elders still actively participate in traditional ritual, domestic, or economic activities and make valued contributions; a control factor which reflects the degree of direct control exercised by the aged over the behavior and welfare of others; and a residual factor which reflects age-graded prestige derived from statuses previously held.

A classic example of the economic role of the aged in agrarian and herding societies is provided by Simmons (1945):

> Old men among the Hopi tended their flocks until feeble and nearly blind. When they could no longer follow the herd, they worked on in their fields and orchards, frequently lying down on the ground to rest. They also made shorter and shorter trips to gather herbs, roots, and fuel. When unable to go to the fields any longer they sat in the house or kiva and spun, knitted, carded wool, or made sandals. . . . Old women would cultivate their garden patches until very feeble and "carried wood and water as long as they were able to move their legs." They prepared milling stones, wove baskets and plaques out of rabbit weed, made pots and bowls from clay, ground corn, darned old clothes, cared for children, and guarded the house; and when there was nothing else to do, they would sit and watch the fruit drying in the sun. The old frequently expressed the desire to keep on working until they died. (pp. 85–86)

MODERNIZATION THEORY: CURRENT CONSIDERATIONS

Ethnographic Data

Since modernization theory was derived from ethnographic records, it is not surprising that most of the ethnographic data are consistent with it. Thus only three portions of this literature will be considered: the work of Cowgill and Holmes (1972); a new perspective gained from the work of Simmons (1945); and some recent criticisms of the earlier literature (Glascock & Feinman, 1980).

In their contribution to the ethnographic evidence, Cowgill and Holmes (1972) invited eight anthropologists, seven sociologists, two psychologists, and one social worker to write chapters on the aged in societies familiar to them. The societies covered the range from the agrarian to the most modern. On the basis of the data yielded by these chapters, Cowgill and Holmes (1972) found evidence for 8 universals and 22 variations in the status of the aged. The 22 variations showed a declining status with increasing modernization. These 22 variations constitute a fruitful set of hypotheses for further cross-cultural testing. Some examples are: a person is classified as old at an earlier chronological age in primitive society; modern societies have higher proportions of women and especially widows; the status of the aged declines as their numbers and proportions increase; the status of the aged is inversely proportional to the rate of social change; stability of residence favors high status of the aged; retirement is a modern invention; the responsibility for the provision of economic security for dependent aged shifts from the family to the state in modern societies; and the individualistic value system of Western society tends to reduce the security and status of older people (Cowgill & Holmes, 1972, pp. 322–323).

Cowgill and Holmes (1972), however, did not take as their starting point the earliest possible point on the modernization continuum. They began with agrarian and herding societies, rather than the more technologically primitive groups of hunters, fishers, and collectors. Simmons (1945), however, believed that the treatment, and perhaps the status, of the aged were different in these two categories of nonindustrial societies. Myerhoff (1978) summarizes Simmons' views succinctly:

> . . . the simplest societies are not the best for the aged, since these societies are often (but certainly not always) found in harsh environments with very rude technological development. These foraging and hunting or fishing societies are often marginal in terms of subsistence and simply cannot afford to support any sizable body of people that can no longer support itself. . . . It is in the middle-range societies—rural, agrarian, peasant communities—that the aged are most useful and sometimes (again, not always) because of this, most secure. There are opportunities in these settings for the aged to perform many tasks requiring manipulative skill rather than heavy manual labor, and thus they are more able to hold their own. (p. 163)

Thus when the technological level is extended downward to the most primitive hunting, gathering, and fishing societies, a curvilinear relationship between modernization and the treatment of the aged emerges. In hunting, fishing, and collecting societies, the aged appear to enjoy a high status, but their survival may be precarious. In stable agrarian and herding societies, by contrast, the role, status, and treatment of the aged reach their zenith, to be followed by precipitous declines with further modernization.

Although the above patterns emerge from the writings of Cowgill and Holmes (1972) and Simmons (1945), greater complexity is suggested by more recent studies. For example, Glascock and Feinman (1980) have suggested that some of the above conclusions may be in error due to a failure to make the following four distinctions: the status of the aged versus the actual treatment of the aged; attitudes toward the aged versus the actual treatment of the aged; the intact aged versus the decrepit aged; and, finally, there may well be fluctuations in the treatment of the aged within the same society under different social, economic, or nutritional conditions.

In short, while there appears to be a consensus that the aged enjoy a high status in nonindustrial societies, there has yet to emerge a consensus regarding the treatment of the aged in nonindustrial societies. For example, even among agrarian societies, about whom there would appear to be a consensus in the literature, Glascock (personal communication, 1981) suggests that societies which rely on grain crops have a much higher level of death-hastening behavior than societies which raise root crops. For the purposes of this chapter, however, the broad outlines of the Cowgill and Holmes (1972) and Simmons (1945) work will be followed.

Finally, since urbanization is important to modernization theory, it should be noted that most studies *within* modern westernized nations show that the status of the aged declines with increasing urbanization. Summaries of this literature may be found in Burgess (1960), Gutmann (1977, 1980), von Mering and Weniger (1959), and Youmans (1967).

Demographic Data

The work of Simmons (1945) extended the linear decline model of Cowgill (1974) downward to a curvilinear relationship. The work of Palmore suggests that the curvilinear relationship may become a cyclical relationship when the highest levels of modernization are reached.

The issue addressed by Palmore was the *relative* status of the elderly and young adult populations in the same society. His measure is the equality index which is "the proportion of the aged who are similar to the same proportion of non-aged" (Palmore & Whittington, 1971, p. 85) on a variety of socioeconomic dimensions. Status is indexed by socioeconomic position.

Palmore and Whittington (1971) undertook an important test of modernization theory by examining longitudinal trends in the United States for the years 1940–1969. They concluded that "the evidence from the past thirty years in the United States supports the theory that the relative status of the aged tends to decline in industrial society" (p. 90).

Although modernization theory can be tested definitively only by longitudinal studies, Palmore and Manton (1974) undertook a cross-national test with 31 nations with *available data*. Their goal was to determine whether or not the relative status of the aged declined with increasing modernization across these 31 nations and, if so, which indices of modernization best predicted the outcome.

One index of modernization, percentage of the labor force engaged in agriculture, yielded a linear decline in the equality index for employment across these 31 nations. With fewer people engaged in agriculture, the rates of retirement and unemployment among the elderly were higher. It is noteworthy that this conclusion meshes nicely with that of Simmons (1945) who suggested that the treatment of the aged was best in agrarian societies as well as with that of Cowgill and Holmes (1972) who took agricultural societies as the zenith from whence came the fall with increasing modernization. Clearly, the centrality of agriculture to the total society is a major determinant of the role and status of the aged.

The most interesting finding of Palmore and Manton (1974), however, was the suggestion that at the most advanced levels of modernization, the relative status of the elderly would begin to rise again after the nadir reached either during or following the most rapid rates of modernization. Using indices such as gross national product per capita, percentage of the population in higher education, and percentage of youth in school, they found that the employment equality index, the education equality index, and the occupation equality index, respectively, declined across these 31 nations in the early stages of modernization. However, at the highest levels of modernization, there was evidence that the equality indices began to rise again.

Further evidence for a final rise in the equality indices was obtained by Palmore (1976) in his continuation of the Palmore and Whittington (1971) longitudinal analyses and his projections to the year 1990. Palmore suggests that the period around 1967 appears to have been a turning point in the United States and that "the relative status of the aged in health, income, occupation, and education is rising and probably will continue to rise for the rest of this century" (p. 301). This rise is attributed to a slowing rate of change at the highest levels of modernization.

> The dramatic rate of improvement in health and longevity achieved in the first half of this century has decreased. Increases in purchasing power of income has leveled off. The massive shifts from rural to urban occupations, and from lower status to higher status occupations have been reduced. Educational attainment has begun to level off. All of these diminished rates of change have the effect of reducing the discrepancies between older and younger generations so that the future older generation will not lag so far behind the younger. As a result, the status of the aged relative to younger people begins to rise. (Palmore, 1976, p. 301)

Palmore emphasizes the necessary role of governmental programs involving income transfer, health care, housing, and food sharing which have helped to create these changes in both the absolute and the relative status of the elderly. It is important to note, however, that Palmore argues only that the elderly begin to move toward equality with the young, not that they return to the high statuses enjoyed in simpler agrarian societies.

Palmore's emphasis on rate of change as a crucial variable receives support from a cross-ethnic survey of psychological well-being. Datan, Antonovsky, and Maoz (1981) surveyed attitudes toward middle age and menopause in five ethnic groups in Israel. A central finding was a curvilinear relationship between self-reported psychological well-being during the midlife transition and cultural stability. Psychological well-being was highest for the two extreme and most stable samples, the most traditional Arab women and the most modern Central European women. It was lowest for the Persian women who both were at the midpoint of the modernization continuum and who experienced the most change. Psychological well-being was intermediate for the two remaining intermediate samples, the Turks and the North Africans. Gutmann (1980) provides further material on the relationship between modernization and the mental health of the aged.

In summary, if the cross-national results of Palmore and Manton (1974) and the longitudinal results of Palmore and Whittington (1971) and Palmore (1976) are replicated with longitudinal studies in other nations, then a cyclical model of modernization may come to replace the decline model of modernization outlined by Cowgill (1974). What will happen in the future presumably will depend on future rates of change.

Cross-Cultural Attitudinal Data

One of the most direct measures of the status and prestige of the aged are attitudes held by younger members of the society toward them. Regrettably, only two cross-cultural surveys of attitudes toward the aged were found. Given their potential importance, however, they will be examined in detail.

Bengtson, Dowd, Smith, and Inkeles (1975) undertook a secondary analysis of 3 items on aging from a 438-item modernity questionnaire used in an earlier study (Inkeles and Smith, 1974). The sample consisted of 5450 males between the ages of 18 and 32 from 6 developing countries (Argentina, Bangladesh, Chile, India, Israel, and Nigeria). The present reanalysis focuses on the two items used across all six nations: "Some people say that a boy learns the deepest and most profound truth from old people; others say that a boy learns most from books and in school. What is your opinion?" and "Some people look forward to old age with pleasure, while others dread (fear) the coming of old age. How do you personally feel about the coming of old age?" The results to be presented consist of the percentage of the sample in each nation which agreed with the most traditional or proaging responses: "a boy learns most from old people" and "look forward to old age." Since the responses to both questions were similar across nations, they were averaged. Thus the data to be presented in Figure 30.2 consist of the average percentage of the total sample in each nation which agreed with the traditional or proaging response.

Arnhoff, Leon, and Lorge (1964) selected 100 items reflecting generally negative stereotypes, misconceptions, and overgeneralized beliefs about the elderly from the widely used attitude scale of 137 items developed by Tuckman and Lorge (1953). Respondents check either yes or no to indicate whether or not the statements generally are true of "old people." The samples ranged from about 180 to 420 college students in each of the following nations: Greece, India, Japan, Puerto Rico, Sweden, the United Kingdom, and the United States. An additional study by Sharma (1971) replicated this study with 200 college students in India and will be grouped with it. The data originally were scored such that a high overall percentage agreement score reflected negative attitudes toward old people. To obtain an index similar to the Bengtson, et al. index, such that a high score would represent a positive attitude, the original results were subtracted from 100% to obtain the revised overall percent agreement score shown in Figure 30.2.

Following Palmore, the present analysis sought an index of modernization which would order the cross-national attitudinal results in a meaningful way. Reviewing the indicators of economic development for the period shortly before both surveys were undertaken (given in Table 4-1 of the original Inkeles & Smith, 1974, study) it was found that "wage and salary earners as a percentage of the working-age population" best ordered the Bengtson et al. results. Similar data for the Arnhoff et al. nations then were obtained. The result of these endeavors, relating attitudes toward the aged to societal modernization, is presented in Figure 30.2.

The most striking conclusions to be drawn from Figure 30.2 are twofold: first, the modernization index meaningfully and regularly organizes the attitudinal data for both studies; and second, the results of the two studies are diametrically opposed. The Bengtson et al. results (dashed line) show that favorable or traditional attitudes toward the aged decline dramatically with increasing modernization. The results of the Arnhoff et al. study (solid line) show that attitudes toward the elderly improve markedly with increasing modernization. Both studies have broad ranges on the modernization index as well as on the percent agreement scale. However, the studies still yield opposite conclusions within the range of overlap in the middle of the modernization index.

How then to resolve the apparent disagreement? The most likely explanation lies with the samples or the instruments. Bengtson et al. used rural cultivators (15%), urban nonindustrial workers (15%), and urban industrial workers (70%) in each country. Arnhoff et al. used college students. Since modernization impacts differ-

Figure 30.2. Attitudes toward the Aged as a Function of Modernization.

entially on various segments of the population, different levels of education might well produce different attitudes, even within the same range of societal modernization.

The instruments used in each study also are deficient. Bengtson et al. used only two items. One requires a choice between alternatives on different dimensions: a boy learning the deepest and most profound truth from old people (perhaps wisdom) versus a boy learning most from books and school (perhaps instrumental). Further, the item very likely taps the *fact* that in primitive society a boy *does* learn most from the old, while in modern society he *does* learn most from school. The second item asks how the respondent feels about his *own* old age, not about old people.

Kogan (1979) has criticized the Tuckman and Lorge (1953) instrument used by Arnhoff et al. for containing items of fact and belief as well as attitudes per se. Additionally, some items are phrased positively and some negatively, leading one to wonder about the interpretation of a total score.

Given these shortcomings, it is difficult to know which set of results to believe. One would like to believe the Arnhoff et al. results because they hold hope for the future. They suggest that attitudes toward the aged improve at the most advanced stages of modernization and thus are consistent with the basic conclusions of Palmore. On the other hand, the results of Bengtson et al. are based upon extraordinarily large samples and provide textbook-perfect support for the theory of Cowgill.

It also has been suggested (Palmore, personal communication, 1981) that two of the samples in the Arnhoff et al. study have unique characteristics. If one assumes that the Puerto Rican sample has an unusually positive response, because of their close association with the United States and that the Japanese sample has a special tradition of great respect for the aged, then a curvilinear relationship may yet be seen in the present data. Other nations in the middle of the modernization continuum, however, must be sampled to confirm this. Further, attitudes, like modernization, may be multidimensional. In any case, it is clear that additional cross-cultural studies of attitudes toward the aged are needed.

The attitudinal issues discussed so far, however, were not the main issue addressed by the Bengtson, et al. paper. Their principal interest was in the distinction between analyses at the societal level (modernization) and at the individual level (modernity). They sought to determine whether or not modernization could predict attitudes both in terms of national differences and in terms of individual differences within nations. The within-nation index of personal modernity was one's exposure to modernizing influences as a function of occupation (urban industrial workers; urban nonindustrial workers; rural cultivators). As outlined earlier, they found clear relationships between increasing levels of societal modernization and decreasingly favorable attitudes toward the aged. However, and crucially, they did *not* find the same relationship *within* each of the six nations.

While the issue they raise is important, their results are not conclusive. Not only was the within-nation range of variability extremely restricted compared to the between-nation variability, but individual modernity was indexed indirectly by occupation.

In summary, our analysis has suggested a number of issues which should be considered in future cross-national surveys of attitudes toward the aged: (1) the targets of the attitudes must be specified (young–old vs. old–old or intact vs. decrepit); (2) attitudes must be distinguished from, but related to, behaviors toward the aged; (3) two or more reliable and valid instruments should be used (Kogan, 1979); (4) attitudes of the aged themselves ought to be sampled; and (5) recent rate of change in both individuals and nations should be examined. Perhaps the latter might fruitfully be studied with groups currently acculturating to the United States (e.g., Finley & Delgado, 1979).

Personal Note

While preparing this chapter, I had the opportunity to speak to a senior-citizens group on Miami Beach. The members mostly were Jewish, and many had spent their childhoods in Russia or Eastern Europe. My topic was the gist of this chapter. Afterward, I asked the audience to tell me what it was like to be old when they were children compared to what it is like to be old today.

I was surprised by the reponses. Contrary to the content of the current theories and data on modernization, virtually all the comments focused on changes in *affective* relationships among family members then and now. *Then* consisted of feelings of belonging, warmth toward parents and relatives, and emotional closeness. *Now* consists of the absence of these. Perhaps future research would benefit from a consideration of changes in affective relationships as a function of modernization. This seems to be uppermost in the minds of at least one group of today's aged.

Future Research

A number of issues have arisen in the present review which should be taken into consideration if future research is to benefit from the mistakes of the past. Foremost among these is the notion that neither modernization nor aging are monolithic, holistic, uniform, or unidimensional entities. Rather, both are multidimensional. Thus if modernization theory is to be retained as a functional theory for cross-cultural research, future research must seek to relate the many dimensions of modernization, independently, to the manifold dimensions of aging. Such a task is going to be extraordinarily complex, and at some future time, modernization theory may well be replaced by independent variables comprising some subset of the dimensions of modernization. Historians have been particularly critical of the breadth and inclusiveness of modernization theory (Achenbaum & Stearns, 1978; Laslett, 1976).

Second, even at the level of the present review, the relationship between modernization and the status and the treatment of the aged appears to be cyclical rather than linear. Further, both the magnitude and timing of the cycles would appear to be determined by a host of yet-to-be-specified historical, cultural, and technological factors. Presumably, the rate of change in modernization will determine whether the cyclic pattern continues in the future or whether it levels out.

Third, it has become clear that at least four distinctions must be made in future research: the status of the aged versus the treatment actually accorded them; attitudes toward the aged versus the treatment actually accorded them; the intact aged versus the decrepit aged; and the possibility that multiple treatments may be afforded the aged within the same society under differing social, economic, and nutritional conditions. These distinctions likely will increase the complexity of future conclusions.

Fourth, it is clear that comparative psychological data are so scarce that virtually any sound cross-cultural study of any topic reviewed in Birren and Schaie's (1977) *Handbook of the Psychology of Aging* would constitute a contribution. Within the past decade, psychologists have come to recognize the impact of generational or cohort factors on aging processes. Perhaps the decade of the 1980s will see this perspective expanded to include cultural factors. Without such a perspective, it will be impossible to know whether or not the findings obtained in Western societies are universal or particular.

Finally, the high level of interest in gerontology in the developed nations today is one consequence of the process of modernization. This has come about for many reasons including: changes in the shape of the population pyramid, dramatic improvements in health status and longevity, increases in educational and economic levels, and, importantly, increases in the political power of the elderly. As these changes occur also in the less developed regions of the world, so too will their interest in gerontology increase. Having experienced these changes ourselves, we are in a unique and fortunate position to alert and assist gerontologists and social planners in developing nations as they, like us, modernize (Finley & Delgado, 1981).

REFERENCES

Achenbaum, W. A., & Stearns, P. N. Essay: Old age and modernization. *Gerontologist,* 1978, **18** (3), 307–312.

Arnhoff, F. N., Leon, H. V., & Lorge, I. Cross-cultural acceptance of stereotypes towards aging. *Journal of Social Psychology,* 1964, **63**, 41–58.

Bengtson, V. L., Dowd, J. J., Smith, D. H., & Inkeles, A. Modernization, modernity, and perceptions of aging: A cross-cultural study. *Journal of Gerontology,* 1975, **30** (6), 688–695.

Birren, J. E., & Schaie, K. W. (Eds.) *Handbook of the psychology of aging.* New York: Van Nostrand, Reinhold, 1977.

Burgess, E. W. (Ed.) *Aging in western societies.* Chicago: University of Chicago Press, 1960.

Cowgill, D. O. Aging and modernization: A revision of the theory. In J. F. Gubrium (Ed.), *Late life: Communities and environmental policy.* Springfield, Ill.: Thomas, 1974, pp. 123–146.

Cowgill, D. O., & Holmes, L. D. (Eds.) *Aging and modernization.* New York: Appleton-Century-Crofts, 1972.

Datan, N., Antonovsky, A., & Maoz, B. *A time to reap: The middle age of women in five Israeli sub-cultures.* Baltimore: Johns Hopkins University Press, 1981.

Delgado, M., & Finley, G. E. The Spanish-speaking elderly: A bibliography. *Gerontologist,* 1978, **18** (4), 387–394.

Finley, G. E. Aging in Latin America. *Spanish Language Psychology,* 1981, **1** (3), 223–248.

Finley, G. E., & Delgado, M. Formal education and intellectual functioning in the immigrant Cuban elderly. *Experimental Aging Research,* 1979, **5** (2), 149–154.

Finley, G. E., & Delgado, M. La psicologia del envejecimiento. *Revista Latinoamericana de Psicologia,* 1981, **13** (3), 415–432.

Fry, C. L. (Ed.) *Aging in culture and society.* New York: J. F. Bergin, 1980.

Glascock, A. P., & Feinman, S. L. Toward a comparative framework: Propositions concerning the treatment of the aged in non-industrial societies. In C. L. Fry & J. Keith (Eds.), *New methods for old age research: Anthropological alternatives.* Chicago: Center for Urban Policy, Loyola University, 1980, pp. 204–222.

Gutmann, D. The cross-cultural perspective: Notes toward a comparative psychology of aging. In J. E. Birren & K. W. Schaie (Eds.), *Handbook of the psychology of aging.* New York: Van Nostrand, Reinhold, 1977, pp. 302–326.

Gutmann, D. Observations on culture and mental health in later life. In J. E. Birren & R. B. Sloane (Eds.), *Handbook of mental health and aging.* Englewood Cliffs, N.J.: Prentice-Hall, 1980, pp. 429–447.

Hauser, P. M. Aging and world-wide population change. In R. H. Binstock & E. Shanas (Eds.), *Handbook of aging and the social sciences.* New York: Van Nostrand, Reinhold, 1976, pp. 58–86.

Heron, A., & Kroeger, E. Developmental perspectives. In H. C. Triandis & A. Heron (Eds.), *Handbook of cross-cultural psychology, developmental psychology* (Vol. 4). Boston: Allyn and Bacon, 1981, pp. 1–15.

Inkeles, A., & Smith, D. H. *Becoming modern.* Cambridge: Harvard University Press, 1974.

Katz, S. H. Anthropological perspectives on aging. *Annals of the American Academy of Political and Social Sciences,* 1978, **438**, 1–12.

Keith, J. The best is yet to be: Toward an anthropology of age. In B. J. Siegel, A. R. Beals, & S. A. Tyler (Eds.), *Annual Review of Anthropology* (Vol. 9). Palo Alto: Annual Reviews, 1980, pp. 339–364.

Kogan, N. Beliefs, attitudes, and stereotypes about old people: A new look at some old issues. *Research on Aging,* 1979, **1** (1), 11–36.

Laslett, P. Societal development and aging. In R. H. Binstock & E. Shanas (Eds.), *Handbook of aging and the social sciences.* New York: Van Nostrand, Reinhold, 1976, pp. 87–116.

LeVine, R. A. Adulthood and aging in cross-cultural perspective. *Items,* 1978, **31/32** (4/1).

Maddox, G. L. Sociology of later life. In A. Inkeles, J. Coleman, & R. H. Turner (Eds.), *Annual Review of Sociology* (Vol. 5). Palo Alto: Annual Reviews, 1979, pp. 113–135.

Missinne, L., & Seem, G. *Comparative gerontology: A selected annotated bibliography.* Washington, D.C.: International Federation on Ageing, 1979.

Myerhoff, B. Aging and the aged in other cultures: An anthropological perspective. In E. E. Bauwens (Ed.), *The anthropology of health.* Saint Louis: C. V. Mosby, 1978, pp. 151–166.

Palmore, E. The future status of the aged. *Gerontologist,* 1976, **16** (4), 297–302.

Palmore, E., & Manton, K. Modernization and status of the aged: International correlations. *Journal of Gerontology,* 1974, **29** (2), 205–210.

Palmore, E., & Whittington, F. Trends in the relative status of the aged. *Social Forces,* 1971, **50**, 84–91.

Press, I., & McKool M. Social structure and status of the aged: Toward some valid cross-cultural generalizations. *Aging and Human Development,* 1972, **3** (4), 297–306.

Sharma, K. L. A cross-cultural comparison of stereotypes towards older persons. *Indian Journal of Social Work,* 1971, **32** (3), 315–320.

Simmons, L. W. *The role of the aged in primitive society.* New Haven: Yale University Press, 1945.

Tuckman, J., & Lorge, I. Attitudes toward old people. *Journal of Social Psychology,* 1953, **37**, 249–260.

von Mering, O., & Weniger, F. L. Social-cultural background of the aging individual. In J. E. Birren (Ed.), *Handbook of aging and the individual.* Chicago: University of Chicago Press, 1959, pp. 279–335.

Youmans, E. G. (Ed.) *Older rural Americans.* Lexington: University of Kentucky Press, 1967.

31

MICROANALYTICAL RESEARCH ON ENVIRONMENTAL FACTORS AND PLASTICITY IN PSYCHOLOGICAL AGING

Margret M. Baltes
Visiting Associate Professor of Psychological Gerontology
Department of Gerontopsychiatry
Free University, Berlin, Germany

Paul B. Baltes
Director and Professor of Psychology
Max Planck Institute on Human Development and Education
Berlin, Germany

The purpose of this chapter is to review the role of research on plasticity in the study of psychological aging. Much of gerontological research deals with the establishment of knowledge about average, or "normative," functions of aging. This is a valid enterprise. At the same time, we argue that knowledge about variability and plasticity is equally important.

In this chapter we focus first on the definition of plasticity and then on the description of two research programs that emphasize a microanalytic treatment of environmental factors as they affect the course of psychological aging. The conceptions and findings on plasticity contribute to a body of knowledge that includes not only information on what aging is like in the average but what it could be like if the environmental conditions for aging were different.

BEHAVIORAL PLASTICITY: ITS ROLE IN THE STUDY OF PSYCHOLOGICAL AGING

We conceive of aging as a process that begins at conception and extends through the entire human life span (Baltes, Reese, & Lipsitt, 1980; Riley, 1979). A concern with process—its temporal characteristics, determinants, degree of generality, and invariance vs. plasticity—is at the core of our thinking about psychological aging.

Historically, the field of gerontology began with a primary focus on the descriptive study of "normative" aging. There was much interest in the search for invariant (robust) and fairly universal (general) processes. During this phase of gerontological research, a fairly strong emphasis on aging as decline or deterioration appeared to be a hallmark of the descriptive evidence.

Such a general stereotype of aging as decline, however, has undergone major revision, particularly during the last decade. While decline may be one of the salient features of aging, at least for the present historical era (Fries, 1980), life-span development from conception to death can take rather varied forms, shapes, and directions (Baltes et al., 1980; Labouvie-Vief & Chandler, 1978). Without denying the possibility and existence of decline phenomena, decline needs to be seen in the context of other information. Among the features attracting attention are such terms as multidirectionality, multilinearity, variability, and plasticity. Of these concepts, plasticity is central to this chapter (see also Baltes & Baltes, 1980; Baltes & Willis, 1982; Willis & Baltes, 1980). It is a main characteristic of variability and plasticity that psychological aging does not always follow a particular course which is largely fixed (invariant) and robust. On the contrary, psychological aging can vary rather markedly depending on both biological and environmental conditions.

What is the meaning of plasticity in the present context? While variability usually refers to differences between individuals, plasticity is used here to denote variability within the same individual. It is measured by the degree to which the same individual exhibits different behaviors if exposed to differing life conditions. Plasticity involves both concurrent and developmental considerations. As to a concurrent orientation, the question "How variable are individuals at any particular moment of their lives if exposed to specific conditions?" is an example, whereas a developmental orientation would be expressed in the question "How variable is the course of development if individuals were exposed to differing life conditions?" The full extent of concurrent or developmental plasticity is, of course, as yet unknown. In principle we might deal here with an unknown "dialectical" quantity that can be approximated only.

In our view, and also that of others (Baltes & Lerner, 1980; Cairns, 1979; Lerner & Busch-Rossnagel, 1981), the study of the extent of and conditions for plasticity is one of the most fundamental questions of human development and aging. It is important to recognize that we do not claim psychological aging to be completely variable and plastic—such an extreme view would seriously be at variance with the basics of the field. It is the study of the conditions and the range of plasticity which is at the core of our interest. In fact, even if one chooses to minimize the role of plasticity, because one opts for normative or universal theories of aging, it would be important, nevertheless, to have information about plasticity. In that instance, only those behaviors that, in the long run, exhibit a fair degree of invariance or robustness (i.e., little plasticity) across a wide range of conditions would become the salient components of theories of normative development and aging, as would be true for Wohlwill's (1973) position. Thus the study of plasticity is a cornerstone of research in psychological aging, irrespective of one's subsequent use of the information obtained.

There are a number of historical trends in the study of development and aging that have contributed to the emerging concern with plasticity specifically and variability, multidirectionality, and multilinearity generally. There is, for example, the evolvement of a life-span view of aging. Life-span research, in part because of its concern with the cumulative effect of differential life histories (e.g., Runyan, 1978) and the process of social change (Elder, 1979; Riegel, 1976; Riley, 1979; Schaie, 1979), has resulted in clear evidence of discontinuity, multidirectionality, and variability in human development (Baltes et al., 1980). Furthermore, there is the increasing use of experimental intervention paradigms in the study of both psychological and biological aging (Birren, 1970; Fries, 1980; Strehler, 1977). Such experimental intervention research has shown that there is much more modifiability in both behavioral and biological aspects of development than traditionally assumed.

Finally, there is the advent of ecological and environmental perspectives in psychology stressing aspects of cultural–ecopsychological relativism and determinism. Similarly, the social sciences in general (e.g., Binstock & Shanas, 1976) have a long tradition in emphasizing the role of social structure and social processes in codetermining the nature of aging.

Role of Environmental Factors

For a comprehensive theory of psychological aging, it is imperative to view aging as the joint and interactive outcome of biological and environmental factors. Although we focus here on the role of ecological factors as codeterminants of plasticity, we do not take an extreme environmental position, which would represent a narrow-minded view. The role of environmental factors may be larger in old age, however, than is traditionally assumed in gerontology, or than in infancy, for a number of reasons. Among them are the increased biological vulnerability or reduced biological reserve of aging organisms (Baltes & Willis, 1982), as well as the fact that for evolutionary arguments (Baltes et al., 1980) it is somewhat difficult to infer a "strong" genetic, universally shared program of aging. Understanding the role of the environment, thus, is one of the building blocks of theory construction in aging.

As research on mechanisms of environmental and biological processes has progressed, the importance of the interplay between organism and environment has come to the foreground. Thus the more recent formulations of theories of development and aging have moved toward interactionistic models in the attempt to represent both environmental and biological factors in the process of aging (e.g., Lerner, 1978; Strehler, 1977). The paradigmatic nature of such interactionistic conceptions, of course, varies widely.

Plasticity: Two Research Programs

Two programs of research will be presented that have been conducted with a primary concern for plasticity and the potential role of environmental factors. They are illustrative only and reflect the primary interest of both authors and their colleagues. Other research programs could have been used as well to communicate similar perspectives.

The first research program conducted by M. M. Baltes and her colleagues involves the application of operant-psychological principles to the study of dependency in aging. The second research program focuses on the effect of cognitive training on intellectual functioning in aging. This work has been done by P. B. Baltes, S. L. Willis, and their coinvestigators.

OPERANT RESEARCH ON DEPENDENCY IN AGING: CONVERGENT OPERATIONS

Operant Psychology

The operant paradigm yields one example of an interactionistic model concerned with plasticity in human development (Baer, 1973; Baltes & Barton, 1977; Baltes & Lerner, 1980; Hoyer, 1974). The analysis of dynamic reciprocal interactions between a behaving organism and an environmental context is at the core of this paradigm. The unit of analysis is privately or publicly observable behaviors of the organism that are controlling and are controlled by environmental events.

The fundamental three-way contingency, S^D–R–S^R, is the basis for the operant analysis of the reciprocal influences between organism and environment. The environment thus plays an important role in the acquisition, maintenance, and extinction of behaviors. Contrary to frequent interpretations, however, the operant paradigm is not necessarily postulating that the organism is passive. How active the organism is conceptualized and how active the organism has to be in order to call organism–environment interchanges a truly reciprocal interaction is a matter of debate between psychologists holding different theoretical world views (see Baltes & Reese, 1977). This debate does not concern us here. What is important is that in the operant model, because of its particular paradigmatic posture, there is a great emphasis on changeability, modifiability, and reversibility of behaviors, including behaviors of elderly.

In addition to the convergence between the theoretical features of the operant model and the interest in plasticity expressed by modern gerontologists, there is a methodological argument for the viability of the operant model. The operant paradigm provides three different types of research strategies which, in concert, can yield a more complete picture of the etiology of a behavior (Baer, 1973). Thus the operant paradigm provides one model not only for the descriptive identification of plasticity but also for its causal analysis.

The first of these three research strategies is *experimental-operant* research which requires the manipulation of environmental conditions in order to produce change in the behavior(s) of interest. This research is aimed at the establishment of the modifiability of behaviors and thus provides one with information about the plasticity of behavior and the sufficient conditions under which it can come about. The second research strategy is *observational-operant*. This strategy requires observations of behavior–environment contingencies in terms of naturally occurring behavior sequences. This research is aimed at the identification of existing conditions in the natural ecology surrounding the emittance of a given behavior.

The third research strategy is *operant-ecological intervention* work in which findings from experimental- and observational-operant research are combined in a hypothesis-directed intervention program delivered in and by the natural ecology.

What is the evidence, to date, of operant findings on environment-related plasticity in the acquisition and maintenance of dependence in aging?

Experimental-Operant Research

To reiterate, experimental-operant work is aimed at examining sufficient conditions for causal control. Thus in the case of dependence—defined as overt acts related to personal maintenance—the focus is on which dependent and independent behaviors are modifiable in the elderly; to what degree are they changeable; and which environmental events are and are not able to produce change. This type of experimental work, requiring high control, is conducted under strict laboratorylike conditions using mostly single-subject designs. Such experimental studies are obviously restricted in their generalizability across people and settings. For example, many of these studies have been done with institutionalized elderly in nursing homes or other hospitallike settings.

Most of the relevant procedures and findings have been reviewed elsewhere in more detail (see Baltes & Barton, 1977, 1979). Suffice it here to say that the findings show consistently a great potential for change. As it is true for much operant work with other behaviors and other populations, operant work on dependency in aging illustrates that deficits can be changed. Classes of behaviors, which have been successfully modified, include self-care, eating, locomotion, communication skills, and socially obstructive behaviors. Operant treatments shown to be effective run the entire gamut of operant strategies classifiable under stimulus control and reinforcement procedures. Plasticity as a function of environmental (treatment) changes is a feature of psychological aging. The kind of overt behaviors we observe in older persons is but one of the possible behavioral outcomes associated with organism–environment interactions during old age.

Observational-Operant Research

The experimental-operant studies with their emphasis on the modification of behaviors through environmental manipulation, however, give us little direct information about how the deficits are acquired and maintained in the natural ecology. To answer this, it is necessary to discover the naturally extant contingencies of existing deficits.

Consequently, in this second type of research, we use the operant paradigm in observing, rather than changing, the behavioral contingencies that exist in the natural ecology in which dependent behaviors occur. The argument is made that if convergence between experimental-operant research and observational-operant research is found, the conclusion that environmental factors play a "causal" role in the development and maintenance of dependent behaviors in the elderly is strengthened. Converging information from both types of operant research strategies is crucial for theory development and treatment programs since it will assure the ecological validity (Bronfenbrenner, 1977) of the knowledge obtained.

Few observational-operant studies, specifically in psychological aging, exist to date. They have been described in detail elsewhere (Baltes & Baltes, 1980; Baltes, Burgess, & Stewart, 1980; Barton, Baltes, & Orzech, 1980). As was true for the experimental work, the observational-operant studies all have been conducted so far with nursing home residents and with a rather general behavior observation code. Behaviors are classified into dependence and independence related to personal maintenance or self-care and nonengagement and independence referring to all other solitary or social activities. The main emphasis was to identify behavior

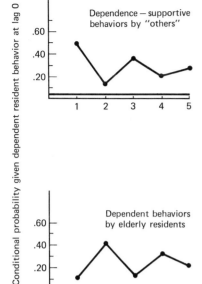

Figure 31.1. Sequential lags for two criterion behaviors, resident independent and dependent behaviors, related to personal maintenance. (Probabilities do not add to 100 per lag since only significantly increased consequent events are shown.)

patterns in the form of sequences of behaviors between residents and all social agents present in a nursing home situation. Sequential lag analyses (Sackett, 1977)[1] were used in identifying the behavior patterns between residents and others.

What are some of the highlights of this observational-operant research? The most consistent finding shows the existence of discrepant social ecologies for dependent and independent behaviors. Specifically, we find a highly consistent and immediately supportive social environment for dependent behaviors related to personal maintenance, whereas independent behaviors in the context of self-care are not associated with observable responses by the social environment. There are other independent pro-social behaviors (such as writing a letter, assisting a fellow resident in walking) of the elderly resident, however, for which we find attention by the social ecology; albeit this attention consists of inconsistent, irregular responses.

Using the operant paradigm as the basis for the interpretation of these naturalistic findings on the interactions between elderly nursing home residents and their social environment, the following conclusions can be offered. Dependent behaviors related to personal maintenance exhibited by the elderly resident are followed by a continuous, immediate reinforcement schedule. Such a schedule typically leads to increments or maintenance of dependency. For independent behaviors related to personal maintenance, in contrast, no social responses are observed. Between these two extremely discrepant schedules we find for those independent behaviors that are related to social or solitary activities other than self-care a very thin intermittent reinforcement schedule.

Figure 31.1 illustrates the most probable sequential behavior patterns for the criterion behavior dependence. Only the statistically significant behavioral consequences are shown. Statistical significance refers to the fact that the "conditional"

---

[1] We are aware of the criticisms directed toward the statistical significance testing of behavior sequences (Allison & Liker, 1981). The equation proposed by Sackett and used by us in the present analyses produces, however, conservative estimates of statistical significance.

probability of a behavior, when following a criterion behavior, deviates from its base probability. The base probability of a consequent behavior—shown as horizontal band in the figure—is an index for the overall probability of that behavior in the total pool of observations. The figure illustrates that dependence-supportive behavior by "others" (social partners) follows dependence of the elderly with a clearly elevated probability and seems to take turns with dependent behavior of residents. This is evidence for resident-social partner congruence. The counterpart "congruent" relationship was not found for independence related to self-care.

Implications of Findings for the Conception of Dependency and Independence in Aging

Dependency and Independence. The findings describing and explaining dependency as a function of direct social consequences demonstrate that dependency is not as such a necessary complement to aging as has been assumed. While some dependency may be correlated with physical dysfunctions and impairment, a certain component can be contributed to environmental factors fostering dependent behaviors in the elderly.

Moreover, the description and explanation of dependent behaviors in terms of social contingencies not only suggests a certain theoretical posture within the learning tradition but also a reinterpretation of dependency. For example, learned helplessness has been proposed as a central mechanism in accounting for dependency. Learned helplessness as defined by its originator Seligman (1975) is the product of a lack of, rather than presence of, consequences upon the emittance of behaviors in aversive situations. Our findings, suggesting presence of direct consequences, are thus squarely opposed to the notion of learned helplessness. Our data on dependency also simplify explanations promoted by Rodin and Langer (1977). The former, using a cognitive approach, have argued that experience of physical deficiencies and of social prejudices leads to perceived lack of control and, in turn, to dependency. Langer (1979) more recently has introduced the notion of self-induced dependency. This concept supposedly refers to the relationship between interpersonal situational factors—such as allowing someone else to do something for you—and the subsequent illusion of incompetence or erroneous inference of incompetence which again results in dependency. Our findings on dependence do not exclude the operation of such cognitive processes. We would argue, however, that given the evidence of a functional relationship between dependency and social consequences, a simple learning process appears to be a more parsimonious explanation of dependency in the elderly resident. We would like to add that different processes might be in operation contingent on the learning stage, that is, the acquisition or maintenance phase, of a behavior.

As to our findings on independent behavior, the outcome is different. First, we do not find evidence for the operation of direct, continuous social contingencies in the nursing home setting that would maintain or increase independent behaviors. Second, despite the complete lack of supportive conditions for independent behaviors related to self-care in the social ecology (the nursing home setting), there is, however, a fair level of independent behavior present, much more so than dependent behavior. Third, independent behaviors related to other than self-care activities are followed only occasionally by reinforcement from others. Several post hoc explanations of the discrepancy between observed behavior frequency and predicted behavior frequency are possible. One refers to the differential consequences effective during the acquisition or maintenance phase of a behavior. Several authors working with different behaviors have shown that during the acquisition phase, direct and immediate positive reinforcement is necessary for the behavior to be learned, whereas different forms of reinforcement (e.g., intermittent or delayed positive reinforcement, negative reinforcement) are able to maintain the behavior (e.g., Patterson's work, 1979, on aggressive behavior in children). Second, one can argue that independence, during its long learning history through life, has become associated with a host of secondary reinforcements, including intrinsic reinforcement patterns (e.g., the self-efficacy concept of Bandura, 1977). As a consequence, the present social ecology alone is not the primary agent in the control of independence in the elderly.

Social Ecology as a System. So far, our data analysis has been straightforward but unidirectional. When the operant paradigm is applied to social episodes, each behavior event is at the same time consequence for one and the antecedent or discriminative stimulus for another behavior. So far, resident behavior has been our target behavior, for which subsequent contingencies were sought. This behavior sequence resident–others (rather than others–resident) has been our stated goal. Nevertheless, it remains only one side of the reciprocal relationship between residents and their social ecology. Furthermore, when we look at resident behaviors as antecedents for others' behaviors, it becomes obvious that dependent resident behaviors function as strong discriminative stimuli setting the occasion for others to behave in a very specific way, namely, supportive of dependence, whereas independent resident behaviors do not exert similar control over others' behavior. Since we have not looked at the consequences for others' behavior, we do not know why the two resident behaviors have such differential effect. In future analyses, it is planned to focus more on the reciprocal nature and systems character of the social ecology of the elderly.

In addition, it would be desirable to observe interactions of elderly persons not only in nursing homes but also in other settings such as families. Moreover, it will be necessary to study the interaction patterns of elderly persons as they move from one social ecology (e.g., family) to another (e.g., nursing home) in order to transcend the context of institutional environments. Such environments as described, for instance, by Goffman (1960) may have their peculiar characteristics. For example, in an institutional system, like nursing homes, dependency of the elderly and dependence-supportive behavior by others may be role congruent for both actors. Thus the interactions observed in our past research would be supported by the administrative system governing the nursing home. To study these issues, however, longitudinal pre- and postinstitutionalization data as well as comparative data from community-dwelling elderly are required. Such studies are the target of the next research project of the first author.

In summary, operant research on dependency in aging, to date, has shown that increased dependency in older nursing home residents is not solely regulated by biological conditions of the aging organism. First, as shown in operant-experimental work, dependency is modifiable. Second, as demonstrated in operant-observational work, dependency is acquired through and maintained by the nature of the existing social contingencies with which the elderly live. Intraindividual plasticity is indeed paramount.

COGNITIVE TRAINING RESEARCH IN INTELLECTUAL AGING

When it comes to questions of variability and plasticity in psychological aging, research on psychometric intelligence has been at the forefront. This is certainly true for variability. Historically, this is illustrated in the recurring debate about whether there is decline in intellectual functioning during aging (Botwinick, 1977; Horn & Donaldson, 1976, 1977; vs. Baltes & Schaie, 1976; Schaie & Baltes, 1977).

In the meantime, the debate has gone beyond the state of a simple dichotomy (Is there decline or not?) to a more comprehensive and dynamic view. It is recognized by most everyone that some decline in intellectual functioning is part of the picture. However, because of a lack of generality across persons and abilities, it is necessary to specify the conditions of occurrence (e.g., interability variation in onset, patterning, and trajectory).

During the last years, research on variability in intellectual aging has been supplemented with questions about plasticity at the level of individual analysis. One way of viewing this issue is to ask: "How much of interindividual variability observed between age and cohort groups can be seen intraindividually in a single individual of a given age?" Another way is to pose the more general question: "Irrespective of the magnitude of variability between persons, what is the extent of and what are the conditions for intraindividual plasticity?" As was true for operant research, such a question requires a particular methodology. First, the methodology is aimed at examining modifiability and at the causal analysis of cognitive performance (Willis & Baltes, 1980). Second, such a methodology tends to emphasize the contextual relativity of performance. In line with the performance–competence distinction, it is emphasized that individuals might be able to perform differently if conditions were different. A given performance is but one index of what individuals can do.

A number of recent survey articles report information on the conditions for and extent of cognitive modifiability in old age (e.g., Denney, 1979; Sterns & Sanders, 1980; Willis & Baltes, 1980). Findings from one such research program conducted by Baltes and Willis (1981, in press) are summarized here. This research program consists of a series of interrelated studies. In terms of intervention strategy, environmental factors such as test familiarity and practice in problem solving are considered.

The Pennsylvania State University's ADEPT Project

The research program ADEPT (Adult Development and Enrichment Project) can be described as an intervention project whose primary aim is the examination of the limits of intraindividual plasticity in intellectual performance in older individuals. More extensive information on rationale and procedure of the present, as well as additional, studies is presented in Baltes and Willis (1982).

Participants in ADEPT are "healthy" community residents aged 60 to 85 from central Pennsylvania. By and large, these elderly persons are somewhat positively selected in terms of health and education when compared with cohort-specific census data. Dependent variables include a multitude of measures covering the major dimensions of psychometric intelligence as formulated in the theory of fluid and crystallized intelligence (Baltes, Cornelius, Spiro, Nesselroade, & Willis, 1980; Horn, 1978). The framework of the fluid–crystallized intelligence theory was chosen because it permits a systematic approach to measurement when assessing the effectiveness of intervention in terms of scope and transfer. In addition, it is fluid intelligence that is consistently identified as the central dimension associated with intellectual decline (Horn, 1978).

Two intervention studies from ADEPT are presented here. The first intervention strategy deals with the role of test familiarity (retesting) in the assessment of plasticity in intellectual performance (Hofland, Willis, & Baltes, 1981). The second study (Willis, Blieszner, & Baltes, 1981) focuses on the role of practice in problem-solving skills related to one factor of fluid intelligence (figural relations) as strategy of intervention.

Environmental Factor: Test Familiarity (Retesting). The first study (Hofland et al., 1981) was aimed at examining the range of intraindividual variability as a function of experience associated with retesting (eight 1-hour practice [retest] sessions distributed over approximately 1 month). The same two measures of fluid intelligence (figural relations and induction) were given at each session without feedback on performance.

Our hypothesis was that older individuals would improve in performance with retesting and probably reach an asymptote (ceiling) after three or four retest trials. Figure 31.2 shows that older individuals did not only improve during the initial sessions but throughout the entire period of retesting. No apparent asymptote was reached. Total improvement in mean scores on both tests is slightly more than one standard deviation.

One may wonder whether the observed increments index the same ability or test-specific skills. An examination of the correlated measurement validity of the two tests used yielded no information on changing validity with retesting. This examination was accomplished by correlating the retest measures, separately by retest occasion, with a set of external criterion tests. These tests represent the measurement space of abilities provided by the theory of fluid–crystallized intelligence. In addition, retest stability coefficients were computed to be exceedingly high and close to the tests' own reliability. Such a finding suggests that subjects shared fairly equally in retest gains, independent of their initial level of functioning.

In sum, the evidence of intraindividual plasticity in this study appears remarkable. This is particularly significant because the tests chosen (indicators of fluid

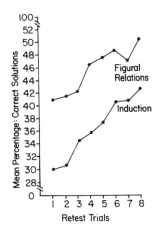

Figure 31.2. Mean percentage of correct solutions across retest trials (from Hofland, Willis, & Baltes, 1981, in press).

intelligence) are among those of psychometric intelligence which, according to the Cattel-Horn theory, should be among the most robust (invariant) when it comes to effects due to *short-term* experimental intervention. In other words, the tests are putative traits rather than states. We need to note, however, that the Hofland et al. study does not permit examination of two questions: (a) whether retesting itself leads to transfer of training to other ability tests and (b) whether older adults differ from younger adult age groups in the magnitude of retest effect. Studies aimed at answering some of these questions are in progress. What we do observe in the study is that elderly subjects evidence a substantial amount of plasticity as a function of retest experience.

Environmental Factor: Practice in Problem-Solving Skills. Figure 31.3 summarizes the outcome of one of the ADEPT studies that focused directly on the training of cognitive skills (Willis, Blieszner, & Baltes, 1981).

Specifically, the intervention focus was on problem-solving skills that are requisites for effective performance on a given factor of fluid intelligence, in this case, "figural relations." In addition, intervention effectiveness was not only measured by performance on the training tests themselves. It was assessed also whether cognitive training generalized to other dimensions of intellectual performance than the target ability (figural relations) selected for training. This approach is known as the assessment of transfer of training. Furthermore, training effectiveness was assessed not only at the end of training but also at delayed posttests in order to examine the degree of maintenance.

In order to give participants experience in problem-solving skills requisite for good performance on the fluid ability "figural relations," a cognitive training program was developed. The training program was based on a content analysis of the cognitive rules and format required for solution of test items for the tests indexing the intelligence factor "figural relations." The intelligence tests used for program development were four subtests of the Culture Fair Test (figure series, figure classify, matrices, and topology). Based on the task analysis of these tests, training problems were developed based on the most frequently occurring relational rules. Much emphasis was placed during training on presenting the training material and the rules with concern for individual styles of problem solving rather than focusing on a single set of correct solution strategies. None of the training items was identical to those used in the criterion tests themselves.

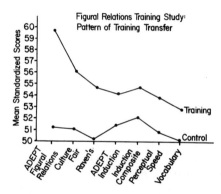

Figure 31.3. Performance on seven transfer tests for training and control groups averaged across three posttests (from Willis, Blieszner, & Baltes, 1981, in press).

Assessment of training effectiveness and of transfer of training to other ability factors (beyond figural relations) was examined by using seven tests as posttests that covered much of the spectrum of the theory of fluid–crystallized intelligence. The tests used for this purpose are shown in Figure 31.3. They are ordered on the basis of their psychometric and conceptual similarity to the training factor (figural relations) along a continuum from near to far transfer.

As can be seen in Figure 31.3, training was effective. Based on statistical analyses, both the predicted training effect on the near transfer measures and the predicted pattern of differential (ordered) transfer to other ability measures were judged to be significant. The pattern of training transfer is maintained across all three posttests spanning a 6-month period. On the nearest transfer measure, the magnitude of training corresponds to approximately one standard deviation if measured in terms of baseline performance of the control group at the first posttest.

The magnitude of the intervention effect shown in this study can be compared to information on aging changes as they are usually observed in longitudinal research over the age span under investigation. Schaie (1979, 1981), for example, finds in his 21-year longitudinal study that average decline on similar measures of intelligence over a 21-year period from age 60 to 80 is in the range from one-half to one standard deviation. The present intervention, then, produced an increment in performance that is at least as large as that obtained in longitudinal research over the age range covered by the ADEPT study participants. It should also be noted that other intervention studies (see Baltes & Willis, 1982) contain additional and supportive evidence. By and large, training effectiveness can be replicated and, moreover, is also present when other factors of intelligence (e.g., induction, attention/memory) are considered. However, the magnitude and scope (breadth) of training varies as does its maintenance over time.

Implications for Conceptions of Intellectual Aging. This evidence for intraindividual plasticity is particularly noteworthy because the target abilities involved are components of fluid intelligence. According to theory and empirical research (Horn, 1978), it is fluid intelligence which exhibits a fairly general pattern of decline with aging. Our own interpretation of the data stresses the role of environmental factors in regulating intellectual performance of older individuals on the measures examined. Without downgrading the role of biological factors, we believe that older individuals generally live in a context of cognitive deprivation (Labouvie-Vief, Hoyer, Baltes, & Baltes, 1974). Consequently, when it comes to psychometric intelligence tests, older persons are likely to function below their optimal level of performance. It follows that relatively short-term behavioral interventions (such as retesting and cognitive training) result in significant improvement.

Whatever the interpretation, however, and there are others, the case for a fair amount of intraindividual plasticity in intellectual performance of aging individuals is real. As a consequence, we argue (see also Baltes & Baltes, 1980; Willis & Baltes, 1980) that the simple establishment of average performance functions is an incomplete description of the behavior of older individuals. Such statements about averages need to be supplemented by statements about potentials and limits, in terms of both floor and ceiling, of performance. Programmatic intervention research is important in gaining such knowledge about intraindividual plasticity, and results from such research place the finding of normative, "average" aging functions (such as presented by Horn, 1978) into a larger context. Conversely, looking only at averages and "untreated" individuals provides a rather narrow and biased view of the world of intellectual aging.

PROSPECTS

The central argument of this chapter is that information about variability and plasticity is imperative for the construction of theories involving psychological aging. We have implied that past traditions in the field of gerontology have deemphasized such a view. To what degree a position stressing plasticity will be supported by empirical evidence is open. The studies summarized here, as well as research by others (for review, e.g., Denney, 1979; Sterns & Sanders, 1980), offer a fair amount of evidence. What is necessary, however, is to include a plasticity perspective in future empirical work in order to obtain the necessary evidence for falsification or verification.

In conclusion, we offer some additional perspectives on what evidence on plasticity might suggest for theories of psychological aging. What general theoretical principles can possibly account for a picture of aging that exhibits less universality than originally expected by many psychologists? How can we deal with the fact that psychological aging exhibits large interindividual differences, multidirectionality, multidimensionality, and intraindividual plasticity? In past work, we have written with colleagues about two possible theoretical interpretations (Baltes & Baltes, 1980; Baltes, Cornelius, & Nesselroade, 1979; Baltes & Willis, 1982).

A first general model is a tricausal model of influences that is formulated on the level of macrostructure and emphasizes variability and plasticity in life-span development resulting from the fact that at least three systems of influences interact in the production of aging: age-graded, history-graded, and nonnormative influences. History-graded and nonnormative influences, in particular, are major contributors to variability and plasticity. In combination with other determinants of differential development (e.g., social class, sex), they are at the basis of an aging process that is primarily one of differentiation.

The second and related model is formulated on the level of individual development and characterizes psychological aging as a process of selective optimization. This approach states that aging individuals invest successively more and more of their time and effort in selected areas of behavior. Such a process of selective optimization corresponds to the perspective that aging can be indexed as a process toward specialization and individualization (Brent, 1978). In other words, depending on individual life histories, individuals engage in a growing commitment to particular trajectories and deemphasize others. At the same time, if aging individuals are prompted to exhibit alternative behaviors, they are often capable of doing so.

The articulation of these two theoretical viewpoints on psychological aging has been enhanced by an approach and by findings that testify to much variability and plasticity. In this sense, they exemplify a more general concern. Any particular research strategy, such as a descriptive approach, results in selected evidence on what aging is like. It does not tell us, however, what aging could be like if environmental conditions for aging were different.

The growing interest in the conditions of and extent for plasticity, exemplifying a change in methodological orientation away from the focus upon a single strategy toward "convergent operations," contributes to a possibly more balanced and surely more encompassing understanding of aging. In this sense, research on plasticity is a necessary part of future work in gerontology.

REFERENCES

Allison, P. D., & Liker, J. K. Analyzing sequential interaction data in two-person systems. Unpublished Paper, Cornell University, 1981.

Baer, D. M. The control of developmental process: Why wait? In J. R. Nesselroade & H. W. Reese (Eds.), *Life-span developmental psychology: Methodological issues.* New York: Academic, 1973.

Baltes, M. M., & Barton, E. M. New approaches toward aging: A case for the operant model. *Educational Gerontology: An International Quarterly,* 1977, **2**, 383–405.

Baltes, M. M., & Barton, E. M. Behavioral analysis of aging: A review of the operant model and research. *International Journal of Behavior Development,* 1979, **2**, 297–320.

Baltes, M. M., Burgess, R. L., & Stewart, R. B. Independence and dependence in self-care behaviors in nursing home residents: An operant observational study. *International Journal of Behavior Development,* 1980, **3**, 489–500.

Baltes, M. M., & Lerner, R. M. Roles of the operant model and its methods in the life-span approach to human development. *Human Development,* 1980, **23**, 362–367.

Baltes, M. M., & Reese, H. W. Operant research in violation of the operant paradigm? In B. C. Etzel, J. M. Leblanc, & D. M. Baer (Eds.), *Contributions to behavioral research: Festschrift in honor of Sidney Bijou.* New York: Erlbaum, 1977.

Baltes, P. B., & Baltes, M. M. Plasticity and variability in psychological aging: Methodological issues. In G. Gurski (Ed.), *Determining the effects of aging on the central nervous system.* Berlin: Schering, 1980.

Baltes, P. B., Cornelius, S. W., & Nesselroade, J. R. Cohort effects in developmental psychology. In J. R. Nesselroade & P. B. Baltes (Eds.), *Longitudinal research in the study of behavior and development.* New York: Academic, 1979.

Baltes, P. B., Cornelius, S. W., Spiro, A., Nesselroade, J. R., & Willis, S. L. Integration versus differentiation of fluid crystallized intelligence in old age. *Developmental Psychology,* 1980, **16**, 625–635.

Baltes, P. B., Reese, H. W., & Lipsitt, L. P. Life-span developmental psychology. *Annual Review of Psychology,* 1980, **31**, 65–100.

Baltes, P. B., & Schaie, K. W. On the plasticity of intelligence in adulthood and old age: Where Horn and Donaldson fail. *American Psychologist,* 1976, **31**, 720–725.

Baltes, P. B., & Willis, S. L. Enhancement (plasticity) of intellectual functioning in old age: Penn State's Adult Development and Enrichment Project (ADEPT). In F. I. M. Craik & S. E. Trehub (Eds.), *Aging and cognitive processes.* New York: Plenum, 1982.

Bandura, A. Self-efficacy: Toward a unifying theory of behavioral change. *Psychological Review,* 1977, **84**, 191–215.

Barton, E. M., Baltes, M. M., & Orzech, M. J. On the etiology of dependence in older nursing home residents during morning care: The role of staff behavior. *Journal of Personality and Social Psychology,* 1980, **38**, 423–431.

Binstock, R., & Shanas, E. (Eds.) *Handbook of aging and the social sciences.* New York: Van Nostrand-Reinhold, 1976.

Birren, J. E. Toward an experimental psychology of aging. *American Psychologist,* 1970, **20**, 124–135.

Botwinick, J. Aging and intelligence. In J. E. Birren & K. W. Schaie (Eds.), *Handbook of the psychology of aging.* New York: Van Nostrand-Reinhold, 1977.

Brent, S. B. Individual specialization, collective adaptation, and rate of environmental change. *Human Development,* 1978, **21**, 21–33.

Bronfenbrenner, U. Towards an experimental ecology of human development. *American Psychologist,* 1977, **32**, 513–531.

Cairns, R. B. *Social development: The origins and plasticity.* San Francisco: Freeman, 1979.

Denney, N. W. Problem solving in later adulthood: Intervention research. In P. B. Baltes & O. G. Brim, Jr. (Eds.), *Life-span development and behavior* (Vol. 2). New York: Academic, 1979.

Elder, G. H. Historical change in life patterns and personality. In P. B. Baltes & O. G. Brim, Jr. (Eds.), *Life-span development and behavior* (Vol. 2). New York: Academic, 1979.

Fries, J. F. Aging, natural death, and the compression of morbidity. *New England Journal of Medicine,* 1980, **303**, 130–135.

Goffman, E. Characteristics of total institutions. In M. R. Stein, A. J. Violich, & D. M. White (Eds.), *Identity and anxiety: Survival of the person in mass society.* New York: Free Press, 1960.

Hofland, B., Willis, S. L., & Baltes, P. B. Fluid intelligence performance in the elderly: Intraindividual variability and conditions of assessment. *Journal of Educational Psychology,* 1981, in press.

Horn, J. L. Human ability systems. In P. B. Baltes (Ed.), *Life-span development and behavior* (Vol. 1). New York: Academic, 1978.

Horn, J. L., & Donaldson, G. On the myth of intellectual decline in adulthood. *American Psychologist,* 1976, **31**, 701–719.

Horn, J. L., & Donaldson, G. Faith is not enough: A response to the Baltes-Schaie claim that intelligence does not wane. *American Psychologist,* 1977, **32**, 369–373.

Hoyer, W. J. Aging as intraindividual change. *Developmental Psychology,* 1974, **10**, 821–826.

Labouvie-Vief, G., & Chandler, M. Cognitive development and life-span developmental theories: Idealistic versus contextual perspectives. In P. B. Baltes (Ed.), *Life-span development and behavior* (Vol. 1). New York: Academic, 1978.

Labouvie-Vief, G., Hoyer, W. J., Baltes, M. M., & Baltes, P. B. Operant analysis of intellectual behavior in old age. *Human Development,* 1974, **17**, 259–272.

Langer, E. J. The illusion of incompetence. In L. C. Perlmutter & R. A. Monti (Eds.), *Choice and perceived control.* Hillsdale, N.J.: Erlbaum, 1979.

Lerner, R. M. Nature, nurture, and dynamic interactionism. *Human Development,* 1978, **21**, 1–20.

Lerner, R. M., & Busch-Rossnagel, N. A. Individuals as producers of their development: Conceptual and empirical bases. In R. M. Lerner & N. A. Busch-Rossnagel (Eds.), *Individuals as producers of their development: A life-span perspective.* New York: Academic, 1981.

Patterson, G. R. A performance theory for coercive family interaction. In R. B. Cairns (Ed.), *The analysis of social interactions: Methods, issues, and illustrations.* Hillsdale, N.J.: Erlbaum, 1979.

Riegel, K. F. The dialectics of human development. *American Psychologist,* 1976, **31**, 689–700.

Riley, M. W. (Ed.) *Aging from birth to death.* Boulder: Westview, 1979.

Rodin, J., & Langer, E. J. Long-term effects of a control-relevant intervention with the institutionalized aged. *Journal of Personality and Social Psychology,* 1977, **35**, 897–902.

Runyan, W. M. The life-course as a theoretical orientation: Sequences of person-situation interactions. *Journal of Personality,* 1978, **46**, 569–593.

Sackett, G. T. The lag sequential analysis of contingency and cyclicity in behavioral interaction research. In J. Osofsky (Ed.), *Handbook of infant development.* New York: Wiley, 1977.

Schaie, K. W. The primary mental abilities in adulthood: An exploration in the development of psychometric intelligence. In P. B. Baltes & O. G. Brim, Jr. (Eds.), *Life-span development and behavior* (Vol. 2). New York: Academic, 1979.

Schaie, K. W. The Seattle Longitudinal Study: A twenty-one year exploration of psychometric intelligence in adulthood. In K. W. Schaie (Ed.), *Longitudinal studies of adult psychological development*. New York: Guilford, in press.

Schaie, K. W., & Baltes, P. B. Some faith helps to see the forest: A final comment on the Horn and Donaldson myth of the Baltes-Schaie position on adult intelligence. *American Psychologist*, 1977, **32**, 1118–1120.

Seligman, M. E. P. *Helplessness: On depression, development, and death*. San Francisco: Freeman, 1975.

Sterns, H. L., & Sanders, R. E. Training and education of the elderly. In R. R. Turner & H. W. Reese (Eds.), *Life-span developmental psychology: Intervention*. New York: Academic, 1980.

Strehler, B. L. *Time, cells, and aging*. New York: Academic, 1977.

Willis, S. L., & Baltes, P. B. Intelligence and cognitive ability. In L. Poon (Ed.), *Aging in the 1980s: Psychological issues*. Washington: American Psychological Association, 1980.

Willis, S. L., Blieszner, R., & Baltes, P. B. Training research in aging: Modification of performance on the fluid ability of figural relations. *Journal of Educational Psychology*, 1981, in press.

Wohlwill, J. F. *The study of behavioral development*. New York: Academic, 1973.

32

BREAKDOWN OF CONTROL PROCESSES IN OLD AGE

Patrick Rabbitt
Fellow, The Queen's College, Oxford

Developmental psychologists and gerontologists face the same problems: they are concerned with descriptions of processes in a state of change. Models for change cannot be simple. For example, we cannot suppose that young children learn to walk or talk merely by practicing the same, unaltered sequences of movements until they can perform them swiftly and flawlessly, nor can we assume that these skills deteriorate in old age in precisely the inverse order to that in which they developed. As children grow up they do not merely learn to do the same things better and faster but rather actively seek and discover new and more efficient ways of doing them. As people grow old they remain innovative, finding complex ways to circumvent their growing disabilities.

Because models for change are so necessary, it has been unfortunate that both developmental psychologists and gerontologists have borrowed their theoretical models of performance, and the paradigms which they use to test them, mainly from human experimental psychology and cognitive psychology. It is arguable that the most remarkable limitation of these disciplines has been their failure to generate any single, useful model for *change*. All experimental paradigms have been tested at particular, arbitrary levels of practice on particular groups of highly selected, fit, intelligent, and motivated young adults. Current models for human performance are essentially models of systems locked in some imaginary "steady state" (Rabbitt, 1979; 1981a,b,c,d). This is emphasized by the remarkable fact that we have no models which explain how human beings learn to perform better when they are practiced at very simple tasks. We only have a tacit—and blatantly invalid—assumption that as people practice a task they learn to go through the same sequence of functional operations ever faster and more accurately. This can easily be shown to be wrong. As young adults practice, they learn new and better ways to do the same things (Rabbitt, 1981a,b,c). But the models we have imported from human experimental psychology and cognitive psychology ignore this obvious fact. They cannot be adapted to describe how people get better with practice. It follows that they also cannot be used to describe whether young and old people improve with practice in different ways, why people gradually perform worse as they get old, or how their performance changes when they contract conditions like schizophrenia. Still less can these models describe how individual differences in strategies of performance may develop. Human beings do not all, and do not always, do the same things in the same ways. Models for performance which neglect the extent of this intersubject and intrasubject variance will not get us very far.

However justified these complaints may be, it is trivial to rehearse them here unless they can be documented by concrete examples and unless alternative, more promising approaches can be suggested. The purpose of this chapter is to make two points. First, that models recently developed from assumptions underlying elementary control theory are specifically developed to describe changing processes, to suggest how self-controlling systems actively adapt to alterations in task demands and to describe individual differences in strategies of control. Recent experiments show that such models may have very wide generality in their application to performance of simple tasks. These models have also suggested new ways of looking at particular functional changes in performance in old age and now allow us to consider whether, and how, old people alter their strategies of control to circumvent or to mitigate the effects of these changes. Second, models which assume that people can exercise active, predictive control over their behavior allow us to progress beyond analyses of very simple laboratory tasks to interpret data from complex real-life situations. Hopefully this may allow us to understand, and even perhaps to mitigate, the difficulties which old people encounter in everyday life.

It is hard to illustrate any universal principles of breakdown of cognitive control in old age because examples of breakdowns of control processes in particular skills are so numerous and so varied. The best-developed formal applications of control theory have been to physiological regulation systems and the control of appetitive behavior (McFarland, 1971; Toates, 1976). Bernstein (1967) has given descriptions of control systems used in skills, such as walking, and some account of their breakdown in old age. Moray (1976) has reviewed applications of control systems theory to human visual search and selective information processing. Rabbitt (1981a) has shown how the efficiency of control of attentional selectivity breaks down in old age. Descriptions of breakdowns in control of rapid sequences of hand movements have been given by Rabbitt and Rodgers (1965) and by Rabbitt and Birren (1967). In this chapter we describe examples of a different kind of cognitive control, and of its breakdown, examining the ways in which people access and use information stored in short- and in long-term memory to control their behavior from moment to moment, and considering why the efficiency with which they can do this declines as they grow old.

LABORATORY EXPERIMENTS SHOWING BREAKDOWN OF CONTROL OF PERFORMANCE BY INFORMATION HELD IN MEMORY

Loss of Control in Continuous Arithmetic Tasks

Gronwall and Sampson (1974) describe a paced, serial addition task (PASAT) which they used in order to detect performance changes in patients recovering from mild, "closed" head injuries. Sequences of 50-tape-recorded digits are played at speeds varying from $1/1.5$ seconds to $1/3.0$ seconds. Subjects have to keep up with the sequence, adding each digit to its predecessor and reporting the sum before the next digit occurs:

Digits presented	4	8	9	5	3	6	2	and so forth
Correct response	–	12	17	14	8	9	8	and so forth

Errors become more frequent as sequences are presented faster. At all speeds, most errors are omissions due to failures to keep up with the speed of the sequence or are due to random lapses in addition. Thomas (1977), however, in attempting to replicate Gronwall and Sampson's (1974) experiments noted that other types of errors occur. There seems to be cases where subjects correctly add together inappropriate numbers, sometimes adding a current digit to the one two-back rather than to the last one in the sequence:

Digits presented	4	8	9	5	3	6	2
Subjects' responses	–	12	13	14	8	9	5

They also sometimes appear to correctly add a current digit to a previous total:

Digits presented	4	8	9	5	3	6	2
Subjects' responses	–	12	21	14	8	14	8

It is hard to ascribe such errors to simple forgetting, since digits two-back in a sequence would be more likely to suffer from interference or trace-decay than digits one-back in a sequence. These errors seem rather to be failures to correctly index and retrieve the correct digit from among others simultaneously available in immediate memory or "working memory"—the running log of information about immediately past events which subjects need in order to correctly carry out the task (Baddeley & Hitch, 1974). It would evidently be more efficient for subjects to be able to forget two-back digits entirely and for them never to enter into memory any information about any totals which they announce. Apparently they cannot do this, and their memories are cluttered with unwanted information which can only cause errors when it is indexed by mistake.

Rabbitt and Heptinstall (described in Rabbitt, 1981a) found that if the basic PASAT task was made harder for young subjects by giving them another secondary task to perform at the same time (serial self-paced choice responses to signal lamps), they made more errors of all kinds, but errors involving retrieval of unwanted digits increased proportionately more than others. Rabbitt and Vyas (in preparation) compared groups of young (aged 18 to 32) and elderly (aged 65 to 81) subjects who had been matched for Mill Hill vocabulary scores both on the PASAT task alone and in the joint-task condition. As Table 32.1 shows, elderly subjects made more PASAT errors of all kinds in the joint-task condition, but the proportion of these rare extra errors, apparently due to failures of accurate retrieval from working memory, was greater for the older group. It is certainly true that older people cannot perform serial addition as fast as the young, and it is also true that they are more prone to other arithmetical errors of untraceable provenance

TABLE 32.1 Errors Made in a Paced Serial Addition Task (PASAT) by Groups of Young and Elderly Subjects under Normal Presentation and Distraction from a Simultaneous, Serial, Self-Paced Choice Reaction Task. Errors as Percentages of all Responses.

Group	Error	No Distraction	Distraction
Young group	Total errors	6.0	14.0
	Errors caused by addition of current digit to inappropriately remembered previous digit (memory slips).	0.02	2.5
Elderly group	Total errors	10.3	28.4
	Memory slips	1.4	7.2

and that they suffer more from distraction by a secondary task. But over and above all these other limitations, they seem to be particularly handicapped in the efficiency with which they can retrieve, from among other data held in working memory, information necessary to control their performance during a fast paced task.

Knowing What to Do Next

The experiment which we have described above may be taken as a particular, concrete instance of a very general situation in everyday life. At each stage during a long continuous task, a person may have to depend, for guidance, on information about what he last did. As we have seen, his efficiency will neither merely depend on the *capacity* of his immediate memory nor just on the *time* for which he can retain data in memory. It will also depend on the precision with which he can exactly recover the data which he requires and on the efficiency with which he can disregard other information about his immediate past, which is available in his working memory but which is irrelevant to a decision that he must currently make. He must not merely be able to remember some things which have recently happened to him or some of the decisions which he last made. He must be able to remember precisely the right things.

Let us consider other cases. Most of us can flawlessly and rapidly recite the English alphabet in the order in which we learned it at school. We can also quickly and accurately decide which of two letters occurs earlier in the alphabet, either in artificial laboratory tasks (Parkman, 1971; Hamilton & Sandford, 1978) or when we are looking up words in a dictionary. It follows that we must have available in our long-term memories what we may loosely call a "representation" of an ordered alphabet that we can also use as a linear "program" to guide our recitation of it when necessary. However, it is one thing to have such a program, but, as we shall see, quite another to use it adequately. When reciting the alphabet, we must know where we are in the sequence at any moment. In other words, in order to know what to say next, we must know what we last said.

Using simple analogies derived from the logic of such devices as programmable hand calculators, we might conceive that we might keep and update our place in such sequences in one of two possible ways. We might have an automatic program instruction counter so that as we read off one letter from the program sequence (e.g., R), the instruction counter automatically moves on to the next "instruction" for the following letter (e.g., S). Alternatively we must use a system by which we remember each letter we have just uttered and use it as an address to index the program at the correct point to retrieve the next. Thus we might remember we have just said "R" (holding R in working memory) and then use this information in working memory in order to index the main program in long-term memory and to retrieve the next instruction. The information in the working memory buffer store would be updated as we find, and say, "S," and the process would continue, for each letter in turn, until we complete the alphabet.

It is likely that both these descriptions are valid for the ways in which people go through various different well-learned routines in everyday life. However work discussed by Rabbitt (1980b; 1981a) suggests that under conditions of distraction from a secondary task people may fall back on the second technique and that in order to decide what to do next, they try to remember more or less about what they have just done.

A test for this arises out of the fact that sequences such as the English alphabet may be particularly easy for subjects to follow because each element in them occurs only once. Thus at any juncture when a person is briefly distracted, he need only remember the single, last item which he spoke aloud in order to locate himself at the unique correct point in the sequence. However, consider other kinds of sequences in which particular items may be repeated:

A   B   C   D   E   C   D   F   B   C   D   G   H   I   and so forth

Here, in order to know what to say next after either of the two B's, a person must remember at least one previous letter, but in order to know what to say after each C, he must remember at least two previous letters and to know what to say after D, he must remember at least three previous letters. If a person is equipped with an automatic program instruction counter-index, repetitions of items in the program will not affect his performance since the index will simply move along to the next instruction, designating it equally clearly whether it is a unique or a repeated item.

Rabbitt (1980; 1981a,b) found that, in fact, even when subjects learn sequences to a very strict criterion of 100 successive correct repetitions, people who are distracted by a secondary task (serial CRT) find sequences with repeats much more difficult to execute properly than sequences without repeats. Their errors most frequently occur at points of repetition and the difficulty of correctly repeating sequences increases with the estimated number of items which people have to remember to resolve potential ambiguities. All these difficulties affect elderly people more than their carefully matched young controls. Rabbitt and Vyas (described in Rabbitt, 1981a) have also tested this in visual search tasks in which young and old people have to scan successive displays looking for each of a sequence of target letters in turn. When they find the letter for which they are currently searching they must look for the next and so on through a learned sequence. Elderly subjects find this task very difficult, search relatively slowly, and frequently forget what they are looking for. They become especially confused when lists contain recurrent items (as described above) and in which, presumably, they always have to remember the last two or three letters that they found in order to decide which letter to search for next.

People Use Variable Samples of Past Events in Order to Predict Future Events

Continuous, serial self-paced choice response tasks (CRT tasks) may seem very trivial and artificial topics for laboratory investigations, but in fact they reveal many control problems which people encounter in everyday life. Consider a simple task in which a man has to make a different response to each of three signals which appear, in long sequences, one at a time in unpredictable order. As soon as he responds to one signal, it disappears and is immediately replaced by the next. Suppose that there are slight differences in event probability so that one signal occurs more often than the others. We know that a young subject can detect even small differences in signal probability and use this information to improve his performance, since he will begin to respond faster to the more frequent signals. In order to decide which signals are most frequent he will have had to consider samples of n successive signals during a sequence and to analyze them in order to discover whether all signals are equiprobable or not. By elementary sampling theory, the accuracy with which he can detect small variations in probability will increase linearly as $\sqrt{n}$; that is, the larger the sample of past events which he can remember and assess, the more accurate his predictions of future events will be.

Remington (1969) and Kirby (1972) have shown that young people use samples of from 3 to 10 successive signals and responses during CRT tasks in order to adjust their running expectancies for current signals. Rabbitt and Vyas (1981) found that elderly people base their running estimates of the probability of successive events on smaller sequences of past events than do the young. This might be expected to have two consequences for age changes in the efficiency of probability estimation—one obvious and one paradoxical. The obvious consequence is that we would expect elderly people to become less able to detect small, stable differences in event probability. Rabbitt and Vyas (1980) found that this was the case. The paradoxical consequence is that while a person who bases his running estimates of event probability on a long-running sample of past events may be expected to detect subtle, stable differences in event probability, he will only be able to readjust his expectancies rather slowly if sudden changes in event probability occur. In contrast a person who works on a short-running sample will only be able to detect relatively coarse differences in event probability, but if sudden changes in event probability occur, he will either detect them rapidly or not at all. Thus we have the paradox that elderly people might actually detect sudden, gross changes in event probability faster than the young because they base their running estimates on relatively small n's.

With this hindsight, results reported by Griew (1962; 1968) suggest that this is indeed the case. Griew found that when sudden changes in signal probability occurred during runs, elderly subjects detected them sooner than young people did. Note that this does not imply that we should make the counterintuitive generalization that elderly people are impulsive, while the young are cautious in all situations. We are not, of course, dealing with long-term memories involving entire life experiences but merely with the effects of changes in short-term, or "working," memory on the ability to carry out simple tasks. Moreover the fact that, in some situations, the young can and do use longer event samples than the old does not mean that they *cannot* also use shorter samples if they wish to do so. If experimental instructions emphasize the need to detect sudden changes, the young no doubt can bring themselves to use as short event samples as the old, but, as a consequence, they will then, like the old, be liable to be misled by random local probability perturbations into believing that probability changes have taken place when in fact they have not. The young can vary their sample sizes and so retain options which the old have lost.

The Gap between Acquiring and Using Information about Event Probability

We have seen that elderly people may lose the ability to hold enough information in working memory to compute accurate predictions of future events. They may thus lack the *information* necessary to exercise predictive control over their behavior. But it seems that possession of this information is a necessary, but not a sufficient, condition for the exercise of control.

This point is brought out by a clever series of experiments by Sandford and Maule (1971; 1973a,b) who compared young and elderly people on simulated industrial inspection tasks. Subjects sat before three sources which they had to interrogate by pressing appropriate response buttons to find out whether signals indicating events with which they were supposed to deal had occurred on any source. The experimenters systematically varied the probabilities of signals among these sources during their various experiments. Young people rapidly detected these probability differences and used this information to guide their interrogations. Elderly people seemed to detect probability differences as efficiently as the young since they could estimate and report them accurately after each experiment. But they apparently could not use their knowledge of probability differences during the tasks, since they interrogated all sources equally often.

Rabbitt (1979) reports a similar experiment in which young subjects readily learned that target symbols in a visual search task occurred more often at some display locations than at others and used this information to guide their search, recognizing targets faster when they occurred at frequent than at infrequent locations. Elderly subjects could also accurately *report* differences in target probability across display locations. However, they detected targets no faster when they occurred at probable than at rare, locations. Other experiments reported by Rabbitt (1981a,b) also emphasize that in old age people may lose the ability to retain and analyze information about their immediate past experience on which predictive control depends. But even when they appear to have deduced the necessary information, and can give a clear and accurate verbal account of it, they seem, quite independently, to lose the ability to *use* this information to maintain active control over their behavior.

APPLYING CONTROL SYSTEM MODELS TO ANALYZE CHANGES IN EFFICIENCY IN EVERYDAY LIFE

Clumsiness and Loss of Motor Control

The failures in control noted in experiments on hand movements and serial self-paced reaction-time tasks (Rabbitt, 1980c; Rabbitt & Birren, 1967; Rabbitt &

Rodgers, 1965) may be considered as part of a more general syndrome of deterioration of motor control in old age (Bernstein, 1967, pp. 92–93). Analyzing changes in efficiency in walking in old age points out that fit, young adults do not walk by reflex, waiting for such cues as the moment of heel contact with the ground, changes of pressure on the soles of the feet, or changes in joint position and muscle tension to be perceptible *before* they make appropriate corrective movements. They rather exercise *predictive* control, initiating complex motor patterns *anticipating* changes which are about to occur. If these changes do not occur, feedback is *later* used to initiate complex patterns of corrective movements.

In contrast to the smooth predictive control exercised by adults, very young children and elderly people walk in characteristically clumsy ways, initiating each movement sequence only in response to a particular feedback cue, rather than playing through motor programs which anticipate what happens next. Note that normal, young adults retain the option of *either* sort of control. A young adult will very rapidly adapt to walking with one leg immobilized by plaster, or can manage to hobble when peripherally disabled by an attack of "pins and needles," or to stagger on when centrally disabled by alcohol. In all these cases there is some regression from anticipatory predictive control to post hoc reactive control of walking. This change may be deliberate, and adaptive, even when normal central and peripheral function is retained if novel, unfamiliar movement patterns are required (e.g., picking one's way down an icy hillside). Again, the young retain the option of exercising *either* predictive *or* reactive control as necessary. The old gradually lose the option of predictive control and have to struggle on as best they can with reactive control. They thus have a smaller range of possible control strategies available to them.

Nearly all everyday motor skills require the use of predictive control if they are to be carried out with maximum efficiency. Typing, playing the piano, using a lathe, driving a car, or playing golf can be done well only if control is anticipatory and does not permanently lag at least one reaction time behind events. Very little is known of the extent or the pattern of age changes in such complex skills. We do not even know how far such skills can be retained if they continue to be practiced into old age. Nor do we know what kind of practice best helps the elderly to retain their skills, or whether the elderly can be helped to mitigate losses of control if they are taught new techniques. These seem very urgent and fruitful topics for further investigation.

Keeping Track of Everyday Conversations

We have seen that when elderly people are distracted while performing a simple laboratory task (e.g., PASAT), they can become confused so that even if they happen to hold necessary information in working memory they may fail to correctly access it and may retrieve unwanted data instead. Rabbitt (1981c) has found that similar difficulties in indexing and retrieving relevant items of information from immediate memory make it hard for elderly people to share in conversations involving groups of two or more other participants. People were asked to listen to series of four sentences which were all spoken by a single speaker or spoken in random order by two or four speakers. Young and elderly people recalled the sentences equally well when they were told to disregard which speaker had read which sentence. However, when they were told not only to remember what had been said but also who had said it, elderly people found the additional demand on memory so severe that they forgot the content as well as the sources of the remarks they had heard. Both young and elderly subjects were particularly prone to forget what had been said when they were required to interject brief remarks of their own into a taped conversation during a pause left for this purpose. This difficulty was most pronounced for the elderly, who could remember very little of what other speakers had said once they themselves had spoken.

These difficulties which elderly people seem to have in keeping track of laboratory analogues of everyday conversations allow us to interpret in more charitable ways some conversational mannerisms which we might otherwise attribute to their egotism or self-absorption. A tendency to interminable monologue may be merely a defense against embarrassment at confusion when trying to follow interchanges between several speakers. A tendency to continually make irrelevant interruptions and to guide a conversation into disjunctive sidetracks may in fact be a desperate attempt to disguise the fact that the drift of remarks has been lost. A failure to precisely acknowledge points of view or feelings which have been overtly or tacitly expressed may be the result of loss of information under conditions of memory overload rather than the result of personal indifference. The difficulties which old people face can be simply, and tactfully, mitigated once they are understood. If elderly people are helped to structure a conversational exchange, if exchanges are made easy for them to follow by providing enough redundancy for them to keep up, they can experience and share the modest daily pleasures of contact rather than becoming bored by, and boring because of, their growing incapacities.

Working Out What Happens Next

As we have seen, people often have to follow through familiar, learned sequence of actions to guide themselves through everyday tasks. Elderly people are vulnerable to distraction, forget what they have just done, and so may lose their place and garble sequences of actions which they are trying to follow, even when they can correctly remember them. This has many obvious implications for efficiency in everyday life. One daily task in which these are evident is the verbal description of familiar routes as compared to the task of following the same routes on a map or plan, or in the real world. If one follows a well-known route through a town, one does not lose one's way, even when distracted, because at each stage of the route one has merely to look around in order to locate oneself. Similarly when tracing a route on a map, one has an *aide memoire* which allows one to locate oneself at any point along it and continue toward a conclusion. Rabbitt (study, in progress) has interviewed 123 people aged between 50 and 85 years who had all lived in Oxford for 20 years or longer and who all, consequently, knew the center of the city very well. Each person was asked to describe simple routes in the city center which led between prominent landmarks along streets with which they were very familiar. The instructions were "How would you explain to a tourist how to get from X to Y?" (a common hazard for elderly people who are too slow moving, and of too polite a generation to easily evade such importunities). A control instruction was "How would you point out the correct route to the tourist using this city plan?"

Subjects aged from 50 to 70 years were very efficient at both tasks, giving concise verbal descriptions of routes. Subjects older than 70 did not make errors in selecting routes or forget parts of routes and give inaccurate directions. But their verbal descriptions were often inadequate because they frequently forgot which section of a route they had just described and either backtracked to repeat information which they had already given or skipped forward neglecting essential instructions for some subsection. Obviously they would not experience such difficulties when traversing a route in reality, since if they were ever momentarily distracted and forgot where they were, they could immediately resolve their uncertainty by looking about them. Many of these older subjects, who could not easily give consecutive descriptions of routes, were greatly helped by being allowed to refer to a plan on which, by moving a finger to trace a route, they found themselves cued in appropriate order to give each instruction necessary for a hypothetical stranger to find his way.

In brief there seemed to be a distinction between retention in long-term memory of an adequate "mental representation" of routes in a familiar city and retention of the ability to refer to this body of information in order to give consecutive instructions. As a test of how detailed our subjects' memories of the routes actually could be, we asked them to describe "mental walks" along the routes after they had described them, "for a tourist," now giving as much detail as possible of all the shops and buildings which they would expect to see. Ever our oldest subjects, who were often unhelpful at describing routes because they seemed unable to specify landmarks which a strange might conveniently use, would include these landmarks among a wealth of other detail when describing a mental walk. Once again, the distinction between the possession and the use of necessary information seemed to be the important factor.

CONCLUSIONS

Both developmental psychology and gerontology have suffered because they have borrowed models for human performance almost exclusively from cognitive and human experimental psychology. These models have proved inadequate because they have not been designed to describe *changes,* within or between individuals. Recent models based on elementary control systems theory provide better descriptions of change. Five examples are given of cases in which such models can be used to understand some of the difficulties which elderly people face in coping with particular situations in everyday life.

REFERENCES

Baddeley, A. D., & Hitch, G. Working memory. In G. H. Bower (Ed.), *The psychology of learning and motivation* (Vol. VII). New York: Academic, 1974, pp. 47–89.

Bernstein, N. *The co-ordination and regulation of movements.* Oxford: Pergamon, 1967, pp. 92–93.

Griew, S. Learning of statistical structure. A preliminary study in relation to age. In C. Tibbets & K. Donahue (Eds.), *Social and psychological aspects of aging.* New York: Columbia University Press, 1962.

Griew, S. Age and the matching of signal frequency in a two-channel detection task. *Journal of Gerontology,* 1968, **23**, 93–96.

Gronwall, D. M. A., & Sampson, H. *The Psychological effects of concussion.* Oxford, England: Auckland University Press/Oxford University Press, 1974.

Hamilton, W. M. E., Sandford, A. J. The symbolic distance effect for alphabetic order judgements: A subjective report and reaction-time analysis. *Quarterly Journal of Experimental Psychology,* 1978, **39**, 33–43.

Kirby, N. H. Sequential effects in reaction time. *Journal of Experimental Psychology,* 1972, **96**, 32–36.

McFarland, D. J. *Feedback mechanisms in animal behavior.* London: Academic, 1971.

Moray, N. The strategic control of information processing. In G. Underwood (Ed.), *Strategies of information processing.* London: Academic, 1976.

Parkman, J. M. Temporal aspects of digit and letter inequality judgements. *Journal of Experimental Psychology,* 1971, **91**, 191–205.

Rabbitt, P. M. A. Some experiments and a model for changes in attentional selectivity with old age. In F. Hoffmeister & C. Muller (Eds.), *Brain function in old age.* Bayer Symposium No. VII. Berlin: Springer-Verlag, 1979.

Rabbitt, P. M. A. Cognitive Psychology needs models for change in old age. In A. D. Baddeley & J. Long (Eds.), *Attention and performance IX.* Hillsdale, N.J.: Erlbaum, in press. (a)

Rabbitt, P. M. A. How do old people know what to do next? In F. I. M. Craik & S. Trehub (Eds.), *Erindale symposium on aging.* New York: Plenum, in press. (b)

Rabbitt, P. M. A. Age and conversational skill. In *New Society.* London: Macmillan, in press, January 1981. (c)

Rabbitt, P. M. A. A fresh look at changes in reaction times in old age. In D. Stein (Ed.), *The psychobiology of aging: problems and perspectives.* New York: Elsevier, North Holland, 1981. (d)

Rabbitt, P. M. A., & Birren, J. E. Age and responses to sequences of repetitive and interruptive signals. *Journal of Gerontology,* 1967, **22**, 143–150.

Rabbitt, P. M. A., & Rodgers, M. Age and choice between responses in a self-paced repetitive task. *Ergonomics,* 1965, **8**, 435–444.

Rabbit, P. M. A., & Vyas, S. M. Selective anticipation for events in old age. *Journal of Gerontology,* 1980, **35**, 913–919.

Remington, R. J. Analysis of sequential effects in choice reaction times. *Journal of Experimental Psychology,* 1969, **82**, 250–257.

Sandford, A. J., & Maule, A. J. Age and the distribution of observing responses. *Psychonomic Science,* 1971, **23**, 419–420.

Sandford, A. J., & Maule, A. J. The allocation of attention in multisource monitoring behaviour. Adult age differences. *Perception,* 1973, **2**, 91–100. (a)

Sandford, A. J., & Maule, A. J. The concept of general experience: Age and strategies in guessing future events. *Journal of Gerontology,* 1973, **28**, 81–88. (b)

Thomas, C. Deficits of memory and attention following closed head injury. Unpublished M.Sc. thesis, University of Oxford, 1977.

Toates, F. *Control theory in biology and experimental psychology.* London: Hutchinson's Educational, 1976.

33

LEARNING IN LATER ADULTHOOD

Marion Perlmutter
Judith A. List
Institute of Child Development
University of Minnesota

The purpose of this chapter is to examine learning in later adulthood, with particular sensitivity to the questions of whether there are age-related changes in learning of older adults and, if so, how these changes can best be characterized and explained. In an introductory section, an attempt is made to define learning and delineate variables relevant to it. In subsequent sections, findings on learning in later adulthood from both laboratory and nonlaboratory research are considered. In the section on learning in laboratory situations, studies of conditioning, verbal learning, and cognitive skill training are reviewed. In the section on learning in naturally occurring situations, physical, cognitive, and social acquisitions of later adulthood are considered. In a final section, some theoretical and practical implications of these findings are discussed.

NATURE OF LEARNING

The nature of learning has been of interest to scholars from a diversity of disciplines and has been of central concern to psychologists attempting to understand behavior and its development. While experimental psychologists have not always agreed on an appropriate definition of learning, one offered by Hilgard and Bower (1975) has often been accepted. They state that:

> Learning refers to the change in a subject's behavior to a given situation brought about by his repeated experiences in that situation, provided that the behavior change cannot be explained on the basis of native response tendencies, maturation, or temporary states of the subject. (p. 17)

Similarly, Botwinick's (1967) definition of learning (also cited by Arenberg & Robertson-Tchabo, 1977) seems to capture the essentials of this construct. He indicates that "learning is the acquisition of information or skills measured by an improvement in some overt response" (p. 48). A slightly different but compatible perspective on learning is taken by more biologically attuned scholars. They consider learning to be a process whereby organisms become adapted to their particular environment. In some respects learning is similar to genetic evolution. Both involve adaptation of organisms to environmental demands, and both can be transmitted across generations. However, the time frame and unit of analysis of the two processes differ. The primary adaptive function of learning, which occurs relatively rapidly, is specific to individuals, while the adaptive function of genetic evolution, which is slow to occur, is believed to be relevant to the species as a whole.

Traditionally it had been assumed that general "laws of learning" were discoverable and that discovering them would be accomplished within the confines of relatively simple, carefully controlled experimental procedures. Now, however, it is more commonly believed that learning phenomena are diverse and contextually determined (Estes, 1975; Jenkins, 1979), and perhaps should be investigated from a broader perspective (Charlesworth, 1979; Neisser, 1978; Perlmutter, 1980). Some of the most salient variables to affect learning in later adulthood are discussed in the sections that follow.

Learner Characteristics

Learner characteristics are known to influence learning, and perhaps the most obvious learner characteristic relevant to the present analysis is age. While age is a convenient and exact index of passage of time, it also is an inexact index of numerous other confounded, and often unrecognized, variables. For example, age is quite predictive of biological state and somewhat predictive of education, income, and lifestyle. Thus even if age was found to differentiate individuals in terms of learning skill, it would still be necessary to determine which factors in the age variable actually were relevant.

A related issue concerns the cohort-specific experiences that are completely confounded with passage of time. That is, since passage of time is uniquely experienced by each generation, it probably is impossible to draw conclusions about effects of passage of time that are totally generalizable across generations. On the other hand, since cohort factors that are most likely to contribute to age differences in learning are sometimes known, at least some attempt can be made to take account of them.

Learners' expertise and prior knowledge of the content to be learned, or content relevant to it, also has been shown to enhance learning (Chi, 1978; Larkin, Heller, & Greeno, 1980). Experts not only have more information within their knowledge structures, but have the information more tightly organized and possibly more easily accessible. In addition, with experience learners become more efficient at assessing the demands of a task, as well as at assessing their own skills, and the learning activities available to them. At first this learning how to learn is achieved within fairly specific domains, but subsequently more general learning strategies are acquired. Since older learners typically have more prior experiences to draw upon, they might be considered generally more "expert" than the younger learners. However, there is still some question as to whether this prior learning facilitates or interferes with older adults' new learning (Postman, 1961).

Learning Contexts

Another possibly important factor in learning is the context within which it occurs. Relevant questions are: How important are differences in learning when it is completed on one's own versus in a social situation? Are there differential outcomes from being tutored by a peer, a parent, or an instructor? What are the unique consequences of learning from being taught versus teaching? How are informal and formal educational experiences distinct? These issues may contribute to age differences in learning. For example, across adulthood the predominance of learning contexts tends to shift (from formal to informal). Likewise, the changing social role that learners bring into learning situations (as youngsters they typically are subservient to teachers, while as adults they generally are peers) may be an overlooked factor in age differences in adult learning.

Learning Contents

Humans learn a variety of content, from fine-grain motor responses, to complex cultural practices, symbols, and conceptual abstractions. Moreover, the nature of representation of these acquisitions may shift. For example, one might learn how to manually operate a machine through physical experience with it, but then transfer this skill to a cognitive domain in order to communicate it to fellow workers. Conversely, one might be taught about general rules of appropriate behavior in a particular culture, but then be left to transform these prescriptions into specific actions. The distinction being made here relates to differences between "learning how" versus "learning about." While some general statements about learning in later adulthood might pervade each of these domains, it is likely that a more specific analysis will point to age-related differences.

Even within a plane of representation, other factors concerning the content of what is to be learned are relevant. For example, the complexity and/or detail of material is known to affect learning, although probably not consistently across age. Likewise, organizational and sequencing factors have known effects on learning, although, to some extent, they, too, interact with age. It will be extremely important to understand these interactions, for they hold both theoretical and practical significance.

Learning Activities

Much learning occurs simply in going through day-to-day life, even when experiences are not directed specifically toward learning. Moreover, this extensive informal education is supplemented by formal learning, although formal education occurs mainly during early life. While the formality of a learning experience, in itself, may not be critical to learning, it often dictates activities that are relevant. That is, the structure inherent in formal educational situations often is designed to promote optimal learning activities. Moreover, a goal of formal education is to teach people the activities that lead to effective learning.

It is interesting to note that at least early development is associated with increasing use of effective learning strategies. That is, much of the early change observed in learning appears to be related to increases in the spontaneous use of efficient learning activities, rather than to increases in their availability. It has been speculated, therefore, that many age changes in learning are the result of acquiring knowledge about learning or metacognition (Cavanaugh & Perlmutter, 1980). This acquisition is likely to accrue throughout life.

Learning Goals

Much learning occurs incidentally, without a specific goal for learning. On the other hand, some things are learned intentionally, after having set very deliberate goals. The most important function of learning goals may be that they provoke the use of particular learning activities. Thus establishing a goal to learn a particular body of knowledge demands exposure to it, as well as work to comprehend and retain it. Similarly, learning to play a musical instrument requires practice. In attempting to investigate learning, it often becomes difficult to gain adequate knowledge of, or control over, subjects' learning goals. This problem may be particularly pertinent to developmental studies, since age is often related systematically to learning goals. For example, a fairly young or old subject is likely to be less motivated to learn a task specified by a college professor than would a student of that professor.

Learning Assessments

In order to determine whether learning has occurred, performance should be assessed both before and after a learning experience. Unfortunately, however, it is not often obvious how performance should be assessed. That is, it may not be clear how to provide ample opportunity for a learner to demonstrate what has been learned. For example, is learning during college adequately assessed by multiple-choice exams, or even by essays and papers? Are there important, albeit difficult to measure, nonacademic kinds of learning that take place during college? Are particular assessment instruments suited equally well for all learners? It is essential that important, but perhaps subtle and difficult to measure, effects of learning not be overlooked.

LEARNING IN LABORATORY SITUATIONS

Most research on learning has involved experiments carried out in the laboratory. In general, conditions are highly controlled, and subjects are tested on tasks designed to elucidate the effects of particular variables on learning or changes in learning that are associated with age. An assumption inherent in this approach has been that there are basic principles of learning that can be understood best by experimentally dissecting each relevant factor. Unfortunately, while this approach allows investigators to gain knowledge of the particular variables studied, within the confines of the particular situations in which they have been investigated, it may not help one assess the relative importance of these variables (variance accounted for) in different situations. This predicament is especially problematic if the effects of variables are not additive. That is, if there are interactions, effects from separate experiments cannot simply be summed, since estimates of all the interactions needed to account for performance are not available. In essence, then, confidence in the value of findings from laboratory experiments must rest on subjective assessments of the importance of the variables that have been investigated and on faith that interaction effects have been adequately taken account of. Since it is not yet known how relevant experimental tasks are to adult life, the significance of research on older adults' learning in laboratory situations remains somewhat suspect. Nevertheless, there has been considerable research involving comparisons of younger (generally college age) and older (generally over age 60) adults' learning, and several seemingly important age differences emerge. A summary of the major findings is provided below.

Conditioning

Conditioning is often viewed as the simplest form of learning. In classical conditioning the learner is trained to make a generalized response to a signal, and in operant conditioning the learner acquires an instrumental response to a discriminated stimulus. During recent years there have been few human-aging studies of conditioning. Still, there are some important issues to be resolved within these paradigms. For example, in two studies of classical conditioning of the eye-blink response, Braun and Geiselhart (1959) and Kimble and Pennypacker (1963) found that older adults had greater difficulty acquiring a conditioned eye blink than did younger adults. Moreover, Kimble and Pennypacker (1963) found that the magnitude of the unconditioned response was significantly correlated with the frequency of conditioned responses. Thus it was hypothesized that over time the eye-blink response becomes habituated or partially adapted out and, therefore, is less susceptible to modification by conditioning. In order to draw firm conclusions about possible age differences in susceptibility to classical conditioning, additional research, involving different responses, is needed. The evidence that is available (Shavonian, Miller, & Cohen, 1968, 1970) seems to indicate that on a variety of autonomic measures there are weaker conditioned and unconditioned responses in older adults. In addition, the strength of the conditioned and unconditioned stimuli are known to influence conditioning and are likely to differ across age, since older adults' sensory functioning is reduced. Similarly, the time parameters chosen in particular experiments might contribute to conclusions about age differences in conditioning, since older adults require longer time to encode and respond to stimuli than do younger adults.

Operant conditioning also has received scant experimental attention in recent research on human aging, although these techniques have begun to be used extensively in a variety of applied settings. Research that is available points to the efficacy of reinforcement procedures, even for adults well into their 70s (Ayllon & Azrin, 1965; Baltes & Zerbe, 1976). Thus although all the parameters of operant conditioning have not been documented for older adults, it is clear that these procedures can be effective controllers of behavior in later life. It should be noted, however, that reinforcers that are optimally effective for young adults probably are not the same as those most effective for older adults.

Verbal Learning

There has been much developmental research on verbal learning. Many investigators have been interested in whether there are age differences in learning of lists of words and whether the magnitude of age differences are affected by specific variables. The overwhelming evidence points to age differences in verbal learning performance; groups of older adults (60+) typically perform statistically worse than groups of younger adults (20s). It should be noted, however, that even the performance of the oldest subjects almost always is well above chance level, and performance of the different age groups almost always overlap. Moreover, the pattern of age differences in various experimental conditions differs; several manipulations attenuate age-related declines. Indeed, the patterns of results in various experiments point to performance factors that contribute to age-related deficits and leave open to question whether there is aging of learning ability per se (Agruso, 1978; Arenberg & Robertson-Tchabo, 1977; Botwinick, 1978; Poon, Fozard, Cermak, Arenberg, & Thompson, 1980).

One of the most clearly demonstrated factors contributing to age differences in learning performance is pacing. There have been a number of studies in which amount of time stimuli are available for study and/or amount of time available for response are varied. In general, it has been found that older adults are especially disadvantaged when time is limited and that this effect is particularly strong when there are limitations in response time (Arenberg, 1965; Canestrari, 1963; Eisdorfer, Axelrod, & Wilkies, 1963; Monge & Hultsch, 1971; Taub, 1967). Thus when sufficient time for response is available, performance of older adults is only slightly worse than that of younger adults. This finding indicates that older adults need more time than younger adults to show what they have learned. However, even when given a long time for response, older adults perform relatively poorly if they have been rushed during study. Thus older adults also need more time to learn material than do younger adults.

Older adults' need for extra learning time may be related to limitations in their spontaneous use of effective coding strategies. In several experiments subjects' performance has been compared after standard intentional learning instructions and experimenter-directed learning instructions designed to enhance acquisition (image, associate, categorize). In general, it has been found that age differences are attenuated when subjects are provided specific learning instructions (Canestrari, 1968; Hulicka & Grossman, 1967; Hultsch, 1971; Perlmutter, 1978, 1979). These findings suggest that older adults spontaneously use effective acquisition strategies less than younger adults, even though such strategies are available to them.

Age differences in motivation also have been hypothesized to contribute to the often observed age decrements on experimental learning tasks. Typically, it has been assumed that older adults are not as motivated as younger adults, since laboratory experiments are not meaningful to them. However, Botwinick (1978) has argued that older adults actually are more involved in experimental situations than are younger adults and, in fact, that they are sometimes inappropriately involved to an extent that depresses their performance. Evidence for this view comes from psychophysiological studies of arousal state that show older adults often are overly aroused during experimental sessions (Furchgott & Busemeyer, 1976; Powell, Eisdorfer, & Bogdonoff, 1964). Moreover, when these arousal states have been reduced by drugs, or adaptation to the laboratory, performance has been found to improve (Eisdorfer, 1968; Eisdorfer, Nowlin, & Wilkie, 1970).

Older adults are thought to be more cautious than younger adults, as well. This tendency toward cautiousness often makes them appear to have learned less than they actually have. Evidence for this view comes from a fairly consistent finding that older adults make many errors of omission but rarely errors of incorrect responding (Eisdorfer, Axelrod, & Wilkie, 1963; Korchin & Basowitz, 1957). Moreover, in one study in which omission errors were discouraged by rewarding all responses, regardless of whether they were correct or incorrect, older adults' learning performance disproportionately improved (Leech & Witte, 1971).

Another hypothesized explanation of age deficits in laboratory learning performance is that older adults are more prone to interference than are younger adults. Unfortunately, as noted by Arenberg and Robertson-Tchabo (1977), methodological difficulties presently make it impossible to evaluate this hypothesis.

Cognitive Skill Training

A few researchers have attempted to train cognitive skills in older adults. Modifiability of intellectual functioning has been demonstrated by Hoyer, Labouvie, and Baltes (1973) and Labouvie-Vief and Gonda (1976), although transfer of training was nonexistent or limited to closely related tasks. On the other hand, Plemons, Willis, and Baltes (1978) demonstrated modifiability of fluid intelligence skills, as well as transfer of training. However, on a 6-month posttest, they found attenuation of the difference between their training and control group that was attributable to practice gains in the control group. Thus the performance gains may not have been dependent upon the training program.

The durability and transfer of enhanced performance of older adults due to training has been studied in several other domains, as well. Sanders and Sanders (1978), for example, found that elderly individuals trained to use an efficient strategy to solve unidimensional concept identification problems demonstrated better performance on a bidimensional conjunctive concept identification problem administered 1 year later. Sterns and his associates (presented in Sterns & Sanders, 1980) have demonstrated that sequenced training of older adults is effective in improving information-processing skills needed for effective driving for at least a 6-month period. Hornblum and Overton (1976) reported rapid improvement in performance on a conservation-of-surface task for older adults trained with a feedback procedure. This training effect was much stronger than that from a nonfeedback procedure, and it resulted in transfer of training on 4 out of 5 tasks. Schultz and Hoyer (1976) found performance gains on both an immediate and delayed test by older adults given visual-auditory feedback on a perspective-taking task; this improvement was greater than that of adults in both nonfeedback practice and untrained control groups. However, no transfer was evident for any of the groups. Zaks and Labouvie-Vief (1980) attempted to train social cognitive functioning in older adults. A group of elderly adults was given a 6-week training program consisting of discussion and role-playing of problems, as well as practice in taking the listener and speaker roles in a referential communication task; a placebo group participated in separate discussion groups, but was not given the opportunity to engage in role-playing, and received practice in only one role of the referential communications task; and a control group received no training. On a spatial perspective-taking task the training group performed better than the placebo group, which in turn outperformed the control group. On a referential communication task the training group outperformed both the placebo and control groups, which were not different from one another. The authors concluded that nonspecific training may facilitate performance on some tasks but that the opportunity to engage in role-playing facilitates social cognitive skills above and beyond the effects of less circumscribed interaction.

LEARNING IN NATURALLY OCCURRING SITUATIONS

Relatively little research on learning in naturally occurring situations has been carried out, and this paucity of research is especially evident for learning in informal situations (nonclassroom learning). Moreover, the little research of this kind that has been carried out has not had a developmental focus. Yet systematic evaluation of learning in naturally occurring contexts, and of age differences in such learning, would be useful. At the very least, such an analysis would permit the extension and validation of current understanding of learning and development derived from laboratory studies. Since the learning that takes place in everyday life tends to be considerably more complex than that tested in most experimental studies, there is some reason to believe that important factors may be missing from current understanding. Moreover, while it is possible that aging detrimentally influences the micromechanisms of learning most heavily relied upon to perform most experimental tasks, aging may have a more limited adverse effect, or perhaps even a positive effect, on some of the macrolearning styles used in everyday life. Thus information about older adults' everyday learning poses a potentially unique data source for those interested in aging and learning. The scant literature on such learning, in a sample of domains pertaining to physical skills, cognitive skills, and social skills, is summarized below.

Physical Skills

The biological deterioration that accompanies aging may provide impetus for learning and compensation. Just as the developing child must learn to cope with rapid physical growth and hormonal change, the aging adult must learn to adapt to sensory loss and skeletal, muscular, and other organic deficits. In this section the impact of physical training on physiological and psychological functioning of older adults and their adaptation to loss of sensory functioning are discussed.

Physical Training. Age-related losses in physical skills are well-documented (Asmussen et al., 1975; Atomi & Miyashita, 1974; Dehn & Bruce, 1972; Drinkwater et al., 1975; Kilbom, 1974; Profant et al., 1972; Robinson et al., 1976). While these losses are at least partially attributable to physiological aging per se, they probably also reflect undiagnosed disease and loss of fitness associated with reduced activity. Of concern for the present purposes is evidence about whether older adults profit from physical training.

Until quite recently it generally was believed impossible to improve older adults' physiological functioning (DeVries, 1975). However, since the mid-1960s numerous studies have demonstrated enhancement of aerobic capacity and cardiac functioning following training in both elderly men and women (Adams & DeVries, 1973; DeVries, 1970; Hartley et al., 1969; Kilbom et al., 1969; Saltin et al., 1969; Sidney & Shephard, 1977). Several additional points also emerge from this more recent research. First, those who begin training programs in poorer physical condition show the greatest gains, regardless of age (DeVries, 1970; Kasch & Wallace, 1976). Second, if compared on a percentage basis, older adults demonstrate gains comparable to younger adults (Adams & DeVries, 1973; DeVries, 1970; Hartley et al., 1969; Saltin et al., 1969; Suominen, Heikkinen, & Parkatti, 1977). Third, while it may not be necessary to require maximal exercise to obtain improvement in aerobic capacity (DeVries, 1970; Suominen, Heikkinen, & Parkatti, 1977), type of exercise probably is important. For example, it has been suggested (DeVries, 1975) that the rhythmic exercises utilized in most training programs (running, jogging, and swimming) yield optimal results.

Other areas of elderly functioning also have shown improvement following physical training. Suominen, Heikkinen, and Parkatti (1977) found improved metabolism in skeletal muscle and connective tissue following 8 weeks of various exercise activities (walking, swimming, gymnastics). Elsayed, Ismail, and Young (1980) documented significant increases on measures of fluid, but not crystallized, intelligence for young and old men in a physical training program. In some studies changes in personality variables related to physical training also have been reported. For example, Young and Ismail (1976) and Sharp and Reilly (1975) have noted increases in conscientiousness and persistence in older males participating in a physical fitness program. Hartung and Farge (1977) have documented specific personality traits of middle-aged runners and joggers, although the directionality and causality of their findings remain unclear, and participation in exercise programs has been found to increase health and exercise consciousness (Gutman, Herbert, & Brown, 1977; Sidney & Shephard, 1977; Thomas, 1979). Finally, DeVries (1975) has suggested that physical training may have a greater sedative effect than popular tranquilizers for anxious adults.

It should be evident that the question posed at the beginning of this section has been answered with a definitive "Yes." In virtually every study reviewed, older adults have shown significant improvements in physiological and intellectual performance following training, and this improvement has been quite comparable to that observed in younger adults. However, to date there have been few studies in which the differential efficacy of various types of exercise on elderly functioning has been investigated (a study by Gutman, Herbert, & Brown, 1977, is an exception). Likewise, the issues of optimal length of time of training, individual differences in response to exercise, and effectiveness of various presentation methods have not been addressed.

Sensory Adaptation. Detrimental effects of aging on sensory functioning are nearly universal (Corso, 1977; Fozard et al., 1977). For this discussion only the adaptations that older adults make to deteriorations in sight and hearing are considered. It will be argued that although there are substantial declines with age in the quality of information received from sense organs, older adults learn to compensate for these deficits, at least to some extent.

One of the most common and clearly established age-related decrements in visual function concerns dark adaptation. It has been found that rate of dark adaptation slows with age and that final level of dark adaptation is lower in older than younger adults. An important implication of these changes is reduction of vision during night driving. Yet there has been little research on the adaptation of older adults to night driving, and vision tests for driver's licenses do not include assessments of such skills. Nevertheless, there are several ways in which older adults may accommodate to diminished vision associated with reduced dark adaptation. For example, they can learn to look to the lower-right corner of the windshield when oncoming headlights are detected, rather than looking directly at them. In addition, they can learn to anticipate problems while driving and to take advantage of environmental context.

Auditory losses also are common in later adulthood, with most older adults experiencing presbycusis. This auditory change is characterized by a loss of hearing for high-frequency sounds. A person suffering from presbycusis may not hear telephones ring, may have difficulty hearing human voices, particularly those of women, and may have problems intelligibly interpreting speech, since consonants such as f, s, and z are particularly difficult to hear. Thus loss of hearing associated with presbycusis has implications for both the safety (detection of warning signals) and social communication of older adults. While environmental (amplification and frequency of signals) and mechanical (hearing aids) adjustments can reduce the negative impact of these hearing losses, such aids are not entirely effective. Over time, however, older adults learn to adapt to some of the limitations of these aids (overamplification of sudden low-frequency noises). Moreover, older adults who experience hearing loss learn to compensate by depending more on lipreading and contextual cues.

Cognitive Skills

Cognitive development is recognized as an important aspect of early childhood. The advent of compulsory education and the theoretical views of Piaget have contributed to this emphasis. However, until recently interest in cognitive development beyond adolescence has been meager. Adults engage in formal education far less frequently than children. In addition, Piagetian theory, as well as most other perspectives on cognition, assumes that cognitive development ceases at the time of maturity. Yet there is no evidence to refute the possibility that cognitive functioning continues to develop beyond adolescence. The prevalence of second careers, job-related retraining, and continuing education attest to the ability of adults to acquire new information, refine acquisition skills, and develop strategies for coping with cognitively complex and changing environments. In this section attention is directed to the role of media exposure and formal educational experiences on cognitive learning in later adulthood.

Media Use. Adults make extensive use of media in today's society, and although much of this use of newspapers, magazines, books, radio, film, and television is geared toward entertainment and recreation, some is motivated toward a desire to increase understanding and competence. For example, about 60% of adults in the United States report regular use of newspapers to acquire new information, 40% report such use of magazines, and about 35% report such use of books (Knox, 1978). While the extent to which adults read is highly associated with level of formal education, it is only slightly associated with age (Parker & Paisley, 1966). Even so, adults who continue to read beyond the age of 60 tend to read material that is readily available; there is a decline in their use of printed material that must be obtained outside of the home (books from libraries and bookstores).

More than half the adults in the United States report regular use of electronic media to acquire new information (Knox, 1978). When entertainment and informational use of electronic media are combined, there is a slight decrease in use in old age (Knox, 1978). However, when informational use of electronic media is analyzed separately, a different pattern emerges; informational use of both radio and television increases in later adulthood (Parker & Paisley, 1966). It appears then that educationally oriented use of media is extensive in today's society and that considerable use of television extends to cohorts that have not experienced this medium during their early years. Research that focuses on the educational impact of television, particularly for adults, still needs to be carried out. Since much adult learning may be accrued through television, appropriate programming should be useful in providing information to the elderly about important topics, including consumerism, health, nutrition, and politics.

Formal Education. Lifelong education recently has received attention by psychologists, sociologists, and educators (Birren & Woodruff, 1973; Huberman, 1974; Havinghurst, 1976; Schaie & Willis, 1978; Sterns & Sanders, 1980). Moreover, each year during the mid-1970s, about 25% of adults in the United States engaged in at least one major continuing education activity. This percentage represents about a 5% increase over a single decade (Knox, 1978). This increased interest in continuing education probably derives from the changing American age and educational structure (Knox, 1978), rapid technological advancement and obsolescence of knowledge (Dubin, 1972; Wrocznski, 1974), increases in multiple careers (Birren & Woodruff, 1973), and changing roles of women in the labor force (Troll, 1975). While there has been much discussion of the underlying assumptions (educational rights, societal necessity, and developmental imperatives) and goals (educational equality, prevention, and intervention) of continuing education, there has been little research to identify methods and strategies that optimize adult learning in formal settings. Moreover, little is known about age differences in the effects of continuing education.

It is known, however, that participation in continuing education declines with age. The highest participation rate is 29% in the 20s, followed by 26% in the 30s, 21% in the 40s, 16% in the 50s, 10% in the 60s, and 4% for persons 70 or older (Knox, 1978). Part of this age-related decline undoubtedly is due to location, topic, schedule, and cost. In addition, it is well-established that level of prior education is highly associated with extent of participation in continuing education (Johnstone & Rivera, 1965; Knox & Videbeck, 1963; London, Wenkert, & Hagstrom, 1963; Parker & Paisley, 1966) and, of course, level of formal education is known to be lower for today's older population. The percentage of adults with only some grade-school experience who participate in continuing education during a year is only 6%, while it increases to 9% for those who have completed eighth grade, 15% for those with some high school, 24% for high school graduates, 36% for those with some college, 39% for college graduates, and 47% for those who have completed a year or more of graduate education (Knox, 1978).

The goals of adults who engage in continuing educational activities vary greatly, but there seem to be some systematic shifts with age. More specifically, occupational considerations are most influential for young adults, leisure interests are more central during middle adulthood, and general information and social contact increase in importance in later adulthood (Hendrickson & Barnes, 1967; Johnstone & Rivera, 1965). Although the needs of older learners have not yet been assessed adequately, many colleges and universities have shown their receptivity to this segment of the student body. For example, they have instituted, modified, and individualized curricula and offered courses at a variety of geographic centers. All these changes serve to increase the availability and flexibility of accredited education and make it more attractive to older adults.

While continuing education appears to be gaining an increasing role in American society, at least for the foreseeable future, its role is likely to remain relatively modest. Nevertheless, its potential for reducing age differences is appreciable. To date, however, little evidence is available concerning how existing programs enhance the lives of participants, nor what the relative merits are of various programs for different segments of the population. Such data will be valuable for assessing developmental and cohort changes in learning and adult cognitive functioning.

Social Skills

Learning social skills and roles is a lifelong process. Acquisition of these competencies typically is an informal learning process, whereby the ritualization of interactions, adjustment to new relationships and roles, and fine tuning of interpersonal skills proceeds in a trial-and-error fashion, with peers, parents, and teachers serving as corrective guides. By adulthood it is often assumed that styles of interaction and patterns of adjustment are well-established. However, adults are often faced with new social demands for which they may not have the requisite skills. Entry into a career, marriage, parenthood, and widowhood are examples of social transitions. In this section retirement is considered as another such transition; the older adult must define and learn to meet the demands of this major role change. In addition, the efficacy of psychotherapy as an avenue for older adults to acquire new patterns of adjustment and to learn or relearn social skills is considered.

Retirement Adjustment. Much research on retirement has been designed to relate retiree's adjustment, satisfaction, and attitudes to a multitude of variables, such as health, income, education, occupational status, and work orientation. Until recently researchers have not considered retirement planning as a potential influence on retirement success. Retirement planning, either informally (through discussions with friends, colleagues, and family), or more formally (through participation in community- or company-based preretirement programs), is an active process of learning about the social role of a retiree, the benefits and options of retirement, and the process of adjusting to it.

Little is known about the impacts of informal methods of learning about retirement, although the vast majority of older workers rely exclusively on this process of information acquisition. The only available description of this learning mode comes from a study carried out by Simpson, Back, and McKinney (1966). They assessed the extent to which a population of workers sought information about retirement from a variety of sources. They found that most preretirees, at all occupational levels, received such information from retired people, fellow workers, and the media, rather than from company officials or social security personnel.

Still, relatively formal retirement planning programs are increasing in number. The main goals of such planning programs have been viewed as either counseling or planning (Kasschau, 1974). Programs that focus on counseling seek to reduce worker anxiety about retirement, dispel myths and stereotypes about aging and retirement, and change workers' attitudes toward loss of the work role. Programs that focus on planning seek to provide the older worker with information about concrete steps that may be taken to ensure a comfortable transition into retirement, such as information about pension benefits, financial planning, legal planning, medical benefits, health care, nutrition, exercise, and activity planning.

An overwhelming majority of studies have indicated that older workers are less interested in the topics emphasized by counseling programs (Kalt & Kohn, 1975) and that they engage in fewer such programs (Kimmel, Price, & Walker, 1978). In addition, research has shown relatively few measurable changes following counseling programs, perhaps because they are designed to modify attitudes and morale, which may be relatively difficult to influence (Glasmer & DeJong, 1975; Tiberi, Boyack, & Kerschner, 1979). On the other hand, evaluation of planning programs has demonstrated that they are quite successful. Stimulation of planning and acquisition of information has been documented following many retirement planning programs (Glasmer & DeJong, 1975; Tiberi, Boyack, & Kerschner, 1979).

The effectiveness of different types of planning programs has received only minimal evaluation. Glasmer and DeJong (1975) exposed workers to retirement information in groups or individually. While the lecture-discussion format appeared most effective, as the authors noted, the greater amount of time spent in group discussion precludes confidence in this finding. Sieman (1976) found that older adults learned as much as younger adults when the information was presented in a self-paced programmed instruction format. Thus, programmed instruction may provide a relatively inexpensive method of transmitting information to elderly adults.

Psychotherapy. It has been estimated that 17% of the elderly currently manifest psychological impairment severe enough to affect their daily functioning (Abrahams & Patterson, 1978–79) and that the prevalence of all forms of psychological disorders in the elderly is about 30% (Whanger & Busse, 1975). Moreover, although only 10% of the population is over 65, this cohort accounts for 25% of all reported suicides (Butler, 1975). When these statistics are juxtaposed against the finding that only about 5% of the people seen in community mental health clinics are over age 65, the incidence of unserved needs becomes real.

Since the writings of Freud, therapy for older adults often has been considered of little value. The emphasis on resolution of early conflicts in psychoanalysis led to the view that persons over 50 would not profit from extended psychotherapy. It was thought that their conflicts would be too rigidly embedded in their personalities to allow sufficient time for change. At present, however, the notion that older adults are unresponsive to therapy generally is considered a myth (Knight, 1978–79; Karpf, 1980).

Still, it remains unclear whether there are age differences in the effectiveness of various therapeutic techniques. Knight (1978–79) argued that therapies typically used with younger adults are appropriate for aged populations as well (dynamic psychotherapy, family and marital counseling, sex therapy, behavioral intervention, and group therapy). Thus according to Knight the selection of a therapeutic technique should depend more on the nature of the problem than on the age of the client. Still, as Knight pointed out, the elderly have concerns that are unique to their cohort. Moreover, Butler and Lewis (1973) considered age-specific therapies for the elderly appropriate. For example, they have suggested that life review therapy is especially suitable for older adults, since it capitalizes on the normal phenomena of reminiscence in old age. Life review therapy facilitates older adults' reorganization of past events to permit them to gain perspective and self-understanding of life-cycle patterns, that in turn, raises self-esteem and decreases anxiety. An exploratory study of the efficacy of behavioral group therapy and life review group therapy indicated greater gains in self-esteem and greater reductions in anxiety and incidence of somatic behavior for participants in life review therapy (Ingersoll & Silverman, 1978). However, the extremely small sample sizes, lack of adequate control groups, short duration of therapy, and inability to control extraneous variables preclude definitive comparison of the two techniques. Clearly, more research on the effectiveness of various therapy techniques for older adults needs to be conducted, and greater attention should be directed to individual differences in responsiveness.

SUMMARY AND CONCLUSIONS

The perspective taken in this chapter has been that effective learning is a lifelong activity. While this view seems hard to dispute, in fact, it has not been central to many theoretical or practical considerations of adulthood. Rather, previous conceptualizations of adulthood have tended to view adults as having fixed or declining modes of functioning. Perhaps it has been the recent advent of large numbers of adults, surviving literally 50 or more years of adulthood, that has called this view into question. Still, the full promise of growth and development during adulthood has yet to be appreciated.

One step toward the appreciation of this perspective may be a better understanding of the learning skills and abilities of older adults. As has been summarized in this chapter, considerable experimental research has been carried out to assess these skills. In general, age-related declines have been noted. However, it is unclear how important these deficits are to learning in everyday life. Some evidence of older adults' effective learning in naturally occurring situations has been documented. Unfortunately, systematic research on possible age limitations, and/or advantages, in such situations is not available. Identification of how individual differences in older adults relate to learning, as well as the context, content, activities, and goals of learning in later adulthood, is of utmost importance. Analysis of these issues should provide a richer and more complete theoretical understanding of learning and should serve practical interests, as well.

REFERENCES

Abrahams, R. B., & Patterson, R. D. Psychological distress among the community elderly: Prevalence, characteristics and implications for service. *International Journal of Aging and Human Development,* 1978–79, **9**, 1–18.

Adams, G. M., & DeVries, H. A. Physiological effects of an exercise training regimen upon women aged 52–79. *Journal of Gerontology,* 1973, **28**, 50–55.

Agruso, V. M., Jr. *Learning in the later years: Principles of educational gerontology.* New York: Academic, 1978.

Arenberg, D. Anticipation interval and age differences in verbal learning. *Journal of Abnormal Psychology,* 1965, **70**, 419–425.

Arenberg, D., & Robertson-Tchabo, E. A. Learning and aging. In J. E. Birren & K. W. Schaie (Eds.), *Handbook of the psychology of aging.* New York: Van Nostrand-Reinhold, 1977.

Asmussen, E., Fruensgaard, K., & Norgaards, S. A follow-up longitudinal study of selected physiological functions in former physical education students—after 40 years. *Journal of the American Geriatrics Society,* 1975, **23**, 442–450.

Atomi, Y., & Miyashita, M. Maximal aerobic power of Japanese active and sedentary adult females of different ages (20–62 years). *Medicine and Science in Sports,* 1974, **6**, 223–225.

Ayllon, T., & Azrin, N. H. The measurement and reinforcement of behavior of psychotics. *Journal of Experimental Analysis of Behavior,* 1965, **8**, 375–383.

Baltes, M. M., & Zerbe, M. B. Re-establishing self-feeding in a nursing home resident. *Nursing Research,* 1976, **25**, 24–26.

Birren, J. E., & Woodruff, D. S. Human development over the lifespan through education. In P. B. Baltes and K. W. Schaie (Eds.), *Life span developmental psychology: Personality and socialization.* New York: Academic Press, 1973.

Botwinick, J. *Aging and behavior.* New York: Springer, 1978.

Botwinick, J. *Cognitive processes in maturity and old age.* New York: Springer, 1967.

Braun, H. W., & Geiselhart. Age differences in the acquisition and extinction of the conditioned eyelid response. *Journal of Experimental Psychology,* 1959, **57**, 386–388.

Butler, R. A. *Why survive? Being old in America.* New York: Harper & Row, 1975.

Butler, R. N., & Lewis, M. I. *Aging and mental health: Positive psychological approaches.* St. Louis: Mosby, 1973.

Canestrari, R. Paced and self-paced learning in young and elderly adults. *Journal of Gerontology,* 1963, **18**, 165–168.

Canestrari, R. E., Jr. Age changes in acquisition. In G. A. Talland (Ed.), *Human aging and behavior: Recent advances in research and theory.* New York: Academic, 1968.

Cavanaugh, J. C., & Perlmutter, M. *Metamemory: A critical examination.* Unpublished manuscript, 1970.

Charlesworth, W. R. An ethological approach to studying intelligence. *Human Development,* 1979, **22**, 212–216.

Chi, M. Knowledge structure and memory development. In R. Siegler (Ed.), *Carnegie-Mellon symposium on cognition.* Hillsdale, N.J. Erlbaum, 1978.

Corso, J. F. Auditory perception and communication. In J. E. Birren & K. W. Schaie (Eds.), *Handbook on the psychology of aging.* New York: Van Nostrand Reinhold, 1977.

Dehn, M. M., & Bruce, R. A. Longitudinal variations in maximal oxygen intake with age and activity. *Journal of Applied Psychology,* 1972, **33**, 805–807.

DeVries, H. A. Physiological effects of an exercise training regimen upon men aged 52–88. *Journal of Gerontology,* 1970, **25**, 325–336.

DeVries, H. A. Physiology of exercise and aging. In D. S. Woodruff & J. E. Birren (Eds.), *Aging: Scientific perspectives and social issues.* New York: D. Van Nostrand, 1975.

Drinkwater, B. L., Horvath, S. M., & Wells, C. L. Aerobic power of females, ages 10 to 68. *Journal of Gerontology,* 1975, **30**, 385–394.

Dubin, S. S. Obsolescence or lifelong education: A choice for the professional. *American Psychologist,* 1972, **27**, 486–498.

Eisdorfer, C. Arousal and performance: Verbal learning. In G. A. Talland (Ed.), *Human aging and behavior.* New York: Academic, 1968.

Eisdorfer, C., Axelrod, S., & Wilkie, F. Stimulus exposure time as a factor in serial learning in an aged sample. *Journal of Abnormal and Social Psychology,* 1973, **67**, 594–600.

Eisdorfer, C., Nowlin, J., & Wilkie, F. Improvement in learning in the aged by modification of autonomic nervous system activity. *Science,* 1970, **170**, 1327–1329.

Elsayed, M., Ismail, A. H., & Young, R. S. Intellectual differences of adult men related to age and physical fitness before and after an exercise program. *Journal of Gerontology,* 1980, **35**, 383–387.

Estes, W. K. The state of the field: General problems and issues of theory and metatheory. In W. K. Estes (Ed.), *Handbook of learning and cognitive processes* (Vol. 1). Hillsdale, N.J.: Erlbaum, 1975.

Fozard, J. L., Wolf, E., Bell, B., MacFarland, R. A., & Podolsky, S. Visual perception and communication. In J. E. Birren and K. W. Schaie (Eds.), *Handbook of the psychology of aging*. New York: Van Nostrand-Reinhold, 1977.

Furchgott, E., & Busemeyer, J. K. Heart rate and skin conductance during cognitive processes as a function of age. Paper presented at the Meetings of the Gerontological Society, 1976.

Glasmer, F. D., & DeJong, G. F. The efficacy of preretirement preparation programs for industrial workers. *Journal of Gerontology*, 1975, **30**, 595–600.

Gutman, G. M., Herbert, C. P., & Brown, S. R. Felden Krais versus conventional exercises for the elderly. *Journal of Gerontology*, 1977, **32**, 562–572.

Hartley, L. H., Grimby, G., Kilbom, A., Nilsson, I. A., Bjure, B. E., & Saltin, B. Physical training in sedentary middle-aged and older men. III. Cardiac output and gas exchange at submaximal and maximal exercise. *Scandinavian Journal of Clinical and Laboratory Investigation*, 1969, **24**, 335–344.

Hartung, G. H., & Farge, E. J. Personality and physiological traits in middle-aged runners and joggers. *Journal of Gerontology*, 1977, **32**, 541–548.

Havighurst, R. J. Education through the adult life span. *Educational Gerontology*, 1976, **1**, 41–51.

Hendrickson, A., and Barnes, R. F. Educational needs of older people. *Adult Leadership*, 1967, **16**, 2 ff.

Hilgard, E. R., & Bower, G. H. *Theories of learning* (4th ed.). Englewood Cliffs, N.J.: Prentice-Hall, 1975.

Hornblum, J. N., & Overton, W. F. Area and volume conservation among the elderly: Assessment and training. *Developmental Psychology*, 1976, **12**, 68–74.

Hoyer, W. J., Labouvie, G., & Baltes, P. B. Modification of response speed deficits and intellectual performance in the elderly. *Human Development*, 1973, **16**, 233–242.

Huberman, M. Looking at adult education from the perspective of the adult life cycle. *International Review of Education*, 1974, **20**, 117–137.

Hulicka, I. M., & Grossman, J. L. Age group comparisons for the use of mediators in paired associate learning. *Journal of Gerontology*, 1967, **22**, 46–51.

Hultsch, D. F. Adult age differences in free classification and free recall. *Developmental Psychology*, 1971, **4**, 338–342.

Ingersoll, B., & Silverman, A. Comparative group psychotherapy for the aged. *Gerontologist*, 1978, **18**, 201–206.

Jenkins, J. J. Four points to remember: A tetrahedral model and memory experiments. In L. S. Cermak & F. I. M. Craik (Eds.), *Levels and processing in human memory*. Hillsdale, N.J.: Erlbaum, 1979.

Johnstone, J. W. C., & Rivera, R. J. *Volunteers for learning*. Chicago: Aldine, 1965.

Kalt, N. C., & Kohn, M. H. Pre-retirement counseling: Characteristics of programs and preferences of retirees. *Gerontologist*, 1975, **15**, 179–181.

Karpf, R. J. Modalities of psychotherapy with the elderly. *Journal of the American Geriatrics Society*, 1980, **28**, 367–373.

Kasch, F. W., & Wallace, J. P. Physiological variables during 10 years of endurance exercises. *Medicine and Science Sports*, 1976, **8**, 5.

Kasschau, P. L. Re-evaluating the need for retirement preparation programs. Industrial Gerontologist, 1974, **1**, 42–59.

Kilbom, A. Physical training with submaximal intensities in women. I. Reaction to exercise and orthostasis. *Scandinavian Journal of Clinical and Laboratory Investigation*, 1971, **28**, 141–161.

Kilbom, A., Hartley, L. H., Saltin, B., Bjure, J., Grimby, G., & Astrand, I. Physical training in sedentary middle-aged and older men: I. Medical evaluation. *Scandinavian Journal of Clinical and Laboratory Investigation*, 1969, **24**, 315–332.

Kimble, G. A., & Pennypacker, H. W. Eyelid conditioning in young and aged subjects. *Journal of Genetic Psychology,* 1963, **103,** 283–289.

Kimmel, D. C., Price, K. F., & Walker, J. W. Retirement choice and retirement satisfaction. *Journal of Gerontology,* 1978, **33,** 575–585.

Knight, B. Psychotherapy and behavior change with the noninstitutionalized age. *International Journal of Aging and Human Development,* **9,** 1978–79, 221–236.

Knox, A. B. *Adult development and learning.* San Francisco: Jossey-Bass, 1978.

Knox, A. B., & Videbeck, R. Adult education and adult life cycle. *Adult Education,* 1963, **13,** 102–121.

Korchin, S. J., & Basowitz, H. Age differences in verbal learning. *Journal of Abnormal and Social Psychology,* 1957, **54,** 64–69.

Labouvie-Vief, G., & Gonda, J. N. Cognitive strategy training and intellectual performance in the elderly. *Journal of Gerontology,* 1976, **31,** 327–331.

Larkin, J. H., Heller, J. I., & Grenno, J. G. Instructional implications of research on problem solving. In W. J. McKeachie (Ed.), *Cognition, college teaching, and student learning.* San Francisco: Jossey-Bass, 1980.

Leech, S., & Witte, K. L. Paired associate learning in elderly adults as related to pacing and incentive conditions. *Developmental Psychology,* 1971, **5,** 180.

London, J. Wenkert, R., & Hagstrom, W. O. *Adult education and social class.* Berkeley: University of California Survey Research Center, 1963.

Monge, R., & Hultsch, D. Paired associate learning as a function of adult age and the length of anticipation and inspection intervals. *Journal of Gerontology,* 1971, **26,** 157–162.

Neisser, U. Memory: What are the important questions? In M. M. Gruneberg, P. E. Morris, & R. N. Sykes (Eds.), *Practical aspects of memory.* London: Academic, 1978.

Parker, E. B., & Paisley, W. J. *Patterns of adult information seeking.* Final Report on USOE Project No. 2583. Stanford, Calif.: Stanford University, 1966.

Perlmutter, M. What is memory aging the aging of? *Developmental Psychology,* 1978, **14,** 330–345.

Perlmutter, M. Age differences in adults' free recall, cued recall, and recognition. *Journal of Gerontology,* 1979, **34,** 533–539.

Perlmutter, M. *New directions in child development: Naturalistic approaches to children's memory.* San Francisco: Jossey-Bass, 1980.

Plemons, J. K., Willis, S. L., & Baltes, P. B. Modifiability of fluid intelligence in aging: A short-term longitudinal training approach. *Journal of Gerontology,* 1978, **33,** 224–231.

Poon, L. W., Fozard, J. L., Cermack, L. S., Arenberg, D., & Thompson, L. W. *New directions in memory and aging: Proceedings of the George A. Talland memorial conference.* Hillsdale, N.J.: Erlbaum, 1980.

Postman, L. The present status of interference theory. In C. N. Cofer (Ed.), *Verbal learning and verbal behavior.* New York: McGraw-Hill, 1961.

Powell, A. H., Eisdorfer, C., & Bogdonoff, M. D. Physiologic response patterns observed in a learning task. *Archives of General Psychiatry,* 1964, **10,** 192–195.

Profant, G. R., Early, R. G., Nilson, K. L., Kusumi, F., Hofer, V., & Bruce, R. A. Responses to maximal exercise in healthy middle-aged women. *Journal of Applied Physiology,* 1972, **33,** 595–599.

Robinson, S., Dill, D. B., Robinson, R. D., Tzankoff, S. P., & Wagner, J. A. Physiological aging of champion runners. *Journal of Applied Physiology,* 1976, **41,** 46–51.

Saltin, L. H., Hartley, A. K., & Astrand, I. Physical training in sedentary middle-aged and older men. II. Oxygen uptake, heart rate and blood lactate concentration at submaximal and maximal exercise. *Scandinavian Journal of Clinical and Laboratory Investigation,* 1969, **24,** 323–334.

Sanders, R. E., & Sanders, J. C. Long-term durability and transfer of enhanced conceptual performance in the elderly. *Journal of Gerontology,* 1978, **33,** 408–412.

Schaie, K. W., & Willis, S. L. Life-span development: Implications for education. *Review of Educational Research,* 1978, **6,** 120–156.

Schultz, W. R., Jr., & Hoyer, W. J. Feedback effects on spatial egocentrism in old age. *Journal of Gerontology,* 1976, **31**, 72–75.

Sharp, M. W., & Reilly, R. R. The relationship of aerobic physical fitness to selected personality traits. *Journal of Clinical Psychology,* 1975, **31**, 428–430.

Shavonian, B. M., Miller, L. H., & Cohen, S. I. Differences among age and sex groups in electrodermal conditioning. *Psycho-physiology,* 1968, **5**, 119–131.

Shavonian, B. M., Miller, L. H., & Cohen, S. I. Differences among age and sex groups with respect to cardiovascular conditioning and reactivity. *Journal of Gerontology,* 1970, **25**, 87–94.

Sidney, K. H., & Shephard, R. J. Activity patterns of elderly men and women. *Journal of Gerontology,* 1977, **32**, 25–32.

Siemen, James R. Programmed material as a training tool for older persons. *Industrial Gerontology,* 1976, **3**, 183–190.

Simpson, I. H., Back, K. W., & McKinney, J. C. Exposure to information on, preparation for, and self-evaluation in retirement. In I. H. Simpson & J. C. McKinney (Eds.), *Social aspects of aging.* Durham, N.C.: Duke University Press, 1966.

Sterns, H. L., & Sanders, R. E. Training and education of the elderly. In R. R. Turner & H. W. Reese (Eds.), *Life-span developmental psychology: Intervention.* New York: Academic Press, 1980.

Suominen, H., Heikkinen, E., & Parkatti, T. Effect of eight weeks' physical training on muscle and connective tissue of the m. vastus lateralis in 69-year-old men and women. *Journal of Gerontology,* 1977, **32**, 33–37.

Taub, H. A. Paired associates learning as a function of age, rate and instructions. *Journal of Genetic Psychology,* 1967, **111**, 41–46.

Thomas, G. S. Physical activity and health: Epidemiologic and clinical evidence and policy implications. *Preventive Medicine,* 1979, **8**, 89–103.

Tiberi, D. M., Boyack, V. L., & Kerschner, P. A. A comparative analysis of four preretirement education models. *Educational Gerontology: An International Quarterly,* 1979, **3**, 355–374.

Troll, L. *Early and middle adulthood.* Monterey, Calif.: Brooks Cole, 1975.

Whanger, A. D., & Busse, E. W. Care in hospital. In J. G. Howells (Ed.), *Modern perspectives in the psychiatry of old age.* New York: Brunner/Mazell, 1975.

Young, R. J., & Ismail, A. H. Personality differences of adult men before and after a physical fitness program. *Research Quarterly,* 1976, **47**, 513–519.

Zaks, P. M., & Labouvie-Vief, G. Spatial perspective taking and referential communication skills in the elderly: A training study. *Journal of Gerontology,* 1980, **35**, 217–224.

34

MEMORY AND AGE

Joan T. Erber
Research Associate
Washington University
Department of Psychology
Aging and Development Program

Memory has been one of the most extensively investigated topics in the aging literature. The quantity and complexity of the information store is not readily condensed into a neat and concise review. Nonetheless, this chapter touches upon many of the issues studied with regard to memory and aging and summarizes the present status of the findings. Studies relating to verbal memory are emphasized, and representative works are cited on the various topics.

Thus far, memory research has largely employed the cross-sectional method, which confounds age and cohort effects. Included among cohort effects are attained educational level and recency of involvement in a formal educational system. Formal education is likely to involve practice in memory skills and is therefore capable of affecting performance on laboratory-type memory tasks. Statistically significant age effects on a memory task must therefore be interpreted cautiously.

Subject selection is an area of consideration in any research dealing with humans, but it is a particularly cogent concern when older adults are involved in memory studies. Recent sophisticated research has attempted to ensure some degree of comparability between young and old subjects in terms of intellectual status, as indexed by a brief measure of verbal intelligence. Matching young and old groups for number of years of education has been less frequent, undoubtedly because of cohort differences in college and even high school completion. Several instances of discrepant findings between studies employing similar memory tasks have led to scrutiny regarding the characteristics of the elderly group. It is often the case that one study used a highly select elderly group (e.g., college alumni), whereas another sampled an elderly group from a much broader range of educational attainment. Some studies (e.g., Rankin & Kausler, 1979) have purposely employed a strategy of comparing highly select older adults with somewhat less select young adults (i.e., older college alumni whose intellectual indices are significantly higher than those of a young undergraduate group). Obtaining an age difference in favor of the young despite the superiority of the older group on both educational and intellectual indices constitutes a compelling case for age deficit rather than age-related cohort

---

Supported in part by NIA program project grant, AG00535 and training grant, AG00030.

differences. The virtues and drawbacks of using a highly select older sample versus using a more representative older sample (who are likely to differ from the young in educational background) are not easily resolved in aging and memory research. The choice ideally depends upon the specific question under investigation.

A highly replicated finding is that older adults in general evidence a greater degree of variability than younger adults on many tasks of cognitive performance. Thus statements regarding either a general memory decline with age or benefits to be realized from memory intervention strategies may not be highly accurate for an individual older adult.

THEORETICAL MODELS IN THE STUDY OF MEMORY

The evolvement of theoretical models in general learning and memory research over the past 25 years has, in turn, guided the paradigms and content of aging research. The associative approach, dominant in the late 1950s and early 1960s, emphasized learning rather than memory. The learning process involved the quantitative formulation and dissolution of stimulus–response bonds, with the subject playing a passive part. Hultsch (1977) describes this as an irreversible decrement model of aging.

The information-processing approach became dominant in the 1960s and is at present a highly influential metamodel guiding research on memory and aging. The subject is viewed as an active attender and organizer of incoming information, capable of altering the quality and thereby the durability of a memory trace. Age-related deficits have indeed been reported, but the focus is often on the potential modifiability of such decrements. This approach has therefore been designated as a decrement with compensation view of aging (Hultsch, 1977).

The most recent theoretical metamodel, the contextual approach, regards learning and memory as dynamic processes. Events or items have meaning only within the context experienced by an individual. Memory for circumstances surrounding the exposure of items or events (i.e., the book that the information was in; the person who gave the instruction) is considered relevant. Research inspired by this approach tends to employ "ecologically valid" measures; memory for prose tends to be favored over memory for unconnected item lists, and memory for the gist of prose passages is favored over verbatim memory. More extensive aging research inspired by this view is undoubtedly forthcoming. The contextual approach holds a nondecremental view of aging (Hultsch, 1977). However, conflicting findings have been reported. For example, Taub (1979) found no age difference in prose comprehension with subjects of high-vocabulary level. In contrast, Gordon and Clark (1974a) obtained a significant age difference with prose materials and high-vocabulary subjects. Further research is essential on the prior experience and ability of young and older adult samples relative to the type and difficulty of prose material. Such investigation is particularly important if disuse, rather than cognitive decline, is postulated as an explanation of any obtained age deficits.

MEMORY STORES

Returning now to the information-processing approach, memory is conceptualized not as a unidimensional entity but, rather, as a series of four discrete stores. A modality-specific sensory memory store ("iconic" for visual and "echoic" for auditory) is followed by a primary, or short-term, store, which is followed by a secondary, or long-term store, and, finally (for some information), by a tertiary, or very long term store.

Sensory Memory

Information initially received peripherally from the senses is held in the sensory store very briefly (¼ to ½ second). This process more closely resembles a perceptual afterimage than what is typically regarded as memory. Some studies (e.g., Walsh & Thompson, 1978) have reported moderate age deficits in visual sensory memory, but others have not (e.g., Kline & Orme-Rogers, 1978). The few studies directly testing echoic memory have typically employed a dichotic listening task. The modest age deficits in sensory memory obtained by some studies are insufficient to account for the cognitive difficulties evidenced by the elderly, the root cause of which most likely lies deeper in the system (Craik, 1977).

Primary Memory

Primary memory (PM) includes material still being rehearsed, still in the focus of conscious attention, and within the span of immediate memory. PM is more akin to a temporary holding and organizing system than to a structured memory store (Craik, 1977). The capacity of PM is limited, with anything beyond approximately five items necessitating retrieval from secondary memory (SM). Memory for the last few items from a long list is one measure of PM, and typically no age difference is obtained, provided the number of items does not exceed PM capacity. Age differences become apparent only when older people must manipulate or reorganize materials prior to recall; for example, repeating a string of digits back in reverse order (Botwinick & Storandt, 1974). Age differences may also appear when older people must divide their attention between two or more simultaneous tasks (Arenberg, 1980a,b). Auditory augmentation of visually presented materials seems to be equally helpful for young and old for items in PM, while it is actually detrimental for both age groups for items in secondary memory (Arenberg, 1976).

Secondary Memory

Secondary memory (SM) represents the permanent, unlimited capacity store, the duration of which is lengthy if not permanent. A large body of research on SM concurs that older adults have difficulty compared with young adults (Craik, 1977). The locus of this difficulty and the means of ameliorating or eliminating it are topics presently under investigation. Three different stages of SM have been suggested: encoding, storage, and retrieval. First, information must be encoded, learned, or put into the system; second, the information is stored until needed; finally, the information must be retrieved from storage.

Storage. While in storage, material may be vulnerable to interference from other items. The idea of increased susceptibility to interference with age was popular with early investigators (e.g., Welford, 1958). This view was in accord with the associative metamodel, wherein new responses replace old ones in a stimulus–response bond. Unfortunately early studies often failed to equate degree of learning of young and old groups. Age-related forgetting over the retention interval was therefore just as likely related to differences in degree of learning as to differential interference in storage (Kausler, 1970). A number of recent attempts with free recall (Smith, 1979), cued recall (Smith, 1974), continuous recognition (Erber, 1978; Wickelgren, 1975), and paired–associate learning (Smith, 1975) failed to obtain evidence for increased interference with age. When retention intervals are very long and/or a great deal of material is interpolated between encoding and retrieval, memory does indeed suffer, but equally so for young and old. Arenberg (1980a,b) suggested that age-related susceptibility to interference may occur not under conditions of interpolated distraction, but under conditions of concurrent distraction (e.g., divided attention), well known as disadvantageous to older adults (Craik & Simon, 1980).

Retrieval. By the late 1960s the interference (storage) hypothesis was overtaken in popularity by the retrieval hypothesis. The retrieval hypothesis attributed age-related memory deficits to inaccessibility of the material at the time of test—a cue-dependent rather than a trace-dependent forgetting (Tulving, 1974). Smith (1980) has enumerated experimental techniques that have been used to assess the retrieval process. A popular one has assumed that recall involves both storage and retrieval of information, whereas recognition involves storage but bypasses any need for retrieval. Several studies (e.g., Schonfield & Robertson, 1966) have reported the old as disadvantaged on recall but not on recognition tasks. Retrieval, rather than storage, was therefore implicated as the locus of the age deficit. More recently, the assumption that recognition bypasses retrieval has been seriously questioned (Tulving & Thomson, 1971). A current view is that recognition provides retrieval support (Perlmutter, 1979a) rather than eliminating the need for retrieval altogether. The research question regarding both recall and recognition memory is moving away from the broad, "Are there age differences?" to the more differentiated, "Under what conditions are there age differences?" Erber (1974), using a difficult recognition task with a five-alternative, forced-choice format, obtained age differences, although they were not of the same magnitude as recall age differences. Kausler and Kleim (1978) found an age deficit on a multiple-item recognition task when target words were embedded in three distractors, but not when they were embedded in only one.

Another technique for examining potential age deficits in retrieval has been the comparison of free-recall with cued-recall performance. If cuing does not differentially aid the recall of the old, then storage, rather than retrieval, is presumed to be the primary locus of any age-related memory impairment. Initial letter cuing at time of recall has not reduced age differences (Drachman & Leavitt, 1972; Smith, 1977). However, semantic cues (e.g., category labels) have been found to reduce (Hultsch, 1975) and in some cases to actually eliminate age differences (Smith, 1977). The type of cues as well as the complexity of the task are clearly relevant to any potential age by cuing interaction.

Encoding. Encoding (or processing) refers to the manner in which material is entered into the memory system. The characteristics of the material that are stored based on initial encoding determine the stability of the memory during the retention interval, as well as the potential effectiveness of retrieval cues. Cues encoded along with items at input have been found highly effective at time of retrieval (Tulving, 1974). In fact, presenting a semantic cue at time of encoding may actually eliminate age differences on a subsequent recall test, even when the cues are not made available at actual time of recall (Smith, 1977). Many recent research studies have concentrated on the analysis of encoding processes, undoubtedly due to the difficulty in independently evaluating retrieval (Smith, 1980).

Encoding processes have been investigated by examining the relationship of memory to both the organization and the length of word lists. Failure to obtain any age-differential ameliorative effect on recall with experimenter-determined subgroupings of words has been a typical outcome (e.g., Mueller, Rankin, & Carlomusto, 1979). With regard to unorganized word lists, Smith (1979) found that young adults recalled a greater proportion of words on both short and long lists than older adults. At the same time, however, the young were more adversely affected than the old as list length increased. This age by list length interaction suggests that young adults engage in spontaneous organizational processing, which most likely loses effectiveness with increasing list length. Older adults are presumably less likely to use spontaneous organizational strategies and are therefore less affected by list length.

Attempts to improve encoding, or processing, in the elderly have included the provision of specific instructions to perform activities with items to be subsequently recalled. Requiring subjects to sort words into categories prior to recall has resulted in a reduced age decrement relative to recall with no prior sorting (Hultsch, 1971). On this basis it seems likely that any age deficit in organization is a "production deficit" rather than a "mediational deficit." While the elderly may not organize spontaneously, they may be induced to do so under the proper conditions, thereby improving memory performance.

Specific instructions to mediate (verbally and/or visually) have been another effort to enhance the quality of encoding. Erber (1976) noted that young adults were more likely than older adults to report the spontaneous use of mediators in learning digit–symbol pairs. Other studies (e.g., Hulicka & Grossman, 1967) found that the elderly improved more than the young on paired–associate learning when provided with imagery-type mediators, even though they did not reach the same level of performance as the young. This again suggests that the old may not use mediators unless specifically instructed, while the young may employ them spontaneously. Treat and Reese (1976) reported that young and old both benefited in paired–associate learning from imagery instructions. The performance level of the old reached that of the young when self-generated imagery instructions were coupled with long retrieval time. Mason and Smith (1977) found neither age group benefited on a free-recall task from instructions to use visual imagery (i.e., form a mental picture of the words as they are presented at a 3-second rate). In light of conflicting results on the effectiveness of visual mediators across various memory paradigms, as well as some research findings indicating age-related decline in visual memory, Winograd and Simon (1980) have cautioned against premature commitment to imagery training for the maintenance of memory skills in the elderly.

Robertson-Tchabo, Hausman, and Arenberg (1976) employed another mnemonic procedure, the method of loci. Older adults were instructed to take an imaginary trip through a familiar spatial location, their residence, stopping at 16 familiar places. At each place they were to associate a high-imagery noun. This procedure led to significant gains in serial recall over a 4-day period relative to a control group who received practice without the method of loci. The apparent effectiveness of this method illustrates well the operation of encoding specificity (Tulving & Thomson, 1973); the same cues (loci) stored during the encoding operation served as effective retrieval cues. Unfortunately, on a fifth-posttest day, when no explicit loci instructions were given, declines in memory performance were related to the failure of many older adults to spontaneously employ this technique. Other studies have also reported that benefits derived from mnemonic devices were not maintained beyond immediate recall. Techniques designed to foster the spontaneous use of mnemonics must be the ultimate goal if such strategies are to aid the elderly in real-world memory demands. Further exploration of individual differences in imagery abilities, as well as of specific conditions under which the elderly can make effective use of verbal and/or visual mnemonics, is certainly in order. A comprehensive memory battery designed through a joint effort of experimental and clinical psychologists could provide a basis for practical recommendations regarding remediation of memory functions (Erickson & Scott, 1977).

Speed. Salthouse (1980) proposed a departure from the encoding, storage, and retrieval framework for investigating memory processes. Birren (1974) has postulated that loss of speed with age represents a fundamental change in central nervous system activity. Salthouse suggests a central speed loss interpretation for a number of memory research findings, and he recommends rate and speed as the best presently available indices for measuring processing capacity. Kausler (1970) proposed that older adults are less likely than young ones to engage in spontaneous rehearsal during an unstructured retention interval. Salthouse suggested slower speed, rather than lack of rehearsal, as the basis of the age-related memory deficit; that is, older adults may be incapable of rehearsing items the same number of times. Several studies (e.g., Waugh, Thomas, & Fozard, 1978) have indeed found that retrieval time from secondary memory increases with age. The balance between speed of retrieval and speed of forgetting could be critical if additional time required by older people to search for and retrieve the first few items in memory increases the chance for unsearched items to be forgotten. On the other hand, Arenberg (1980a) points out that speed is as likely an outcome as an explanation of age-related memory problems. A procedure that aids in memory performance may result *in,* rather than *from,* faster encoding and retrieval processes. Hartley, Harker, and Walsh (1980) have also questioned the adequacy of the speed-of-processing hypothesis in articulating the specific secondary memory mechanism presumed to be adversely affected by slower processing. Also, the speed hypothesis has not yet explained why age differences have been found in speed of retrieval from PM but not in overall PM performance.

Levels of Processing. Craik and Lockhart (1972) have conceptualized memory not as a series of different stores, but rather as a continuum. The durability of a memory trace is a function of the "depth of processing" carried out on the material at the time of encoding. Processing may vary along a qualitative dimension of depth (sensory or structural = shallow; semantic = deep), as well as along the quantitative dimension of elaboration (How fully?) within any given depth domain. The critical factor for memory is the encoding operation carried out on the material; specific intention to learn is not essential (Craik & Tulving, 1975). Several levels of processing studies have employed incidental learning paradigms. Various depths of processing are induced via instructions to perform some specific task while the materials are being exposed. A production deficiency hypothesis would predict that older adults may not spontaneously process deeply. However, they can be induced to do so via performing a semantic orienting task, with a consequent reduction or even elimination of age differences in memory performance. Eysenck (1974) found that young and old performed at a similar low level of recall following a shallow task (letter counting or rhyming). Surprisingly, however, the old were notably inferior to the young following a deep, or semantic, task. To explain this result, Eysenck (1974; 1977) invoked a processing-deficit hypothesis, which predicts that age-related memory decrements increase as depth of processing required by the orienting task increases. Not all research findings have concurred with this prediction, however. Perlmutter (1979a) attenuated a free-recall age difference with a semantic orienting task followed by cued recall. Both Perlmutter (1979a) and White (as cited by Craik, 1977) actually eliminated age differences when a semantic orienting task was followed by recognition. These results suggest older adults are capable of deriving benefit from effective encoding operations if adequate retrieval support is provided.

Craik (1980) reported that older adults performed well on a memory task when the same strong-word associate present at time of encoding was also used as a retrieval cue. When the same *weak* associate present at encoding was the retrieval cue, however, older adults were differentially disadvantaged. With strong and compatible encoding/retrieval cues, older people seem to be guided, or "driven," by the task stimuli to perform well. They may be less prone than young adults to spontaneously transform weak cues presented at encoding into potentially useful retrieval cues, probably due to inefficiency rather than to true loss. In an unstructured situation, older adults are less consistent than young adults in semantic elaboration across trials (Perlmutter, 1979b). A mismatch between encoding and retrieval cues could easily occur, with a resultant negative effect on memory performance. Perlmutter (1979a) also found that for young adults, self-generated associations were more effective as recall retrieval cues than experimenter-provided associations. Older adults were not differentially advantaged by self-generated cues, another possible indication of their inefficiency in producing stable retrieval cues. Taken together, these findings present a picture of the older adult as performing most efficiently on recall tasks when guidelines are clearly delineated and when structure is provided in terms of specific encoding tasks along with retrieval support.

Release from proactive inhibition (PI) is another research paradigm used to investigate whether older adults spontaneously encode on semantic dimensions. Community-living elderly are similar to young adults in evidencing a significant degree of release from PI on shift trials (e.g., Mistler-Lachman, 1977). Even so, the young are consistently superior in memory performance. It seems likely that young adults may engage in more extensive, richer, rehearsal or elaboration within a processing dimension, thereby ensuring a more durable memory trace (Rankin & Kausler, 1979).

As has been suggested previously, older adults may be penalized when attention must be divided between two input sources. Erber, Herman, and Botwinick (1980) obtained evidence in this direction using an incidental levels-of-processing-type paradigm. The young improved significantly in their performance when an intentional component was added to a shallow-task instruction. However, the old remained at the same low level as they had with the shallow instruction alone. Similarly, the young recalled more when a semantic instruction was added to an intentional one, whereas the old did not improve with the additional component. Older adults seem to have difficulty in dealing simultaneously with two instructional components or when items to be remembered are embedded in too much distracting or irrelevant information. Kausler and Kleim (1978) proposed that distractors are most likely processed at a shallow level such that they interfere with target items without actually being retained in memory.

The levels-of-processing incidental paradigm has been extended from verbal materials to face recognition. Smith and Winograd (1978) found that judgments of friendliness of unfamiliar faces (deep) led to better recognition than judgments of nose size (shallow) for both young and elderly adults. This parallel effect of orienting conditions on the two age groups does not support Eysenck's (1974) processing-deficit hypothesis; age differences in face recognition did not increase with the deep task. Neither was there evidence for compensation, since the deep task did not reduce the age differences. Faces apparently necessitate different encoding operations than words; thus the effect of the orienting tasks is not identical for the two types of stimuli. It is important to note that age-related recognition deficits have not been obtained when faces are very familiar, even when the memory has been maintained over a long period of time (Bahrick, Bahrick, & Wittlinger, 1975).

Automatic Processing. A recent distinction has been made between effortful and automatic processing. Effortful processing involves attention-demanding active encoding and retrieval strategies. Automatic processing presumably occurs spontaneously, unintentionally, without interfering or competing with the limited human capacity for effortful processing. Effortful processes decline with age, but automatic ones (e.g., frequency judgments) have been postulated as impervious to age decrement (Hasher & Zacks, 1979). Research on age differences in accuracy of word frequency judgments have produced conflicting results. Attig and Hasher (1980) employed a two-alternative, forced-choice recognition test (Which was more frequent?) and found no age difference. Hasher and Zacks (1979) used a frequency estimation test (similar to recall) and found the older group performed at a lower level. Recall and recognition are well known for their differential age sensitivity (e.g., Schonfield & Robertson, 1966). A definitive statement regarding age sensitivity to frequency information therefore awaits further investigation. Determination of the automaticity of frequency judgments might be made by introducing a frequency judgment task concurrently with an effortful one. If the effortful task is unaffected by the introduction of the frequency task, then frequency judgments may be deemed automatic. A subsequent introduction of the age variable would allow for an unconfounded determination regarding the existence, or lack of, an age difference on an "automatic process."

Metamemory. Metamemory refers to a constellation of capacities concerned with knowledge about one's own memory processes and abilities. Several studies have found young children less accurate than older children in the prediction of recall span as well as in judgments of "feeling of knowing." Thus far, studies comparing young and elderly adults have produced conflicting results. Some have found young and elderly adults comparable in prediction of recall ability (Perlmutter, 1978) and on recall efficiency (Lachman, Lachman, & Thronesbery, 1979). Zelinski, Gilewski, and Thompson (1980) found an even closer correspondence for the elderly than for the young between self-evaluation of memory functioning and actual laboratory task performance. In contrast, Kahn, Zarit, Hilbert, and Niederehe (1975), using a somewhat more open-ended approach, found that memory complaint in the elderly was not systematically correlated with memory performance. Bruce, Coyne, and Botwinick (1980) reported the young and elderly as equally inaccurate in predicting the number of words they could recall from a 20-item list; the young underestimated and the elderly overestimated the number they actually did recall, even though the elderly spent less time in studying the list. More research is definitely necessary in order to determine which aspects of metamemory are maintained and which ones decline with age.

Tertiary Memory

Ribot's Law (1882) postulated that material is forgotten in the reverse order in which it was acquired. This law has become a stereotype regarding the pattern of memory decline with age, perhaps via its assertion by the aged themselves (Kahn & Miller, 1978). A number of recent research efforts have been directed toward investigating the validity of this law, as well as toward a more general exploration of the accuracy of remote memory in the elderly.

Public Events. The study of tertiary memory is plagued by methodological difficulties. An important concern is ensuring the equivalence of the questions sampled from recent to remote times. Questionnaires have been devised in which items deal solely with notable public events across decades (e.g., Botwinick & Storandt, 1974; Squire, 1974; Warrington & Silberstein, 1970). Squire and Slater (1975) have further restricted item content to questions on race horses or television shows. Even so, the equivalence of sampled items is still not assured, and absolute comparison of memory performance across the decades cannot strictly be made.

Another concern in remote public events research is in determining whether the information was actually acquired during the specified time or whether it was learned at a more recent time via less direct sources. The latter would represent historical knowledge rather than remote memory. Poor performance by a group too young to have directly experienced the events has been used to verify the contribution actual memory, rather than historical knowledge, makes to performance. Related to this issue is the decision of how far back in time to go in order to test for Ribot's Law. Warrington and Sanders (1971) went back 25 years in the life of each subject and found that older people displayed less accurate memory for items sampled from all time periods. Botwinick and Storandt (1980) went back 60 years, testing for memories occurring very early in the lives of the old. They found no memory decline up to the 60s. However, they acknowledge that mixing remote memory with historical knowledge for the younger subjects is a methodological price to be paid for going back 60 years.

A basic consideration is that remote knowledge was acquired under uncontrolled conditions. Failure to remember could therefore represent a lack of original learning, an impairment in retrieval, or both. Nevertheless, the disadvantages of uncontrolled encoding and storage conditions are certainly no greater with the public events questionnaire method than with other methods thus far devised for investigating remote memory.

Some attention has been directed toward ascertaining whether people best remember items or events from specific periods in their lives. Conflicting findings have been reported regarding such an interaction. Some studies have obtained no differential memory performance for particular decades (Warrington & Sanders, 1971; Squire, 1974). Others (Botwinick & Storandt, 1974; Botwinick & Storandt, 1980; Storandt, Grant, & Cooper, 1978; Perlmutter et al., 1980) have reported interactions, but not all in the same direction.

Naming Latency. Poon and Fozard (1978) instructed people of different ages to view slides of objects and to name them. Unique dated objects were named more quickly by the old than the young, while the reverse was true for unique contemporary objects. It was concluded that retrieval from long-term lexical memory is related to familiarity of the information.

Method of Cross-Sectional Adjustment. Bahrick, Bahrick, and Wittlinger (1975) and Bahrick (1979) have employed the cross-sectional adjustment technique to investigate memory for names and faces of former high school classmates as well as for the geography of a former college town. They initially obtained unadjusted retention curves and then made subsequent statistical adjustments for variables such as attendance at class reunions. Results with this technique indicate that material learned and retained over a longer period of time in real-life situations is more resistant to forgetting than laboratory-learned material.

Personal Recollection. Inability to check the accuracy of personal recollections is not an overriding concern when memory is viewed as a belief system. A handful of studies on personal recollection have included people's ratings of the clarity of personal recollections surrounding the Kennedy assassination (Yarmey & Bull, 1978). Also, latencies and time placements of personal experiences reported as associative responses to common words have been measured (Franklin & Holding, 1977; Robinson, 1976). The few available studies indicate neither selective sparing nor any notable decline in memory for remote personal events. Further work, analogous to that on the public events research, is needed for ascertaining any existent sampling bias in the items recollected from various time periods.

Actualization of World Knowledge. Tulving (1972) distinguished between two types of memory: episodic and semantic. The episodic category includes memory for specific items and events and encompasses most laboratory tasks. Semantic memory, which has lately received increased attention, includes meaningful material acquired over an extended period of time.

World knowledge has been conceptualized as a vast information store, both factual and inferential, typically not acquired for the express purpose of laboratory performance. It tends to have high salience for everyday life and tends toward the realm of semantic memory. Lachman and Lachman (1980) have evolved an "efficiency statistic," which makes allowance for the size of the data base for total knowledge. Memory performance of any particular age group can be weighted for generational differences in acquired information. Thus far, there is little reason to believe that retrieval efficiency for very long term memory of this type declines with age.

In general, no substantial evidence for Ribot's Law has been reported. Some studies have reported that older people have somewhat poorer memory relative to the young, but this is typically found for both recent and remote events. Other studies have found that older people's memory for remote events or recollections holds up well, even though it is not usually superior to memory for more recent information. It does seem quite certain, in any case, that age deficits are much less or even nonexistent when material learned outside the laboratory is the basis of concern.

NONCOGNITIVE FACTORS AFFECTING MEMORY

It is difficult to make an operational distinction between memory competence and memory performance. Even so, considerable research attention has been devoted to ascertaining the influence of emotional context on cognitive performance of the elderly. The hypothesis directing much of this study is that older people are differentially affected, presumably in a negative direction, by the emotional context surrounding a cognitive task, such that they are prevented from outwardly demonstrating their true competence.

Motivation

Opposing motivational hypotheses have been advanced to explain the lower performance level of older adults on many secondary memory tasks. One is that older adults are less motivated (i.e., underaroused) because the test materials are not personally relevant to them. The elderly are further removed than the young from the formal educational system, in which abstract materials are the norm. Indirect support for this hypothesis has been the observed enhancement in the performance of older adults when concrete, rather than abstract, materials are used (e.g., Hulicka, 1967). A problem in interpreting this finding is that personal relevance (meaningfulness) and difficulty of the material are confounded; difficult tasks generally distinguish between the young and old to a greater degree than easy tasks (Botwinick, 1978, p. 274).

The other motivational hypothesis is that older adults are overaroused, to the detriment of their learning/memory performance. Powell, Eisdorfer, and Bogdonoff (1964) reported that older adults had higher levels of free fatty acids (FFA) in their blood samples than young adults during a serial learning task. Also, their FFA levels took longer than the young group's to return to normal upon task completion. FFA level is one physiological indicator of autonomic nervous system activity. Eisdorfer, Nowlin, and Wilkie (1970) directly tested the overarousal explanation by administering the drug Propanol (which partially blocks autonomic end-organ response) to older adults prior to a serial learning task. They performed at a higher level than an older adult placebo group. A young group was not tested in either condition, so the effect of the drug on their performance is unknown. Furchtgott and Busemeyer (1979) simultaneously measured heart rate and skin conductance during several cognitive tasks. Skin conductance showed no significant change for either age group during task performance, although in general it was lower for the old. It was not significantly correlated with heart rate, which actually showed greater increases in the young during task performance. The only evidence pointing to overarousal in the old was that their heart rates showed less recovery during rest periods than did those of the young. Support for the overarousal hypothesis seems to be highly dependent upon the type of task and the physiological indicator used.

Anxiety

The overarousal hypothesis has been translated by some into a view of the older adult as experiencing a high anxiety level when performing cognitive tasks. Task instructions have been manipulated in an attempt to vary emotional context and, indirectly, anxiety level. Ross (1968) found that young adults performed at the same level on a paired–associate task regardless of instructions (challenging, neutral, or supportive). Older adults, however, did best with supportive instructions and worst with challenging ones. Lair and Moon's (1972) results concurred with those of Ross; elderly adults showed decrement on a substitution task with reproof instructions but immediate gains with praise. Belluci and Hoyer (1975) found the old benefited to a larger extent than the young on a psychomotor task with noncontingent praise.

Researchers employing self-report scales to measure anxiety have reported conflicting outcomes. Whitbourne (1976) found that a high level of self-reported-debilitating anxiety in the old correlated with a greater degree of behavioral manifestation of test anxiety (e.g., refusals, expressions of feelings of inadequacy, etc.). More interesting, however, was a significant correspondence in the old between low-recall performance and self-reported-debilitating anxiety. Others (Mueller, Rankin, & Carlomusto, 1979; Perlmutter, 1978) have not obtained age differences on posttask self-report anxiety questionnaires. Costa, Fozard, McCrae, and Bosse (1976) found that the old performed at a lower level than the young on tests of cognitive ability and, also, that anxious subjects scored lower than nonanxious ones (anxiety was measured by the Cattell Sixteen Personality Factor Questionnaire). Lack of a significant age by anxiety interaction, however, indicated that cognitive decline in the elderly was not attributable to anxiety. The results from these studies do not allow for any clear statement regarding the unique role anxiety might play in the cognitive performance of the elderly. Wallach, Riege, and Cohen (1980) reported that an elderly group recognized as many neutral words, but fewer emotional ones, than a young group. One interpretation was that the elderly had a greater degree of perceptual defense against words connoting emotional intensity. Perceptual defense could well play a part in their responses on self-report questionnaires as well as on other types of personality tests.

Cautiousness

When payoffs are uncertain, older adults have been found more likely to choose easier tasks, on which likelihood of success is great. They are also less likely to raise their level of aspiration following success (Okun & DiVesta, 1976). They are more prone to omission errors (no response) than commission errors (wrong response) on learning tasks. This tendency is interpreted by some as cautiousness, or reluctance to venture a response that might be wrong. Leech and Witte (1971) induced older adults via monetary rewards to make commission errors on a paired–associate task. By so doing they reduced the omission error rate as well as the trials to criterion. Erber, Feely, and Botwinick (1980), using a similar task and monetary reward format, found that only the low-socioeconomic-status older adults showed a trend in the direction reported by Leech and Witte.

Cautiousness has also been investigated with Signal Detection Theory (SDT) methodology, which allows for the separate analysis of memory and response criterion. Recognition of words, nonsense syllables, prose, and faces have been studied (e.g., Ferris, Crook, Clark, McCarthy, & Rae, 1980; Gordon & Clark, 1974a,b; Harkins, Chapman, & Eisdorfer, 1979). Inconsistent results have been obtained. In some cases the elderly are more cautious as indicated by lower false alarm rate, while in others they are actually more lax than a younger group. Hilbert, Niederehe, and Kahn (1976) found that older adults rated as depressed evidenced a laxer response criterion (greater tolerance for false–positive errors) on digit recognition, possibly indicating an acquiescent response bias. Kahn, Zarit, Hilbert, and Niederehe (1975) reported that a nonorganic but depressed elderly group displayed exaggerated memory complaints relative to their actual memory performance on varied materials. If such complaints are also a form of cautiousness, then the conflicting outcomes of these two studies on depressed elderly further serve to demonstrate that the construct of cautiousness is highly situation and index specific.

CONCLUSION

Based on the present memory literature, the difficulty older people have seems to lie mainly within the realm of secondary memory. It is most apparent in laboratory-type episodic memory tasks when structure is not provided, when little or no retrieval support is available, and in some cases when several elements must be attended to simultaneously. Ameliorating any or all of these conditions often results in a marked improvement in performance.

Further study is needed for a more precise determination of the type of structure or mnemonic likely to optimize the older adults' memory performance over an extended time period. A first step might be an effort to match individual differences in spontaneous memory strategy with appropriate techniques for enhancing performance. Research tying metamemory skills to actual memory performance might supply some notion of how to approach this goal.

More work is likely forthcoming on memory for connected items and for contextual cues. Studies of tertiary memory for geographical areas and familiar faces are also increasing the knowledge base for semantic-type memory.

Further examination of the relationship between emotional factors and memory performance will, hopefully, resolve the present discrepant findings. Using a group of indices on the same people as they perform different tasks may elucidate any existent lawful relationships. A clear statement regarding differential influence of emotional factors on memory performance of the elderly awaits such investigation.

REFERENCES

Arenberg, D. The effects of input condition on free recall in young and old adults. *Journal of Gerontology,* 1976, **31**, 551–555.

Arenberg, D. Comments on the processes which account for memory decline with age. In L. W. Poon et al. (Eds.), *New directions in memory and aging.* Hillsdale, N.J.: Erlbaum, 1980, pp. 67–72. (a)

Arenberg, D. Localization of decline and the role of attention in memory. In L. W. Poon et al. (Eds.), *New directions in memory and aging.* Hillsdale, N.J.: Erlbaum, 1980, pp. 19–22. (b)

Attig, M., & Hasher, L. The processing of frequency of occurrence information by adults. *Journal of Gerontology,* 1980, **35**, 66–69.

Bahrick, H. P. Maintenance of knowledge: Questions about memory we forgot to ask. *Journal of Experimental Psychology: General,* 1979, **108**, 296–308.

Bahrick, H. P., Bahrick, P. O., & Wittlinger, R. P. Fifty years of memory for names and faces: A cross-sectional approach. *Journal of Experimental Psychology: General,* 1975, **104**, 54–75.

Belluci, G., & Hoyer, W. Feedback effects on the performance and self-reinforcing behavior of elderly and young adult women. *Journal of Gerontology,* 1975, **30**, 456–460.

Birren, J. E. Translations in gerontology—from lab to life: Psychophysiology and speed of response. *American Psychologist,* 1974, **29**, 808–815.

Botwinick, J. *Aging and behavior* (2nd ed.). New York: Springer, 1978.

Botwinick, J., & Storandt, M. *Memory, related functions, and age.* Springfield, Ill.: Thomas, 1974.

Botwinick, J., & Storandt, M. Recall and recognition of old information in relation to age and sex. *Journal of Gerontology,* 1980, **35**, 70–76.

Bruce, P. R., Coyne, A. C., & Botwinick, J. Adult age differences in metamemory. Paper presented at American Psychological Association Meeting, Montreal, 1980.

Costa, P. T., Jr., Fozard, J. L., McCrae, R. R., & Bosse, R. Relations of age and personality dimensions to cognitive ability factors. *Journal of Gerontology,* 1976, **31**, 663–669.

Craik, F. I. M. Age differences in human memory. In J. E. Birren & K. W. Schaie (Eds.), *Handbook of the psychology of aging.* New York: Van Nostrand-Reinhold, 1977.

Craik, F. I. M. Aging and memory: Downhill all the way? Invited address, American Psychological Association Meeting, Montreal, 1980.

Craik, F. I. M., & Lockhart, R. S. Levels of processing: A framework for memory research. *Journal of Verbal Learning and Verbal Behavior,* 1972, **11**, 671–684.

Craik, F. I. M., & Simon, E. Age differences in memory: The roles of attention and depth of processing. In L. W. Poon et al. (Eds.), *New directions in memory and aging.* Hillsdale, N.J.: Erlbaum, 1980, pp. 95–112.

Craik, F. I. M., & Tulving, E. Depth of processing and the retention of words in episodic memory. *Journal of Experimental Psychology: General,* 1975, **104**, 268–294.

Drachman, D. A., & Leavitt, J. Memory impairment in the aged: Storage versus retrieval deficit. *Journal of Experimental Psychology,* 1972, **93**, 302–308.

Eisdorfer, C., Nowlin, J., & Wilkie, F. Improvement in learning in the aged by modification of autonomic nervous system activity. *Science,* 1970, **170**, 1327–1329.

Erber, J. T. Age differences in recognition memory. *Journal of Gerontology,* 1974, **29**, 177–181.

Erber, J. T. Age differences in learning and memory on a digit–symbol substitution task. *Experimental Aging Research,* 1976, **2**, 42–53.

Erber, J. T. Age differences in a controlled-lag recognition memory task. *Experimental Aging Research,* 1978, **4**, 195–205.

Erber, J. T., Feely, C., & Botwinick, J. Reward conditions and socioeconomic status in the learning of older adults. *Journal of Gerontology,* 1980, **35**, 565–570.

Erber, J. T., Herman, T., & Botwinick, J. Age differences in memory as a function of depth of processing. *Experimental Aging Research*, 1980, **6**, 341–348.

Erickson, R. C., & Scott, M. Clinical memory testing: A review. *Psychological Bulletin*, 1977, **84**, 1130–1149.

Eysenck, M. W. Age differences in incidental learning. *Developmental Psychology*, 1974, **10**, 936–940.

Eysenck, M. W. *Human memory: Theory, research, and individual differences.* Oxford: Pergamon, 1977.

Ferris, S. H., Crook, T., Clark, E., McCarthy, M., & Rae, D. Facial recognition memory deficits in normal aging and senile dementia. *Journal of Gerontology*, 1980, **35**, 707–714.

Franklin, H. C., & Holding, D. H. Personal memories at different ages. *Quarterly Journal of Experimental Psychology*, 1977, **29**, 527–532.

Furchtgott, E., & Busemeyer, J. K. Heart rate and skin conductance during cognitive processes as a function of age. *Journal of Gerontology*, 1979, **34**, 183–190.

Gordon, S. K., & Clark, W. C. Application of signal detection theory to prose recall and recognition in elderly and young subjects. *Journal of Gerontology*, 1974, **29**, 64–72. (a)

Gordon, S. K., & Clark, W. C. Adult age differences in word and nonsense syllable recognition memory and response criterion. *Journal of Gerontology*, 1974, **29**, 659–665. (b)

Harkins, S. W., Chapman, C. R., & Eisdorfer, C. Memory loss and response bias in senescence. *Journal of Gerontology*, 1979, **34**, 66–72.

Hartley, J. T., Harker, J. O., & Walsh, D. A. Contemporary issues and new directions in adult development of learning and memory. In L. W. Poon (Ed.), *Aging in the 1980's: Psychological issues.* Washington, D.C.: American Psychological Association, 1980.

Hasher, L., & Zacks, R. T. Automatic and effortful processes in memory. *Journal of Experimental Psychology: General*, 1979, **108**, 356–388.

Hilbert, N. M., Niederehe, G., & Kahn, R. L. Accuracy and speed of memory in depressed and organic aged. *Educational Gerontology*, 1976, **1**, 131–146.

Hulicka, I. M. Age differences in retention as a function of interference. *Journal of Gerontology*, 1967, **22**, 180–184.

Hulicka, I. M., & Grossman, J. L. Age group comparisons for the use of mediators in paired-associate learning. *Journal of Gerontology*, 1967, **22**, 46–51.

Hultsch, D. Adult age differences in free classification and free recall. *Developmental Psychology*, 1971, **4**, 338–342.

Hultsch, D. F. Adult age differences in retrieval: Trace-dependent and cue-dependent forgetting. *Developmental Psychology*, 1975, **11**, 197–201.

Hultsch, D. F. Changing perspectives on basic research in adult learning and memory. *Educational Gerontology*, 1977, **2**, 367–382.

Kahn, R. L., & Miller, N. E. Adaptational factors in memory function in the aged. *Experimental Aging Research*, 1978, **4**, 273–289.

Kahn, R. L., Zarit, S. H., Hilbert, N. M., & Niederehe, G. Memory complaint and impairment in the aged: The effect of depression and altered brain function. *Archives of General Psychiatry*, 1975, **32**, 1569–1573.

Kausler, D. H. Retention-forgetting as a nomological network for developmental research. In L. R. Goulet & P. B. Baltes (Eds.), *Life-span developmental psychology.* New York: Academic, 1970.

Kausler, D. H., & Kleim, D. M. Age differences in processing relevant versus irrelevant stimuli in multiple-item recognition learning. *Journal of Gerontology*, 1978, **33**, 87–93.

Kline, D. W., & Orme-Rogers, C. Examination of stimulus persistence as a basis for superior visual identification performance among older adults. *Journal of Gerontology*, 1978, **33**, 76–81.

Lachman, J. L., & Lachman, R. Age and the actualization of world knowledge. In L. W. Poon et al. (Eds.), *New directions in memory and aging.* Hillsdale, N.J.: Erlbaum, 1980, pp. 285–312.

Lachman, J. L., Lachman, R., & Thronesbery, C. Metamemory through the adult life span. *Developmental Psychology,* 1979, **15**, 543–551.

Lair, C. V., & Moon, W. H. The effects of praise and reproof on the performance of middle aged and older subjects. *Aging and Human Development,* 1972, **3**, 279–284.

Leech, S., & Witte, K. L. Paired-associate learning in elderly adults as related to pacing and incentive conditions. *Developmental Psychology,* 1971, **5**, 180.

Mason, S. E., & Smith, A. D. Imagery in the aged. *Experimental Aging Research,* 1977, **3**, 17–32.

Mistler-Lachman, J. L. Spontaneous shift in encoding dimensions among elderly subjects. *Journal of Gerontology,* 1977, **32**, 68–72.

Mueller, J. H., Rankin, J. L., & Carlomusto, M. Adult age differences in free recall as a function of basis of organization and method of presentation. *Journal of Gerontology,* 1979, **34**, 375–380.

Okun, M. A., & Di Vesta, F. J. Cautiousness in adulthood as a function of age and instructions. *Journal of Gerontology,* 1976, **31**, 571–576.

Perlmutter, M. What is memory aging the aging of? *Developmental Psychology,* 1978, **14**, 330–345.

Perlmutter, M. Age differences in adults' free recall, cued recall, and recognition. *Journal of Gerontology,* 1979, **34**, 533–539. (a)

Perlmutter, M. Age differences in the consistency of adults' associative responses. *Experimental Aging Research,* 1979, **5**, 549–553. (b)

Perlmutter, M., Metzger, R., Miller, K., & Nezworski, T. Memory of historical events. *Experimental Aging Research,* 1980, **6**, 46–60.

Poon, L. W., & Fozard, J. L. Speed of retrieval from long-term memory in relation to age, familiarity, and datedness of information. *Journal of Gerontology,* 1978, **33**, 711–717.

Poon, L. W., Fozard, J. L., Cermak, L. S., Arenberg, D., & Thompson, L. W. (Eds.), *New directions in memory and aging: Proceedings of the George A. Talland Memorial Conference.* Hillsdale, N.J.: Erlbaum, 1980.

Powell, A. H., Jr., Eisdorfer, C., & Bogdonoff, M. D. Physiologic response patterns observed in a learning task. *Archives of General Psychiatry,* 1964, **10**, 192–195.

Rankin, J. L., & Kausler, D. H. Adult age differences in false recognitions. *Journal of Gerontology,* 1979, **34**, 58–65.

Ribot, T. *Diseases of memory.* New York: Appleton, 1882.

Robertson-Tchabo, E. A., Hausman, C. P., & Arenberg, D. A classical mnemonic for older learners: A trip that works! *Educational Gerontology,* 1976, **1**, 215–226.

Robinson, J. A. Sampling autobiographical memory. *Cognitive Psychology,* 1976, **8**, 578–595.

Ross, E. Effects of challenging and supportive instructions in verbal learning in older persons. *Journal of Educational Psychology,* 1968, **59**, 261–266.

Salthouse, T. A. Age and memory: Strategies for localizing the loss. In L. W. Poon et al. (Eds.), *New directions in memory and aging.* Hillsdale, N.J.: Erlbaum, 1980, pp. 47–66.

Schonfield, D., & Robertson, E. A. Memory storage and aging. *Canadian Journal of Psychology,* 1966, **20**, 228–236.

Simon, E. Depth and elaboration of processing in relation to age. *Journal of Experimental Psychology: Human Learning and Memory,* 1979, **5**, 115–124.

Smith, A. D. Response interference with organized recall in the aged. *Developmental Psychology,* 1974, **10**, 867–870.

Smith, A. D. Aging and interference with memory. *Journal of Gerontology,* 1975, **30**, 316–325.

Smith, A. D. Adult age differences in cued recall. *Developmental Psychology,* 1977, **13**, 326–331.

Smith, A. D. The interaction between age and list length in free recall. *Journal of Gerontology,* 1979, **34**, 381–387.

Smith, A. D. Age differences in encoding, storage, and retrieval. In L. W. Poon et al. (Eds.), *New directions in memory and aging.* Hillsdale, N.J.: Erlbaum, 1980, pp. 23–46.

Smith, A. D., & Winograd, E. Adult age differences in remembering faces. *Developmental Psychology,* 1978, **14,** 443–444.

Squire, L. R. Remote memory as affected by aging. *Neuropsychologia,* 1974, **12,** 429–435.

Squire, L. R., & Slater, P. C. Forgetting in very long-term memory as assessed by an improved questionnaire technique. *Journal of Experimental Psychology: Human Learning and Memory,* 1975, **104,** 50–54.

Storandt, M., Grant, E. A., & Cooper, B. C. Remote memory as a function of age and sex. *Experimental Aging Research,* 1978, **4,** 365–375.

Taub, H. A. Comprehension and memory of prose materials by young and old adults. *Experimental Aging Research,* 1979, **5,** 3–13.

Treat, N. J., & Reese, H. W. Age, pacing, and imagery in paired-associate learning. *Developmental Psychology,* 1976, **12,** 119–124.

Tulving, E. Episodic and semantic memory. In E. Tulving & W. Donaldson (Eds.), *Organization of memory.* New York: Academic, 1972.

Tulving, E. Cue-dependent forgetting. *American Scientist,* 1974, **62,** 74–82.

Tulving, E., & Thomson, D. M. Retrieval processes in recognition memory: Effects of associative context. *Journal of Experimental Psychology,* 1971, **87,** 116–124.

Tulving, E., & Thomson, D. M. Encoding specificity and retrieval processes in episodic memory. *Psychological Review,* 1973, **80,** 352–373.

Wallach, H. F., Riege, W. H., & Cohen, M. J. Recognition memory for emotional words: A comparative study of young, middle-aged and older persons. *Journal of Gerontology,* 1980, **35,** 371–375.

Walsh, D. A., & Thompson, L. W. Age differences in visual sensory memory. *Journal of Gerontology,* 1978, **33,** 383–387.

Warrington, E. K., & Sanders, H. I. The fate of old memories. *Quarterly Journal of Experimental Psychology,* 1971, **23,** 432–442.

Warrington, E. K., & Silberstein, M. A questionnaire technique of investigating very long term memory. *Quarterly Journal of Experimental Psychology,* 1970, **22,** 508–512.

Waugh, N. C., Thomas, J. C., & Fozard, J. L. Retrieval times from different memory stores. *Journal of Gerontology,* 1978, **33,** 718–724.

Welford, A. T. *Aging and human skill.* London: Oxford University Press, 1958.

Whitbourne, S. K. Test anxiety in elderly and young adults. *International Journal of Aging and Human Development,* 1976, **7,** 201–210.

Wickelgren, W. A. Age and storage dynamics in continuous recognition memory. *Developmental Psychology,* 1975, **11,** 165–169.

Winograd, E., & Simon, E. W. Visual memory and imagery in the aged. In L. W. Poon et al. (Eds.), *New directions in memory and aging.* Hillsdale, N.J.: Erlbaum, 1980, pp. 485–506.

Yarmey, A. D., & Bull, M. P. Where were you when President Kennedy was assassinated? *Bulletin of the Psychonomic Society,* 1978, **11,** 133–135.

Zelinski, E. M., Gilewski, M. J., & Thompson, L. W. Do laboratory tests relate to self-assessment of memory ability in the young and old? In L. W. Poon et al. (Eds.), *New directions in memory and aging.* Hillsdale, N.J.: Erlbaum, 1980, pp. 519–538.

35

COGNITIVE STYLES IN OLDER ADULTS

Nathan Kogan
Professor of Psychology
Graduate Faculty
New School for Social Research

In 1972, this author presented a paper at the third West Virginia Conference on Life-Span Developmental Psychology. The paper was entitled "Creativity and Cognitive Style: A Life-Span Perspective," and in expanded and modified form, it subsequently appeared as a chapter (Kogan, 1973) in the book based on the conference proceedings (Baltes & Schaie, 1973). Approximately 8 years have passed since the appearance of that chapter, and hence it would seem timely to inquire how much progress has occurred over this time span. This chapter, then, emphasizes research published since 1972 with occasional reference made to earlier work where completeness and clarity require it. Though a number of relevant publications have appeared since 1972, it would be fair to say that cognitive styles do not represent one of the "hot" topics in the psychology of aging. For example, one searches in vain for relevant index entries in a recently published *Handbook* (Birren & Schaie, 1977). It is important to note, however, that the term *cognitive style* represents a generic category, and one often finds relevant research conducted within this domain that does not carry the foregoing label. It is the author's hope that the reader will consult his 1973 chapter as a prelude to this one. Given space limitations, it will not be feasible to attempt detailed discussion of earlier studies and background issues before turning to the more contemporary work.

DEFINITION AND CLASSIFICATION

Cognitive styles refer to individual variation in *modes* of attending, perceiving, remembering, and thinking. Four cognitive styles received detailed discussion in Kogan (1973)—constricted versus flexible control, field dependence–independence, conceptualizing styles, and reflection–impulsivity. The first of these is omitted from

---

The author would like to acknowledge the assistance of Mindy Chadrow for the bibliographic work involved in the preparation of this chapter.

consideration in this chapter in the absence of relevant published research since 1972. Approximately half the 1973 chapter was devoted to the topic of creativity. Again, there is a dearth of relevant material published over the past 8 years, but one highly relevant paper has appeared and is discussed in the section titled "Divergent Thinking." There has been a resurgence of interest in the area of metaphor recently, and the limited research on that topic relevant to aging is described in the section titled "Metaphoric Style."

SPECIFIC STYLES

Reflection–Impulsivity (R–I)

Conceptual and Empirical Background. This cognitive style concerns the extent to which an individual reflects on the potential accuracy of his or her hypothesis when solving problems containing response uncertainty. Originally proposed by Kagan and his associates (Kagan, Rosman, Day, Albert, & Phillips, 1964), R–I has generated an extensive body of research, a large portion of which has been reviewed by Messer (1976) and Kogan (1976).

The R–I dimension contains a latency component—the speed with which an initial hypothesis is ventured on matching-to-sample tasks such as the Matching Familiar Figures (MFF) test—and an error component—the number of inaccurate matches made. Considerations of response speed and cautiousness of response in cognitive tasks have been dominant in the aging literature for many years (see Botwinick, 1978, for a review). On the basis of that literature, Kogan (1973) ventured the hypothesis that older adults on the MFF test would exhibit the prototypical reflective pattern: long latencies and few errors. In other words, given the demonstrated decline in response time and enhanced cautiousness observed in older adults, there did not seem to be any basis for anticipating greater impulsivity or lesser reflection in older subjects than would be observed in younger adult samples.

Recent Research. Two relevant studies have been published (Coyne, Whitbourne, & Glenwick, 1978; Denney & List, 1979), neither of which confirmed this author's 1973 hypothesis. Regrettably, the two investigations also failed to agree with each other. The Denney and List study used five age groups of males and females ranging from the 30s to the 70s. Both latencies and errors showed significant increases with increasing age of subject when both ANOVA and regression analyses were applied to the data. Of further interest is the evidence that critical demographic variables—sex, education, occupation, and retirement—did not make a significant contribution to MFF performance. Though Denney and List did not report use of the median-split procedure, the indication that the older adults manifested the highest latency and error means strongly suggests that these individuals would be classified neither as reflectives nor impulsives. Rather, they would likely be assigned to the slow-inaccurate quadrant. The Coyne et al. (1978) investigation is based on a comparison of two age groups—18- to 27-year-olds versus 61- to 87-year-olds. Application of median splits to the latency and error data yielded significant age differences: older subjects were disproportionately represented in the impulsive quadrant, younger subjects in the reflective quadrant. ANCOVA controlling for level of education did not alter these outcomes.

Discussion. Both the foregoing researches agree that older adults make more errors on the MFF than do middle-aged and younger adults. In both studies, these age effects cannot be attributed to educational differences, though both sets of authors acknowledge the possible role of other types of cohort influences that could not be assessed with a cross-sectional design. The major discrepancy between the two investigations concerns the latency outcomes. The older adults of the Denney and List work are markedly inefficient (Salkind & Wright, 1977). They make relatively more errors than the younger subjects despite their slowness. In the Coyne et al. work, the older sample responds more quickly than the younger group. Thus elderly adults emerge as either slow-inaccurate or impulsive on the MFF Test. Further research will clearly be required to resolve this discrepancy.

Missing from either study is an empirical examination of the cognitive strategies employed by younger and older adults that might help explain the observed age differences. An abundance of such research has been conducted with children with particular emphasis on eye-movement patterns (see Messer, 1976, for a review), but no comparable research exists for older samples. Possibly the most relevant work has been carried out by Kleinman and Brodzinsky (1978) in the area of haptic exploration. The visual MFF requires extensive perceptual scanning behavior, and in that sense the task has much in common with haptic matching in which geometric forms are palpated. Working with young, middle-aged, and elderly adults reasonably matched for education, Kleinman and Brodzinsky observed that the last of these age groups was least accurate on haptic matching, results consistent with those reported earlier for the MFF. Analysis of the strategies employed by the elderly adults reinforces earlier evidence indicating the difficulties these subjects experience with problems of selective attention (Rabbitt, 1968). Thus the older respondents scanned the standard and comparison stimuli more cursorily. Further, their feature-analyzing strategies were less efficient than those of the younger participants.

We presently do not know whether these deficits in selective attention observed in elderly adults reflect central nervous system changes with age or are of an experiential or motivational nature. On the motivational side, Block and Peterson (1955) and Smock (1955) long ago observed that an impulsive performance (fast and inaccurate) can serve the function of removing the subject (doubting his or her competence) from an unfamiliar and potentially threatening testing situation. Scanning tasks of the sort described may well place an excessive information-processing burden on the older adult; hence there may well be a wish to have the task over with as quickly as possible. Unfortunately, the Denney and List latency data described above fail to fit the foregoing hypothesis. More research on this issue is clearly required.

A short time ago, Zelniker and Jeffrey (1976) constructed an alternate version of the MFF that called upon global–holistic, rather than feature–detail, processing. Children classified as impulsive on the standard MFF performed at the same level as their reflective peers on the global form of the MFF. Given our current knowledge about older adults' inefficient processing of detail in visual and haptic arrays, it would be of considerable interest to find out whether they would produce a more accurate and efficient performance when the stimulus material lends itself to a more global processing style.

Field Dependence–Independence (FD–FI)

Conceptual and Empirical Background. The current theoretical position in regard to the present cognitive style has been recently elucidated by Witkin (1978). In the briefest terms, FD–FI concerns a capacity to overcome embedded contexts in perception; that is, to separate a part from an organized whole. As the theory has evolved over the past 30 years, the value-laden aspects of the style have been muted. The current view stresses the bipolarity of the style: field independents excel in the solution of analytic restructuring tasks and field dependents are distinguished by their social sensitivity. It should be noted that virtually all the research relating FI–FD to aging has not been framed within this bipolar conception but, rather, has been based on an earlier theoretical perspective (Witkin, Dyk, Faterson, Goodenough, & Karp, 1962). This applies to the research reviewed in the author's earlier chapter (Kogan, 1973) as well as to the later research reported here.

The relevant studies described in Kogan (1973) uniformly found significant declines in FI with increasing age beyond young adulthood. All employed a cross-sectional design, and most could be faulted for failing to control for education and other critical demographic variables. The more recent research reviewed below, though again based exclusively on cross-sectional designs, has at least made a serious effort to control demographic variables.

Recent Research. Some of the studies reviewed in this author's 1973 chapter were carried out within the theoretical framework of Heinz Werner (1957). According to this view, development over the life span is distinguished by increases in differentiation and hierarchic integration up to some midpoint after which dedifferentiation commences. Though a reasonable case can be made for this "orthogenetic principle," it is dubious whether it can be effectively tested with cross-sectional designs, particularly when these lack any control for educational level. In this regard, Eisner (1972) obtained the expected inverted U-shaped curve for performance on the Embedded Figures Test (EFT) in a sample ranging from 10 to 83 years of age. It is difficult to interpret this finding, however, in the absence of any controls for demographic differences. Thus the young adults were college students, whereas middle-aged and older adults were community volunteers whose level of education was left unspecified.

The necessity for an education control is well illustrated in the research of Gruenfeld and MacEachron (1975). The subjects in that study—adult females on the nursing staff of a pediatric hospital—were divided into five age groups extending from the 20s to the 60s and beyond. They were administered the portable Rod-and-Frame Test (RFT; Oltman, 1968). Age and RFT performance were significantly and inversely related. With educational background controlled, however, the partial correlation between age and RFT performance dropped to a negligible nonsignificant value. Similar suggestive findings were reported by Tramer and Schludermann (1974) based on a sample of male volunteers from a veterans hospital. Field dependence (children's form of the EFT) was associated with both increasing age and lesser education. No partial correlational analysis was reported, however.

Within the limitations of the cross-sectional design, one of the better controlled life-span studies in the FI–FD domain has been carried out by Lee and Pollack (1978). Those authors selected female subjects in their 20s, 30s, 40s, 50s, 60s, and 70s who were closely comparable in visual acuity and IQ. A significant age effect in mean time to solution on the EFT was obtained, with older adults taking more time. The major point of transition was found to occur between the 40s and 50s. Virtually no differences were observed between the 20s and 40s, and negligible differences were obtained between the 50s and 70s. The outcome for mean number of EFT items correct corresponded closely to that reported above for time scores.

The Lee and Pollack findings are intriguing, and it would be of considerable value to be able to understand why the sharp decline was manifested between the 40s and 50s. In this regard, the foregoing authors conducted a qualitative analysis of solution strategies but found that these were not related to subject age. For lack of any alternative, they speculated about a lower-order perceptual difficulty emerging in the 50s and possibly linked to hormonal change. Recall that the subjects were exclusively female. Clearly, a replication with a male sample is called for as is an expansion of the number of perceptual tests. Is the observed difficulty peculiar to embedded figures, or will it also be found for other FI–FD measures? In a subsequent study (Lee & Pollack, 1980), the portable RFT was administered to females in their 40s, 50s, and 60s. No significant age effects were found, suggesting that aging affects the perception of figure–ground embeddedness but has little influence on the ability to estimate the true vertical. Such results call into question the unitary trait conceptualization of FI–FD in the latter portion of the life span.

Results partially inconsistent with those reported above are offered by Panek, Barrett, Sterns, and Alexander (1978). Those authors tested 175 females ranging in age from 17 to 72 (divided into seven age groups) on the group EFT, the RFT, and a set of selective attention and reaction time measures. Subjects in each age group had the same modal educational level, reported their health as good to excellent, and were screened for normal visual and auditory acuity. Highly significant linear trends were obtained for *both* the EFT and RFT—increasing errors and degrees of deviation from the vertical, respectively, with increasing age. Decrements across the age span appeared to be more gradual than those observed in the Lee and Pollack (1978) research. On the whole, however, the performance of the groups in the middle range of age (33 to 56) was closer to that of the older groups (57 to 72) than to that of the younger subjects (17 to 32). Of particular importance is the relatively close correspondence in the age trends for *both* of the FI–FD indices. Again, one finds a pattern of consistent decline that apparently cannot be attributed to education, health, or acuity factors.

Here then is one of the unresolved issues in the literature. Does aging have a general impact on FI–FD, as Panek et al. have demonstrated, or a more specific impact on EFT performance alone, as Lee and Pollack have claimed? Only further research specifically directed to this issue can provide an answer.

In the Kogan (1973) chapter, studies were described in which comparisons were drawn between different types of elderly persons (e.g., employed vs. retired). Research in this tradition has continued. Markus and Nielson (1973) administered the Children's Embedded Figures Test (CEFT) (Karp & Konstadt, 1963) to five mixed-sex samples of older adults: residents of two homes for the aged, recent hospital dischargees, and two samples of community residents (one affiliated with a senior center). Within each setting, subjects were divided by sex into those less than and more than 75 years of age. As might be expected, institutionalized subjects performed more poorly than home residents, and those over 75 did less well than those younger than 75. The Markus and Nielson research points dramatically to the wide individual differences in FI–FD among older adults, though it is to be regretted that no statistical tests were reported. Further testimony to the broad range of individual differences among the elderly comes from a study by Rotella and Bunker (1978) indicating that "supersenior" tennis players (average age of 72.5) were more field independent on the RFT than a "nonrandom sample" of community volunteers matched for chronological age.

Some of the publications cited above report correlations between FI–FD and other variables in elderly samples. Eisner (1972) found significant positive correlations between EFT and Stroop Test performance in all age groups tested. Tramer and Schludermann (1974) report a significant positive correlation between FI and intellectual ability (Raven's Progressive Matrices) of a magnitude comparable to that found in younger groups. These findings suggest that the meaning of the FI–FD construct may not undergo any fundamental change from young to elderly adulthood.

Let us inquire, finally, into the predictive validity of FI–FD for older adults. In a study of postrelocation mortality among the institutionalized aged (Markus, Blenkner, Bloom, & Downs, 1972), the CEFT was included as a possible predictor of survival. The sample consisted of the entire populations of two homes for the aged that experienced relocation from the downtown area of a large city to new suburban facilities. The CEFT failed as a consistent mortality predictor. FI and FD individuals manifested similar mortality rates.

Discussion. On the whole, the increment in knowledge gained over the past 8 years in respect to FI–FD and aging is disappointing in its scope. It will be granted that there is now greater sensitivity to the role of cohort effects, and hence more of an effort is now made to match age groups on critical demographic variables or to control the latter statistically. The most provocative evidence is offered by the Panek et al. (1978) and Lee and Pollack (1978, 1980) studies. For the first time, there is an indication of decline in FI with age that may not be directly attributable to IQ, education, and visual acuity effects. Unfortunately, it is exceedingly difficult at present to pinpoint the generality of and the basis for the shift toward greater FD in middle age. Lee and Pollack's (1978) suggestion of hormonal changes during middle age as a causal influence in female subjects received no support in their subsequent study (Lee & Pollack, 1980).

As indicated earlier, almost all the research on FI–FD in relation to age subscribes to the earlier value-laden theory (Witkin et al., 1962). The revised version of the theory (Witkin, 1978) asserts that FD individuals possess various social sensitivities. With one possible exception (Tramer & Schludermann, 1974), no one has explored the positive FD side of the coin in elderly samples. The foregoing investigators employed a "socializing scale" that attempted to assess the older person's frequency of social contacts. The results were directly contrary to expectations: a marginally significant correlation in which FI was associated with greater socializing. Frequency of social contact may not be the best variable to employ in the present context, however, for it must substantially reflect environmental factors over which the elderly person would have little control. The examination of the positive side of FD presents formidable conceptual and empirical difficulties (see Kogan, 1980), but our understanding of the role of FI–FD in aging will remain incomplete until these difficulties are overcome.

Styles of Conceptualization

Conceptual and Empirical Background. When individuals are presented with a set of objects for classification, the bases for grouping can vary. Though different categorical schemes have been employed to characterize these bases for grouping, a gross categorization common to all concerns the distinction between complementarity and similarity. Within the latter, finer distinctions are sometimes drawn (e.g., grouping by perceptual similarity vs. grouping into nominal classes). Research reported by Kogan (1974) indicated a predominance of categorical–inferential (similarity) groupings in both younger and older adults (though slightly higher in the younger sample) and approximately twice as many relational–thematic (complementary) groupings in the older relative to the younger subjects. Since complementary groupings were observed to decline with development in childhood, their greater frequency in late adulthood could be viewed as evidence for cognitive regression. This author, however, advanced the alternative proposal that complementary responses in an object-sorting task might represent a more imaginative choice than is typified by the more dominant similarity preferences.

Recent Research. Several replications of the increase in complementarity in older adults are now available in the published literature. Denney and Lennon (1972) and Denney (1974a) compared middle-aged and elderly individuals in grouping geometric forms and observed that the latter were more likely to construct graphic designs, whereas the former grouped the stimuli by color, form, or size. It should be noted that the foregoing effects were much weaker in the Denney (1974a) study, where age groups were reasonably equivalent in SES and other demographic indicators.

Cicirelli (1976) replicated the Kogan (1974) investigation of object-sorting behavior with an expanded number of age groups. Five- to seven-year-olds, as well as college students, were included, and elderly subjects were divided into those in their 60s, 70s, and 80s for purposes of analysis. As in the earlier Kogan (1974) work, categorical–inferential (similarity) groupings were predominant for both college students and elderly adults, but relational-thematic (complementary) groupings were considerably more popular in the older subsamples. In this latter respect, the performance of the older subjects resembled that of the young children in the study. Within the three older subsamples, there were weak trends indicating a decline in similarity and an increase in complementarity from the 60s to the 80s. Worthy of note is the evidence that education was only marginally correlated with age in the three elderly groups and hence could exert little influence on the principal outcomes.

All the foregoing research is based on the grouping of a large array of geometric forms or common objects. In a further study by Denney (1974b), a Conceptual

Styles Test, constructed by D. R. Denney (1971), was employed. The test is comprised of 30 pictorial triads, in each of which a complementary and a similarity pairing were feasible. Subjects (30- to 50-year-olds and 70- to 90-year-olds) were asked to select their preferred pairings in each item and to specify the basis for the pairing. A significant difference in the direction of greater complementarity for the elderly adults was observed in the male subjects, but not for the females. Middle-aged females manifested just as much complementarity as their elderly counterparts. Denney (1974b) explained the sex difference as due to the absence of occupational and educational pressures for the middle-aged women in her sample. Thus a younger group of women (15 to 29) exhibited less complementarity (about the same level as the middle-aged men), and a second investigation demonstrated that (for both males and females) complementary responses are significantly lower for those in professional, as opposed to those in nonprofessional, occupations.

Finally, let us consider a recent study by Smiley and Brown (1979). This research used a verbal triads procedure in which one word was a standard and the remaining two words were thematically and taxonomically related to it. A typical item was as follows: river, boat, lake (standard, thematic connection, and taxonomic connection, respectively). Subjects (ranging from preschool children to elderly adults) expressed their pairing preference and explained the basis for it. In addition, subjects were also asked to explain the basis for the nonpreferred pairing. Preschoolers, first graders, and elderly adults overwhelmingly preferred the thematic classification; fifth graders and college students showed a predominant preference for the taxonomic classification. Of particular interest is the evidence that all groups (exclusive of preschoolers) were generally able to explain the basis for the nonpreferred pairing. Clearly, subjects' performance points away from capacities to the manifestation of stylistic preferences. Note, finally, that Smiley and Brown (1979) maintain that the findings for the elderly adults were independent of level of education.

Discussion. Research on styles of conceptualization offers one of the more robust sets of findings in the aging literature. Across a diverse set of tasks—sorting of geometric forms and common objects, preferential responses in pictorial and verbal triads—elderly adults (relative to young adults) manifest a stronger disposition toward complementary responding. Regrettably, only Denney's research has made systematic use of middle-aged subjects, and here the findings are somewhat equivocal. It appears that a preference for complementarity may have less to do with age, as such, and more to do with one's proximity to formal education and higher-level occupations. The claim of Smiley and Brown (1979) that educational level is independent of preferred mode of classification in elderly adults suggests that the type of occupational involvement may be the more critical variable for adults beyond the years of formal education. More systematic research directed toward this issue is clearly needed.

There is little basis any longer for linking complementary responding in older adults to cognitive regression. At the same time, it is dubious whether this author's earlier claim (Kogan, 1973, 1974) linking complementarity to a more freewheeling cognitive approach can any longer be justified in the absence of directly relevant empirical evidence. The opportunity for creativity on object-sorting and triads procedures is decidedly limited. Complementarity in older adults may well be adaptive, for the relation between objects in the natural world is generally based on function and contiguity. Outside of formal educational and higher-level occupational contexts, the abstraction of similarities has no apparent adaptive purpose. Explanation along these lines would seem to have greater credibility than is the case for a "creativity" interpretation.

Other Cognitive Styles

Conceptual Differentiation and Category Breadth versus Narrowness. Use of an object-grouping procedure for the study of conceptualization styles allows for the assessment of a variety of other indices. Thus conceptual differentiation–the number of groups formed (Gardner & Schoen, 1962)–and compartmentalization–number of objects left ungrouped (Messick & Kogan, 1963)–were examined in older and younger adults in Kogan's (1974) research. Older adults in the object-sorting procedure produced a smaller number of groupings (low conceptual differentiation), but showed only a negligible tendency toward compartmentalization. Subjects were also given a photo-sorting task consisting of 27 male or female photos for grouping. One-half of the subjects were assigned the male photo set, the other half responded to the female photo set. In both conditions, the older adults formed significantly fewer groupings than the younger adults, results completely consistent with the object-sorting data. In the case of the photo sorts, however, the older adults manifested significantly *less* compartmentalization (fewer ungrouped singles) than did the younger subjects.

In his replication of the object-sorting portion of the Kogan (1974) study, Cicirelli (1976) also found lower levels of conceptual differentiation in the elderly subjects of his sample. Note that forming fewer groups is linked to greater abstraction by Gardner and Schoen (1962), suggesting higher levels of performance on the part of the older relative to the younger adults in the sample under study. Such an inference does not seem to be warranted in the present case, for the reduced number of groupings in the elderly appears to be related to the larger number of objects left ungrouped by those subjects.

In a further replication based on the same object-sorting test administered to college students (averaging 25 years of age) and community volunteers (averaging 73 years of age), Finley and Delgado (1979) found no significant age differences in either conceptual differentiation or compartmentalization. Note that the older adults in the foregoing study had significantly less formal education than the college students, though the two age groups did not differ in WAIS vocabulary level.

In addition to conceptual differentiation and compartmentalization scoring, object-sorting data can be examined for the average number of items that subjects assign to each category formed. Such an index is relevant to *breadth* of categorizing, for larger scores imply more extended boundaries for including exemplars in a category. Both Cicirelli (1976) and Finley and Delgado (1979) obtained such a category-breadth index, but neither found significant age differences. The latter authors also administered a "bandwidth" measure of category breadth–the Pettigrew (1958) multiple-choice instrument. The items of that instrument state the central-tendency value of a category (e.g., length of whales), and subjects are required to select the high and low boundary values from the alternatives provided. Choices further from the central tendency imply greater category breadth.

Correlations between the Pettigrew index and the mean objects-per-group score were negligible in both the older and younger samples, pointing to the previously observed inconsistencies manifest in assessments of category breadth with bandwidth and object-sorting instruments (Kogan, 1971). Finley and Delgado (1979), however, did obtain a significant age difference in the Pettigrew bandwidth task, older adults exhibiting greater breadth of categorization relative to the younger adults of the study. It can be argued that these data support other evidence indicative of greater caution in the elderly (see Botwinick, 1978). Since there is likely to be at least one highly deviant instance for most events, an effort to assimilate most instances to a category (breadth) would seem to represent a more cautious approach than one that excludes potentially relevant exemplars (narrowness). Kogan and Wallach (1964), in fact, observed that greater breadth on the Pettigrew index was associated (at least in females) with more cautious strategies in explicit decision-making tasks. Finley and Delgado (1979) also included several memory procedures in their investigation, and observed that category breadth of the bandwidth type was significantly (though marginally) related to superior delayed story recall in the older adults of the sample. On the other hand, the object-sorting form of breadth (mean objects per group) proved to be significantly related to both immediate and delayed story recall in the college student sample. Partial correlations controlling for vocabulary and years of formal education were employed.

Given the importance of memory phenomena for older adults, the evidence of possible linkages between categorization style and extent of recall deserves further theoretical and empirical consideration. At the present time, however, the general pattern of the findings across the studies reviewed above offers too many inconsistencies to permit any straightforward generalizations regarding possible life-span developmental implications of conceptual differentiation, compartmentalization, and breadth of categorization.

Divergent Thinking. Half the Kogan (1973) chapter was devoted to creativity, with a strong emphasis on divergent-thinking performance. At that time, there was no relevant research to report on divergent thinking in older adults. This gap has been filled by the work of Alpaugh and Birren (1977). Subjects (public school teachers) ranging in age from the 20s to the 70s were administered the WAIS, six Guilford (1967) divergent-thinking tests, and the Barron-Welsh Art Scale (Barron, 1963)—a preference-for-complexity measure. Over the age range examined, WAIS intelligence level remained constant (all subjects had at least 4 years of college), but a composite of the Guilford tests and the Barron-Welsh Art Scale both showed steady age-based declines. Alpaugh and Birren speculate that the divergent-thinking decline with age may derive from CNS slowing of information processing, whereas the decline in preference for complexity may reflect age-based losses in interest or motivation. Those authors also raise the possibility that the processes tapped by their study might account for the age-based declines in real-world creativity documented by Lehman (1953). These interpretations and extrapolations unfortunately lose much of their force in the context of strong cohort effects. Since the background of the older subjects was rural and that of the younger subjects was urban, differential cultural experience of the diverse age groups might well be responsible for the observed age differences.

Metaphoric Style. A promising new development in the cognitive-style domain is reflected in a recent article by Boswell (1979) on metaphoric processing in older adults. Thirty retired adults, with an average age of 70, and 31 high school students, with a mean vocabulary level comparable to that of the older subjects, were given four verbal metaphors to interpret (e.g., "A nation is a warm ocean"). These were scored by five judges (graduate students in English) along a 5-point scale with poetic-synthesizing interpretations at one extreme and literal-analytic interpretations at the other. Results indicated a significant difference between the elderly adults and the high schoolers, the former yielding ratings near the middle of the scale, whereas the latter's interpretations were closer to the literal-analytic pole. Level of education was unrelated to metaphor score in the older group. It would be premature, of course, to speak of a new cognitive style at this point in time. Other measures of a metaphoric style have to be developed and applied in order to know whether a coherent construct exists. A promising lead in this direction is offered by a recent monograph (Kogan, Connor, Gross, & Fava, 1980) describing a visual metaphoric triads task. Until that task, as well as others, is examined in samples of elderly adults, our knowledge of the relation between aging and metaphoric processing will remain limited.

ISSUES OF MODIFIABILITY AND ECOLOGICAL VALIDITY

Though modifiability and ecological validity are separate issues, they are discussed in a single section of this chapter in the light of their obvious connection in aging research. If we should seek to modify the cognitive styles and strategies of older adults, it should be with the conviction that a change would be in the interests of the older persons concerned. Most modification or intervention research with older adults has not been governed by this pragmatic goal. We do not seek to alter styles (or abilities) with the knowledge that these new modes of responding will necessarily facilitate the older person's functioning in the real world. Rather, we seek to modify in order to test the immutability of our favorite constructs. When these constructs are value-laden (as evidenced by inferior performance on the part of older relative to younger adults), further justification is offered for our modification efforts. For if we can point to evidence that our interventions eliminate or markedly reduce the disparity in cognitive performances between those older and younger, it will be much more difficult to sustain stereotypes about old age-related decrements and inflexibility. The dissemination of such information can have practical consequences, of course. The paradox is that the reduction of stereotyping can occur in the absence of information that the changes produced by our interventions have any impact whatever on the way the older person functions in his or her everyday world.

Let us next consider the aspects of modifiability and ecological validity for the major cognitive styles and strategies treated in this chapter. The reflection–impulsivity dimension has just recently been introduced to the aging field, and hence it is a bit soon to expect intervention research with the R–I construct. There is extensive research of this kind with children (see Messer, 1976), where it has been found that both latencies and errors can be modified in the reflective direction through strategy training. There is no inherent reason why such training would not be effective with older adults. The critical question, of course, is the worthwhileness of such intervention efforts. Reflection does appear to have adaptive value for a diversity of school-related behaviors in children. Conceivably, R–I in older adults might be of relevance for jobs that entail scanning of complex visual arrays. We simply do not know because the critical research has not yet been done.

In respect to field independence–dependence, there is again no aging-relevant modification research. Though Witkin (1978) has discussed the possibility of modifying FI–FD in children, it is difficult to comprehend the purpose of so doing if the current theoretical position is as bipolar and value-free as Witkin claims. Modification efforts would presumably be directed toward making subjects more FI, that is, performing better on the EFT and RFT. If however, the older person's adaptation is more closely linked to interpersonal as opposed to restructuring skills, it is FD rather than FI that should be fostered. Viewed from this perspective, one can begin to understand why FI did not predict mortality after relocation in institutionalized older persons (Markus et al., 1972). With a bipolar value-free theory, there is considerable ambiguity as to whether FI or FD would be more conducive to the older person's morale, adaptation, and survival. A number of years ago, Neugarten, Havighurst, and Tobin (1968) observed that both "activity" and "disengagement" could be associated with life satisfaction in older individuals. One might speculate that those persons who maintain an active social life well into old age are distinguished by FD, whereas those who disengage and fall back on their own inner resources are more likely to be FI. On the other hand, FI and FD may simply be associated with the direction of one's orientation—inward or outward—rather than with its adaptive success. All of this is conjectural, of course, and hopefully will eventually be confirmed or rejected in the face of empirical evidence.

The efficacy of modification with older adults is most clear in the realm of conceptualization strategies. There does not appear to be the slightest doubt that initial preferences for complementarity can be shifted toward similarity with a modest amount of training. Denney (1974a) took elderly subjects who did not group a set of geometric stimuli by similarity and assigned them randomly to a modeling and a control condition. In a first posttest, all the subjects in the modeling condition grouped by similarity. In a second posttest with a different set of geometric stimuli, subjects in the modeling condition maintained their gains but control subjects manifested a strong shift to similarity grouping, as well. Apparently, the mere opportunity to inspect and repeatedly group the stimuli can evoke similarity responding in elderly adults even in the absence of explicit modeling.

It is evident that the capacity to group by similarity, if not overtly expressed, is nevertheless latent in most healthy older adults. Given the general availability of both major modes of conceptualization, it makes little sense to treat these in a predictive or ecological validity framework. Conceptualization preferences are clearly a consequent rather than an antecedent of the individual's location in an educational and/or occupational hierarchy.

CONCLUSIONS AND IMPLICATIONS FOR FURTHER RESEARCH

If the reader's overall impression of this chapter is that of a parade of unintegrated constructs, that impression is veridical in most respects. It is apparent that each construct has been studied in isolation from each of the others. This state of affairs is not peculiar to research on aging, for the large proportion of cognitive-style studies with individuals of all ages has relied on univariate designs. For constructs of recent vintage, such studies serve a useful conceptual purpose. Further, they can be pursued by investigators with limited resources. Most of the constructs reviewed here have now been with us for a good many years, however, and, in this author's judgment, the time may be ripe for broad-scope multivariate investigations of cognitive styles and related variables. If such projected research is to include age comparisons, their value would obviously be enhanced if sequential designs of the sort advocated by Schaie (1977) were to be used. Such a proposal may strike a utopian note in an era of shrinking research funding, but one must nevertheless hope that a broad-gauged approach to the problems examined in this chapter would strike a responsive chord in the proper quarters.

None of the foregoing is intended to imply that more circumscribed work of a particular character would not be valuable. What we clearly do *not* need are more cross-sectional young versus old comparisons on the older constructs reviewed in this chapter. If cross-sectional research is to be done, however, sampling subjects demographically comparable across the entire adult life span is essential. Consistent with recommendations advanced by Krauss (1980), there is much to be said for a shift in emphasis toward the examination of differences within the elderly population. In many age-comparative studies, variability among the old exceeds the variability observed in the young. Ideally, we should like to understand the roots of these striking individual differences. Within the cognitive-style domain, examination of demographic, health, ability, and personality correlates represents one possible route to such understanding.

Most of the cognitive styles discussed in this chapter could clearly profit from better integration with related work deriving from other theoretical traditions. Research on reflection–impulsivity is closely allied with experimental analyses of selective attention in older adults. As Hoyer and Plude (1980) have noted, the experimental literature on attentional processes has been concerned with the issue of whether older adults experience difficulties in discriminating relevant from irrelevant information. The MFF measure of R–I requires precisely this kind of discrimination, and hence the articulation between these separate traditions is deserving of encouragement. Similarly, the EFT measure of FI–FD has a strong selective-attention component, which again calls for research cutting across topical boundaries. There is an overlap between cognitive styles and problem-solving research in older samples (Rabbitt, 1977). Thus styles of conceptualization necessarily involve issues of abstraction, whereas FI–FD is concerned with restructuring skill. Finally, additional research is needed on the linkages between cognitive styles and intellective abilities, one of the most thoroughly studied topics in the literature on aging (e.g., Willis & Baltes, 1980; Botwinick, 1977; Horn, 1978). The style–ability distinction is of importance, not only in the sense of discriminant validity but also from the perspective of process similarities. Is the meaning of FI–FD in older adults, for example, exhausted by reference to a spatial visualization factor (Horn, 1978)?

Much of the foregoing discussion has urged that cognitive-style researchers make a greater effort to integrate their work with "mainstream" influences in the field of cognition and aging. In closing, it is only fair to note that influence can be bidirectional, and it is my hope that this chapter will find readers outside of the cognitive-style tradition for whom this approach might offer something of value. As stated earlier, cognitive styles are not among the "hot" topics in the psychology of aging.

If this chapter should generate an extra bit of heat, it will have accomplished its objective.

REFERENCES

Alpaugh, P. K., & Birren, J. E. Variables affecting creative contributions across the adult life span. *Human Development,* 1977, **20**, 240–248.

Baltes, P. B., & Schaie, K. W. (Eds.) *Life-span developmental psychology: Personality and socialization.* New York: Academic, 1973.

Barron, F. *Creativity and psychological health.* New York: Van Nostrand, 1963.

Birren, J. E., & Schaie, K. W. (Eds.) *Handbook of the psychology of aging.* New York: Van Nostrand-Reinhold, 1977.

Block, J., & Petersen, P. Some personality correlates of confidence, caution, and speed in a decision situation. *Journal of Abnormal and Social Psychology,* 1955, **51**, 34–41.

Boswell, D. A. Metaphoric processing in the mature years. *Human Development,* 1979, **22**, 373–384.

Botwinick, J. Intellectual abilities. In J. E. Birren & K. W. Schaie (Eds.), *Handbook of the psychology of aging.* New York: Van Nostrand-Reinhold, 1977.

Botwinick, J. *Aging and behavior* (2nd ed.). New York: Springer, 1978.

Cicirelli, V. G. Categorization behavior in aging subjects. *Journal of Gerontology,* 1976, **31**, 676–680.

Coyne, A. C., Whitbourne, S. K., & Glenwick, D. S. Adult age differences in reflection–impulsivity. *Journal of Gerontology,* 1978, **33**, 402–407.

Denney, D. R. The assessment of differences in conceptual style. *Child Study Journal,* 1971, **1**, 142–155.

Denney, N. W. Classification abilities in the elderly. *Journal of Gerontology,* 1974, **29**, 309–314. (a)

Denney, N. W. Classification criteria in middle and old age. *Developmental Psychology,* 1974, **10**, 901–906. (b)

Denney, N. W., & Lennon, M. L. Classification: A comparison of middle and old age. *Developmental Psychology,* 1972, **7**, 210–213.

Denney, N. W., & List, J. A. Adult age differences in performance on the Matching Familiar Figures test. *Human Development,* 1979, **22**, 137–144.

Eisner, D. A. Developmental relationships between field independence and fixity-mobility. *Perceptual and Motor Skills,* 1972, **34**, 767–770.

Finley, G. E., & Delgado, M. Category width and memory in younger and older adults. Paper presented at the 32nd Annual Scientific Meeting of the Gerontological Society, Washington, D.C., 1979.

Gardner, R. W., & Schoen, R. A. Differentiation and abstraction in concept formation. *Psychological Monographs,* 1962, **76** (41, Whole No. 560).

Gruenfeld, L. W., & MacEachron, A. E. Relationship between age, socioeconomic status, and field independence. *Perceptual and Motor Skills,* 1975, **41**, 449–450.

Guilford, J. P. *The nature of human intelligence.* New York: McGraw-Hill, 1967.

Horn, J. L. Human ability systems. In P. B. Baltes (Ed.), *Life-span development and behavior.* New York: Academic, 1978.

Hoyer, W. J., & Plude, D. J. Attentional and perceptual processes in the study of cognitive aging. In L. W. Poon (Ed.), *Aging in the 1980s: Psychological issues.* Washington, D.C.: American Psychological Association, 1980.

Kagan, J., Rosman, B. L., Day, D., Albert, J., & Phillips, W. Information processing in the child: Significance of analytic and reflective attitudes. *Psychological Monographs,* 1964, **78** (1, Whole No. 578).

Karp, S. A., & Konstadt, N. L. *Manual for the children's embedded figures test.* Brooklyn, N.Y.: Cognitive Tests, 1963.

Kleinman, J. M., & Brodzinsky, D. M. Haptic exploration in young, middle-aged and elderly adults. *Journal of Gerontology,* 1978, **33,** 521–527.

Kogan, N. Educational implications of cognitive styles. In G. S. Lesser (Ed.), *Psychology and educational practice.* Glenview, Ill.: Scott, Foresman, 1971.

Kogan, N. Creativity and cognitive style: A life span perspective. In P. B. Baltes & K. W. Schaie (Eds.), *Life-span developmental psychology: Personality and socialization.* New York: Academic, 1973.

Kogan, N. Categorizing and conceptualizing styles in younger and older adults. *Human Development,* 1974, **17,** 218–230.

Kogan, N. *Cognitive styles in infancy and early childhood.* Hillsdale, N. J.: Erlbaum, 1976.

Kogan, N. A style of life, a life of style. *Contemporary Psychology,* 1980, **25,** 595–598.

Kogan, N., Connor, K., Gross, A., & Fava, D. Understanding visual metaphor: Developmental and individual differences. *Monographs of the Society for Research in Child Development,* 1980, **45** (1, Serial No. 183).

Kogan, N., & Wallach, M. A. (1964). *Risk taking: A study in cognition and personality.* New York: Holt, Rinehart and Winston, 1964.

Krauss, I. K. Between- and within-group comparisons in aging research. In L. W. Poon (Ed.), *Aging in the 1980s: Psychological issues.* Washington, D.C.: American Psychological Association, 1980.

Lee, J. A., & Pollack, R. H. The effects of age on perceptual problem-solving strategies. *Experimental Aging Research,* 1978, **4,** 37–54.

Lee, J. A., & Pollack, R. H. The effects of age on perceptual field dependence. *Bulletin of the Psychonomic Society,* 1980, **15,** 239–241.

Lehman, H. *Age and achievement.* Princeton, N.J.: Princeton University Press, 1953.

Markus, E., Blenkner, M., Bloom, M., & Downs, T. Some factors and their association with post-relocation mortality among institutionalized aged persons. *Journal of Gerontology,* 1972, **27,** 376–382.

Markus, E., & Nielsen, M. Embedded-figures test scores among five samples of aged persons. *Perceptual and Motor Skills,* 1973, **36,** 455–459.

Messer, S. Reflection–impulsivity: A review. *Psychological Bulletin,* 1976, **83,** 1026–1053.

Messick, S., & Kogan, N. Differentiation and compartmentalization in object-sorting measures of categorizing style. *Perceptual and Motor Skills,* 1963, **16,** 47–51.

Neugarten, B. L., Havighurst, R. J., & Tobin, S. S. Personality and patterns of aging. In B. L. Neugarten (Ed.), *Middle age and aging.* Chicago: University of Chicago Press, 1968.

Oltman, P. K. A portable rod-and-frame apparatus. *Perceptual and Motor Skills,* 1968, **26,** 503–506.

Panek, P. E., Barrett, G. V., Sterns, H. L., & Alexander, R. A. Age differences in perceptual style, selective attention, and perceptual-motor reaction time. *Experimental Aging Research,* 1978, **4,** 377–387.

Pettigrew, T. F. The measurement and correlates of category width as a cognitive variable. *Journal of Personality,* 1958, **26,** 532–544.

Rabbitt, P. Age and the use of structure in transmitted information. In G. A. Talland (Ed.), *Human aging and behavior: Recent advances in research and theory.* New York: Academic, 1968.

Rabbitt, P. Changes in problem solving ability in old age. In J. E. Birren & K. W. Schaie (Eds.), *Handbook of the psychology of aging.* New York: Van Nostrand-Reinhold, 1977.

Rotella, R. J., & Bunker, L. K. Field dependence and reaction time in senior tennis players (65 and over). *Perceptual and Motor Skills,* 1978, **46,** 585–586.

Salkind, N. J., & Wright, J. C. The development of reflection-impulsivity and cognitive efficiency: An integrated model. *Human Development,* 1977, **20,** 377–387.

Schaie, K. W. Quasi-experimental research designs in the psychology of aging. In J. E. Birren & K. W. Schaie (Eds.), *Handbook of the psychology of aging*. New York: Van Nostrand-Reinhold, 1977.

Smiley, S. S., & Brown, A. L. Conceptual preference for thematic or taxonomic relations: A nonmonotonic age trend from preschool to old age. *Journal of Experimental Child Psychology,* 1979, **28**, 249–257.

Smock, C. D. The influence of psychological stress on the "intolerance of ambiguity." *Journal of Abnormal and Social Psychology,* 1955, **50**, 177–182.

Tramer, R. R., & Schludermann, E. H. Cognitive differentiation in a geriatric population. *Perceptual and Motor Skills,* 1974, **39**, 1071–1075.

Werner, H. The concept of development from a comparative and organismic point of view. In D. B. Harris (Ed.), *The Concept of development: An issue in the study of human behavior*. Minneapolis: University of Minnesota Press, 1957.

Willis, S. L., & Baltes, P. B. Intelligence in adulthood and aging: Contemporary issues. In L. W. Poon (Ed.), *Aging in the 1980s: Psychological issues*. Washington, D.C.: American Psychological Association, 1980.

Witkin, H. A. *Cognitive styles in personal and cultural adaptation*. Worcester, Mass.: Clark University Press, 1978.

Witkin, H. A., Dyk, R. B., Faterson, H. F., Goodenough, D. R., & Karp, S. A. *Psychological differentiation*. New York: Wiley, 1962.

Zelniker, T., & Jeffrey, W. E. Reflective and impulsive children: Strategies cf information processing underlying differences in problem solving. *Monographs of the Society for Research in Child Development,* 1976, **41** (5, Serial No. 168).

36

AGING, THE LIFE COURSE, AND MODELS OF PERSONALITY

Robert R. McCrae
Senior Staff Fellow

Paul T. Costa, Jr.
Chief

*Section on Stress and Coping
Laboratory of Behavioral Sciences
Gerontology Research Center
National Institute on Aging, NIH*

As researchers in the field of personality and aging, we are usually asked only one question: What happens to personality with age? When we answer that the best evidence to date suggests that personality is basically stable and unchanging across the adult portion of the life span, we immediately lose the interest and sometimes the goodwill of the questioner. In our view, such a reaction is prompted by a rather limited conception of the many ways in which the relations between age and personality can be profitably studied. This chapter is an attempt to review briefly a number of alternative models of personality in relation to life-span development and to indicate some of the directions in which one of them, the dimensional continuity model, can be taken.

ALTERNATIVE MODELS OF PERSONALITY AND AGING

The Growth/Decline Model

The most common conception of aging research can be labeled the *growth/decline model*. In this approach, researchers attempt to chart the increase or decrease of a variable as a function of chronological age. Field independence (or spatial ability), for example, is known to increase with age during childhood and adolescence and to decrease thereafter, especially in old age (Schwartz & Karp, 1967). Many physical functions, such as height, weight, and pulmonary capacity, show the same pattern. The major concerns within this model are the identification of variables which show some age-related pattern of growth or decline; the separation of "maturational" changes from generational differences or historical changes; and, ideally, the discovery of the mechanisms which account for the life-span changes in the variable.

A welter of cross-sectional studies of personality variables have implicitly followed this model (Neugarten, 1977), and a small but growing number of longitudinal studies have used repeated measures and cross-sequential techniques to confirm or reject the cross-sectional findings. A recent review (Costa & McCrae, 1980a) argued that the general conclusion to be drawn from these studies was that there are few, if any, meaningful age-related changes in the level of personality traits in the adult years. More precisely, there is no replicated longitudinal evidence of change in such traits as emotional stability, hostility, ascendance, sociability, or imaginativeness. A few personality variables have shown a pattern which can be interpreted as maturational change, but the amount of change is modest, and the range of individual differences at any single age is larger than the variation across ages.

Consider the trait of "masculinity." Studies from projective methods (TAT) have suggested the hypothesis that men decline in some masculine characteristics as they age, while women show an increase in masculine characteristics (Neugarten & Gutmann, 1968). In a study using the Guilford-Zimmerman Temperament Survey on a large sample of men ranging in age from 20 to 90, Douglas and Arenberg (1978) found statistically significant decreases in masculinity which were present in cross-sectional, longitudinal, and cross-sequential analyses. However, the conclusion that aging men become "feminized" (based upon the decreases in the masculinity scores) is premature. In a somewhat whimsical vein, we calculated from the cross-sectional differences that the average man would score the same as the average female college student only if he lived to the age of 269. Extrapolating from longitudinal rates of change, the estimate is 211 years.

To be sure, research to date has not exhausted the personality variables which might be substantially related to age. Within the growth/decline model, the search for personality traits, processes, or structures which are ordered by age continues to remain one direction for research, as does the effort to explain the mechanisms which underlie the small age changes which have already been documented. But the most pervasive domains of personality—neuroticism, extraversion, and openness to experience—show little evidence of age-related change; and an approach which disregards these domains because they do not show such change can hardly claim to provide a comprehensive psychology of aging and personality.

The Life Stage Model

Researchers who employ the growth/decline model have generally been empiricists who prefer to gather facts before propounding theories. But the most elegant models of the interaction of personality and aging have been put forward by stage theorists like Erikson (1950) and, more recently, Levinson et al. (1978). These writers reject the idea that a psychology of adult development can be based on tracing the rise and fall of discrete variables. Instead, they hold that there are qualitative shifts in the nature and relevance of personality variables and syndromes as a function of the stage of adult development. Typically these theorists posit an interaction of social and intrapsychic factors in which personality must be considered both cause and effect in the shaping of the life course.

Erikson describes the life cycle in terms of the succession of stages of psychosocial development. For Erikson, man and society have evolved to a mutual accommodation in which the well-adjusted, mature individual is one whose psychological organization meshes with the social age-grading requirements of his society. Correspondingly, culture has evolved institutions to complement the capacities of the individual at each stage, educating the child concerned with developing industry and revering the elder who has attained integrity.

Levinson et al. (1978) have proposed a complex stage model of adult development in which the life structure, rather than personality itself, is the variable to be explained. Nevertheless, certain personality characteristics are inextricable elements of the life structure and its changes. At the midlife transition, hypothesized to occur universally around age 40, the individual may go through a period of inner turmoil which may resemble neurosis. There is also a reemergence of the repressed wishes and dreams of youth. These intrapsychic changes are instrumental in the reshaping of the life structure for the midlife period.

Such theories are rich and appealing, but their commendable complexity often makes them difficult to test empirically. So interlinked are the elements of individual, society, and history that almost any phenomenon can be explained, though few can be predicted. There are a number of studies which offer support for some of the propositions of Erikson's theory (e.g., Whitbourne & Waterman, 1979), particularly for the period of adolescence and young adulthood, but the theory as a whole cannot be considered empirically established. Furthermore, there have been empirical studies which directly contradict the premises of some stage theories of adult development. An attempt to locate the universal "midlife crisis" (Cooper, 1977) found only a small group of men with signs of a crisis, and these were found to vary in age from 30 to 60, the effective range of the study population. These negative results were confirmed in a second study (Costa & McCrae, 1978). Another recent attempt to detect age-related life stages in respondent's life orientation (futurity) and satisfaction has also found little support for uniform and universal age-related stages (Lacy & Hendricks, 1980).

Critics could question the sensitivity of the measures used in investigating so complex a phenomenon, but such a criticism is itself an admission that the life stage model is at best a difficult basis for the empirical investigation of adult development. More elaborate attempts to address the question of qualitative change in personality at different stages of life using standard personality instruments have occasionally been made. Factor analyses of personality scales within age groups and at different points in time have been used for this purpose but have yielded no evidence of theoretically relevant change in the interrelation of traits. Quite the contrary: Data from large samples of men showed striking invariance of personality structure across age and time (Costa & McCrae, 1980a; McCrae, Costa, & Arenberg, 1980). Again, it must be noted that this is only one of several ways in which qualitative changes might be sought; but there is little in the existing literature to encourage research in this direction.

The basic insight which distinguishes the life stage model theorists from others is their recognition that personality, cultural age norms, and expectations interact in determining the life course of the individual. What is needed is a model which incorporates this conception into a psychology of aging built on sound empirical findings about personality.

The Typological Model

A third model, which makes certain contributions along these lines, might be called the typological model. In this approach, individuals of a given age group are classified into types which are based on the life-style, personality, and adjustment of the individuals. The use of personality measures or ratings and the statistical classification of individuals adds an empirical element to these approaches which is laudable. At the same time, the labels chosen to characterize the groups often summarize holistically the life structure, as well as the personality, of the individual. For example, the "rocking chair men," one of five types identified by Reichard, Livson, and Peterson, (1962) are described as a "passive-dependent group [that] tended to lean on others for material and emotional support. Unambitious men who found little satisfaction in work, they were glad to take it easy when retirement came" (p. 129).

Neugarten, Crotty, and Tobin (1964) used a similar technique to identify six personality types in an aged population. More recently, Maas and Kuypers (1974) again used a typological approach in order to investigate the 40-year predictors of personality and life-style in 70-year-old men and women. They described four life-style clusters for men and six for women, as well as three personality types (based upon the California Q-sort method) for men and four for women. However, Maas and Kuypers found "a relatively random association between personality Q-groups and life style clusters" (p. 156), and thus concluded that personality organization and patterns of life-style are quite independent of each other.

It is the multiplicity of distinct types that is the major shortcoming of the typological approach. If the three studies cited above had concurred in the number and nature of personality or life-style types, there would be a sounder basis for employing this model. As it is, the clustering seems to depend on the particular set of variables included and seems not to be particularly robust across studies. Another point needs to be considered: in all three of these studies, subjects were restricted to older individuals. A cluster analysis of personality and life-style variables which showed that age ordered the clustering of subjects better than gender, occupation, or some other variable would materially strengthen the claim that the various typologies capture the interrelation of age and personality. But within the age range involved, Neugarten et al. (1964) report that "an important finding is that the personality types described here are not, on the whole, related to age. . . . over the wide age ranges from the early fifties to the late eighties" (p. 186).

That different types emerge in different studies should not obscure the fact that all three reach a similar conclusion with regard to the continuity of personality in adulthood. Although they are careful to point out that there is notable change in life-styles, Maas and Kuypers (1974) also show longitudinal evidence of continuity. The 70-year-old men they classify as "unwell-disengaged" were the most explosive, tense, and nervously unstable of the groups 40 years earlier. Likewise, "the fearful-ordering mothers remain, over their adult life course, depressed in mood and activity level, low in adaptive capacity, and low in self worth" (p. 203). Continuity is also a conclusion from the other studies: "The histories of our aging workers suggest that their personality characteristics changed very little throughout their lives"

(Reichard, Livson, & Peterson, 1962, p. 163). "The implication . . . is that personalities maintain their characteristic patterns of organization as individuals move from middle into old age" (Neugarten, Crotty, & Tobin, 1964, p. 187).

The Dimensional Continuity Model

The dimensional continuity model of aging and personality, our preferred model, argues that the cardinal feature of personality in adulthood is the stability of a number of its major dimensions. This premise is based on a growing body of data which show that the retest correlations of personality measures administered over a period of many years are extremely high; in some cases rivaling the short-term retest reliabilities of the measures (Costa & McCrae, 1977; Costa, McCrae, & Arenberg, 1980). Correlations ranging from .59 to .87 over a 12-year interval statistically confirm the impressions of stability which retrospective accounts give and lay the basis for an entirely new approach to the relation of personality to the life course.

It might be useful to compare the present stability model of personality with the epigenetic model of Erikson. Erikson proposes eight stages, each with a better and worse resolution. He argues that success in preceding stages is the best preparation for success in the next but views each new conflict as a potential for change, either positive or negative. Our longitudinal data lead to the simpler, if less optimistic, conclusion that success in any single developmental task is likely to be a reflection of stable personality traits. Rather than the branching life patterns that a series of developmental successes and failures could provide in theory, we would hypothesize a preponderance of straight lines: some individuals would show a constellation of mistrust, doubt, guilt, inferiority, identity diffusion, isolation, stagnation, and despair throughout their lives; others would show a constellation of trust, autonomy, initiative, industry, identity, intimacy, generativity, and integrity; and most people would show a lifelong pattern of moderate adjustment.

The point is that whereas the Eriksonian view leads one to look for the possible emergent changes in personality at each stage, our position emphasizes the continuity in outcomes across the stages. Radical changes in personality may perhaps result from effective therapeutic interventions or catastrophic changes in health or social status, but our data suggest that age, per se, does not bring about stage changes in adult personality.

Unlike typologies, which have not proven easy to replicate, a large literature recently has concurred in the identification of a few basic dimensions of personality traits. Neuroticism and extraversion are found in the theories and measures of Cattell (1973), Guilford (1976), and Eysenck (1960). Another domain, openness to experience (McCrae & Costa, 1980) is beginning to be recognized as a third pervasive dimension of personality (e.g., Tellegen & Atkinson, 1974). Within each of these domains, a number of distinct but covarying traits can be enumerated. Neuroticism includes anxiety, hostility, depression, self-consciousness, impulsiveness, and vulnerability to stress. Extraversion includes warmth, gregariousness, assertiveness, activity, excitement seeking, and positive emotions. Openness is manifested in the areas of fantasy, aesthetics, feelings, actions, ideas, and values. Grouping these traits into three domains provides a model which is reasonably comprehensive (though certainly not exhaustive of personality dimensions) while being conceptually manageable. The demonstrated empirical relations between alternate measures of traits in these domains (Costa & McCrae, 1980a) makes it possible to compare studies using different instruments. The internal consistency, retest reliability, and discriminant validity of these well-constructed objective personality tests also contribute to the conclusion that the dimensional continuity approach offers a sound empirical basis for the study of aging and personality.

What the dimensional continuity approach lacks, in the eyes of most researchers, is any application. Having said that personality is stable, what more can be said? Is there any future to research on personality and adult development, other than to refute or confirm, to qualify or delimit the central claim of stability?

In fact, there are a number of important directions for future research. Elsewhere (Costa & McCrae, 1980a) we have argued that one such direction is the search for an explanation of personality stability. Little is known about the mechanisms which maintain characteristic levels of various dispositions, although a number of theoretically relevant possibilities can be cited, from genetics to self-image. The institutionalization of the self in the interlocking obligations and expectations of the life structure is doubtless another source of personality stability.

This chapter, however, is directed at another way in which personality and aging can fruitfully be studied. Here we will argue that the enduring dimensions of personality can be viewed as a framework for understanding the life course of the aging person. Rather than taking personality as the dependent variable to be understood in terms of age, or stage in some career, we view personality dimensions as independent variables which function jointly with age and stage to influence some of the outcomes of life. In looking for these interrelations of personality and the life course, we will be following in the tradition of major personality theorists like Murray (1938) who held that the phenomenon to be explained is not a single process or specific behavior, but the complete life of the individual (White, 1963).

THE LIFE COURSE AS A FUNCTION OF AGE AND PERSONALITY

Life-span developmentalists share a concern for the course of life as a major element in their theories and research. They differ, however, in their goals. Some researchers are primarily interested in psychological or biological processes and view aging as a quasi manipulation, the effects of which may help elucidate the mechanisms underlying the process. Sociologists (e.g., Elder & Rockwell, 1974) take social, historical, and developmental "ages" into account in attempting to understand career development or such social phenomena as the unwed mother. Personality researchers have historically considered their primary concern to be an understanding of the whole person, both at a given time and across the life span; and some, like Erikson (1962), have written biographies. The relevance of enduring personality dispositions to the work of life-span developmentalists will necessarily vary with their goals. Our purpose here is to describe one of the several approaches to the study of the life span, the approach we find valuable for the study of aging and personality.

Determinants of the Life Course

It is possible to view a person's life in cross section and to describe what Levinson et al. have called a "life structure." At any given time the life structure can be described in terms of occupational, social, and family roles; intimate, personal, and professional relationships; and the goals, values, motives, and memories which constitute the inner aspect of the life structure. One of the tasks of the individual is to manage all these elements at one time, to avoid role conflict, and to accommodate all the intrapsychic needs, values, and preferred styles. Murray and Kluckhohn (1953) refer to this process as "scheduling."

Viewed longitudinally, the sequence of any one of these elements is often called a *career,* and it is possible to speak of social, leisure time, or family as well as occupational careers. From the psychological side, Murray calls these temporal sequences "serials." Taken as a whole, the more-or-less coherent progress of all these aspects of life can be called a "life course," which is the proper subject of a complete biography.

The life-span developmentalist is likely to view the life course as a series of *changes* which are shaped by the succession of age-related roles prescribed by the culture, by the biological and cognitive development of the individual, and by the particular historical events, shared and idiosyncratic, that define the context in which the individual ages. The choice of retirement, for example, can be seen to be influenced by social policies of mandatory retirement, by the physical health of the individual, and by prevailing economic conditions. As an element in the life course, retirement then influences a host of other events and provides the individual with new choices for the use of leisure time, new residence, and so on.

By and large, developmentalists have concentrated on the explanation for changes in the life course and have given less thought to factors which provide continuity. For the individual, however, continuity is as important as change, for it provides the basis for a sense of identity. A number of sources of continuity in the life course can be easily recognized. The social structures of the family and social class provide an enduring set of opportunities and expectations while biological and cognitive abilities of the individual set certain stable limits to achievement. This chapter will argue that the stability of dimensions of personality is yet another powerful source of continuity in the life course of the individual, although it has been perhaps less widely recognized as such by life-span theorists.

Both continuity and ordered change are necessary for the smooth functioning of society, and mechanisms for preserving continuity (such as marriage contracts and seniority systems) have been institutionalized. But the student of personality must also point out that much of life's continuity is the result of the individual's own action. Most people have considerable say in the shaping of their own lives, and the successful management of the life structure at any one time depends on a history of preparation and planning. In making these decisions, one of the major considerations of the individual is his or her own personality dispositions. Indeed, if personality were not stable, our ability to make wise choices about our future lives would be severely limited.

Personality and Critical Life Choices

The statement that personality and the life course are "interrelated" means that there are a number of phenomena that are best explained in terms of both conceptions jointly. In Runyan's (1980) terms, we could say that personality influences the state of the individual at any given stage. Some of the examples which we will provide are obvious on reflection, and some are empirical discoveries. It is also relatively easy to generate speculations about possible relations to be confirmed or disconfirmed by research. For this reason, we believe this to be a fruitful model for developmental research.

Perhaps the most obvious example of the interaction of aging and personality is to be found in the transitions of the life course. On the basis of biological capacity and social requirements, cultures have dictated that certain roles must be adopted or given up at certain times. Young adults are expected to marry, begin a family, and take up an occupation. Middle-aged persons are supposed to sustain their families and advance their careers. Older individuals, at least until recently, were expected to step down from positions of responsibility and adapt gracefully to a period of relative inactivity. But at each transition there is also choice, and in a culture like ours, where the individual is given wide latitude in the choice of roles, the role of choice becomes more central in the shaping of the life course.

Most individuals start a career in their 20s. But what determines the selection of a particular occupation? Intelligence, education, social class, role models, and a large element of chance go into the choice, but so do personality dispositions. Holland (1966) has developed a theory of occupational development based on personality types, and scales of occupational interest are known to correlate with dimensions of personality, particularly extraversion (Costa, Fozard, & McCrae, 1977). Young people may not know the professions most suited to their temperaments, but if they have made a mistake, they soon discover it. The period of occupational adjustment in the 20s is in part a period of self-discovery in which the tastes, interests, and capacities of the individual become more apparent to him or her. The introvert does not last long as a door-to-door salesperson and is not likely to try that line of work again.

Recently, attention has been drawn to the phenomenon of midlife career shifts (Clopton, 1973). To the extent that these changes are voluntary, they highlight the proactive choice of the individual in shaping the life course. From the viewpoint of personality stability, we might form two hypotheses about the kinds of people who would choose a new career. First, we would expect that individuals whose initial career choice was incompatible with their temperament would be most likely to change, in an attempt to find a more satisfying occupation. There is some evidence from the study of vocational interests that this process does occur (Strong, 1955). Second, we might hypothesize that the need for change or variety is greater in some individuals and that after a number of years in a particular field, they might want to move on. Our own research provides some data consistent with this hypothesis (Costa & McCrae, 1978). Men who had changed their line of work in the previous 10 years were significantly higher in openness to experience than those who had not. It is possible, of course, that the change in vocation led to higher openness instead of the other way around, and longitudinal research is currently in progress which would allow an assessment of that alternative.

Both vocational counseling and industrial selection have capitalized on the association between personality and occupation in the use of interest and attitude scales. Less attention has been paid to the role of personality in determining the use of time after retirement, although there is every reason to believe that personality should figure prominently. Freed from the necessity of working, the older person can spend time as he or she wishes, and an even clearer expression of individual temperament should result.

The continuity of abilities, interests, and values into old age can result in a continuity of activities. Havighurst et al. (1979) report a study on the postretirement publications of a group of scientists. They find clear evidence that retirement per se has little effect and that there is a pattern of continuity between productivity before and after the event. Despite major changes in social-role requirements, the majority of their subjects continued to publish at about the same rate.

Scientists, however, belong to a rather small group who can, if they desire, continue to work after formal retirement. For most people, retirement is a source of discontinuity, but it is also an opportunity to do other things which may have been impossible before. Sociologists have dealt extensively with the variables which affect the age of retirement, including health, income, and job satisfaction. But relatively little research has gone into a specification of the determinants of how retired people employ their own leisure. What kinds of people travel? What kinds retire to a farm? Who joins senior citizen clubs? Who becomes a burden on the local health clinic? Who goes back to college? Any complete theory of aging and the life course must surely address these questions, and it is probable that stable personality traits will be one of the explanatory factors.

Personality and Adaptation

In addition to the choice of roles and relationships, personality influences adjustment to the circumstances—chosen or not—in which the individual finds himself. Any clinical psychologist or psychiatrist who has taken life histories knows that the maladjusted adult typically shows a lifelong pattern of poor adaptation. Prospective studies from deviant children confirm this impression (Robins, 1966), as do recidivism rates for treated patients of all kinds (Moss & Susman, 1980).

Research on the "midlife crisis," also, is instructive here (Costa & McCrae, 1978). A series of questions was asked which addressed the characteristics and concerns of men in the middle portion of their lives: questions about marital satisfaction, declining power and potency, career fulfillment, and problems with children and with aging parents. Although there was no evidence that individuals scoring high on this checklist of problems clustered at any particular age within the range from 30 to 60, there was a strong association ($r = .51$) with concurrent measures of neuroticism. Further, personality scales measuring neuroticism 10 years earlier showed a highly significant predictive correlation. Those individuals with a history of neurotic traits were most likely to suffer during midlife the complaints identified as constituting a midlife crisis. In a similar vein, Lowenthal and Chiriboga (1972) have reported that most women do not experience unhappiness in the "empty nest" period and that those who do have a preexisting history of maladjustment.

The 40-year longitudinal findings of Maas and Kuypers (1974) give a similar impression with regard to old age. They argue that "old age does not usher in or introduce decremental psychological processes. Rather, old age may demonstrate, in perhaps exacerbated forms, problems that have long-term antecedents" (p. 203). Research on health complaints (Costa & McCrae, 1980b) similarly finds no longitudinal increase in symptoms for most body systems, but more neurotic individuals of all ages report more physical problems. The same pattern of invariant relations between personality and outcome variables across the adult life span is seen in studies on psychological well-being or personal adjustment to aging (Costa, Mc-

Crae, & Norris, 1980). Even death anxiety, which might be imagined to be most relevant to older persons, shows little relation to age, but at all ages, it is the characteristically anxious person who shows fear of death (Kastenbaum & Costa, 1977).

Questions and Methods for Future Research

One of the chief obstacles to the study of lives is the sheer quantity of information which must be synthesized in order to make sense of the subject. Biographers adopt some implicit framework, and their task is simplified by the fact that they treat one individual and are allowed post hoc and idiographic interpretations. The life-span developmentalist who desires a scientific theory of the course of human life likewise must account for the similarities and differences of all people. Clearly some organizing principle is necessary, and the continuity of personality dimensions might well provide this principle and form a better basis for an understanding of the life course. Some empirical observations can be profitably understood in this framework, but the extent and limits of the approach are unknown. For the researcher interested in exploring the life course from the viewpoint of the individual, we suggest such questions as these: How do the lives of introverts differ from the lives of extraverts? Which aspects of the life structure are influenced by openness to experience? Does neuroticism or the poor coping styles associated with it interfere in the individual's ability to make effective schedules and serials to order and plan a life? What role do personality dispositions play in adapting to stressful life events?

Another difficulty in the study of lives is the time required to observe the phenomenon. Historical biography is one solution to this problem, as is the analysis of personal documents covering an extended time (Allport, 1965). Prospective longitudinal studies are invaluable for the objectivity and pertinence of the data they provide. But researchers should also utilize the retrospective account of the individual as a source of data. Older persons in particular have a unique perspective on the life course, and through the process of reminiscence, many of them are engaged in making sense of their own lives and of transmitting their insights to others (Butler, 1963). Thus far, retrospective accounts and prospective longitudinal studies have shown substantial agreements in pointing to stability in personality and continuity in life course. The accumulated experience of older men and women should be regarded as a scientific, as well as a social, resource.

CONCLUSIONS

Personality has usually been regarded as an outcome of living, thought to develop or decline with age, or to metamorphose through different life stages. But objective personality data and clinical impressions find more stability than change in the period of adulthood and old age.

This fact encourages a recasting of the fundamental question—instead of asking how personality is changed by aging, we now can ask how the life course is shaped by enduring personality dispositions. The widely recognized and replicated dimensions of neuroticism, extraversion, and openness to experience provide a useful framework in which to organize data on continuity and changes in the life structure. Stable personality dimensions can help explain the choices which are required at certain age-related transition points, the maintenance of life-style across different developmental periods, and the level of adaptation at all ages.

Universal developmental stages cannot account for the continuity of the life course, just as enduring individual differences cannot account for all important life changes. But a model that considers biological, cultural, and historical influences on aging in conjunction with stable dimensions of personality offers promise as a sound basis for an understanding of the life course.

REFERENCES

Allport, G. W. (Ed.) *Letters from Jenny.* New York: Harcourt, Brace & World, 1965.

Butler, R. N. The life review: An interpretation of reminiscence in the aged. *Psychiatry,* 1963, **26**, 65–76.

Cattell, R. B. *Personality and mood by questionnaire.* San Francisco: Jossey-Bass, 1973.

Clopton, W. Personality and career change. *Industrial Gerontology,* 1973, 9–17.

Cooper, M. W. An empirical investigation of the male midlife period: A descriptive, cohort study. Unpublished manuscript, University of Massachusetts at Boston, 1977.

Costa, P. T., Jr., Fozard, J. L., & McCrae, R. R. Personological interpretation of factors from the Strong Vocational Interest Blank scales. *Journal of Vocational Behavior,* 1977, **10**, 231–243.

Costa, P. T., Jr., & McCrae, R. R. Age differences in personality structure revisited: Studies in validity, stability, and change. *Aging and Human Development,* 1977, **8**, 261–275.

Costa, P. T., Jr., & McCrae, R. R. Objective personality assessment. In M. Storandt, I. C. Siegler, & M. F. Elias (Eds.), *The clinical psychology of aging.* New York: Plenum, 1978.

Costa, P. T., Jr., & McCrae, R. R. Still stable after all these years: Personality as a key to some issues in adulthood and old age. In P. B. Baltes & O. G. Brim (Eds.), *Life span development and behavior* (Vol. III). New York: Academic, 1980. (a)

Costa, P. T., Jr., & McCrae, R. R. Somatic complaints in males as a function of age and neuroticism: A longitudinal analysis. *Journal of Behavioral Medicine,* 1980, **3**, 245–257. (b)

Costa, P. T., Jr., McCrae, R. R., & Arenberg, D. Enduring dispositions in adult males. *Journal of Personality and Social Psychology,* 1980, **38**, 793–800.

Costa, P. T., Jr., McCrae, R. R., & Norris, A. H. Personal adjustment to aging: Longitudinal prediction from neuroticism and extraversion. *Journal of Gerontology,* 1980, **36**, 78–85.

Douglas, K., & Arenberg, D. Age changes, cohort differences, and cultural change on the Guilford-Zimmerman Temperament Survey. *Journal of Gerontology,* 1978, **33**, 737–747.

Elder, G. H., & Rockwell, R. C. The life course and human development: An ecological perspective. *International Journal of Behavioral Development,* 1979, **2**, 1–33.

Erikson, E. H. *Childhood and society.* New York: Norton, 1950.

Erikson, E. H. *Young man Luther: A study in psychoanalysis and history.* New York: Norton, 1962.

Eysenck, H. J. *The structure of human personality.* London: Methuen, 1960.

Guilford, J. S., Zimmerman, W. S., & Guilford, J. P. *The Guilford-Zimmerman temperament survey handbook: Twenty-five years of research and application.* San Diego, Calif.: Edits, 1976.

Havighurst, R. J., McDonald, W. J., Maeulen, L., & Mazel, J. Male social scientists: Lives after sixty. *The Gerontologist,* 1979, **19**, 55–60.

Holland, J. L. *The psychology of vocational choice: A theory of personality types and model environments.* Waltham, Mass.: Blaisdell, 1966.

Kastenbaum, R., & Costa, P. T., Jr. Psychological perspectives on death. In M. R. Rosenzweig & L. W. Porter (Eds.), *Annual Review of Psychology,* 1977, **28**, 255–249.

Lacy, W. B., & Hendricks, J. Developmental models of adult life: Myth or reality. *International Journal of Aging and Human Development,* 1980, **11**, 89–110.

Levinson, D. J., Darrow, C. N., Klein, E. B., Levinson, M. H., & McKee, B. *The seasons of a man's life.* New York: Knopf, 1978.

Lowenthal, M. F., & Chiriboga, D. Transition to the empty nest. *Archives of General Psychiatry,* 1972, **26**, 8–14.

Maas, H. S., & Kuypers, J. A. *From thirty to seventy.* San Francisco: Jossey-Bass, 1974.

McCrae, R. R., & Costa, P. T., Jr. Openness to experience and ego level in Loevinger's Sentence Completion Test: Dispositional contributions to developmental models of personality. *Journal of Personality and Social Psychology,* 1980, **38**, 1179–1190.

McCrae, R. R., Costa, P. T., Jr., & Arenberg, D. Constancy of adult personality structure in males: Longitudinal, cross-sectional and times of measurement analyses. *Journal of Gerontology,* 1980, **35**, 877–883.

Moss, H. A., & Susman, E. J. Constancy and change in personality development. In O. G. Brim, Jr., & J. Kagan (Eds.), *Constancy and change in human development.* Cambridge, Mass.: University Press, 1980.

Murray, H. A. *Explorations in personality.* New York: Oxford, 1938.

Murray, H. A., & Kluckhohn, C. Outline of a conception of personality. In C. Kluckhohn & H. A. Murray (Eds.), *Personality in nature, society, and culture* (2nd ed.). New York: Knopf, 1953.

Neugarten, B. L. Personality and aging. In J. E. Birren & K. W. Schaie (Eds.), *Handbook of the psychology of aging.* New York: Van Nostrand Reinhold, 1977.

Neugarten, B. L., Crotty, W. J., & Tobin, S. Personality types in an aged population. In B. L. Neugarten (Ed.), *Personality in middle and later life.* New York: Atherton, 1964.

Neugarten, B. L., & Gutmann, D. L. Age-sex roles and personality in middle age: A thematic apperception study. In B. L. Neugarten (Ed.), *Middle age and aging.* Chicago: University of Chicago Press, 1968.

Reichard, S., Livson, F., & Peterson, P.G. *Aging and personality.* New York: Wiley, 1962.

Robins, L. N. *Deviant children grow up.* Baltimore: Williams and Wilkins, 1966.

Runyan, W. McK. A stage-state analysis of the life course. *Journal of Personality and Social Psychology,* 1980, **38**, 951–962.

Schwartz, D. W., & Karp, S. A. Field dependence in a geriatric population. *Perceptual and Motor Skills,* 1967, **24**, 495–504.

Strong, E. K., Jr. *Vocational interests 18 years after college.* Minneapolis, University of Minnesota, 1955.

Tellegen, A., & Atkinson, G. Openness to absorbing and self-altering experience ("absorption"), a trait related to hypnotic susceptibility. *Journal of Abnormal Psychology,* 1974, **83**, 268–277.

Whitbourne, S. K., & Waterman, A. S. Psychological development during the adult years: Age and cohort comparisons. *Developmental Psychology,* 1979, **15**, 373–378.

White, R. W. (Ed.) *The study of lives: Essays on personality in honor of Henry A. Murray.* New York: Atherton, 1963.

37

THE WELL-BEING AND MENTAL HEALTH OF THE AGED

M. Powell Lawton
Director, Behavioral Research
Philadelphia Geriatric Center

While there has been a great deal of research dealing with the psychological well-being of older people, the state of the art in this area is still poorly developed. Concepts are unclear, and its present status may best be described as one where definitions are still being sought and operationalization is proceeding very slowly. The phase of investigating the discriminant validity of the many separate aspects of psychological well-being that have been identified is hardly started, and even the easier task of simply determining the interrelationships among these aspects has proceeded only a short distance.

This chapter reviews research relevant to the general areas of psychological well-being and mental health of the aged, anchoring it to some extent within a conceptual structure that may add some clarity to the area. The author's view of "well-being" is discussed, followed by a more detailed consideration of research that has led to several definitions of psychological well-being. Factorially based approaches to the definition of psychological well-being are then reviewed, and the chapter is concluded by consideration of the relatively small amount of information relating to the discriminant validity among aspects of psychological well-being.

While "mental health" is a focal construct, an attempt at a definition is delayed until the broader concept "well-being" has been discussed. Well-being, referred to elsewhere by this author as "the good life" (Lawton, 1978), may be viewed as an optimum state, in both an objective and subjective sense, in behavioral, psychological, and environmental spheres. More specifically, the good life was seen as being indicated by well-being in four different sectors, related to one another but with important areas of independence from one another: behavioral competence, objective environmental quality, domain-specific perceived quality of life, and psychological well-being. The next section describes these four sectors of well-being and discusses how they relate to mental health.

SECTORS OF THE GOOD LIFE

Competence has been suggested (Lawton, 1972a) as the theoretical upper limit of performance of which the individual is capable in the important areas of living. The substrata of biological health and cognitive capacity set limits for behavioral competence and may be thought of as the domains of primary competence. As the domains of competence become further removed from the biopsychological levels, external influences upon behavior become more pronounced, and behavioral competence in these secondary domains depends less upon biological and cognitive competence. Lawton, Kleban, and Moss (1980) have suggested that the areas of behavioral competence that most completely represent this behavioral aspect of well-being and are at the same time measurable with relative ease are physical health, activities of daily living, cognition, time use, and social behavior. Competence in each domain may be indexed by psychometric assessments of functioning in each area ("normative competence," Lawton, 1972a). Instruments that have been designed for the systematic multidimensional evaluation of behavioral competence include the OARS (Duke University, 1978), the Comprehensive Assessment and Referral Evaluation (CARE, Gurland et al., 1977–78), and the Multilevel Assessment Instrument (Lawton et al., 1980).

Objective environment is defined either by physical measures in the centimeters-grams-seconds system or by observational consensus (Lawton, in press). Examples of objective physical environment are presence of a spouse, existence of relatives, income and assets, neighborhood or community resources, and dwelling-unit characteristics; the environment as *perceived* by the focal individual is not included (Lawton, in press). Evaluations of objective environmental *quality* are inevitably consensual and therefore value-determined, whether in the social or physical realm. The consensual objective environment is defined by the modal judgments of a number of people evaluating the same environment. While there is inevitably considerable individual variation in assessing environmental qualities such as size, noisiness, or temperature comfort, consensus is relatively high in these domains as compared to other possible qualities such as beauty, complexity, or liking, which probably do not attain sufficient multijudge consensus to be characterized as "objective."

Perceived quality of life is the individual's evaluation of the adequacy of specific domains of life. For example, Campbell, Converse, and Rodgers (1976) represented this sector in terms of expressed satisfaction with 17 domains, including marriage, family life, what one does in spare time, housing, health, and others. The environment as perceived or evaluated by the individual is included in this sector.

Psychological well-being is the individual's net internal experience of well-being compounded from, but not limited to, her perception of her competences, her environment, and the adequacy of the salient domains of everyday life.

The burden of proof is on the person who asserts that the good life may be so neatly subdivided; doubtlessly, a compelling case could be made that the four sectors are so highly interrelated as to make hopeless any attempt to separate them. However, more heuristic approaches have, in fact, shown that each can be measured separately (Lawton et al., 1980). It is also suggested that the factors associated with the well-being of the aged may be understood more easily (at least in the primitive stages of conceptual development) by making such distinctions even if empirical data show them later to be more interdependent than portrayed here.

Acknowledgment of these distinctions, for example, saves us from such reductions to absurdity as that voiced by Taylor (1977), in his otherwise very perceptive critique of the construct "morale": "We must ask ourselves: If a person continues to keep himself clean, wants to see his family, thinks in a reasonably orderly fashion, and relishes any part of his life history, does he have to be happy too?" (p. 30). In my conception, all these indicators other than the internal process of reminiscing belong properly to the domains of behavioral competence. There is no reason to expect total congruity between behavioral competence and psychological well-being. Or later, "If she worries whether her Social Security check will come on time next month, the fact that it makes her unhappy seems almost impertinent" (Taylor, 1977, p. 33). To understand this woman, one surely must know that her income is limited (objective environment) and that she is dissatisfied in the economic domain of life quality. The fact that she may or may not be unhappy as the result of these economic circumstances constitutes one *additional* piece of information relevant to her overall well-being; this feeling is neither irrelevant to general well-being nor does it stand as a proxy for economic deprivation or for dissatisfaction with the level of economic well-being. It is a task for behavioral science to understand the conditions determining the relationships among the sectors of well-being and for applied social science to determine how that knowledge may be used in dealing with clients or designing social programs. We must never forget that the achievement of well-being in any of the four sectors is a legitimate goal in its own right that does not require justification through its contribution to goal achievement in any other sector (Carp, 1977; Rosow, 1977).

"Mental health" is less easily defined. Lip service is paid by all to the concept of "positive mental health" (Jahoda, 1958) as a necessary complement to the more usual definitions of the negative aspects of mental health. Yet operationalizing positive mental health has remained difficult. It is suggested here that mental health may be represented loosely as a combination of three of the four features of the good life—that is, behavioral competence, perceived quality of life, and psychological well-being, excluding the objective environment. However, in common parlance, mental health often denotes simply the absence of psychological symptoms. Since there seems to be a need for such a term, it is suggested here that "poor mental health" be reserved explicitly for the subset of indicators of the "poor life" that refer to psychological symptoms and maladaptive behaviors of classic neurotic or psychotic types. While it would be expected that lowered behavioral competence and poorer perceived quality of life would occur along with indicators of poor mental health, this is not always the case; psychopathology may or may not result in impaired behavior or dissatisfaction with aspects of life. The fourth sector, psychological well-being, deals with subjective experience; psychological symptoms where they are manifested as subjective discomfort are a form of psychological well-being and therefore a subarea of mental health.

DEFINING AND MEASURING PSYCHOLOGICAL WELL-BEING

Reviews of the constructs of psychological well-being and their operationalizations in gerontological research have been written by George (in press), George and

Bearon (1979), Stones and Kozma (1980), and Lawton (1977), while summaries of the correlates of psychological well-being appear in Adams (1971) and Larson (1978). One emerges from this literature and the studies reviewed by these authors with bewilderment over the variety of constructs that fall easily into this general category. The problems of measurement begin with the criterion against which any proposed definition of psychological well-being is to be assessed.

The earlier tradition in mental health research tended to seek criteria of psychological well-being external to the individual: ratings by observers, clinicians' judgments, psychiatric diagnosis, objective behaviors, and so on (Leighton, Leighton, & Danley, 1966; Hollingshead & Redlich, 1958; Srole et al., 1962). These external criteria were likely to be composed primarily of indicators of poor mental health, as defined above. As the construct "psychological well-being" became more explicitly defined in terms of the subjective state of the individual, investigators searched for a criterion that was itself a subjective state. Happiness appears to have been accorded a position of primacy in this respect (Bradburn, 1969; Gurin, Veroff, & Feld, 1960; Stones & Kozma, 1980; Wilson, 1967). The rationale for considering happiness so central has been questioned by many (Rosow, 1977; Taylor, 1977), and in fact, in the aging literature, life satisfaction and morale have been at least as prominent (George, 1979). Further constructs deserving some mention are affect balance (Bradburn, 1969), psychopathology, and self-esteem. Each of these constructs will be discussed in terms of their research origins and measurement.

Life satisfaction is thought of as a cognitive assessment of the extent to which one's life in general is experienced as satisfying, or, as George (1979) puts it, "an assessment of the overall conditions of existence as derived from a comparison of one's aspirations to one's actual achievements" (p. 210). Theoretically, if one could name and measure achieved satisfaction with *all* the domains of perceived quality of life relevant to an individual and weigh them according to their salience to that individual's life goals, a measure of overall life satisfaction should result. Such an ideal accretive index does not seem possible to construct; for example, the multiple correlation between satisfaction with 17 domains and general life satisfaction was .73 in the study by Campbell et al. (1976). Thus "generalized life satisfaction" may be thought of as composed of domain satisfactions in the perceived quality of life sector *plus* other less domain-specific satisfactions.

Life satisfaction among the aged was operationalized as a criterion by Neugarten, Havighurst, and Tobin (1961), who collected definitions of psychological strength in later life from senior researchers and clinicians. A series of successive approximations converged on a staff-consensus definition of life satisfaction along five dimensions: zest versus apathy, resolution and fortitude, goodness of fit, self-concept, and mood tone. From the extended definitions of these rating scales, interview items comprising the widely used Life Satisfaction Indices (LSI) were constructed. In gerontological research the LSI has been a ubiquitous criterion for general psychological well-being.

Life satisfaction was selected by Andrews and Withey (1976) as the most useful criterion for general well-being, as in the question "How do you feel about life as a whole?" with a 7-point scale with named points running from "delighted" to "terrible." In a similar study, Campbell et al. (1976) used the 7-point-rating scale, "How satisfied are you with your life as a whole these days?" as one of two major indicators of well-being. While these simple items have obvious psychometric limitations, at least they afford clear representations of generalized life satisfaction.

Less frequent as a criterion for psychological well-being in gerontological research than in research dealing with people of all ages is *happiness*. Beginning in the research *Americans View Their Mental Health* (Gurin, Veroff, & Feld, 1960) and continuing in the large-sample research by Bradburn (1969), a single item, "Taking things all together, how would you say things are these days? Would you say that you're very happy, pretty happy, or not too happy?" was used as a criterion against which other measures were validated. Happiness has been characterized as an affective response, as contrasted to the cognitive response of life satisfaction, for example, "transitory moods of gaiety or euphoria" (George, 1979, p. 210). Andrews and McKennell (in press), in a linear structural-relation analysis of national social-indicator data, did in fact report findings consistent with the hypothesis that the satisfaction measures were more clearly determined by cognitive factors and the happiness measures by affective factors. However, a "transitory mood," or affect, has a clearly short time referent; when phrased in terms of "these days," a longer-term averaging process, and therefore a significant cognitive element, is implied. Thus a need is still seen for a more completely affective indicator.

Finally, *morale* was one of the earliest constructs to be used in measuring psychological well-being among the aged (Kutner, Fanshel, Togo, & Langner, 1956). Unlike life satisfaction or happiness, morale has never been defined for its measurement purpose in gerontology in a way that highlights its discriminant validity in relation to other indicators of psychological well-being. For example, in defining morale, George (1979) turns to the dictionary definition and then notes that "gerontologists often discuss the wellbeing of older people in terms of morale" (p. 210), correctly implying that "morale" can and does include many diverse aspects of well-being. The more extensively developed morale scales (Lawton, 1972b; Pierce & Clark, 1973; Schooler, 1970) in fact all started with the deliberate intent of representing psychological well-being in its broadest form. It is suggested, then, that despite its frequent use, the term "morale" be abandoned in favor of other more precise terms.

A now-familiar indicator of psychological well-being is the pair of measures of positive affect and negative affect whose difference is called the *Affect Balance Scale* (ABS, Bradburn, 1969). In fact, these scales represent Bradburn's further operational definition of happiness, through their demonstrated correlations with the happiness item. Negative affect is measured by five items best described as typical neurotic symptoms (e.g., depressed, lonely, restless) anchored to the time frame of "the past few weeks." Positive affect is also measured by five items with clearly emotional referents ("proud," "pleased about accomplishment," "excited," and so on). The particularly interesting aspect of the ABS is that while both the positive and the negative affect scales correlate substantially with the happiness item, they are unrelated to one another. This pattern of findings has been replicated in a number of other studies (Andrews & Withey, 1976; Campbell et al., 1976; Costa & McCrae, 1980). The ABS has been used as a measure of psychological well-being in several studies of older people (Gaitz & Scott, 1972; National Council on the Aging, 1975; Wilker, 1975). With its clearly affective meaning and the specific time referent of "the past few weeks," the positive affect scale approaches the criteria for a true affective scale. On the other hand, negative affect is simply a five-item measure of neurotic psychopathology, tied to a recent time frame.

The oldest tradition in the measurement of psychological well-being attempts to measure *psychopathological symptoms* which, as mentioned earlier, lie in both the behavioral competence and psychological well-being sectors. No attempt to survey this vast area will be made here, but a few measures of demonstrated utility with older people will be mentioned. The Minnesota Multiphasic Personality Inventory has been used extensively with the elderly (see Lawton, Whelihan, & Belsky, 1980, for a review of this work), but its 550 items constitute a formidable barrier in many research situations. A shortened version, the Mini-Mult (Kincannon, 1968) was included in the Duke OARS instrument, and some of its psychometric characteristics were reported by Fillenbaum and Pfeiffer (1976). While a full profile of clinical scales is yielded, this report indicates that caution is appropriate from the point of view of the shortened scales' validity with the aged.

In fact, it is worth noting the absence of any survey-method attempt to distinguish through self-report measures the full spectrum of psychiatric syndromes in a large-scale nonclinical elderly population. As Dohrenwend, Yager, Egri, and Mendelson (1978) have shown, psychotic syndromes are particularly difficult to discern in "normal" populations, as compared to the relative ease with which they may be distinguished from other nonpsychotic syndromes, *given* a known clinical population within which the distinction is to be made. Among the aged, too little is known about the effectiveness of self-responding survey-type questions in eliciting admission of major symptoms among older people outside of institutions. In addition, there is the problem of very low base rate (and therefore yield) of such symptoms, and the large number of questions that would, therefore, have to be asked to make a few positive determinations. The CARE (Gurland et al., 1977–78) approaches the latter problem by providing a supplementary in-depth inquiry regarding psychosis, to be used only if some indication is given by the "core" items or clinical judgment of a need for them. For our present purpose, however, the need for expert clinical judgment of this type puts this solution out of range of those available in the typical situation where the instruments are administered by research assistants or nonpsychologically trained service personnel.

The Langner Twenty-Two Item Screening Index (1962) was developed for use in the Midtown Study (Srole et al., 1962) and originally was not used with older Ss. Since that time, however, it has had use in older samples in the 20-year follow-up of the Midtown sample (Srole & Fischer, 1978) and in an all-ages sample of Gaitz and Scott (1972). The items include content related to neurotic symptoms of anxiety, depression, and psychophysiological responses. This item set is closely related to that used in the national Health Interview Survey (National Center for Health Statistics, 1970), whose item scores are available by age for a large national sample. A successor to this item set is the 18-item General Well-Being Schedule (Dupuy, 1978), used in the 1971–1975 Health and Nutrition Examination Survey of the National Center for Health Statistics; findings are still being reported from this study. It should be noted that the age range is truncated at age 74 in this data set.

Since depression is so common in the aged (Gurland, 1976), it is understandable why explicit measures of this symptom should be sought. Instruments constructed by Hamilton (1960) and by Beck et al. (1951) are in general use, but the Zung Depression Scale (1965) has been used explicitly with the aged (Gallagher, McGarvey, Zelinski, & Thompson, 1978; Morris, Wolf, & Klerman, 1975) and in 20 items covers the major clinical symptoms associated with depression. Occasionally one also sees the Cornell Medical Index (Brodman et al., 1949) used; among its 195 items are 51 that are sometimes used as a pool of psychological symptoms. No extensive psychometric analyses of the CMI with older populations have been reported.

Looking back at the array of these and many other measures of psychopathology, one sees that their content is very broad. Some include both psychotic and psychological symptoms, some include behaviors, some have many psychophysiological symptoms, and most overlap significantly in content with all the other domains of well-being discussed previously, especially happiness and morale. A common element, however, is the intrapsychic neurotic symptoms of anxiety and depression.

A measure sometimes used with older people, despite its origin with an adolescent sample, is the Rosenberg Self-Esteem Scale (1965). *Self-esteem,* in every item, does refer to a self-evaluation (e.g., "On the whole, I am satisfied with myself" or "At times I think I am no good at all"), but it is not clearly distinguished from depressive feelings or other kinds of unhappiness. For example, "Compared to other people I get down in the dumps too often" would seem to fit quite well in a measure of depression. Another item, "When I think back over my life, I didn't get most of the important things I wanted," fits clearly into the life satisfaction category. While the scale presently consists of 10 items, in its original form it was scaled according to Guttman criteria in a six-item form, after some item combinations. Reports on its use with older subjects have been done by Atchley (1969; 1976) and Ward (1977).

SOME MORE PROBLEMATIC DOMAINS OF PSYCHOLOGICAL WELL-BEING

Life satisfaction, happiness, positive affect, psychopathological symptoms, and self-esteem (but not morale) thus appear defensible as constructs falling into the sector of psychological well-being, constructs that are also capable of being defined and measured. Other domains that might also be justified as indicators of psychological well-being and have sometimes been used as indicators of psychological well-being are also more problematic: age-related morale, self-rated health, psychophysiological symptoms, and loneliness. Lawton (1977) suggested that *age-related morale* might be a distinguishable aspect of psychological well-being (typical item: "I am as happy now as when I was younger"). For the most part, clusters of items with this content have appeared primarily when using the Philadelphia Geriatric Center Morale Scale (Lawton, 1975). As will be argued in the next section, this cluster may be an instrument-specific subvariety of life satisfaction.

Self-rated health (i.e., an overall estimate of total health; for example, "My health is excellent/good/not so good/poor") while showing relatively high validity as a measure of physical health (e.g., Maddox & Douglass, 1973) is also consistently highly correlated with most measures of psychological well-being (Larson, 1978). However, because it shares so much variance with a construct that is central to the sector of behavioral competence—biological health—it is suggested that measures of psychological well-being be carefully purged of direct health indicators.

Even more problematic are *psychophysiological symptoms* (e.g., palpitations, sweating, headaches), whose legitimacy as indicators of neurotic conditions is unassailable. Nonetheless, not only do they straddle the sectors of behavioral competence (physical health) and psychological well-being; they *may* also be indicators of many serious somatic illnesses of old age, and if for no reason other than as a cautionary measure against masking such a diagnosis, they should be excluded from measures of psychological well-being.

Loneliness and similar feelings form the last of the problematic categories. Theoretically, loneliness is a subjective affect and should therefore be included under the rubric of psychological well-being. If the affective experience itself could be measured in relatively pure form, there would be no problem. However, it is probable that the subjective affect is always mixed with subjective evaluation of the adequacy of one's social interaction (i.e., "I see enough of other people"), which is, according to the territorial subdivisions of the good life proposed above, a domain-specific aspect of perceived quality of life (either family or friends). Thus again it is suggested that the quality of social relationships and affects associated with perceived quality be reserved as one dimension (a particularly central one, as seen in the analysis by Campbell et al., 1976, p. 70) of perceived quality of life.

FACTORIAL ATTEMPTS TO DIMENSIONALIZE PSYCHOLOGICAL WELL-BEING

Exploratory factor analysis is a first-level approach to dimensionalizing a diverse and psychometrically primitive domain. Despite the 25-year or more history of research on the psychological well-being of the elderly, relatively few such studies have been reported, the majority of them dealing with only a few instruments or constructs at a time. The first was the Philadelphia Geriatric Center Morale Scale (Lawton, 1972b), which was constructed so as to represent the widest variety of item content. Later uses of the scale (Lawton, 1975; Morris & Sherwood, 1975; Schooler, 1970) showed convergence on three factors. *Age-related morale* consisted of five items, all with an explicit or implicit comparison between overall quality of life as experienced in the present and some earlier period of life. *Agitation* was very clearly a neurotic symptom or negative affect. *Lonely dissatisfaction* mixed two items on interpersonal relationships with four others dealing with dissatisfaction and unhappiness.

The LSI-A and its shorter versions have been factored by several investigators (Adams, 1969; Bigot, 1974; Dobson, Powers, Keith, & Goudy, 1979; Wilker, 1975). While comparisons are made difficult by the differing item pools used, some convergence in factor definition was observed, within very rough limits. A clear negative-affect factor was found by Dobson et al. and by Wilker, while Adams' first factor seemed to combine happiness and negative affect (he named the factor "mood tone" after the original Neugarten et al., 1961, construct). Somewhat similar factors relating to the root definition of life satisfaction (congruence between expected and achieved goals) were found by all but Bigot, whose British sample showed little similarity to the other three samples in their structure of life satisfaction.

The LSI items have also been combined with those from other scales for component analyses. Dobson et al. (1979) treated the original six-item Rosenberg Self-Esteem Scale (1965) in this fashion and found that their LSI "congruence" factor was fortified by four of the self-evaluative items, while one ("I certainly feel useless at times") clearly belonged to the LSI negative-affect factor. Ward (undated) performed component analyses of the items of the LSI-A and the ABS within age groups for a national sample of all ages (National Council on the Aging, 1975). His older age groupings included the "young-old" (ages 55 to 74) and the "old-old" (75+), each of which was divided into subgroups of males, "work-oriented" females (i.e., either employed or self-characterized as "retired"), and housewives. Within each of these three subgroups, for both ages, in all but one instance an ABS negative-affect factor was clearly evident and positive affect occurred in all six; in five instances an LSI happiness factor occurred. This latter factor was identical to one found by Bigot (1974). Finally, the two female old-old groups exhibited the general life satisfaction (congruence) factor reported by Adams (1969), Dobson et al. (1979), and Wilker (1975).

Andrews and Withey (1976) factored 12 measures of psychological well-being that included ratings of general life satisfaction, happiness, worries, positive affect, and negative affect. Their strongest factor included both satisfaction and happiness (interpreted as cognitions), while clearly separate factors for negative and positive affect also emerged. Their later analysis of some of their same data plus other British data demonstrated, however, that the satisfaction measures tipped toward the cognitive domain while happiness had a stronger affective component (Andrews & McKennell, in press).

The most sophisticated analysis of the well-being domain was performed by Burt, Wiley, Minor, and Murray (1978) on adults of all ages (cumulated NORC General Surveys). The measures included negative affect, positive affect, general satisfaction (a composite of satisfaction with life in general and the happiness rating), and satisfaction with each of eight domains of perceived life quality. The best structural fit to observed data occurred with components of negative affect, positive affect, general satisfaction, and a single overall dimension of domain satisfactions.

Each of the multivariate analyses described above involved only a limited number of the possible items or scales that might have been included. Lawton and Kleban (research in progress) examined the factorial structure of a number of indicators of psychological well-being and perceived quality of life in a sample of 284 older people from groups chosen purposively to represent a range of general competence. The items were assembled from a variety of sources and included the Gurin et al. (1960) happiness item, the Campbell et al. (1976) general satisfaction item, the Bradburn Affect Balance Scale (1969), the Rosenberg Self-Esteem Scale (1965), and the Philadelphia Geriatric Center Multilevel Assessment Instrument personal adjustment subscale in toto. Twelve factors were strong enough to interpret. The largest by far were a happiness factor and a negative-affect factor. Smaller factors included expression/denial of negative affect, self-concept (two separate factors composed primarily of Rosenberg self-esteem items), social ease, generalized life satisfaction (identical to the reappearing LSI factor indicating congruence between expected and attained life goals), and positive affect. Other clearly marked factors outside the realm of psychological well-being were environmental satisfaction, self-rated health, satisfaction with time use, and satisfaction with neighborhood resource access. The negative-affect factor contained many more items than the Bradburn index; as a matter of interest, rather than being uncorrelated with positive affect, this factor showed a correlation of −.38 with the Bradburn positive-affect index.

To summarize, the group of factor-analytic studies of psychological well-being, while not exactly tidy in their overall pattern, shows some recurrent findings. First, positive affect is clearly a cluster of indicators different from all others, including the related construct happiness. Negative affect *is* mild psychopathology, in which both anxiety and depressive feelings figure prominently. As in Adams's "mood-tone" factor, negative affect and (absence of) happiness are statistically related (for example, in the Lawton and Kleban data, these two factors, though nominally orthogonal, when scored for high-loading items, showed a correlation of .74). However, they appear separately often enough to warrant separate use, especially in light of happiness as the posited net of negative and positive affect. Also a frequent, though not invariably present, factor was that aspect of generalized life satisfaction that refers to congruence between past expectations and present reality. No other clusters appeared with any frequency; there is doubt, for example, as to whether self-esteem as operationalized by Rosenberg (1965) forms a homogeneous content area.

THE DISCRIMINANT VALIDITY OF THE VARIETIES OF WELL-BEING

While hundreds of published reports have dealt with predictors or correlates of different measures of well-being, only a few have done so in such a way as to allow the performances of two or more aspects of psychological well-being to be compared. With no attempt at an exhaustive review of indicators of well-being, the task may be somewhat delimited by searching for the correlates of the major domains yielded by the preceding review: positive affect, negative affect, happiness, and generalized life satisfaction.

Across the full age span, Bradburn (1969), Andrews and McKennell (in press), and Gaitz and Scott (1972) reported that positive affect *and* negative affect both decreased with age. Happiness behaved similarly (Bradburn, 1969; Campbell et al., 1976; Gurin et al., 1960; Gaitz & Scott, 1972). On the other hand, generalized life satisfaction increased with age when measured in relatively pure cognitive form (Andrews & McKennell, in press; Campbell et al., 1976; Gurin et al., 1960), but not in the form of the LSI which, as has been seen, is actually multidimensional (National Council on the Aging, 1975). Thus if only with respect to their behavior in relation to chronological age, satisfaction and negative affect–positive affect–happiness are worth distinguishing. The patterning of these findings suggests a diminution of emotional responsiveness with age plus an inclination to be more accepting of the status quo. Almost universally, women have been found to exhibit a greater degree of negative affect, while positive affect is sometimes found to be unrelated to sex (i.e., Wilker, 1975; Bradburn, 1969) and sometimes lower among women (Knapp, 1976). Income was found to be more strongly related to positive affect than to negative affect both for a younger adult sample (Bradburn, 1969) and an older sample (Wilker, 1975).

Of greater interest are the correlates of these several domains of psychological well-being with behaviors and with personality traits. A fairly consistent but by no means inevitable association between positive affect and social behavior (including both formal and informal activities) has been reported (Beiser, 1974; Bradburn, 1969; Lawton & Kleban, research in progress; Wilker, 1975), while negative affect tends to be unrelated to the social behavioral domain. Wilker (1975) also included some environmental satisfaction variables (a perceived quality of life domain) and found them related to positive, but not negative, affect. Among his under-60 Ss, Bradburn (1969) found health related to negative but not to positive affect. In elderly samples, this latter finding occurred in the Lawton and Kleban data, but Wilker (1975) found that poor physical health was associated with negative affect, low positive affect, and low life satisfaction.

In the personality test arena, Costa and McCrae (1980) hypothesized that longstanding "personality dispositions" were differentially responsible for positive and for negative affect. In their large all-ages sample they found, as predicted, that traits named general emotionality, fear, anger, and poor inhibition of impulse were moderately related to negative affect but minimally to positive affect; an opposite pattern was observed for the traits sociability, tempo, and vigor. Costa and McCrae, noting the consistency of this pattern with the classical second-order personality factors of neuroticism (N) and extraversion (E), tested this suggestion directly with measures of N and E from both the Cattell 16 PF Scale (Cattell, Eber, & Tatsuoka, 1970) and the Eysenck Personality Inventory (Eysenck & Eysenck, 1969). In every comparison the predicted pattern was repeated. Furthermore, N and E at baseline significantly predicted negative and positive affect, respectively, 10 years later, attesting to the stable temperamental quality of the relationship. They suggest that rather than the explanation for positive and negative affect's lying in different life experiences associated with each, the explanation may lie in the basic intraindividual characteristics indexed by neuroticism and extraversion.

> Low N introverts and high N extraverts may have similar levels of . . . happiness, but they achieve this result in utterly different ways. The former are seldom depressed but just as seldom elated. The latter are prone to both extremes and reach "average" satisfaction only because there is as much satisfaction as dissatisfaction in their lives. . . . The independence of E and N argues that introverts are no more prone to anxiety, depression, or anger than are extraverts. Whether they should be considered lower in mental health simply because they show less zest, vigor, or enthusiasm is a thorny question. (p. 676)

CONCLUSION

This review of knowledge dealing with one sector of the good life, psychological well-being, suggests that some nascent conceptual order is observable as research data accumulate from both older populations and the general adult population. The evidence suggests that at least four domains of psychological well-being may be conceptually and empirically differentiated: happiness, positive affect, negative affect (mild psychopathology), and generalized life satisfaction based on perceived congruence between desired and achieved life goals.

The discriminant validity of these domains is very incomplete; such evidence as there is seems based primarily on the distinction between negative affect versus positive affect. The question as to whether any of these dimensions should be thought of as a more basic criterion of psychological well-being than any other remains unanswered. The empirical performance of happiness as affected independently by positive and by negative affect perhaps argues that happiness should be seen as an ultimate outcome. However, like Taylor (1977), this author agrees that the search for what in addition to happiness contributes to a good life should be continued. The dimensionalizing of the other sectors of well-being—behavioral competence, objective environment, and perceived quality of life—should proceed in analogous fashion to that reviewed here in the case of the psychological well-being sector.

A need has been suggested for good population data on indicators of mental health other than those in the psychological well-being sector, especially on signs of major distress in the noninstitutionalized older population: organic brain syndromes, psychotic depression, and other psychoses. The technology for such measurement needs development to determine whether such data can be obtained by survey methods, rather than only by prohibitively expensive psychiatric interview.

If not the last word, at least the most provocative question on the agenda today has been provided by Costa and McCrae's (1980) suggestion that positive and negative affect may be the time-honored, two-factor personality traits: extraversion and neuroticism in disguise. Several research challenges have been provided. First, can research provide evidence that domains of psychological well-being, other than positive and negative affect, have differential explanatory value for happiness, for clinical psychopathological syndromes, or other indicators of mental health? Second, is positive affect truly one and the same construct as extraversion? It is clear that the temporal dimension of affective experience and the manner in which affective experiences are subjectively summed to produce cognitions of well-being in general need much further research. Third, if stable, intrapersonal, temperamental factors contribute to the experience of positive and negative affects, how do external experiences modify such long-term propensities?

Finally, this chapter has perforce drawn upon research dealing with many different age ranges. Naturally, research concerned with the well-being of the elderly will frequently have to be limited to older subjects. However, there seems to be no excuse for arbitrarily truncating on "adult population" at age 60, 65, or any other upper limit. It is clear that maximum gain in knowledge accrues when large probability samples are constructed with enough subjects in each broad category to allow both between- and within-age analyses to be performed.

REFERENCES

Adams, D. L. Analysis of a life satisfaction index. *Journal of Gerontology,* 1969, **24**, 470–474.

Adams, D. L. Correlates of satisfaction among the elderly. *Gerontologist,* 1971, **11**, 64–68.

Andrews, F. M., & McKennell, A. C. Measures of self-reported well-being: Their affective, cognitive, and other components. *Social Indicators Research,* in press.

Andrews, F. M., & Withey, S. B. *Social indicators of well-being.* New York: Plenum, 1976.

Atchley, R. C. Respondents vs. refusers in an interview study of retired women: an analysis of selected characteristics. *Journal of Gerontology,* 1969, **24**, 42–47.

Atchley, R. C. Selected social and psychological differences between men and women in later life. *Journal of Gerontology,* 1976, **31**, 204–211.

Beck, A. T., Ward, C. H., Mendelson, M., Mock, J., & Erbaugh, J. An inventory for measuring depression. *Archives of General Psychiatry,* 1951, **4,** 561–571.

Beiser, M. Components and correlates of mental well-being. *Journal of Health and Social Behavior,* 1974, **15,** 320–327.

Bigot, A. The relevance of American life satisfaction indices for research on British subjects before and after retirement. *Age and Ageing,* 1974, **3,** 113–121.

Bradburn, N. M. *The structure of psychological wellbeing.* Chicago: Aldine, 1969.

Brodman, K., Erdmann, A. J., Lorge, I., & Wolff, H. G. The Cornell Medical Index: An adjunct to a medical interview. *Journal of the American Medical Association,* 1949, **140,** 530–534.

Burt, R. S., Wiley, J. A., Minor, M. J., & Murray, J. R. Structure of well-being. *Sociological Methods and Research,* 1978, **6,** 365–406.

Campbell, A., Converse, P. E., & Rodgers, W. L. *The quality of American life: Perceptions, evaluations, and satisfactions.* New York: Russell Sage, 1976.

Carp, F. M. Morale: What questions are we asking of whom? In C. Nydegger (Ed.), *Measuring morale: A guide to effective assessment.* Washington, D.C.: Gerontological Society, 1977.

Cattell, R. B., Eber, H. W., & Tatsuoka, M. M. *Handbook for the 16 Personality Factor Questionnaire.* Champaign, Ill.: Institute for Personality and Ability Testing, 1970.

Costa, P. T., & McCrae, R. R. Influence of extraversion and neuroticism on subjective well-being: Happy and unhappy people. *Journal of Personality and Social Psychology,* 1980, **38,** 668–678.

Dobson, C., Powers, E. A., Keith, P., & Goudy, W. J. Anomia, self-esteem, and life satisfaction: Interrelationships among three scales of well-being. *Journal of Gerontology,* 1979, **34,** 569–572.

Dohrenwend, B. P., Yager, T. J., Egri, G., & Mendelson, F. S. The Psychiatric Status Schedule as a measure of dimensions of psychopathology in the general population. *Archives of General Psychiatry,* 1978, **35,** 731–737.

Duke University Center for the Study of Aging. *Multidimensional functional assessment: The OARS methodology* (2nd ed.). Durham, N.C.: Duke University, 1978.

Dupuy, H. J. Self representations of general psychological well-being of American adults. Paper presented at the annual meeting of the American Public Health Association, Los Angeles, 1978.

Eysenck, H. T., & Eysenck, S. B. *Eysenck Personality Inventory Manual.* San Diego: Educational and Industrial Testing Service, 1964.

Fillenbaum, G., & Pfeiffer, E. The mini-mult: A cautionary note. *Journal of Consulting and Clinical Psychology,* 1976, **44,** 698–703.

Gaitz, C. M., & Scott, J. Age and the measurement of mental health. *Journal of Health and Social Behavior,* 1972, **13,** 55–67.

Gallagher, D., McGarvey, W., Zelinski, E., & Thompson, L. W. Age and factor structure of the Zung Depression Scale. Paper presented at the annual meeting of the Gerontological Society, Dallas, 1978.

George, L. K. The happiness syndrome: Methodological and substantive issues in the study of social-psychological well-being in adulthood. *Gerontologist,* 1979, **19,** 210–216.

George, L. K. Subjective well-being: Conceptual and methodological issues. In C. Eisdorfer (Ed.), *Annual review of gerontology and geriatrics* (Vol. II). New York: Springer, in press.

George, L. K., & Bearon, L. B. *Quality of life in older persons.* New York: Human Sciences Press, 1980.

Gurin, G., Veroff, J., & Feld, S. *Americans view their mental health.* New York: Basic Books, 1960.

Gurland, B. J. The comparative frequency of depression in various adult age groups. *Journal of Gerontology,* 1976, **31,** 283–292.

Gurland, B. J., Kuriansky, J., Sharpe, L., Simon, R., Stiller, P., & Berkett, P. The comprehensive assessment and referral evaluation (CARE)—rationale, development, and reliability. *International Journal of Aging and Human Development,* 1977–78, **8,** 9–41.

Hamilton, M. A rating scale for depression. *Journal of Neurology, Neurosurgery, and Psychiatry,* 1960, **23,** 56–62.

Hollingshead, A., & Redlich, F. *Social class and mental illness,* New York: Wiley, 1958.

Jahoda, M. *Current concepts of positive mental health.* New York: Basic Books, 1958.

Kincannon, J. C. Prediction of the standard MMPI scale scores from 71 items: The mini-mult. *Journal of Consulting and Clinical Psychology,* 1968, **32,** 319–325.

Knapp, M. R. J. Predicting the dimensions of life satisfaction. *Journal of Gerontology,* 1976, **31,** 595–604.

Kozma, A., & Stones, M. J. Some research issues and findings in the assessment of well-being in the elderly. *Canadian Psychological Review,* 1978, **19,** 241–249.

Kutner, B., Fanshel, D., Togo, A., & Langner, T. S. *Five hundred over sixty.* New York: Russell Sage Foundation, 1956.

Langner, T. S. A 22-item screening score of psychiatric symptoms indicating impairment. *Journal of Health and Human Behavior,* 1962, **3,** 269–276.

Larson, R. Thirty years of research on the subjective well-being of older Americans. *Journal of Gerontology,* 1978, **35,** 109–125.

Lawton, M. P. Assessing the competence of older people. In D. Kent, R. Kastenbaum, & S. Sherwood (Eds.), *Research, planning and action for the elderly.* New York: Behavioral Publications, 1972. (a)

Lawton, M. P. The dimensions of morale. In D. P. Kent, R. Kastenbaum, & S. Sherwood (Eds.), *Research, planning and action for the elderly.* New York: Behavioral Publications, Inc., 1972, pp. 144–165. (b)

Lawton, M. P. The Philadelphia Geriatric Center morale scale: A revision. *Journal of Gerontology,* 1975, **30,** 85–89.

Lawton, M. P. Morale: What are we measuring? In C. Nydegger (Ed.), *Measuring morale: A guide to effective assessment.* Washington, D.C.: Gerontological Society, 1977.

Lawton, M. P. What is the good life for the aging? Los Angeles, University of Southern California Ethel Percy Andrus Center, 1978.

Lawton, M. P. Competence, environmental press, and the adaptation of older people. In M. P. Lawton, P. G. Windley, & T. O. Byerts (Eds.), *Aging and the environment: Theoretical approaches.* New York: Springer, in press.

Lawton, M. P., Kleban, M. H., & Moss, M. The Philadelphia Geriatric Center Multilevel Assessment Instrument. Final report to the National Institute of Aging. Philadelphia: Philadelphia Geriatric Center, 1980.

Lawton, M. P., Whelihan, W. M., & Belsky, J. K. Personality tests and their uses with older adults. In J. E. Birren & B. Sloane (Eds.), *Handbook of aging and mental health.* New York: Prentice-Hall, 1979.

Leighton, A. H., Leighton, D. C., & Danley, R. A. Validity in mental health surveys. *Canadian Psychiatric Association Journal,* 1966, **11,** 167–178.

Maddox, G. L., and Douglass, E. Self-assessment of health: A longitudinal study of elderly subjects. *Journal of Health and Social Behavior,* 1973, **14,** 87–93.

Morris, J. N., Wolf, R. S., & Klerman, L. V. Common themes among morale and depression scales. *Journal of Gerontology,* 1975, **30,** 209–215.

National Center for Health Statistics. Selected symptoms of psychiatric distress. *Vital and Health Statistics,* Series 11, No. 37. Washington, D.C.: U.S. Government Printing Office, 1970.

National Council on the Aging. *The myth and reality of aging in America.* Washington, D.C.: National Council on the Aging, 1975.

Neugarten, B. L., Havighurst, R. J., & Tobin, S. S. The measurement of life satisfaction. *Journal of Gerontology,* 1961, **16,** 134–143.

Pierce, R. C., & Clark, M. M. Measurement of morale in the elderly. *International Journal of Aging and Human Development,* 1973, **4**, 83–101.

Rosenberg, M. *Society and the adolescent self-image.* Princeton, N.J.: Princeton University Press, 1965.

Rosow, I. Morale: Concept and measurement. In C. Nydegger (Ed.), *Measuring morale: A guide to effective assessment.* Washington, D.C.: Gerontological Society, 1977.

Schooler, K. K. *Residential physical environment and health of the aged.* Final report, USPHS Grant EC 00191. Waltham, Mass.: Brandeis University, Florence Heller School for Advanced Studies in Social Welfare, 1970.

Srole, L., & Fischer, A. The Midtown Manhattan study: Longitudinal focus on aging, genders, and life transitions. Paper presented at the annual meeting of the Gerontological Society, Dallas, 1978.

Srole, L., Langner, T. S., Michael, S. T., Opler, M. K., & Rennie, T. *Mental health in the metropolis: The Midtown Manhattan study.* New York: McGraw-Hill, 1962.

Stones, M. J., & Kozma, A. Issues relating to the usage and conceptualization of mental health constructs employed by gerontologists. *International Journal of Aging and Human Development,* 1980, **11**, 269–281.

Taylor, C. Why measure morale? In C. Nydegger (Ed.), *Measuring morale: A guide to effective assessment.* Washington, D.C.: Gerontological Society, 1977.

Ward, R. A. The impact of subjective age and stigma on older persons. *Journal of Gerontology,* 1977, **32**, 227–232.

Ward, R. A. Sex differences in adult development and well-being. Department of Sociology, State University of New York at Albany.

Wilker, L. Towards a convergence in the measurement of psychological well-being. Paper presented at the annual meeting of the Gerontological Society, Louisville, Ky., 1975.

Wilson, W. Correlates of avowed happiness. *Psychological Bulletin,* 1967, **67**, 294–306.

Zung, W. W. K. A self-rating depression scale. *Archives of General Psychiatry,* 1965, **12**, 63–70.

AUTHOR INDEX

Pages in *italics* indicate where full references appear.

Abbs, J., 41, *54*
Abelson, H. I., 357, *375*
Abelson, R., 100, 120, *146*
Abelson, R. P., 275, *288*
Abrahams, R. B., 563, *564*
Abrams, D. B., 361, *374, 379*
Abramson, A., 38, 52, *54, 58*
Achenbach, T. M., 383, *391*
Achenbaum, W. A., 521, *522*
Acredolo, L. P., 135, *140*, 280, *283*
Adams, B. N., 487, *496*
Adams, C., 233, *237*
Adams, D. L., 617, 621–623, *625*
Adams, G. M., 558, *564*
Adamson, L., 28, *31*
Adelson, J., 330, 336, *338*
Ades, A. E., 49, 50, *54*
Adler, P. T., 365, 366, *374*
Agar, M. H., 368, *374*
Agruso, V. M., Jr., 556, *565*
Ahammer, I. M., 492, *496*
Ahlgren, A., 362, *374*
Ahr, P. R., 231, *236*
Ahrend, R. A., *212*
Ainsworth, M. D. S., 10, 14, 15, *15, 16*, 151, *159*, 183, 186, 188, 196, *209*, 309, *313*
Akamatsu, T. G., 387, *391*
Albert, J., 587, *599*
Aldous, J., 485, *496, 497*
Alexander, C. N., 405, *413*
Alexander, K., 405, *413*
Alexander, R. A., 590, *600*
Allain, A. N., 356, 358, 361, *379*
Allen, A., 362, *374*
Allen, K. E., 243, *252*
Allik, J., 270, *289*
Allison, P. D., 529, *537*
Allison, T., 7, *15*
Allport, G. W., 611, *612*
Alpaugh, P. K., 595, *599*
Als, H., 28, *31*, 199, *209*
Altemeyer, R., 272, *284*
Andersen, E., 256, *265*
Anderson, B. J., 200, *213*
Anderson, N. H., 225, *240*
Anderson, R. B., 73, 75, *75*
Anderson, R. C., 111, *140*
Andrews, F. M., 617, 618, 622, 623, *625*

Anglin, J., 108, 120, *140*, 255, *265*
Annis, H. M., 360, *377*
Antell, S., 123, *140*
Antonova, T. G., 69, *76*
Antonovsky, A., 517, *522*
Appel, L., 280, *284*
Archer, R. P., 359, 360, 364, *379*
Arenberg, D., 270, *288*, 461, *467*, 551, 556, 557, 564, 567, 571–574, *582, 584*, 603, 604, 606, *612, 613*
Arend, R., 189, 191, *209*
Arendt, H., 500
Arey, S. A., 465, *470*
Aries, E., 480, *482*
Aries, P., 486, *496*
Aristotle, 221
Arkhelin, D. L., *446*
Arlin, P. K., 436, 443, *444*
Arling, G. L., *213*
Arnhoff, F. N., 518, 519, 520, *522*
Aronfreed, J., 316, 322, *323*
Aronson, E., 27, *29*
Arter, J. A., 350, *354*
Asher, S. R., 332, *339*
Asher, S. T., 332, *337*
Aslin, R., 35, 48, 53, *54*
Aslin, R. N., 93, 96, 102, *103*, 135, *145*
Asmussen, E., 558, *565*
Astrand, I., 566, *567*
Atchley, R., 475, 476, *484*
Atchley, R. C., 499, 620, *625*
Atkinson, G., 606, *613*
Atkinson, J., 96, *103*
Atkinson, R., 270, *284*
Atomi, Y., 558, *565*
Attig, M., 235, *238*, 574, *582*
Atz, J. W., 5, *15*
Axelrod, S., 556, *565*
Ayllon, T., 555, *565*
Azrin, N. H., 555, *565*

Babijian, H., 337, *338*
Bachman, J. G., 357, *376*
Bachrach, L. L., 459, *466*
Back, K. W., 562, *568*
Bacon, S. D., 357, *377*
Baddeley, A. D., 542, *549*
Baer, D. M., 242–245, *252*, 527, *537*

Bahrick, H. P., 576, 578, *582*
Bahrick, L., 133, *140*
Bahrick, L. E., *147*
Bahrick, P. O., 576, 578, *582*
Bakeman, R., 165, *178*
Baker, H., 277
Baker, N., 293, 294, *302*
Baldwin, W., 425, *429*
Bales, R. F., 201, *213*
Balijian, H., 190, *209*
Ball, W. N., 137, 138, *140*
Balswick, J., 474, 478, *483*
Balter, M. B., 361, *378*
Baltes, M. M., 525–528, 535, 536, *537*, *538*, 557, *565*
Baltes, P. B., 282, *286*, 524–528, 532–536, *537–539*, 555, *566–567*, 586, 598, *599*, *601*
Ban, P., 185, 196, 200, 202, *209*
Bandura, A., 180, *209*, 218, *236*, 246, *252*, 292, *301*, 531, *537*
Banks, M. S., 23, 24, *29*, *31*, 96, 97, *104*
Barclay, J., 272, *284*
Bardwell, R., 388, *392*
Barenboim, C., 316, *325*
Barker, R. G., 282, *284*
Barnes, R. A., 272, *287*
Barnes, R. F., 561, *566*
Barnet, A. B., 36, *54*, *59*, 69, *75*
Barnett, M. A., 318, 321, 322, *324*, *325*
Barrera, M. E., 100, *104*, 125–127, *140*, *144*
Barrett, G. V., 590, *600*
Barriere, M., 47, *58*, 132, *144*
Barron, F., 595, *599*
Barron, S. B., 438, *446*
Barry, W. A., 199, *209*
Barten, S., 102, *104*
Bartlett, F. C., 111, *140*
Barton, E. M., 527, 528, *537*, *538*
Bartoshuk, A. K., 69, *76*
Basowitz, H., 556, *567*
Bassili, J. N., 95, *104*
Bates, E., 139, *141*, 256, 260, 261, *265*
Bateson, P. P. G., 9, *15*, 232, *236*
Bauer, J. A., 102, *105*
Bauman, K., 420, *431*
Baumrind, D., 304, 313, *313*, 491, *496*
Bausell, R., 272, *287*
Baxter, J., 362, *377*
Bayley, J., 41, *57*
Bayley, N., 304, *313*
Bazell, S., 369, *377*
Beach, D. H., 270, *285*
Bean, L. L., 464, *469*
Bear, G. G., II., 384, *391*
Bearison, D., 235, *236*
Bearon, L. B., 617, *626–627*
Beason-Williams, L., 157, *163*
Beck, A. T., 619, *626*

Beck, R. W., 487, *499*
Becker, H. S., 397, 412, *413*
Becker, J., 172, 173, *178*
Becker, P. T., 157, *163*
Beecher, M., 53, *60*
Beeson, D. L., 460, *466*
Beever, A., 421, *430*
Beilin, H., 227, 228, 230, 231, *236*
Beiser, M., 624, *626*
Belfer, M. L., 367, *374*
Belhar, M. C., *209*
Bell, B., *566*
Bell, R., 481, *482*
Bell, R. A., 125, *143*, 148, 155, *159*
Bell, R. Q., 308, *314*, 387, *391*
Bell, R. Z., 493, *496*
Bell, S., 151, *159*
Bell, S. M., 14, 15, *16*
Belle, D., *467*
Belluci, G., 580, *582*
Bellugi, U., 254, *265*
Belmont, J., 270, *284*
Belmont, L., 425, *429*
Belsky, J. K., 619, *627*
Bem, D. J., 362, *374*
Bem, S. L., 295, *301*
Bench, J., 34, *54*
Bench, R. J., 33, *54*
Bengston, V. L., 397, *413*, 493, 494, *497*, 518–520, *522*
Benigni, L., *265*
Bennett, C., 367, *379*
Benson, F. W., 332, *338*
Benson, N., 125, *141*
Bentler, P. M., 362, *376*
Benward, J., 366, *375*
Berberich, J. P., 244, *252*
Berg, K. M., 36, 48, *54*
Berger, A., 420, *430*
Berger, B. M., 397, 400, 409, 410, *413*
Bergman, T., 23, *30*, 126, *143*
Berkett, P., *626*
Berkowitz, H., *469*
Berman, W., 365, *376*
Bermuth, H. V., *30*
Berndt, T. J., 330, *338*
Berney, K., 272, *284*
Bernbach, H., 270, *284*
Bernstein, N., 541, 547, *549*
Bernuth, M., 35, *57*
Berry, G. J., 359, *374*
Berry, K. L., 372, *374*
Berscheid, E., 333, *338*
Bertenthal, B. I., 124, 125, *141*
Bertoncini, J., 45, 47, *54*, *58*, 132, *144*
Beschner, G. M., 357, 358, *374*
Best, C., 50, *54*, *56*
Beukema, P., *467*

Bierman, K. L., 328, 329, 330, 331, *338*
Bigi, L., 280, *287*
Bigot, A., 621, 622, *626*
Billingsley, A., 422, *429*
Binstock, R., 526, *538*
Birch, H., 309, *315*
Birch, H. G., 100, *105*
Bird, C., *466*
Birns, B., 34, *54*, 102, *104*
Birren, J. E., 456, *466*, 522, *522*, 526, *538*, 541, 546, *550*, 561, *565*, 574, *582*, 586, 595, *599*
Birrer, C., 11, *16*
Bitterman, M. E., 7, *16*
Bjure, B. E., *566*
Bjure, J., *566*
Black, O., 394, *497*
Blackman, S., 442, *445*
Blair, N. L., 357, 358, *378*
Blank, M., 34, *54*, 260, *265*
Blasi, A., 449, *454*
Blauner, R., 486, *497*
Blehar, M. C., 309, *313*
Blenkner, M., 496, *497*, 591, *600*
Blieszner, R., 533, 534, *539*
Block, J., 192, *209*, 305, *314*, 461, 462, *466*, 588, *599*
Block, J. H., 305, 312, *314*
Block, M. R., *497*
Bloom, L., 254, 255, 258, *265*
Bloom, M., 591, *600*
Bloome, D., 257, *265*
Blumstein, S., 38, 53, *54*
Blurton-Jones, N. G., 157, *159*
Bodmer, W. F., 4, *16*
Bogdonoff, M. D., 556, *567*, 579, *584*
Boggs, S. T., 256, *268*
Bohemeier, J. L., 474, *482*
Bohn, C. J., *497*
Bolles, R. C., 7, *16*
Bone, R. N., 363, *380*
Bonnevaux, B., 282, *289*
Boococh, S. S., 398, *413*
Book, W. F., 438, *444*
Boomer, M., 135, *141*
Borke, H., 316, *324*
Borkovec, T. D., 385, *391*
Borkowski, J. G., 281, *284*
Bornstein, M. H., 86, 89, *90*
Bortner, R. W., *497*
Borton, R. W., 21, 29, *30*, 128, 135, 138, *141*, *145*
Bosack, T. N., 64, *76*
Bosse, R., 580, *582*
Boswell, D. A., 595, *599*
Botwinick, J., 532, *538*, 551, 556, *565*, 571, 575, 577–579, 581, *582*, *583*, 587, 595, 598, *599*
Bourne, L. E., 345, *354*

Bovert, M., 222, 228, 231, *237*
Bower, G. H., 279, *286*, 551, *566*
Bower, T. G. R., 21, 28, 29, *29*, 98, 99, *104*, 108, 118, 119, 121, 123, 124, 128, 134, *141*, 185, *209*
Bowerman, M., 254, 260, *265*
Bowers, K. S., 217, *236*
Bowes, W., 69, *76*
Bowker, L. H., 358, 361, 365, *374*
Bowlby, J., 10, *15*, 183, 188, 189, 194, *201*, *209*
Boyack, L. V., 563, *568*
Boyd, E., 46, *55*
Boyd, E. F., 11, 14, 15, *17*
Brachfeld, S., 165, *178*
Brackbill, Y., 11, 12, *16*, 69, *76*
Bradburn, N. M., 617, 618, 622–624, *626*
Braddick, F., 96, *103*
Braddick, O., 96, *103*
Bradford, L. J., 34, *54*
Brady, J. E., 335, *339*
Bragg, H., 476, *484*
Braine, L. G., 81, 82, *90*
Braine, M. D. S., 254, *265*, 344, *352*
Braine, M. S., 226, *236*
Brainerd, C. J., 226, *236*, 351, *352*
Brakarsh, D., 372, *374*
Brake, M., 398, 401, 411, *413*
Brake, S. C., 71, *76*
Brannock, J., 346, *353*
Bransford, J., 272, 278, *284*
Bransford, J. D., 110, 111, *141*
Brantley, P. J., 361, *379*
Braucht, G. N., 359, 372, *374*
Braukmann, C. J., 390, *391*
Braun, H. W., 555, *565*
Braunstein, M. L., 94, *104*
Brawley, E. R., 243, *252*
Brazelton, T. B., 21, 27, 28, *29*, 31, 69, 75, *76*, 165, 167, *178*, 190, 199, *209*
Brehmer, B., 440, *444*
Breland, K., 70, *76*
Breland, M., 70, *76*
Bremner, J. G., 119, *141*
Brenner, J., 172, *178*
Brent, S. B., 451, *454*, 536, *538*
Bretherton, I., *265*
Bridger, I., 166, *178*
Bridger, W. H., 34, *54*, 89, *91*, 92, 128, *143*
Brill, N. Q., 363, 369, *374*, *376*
Brill, S., 40, 42, *56*
Brim, O. G., 460, 493, 494, *497*
Brittain, C. V., 404, *413*
Brodman, K., 620, *626*
Brody, E. M., 491, *497*
Brodzinsky, D. M., 229, 235, *236*, 239, 588, *600*
Broen, P. A., 263, *265*
Bromley, D. B., 330, *339*, 362, *374*

Bronfenbrenner, U., 180, 198, *209*, 282, *284*, 528, *538*
Bronowski, J., 4, *16*
Bronshtein, A. T., 69, *76*
Bronson, G., 36, *54*
Bronson, W. C., 170, *178*
Brook, J. S., 363, *374*
Brook, R., 364, *374*
Brooks, J., *178*, 181, 190, *212, 214*, 329, *339*
Brooks-Gunn, J., 148, *161*, 183, 184, *211*
Broughton, J., 349, 350, *352*, 443, *444*
Broughton, J. M., 128, *141*
Broverman, D. M., 463, *466*
Broverman, I., 463, *466*
Brown, A., 277, 278, *284*, 300, *301*
Brown, A. L., *284*, 593, *601*
Brown, C., 195, 197, *209*
Brown, G., 222, 225, *236*
Brown, J., 37, 40, *42*, 66, *76*, 165, *178*
Brown, R., 253–255, 258, *265, 266*
Brown, S. R., 559, *566*
Bruce, P. R., 577, *582*
Bruce, R. A., 558, *565, 567*
Bruner, J., 279, *284*
Bruner, J. S., 109, *141*, 149, 152, 154, *159*, 336, *338*
Bruno, L. A., 81, *91*
Brustman, B., 363, *380*
Bryan, J. H., 322, *324*
Bryan, W. I., 438, *444*
Bryant, P., 283, *284*
Bryant, P. E., 119, 128, *141*
Buchanan, N., 170, *179*
Buck, R. W., 175, 320, *324*
Buckley, N., 316, *324*
Buda, F., 33, *60*
Buell, J. S., 243, *252*
Buhrmester, D., 336, *338*
Bull, D., 32, 34, *60, 61*
Bull, M. P., 578, *585*
Bullough, V., 464, 465, *467*
Bullough, B., 464, 465, *467*
Bullowa, M., 148, *160*
Bunk, B., 476, 477, *482*
Bunker, L. K., 591, *600*
Burgess, E. W., 516, *522*
Burgess, R. L., 159, *160*, 528, *537*
Burgos, W., 359, *375*
Burke, D., 270, *286*
Burke, E. L., 365, *374*
Burke, P., 73, *76*
Burkhart, B. R., 363, *378*
Burt, R. S., 622, *626*
Busch-Rossngel, N. A., 282, *286*, 525, *538*
Busemeyer, J. K., 556, *566*, 580, *583*
Bushnell, I. W. R., 102, *104*
Busse, E. W., 563, *568*
Butler, M. C., 359, 362, 367, 369, *375*

Butler, R. A., 563, *565*
Butler, R. N., 563, *565*, 6ll, *612*
Butterfield, E., 48, *54*, 270, *284*
Butterfield, E. C., 109, *144*
Butterworth, C. E., 405, *414*
Butterworth, G., 119, 135, *141*
Bystroletova, G. N., 65, *76*

Cabral, G., 328, *338*
Cahalan, D., 366, *374*
Cain, R. L., 200, *213*
Cain, V., 425, *429*
Cairns, G., 48, *54*
Cairns, R. B., 9, 10, *16*, 148, *160*, 308, *314*, 525, *538*
Caldwell, R., 126, *142*
Caldwell, S. B., 424, *430*
Calligan, R. F., 135, *142*
Camaioni, L., 265
Campbell, A., 457, *467*, 477, 482, 615, 617, 618, 621–623, *626*
Campbell, B., 4, 5, *16*
Campbell, C. B. G., 5, *17*
Campbell, D. T., 4, 5, *16*
Campbell, E. Q., 336, *338*, 405, 406, 408, 410, *413*
Campbell, R., 255, *266*
Campos, J. J., 125, *141*, 148, *160*, 180, *209*
Candy, S., 480, *482*
Canestrari, R., 556, *565*
Canestrari, R. E., Jr., 556, *565*
Cannon, W. B., 74, *76*
Caputo, N., 125, *142*
Card, J., 424, 425, *429*
Carden, G., 41, *54*
Carey, S., 255, *266*
Carlomusto, M., 573, 580, *584*
Carlson, D. B., 155, *160*
Carlson, J. E., 365, 366, *377*
Carlson, R., *497*
Carlson, V. R., 99, *104*, 123, *141*
Carlson-Luden, V., 139, *141*
Carman, R. S., 361, *374*
Carmichael, L., 228, *236*
Caron, A. G., 80, *90*, 99, *104*, 123, 126–128, *141, 142*
Caron, R. F., 12, *16*, 80, *90*, 99, *104*, 123, 126, 127, *141, 142*
Carp, F. M., 616, *626*
Carpenter, G. C., 27, *29*
Carpenter, R., 73, *78*
Carr, L. G., 464, *467*
Carrol, E. M., 363, *374*
Carroll, M., 367, *374*
Carter, A., 272, *288*
Carter, D. B., 294, 297, *301, 302*
Carter, D. E., 332, *338*
Carter, H., 475, *482*

Carter, T. M., 405, *416*
Cartwright, D. S., 386, *391*
Cartwright, L. K., 503, *506*
Case, R., 283, *284*, 437, *444*
Casey, J. J., 361, 367, 372, *378*
Castrec, A., 123, *147*
Catania, A. C., 12, *16*
Cattell, R. B., 386, *391*, 606, *612*, 624, *626*
Cauble, A. E., 458, *469*
Caudill, B. D., 365, *374*
Cavalli-Sforza, L. L., 4, *16*
Cavanagh, J. P., 11, *19*
Cavanaugh, J. C., 281, *284*, 553, *565*
Cazden, C., 254, *265*
Ceraso, J., 440, *444*
Cermak, L. S., *288*, 556, *567*, *584*
Chadrow, M., 586
Chadwick, O. F. D., 388, *393*, 404, *416*
Chagnon, N. A., 7, *16*
Chan, S., 136, *143*
Chandler, M., 525, *538*
Chandler, M. J., 316, *324, 325*, 350, *352*, 452, 453, *454*
Chang, H., 48, *61*
Chang, N., 132, *141*
Chap, J., 273, *284*
Chapman, A. J., 335, *338*
Chapman, C. R., 581, *583*
Chapman, J. L., 440, *444*
Chapman, J. P., 440, *444*
Chapman, M., 233, *236*
Chapman, R. H., 235, *236*, 346, *352*, 440, *445*
Charlesworth, W. R., 552, *565*
Chase, W. G., 278, *284*, 438, *444*
Chein, I., 361, 366, *375*
Chess, S., 148, 156, *163*, 309, *315*
Chi, M., 552, *565*
Chi, M. T. H., 270, 278, 279, *284*, 438, *444*
Chiesi, H. L., 278, *284*
Childs, M. K., 334, *339*
Chilman, C., 419, 429, *429*
Chinsky, J. M., 270, *285*
Chiriboga, D. A., 456–460, *467, 469*, 477, *483*, 610, *613*
Chomsky, N., 218, *236*, 253–255, 258, 259, 262, *266*
Christenson, T. E., 307, *314*
Christie, R. L., 369, *374*
Cicchetti, D., 137, *142*
Cicirelli, V. G., 198, *209*, 592, 594, *599*
Cisin, I., 357, 361, 366, *374, 375, 378*
Clark, B. R., 401, *413*
Clark, E., 581, *583*
Clark, E. V., 108, 120, 121, *142*, 254, 255, 258–260, *266*
Clark, H., 254, *266*
Clark, K. B., 465, *467*
Clark, M., 412, *413*

Clark, M. M., 618, *628*
Clark, W. C., 570, 581, *583*
Clarke, J., 398, *413*
Clarke-Stewart, K. A., 196, 198, 199, 202, *209*
Clarkson, F. E., 463, *466*
Claxton, V., 128, *141*
Clayton, R. R., 358, 359, *377*, 474, *482*
Cleary, P. D., 464, *468*
Clifton, R. K., 21, *31*, 68, *76*
Clopton, W., 609, *612*
Coates, D. L., 182, *211*
Cobbs, P. M., 465, *467*
Cohen, A. K., 396, 397, 400, 408, *413*
Cohen, A. Y., 368, *375*
Cohen, C. I., 479, *482*
Cohen, D., 24, *30*
Cohen, F., 460, *467*
Cohen, J., 401
Cohen, L., 24, *29*
Cohen, L. B., 85, 86, 89, *90*, 98, *105*, 108, 125–127, *142*
Cohen, M. J., 580, *585*
Cohen, P., 397, 409, *413*
Cohen, S., 47, *55*
Cohen, S. I., 555, *568*
Cohler, B., 458, *467*, 497
Colby, A., 348, 349, *352*
Cole, M., 270, 282, *284, 285, 289*, 439, *444*
Coleman, J. S., 395–406, 408, 409, 411, 412, *413, 415, 416*
Coles, L. C., 477, *484*
Collis, K. F., 346, *353*
Collmer, C. W., 149, *162*
Comean, J. K., 458, *469*
Comenius, J. A., 153
Cometa, N. S., 350, *352*
Conant, J. B., 397, 402, *413*
Condon, W. S., 22, *29*
Connolly, J., 333, *338*
Connolly, K., 65, *76*
Connor, K., 596, *600*
Contole, J., 157, *160*
Converse, P. E., 457, *467*, 477, *482*, 615, *626*
Conway, E., 69, *76*
Conway, J., 363, *376*
Cook, D., 479, *482*
Cook, M., 99, *104*
Cooley, C. H., 492
Coop, R. H., 149, *161*
Cooper, B. C., 578, *585*
Cooper, F., *58*
Cooper, M. W., 604, *612*
Cooper, R., 280, *284*
Cooper, R. G., 123, *147*
Cooper, W., 38, 49, *55*
Coopersmith, S., 466, *467*
Corballis, M., 50, *61*
Corbit, J., 49, *55, 56*

Cornelius, S. W., 533, 536, *537*
Cornell, E. H., 83, 85, 87, 89, *90*, 123, 124, 126, *142*
Corrigan, R., 260, *266*
Corsale, J., 270, *288*
Corso, J. F., 559, *565*
Cortelyou, A., 133, *146*
Corter, C. M., 135, *142*
Costa, P. T., 618, 624, 625, *626*
Costa, P. T., Jr., 461, *467*, 580, *582*, 603, 604, 606, 607, 609–611, *612, 613*
Cotton, C., 71, *76*
Cowan, N., 37, 39, 46, *55, 56*
Cowan, P. A., 351, *352*
Cowen, E. L., 190, *209*, 337, *338*, 390, *391*
Cowgill, D. O., 513, 515–518, 520, *522*
Coyne, A. C., 577, *582*, 587, 588, *599*
Craik, F. I. M., 270, *285*, 571, 572, 574, 575, *582*
Crain, W. C., 363, 364, *375*
Crawford, A., 427, *429*
Crawford, S., 421, *429*
Cremin, L. A., 397, *413*
Cressey, D. R., 382, *393*
Criswell, J. H., 332, *338*
Croft, K., 230, *237*
Cromer, R. F., 259, *266*
Cronbach, L. J., 220, *236*
Crook, T., 581, *583*
Crossley, H. M., 366, *374*
Crotty, W. J., 605, 606, 606, *616*
Cullen, E., 6, *16*
Culp, R., 46, *55*
Cusick, P. A., 398, 401, 411, *413*
Cutter, H. S. G., 361, 362, *375*
Cutting, J., 51, *57*, 95, *104*
Cvetkovich, G., 420–423, *429*

Dabrowski, K., 451, *454*
Dahlem, N., 272, *285*
Dale, L. G., 234, *236*
Dale, P., 47, *60*, 254, *266*
Damico, S. B., 407, *414*
Danley, R. A., 617, *627*
Dann, S., 197, *210*
Danner, F. N., 226, 233, *236*
Darby, B. L., 135, *145*
Darrow, C. N., *483, 507, 613*
Darwin, C., 4
Dassen, P. R., 439, *444*
Datan, 282, *285*, 460, *469*, 517, *522*
Daum, C., 102, *105*
Davidson, W. S., III., 390, *393*
Davis, K., 403, *414*
Davis, M., 185, *212*
Davison, M. I., 354
Davitz, J. R., 332, *338*
Day, D., 587, *599*

Day, J., 277, *284*
Day, M. C., 102, *104*, 219, 223, 226, 233, *236, 240*
Day, R. H., 89, *92*, 98, 99, *104*, 123, *146*
Dayton, G. O., 102, *104*
DeCasper, A. J., 22, *29*, 47, *55*, 132, *142*
Deen, M., 51, *59*
DeForest, M., 274, *287*
DeFrain, J., 225, *237*
Degen-Horowitz, F., 127, *147*
DeGroot, A., 278, *285*
Dehn, M. M., 558, *565*
DeJong, G. F., 563, *566*
DeLeon, G., 367, *375*
Delgado, M., 511, 521, 522, *522*, 594, 595, *599*
Deloache, J. S., 86, 89, *90*, 271, *285*, 300, *301*
DeLucia, C. A., 49, *60*, 67, *78*, 81, *92*
Dembo, R., 273, 359, 372, *375*
Dempsey, J., 157, *160*
DeMyer-Gapin, S., 384, *391, 394*
Denisova, M. P., 63, *76*
Denney, D. R., 593, *599*
Denney, N. W., 532, 536, *538*, 587, 592, 593, 597, *599*
Densen-Gerber, J., 366, 373, *375, 378*
DePaulo, B. M., 385, *392*
DePaulo, P., 9, 11, *17*
de Ribaupierre, A., 442, *445*
Dericco, D. A., 361, 365, 366, *375*
Desforges, C., 222, 225, *236*
Des Jarlais, D. D., 359, *375*
DeTine-Carter, S. L., 332, *338*
deVilliers, J. G., 254, *266*
deVilliers, P. A., 254, *266*
DeVries, H. A., 558, 559, *564, 565*
Dewey, J., 110, *142*
Dietrich, C., 384, *391*
Dill, D., 465, *467, 567*
Dino, G. A., 318, 321, 322, *324*
Dirks, J., 89, *90*, 100, *104*
DiVesta, F. G., 581, *584*
DiVitto, B., 165, *178*
Dlugokinski, E. L., 321, *324*
Dobbing, J., 72, 75, *76*
Dobson, C., 621, 622, *626*
Dobson, V., 96, *104*
Dodd, B., 133, *142*
Dodds, C., 86, *91*
Doherty, M. E., 347, *353, 446*
Dohrenwend, B. P., 456, 464, 465, *467*, 619, *626*
Dohrenwend, B. S., 456, 464, 465, *467*
Dolecki, P., 225, *240*
Dolinsky, H., 228, *239*
Donaldson, G., 532, *538*
Donovan, J. E., 357, 358, *375*
Donovan, W., 40, 42, *59*
Dooling, D. J., 277, 278, *285, 289*

Doran, L., 421, *429*
Dore, J., 256, *266*
Dorman, M., 38, 40, 42, *55, 58*
Doty, D., 48, *55*
Douglas, J. A., 234, *236*
Douglas, K., 603, *612*
Douglass, E., 620, *627*
Douvan, E., 330, 336, *338*
Dowd, J., 22, *29*
Dowd, J. J., 518, *522*
Dowling, K., 102, *105*
Downs, T., 591, *600*
Doyle, A., 333, *338*
Drachman, D. A., 572, *582*
Drage, J. S., 73, *77*
Dreyer, P. H., 397, *414*
Drinkwater, B. L., 558, *565*
Drozdal, J. G., Jr., 281, *285, 289*
Drury, A., 277, *287*
Dryfoos, J., 425, *429*
Dubery, D. R., 292, 299, *302*
Dubin, 561, *565*
Duck, S. W., 333, *338*
Dumais, S. T., 96, *103, 104*
Duncan, D. F., 367, *375*
Dunham, H. W., 456, 464, *467*
Dunkeld, J., 24, *30*
Dunn, J., 196, 198, 201, 202, *209, 210*
Dunphy, D. C., 330, *338*
Dupuy, H. J., 619, *626*
Dyhdalo, L., 390, *393*
Dyk, R. B., 589, *601*
Dymond, R. F., 316, *324*
Dytrych, Z., 155, *161*

Early, R. G., *567*
Ebbesen, E., 293, 298, *302*
Eber, H. W., 624, *626*
Eckerman, C. O., 164, 166, 170, 171, 172, 176, 177, *178*, 309, *314*
Eckland, B. K., 405, *413*
Edelbrock, C. G., 383, *391*
Eder, D., 329, *338*
Edwards, C. P., 204–206, *210*
Edwards, T., 43, 44, *56*
Egger, G. J., 369, 370, *375*
Egri, G., 619, *626*
Ehrhardt, A. A., 311, *314*
Eibl-Eibesfeldt, I., 5, *16*
Eichberg, R. H., 365, *374*
Eilers, R., 32, 39, 41–44, 48, 52, 53, *55, 59*
Eilers, R. E., 22, *30*
Eimas, P., 21, 22, *30*, 32, 37, 39, 41–45, 49, 51, 52, *55, 56, 58*, 131, *142*
Eimas, P. D., 32, 49, *56*, 68, *76*
Einhorn, H. J., 441, *445*
Einstein, A., 140
Eisdorfer, C., 556, *565, 567*, 579, 581, *582–584*

Eisenberg, R., 34, 35, 48, *56*
Eisenberg-Berg, N., 316, 318–322, *324, 325*
Eisenburg, A. R., 263, *268*
Eisert, D. C., 225, *240*
Eisler, R. M., 372, *375*
Eisner, D. A., 589, 591, *599*
Eitzen, D. S., 406, *414*
Ekman, P., 157, *160*
Elder, G. H., 526, *538*, 607, *612*
Elder, G. H., Jr., 460, *467*
Elkin, F., 403, *414*
Elkind, D., 234, 235, *236*
Ellgring, D., 148
Ellinwood, E. H., Jr., 361, 366, *375*
Elliott, G. B., 33, *56*
Elliott, K. A., 33, *56*
Elmer, E., 159, *160*
Elsayed, M., 559, *565*
Emde, R. N., 11, *16*, 74, *76*, 148, 152, 157, *160*, 185, *212*
Emerson, P., 14, *19*
Emery, J., 73, *78*
Empey, L. T., 382, 386, *391*
Endman, M., 34, *61*
Engen, T., 68, *76*
Ennis, R. H., 343, *352*
Enright, M., 71, *78*
Entus, A., 50, *56*
Epperson, D. C., 403, 404, *414*
Epstein, N., 318, *325*
Erbaugh, J., *626*
Erber, J. T., 572, 573, 575, 581, *582, 583*
Erdman, A. J., 526
Erikson, E., 9, 456, *467*, 496
Erikson, E. H., 448, 453, *454*, 501, 502, *506*, 603, 604, 606, 607, *612*
Erikson, R. C., 574, *583*
Eron, L. D., 348, *352*, 388, *392*
Ertel, D., 363, *375*
Ervin, F., 7, *16*
Ervin-Tripp, S., 255, 256, 261, 263, *266*
Erwin, R., 51, 53, *59*
Escalona, S. K., 34, *54*
Eson, M. E., 350, *352*
Estes, W. K., 552, *566*
Etzel, B. C., 12, *16*
Ensminger, M. E., 370, *376*
Evans, J., 423, *429*
Evans, J. StB., 440, *446*
Eveland, L. K., 369, *376*
Everett, B. A., 230, *237*
Everett, M. W., 359, *379*
Evers, M., 476, *483*
Eysenck, H. J., 606, *612*
Eysenck, H. T., 624, *626*
Eysenck, M. W., 575, 576, *583*
Eysenck, S. B., 624, *626*

Fagan, J. F., 80–83, 85–90, *90–92*, 97, 98, *104*, 108, 123, 126, 127, *142*, 280
Fagen, J. W., 71, *78*
Fagot, B. I., 306–308, 312, *314*
Falmagne, R. J., 344, *352*, 440, *445*
Fanshel, D., 618, *627*
Fantz, R. L., 24, *30*, 69, *76*, 80, 81, 87, *91*, 97, 100, *104*
Farge, E. J., 559, *566*
Faris, R., 456, 464, *467*
Farley, E. C., 371, *376*
Farrington, D. P., 388, *394*
Farudi, P. A., 387, *391*
Faterson, H. F., 589, *601*
Faust, R., 369, *376*
Fava, D., 596, *600*
Fay, B., 320, *324*
Feely, C., 581, *582*
Feinleib, M., 505, *506*
Feinman, S., 199, *210*, 515, 516, *523*
Feiring, C., 180–182, 194, 196, 198–203, 206, *210, 211*
Fejer, D., 358, 366, *379*
Feld, S., 463, *467*, *468*, 617, 618, *626*
Feldbaum, C. L., 307, *314*
Feldman, C. F., 224, 227, *237*
Feltovich, P. J., 278, *284*
Ferris, S. H., 581, *583*
Feshbach, N. D., 316–320, 323, *324*, *325*
Fichner-Rathus, L., 366, *378*
Field, J., 99, *104*, 109, 121, *142*
Field, T., 24, 25, 28, *30*, *31*, 157, *160*, 165–168, 170, 171, 175, *178*, *179*
Fifer, W., 22, *29*, 47, *55*, 132, *142*
Figlio, R. M., 389, *394*
Figurin, N. L., 63, *76*
Fijimura, O., 52, *58*
Fillenbaum, G., 619, *626*
Fillenbaum, S., 449, *454*
Fillmore, C. J., 254, *266*
Fillmore, K. M., 357, 370, *375*, *377*
Findlay, J. M., 23, *31*
Finkelstein, N. W., 170, *179*
Finley, G. E., 282, *286*, 511, 521, 522, *522*, 594, 595, *599*
Firestone, I., 245, *252*
Firestone, I. G., 321, *324*
Firth, R., 487, *497*
Fischbein, E., 346, *352*
Fischer, A., 619, *628*
Fischer, A. K., 463, *469*
Fischer, T., 171, 173, *178*
Fischoff, B., 441, *446*
Fishburne, P. M., 357, 358, *375*
Fisher, J. C., 361, 362, *375*
Fiske, D. W., 360, 368, *375*
Fiske, M., 458, *467*
Fitzgerald, H. E., 11, *16*

Flanders, N. A., 304, *314*
Flavell, E. R., 230, *237*
Flavell, J. H., 108, *142*, 217, 218, 226–228, 230, 232, 233, *237*, *238*, 269–271, 280, 281, *284–287*, 289, 300, *301*, 351, *352*
Flock, H. R., 94, *104*
Floyd, H. H., Jr., 404, *414*
Foard, C., 51, *57*
Fodor, E. M., 384, *391*
Fodor, J., 40, 42, *56*
Fogel, A., 27, 28, *30*, 156, 157, *160*, 166–169, *178*
Foggitt, R. H., 385, *393*
Follingstad, D., 372, *374*
Foner, A., *498*
Fontane, V. J., *160*
Foot, H. C., 335, *338*
Foote, N., *497*
Foppa, K., 148, *163*
Forge, A., *497*
Forrest, J., 423, *429*
Forsyth, R., 360, 363–365, 367, 370, *378*
Foss, D., 411, *414*
Fowles, D. C., 385, *391*
Fox, G., 420, *429*
Fox, R., 96, *104*
Fox, S., 363, *376*
Foy, D. W., 372, *375*
Fozard, J. L., 270, *288*, 556, 559, *566*, *567*, 574, 578, 580, *582*, *584*, *585*, 609, *612*
Fraiberg, S., *160*
Frankel, F., 270, *284*
Franklin, H. C., 578, *583*
Franks, J., 111, *141*, 272, *284*
Frase, L. T., 440, *445*
Fraser, C., 254, *266*
Frazer, J., 197, *210*
Frease, D. E., 407, *414*
Freeman, R., Jr., 50, *59*
Freese, M. P., 73, *78*, 157, *163*
French, R. L., 411, *414*
Freud, A., 197, *210*
Freud, S., 5, 9, *16*, 186, 189, *210*, 305, 448, *454*, *563*
Friedman, A. S., 357, *374*
Friedman, S., 27, *29*
Friedman, S. B., 81, *91*
Friedrich, D., 270, *285*
Friedrich, L. K., 246, *252*
Fries, J. F., 525, 526, *538*
Friis, H., *499*
Fruensgaard, K., *565*
Fry, C. L., 511, *522*
Fujisaki, H., 45, 49, *56*
Fulton, D., 272, *284*
Fulton, R., *56*
Funder, D. C., 362, *374*
Furchgott, E., 556, *566*, 580, *583*

Furman, W., 190, 191, *210,* 328–334, 336, *338, 339*
Furstenberg, F., 420, 424, 425, 427, *429,* 478, *482*
Furth, H. G., 272, *285*

Gabrielson, I., 422, *429*
Gabrielson, M., 422, *429*
Gaensbauer, T., 148, 152, 157, *160*
Gagné, R. M., 273, *285*
Gagnon, J., 420, *430*
Gaitz, C. M., 618, 619, 623, *626*
Galambos, R., 33, 34, *56, 59, 60*
Galanter, E., 437, *445*
Gallagher, D., 619, *626*
Gallagher, J. M., 350, *352*
Gallas, H., 46, *55*
Galligan, R., 198, *213*
Gallistel, C. R., 225, 227, 228, *237*
Garcia, E., 244, 245, *252*
Garcia, J., 7, *16*
Gardner, H., 24, *30,* 350, *352*
Gardner, J., 24, *30*
Gardner, J. M., 21, *30*
Gardner, R. W., 594, *599*
Garlington, W. K., 361, *375*
Garner, W. R., 122, *142*
Garrett, M., 40, 42, *56*
Gavin, W., 41–44, 48, 52, 53, *55, 59*
Gay, J., *444*
Geer, B., 412, *413*
Geerken, M. R., 463, *467*
Geisheker, E., 321, *324*
Geismar, L. L., 466, *467*
Geitman, H., 280, *285*
Gekoski, M. J., 71, *78,* 138, *146*
Gelber, E. R., 24, *29,* 108, *142*
Gelfand, D. E., 488, *497*
Gellert, B., 328, *338*
Gelman, R., 123, *147,* 225, 227, 228, *237,* 263, *268,* 269, *285*
Gentner, D., 255, *266*
George, L. K., 616–618, *626*
Gerald, D. L., 361, *375*
Gerard, R. E., 384, *392*
Gergen, K. J., 220, *237*
Gerhart, V. C., *467*
Gesten, E. L., 390, *391*
Gewirtz, J. L., 4, 9–15, *16, 17*
Geyer, R. F., 506, *506*
Ghiselin, M., 4, *17*
Gibbs, J. C., 282, *285,* 349, *353*
Gibson, E. J., 21, 23, *30,* 89, *90,* 94, 95, 99–101, 103, *104, 105,* 108, 114–118, 121, 124, 129–131, *142, 143, 147*
Gibson, J. J., 93, 94, 101, *105,* 108, 114, 116–118, 120, 121, 124, 127, 129, 130, 137–139, *143*

Gibson, K. R., 156, *162*
Gilbert, J., 52, *61*
Gilchrist, L., 423, *430*
Gilewski, M. G., 577, *585*
Gilligan, C., 348, 349, *353,* 452, 453, *454*
Gilman, E. A., 54, *56*
Gilmore, L. M., 136, *143*
Girton, M. R., 102, *105*
Glanville, B., 50, *54, 56*
Glascock, A. P., 511, 515, 516, *523*
Glaser, R., 278, *284, 444*
Glasmer, F. D., 563, *566*
Gleitman, H., 263, *267*
Gleitman, L. R., 263, *267,* 280, *285*
Glenn, C. G., 274, 275, *289*
Glenwick, D. S., 587, *599*
Glick, J., 231, *240,* 439, 444, *445*
Glick, P. C., 474, 475, *482,* 486, *497*
Glickman, S. E., 7, *17*
Glucksberg, S., 300, *301*
Goetz, E. M., 243, *252*
Goffman, E., 110, *143,* 531, *538*
Golbeck, S. L., 224, 235, *238*
Goldberg, S., 135, *143,* 156, 157, *160,* 165, *178,* 182, *210,* 211
Golden, M. M., 337, *339*
Golden, R. E., 387, *392*
Goldfarb, W., 74, *76*
Goldman, D., 129, *143*
Goldsmith, B., 373, *378*
Goldsmith, S., 422, *429*
Goldstein, G. S., 357, 359, 364, 365, 369, *377*
Goldstein, K., 442, *445*
Golenski, J., 48, *61*
Golinkoff, R. M., 136, *143*
Gollin, E. S., 230, *240*
Gonda, J. N., 557, *567*
Gonzales, M., 282, *289*
Goodenough, D. R., 442, *446,* 589, *601*
Goodsitt, J., 46, *56*
Goodwin, D. W., 364, *375*
Goody, J., 282, *285*
Gordon, C. W., 396, 397, 399, *414*
Gordon, F. R., 121, 128, *143*
Gordon, M. M., 396, *414*
Gordon, S. K., 570, 581, *583*
Goren, C. C., 24, *30*
Gorenstein, E. E., 385, *391*
Gorman, B. S., 363, *375*
Gormezano, I., 36, *56*
Gorsuch, R. L., 359, 362, 367, 369, *375*
Gorzycki, P., *56*
Goslin, D., 493, *497*
Gottfried, A. W., 89, *91, 92,* 128, *143*
Gottlieb, D., 401, *414*
Gottlieb, G., 9, *17,* 232, *237*
Gottman, J., 334, *339*
Gottman, J. M., *212,* 332, *337*

Goudy, W. J., 621, *626*
Gove, F., 189, *209*
Gove, W. R., 463, *467*
Graeven, D. B., 366, *375*
Graham, F. K., 36, 37, 40, 42, 48, *54, 56, 57, 69, 75, 76*
Graham, J., 388, *392*
Graham, P., 388, *392, 393,* 404, *416*
Grant, E. A., 578, *585*
Gratch, G., 108, 121, 135, *143, 145*
Gray, G., 367, *379*
Gray, P. H., 10, *17*
Graziano, W., 333, *338*
Green, J., 358, 359, 365, 369, *375*
Green, J. A., 157, *160*
Green, L. L., 363, *375*
Green, S. B., 363, *378*
Greenbaum, C. W., 199, *210*
Greenberg, R., 24, *30*
Greenfield, P., 120, *143,* 258, *266*
Greenfield, P. M., 282, *285,* 448, 449, *454*
Greeno, J. G., 552, *567*
Greenspan, S., 316, *324, 325*
Greeg, G., 159, *160*
Greif, E. B., 348, *353*
Gribble, C. M., 295, *303*
Grier, W. H., 465, *467*
Griew, S., 545, *549, 550*
Griffiths, K., 99, *104*
Grimby, G., *566*
Grimwade, J., 33, *61*
Grisdela, M., 421, *430*
Grisell, J., 365, *376*
Gronwall, D. M. A., 541, 542, *550*
Gross, A., 596, *600*
Gross, K. A., *446*
Gross, R., 397, *414*
Grossman, J. L., 556, *566,* 573, *583*
Grote, B., 420–423, *429*
Gruber, H., 443, *445*
Gruendel, J., 275, *287*
Gruendel, J. M., 256, 261, *267*
Gruenfeld, L. W., 589, *599*
Guilford, J. P., 595, *599, 612*
Guilford, J. S., 606, *612*
Gunter, M., 75, *76*
Guppy, N., 504, *507*
Gurin, G., 463, *468,* 617, 618, 622, 623, *626*
Gurland, B. J., 619, *626*
Gustafson, G. E., 157, *160*
Gutman, G. M., 559, *566*
Gutmann, D. L., 458, *468, 469,* 472, 476, 478, *482,* 603, *613*
Guttman, D., 503, *507,* 511, 516, 518, *523*
Guze, S. B., 364, *375*
Gyomroi-Ludowyk, E., 197, *210*

Haaf, R., 125, *143*
Haan, N., 461, 462, *468*
Hefferty, D., 406, *416*
Hagen, J. W., 270, 271, *285, 286*
Hager, D. L., 358, *375*
Hager, J. L., 7, *19,* 70, *78*
Hagestad, G. O., 447, *454,* 460, 485, 489, 492, 493, 495, *497, 498*
Hagstrom, W. O., 561, *567*
Hainline, L., 23, *30,* 126, *143*
Haith, M. K., 124, *141*
Haith, M. M., 23, *30,* 68, 69, 77, 126, *143,* 148, *160*
Halford, G. S., 231, *237,* 346, *353*
Halikas, J. A., 364, *375*
Hall, G. S., 403, *414*
Hall, S., 398, 411, *413, 414*
Hall, W. G., 71, *76*
Haller, A. O., 405, *414*
Hallinan, M. T., 329, *338*
Hallock, N., 157, *160*
Halprin, F., 492, *497*
Halstead, D. L., 366, *379*
Halverson, C. F., 329, *339*
Halwes, T., 270, *287*
Hamilton, M., 619, *627*
Hamilton, M. L., 320, *325*
Hamilton, W. M. E., 542, *550*
Hanks, C., 196, *213*
Hansen, D., 270, *284*
Hansen, G., *498*
Harding, C. G., 136, *143*
Hardy-Brown, K., 225, *240,* 350, *353*
Hare, N., 465, *468*
Hareven, T. K., 485, *497*
Hargreaves, D. H., 406, *414*
Harker, J. O., 574, *583*
Harkins, S. W., 581, *583*
Harlow, H. F., 189, 190, *210, 212, 213*
Harlow, M. D., 189, 190, 210, *210*
Harmatz, J. S., 367, *374*
Harmon, R., 148, 152, 157, *160*
Harper, L. V., 148, 156, *159*
Harre, R., 407, *415*
Harris, C., 277
Harris, F. R., 243, *252*
Harris, P. L., 118, 119, 121, 130, 131, 134, 135, *143*
Hart, B. M., 243, *252*
Harter, N., 438, *444*
Hartig, M., 295, *301*
Hartley, A. K., *567*
Hartley, J. T., 574, *583*
Hartley, L. H., 558, *566*
Hartung, G. H., 559, *566*
Hartung, J., 362, *377*
Hartup, W. W., 190, 191, 197, *210,* 327, 332, 333, 335–337, *339,* 492, 493, *497*

Hasher, L., 576, *582, 583*
Hauser, P. M., 511, *523*
Hauser, R. M., 405, *416*
Hausman, C. P., 573, *584*
Havighurst, R. J., 397, *414,* 561, *566,* 597, *600,* 610, *612,* 617, *627*
Hay, J. C., 95, *105*
Hayden, B., 385, *392*
Hayes, D. S., 331, *339*
Haymes, M., 363, *375*
Haynes, N. M., 390, *392*
Haynes, S. G., 505, *506*
Heath, D. H., 199, *210*
Hebb, D. O., 152, *160*
Hebdige, D., 398, 411, *414*
Hecox, K. E., 33, 34, 36, *56*
Heikkinen, E., 558, 559, *568*
Hein, A., 102, *105*
Heinsohn, A. L., 401, *414*
Held, R., 102, *105*
Heller, J. I., 552, *567*
Helmreich, R. L., 464, *469*
Helson, H., 441, *445*
Hendricks, J., 604, *613*
Hendrickson, A., 561, *566*
Henle, M., 440, *445*
Henn, F. A., 388, *392*
Hennessy, B., 35, *54*
Henry, J., 410, *414*
Herbert, C. P., 559, *566*
Herman, J., 270, *289*
Herman, T., 575, *583*
Heron, A., 511, *523*
Heroux, L., 100, *105,* 127, *144*
Hess, B., *498*
Hess, E. H., 8–10, *17, 160*
Hess, R., 51, *59*
Hetherington, E. M., 387, *392*
Hiatt, S., 130, *144*
Hiel, C. T., 474, *482*
Hiel, R., 487, 491, 492, *497*
Higgins, E. T., 300, *301*
Higgins, R. L., 361, *375, 376*
Hilbert, N. M., 577, 581, *583*
Hilgard, E. R., 551, *566*
Hillenbrand, J., 43, 44, *56*
Hillman, K. G., 404, *414*
Hillyard, S. A., 33, *59*
Hinde, R. A., 5, 7, 9, 10, 13, *17,* 70, *77,* 148, 151, *157,* 182, *210*
Hindelang, M. J., 408, *414*
Hirsch, I. J., 35, *56*
Hirtle, *445*
Hitch, G., 542, *549*
Hitler, A., 277
Ho, V., 233, *237*
Hochman, J. S., 363, 369, *376*

Hodos, W., 5, *17*
Hoeffel, E. C., 449, *454*
Hofer, V., *567*
Hofferth, S., 427, *430*
Hoffman, H., 50, *54*
Hoffman, H. S., 9–11, *17, 18*
Hoffman, M. L., 316–321, *325,* 348, *353*
Hofland, B., 533, 534, *538*
Hogan, D., 502, *506*
Hogan, R., 363, *376*
Hogart, R. M., 441, *445*
Holding, 578, *583*
Holland, J. L., 609, *612*
Holland, V. M., 227, *237*
Hollingshead, A. B., 195, *211,* 396, 397, 400, *414,* 456, 464, *468,* 617, *627*
Holmberg, M. C., 309, 310, *314*
Holmberg, T., 41, 43, *56*
Holmes, L. D., 515, 516, 517, *522*
Holmes, T. H., 456, *468*
Holstein, C., 348, *353*
Holzer, C. E., 465, *470*
Homa, D., 129, *143*
Honigmann, J. J., 465, *468*
Hood, S. Q., 298, *302*
Hooper, F. N., 225, *237*
Horman, R. E., 364, *376*
Horn, J. L., 532, 533, 535, 536, *538,* 598, 599
Hornblum, J. N., 227, 237, *237,* 557, *566*
Hornick, J., 421, *429*
Horowitz, F. D., 63, *77,* 132, *143,* 145
Horowitz, M. J., 460, *468*
Horowitz, S., 494, *497*
Horwath, S. M., *565*
Hoving, K. L., 270, *285, 287*
Howard, J. A., 318, 321, 322, *324, 325*
Howard, K. I., 386, *391, 393*
Howes, C., 309, *314,* 329, *339*
Hoyer, W., 580, *582*
Hoyer, W. J., 527, 537, *538,* 557, *566, 568,* 598, 599
Huba, G. J., 362, 363, 365, 367, *376*
Hubel, D. H., 22, 24, *30,* 72, *77*
Huberman, M., 561, *566*
Hubert, J., *497*
Huesmann, L. R., 388, *392*
Hugo, V., 492
Huizinga, J., 153, *160*
Hulicka, I. M., 556, *566,* 573, 579, *583*
Hull, C. L., 218, *237*
Hull, W., *56*
Hulsebus, R. C., 63, *77*
Hultsch, D., *567,* 573, *583*
Hultsch, D. F., 457, *468, 497,* 556, *566,* 570, 572, *583*
Humes, M., 363, *376*
Humphrey, K., 52, *61*

Hundleby, J. D., 368, 370
Hunt, J. McV., 137, *147*, 148, 152, *160*
Hunt, L. G., 371, *376*
Hunt, R. G., 371, *376*
Hurowitz, L., 47, *58*, 134, *144*
Huston, A. C., 242, 257, *267*
Huston-Stein, A., 282, *286*
Hutt, C., *30*
Hutt, S. J., 22, *30*, 35, *57*
Huttenlocher, J., 270, *286*
Huxley, J. S., 4, *18*
Hyman, H. K., 491, *499*
Hymes, D., 255, *266*, *267*

Iannotti, R. J., 316, 318, *325*
Ignatoff, E., 167, *178*
Ingersoll, B., 564, *566*
Ingram, G. L., 384, *392*
Inhelder, B., 152, *162*, 222, 228, 231, 233, *237*, 271, 273, 281, *286*, *288*, 344–346, 351, *353*, 435, 437, *445*
Inkeles, A., 518, *522*, *523*
Irons, W., 7, *16*
Irwin, J., 397, 398, 411, 413, *414*
Isen, A. M., 225, *240*
Ismail, A. H., 559, *565*, *568*
Izard, C. E., 126, *144*, 332, *339*
Izzo, L. D., 190, *209*, 337, *338*

Jacklin, C. N., 305, *314*, 329, *339*
Jackson, J., 474, *482*
Jackson, J. P., 229, *236*
Jacobs, R., 473, 474, 478, *482*
Jacobson, M., 29, *30*
Jacobson, S., 132, *144*
Jacobson, S. W., 24, 25, *30*
Jacquette, D., 331, *339*
Jahoda, M., 456, *468*, 616, *627*
Jamison, W., 235, *237*, *240*
Jargensen, S. R., *497*
Jassik-Gershenfeld, D., 47, *58*, 132, *144*
Jaynes, J., 4, *18*
Jefferson, T., 398, 411, *413*, *414*
Jeffrey, W., 108, *144*
Jeffrey, W. E., 98, *105*, 558, *601*
Jenkins, J., 52, *58*, 552, *566*
Jenkins, R. L., 388, *392*
Jerison, H. J., 4, *18*
Jesness, C. F., 389, *392*
Jessor, R., 357, 358, 363–365, 368–370, *375*, *376*, *378*, 420, *429*
Jessor, S., 357, 364, 365, *368*, 420, *429*
Jewett, D. L., 33, *56*
Johansson, B., 33, *57*
Johansson, G., 95, 99, *105*
Johnson, J. S., 459, 461, *469*
Johnson, J. W., 272, 273, 281, *286*
Johnson, M., 278, *284*

Johnson, M. E., *498*
Johnson, N. S., 274, *286*
Johnson, W., 185, *212*
Johnson-Laird, P. N., 110, 111, *145*, 344, 347, 355, 440, *446*, 449, *455*
Johnston, J., *30*, 100, *105*, 108, *143*
Johnston, J. R., 260, 261, *267*
Johnston, L. D., 357, 358, 368, 369, *376*
Johnstone, J. W. C., 561, *566*
Jolly, A., 336, *338*
Jones, D., 364, *379*
Jones, K., 96, *105*
Jones, M. H., 102, *104*
Jones, O. H. M., 157, *160*
Jones, P., 128, *141*
Jongeward, R., 270, 271, *285*
Jordan, R., 229, *239*
Jorgenson, S., 421, *429*
Jormakka, L., 333, *339*
Joy, C. B., 458, *469*
Judd-Engel, N., 41, 42, *59*
Jung, C., 472, *483*
Jurkovic, G. J., 384, *392*
Jusczyk, P., 32, 37, 39–43, 46, 48, 51, *54*, *56*, *57*, 68, *76*

Kagan, J., 101, *105*, 108, 117, 118, 121, 129, 130, *144*, 282, *286*, 587, *599*
Kahle, L. R., 225, *240*, 362, *376*
Kahn, R., 500, *506*, 577, 581, *583*
Kahneman, D., 441, *446*
Kail, R. V., 271, *286*
Kail, R. V., Jr., 270, 271, *285*
Kalt, N. C., 563, *566*
Kamenetskaya, N. H., 69, *76*
Kamens, D. H., 408, *415*
Kandel, D. B., 332, 357, 360–362, 364–366, 368–370, 372, *376*, *378*, 404, 405, *415*
Kanfer, F. H., 295, *301*
Kanter, R. M., *506*
Kantner, J., 419–421, *429–431*
Kaplan, B., 279, *280*
Kaplan, E., 46, *57*
Kaplan, G. A., 94, *105*
Kardiner, A., 465, *468*
Karmel, B. Z., 36, *57*
Karniol, R., 294, *302*
Karoly, P., 292, *301*
Karp, S. A., 589, 591, *600*, *601*, 602, *613*
Karpf, R. J., 563, *566*
Kasatkin, N. I., 65, *77*
Kasch, F. W., 558, *566*
Kass, J., 37, 43, 45, *60*
Kasschau, P. L., 562, *566*
Kastenbaum, R., 611, *612*
Katz, S. H., 513, *523*
Kaufman, I. C., 193, 194, *213*
Kausler, D. H., 569, 572, 574–576, *583*, *584*

Kavanau, J. L., 73, 77
Kawashima, T., 45, 49, 56
Kaye, H., 64, 65, 77
Kearsley, R. B., 22, 30
Keasey, B., 225, 240
Keating, D., 123, 140
Keating, D. P., 225, 235, 237, 351, 353
Keats, J. A., 346, 353
Keener, E. O., 256, 267
Keith, J., 511, 523
Keith, P., 621, 626
Kellam, S., 370, 376, 425, 427, 429
Kellas, G., 270, 286
Keller, H., 277
Kelly, G., 292, 301, 385, 392
Kelly, J. A., 155, 160
Kelly, M., 281, 286
Kendrick, C., 196, 198, 201, 202, 209, 210
Keniston, K., 397, 415
Kennedy, C. B., 86, 91
Kennedy, G. E., 4, 18
Kennell, H. J., 10, 18
Kennell, J., 148, 155, 159, 160, 192, 211
Kenney, J. L., 350, 353
Kerr, J. C., 136, 143
Kerschner, P. A., 563, 568
Kessen, W., 23, 31, 68, 69, 77
Kessler, R. C., 357, 376, 464, 465, 468
Keyes, R., 410, 415
Khavari, K. A., 363, 376
Kilbom, A., 558, 566
Kilpatrick, D. G., 363, 376
Kimble, G. A., 70, 77, 555, 567
Kimble, K. A., 11, 18
Kimmel, D. C., 563, 567
Kincannon, J. C., 619, 627
King, K., 474, 483
King, L. M., 318, 321, 324
King, R. A., 319, 320, 326
King, S., 421, 429
Kingsley, P., 270, 285, 286
Kintsch, W., 110, 144, 440, 445
Kirby, M. W., 359, 374
Kirby, N. H., 545, 550
Kirigin, K. A., 390, 391
Kirk, R. S., 366, 376
Kirschner, B., 231, 239
Kister, M. C., 300, 302
Klahr, D., 269, 286
Klatt, D., 52, 60
Klaus, M. H., 10, 18, 148, 155, 159, 160, 192, 211
Kleban, M. H., 615, 622–623, 627
Kleim, D. M., 572, 576, 583
Klein, D. M., 493, 495, 497
Klein, E. B., 483, 507, 613
Klein, M., 159, 160
Klein, R., 52, 57

Klein, R. E., 282, 286
Kleiner, R. J., 466, 469
Kleinman, J. M., 588, 600
Klerman, G. L., 456, 470
Klerman, L. V., 619, 627
Kline, D. W., 571, 583
Klinge, V., 359, 365, 366, 376, 377
Klingel, D. M., 365, 376
Kluckholm, C., 608, 613
Knapp, M. R. J., 623, 627
Knight, B., 563, 567
Knipscheer, C. P. M., 487, 498
Knox, A. B., 560, 561, 567
Koch, D., 453, 454
Koelling, R., 7, 16
Koffka, K., 117, 144
Kogan, N., 520, 521, 523, 586, 587, 589, 590, 592, 592–596, 600
Kohlberg, L., 347–349, 353, 453
Kohler, C. J., 128, 146
Kohn, M. H., 563, 566
Kohn, M. L., 502, 507
Kohn, P. M., 360, 377
Kolb, S., 72, 78
Kolton, S., 40, 42, 59
Konner, M., 72, 77, 154, 161, 206, 211
Konstadt, N. L., 591, 600
Koo, H., 475, 483
Korchin, S. J., 556, 567
Korman, M., 365, 377
Korner, A. F., 21, 30, 157, 163
Koslowski, B., 165, 178, 198, 209
Koslowzki, L. T., 95, 104
Kosslyn, S. M., 279, 286
Kotsonis, M. E., 300, 301
Kozma, A., 617, 627, 628
Krachkovskaia, M. V., 65, 77
Kramer, D., 235, 238
Kramer, R., 349, 353
Krang, H. I., 33, 59
Krasnogorskii, N. I., 63, 77
Krauss, I. K., 598, 600
Krauss, N., 464, 467
Krauss, R. M., 300, 301
Krebs, D. L., 320, 325
Kreitzberg, V., 196, 212
Kremenitzer, J. P., 102, 105
Kreutzer, M., 280, 281, 286
Kreyberg, P. C., 337, 339
Kristensen, T. S., 505, 506, 507
Kristofferson, M. W., 272, 287
Kroeger, E., 511, 523
Kroy, M., 347, 354
Kuhl, P., 32, 38, 41, 43, 45, 47, 53, 56, 57, 131, 144
Kuhn, D., 233, 237, 346, 347, 351, 353
Kulik, J. A., 363, 377
Kulka, R. A., 362, 376

Kuriansky, J., *626*
Kurth, S. B., *339*
Kurtines, W., 348, *353*
Kurtzberg, D., 101, 102, *105*
Kusumi, F., *567*
Kutner, B., 618, *627*
Kutz, S., 164, 172, *178,* 309, *314*
Kuypers, J. A., 451, *454,* 494, *497,* 605, 610, *613*

LaBarba, R. C., 155, *160*
LaBarbera, J. D., 126, *144*
Labouvie, G., 557, *566*
Labouvie-Vief, G., 438, 443, *445, 446,* 449–451, 454, *454, 455,* 525, 535, *538,* 557, *567, 568*
Labov, W., 110, *144*
Lacey, C., 406, 407, *415*
Lachar, D., 365, *376*
Lachman, J. L., 109, *144,* 577, 578, *583, 584*
Lachman, R., 109, *144,* 278, 287, 577, 578, *583, 584*
Lacy, W. B., 604, *613*
Ladis, B., 73, *77*
Ladner, J., 421, 422, *430*
LaFrance, M., 157, *161*
LaGaipa, J. J., 329, 330, 336, *338*
Lahey, M., 254, *265*
Laing, R. D., 185, *211*
Lair, C. V., 580, *584*
Lamb, M. E., 9, *18,* 148, 149, *161,* 195, 196, 198, 202, *211*
Lamson, G., 121, *143*
Lanaro, P., 230, *241*
Landau, R., 199, *210*
Landis, T. Y., 277, 278, *286*
Langer, E. J., 530, *538*
Langer, J., 230, 231, *237,* 240
Langner, T. S., 456, *468,* 618, *627, 628*
Langworthy, R. L., 336, *339*
Lansky, D., 359, *377*
Lapata, H., 479, *483*
Larkin, J. H., 552, *567*
Larkin, R. W., 398, 401, 410, 411, *414*
Larsen, G. Y., 226, *238*
Larson, L. E., 404, *415*
Larson, R., 717, 620, *627*
Lashley, K. S., 107, *144*
Lasker, G. W., 4, 5, *18*
Lasky, R., 52, *57,* 83, *91*
Laslett, P., 521, *523*
Lassey, M. L., 365, 366, *377*
Lave, C., 282, *289*
Lavenkar, M. A., 376, *377*
Lavik, N. J., 388, *392*
Lawson, A. E., 234, *238*
Lawson, D. M., 358, *378*
Lawson, K. R., 93, 102, *105*
Lawton, M. P., 614, 615, 617–624, *627*

Lawton, S., 277, *284*
Lazarus, R. S., 460, *468*
Leahy, R. L., 101, 102, *105*
Leavitt, J., 572, *582*
Leavitt, L., 37, 40, 42–45, 48, 49, *57, 59–61*
Leavitt, L. A., 24, *31,* 69, 76, 86, *91*
Lecours, A., 33, *61*
Lee, C-F., 363, 364, *378*
Lee, G., *483*
Lee, J. A., 590, 591, *600*
Lee, R., 390, *392*
Lee, R. S., 361, *375*
Leech, S., 557, *567,* 581, *584*
Lefkowitz, M. M., 388, *392*
Lehman, H., 595, *600*
Leiderman, A. F., 193, *211*
Leiderman, P. H., 148, 159, *161*
Leighton, A. H., 617, *627*
Leighton, D. C., 617, *627*
Leiman, B., 316, 320, 321, *325*
Lempers, J. A., 492, *497*
Lenard, H. G., *30,* 35, *57*
Lennon, M. L., 592, *600*
Lennon, R., 320, *324*
Lennox, K., 359, *377*
Leon, H. V., 518, *522*
Leonard, C., 280, *286*
Leontiev, A. N., 450, *454*
Lepenies, W., 148, *163*
Lepper, M. R., 299, *302*
Lerner, R. M., 282, *286,* 493, *498,* 525–527, *537, 538*
Leslie, S. A., 388, *392*
Lesser, G. S., 404, 405, *415*
Levenson, R., 50, *56*
Levikova, A. M., 65, *77*
Levin, B., 235, *238*
Levin, H., 115, *142,* 305, *315*
Levine, D. I., 346, *353*
Levine, L. E., 320, *325*
LeVine, R. A., 511, *523*
Levinson, D. J., 472, *483,* 603, 604, *613*
Levinson, D. L., 502, *507*
Levinson, M. H., *483, 507, 613*
Levinson, R. B., 384, *392*
Levitin, T. E., 495, *498*
Levitt, A., 41, *54*
Levy, B. A., 272, *287*
Levy, S. J., 373, *379*
Lewin, K., 218, *238*
Lewis, A., *468*
Lewis, B. C., 295, *303*
Lewis, M., 47, *58,* 69, *77,* 134, *144,* 148, 157, *161,* 170, 172, 177, *178, 179,* 180–188, 190–194, 196–206, *209–212, 214,* 329, *339,* 493, *498*
Lewis, M. I., 563, *565*
Lewis, R. A., 477, *484*

Lewis, T. L., 272, *287*
Liben, L. S., 224, 234, 235, *238*, 272, 273, 277, *286, 287*
Liberale, M., 397, *415*
Liberman, A., 38, 44, 52, *55, 58*
Lichenstein, S., 441, *446*
Lieberman, G. L., 491, *498*
Lieberman, M. A., *468*
Lied, E. R., 361, 366, *377*
Liker, J. K., 529, *537*
Lincoln, A., 277, 278
Lindeman, C., 421, *430*
Lindenthal, J. J., 464, 465, *469, 470*
Lindgren, S. D., 384, *393*
Linn, M., 234, *238*, 346, *353*
Linn, S., 130, *144*
Lintz, L. M., 11, *16*
Lipman, A., 395
Lipset, S. M., 397, *415*
Lipsitt, L., 185, *212*
Lipsitt, L. P., 11, 12, *19*, 63–66, 68, 70, 72–74, 76–78, 159, *161*, 524, *537*
Lisker, L., 38, 52, *54, 58*
List, J. A., 587, *599*
Little, A. M., 36, *58*
Livesley, W. J., 330, *339*
Livson, F., 458, 46–463, *468*, 478, *483*, 605, *613*
Lockard, R. B., 7, *18*
Lockhart, R. S., 270, *285*, 574, *582*
London, J., 561, *567*
Longino, C. F., 395, 396, *415*
Lopata, H. Z., 487, *498*
Lorence, J., 502, *507*
Lorenz, K., 5, 6, 8, 9, *18*
Lorge, I., 518, 520, *522, 523, 626*
Lotecka, L., 365, 366, *374*
Lott, F. D., 7, *18*
Lovaas, O. L., 244, *252*
Love, C. T., 390, *393*
Lowenthal, M., 477, *483*
Lowenthal, M. F., 456–460, 462–464, *467–469*, 610, *613*
Lowenthal, M. J., 466
Lucas, D., 71, *78*
Lueger, R. J., 385, 387, *392*
Luker, K., 421, *430*
Lukoff, I. F., 363, *374*
Lunzer, E. A., 352, *353*
Luppova, V. A., 69, *76*
Luria, A., 196, *212*, 292, *301*, 438, 439, *445*
Lyons, J., 253, *267*
Lyons-Ruth, K., 134, *144*

Maas, H. S., 451, *454*, 605, 610, *613*
Mabry, E., 363, *376*
McArthur, C., 367, *379*
McBride, D. C., 358, *377*

McCall, R. B., 86, *91*, 108, 125, *144*
McCandless, B. R., 149, *161*
McCarrell, N., 280, *284*
McCarrell, N. S., 110, 111, *141*
MacCarthy, D., 159, *161*
McCarthy, J., 427, *430*
McCarthy, M., 581, *583*
McCartney, K. A., 275, 278, *287*
McCauley, C., 270, *286*
McCawley, J. D., 254, 259, *267*
McClelland, K. A., 396, *415*
McCluskey, K. A., 230, *238*
Maccoby, E. E., 304, 305, *314, 315*, 329, *339*
McCoy, C., 358, 369, *377*
McCrae, R. R., 461, *467*, 580, *582*, 603, 604, 606, 607, 609–611, *612, 613*, 618, 624, *625, 626*
McCron, R., 402, *415*
McCubbin, H. I., 458, *469*
McDill, E., 402, 405, *415, 416*
MacDonald, R., 497
McDonald, W. J., *612*
MacEachron, A. E., 589, *599*
McEwen, C., 395
McFadden, M., 358, *379*
McFarland, C. E., 270, *286*
McFarland, D. J., 541, *550*
MacFarland, R. A., 566
McGarvey, W., 619, *626*
McGee, B., *507*
McGee, J., 476, *483*
McGrath, S. K., 88, *91*
McGraw, M. B., 72, 74, *77*
McGuire, I., 156, *163*
McGurk, H., 82, 89, *91*
McIntyre, C. W., 230, *238*
Macionis, J., 472, *483*
McKean, C. M., 33, *60*
McKee, B., *483, 613*
McKeithen, K. B., 438, *445*
McKenna-Hartung, S., 362, *377*
McKennell, A. C., 618, 622, 623, *625*
McKenzie, B. E., 98, 99, *104, 144*
McKinney, J. C., 562, *568*
Macklin, E., 477, 481, *483*
McKool, M., 514, *523*
McLanahan, A. G., 225, *240*
McLaughlin, J., 427, *430*
MacNamara, 120, *144*, 258, *267*
McPherson, B., 504, *507*
McWeeny, P., 73, *78*
Maddox, G. L., 511, *523*, 620, *627*
Maeulen, L., *612*
Main, M., 165, *178*, 199, *209*
Maisel, E. B., 36, *57*
Malatesta, V. J., 363, *377*
Malpass, R. S., 282, *287*
Mandler, J., 110, 111, *144*

Mandler, J. M., 274, 279, *286, 287*
Mangelsdorff, D., 363, *380*
Mangione, P., 148
Manheimer, D. L., 361, 369, *377, 378*
Mankin, D., 363, *376*
Mann, C., 476, 478, *483*
Manton, K., 516–518, 523
Maoz, B., 517, *522*
Maracek, J., 427, *430*
Maratos, O., 24, *30*
Maratsos, M., 256, *267*
Marcus, R. F., 320, *325*
Marden, P., 357, *377*
Margulies, R. Z., 357, *376*
Markman, E., 280, *287*, 300, *301*, 344, *354*
Markus, E., 590, 591, 597, *600*
Markus, G. B., 196, *214*
Marlatt, G. A., 361, 365, 366, *374–377*
Marler, P., 7, *18*
Marquis, D., *58*
Marquis, D. P., 65, *77*
Marsh, P., 407, *415*
Marshall, E. R., 158, *161*
Marshall, J. T., Jr., 158, *161*
Martin, B., 387, *392*
Martin, G., 277
Martin, J., *467*
Martin, R. M., 85, *91*
Martindale, L., 421, *430*
Martorano, S. C., 233, 235, *238*, 346, *353*
Marx, K., 500
Masangkay, Z. S., 230, *238*
Maslow, A. H., 456, 462, 465, *469*
Mason, S. E., 573, *584*
Mason, W. A., 7, *18*
Massaro, D. W., 36, 37, *58*
Masters, J. C., 24, *30*, 292, *302*, 328, 329, 332, 333, *339*
Matas, L., 189, 191, 193, *212*
Matchett, W. F., 358, *377*
Matějček, Z., 155, *161*
Matza, D., 397, 400, 401, 409, *415*
Maudry, M., 166, 173, *178*
Maugham, B., 407, *416*
Maule, A. J., 546, *550*
Maurer, D., 23, *30*, 100, 101, *104, 105*, 125–127, 140, *144*, 272, *287*
Maynard, J., 89, *90*
Mayo, C., 157, *161*
Mazel, J., *612*
Meacham, J., 282, *287*
Mead, G. H., 158, *161*
Mead, M., 154, *161*, 397, *415*, 447, *454*
Mechanic, D., 457, 461, *469, 470*
Meehan, A., 226, 235, *239*
Megargee, E. I., 387, *392*
Megaw-Nyce, J., 99, *105*, 124, *143*
Mehler, J., 45, 47, *54, 58*, 132

Mehrabian, A., 318, *325*
Meichenbaum, D., 292, 295, 298, *302*
Meicler, M., 135, *145*
Mellinger, G. D., 361, 369, *377, 378*
Meluish, E., 47, *58*
Melton, E. M., 322, *324*
Meltzoff, A. N., 21, 24, 25, 29, *30*, 118, 119, 121, 128, 132, 134, *145*, 185, *212*
Mendelson, F. S., 619, *626*
Mendelson, M., *626*
Menken, J., 427, *430*
Mentz, L., 34, *54*
Merritt, F., 74, *77*
Messer, D. J., 156, *161*
Messer, S., 587, 588, 596, *600*
Messick, S., 594, *600*
Messmer, J., III, 74, *77*
Metzger, R., *584*
Meyer, B., 311, *314*
Miano, V. N., 73, *78*
Michael, S. T., 456, *468, 628*
Michalson, L., *178*, 190, *212*, 329, *339*
Michotte, A., 117, 137, 138, *145*
Milewski, A. E., 67, 68, *77*, 81, *91, 145*
Milhoj, P., *499*
Millar, W. S., 11, 12, *18*, 138, *145*
Miller, B., *497*
Miller, B. C., 475, *483*
Miller, C., 40–42, 46, 47, 49, 50, *58, 59*
Miller, C. L., 132, *145*
Miller, D. J., 124, *145*
Miller, D. T., 294, *302*
Miller, G. A., 110, 111, *145*, 219, *238*, 437, *445*
Miller, J., 39, 41–45, 47, 49, 53, *56–58*
Miller, J. D., 131, *144*
Miller, J. G., 150, *161*
Miller, K., *584*
Miller, L. H., 555, *568*
Miller, N. E., 577, *583*
Miller, P. H., 226, 230, *238*, 280, *287*
Miller, P. M., 372, *375, 377*
Miller, S., 475, 476, *484*
Miller, S. J., *499*
Miller, W., 421, 422, *430*
Miller, W. B., 409, *415*
Miller, W. C., 363, *376*
Mills, M., 47, *58*
Millsom, C., 282, *288*
Milner, P., 152, *161*
Minifie, F., 41, 43, 44, *54–56*
Minor, M. J., 622, *626*
Minsky, M. L., 110, *145*
Miran, M., 372, 373, *375*
Miranda, S. B., 24, *30*, 81, 82, 87, *91*, 97, 100, *104*
Mirandé, A., 478, *484*
Mischel, H. N., 298, *302*

Mischel, W., 218, *238*, 292–300, *302, 303,* 360, 362, *377*
Missinne, L., 511, *523*
Mistler-Lachman, J. L., 575, *584*
Mitchell, S., 388, *392*
Mitteness, L., 496, *498*
Miyashita, M., 558, *565*
Miyawaki, K., 52, *58*
Mock, J., *626*
Moely, B., 270, *287*
Moffitt, A., 39, 40, 42, *58*
Molfese, D., 50, 51, 53, *58, 59*
Molfese, V. G., 50, 51, *59*, 225, *238*
Monane, J. H., 198, *212*
Monay, N., 541, *550*
Money, J., 311, *314*
Monge, R., 556, *567*
Moody, D., 53, *60*
Moon, W. H., 580, *584*
Moore, B., 293, 294, *302*
Moore, J., 39, 41, 43, 52, *55*
Moore, J. M., 22, *30*, 34, *59*
Moore, J. W., 36, *56*, 486, *498*
Moore, K., 424, 427, *430*
Moore, K. M., 132, *145*
Moore, M., 185, *212*
Moore, M. E., *498*
Moore, M. J., 23, *30*, 126, *143*, 474, *483*
Moore, M. K., 21, 24, 25, *30*, 118, 119, 121, 128, 134, 135, 137, *145*
Moos, B. S., 363, *377*
Moos, R. H., 363, 369, *377*
Morgan, K., 41, 43, *56*
Morganstern, O., 441, *446*
Morris, J. N., 619, 621, *627*
Morris, N., 420, *431*
Morrison, P., 427, *430*
Morse, P., 32, 37, 40–50, 53, *55–60*
Morse, P. A., 22, 24, *31*, 37, *61*, 86, *91*, 132, *145*
Mortimer, J. T., 502, *507*
Mortimore, P., 407, *416*
Moshman, D., 232, *238*, 345–347, *353*, 440, *445*
Moss, H. A., 610, *613*
Moss, M., 615, *627*
Mount, R., 130, *144*
Mueller, E., 164, 168, 170, 172, *178*, 197, *212*, 329, *339*
Mueller, J. H., 573, 580, *584*
Muller, A. A., 135, *145*
Mungham, G., 398, 411, *415*
Muntjewerff, W. J., *30*
Murdoch, G., 402, 407, 409, *415*
Murdy, W. H., 4, 5, *18*
Murphy, J. M., 349, *353*, 452, 453, *454*
Murphy, J. P., 373, *378*
Murphy, K. P., 33, *59*
Murphy, L. B., 320, *325*

Murray, F. B., 272, *287*
Murray, H., 471, *483*
Murray, H. A., 201, *212*, 607, 608, *613*
Murray, J., 41, 51, *57*
Murray, J. R., 622, *626*
Murstein, B. I., 473, 474, *483*
Mussen, P., 316, 318, 319, 322, *324, 325*
Musto, D. G., 465, *469*
Muyerhoff, B., 515, *523*
Myers, J. K., 464, *469*
Myers, N. A., 279, *288*
Myers, R. S., 127, *142*
Mynatt, C. R., 347, *353*, 446

Naditch, M. P., 361, 364, *377*
Naeye, R., 73, 74, *77*
Naroll, F., 4, *18*
Naroll, R., 4, *18*
Nasby, W., 385, *392*
Nathan, P. E., 358, 359, *377, 378*
Naus, M. J., 270, 271, 276, *287, 288*
Neal, C., 320, 322, *324*
Neary, R., 363, *380*
Needle, R. H., 458, *469*
Neimark, E. D., 225, 226, 234, *238*, 346, 351, 352, *353, 354*, 436, 440, 442, *445*, 448, *454*
Neisser, U., 110, 133, 140, *145, 147*, 271, *287*, 449, *454*, 552, *567*
Nekula, M., 166, 173, *178*
Nelson, C. A., 24, *31*, 86, 87, 89, *91*
Nelson, E. A., 412, *415*
Nelson, K., 108, 120, 121, 135, 136, *145*, 255, 256, 258–262, *266, 267*, 275, 280, *287, 288*
Ness, S., 316, *324*
Nesselroade, J. R., 533, 536, *537*
Neugarten, B. L., 447, *454*, 456, 458, 460, *469*, 458, 486, 491, 495, *497, 498*, 503, *507*, 597, *600*, 603–606, *613*, 617, 621, *627*
Neumann, P. G., 129, *146*
Nevis, S., 87, *91*, 100, *104*
Newcomb, A. F., 335, *339*
Newman, B. M., 149, *161*
Newman, J. P., 385, *391*
Newman, P. R., 149, *161*
Newman-Hornblum, J., 227, 235, *238, 239*
Newport, E. L., 263, *267*
Newson, M. N., 68, *76*
Newton, I., 140
Nezworski, M. T., 274, *289*
Nezworski, T., *584*
Nicol, A. R., 385, *393*
Nida, S., 276, *288*
Niebuhr, V. N., 225, *238*
Niederehe, G., 577, 581, *583*
Nielsen, D. W., 35, *61*
Nielsen, M., 590, *600*
Nilson, K. L., *567*
Nilsson, I. A., *566*

Nisbett, R., 441, *445*
Nitsch, K., 111, *141*
Noblit, G. W., 407, 408, *415*
Nodine, C. F., 350, *353*
Nolan, E., 282, *286*
Norcia, A., 121, *147*
Nordquist, V. M., 244, *252*
Norem-Hebeisen, A. A., 262, *274*
Norgaards, S., *565*
Norris, A. H., 611, *612*
Notman, M. T., 462, *469*
Nottebohm, F., 7, *18*
Novak, M. A., 190, *212*
Nowlin, J., 556, *565*, 579, *582*
Nydegger, C. N., 496, *498*
Nylander, I., 369, 370, *377*

Oberg, C., 121, *147*
O'Berien, P. C., 343, *354*
Obmascher, P., 9, *18*
O'Brien, C., 170, *179*
O'Brien, D., 232, 233, *238*
O'Connell, M., 474, *483*
O'Connor, M., 149, *161*
Oden, S. L., 332, *337*
Odom, R. D., 229, *239*
O'Donnell, J. A., 359, *377*
Oetting, E. R., 357, 359, 364, 365, 369, *377*
Offer, D., 456, *469*
Ohlrich, E. S., 36, *59*
Okun, M. A., 581, *584*
Olds, J., 152, *161*
O'Leary, S., 196, 200, *212*, 292, 298, *302*
Oller, D. K., 41, 42, 52, 53, *55*
Olrich, E. S., *54*, 69, *75*
Olsen, H., 487, *498*
Olsen, J. K., *497*
Olson, D., 257, *267*, 449, *454*, 475, *483*
Olson, F., 270, *287*
Olson, G. M., 79, 86, 87, *91*, *92*
Oltman, P. K., 442, *446*, 589, *600*
Olum, V., 138, *146*
Olweus, D., 387, 388, *392*
O'Malley, P. M., 357, 362, 364, 369, *376*, *377*
O'Neal, E., 307, *314*
O'Neill, G., 476, *483*
O'Neill, N., 476, *483*
Opler, M. K., *628*
Oppenheimer, V. K., 491, *498*
Orgel, A. R., 390, *391*
Orme-Rogers, C., 571, *583*
Ornstein, P. A., 270, 271, 276, *287*, *288*
Orris, J. B., 384, *392*
Ortony, A., 110, 111, *146*, 350, *354*
Orzech, M. J., 528, *538*
Osherson, D. N., 344, *354*
Osofsky, H. J., 149, *161*, 423, *430*
Osofsky, J. D., 148, 149, *161*, 423, *430*

Oster, H., 157, *160*
Osterman, P., 397, *414*
O'Toole, D. H., 372, *375*
Ouston, J., 407, *416*
Over, R., 157, *160*
Overton, W. F., 217, 219, 220, 223, 224, 226–229, 231–235, *236–239*, 435, 557, *566*
Ovesey, L., 465, *468*
Owsley, C., 30, 47, *60*, 99, 100, *105*, 108, 124, 131, 134, *143*, *146*, *147*

Padden, D., 53, *57*
Page, J. B., 358, *377*
Paget, K. F., 453, *454*
Paisley, W. J., 560, 561, *567*
Paivio, A., 280, *288*
Palermo, D. S., 50, *59*, 109, *147*, 227, *237*
Palmer, T., 389, 390, *392*
Palmore, E., 511, 516–518, 520, *523*
Palonsky, S. B., 401, 411, *415*
Panek, P. E., 590, 591, *600*
Pannabecker, B. J., 185, *212*
Papoušek, H., 11, 12, *18*, 66, *77*, *78*, 149, 151, 152, 154–159, *161*, *162*, 185, *212*
Papoušek, M., 149, 151, 152, 154–159, *161*, 162, 185, *212*
Paris, S. G., 272, 277, *288*
Parisi, S. A., 126, *144*
Parkatti, T., 558, 559, *568*
Parke, F., 198, *212*
Parke, R. D., 149, *162*, 196, 198, 200, 202, *212*, 295, *302*
Parker, E. B., 560, 561, *567*
Parker, R., 231, *239*
Parker, S., 156, *162*, 466, *469*
Parker, T., 282, *289*
Parkhurst, J., 334, *339*
Parkman, J. M., 542, *550*
Parmelee, A. H., 82, *92*
Parry, H. J., 361, *378*
Parsons, T., 201, *213*, 395, 410, 416, 487, *498*
Pascale, P. J., 366, *379*
Pascual-Leone, J., 221, 226, 233, *239*, 283, *288*, 351, 352, *354*, 435, 437, 438, 442, *445*
Paton, S. M., 363, *378*
Patsiokas, A. T., 356
Patterson, C. J., 292, 294–297, 299, 300, *301*, *302*
Patterson, G. R., 196, *213*, 306, 307, 309, *314*, 531, *538*
Patterson, J. M., 458, *469*
Patterson, R. D., 563, *564*
Paulus, D. H., 344, *354*
Pearl, R. A., 86, *90*
Pearlin, L. I., 459, 461, 463, 464, 466, *469*
Pearlstone, Z., 218, *240*
Pearson, G., 398, 411, *415*
Peck, R. F., *469*

Pedersen, A., 190, *209*, 337, *338*
Pedersen, F. A., 195, 196, 198, 200, *213*
Peel, E. A., 346, *354*
Pennypacker, H. W., 555, *567*
Peplau, L. A., 474, *482*
Pepper, M. P., 464, *469*
Perkins, J., 128, *141*
Perlmutter, M., 79, *92*, 271, 279, 280, *288*, 552, 553, 556, *565, 567*, 572, 575, 577, 578, 580, *584*
Perloff, B. F., 244, *252*
Perry, A., 35, *54*
Perry, W. I., 452, *454*
Perry, W. L., 349, *354*
Pervin, L. A., 191, *213*
Petersen, P., 588, *599*
Peterson, G. W., 225, *237*
Peterson, P. G., 605, 606, *613*
Peterson, R. F., 244, *252*
Peterson, W. A., 396, *415*
Petrovich, S. B., 5, 9, *17, 18*
Pettigrew, T. F., *469*, 594, *600*
Pfeiffer, E., 619, *626*
Phelps, E. A., 346, 347, *353*
Phelps, G., 407, *415*
Phillips, W., 587, *599*
Philp, A. G., 335, *339*
Piaget, J., 5, 9, 11, *18*, 71, 108, 112–119, 121, 128, 137–140, *146*, 152, 158, *162*, 185, *213*, 218–222, 224, 226, 227, 229, 230, 233, 235, *237, 239*, 258, 267, 271, 273, 279, 280, *286, 288*, 336, *339*, 344–346, 351, *353*, 435, 436, 438, *443, 445*, 448–450, 452, *454*
Pichel, J., 389, *392*
Pick, H. L., Jr., 121, *147*
Picou, J. S., 405, *416*
Picton, T. W., 33, *59*
Piechowski, M. M., 451, *455*
Pierce, R. C., 618, *628*
Pink, W., 407, *416*
Pisoni, D., 35, 45, 48, 49, 51, 53, *54, 57, 60*
Pittendrigh, C. S., 6, *18*
Pittman, L., 276, *288*
Pizarro, M., 511
Plath, D. W., 492, 495, *498*
Plemons, J. K., 457, *468*, 557, *567*
Pliske, D. B., *446*
Ploog, D., 148, 152, *162, 163*
Plude, D. J., 598, *599*
Podolsky, E., 367, *378*
Podolsky, S., *566*
Polk, K., 406, 407, *416*
Pollack, R. H., 590, 591, *600*
Pomerleau-Malcuit, A., 21, *31*
Poon, L. W., 270, *288*, 556, *567*, 578, *584*
Popper, K. R., 347, *354*
Porter, J. R., 466, *469*
Posnansky, C. J., 272, *287*

Postman, L., 218, *239*, 552, *567*
Potvin, R. H., 363, 364, *378*
Powell, A. H., 556, *567*
Powell, A. H., Jr., 579, *584*
Powers, E. A., 621, *626*
Preble, E., 361, 367, 372, *378*
Premack, D., 66, *78*
Prendergast, T. J., 366, *378*
Prentice, N. M., 384, *392*
Press, I., 514, *523*
Presser, H., 420, 421, 427, *430*
Pressley, G. M., 299, *302*
Pribram, K., 437, *445*
Price, K. F., 563, *567*
Price-Bohem, S., 478, *483*
Profant, G. R., 558, *567*
Profitt, D. R., 95, *104*
Protestos, C., 73, *78*
Provitera, A., 440, *444*
Prudhomme, C., 465, *469*
Puff, R. C., 102, *146*
Purpura, D. P., 72, 75, *78*
Pylyshyn, Z. W., 280, *288*

Quasebarth, S. J., 297, *301*
Quay, H. C., 382–385, 389, 390, *392, 393*

Rabbitt, P. M. A., 540–542, 544–548, *550*, 588, 598, *600*
Rabinovitch, S., 41, 42, *61*
Radke-Yarrow, M., 319, 320, 323, *326*
Rae, D., 581, *583*
Rainwater, L., 422, *430*
Rajecki, D. W., 9, *18*
Rake, D. F., 190, 191, *210*
Rake, R. H., 456, *468*
Ramey, C. J., 170, *179*
Ramey, J. W., 476, *483*
Randall, C. L., 361, *379*
Rankin, J. L., 569, 573, 575, 580, *584*
Raphael, L., 38, *55, 60*
Rappaport, J., 390, *393*
Rathus, S. A., 366, *378*
Ratner, A. M., 9–11, *18*
Raykowski, H., 479, *482*
Rebelsky, F., 196, *213*
Redlich, F., 456, 464, *468*, 617, *627*
Rees, E., *444*
Reese, H. W., 218, 220, 223, 228, *239*, 279, 282, *285, 288*, 524, 527, *537*, 573, *585*
Rehberg, R. A., 406, *416*
Reichard, S., 605, 606, *613*
Reid, J. B., 366, *378*
Reilly, R. R., 559, *568*
Reitman, J. S., *445*
Remington, R. J., 545, *550*
Renner, V. J., 456, *466*
Rennie, D. L., 246, *252*

Rennie, T., *628*
Rescorla, L., 261, *267*
Rest, J. R., 348, *354*
Reuter, H. H., *445*
Reuterman, N. A., 386, *391*
Revlis, R., 440, *445*
Rewey, H., 48, *60*
Reynolds, D., 407, *416*
Reynolds, H. N., 94, *105*
Reynolds, I., 369, *375*
Reynolds, N. J., 243, *252*
Reynolds, R. E., 350, *354*
Rezba, C., 121, *146*
Reznick, J., 120, *144*
Rheingold, H. L., 11, *18,* 493, *498*
Ribot, T., 577, 579, *584*
Rice, M. L., 257, 260–262, 264, *267*
Rich, A., 170, *178*
Richards, D. D., 225, *240*
Richards, M. P. M., 148, *162*
Richman, C. L., 276, *288,* 384, *393*
Richter, C., 74, *78*
Rickel, A. U., 390, *393*
Ridberg, E. H., 387, *392*
Ridgeway, D., 363, *378*
Rieber, R. W., 253, *267*
Riebman, B., 234, 235, *239*
Rieff, M., 231, *239*
Riege, W. H., 580, *585*
Riegel, K. F., 282, *288,* 443, 445, 450, 453, *455,* 526, *538*
Rifkin, B., 350, *354*
Riggs, H. E., 33, *60*
Rigsby, L. C., 402, *416*
Riley, C. A., 225, *240*
Riley, M. W., 485, 494, *498,* 524, 526, *538*
Ringel, B. A., 271, *288*
Ripich, D., 257, *265*
Rittenhouse, J. D., 357
Ritter, K., 281, *289*
Rivera, R. J., 561, *566*
Rivest, L., 333, *338*
Robb, K., 270, *285*
Robbins, S., *354*
Roberge, J. J., 344, *354*
Roberts, B., 398, *413*
Robertson, E. A., 572, 576, *584*
Robertson, J., 194, *213*
Robertson, J. F., *498*
Robertson-Tchabo, E. A., 551, 556, 557, *564,* 573, *584*
Robins, L. N., 369, 370, *378,* 388, *393,* 610, *613*
Robinson, I. E., 474, *483*
Robinson, J., 11, *16,* 74, *76*
Robinson, J. A., 578, *584*
Robinson, R. D., *567*
Robinson, S., 558, *567*
Robson, K., 156, 157, *162*

Rockwell, R. C., 607, *612*
Rodgers, M., 541, 547, *550*
Rodgers, R. H., 489, *498*
Rodgers, W. L., 457, *467,* 615, *626*
Rodin, J., 530, *538*
Roe, K., 319, 320, *325*
Roff, J., 337, *339*
Rogers, W. L., 477, *482*
Rogoff, B., 282, *286*
Rohrbaugj, J., 369, *378*
Rohrs, C. C., 373, *378*
Rohwer, W. D., 279, *288*
Roitzsch, J. C., 363, *376*
Roke, E. G., 320, *325*
Rollins, B. C., 198, *213*
Romano, M. N., 33, *56*
Ronch, J., 102, *104*
Room, R. G. W., 359, *377*
Roopnarine, J., 175, *179*
Rorke, L. B., 33, *60*
Rosa, P., 388, *392*
Rosch, E., 110, *146*
Rose, A. M., 396, *416*
Rose, S. A., 82, 83, 89, *91, 92,* 128, *143*
Rosen, R. A., 421, *430*
Rosenbaum, E., 412, *415*
Rosenberg, M., 623, *628*
Rosenblatt, J. S., 156, *162*
Rosenblatt, P. C., 200, *213*
Rosenblith, J. F., 73, 75, *75*
Rosenbloom, S., 27, *29*
Rosenbloom, L. A., 148, 157, *161,* 185–187, 193, 194, 197, 198, *212, 213,* 494, *498*
Rosenbluth, J., 358, 368, *378*
Rosenfeld, A., 148, *161*
Rosenfeld, E., 361, *375*
Rosenfeld, H. M., 127, *147*
Rosenkrantz, P. S., 463, *466*
Rosenmayr, L., 487, 492, *498*
Rosman, B. L., 587, *599*
Rosner, B., 51, *57*
Rosow, I., 616, 617, *628*
Ross, B. M., 272, 273, 282, *284, 285, 288*
Ross, E., 580, *584*
Ross, G., 280, *288*
Ross, G. S., 125, 137, *146*
Ross, H., 424, *430*
Ross, L., 441, *445*
Ross, R. R., 390, *393*
Rossner, E., 407, *415*
Roszak, T., 397, *416*
Rotella, R. J., 591, *600*
Roth, B. K., *498*
Rovee-Collier, C. K., 70, 71, *78,* 138, *146*
Rowburry, T. G., 244, *252*
Roy, M. A., 9, 10, *19*
Rubenstein, J. C., 309, *314*
Rubin, Z., 472, 474, *482, 483*

Ruel, J., 123, *147*
Ruff, H. A., 86, 87, *92*, 99–102, *105*, 108, 115, 125, 126, 128, *146*
Rummelhart, D. E., 110, 111, *146*
Runyan, W. M., 526, *539*
Runyan, W. McK., 608, *613*
Rupenthal, G. C., 192, *213*
Ruse, B. R., 358, *377*
Russ-Eft, D., 421, 427, *430*
Russell, J. A., 363, *378*
Rutter, M., 194, *213*, 388, *392*, *393*, 404, 407, *416*
Ryan, E. B., 124, *145*
Ryan, J. F., 156, *162*
Ryder, N., 489, *498*

Saarmi, C. I., 234, *240*
Sabatini, P., 438, *446*, 449, *455*
Sabshin, M., 456, *569*
Sachs, J., 256, *268*
Sackett, G. P., *213*
Sackett, G. T., 529, *539*
Sadava, S. W., 360, 363–365, 367, 370, *378*
Sagi, A., 317, *325*
Sagotsky, G., 299, *302*
Salamy, A., 33, *60*
Salapatek, P., 23, 24, *29–31*, 68, 69, *77*, 96, 97, 101, *104*, *105*
Salatas, H., 281, *289*
Salkind, N. J., 588, *600*
Salthouse, T. A., 574, *584*
Saltin, B., *566*
Saltin, L. H., 558, *567*
Saltzstein, H. D., 321, *325*
Salzman, E., 258, *266*
Sameroff, A. J., 11, *19*, 63, 66, 70, *78*
Sampson, H., 541, 542, *550*
Sanborn, M. D., 369, *378*
Sandberg, S. T., *393*
Sander, L. W., 22, *29*
Sanders, H. I., 577, 578, *585*
Sanders, J. C., 557, *567*
Sanders, R. E., 532, 536, *539*, 557, 561, *567*, *568*
Sandford, A. J., 542, 546, *550*
Sandler, H., 425, 427, *430*
Santo, Y., 368, *375*
Sarty, M., 24, *30*
Sawhill, I., 424, *430*
Sawin, D., 196, 202, *212*, 295, *302*, 316, 320, *325*
Scanzoni, J., 475, *483*
Scanzoni, L., 475, *483*
Scardamalia, M., 226, *240*, 346, *354*
Schaef, R. D., 366, *375*
Schaefer, E. S., 304, *313*, 366, *378*
Schaeffer, B., 244, *252*
Schaeffer, R. A., 235, *237*

Schaeffer, S., 191–193, *212*
Schafer, W. E., 406, 407, *416*
Schaffer, H. R., 14, *19*, 138, *145*, 148, 149, *162*
Schaie, K. W., 449, 453, *455*, 522, *522*, 526, 532, 535, *537*, *539*, 561, *567*, 586, 598, *599*, *601*
Schank, R., 110, 120, *146*, 275, *288*
Scheflen, A. E., 151, 158, *162*
Schiefelin, B. B., 263, *268*
Schiff, N. B., 264, *268*
Schinke, S., 423, *430*
Schlegel, R. P., 368, 369, *378*
Schlesinger, I. M., 259, 260, *268*
Schludermann, E. H., 589, 591, *601*
Schmeidler, J., 359, *375*
Schneider, B., 32, 34, 35, *60*, *61*
Schoen, R. A., 594, *599*
Schoetzan, A., 157, *162*
Scholnick, E. K., 272, 273, *286*
Schonfield, D., 572, 576, *584*
Schooler, C., 461, 464, 466, *469*, 502, *507*
Schooler, K. K., 618, 621, *628*
Schraf, B., 35, *60*
Schram, R. W., 477, *484*
Schuessler, K. R., 383, *393*
Schüller, V., 155, *161*
Schulman, N., 476, *484*
Schulman-Galambos, C., 34, *60*
Schulte, D., 270, *285*
Schultz, W. R., Jr., 557, *568*
Schwartz, D. W., 602, *613*
Schwartz, M., 89, *92*, 123, *146*
Schwarz, R. M., 363, *378*
Scott, J., 618, 619, 623, *626*
Scott, J. P., 9, *19*
Scott, J. W., 475, *484*
Scott, M., 574, *583*
Scott, T. G., 384, *391*, *394*
Scribner, S., 282, *285*, *288*, 439, *446*, 448, *455*
Sears, R. R., 305, 308, *314*, *315*
Sears, R. R. A., 495, *499*
Seashore, M. J., 159, *161*
Seefeldt, F. M., 350, *354*
Seem, G., 511, *523*
Segal, B., 363, *378*
Seidman, E., 390, *393*
Seligman, M. E. P., 7, 13, *19*, 70, 74, *78*, 530, *539*
Seligson, T., 397, *415*
Sellin, T., 389, *394*
Sells, S. B., 337, *339*, 440, *446*
Selman, R. L., 331, *339*
Selstad, G., 423, *429*
Semler, I., 365, *377*
Sewell, W. H., 405, *416*
Shader, R. I., 367, *374*
Shah, F., 421, *430*
Shanas, E., 487, *499*, 526, *538*

Shanks, B., 36, *54, 59,* 69, *75*
Shankweiler, D., 38, *58*
Shapiro, B. J., 343, *354*
Sharma, K. L., 518, *523*
Sharp, D., 270, 282, *285, 289, 444*
Sharp, M. W., 559, *568*
Sharpe, L., *626*
Shatz, M., 263, *268*
Shavonian, B. M., 555, *568*
Shayer, M., 225, *240*
Shea, S. L., 96, *104*
Sheldon-Wildgen, J., 242
Shephard, R. J., 558, 559, *568*
Shepherd, P. A., 81, 82, 89, 90, *91, 92*
Sherman, J. A., 244, *252*
Sherrod, K. F., 427, *430*
Shields, M. M., 256, 261, *268*
Shiffrin, R., 270, *284*
Shinar, E. H., 505, *507*
Shipley, E. F., 280, *285*
Short, J. F., Jr., 386, *393*
Shuman, H. H., 157, *160*
Siddle, D. A. T., 385, *393*
Sidney, K. H., 558, 559, *568*
Siegel, A., 270, *289*
Siegel, J. S., 488, *499*
Siegel, L. J., 366, *378*
Siegel, L. S., 277, *287*, 316, *324*
Siegler, R. S., 225, 235, *240*, 269, 273, *289*, 344, 346, 351, *354*
Siemen, J. R., 563, *568*
Sigman, M., 82, *92*
Signorella, M. L., 235, *237, 240*
Silberstein, M., 577, *585*
Silverman, A., 564, *566*
Silverstone, B., 491, *499*
Simmel, E. C., 148, *163*
Simmel, M., 438, *446*
Simmer, M. L., 317, *325*
Simmons, L. W., 514–517, *523*
Simon, D. P., 438, *446*
Simon, E., 572, 573, *582, 584, 585*
Simon, H. A., 278, *284*, 438, 441, *444, 446*
Simon, M. B., 370, *376*
Simon, R., *626*
Simon, W., 420, *430*
Simpson, I. H., 562, *568*
Sims-Knight, J. E., 230, 231, *240*, 280, *284*
Sinclair, H., 222, 228, 231, *237*, 258, *268*
Singer, L. T., 89, *91*, 126, *142*
Singleton, L. C., 332, *339*
Sinnott, J., 53, *60*, 124, *145*
Sipple, T. S., 225, *237*
Siqueland, E. R., 11, 12, *19*, 37, 39, 41, 42, 49, *56, 60*, 66–68, *76–78*, 81, *91, 92*
Skard, A. G., 494, *499*
Skinner, B. F., *19*, 71, *78*
Skolnick, A., 486, *499*

Skrzypek, G. J., 384, *393*
Slater, A. M., 23, *31*
Slater, P., 411, *416*, 577, *585*
Slatin, 359, *377*
Slobin, D. I., 254, 258, 259, 263, *268*
Sloman, J., 165, *178*
Slovic, P., 441, *446*
Smart, R. G., 357, 358, 364, 366, 367, *378, 379*
Smiley, S., 277, *284*, 593, *601*
Smith, A. D., 572, 573, 576, *584, 585*
Smith, D., 373, *379*
Smith, D. H., 518, *522, 523*
Smith, D. M., 410, *416*
Smith, J., *499*
Smith, J. H., 120, *143*
Smith, J. T., 335, *338*
Smith, L., 51, *57*, 491, *499*
Smith, R. A., 295, *303*
Smith, W. G., 361, *375*
Smock, C. D., *601*
Smyer, M. A., 495, *497*
Smyth, C. N., 33, *59*
Snow, C. E., 157, *163*, 263, *268*
Snowdon, C., 53, *59*
Snyder, E. E., 404, 406, *416*
Sobowale, N. C., 367, *379*
Sokol, S., 96, *105*
Solkoff, N., 71, *78*
Somers, R. H., 369, *377*
Sorenson, R., 420, *430*
South, D. R., 404, *414*
Souther, A. F., 23, *31*
Spady, W. G., 406, *416*
Spanier, G. B., 477, *484*, 493, *498*
Sparkman, E., 221, 233, *239*
Spearman, C., 437, *446*
Specht, T., 74, *77*
Spelke, E., 47, *60*, 100, *106*, 127, 132–134, *146, 147*
Spence, J. T., 464, *469*
Spencer, F. W., 385, *393*
Spencer, T., 270, *285*
Sperr, S., 231, *239*
Spilich, G. J., 278, *284*
Spiro, A., 533, *537*
Spitz, R., 74, *78*
Spock, B., *60*
Spring, D., 47, *60*
Springer, C. J., 271, *288*
Springer, M., 421, *430*
Squire, L. R., 577, 578, *585*
Sroges, W. R., 7, *17*
Srole, L., 463, 464, *469*, 617, 619, *628*
Sroufe, L., 137, *142*, 147, 152, 157, *163*, 185, 189, 192, 209, *212–214*
Stack, C., 422, *430*
Stanovich, K. E., 271, *285*
Staples, R., 422, *430*, 478, *484*

Starkey, P., 123, *147*
Starr, M., 199, *210*
Staub, E., 316, 318, *325*
Staudenmeyer, H., 344, 345, *354*, 440, *446*
Stayton, D. J., 151, 152, *159*
Stearns, P. N., 521, *522*
Stebbins, W., 53, *60*
Stechler, G., 27, *29*
Stehouwer, J., *499*
Stein, A. H., 246, *252*
Stein, N. L., 274, 275, *289*
Steinberg, B., 85, *92*
Steinschneider, A., 69, 73, *76, 78*
Stenberg, C., 180, 185, *209, 212*
Stephenson, G. R., 149, *161*
Stern, C., 5, *19*
Stern, D., 155, *163*
Stern, D. N., 157, *163*, 165, 167, *179*, 198, *213*
Stern, L., 159, *160*
Sternberg, R. J., 345, 350, 351, *354*
Sterns, 532, 536, *539*, 557, 561, *568*, 590, *600*
Stevens, K., 38, 52, 53, *60*
Stevenson, H. W., 282, *289*
Stevenson-Hinde, J., 7, 13, *17*
Stewart, C., 422, *431*
Stewart, C. S., 358, *375*
Stewart, D. J., 384, *393*
Stewart, R. B., 528, *537*
Stiller, P., *626*
St. James-Roberts, I., 71, *78*
Stokes, J. P., 362, 363, *379*
Stoller, S., 28, *31*
Stone, C. A., 219, 223, 226, 233, *240*
Stones, M. J., 617, *627, 628*
Storandt, M., 571, 577, 578, *582, 585*
Stotland, E., 321, *326*
Stouwie, R., 387, *392*
Strange, W., 52, *58*
Stratton, T., 65, *76*
Strauss, M. S., 85, 89, *90, 92*, 124, 126, 127, 129, 130, *142, 147*
Strauss, S., 228, 230, 231, *240*, 347, *354*
Strayer, J. A., 320, 323, *326*
Streeter, L., 52, *60*
Strehler, B. L., 526, *539*
Streib, G. F., 487, 491, *499*
Streit, F., 366, *379*
Striar, D. E., 373, *379*
Stricker, G., 462
Strock, B. D., 36, *56*, 69, *76*
Stroh, V., *148*
Strong, E. K., Jr., 609, *613*
Studdert-Kennedy, M., 38, *58*
Sturner, W. Q., 73, *77*
Suchindran, C. M., 475, *483*
Suci, G., 136, *143*
Sugarman, B., 406, 407, *416*
Sulin, R. A., 277, *289*

Sullivan, E., 423, *429*
Sullivan, H., 184, *213*, 317, *326*, 329, 336, *339*, 492, *499*
Sullivan, M. W., 71, *78*
Sundy, M. S., 337, *339*
Suomi, K., 37, 41, 42, *55, 59*
Suomi, S. J., 149, *161*, 190, 192, 202, *213*
Suominen, H., 558, 559, *568*
Susman, E. J., 610, *613*
Suther, P. B., 359–361, 363, 364, 368, *376, 377, 379*
Sutherland, E. H., 396, *416*
Swoboda, P., 37, 43–45, 48, 49, *60*
Sykes, G. M., 409, *415*
Sylva, K., 336, *338*
Syrdal-Lasky, A., 52, *57*
Sytova, V. A., 69, *76*
Szandorowska, B., 364, *374*

Taddonio, J. L., 344, *354*, 440, *446*
Tanner, J., 398, *416*
Taplin, J. E., 344, *354*, 440, *446*
Tartter, V. C., 32, 49, *56*
Tatsuoka, M. M., 624, *626*
Taub, H. A., 556, *568*, 570, *585*
Taylor, C., 616, 617, 625, *628*
Taylor, E., *393*
Taylor, J., 53, *59*, 199, *210*
Tecce, J. J., 27, *29*
Teeland, L., 487, 494, *499*
Tees, R., 52, *61*
Telleen, S., 320, *325*
Tellegen, A., 606, *613*
Teller, D., 96, *104*
Tenney, Y. J., 271, *289*
Tennov, D., 472, *484*
Tennyson, R. A., 386, *393*
Tesi, G., 364, *376*
Tessler, R., 457, *470*
Testa, T., 138, *147*
Thelen, M. H., 246, *252*
Thiessen, D. D., 6, *19*
Thistle, R., 367, *378*
Thoman, E. B., 73, *78*, 148, 157, *163*
Thoman, E. P., 21, *30*
Thomas, A., 148, 156, *163*, 309, *315*
Thomas, C., 542, *550*
Thomas, C. S., 465, *470*
Thomas, G. S., 559, *568*
Thomas, J. C., 574, *585*
Thompson, E., 40, 42, 46, *57*
Thompson, G., 34, *59*
Thompson, L. W., 270, *288*, 556, 567, 571, 577, *584, 585*, 619, *626*
Thompson, M., 34, *59*
Thompson, P. J., 357
Thomson, D. M., 572, 574, *585*
Thoreau, H. D., 290, 300, 301, *302*

Thorpe, W. H., 11, *19*
Thronesbery, C., 577, *584*
Thum, D., 365, 366, *379*
Thurnher, M., *469*, 477, *483*
Tiberi, D. M., 563, *568*
Tietze, C., 423, *429*
Till, J., 40, 42, *60*
Tinbergen, N., 6, *19*, 21, *31*
Tizard, B., 194, *213*
Tizard, J., 194, *213*
Toates, F., 541, *550*
Tobin, S., 597, *600*, 605, 606, *613*, 617, *627*
Todd, D. M., 402, *416*
Tognoli, J., 480, *484*
Togo, A., 618, *627*
Tomlinson-Keasey, C., 225, 226, *240*
Toner, I. J., 295, *303*
Toniolo, T. A., 225, *237*
Topinka, C. V., 85, *92*
Topper, M. D., 359, *379*
Torrey, B., 421, *429*
Toulnim, S., 224, 227, *237*
Townsend, M., 277, *284*
Townsend, P., 487, *499*
Trabasso, T., 225, *240*, 283, *284*, *289*
Tramer, R. R., 589, 591, *601*
Tramontana, M. G., 389, *393*
Trampe, J. P., *498*
Trasler, G., 385, *393*
Travers, S. H., 281, *286*
Treas, J., 478, *484*, 488, *499*
Treasure, K. G., 357, 358, *374*
Treat, N. J., 573, *585*
Trehub, S., 32, 34, 35, 41–44, 46, 48, 51, *60*, *61*, 132, *143*
Treiber, F., 23, *31*, 363, *377*
Trevarthen, C., 28, *31*, 109, 147, 167, 169, *179*
Trimboli, F., 365, *377*
Troll, L., 475, 476, 478, *484*, 486–489, 492, 493, *499*, 561, *568*
Tronick, E., 28, *31*, 199, *209*
Trost, M. A., 190, *209*, 337, *338*
Trussel, J., 427, *430*
Tschirgi, J. E., 346, *355*
Tucker, P., 124, *141*
Tucker, T., 225, *240*
Tuckman, J., 518, 520, *523*
Tulkin, S. R., 148, *161*
Tulving, E., 109, *147*, 218, *240*, 572–575, 578, *582*, *585*
Turiel, E., 349, *355*
Turkewitz, G., 21, *30*, 93, 156, *163*
Turner, B. F., 462, 463, *466*, *470*, 476–478, *484*
Turner, C. B., *466*, *470*
Turner, N., 476, *484*
Turner, R. H., 459, *499*
Turnure, C., 46, *61*
Tversky, A., 441, *446*

Tweney, R. D., 347, *353*, 440, *446*
Tzankoff, S. P., *567*

Udry, J., 420, *431*
Uhlenberg, P., 486, *499*
Upton, L. R., 277, *288*
Uzgiris, I., 137, *147*, 230, *240*

Vago, S., 346, *354*
Vaillant, G. E., 361, 367, 369, *375*, *379*, 456, 459, 461, *470*, 502, *507*
Valdes-Dapena, M., 73, 74, *78*
Vandell, D., 164, 168, 169, 170, 177, *178*, *179*, 197, *212*, 309, 310, *315*, 329
van Hilst, A., 478, *484*
Van Twyver, H. B., 7, *15*
Vargha-Khadem, F., 50, *61*
Vaughan, H. G., 93, 101, 102, *105*
Vaughn, B. E., 230, *238*
Vaziri, H., 359, *377*
Vener, A., 358, *375*, 422, *431*
Verbrugge, R., 52, *58*
VerHoeve, J., 37, 46, *56*, *61*
Veroff, J., 463, *468*, 617, 618, *626*
Vetter, H., 253, *267*
Videbeck, R., 561, *567*
Vietze, P., 81, *91*, 126, *144*, 427, *430*
Vigorito, J., 37, 39, 41, *56*, 68, *76*
Vinick, B., 473, 474, 478, *482*
Vogel, S. R., 463, *466*
Volpe, J., 328, *338*
Volterra, V., *265*
von Bertalanffy, L., 150, 151, *163*, 180, *214*
von Granach, M., 148, *163*
Von Derlippe, A., 305, *314*
Vonèche, J., 443, *445*
von Mering, O., 516, *523*
von Neumann, J., 441, *446*
Voss, H. L., 359, *377*
Voss, J. F., 278, *284*
Vurpillot, E., 123, *147*
Vyas, S. M., 542, 544, 545, *550*
Vygotsky, L. S., 292, *303*, 438, *446*, 450, 453, *455*

Wachowski, D., 476, *484*
Waddell, J. O., 359, *379*
Wagner, D. A., 282, *289*
Wagner, J., 228, *239*, *567*
Wahler, R. G., 244, *252*
Waite, L. H., 185, *214*
Walder, L. O., 388, *392*
Waldrop, M. F., 329, *339*
Wales, R., 255, *266*
Walk, R. D., 102, *106*
Walker, A., 99, *105*, 124, 131, 133, *140*, *143*, *147*
Walker, A. S., 127, *147*

Walker, D., 33, *61*
Walker, J. W., 563, *567*
Walker, L. D., 230, *240*
Wall, S., *209,* 309, *313*
Wallace, A. R., 4
Wallace, J. P., 558, *566*
Wallach, H. F., 580, *585*
Wallach, M. A., 595, *600*
Walley, A., 51, *57*
Walsh, D. A., 571, 574, *583, 585*
Walters, J., 487, *499*
Walters, L. H., 487, *499*
Wanska, S. K., 225, *237*
Ward, R. A., 620, *628*
Warden, C., 54, *61*
Warheit, G. J., 465, *470*
Warner, L., 54, *61*
Warren, M. Q., 389, *393*
Warrington, E. K., 577, 578, *585*
Washington, R. E., 466, *469*
Wason, P. C., *240,* 344, 347, *355,* 440, *446,* 449, *455*
Waterman, A. S., 604, *613*
Waters, E., 191, 193, *209, 214,* 309, *313*
Waters, R., 53, *61*
Watson, J. S., 82, *92,* 108, 138, *147*
Watson-Gegeo, K. A., 256, *268*
Waugh, N. C., 574, *585*
Webb, R. A. J., 369, *375*
Wechsler, H., 358, 365, 366, *379*
Wedderburn, D., *499*
Wedenberg, E., 33, *57*
Weimer, W. B., 109, 140, *147*
Weingarten, H., 478, *484*
Weinraub, M., 181, 182, 194, 196, 198, 202, *211, 212, 214*
Weinstein, R. M., 361, *379*
Weir, C., 35, *61*
Weiss, I. P., 36, *54, 59*
Weiss, S., 126, *142*
Weissman, M. M., 456, *470*
Welcher, W., 423, *429*
Welford, A. T., 571, *585*
Wellman, H., 280, 281, *285, 289,* 300, *301*
Wells, C. L., *565*
Weniger, F. L., 516, *523*
Wenkert, R., 561, *567*
Werker, J., 52, *61*
Werner, H., 218, *240,* 589, *601*
Werner, J., 279, *289*
Werner, J. S., 67, *78,* 79, 81, *92*
Wertheimer, M., 121, *147*
West, D. J., 388, *394*
West, M. J., 157, *160*
Westin, B., 33, *57*
Westley, W. A., 403, *414*
Wexler, H. K., 367, *375*
Whanger, A. D., 563, *568*

Whatley, J. C., 309, *314*
Whatley, J. L., 164, 166, 170, 171, 172, *178*
Wheeler, K., 94, *105*
Whelihan, W. M., 619, *627*
Whitbourne, S. K., 580, *585,* 587, *599,* 604, *613*
White, D. E., 231, *240*
White, R. W., 156, *163,* 607, *613*
Whitehead, P. C., 364, *374*
Whitehill, M., 384, *494*
Whiteman, M., 363, *374*
Whiting, B. B., 154, *163,* 189, 204, 206, *210, 214*
Whiting, J. W. M., 154, *163,* 189, 197, 204, *214*
Whittington, F., 516–518, *523*
Wickelgren, W. A., 572, *585*
Wiesel, T. N., 22, 24, *30,* 77
Wilcox, P., 465, *470*
Wilcox, S., 23, *31*
Wilen, J. B., 495, *499*
Wiley, J. A., 622, *626*
Wilkening, F., 225, *240*
Wilker, L., 618, 621–624, *628*
Wilkie, F., 556, *565,* 579, *582*
Wilkinson, A., 282, *289*
Williams, G. C., 6, *19*
Williams, L., 40, 42, 46, 48, *61*
Williams, T. H., 405, *416*
Willis, P. E., 398, 407, 409, 411, *417*
Willis, S. L., 525–527, 532–536, *537–539,* 557, 561, *567,* 598, *601*
Williston, J. S., 33, *56*
Wilner, N., 460, *468*
Wilson, A. B., 390, *391*
Wilson, E., 471, 481, *484*
Wilson, E. O., 7, *19,* 148, *163,* 201, *214*
Wilson, F., 419, *431*
Wilson, G. T., 361, *374,* 379
Wilson, K., 170, *179*
Wilson, M. A., 124, *145*
Wilson, W., 39, 41, 43, 48, 52, 53, *55, 61,* 617, *628*
Wilson, W. R., 22, *30*
Winer, G. A., 225–227, 230, *240*
Wingard, J. A., 262, *376*
Winnicott, D. W., 155, *163*
Winograd, E., 573, 576, *585*
Winslow, R. W., 373, *379*
Wise, S., 28, *31*
Wishart, J. G., 21, 28, *29,* 118, *141*
Withey, S. B., 617, 618, 622, *625*
Witkin, H. A., 442, *446,* 589, 591, 597, *601*
Witte, K. L., 557, *567,* 581, *584*
Witthinger, R. P., 576, 578, *582*
Wohlwill, J. F., 217, 218, 225, 227, 228, 230, 232, 233, *237, 240, 241,* 525, *539*
Wolf, E., *566*
Wolf, M. M., 243, 244, 251, 252, 390, *391*
Wolf, R. S., 619, *627*

Wolff, H. G., *626*
Wolff, P. H., 69, *78*, 154, *163*
Wolfgang, M. E., 389, *394*
Wolman, B. B., 462
Wong, A. C., 234, *236*
Wood, C., 33, 61, 331
Woodruff, D. S., 561, *565*
Woods, J. H., 367, *380*
Woodson, R., 24, *30*
Woodworth, R. S., 440, *446*
Worel, L., 155, *160*
Worner, W. J., *446*
Wright, H. F., 282, *284*
Wright, J. C., 257, *267*, 588, *600*
Wu, P. K., 24, *30*
Wunsch, J., 137, *147*

Yager, T. J., 619, *626*
Yakovlev, P. I., 33, *61*
Yarmey, A. D., 578, *585*
Yarrow, L. J., 157, *163*, 213
Yates, B. T., 298, *303*
Yinger, M., *417*
Yogman, M., 199, *209*
Yonas, A., 101, *106*, 121, 128, 143, *147*
Yost, W. A., 35, *61*
Youmans, E. G., 516, *523*
Young, E. B., 364, *376*
Young, G., 172, 177, *178*, *179*, 190, *212*, 329, *339*
Young, H. B., 363, *376*
Young, J., 401, 410, *417*

Young, R. J., 559, *568*
Young, R. S., *565*
Young-Browne, G., 127, *147*
Younger, B., 46, *58*, 132, *145*
Youniss, J., 184, *214*, 231, *236*, *241*, 272, *285*, 328, *338*
Yudovich, J., 196, *212*
Yule, W., 388, *393*, 404, *416*
Yussen, S., 280, *284*

Zacks, R. T., 576, *583*
Zahn-Waxler, C., 319, 320, 323, *326*
Zajonc, R. B., 196, *214*
Zaks, P. M., 557, *568*
Zarit, S. H., 577, 581, *583*
Zazzo, R., 196, *214*
Zeigler, B. L., 36, *56*
Zeiss, A., 293, 294, *302*
Zelazo, N. A., 72, *78*
Zelazo, P. R., 72, 78
Zelinski, E., 577, *585*, 619, *626*
Zelnik, M., 419, 420, 421, *429–431*
Zelniker, T., 588, *601*
Zentale, T. R., 233, 235, *238*
Zerbe, M. B., 555, *565*
Zich, J., 295, *301*
Zimmerman, B. J., 230, *241*
Zimmerman, W. S., *612*
Zivin, G., 295, *303*
Zucker, K. J., 135, *142*
Zuckerman, M., 363, *374*, *380*
Zung, W. W. K., *628*
Zylman, R., 357, *377*

SUBJECT INDEX

Abortion, in adolescence, 423
Abstraction, 114–116, 345–347
Academic achievement, and adolescent subculture, 399–408
Accommodation, 111–114
Acquaintance process, 333–335
Activation, 228, 229
  age by condition designs, 228
  conditions design, 229
  law of anticipatory function, 228
Activities of daily living, 615
Adjustment, 604, 611, 612
Adolescence:
  abortion, 423
  academic achievement, 399–408
  aggression, 381–394
  alcohol use, 356–379
  artistic understanding, 350
  athletics, 399, 402, 404
  birthrates of, 417
  births, nonmarital, 423–425
  childbearing, 418–431
    causes of, 418–425
    consequences of, 425, 427
  cognitive development, 343–355
  conduct disorder, 383–391
  conflict with adults, 395–399, 402–411
  contraceptive use, 421–423
  counterculture, 397
  delinquency, 381, 382, 396, 407–409
  dimensions of deviance, 381, 382
  drug use, 356–379
  educational aspirations, 395
  family interaction, 366, 367, 387
  generation gap, 397
  logical reasoning, 343–347, 349, 350
  moral development, 347–349, 384
  parental interaction, 398, 399, 402–406
  peer influences, 398, 403–406
  philosophical understanding, 347–350
  popularity, 399, 400, 403
  popular music, 398, 400, 407, 411
  schools, 395–417
  sexual intercourse, nonmarital, 419–421
  social groups, 399–402, 406–408
  socialized aggressive disorder, 383, 386–391
  stress, psychological, 403
  subcultures, 395–417
Adults:
  attitudes and behaviors, 306, 309
  change over time, 310
  control over childrens' development, 307, 309
  limitations on influence, 308, 310
  socializing agents, 304, 313
Affect, 316–323, 618, 621
  negative, 618, 621–624
  positive, 618, 620–624
Affect balance, 611
Affect Balance Scale (ABS), 618, 622
Affiliation need, 471
Affordance theory, 116–118
Age-related morale, 620, 621
Aggression:
  adolescence, 381–394
  arrest rates, 381
  conduct disorder, 383–386
  delinquency, 381, 382
  family, dynamics of, 387
  persistence of, 388, 389
  prevalence of, 387, 388
  prevention of, 390
  sex differences in, 381
  socialized type of, 386–387
  treatment of, 389, 390
Aging, 234, 511–536
  anthropology of, 511, 515
  attitudes toward, 518–522
  average aging, 524–537
  biological factors, 524, 535
  cross-cultural, 511–522
  decline, 526
  demography of, 511, 516–518
  ecological perspectives, 526
  environmental factors, 526
  experimental intervention paradigm, 526
  life-span view, 526
  modernization and, 511–522
  normative aging, 525, 536
  psychological aging, 524
  psychology of, 511, 518–522
  selective optimization, 536
  sociology of, 511
  theories of psychological aging, 536
Agitation, 621
Alcohol use, in adolescence:
  age and use, 358, 359
  delinquency and use, 370
  ethnicity and use, 358, 359
  family interaction and use, 366, 367
  frequency of use, 357, 358
  motives for use, 361, 362
  peer influence and use, 365, 366

sex and use, 358, 359
situational factors and use, 367, 368
social deviance and use, 363
social isolation and use, 364
stress and use, 367
Alpha and omega view, 487–489, 492, 494
Altruism, 316–323
Amodal information, 114–116
Anesthesia, 69
Animacy, 135–137
Anoxia, 68
Anxiety, 620, 623
Anxiety, withdrawal disorder, 383
Apgar scores, 69
Arrest rates, in adolescence, 381
Artistic understanding, 350
Asphyxia, 69
Assimilation, 111–114
Attachment:
  acquisition, 7–12
  animal models of, 7–10
  birth order, 187–190
  child abuse, 190–195
  conditioning approach in, 10–12
  critical period, 190–192
  father-infant, 190–192, 195–197
  fearfulness, 192–195
  imprinting, 7–12
  interactions, 182–187
  mother-infant, 182–185, 187–192
  others, 187–190
  peer relationships, 187–192, 197–199
  peer vs. maternal, 190–195
  quality, 185–190
  siblings, 185–187, 195–197
  social isolation, 187–192
  social relationships, 182–185
  traits, 190–192
Attention, 114–116, 226, 230, 234
Attentional processes, relation to, 598
Attitudes toward aging, 518–522
Audiovisual analysis, 149
Auditory perception, 32, 36–38

Babkin reflex, 65
Behavior:
  of children, 242
  control, 244
  experimental analysis, 250
  prediction of, 250
  production of, 250
  selection for analysis, 250
Behavioral development:
  of bird song acquisition, 7–10
  constraints on, 3–7
Behavioral trap, 244
Behavior pattern, sequential, 529

Bidirectionality, 308, 312
  child effects, 308
  interchanges, development of, 309
  synchrony, 308
Biological changes, 450, 456, 457
Birthrates, in adolescence, 417
Births, nonmarital, in adolescence, 423–425
Brazelton Neonatal Behavior Assessment, 27
Breadth of categorization, 594
  age, relation to, 594
  caution, relation to, 595
  memory, relation to, 595

Cardiac activity, 28
Careers, 502, 608, 609
Categorical perception, 22
Categories, 107–109
  age, 126–128
  coding of, 128–130
  facial expressions, 126–128
  object, 123–126
  sex, 126–128
Cattell 16 PF Scale, 624
Causality, 111–114
  assessment of, 135–137
  mechanical, 135–137
  social, 137–140
Childbearing, in adolescence, 418–431
Childrearing:
  and adolescence, 425–427
  consequences, for adolescent mothers, 425
  consequences, for children of adolescent mothers, 425–427
Choice reaction time tasks (CRT tasks), 544, 545
Classical conditioning, 63, 64
Class inclusion, 225, 230, 231
"Closed" head injuries, 541
Cognition:
  adolescence, 343–355
  adult, 107–109
  role in language acquisition, 258
    cognition hypothesis, 259
    interaction hypothesis, 259
    local homologies, 260
    measurement problem, 260
    production and comprehension differences, 261
    social cognition, 261
Cognitive development:
  in adolescence, 343–355
  and conduct disorder, 384
Cognitive skill training, 557
Cognitive strategies, 588
  attention, 588
  deficits in selective attention, 588
  haptic exploration, 588
  visual scanning patterns, 588

Cognitive structures, 344–345
Cognitive styles, 226, 234, 349, 350, 442, 586, 587, 596, 598
  abilities, relation to, 587
  definition of, 586
  ecological validity, 596
  methodological issues, 598
  modifiability, 596
Competence, 218, 232
  behavioral, 614, 616, 619, 625
  communicative, 254, 255, 258
  concrete operational, 231
  construct validity, 225
  definition of, 218, 232
  diagnosis, 226, 227
  formal operational, 231, 234
  level of generality, 224
  normative, 615
  Piaget's theory, 218
  primary, 615
Competence-performance, distinction, 351–352
Complexity, 451, 458
Comprehensive Assessment and Referral Evaluation (CARE), 615, 619
Concepts, 109–111
  event, 118–121
  object, 118–121
  relational, 109–111, 118–121
Conceptual differentiation, 594
  age, relation to, 594
  modification, 596
  object-grouping procedure, 594
Conceptualization, styles of, 592
  age, relation to, 592
  assessment, 592–593
    Conceptual Styles Test, 593
    verbal triads procedure, 593
  capacity-style distinction, 593
  education, relation to, 592, 593
  sex differences, 593
Concrete operations, 344–346
Conditional reasoning, 232
Conditioning, 10–15, 62–75, 555
Conduct disorder:
  adolescence, 382–386
  cognitive development, 384
  correlates of, 385, 386
  family interaction and, 387
  frequency of, 387, 388
  passive avoidance, 384
  persistence of, 388, 389
  physiological factors in, 384
  prevention of, 390
  response to social reward, 384
  stimulation seeking, 384
  treatment of, 389, 390
Congruence, 621, 622, 623, 624

Conjugate reinforcement, 68, 71
Conservation, 230, 234
Constraints on learning, 3–5, 12–14
Context, 437, 448, 449, 462
Contextual-setting conditions, 12–14
  deprivation, 12–14
Contingent responsivity, 166–168
Continuity, 447, 492, 793
Contraceptive use:
  in adolescence, 421–425
  and ethnicity, 421
  failure to use, 421–422
Control problems in everyday life, 545
Convergent operations, 537
Conversational skills of elderly, 547, 548
Coping strategies, 460, 461, 465
Cornell Medical Index, 620
Counterculture, in adolescence, 399
Courtship, 473, 474
Crib death, 72, 73, 74
Critical period, 70
Cross-cultural, 439, 465, 474, 478, 479
  aging, 511–522
  attitudes towards aging, 518–522
  demography of aging, 511, 516–518
Cross-modal integration, 126–128, 133–135
Crying, infant, 12–14
  conditioning of, 12–15
  indicators of attachment, 14, 15
  mother's departure, 12–14
  operant learning of, 12–14
  outcome, 12–14
  process, 12–14
  shaping of, 12–14
  stranger's approach, 12–14
Cultural conventions, 438, 439
Cultural factors, 226

Deductive reasoning, 343, 345
Defensive behavior, 75
Delinquency:
  alcohol use, 370
  drug use, 369–370
  problems in study of, 382
  rates of, in adolescence, 381
  sex differences in, 381
  subcultures of, 381–382, 396, 407–409
Demography of aging, 511, 516–518
Dependency:
  experimental-operant work, 528–532
  observational-operant work, 528
  reinforcement schedule, 528
  self-induced dependency, 529
  social environment, 530
Depression, 620, 623, 625
Developmental neurology, 33
Developmental reciprocities, 495–496

Developmental theory, 111–114
  Gibsonian, 114–116
  identity, 116–118
  neo-constructivist, 116–118
  Piagetian, 111–114
  psycholinguistic, 118–121
Differentiation theory, 114–116
Discontinuity, 447
Disequilibrium, 450, 451
Dishabituation, 69
Distinctive features, 114–116
Distraction:
  resistance to, 295
  tasks, 543, 544, 547
Divergent thinking, 595
  age, relation to, 595
Double storage system, 128–130
Down's syndrome, 69
Drug use, in adolescence:
  age and use, 358, 359
  attitudes toward use, 360, 361
  behavior systems and use, 368
  delinquency and use, 369, 370
  ethnicity and use, 358, 359
  factors associated with use, 359–371
  family interaction and use, 366, 367
  frequency of use, 357, 358
  motives for use, 361, 362
  patterns of, 356, 357
  peer influences and use, 365, 366
  prevalence of, 357
  prevention of, 371, 372
  school performance and use, 369, 370
  self-esteem and use, 362
  sensation-seeking and use, 363
  sex and use, 358, 359
  situational factors and use, 367, 368
  social deviance and use, 362–364
  social isolation and use, 371, 372
  states and traits associated with use, 362–365
  stress and use, 367
  treatment of users, 373
Dynamic configurations, 130–133
  temporal, 130–133

Early life history, 459
Ecological optics, 107–109
Ecological perspective, 21
Ecological variables, 171–174, 177
Economic well-being, 616
Edge, perception of, 93–97
  contrast, 93–95
    brightness, 93–97
    color, 93–95
    texture, 93–95
  reversible occlusion, 93–95
  stereopsis, 93–95

Education, formal, 361
Elderly people, 545
Empathy, 316–323
  antecedents of, 317–319
  conceptualizations of, 316
  Feshbach Affective Situations Test of, 319
  individual differences in, 323
  and induction socialization technique, 318, 321
  and interpersonal behavior, 323
  measurement of, 319
  and prosocial behavior, 316–323
  sex differences in, 319
Empiricism, 111–114, 220, 223, 224
Environment:
  consensual, 615
  objective, 615, 616, 625
  role in language acquisition, 262
    observational learning, 263
    speech to young children, 262
Environmental:
  quality, 614
  satisfaction, 622, 624
Event perception, 114–116
Event probability, 546
  running estimates of, 545
  variations in, 545, 546
Event roles, 135–137
Everyday life efficiency, 548
Evolution, 3–5
  biological, 3–5, 14, 15
    genetic change, 3–5
    genetic variation, 3–5
    Lamarckian, 3–5, 10–12
    Mendelian, 3–5
    selective retention, 3–5
  cultural, 3–5, 14, 15
    ecological adaptation of behavior, 5–7
    Lamarckian, 3–5
    learning, 3–5
    mechanisms of, 3–5
  Hominid, 3–5
  mechanisms of:
    co-evolution, 5–7
    cultural, 3–7
    organic, 3–7
  organic, 3–5
  paleoanthropology, 3–5
    natural selection, 3–5
  teleonomy, 5–7
    clinical variations, 5–7
Expertise, 438
Explanation of behavior, 218
  causal, 220
  formal, 219, 223, 224
  Piaget's, 218, 219
Extraversion, 624, 625
Eye-blink response, 64
Eysenck Personality Inventory, 624

Face-to-face interaction, 27, 28
Face perception, 23, 123–126
Factor analysis, 621, 623
Family interaction:
 adolescent subcultures, 398–399, 402–406
 aggression, 387
 alcohol use, 366–367
 drug use, 366–367
Father, infant interaction, 28, 164–171
Feeding schedules, 65
Field Dependence-Independence, definition of, 589
 age, relation to, 589–592
 assessment, 589–592
  Children's Embedded Figures Test, 591
  Embedded Figures Test, 589
  Rod-and-Frame Test, 589
 attentional processes, relation to, 598
 bipolarity of construct, 589
  social competence, 591
 modification, 597
 other tests, relation to, 591
  Raven's Progressive Matrices, 591
  Stroop Test, 591
 predictive validity, 591
 sex differences, 590
Field factors, 442, 443
Figurative solutions, 227
Filtering, 114–116
Fixed action patterns, 21
Formal operations, 344–346, 349–350
Frames, 109–111
Friends, 473, 479–480
Friendship, 327
 behavioral features of, 330–331
 conceptions, 330–331
 developmental course, 328–330
 sex differences, 329
 vs. popularity, 327
Friendship process, 332–336
 ending relationship, 335–336
 maintaining friendship, 335
 perceived similarity, 332–333
 selection of friends, 332–333
Fundamental adaptive system, 151, 152

General Well-Being Schedule, 619
Generational squeeze, 491
Generation gap, in adolescence, 397
Good life, 614, 615
Goodness of fit, 617
Grammar, 253
Grasping reflex, 74
Gratification, delay of, 293
Groups, social:
 adolescence, 399–402, 406–408
 aggression, 383, 386–391
Growth, 450
Growth orientation, 457

Habituation, 25, 68, 121–123
Happiness, 617, 618, 620–624
Health, 615, 620, 622, 624
 and nutrition examination survey, 619
 interview survey, 619
Hearing, 33
Heart rate, 69
Hedonic stimulation, 62
Helping, 316–323
Historical events, 448, 459, 486
Homology, 5–7
 phylogenetic relatedness, 5–7
Homosexual relationships, 480–481

Identity rules, 118–121
Illegitimate birth, rates of:
 adoption, 417
 adolescence, 417
 ethnicity, 417
Imitation, 25, 111–114, 244
 deferred, 111–114
Immaturity disorder, 383
Imprinting, 7–10
 altricial special, 7–10
 critical period in, 7–12
 humans, 10–12
 model for attachment, 7–12
 precocial species, 7–10
 process, 7–10
Independence, 529
 social environment, 529
Industrial inspection tasks, 546
Infant-adult interaction, 154
Infant learning, 62–75
Influence:
 child on mother, 10–15
 parent on child, 10–15
Information processing, 166–168, 344, 351
Institutional environment, 531
Institutional treatment, aggression in adolescence, 389–390
Integrative processes, 152
Intellectual aging, 532
 cognitive training, 534
 environmental factors, 533, 534
 intervention studies, 532
 plasticity, 532
 variability, 532
Intellectual functioning, 526
 cognitive training, 526
 fluid, crystallized intelligence, 532
 theory, 532
Intentionality, 290
Interactional synchrony, 29
Interactions, social:
 biological basis, 180–182
 cognitive structures, 182–187
 context of, 164–166

development of self, 182–187
direct, 197–199
dyadic, 197–202
face-to-face, 164–171
father-child, 180–182, 195–197
father-infant, 164–171
functions, 185–187
habits, 182–187
imitation, 182–187, 197–199
indirect, 197–199
mother-child, 180–182, 195–199
mother-infant, 164–171
others, 195–197
parental, 197–199
peer, 164–171, 190–195, 197–199
relationships, 180–185
sibling, 195–197
social reflexive behavior, 182–185
triadic, 197–202
Interpersonal, 457, 458, 471–473
Intrinsic motivation, 152
Intimacy, 472–474
Invariant structure, 114–116

Judgment and decision-making, 440, 441

Knowledge, 107–109
lexical, 109–111
organization of, 107–109
permanent, 109–111
Kohlberg's Theory, 347–349

Lack of control, 530
Langner Twenty-Two Item Screening Index, 619
Language acquisition, 253
Language development, in birds, 7–10
Learned helplessness, 530
Learning:
attachment, 7–12
behavioral approaches to, 243–252
comparative studies of, 5–7
constraints on, 3–5
biological preparedness, 5–10
bird song acquisition, 7–10
ontogenetic, 3–5, 7–10
phylogenetic, 3–5
crying, 12–15
ecology of, 12–14
environmental effects, 243–252
facilitators of, 12–14
imprinting, 7–10
laboratory situations, 554
media use, 560
naturally occurring situations, 558
cognitive skills, 560
physical skills, 558
social skills, 562

nature of, 551
activities, 553
assessments, 554
characteristics, learner, 552
contents, 553
contexts, 552
goals, 554
operant learning, 3–5, 10–15
acquisition, 10–12
definition, 10–12
extinction, 10–12
functional analysis, 10–12, 14, 15
stimulus, definition, 10–12
response, definition, 10–12
process, 10–12
role of environment in, 12–14
social, 3–5, 12–14
verbal, 555
Life course, 606, 607, 611
determinants of, 604, 607–612
life structure and, 608
Life satisfaction, 617–620, 622–624
Life Satisfaction Index (LSI), 617, 621, 623
Life-span development, 524, 536
Liking and loving, 471, 472
Linear gradient perception, 23
Living systems, 150
Logical operations, relation to memory, 271, 272
Logical reasoning, 343–347, 349, 350, 439, 440, 452
Loneliness, 620
Lonely dissatisfaction, 621
Long-term store, 270

Maladaptation, 451, 456
Marriage, 473–477
marriage satisfaction, 477
marriage styles, 475–476
remarriage, 478
Marxism, sociological, and adolescence, 398, 410
Matching Familiar Figures Test, 587–588
error component, 587
global-holistic version, 588
latency component, 587
Maternal:
preoccupation, 155
rejection, 159
Maturation, 12–14
Meaning, 109–111
extensional, 109–111
intensional, 109–111
Mediation deficiency, 271
Memory, 63, 67, 69, 109–111, 226, 234, 270–282, 542
capacity, 270, 272
contextual effects, 282
control processes, 270
cross-cultural, 282

encoding, 273
episodic, 109–111
experts and novices, in, 278
indexing, 542
inferential, 272
knowledge, role of, 273–279
long-term improvement, 272
Piagetian approach to, 271
semantic, 109–111
strategies, 270
structural components, 270
Memory stores, 570
  primary memory, 571
  secondary memory, 571
    automatic processing, 576
    encoding, 572
    levels of processing, 574
    metamemory, 576
    retrieval, 572
    speed, 574
    storage, 571
  sensory memory, 571
  tertiary memory, 577
    cross-sectional adjustment, 578
    naming latency, 578
    personal recollection, 578
    public events, 577
    world knowledge, 578
Mental health, 614, 616, 617, 625
Metacognition, 349–350
Metalogic, 343–350
Metamemory, 280–281
Metaphoric style, 595
  age, relation to, 595
Metatheory, 345–350
"Mid-life crisis," 604, 610
Midtown Scale, 619
Minnesota Multiphasic Personality Inventory, 619
Modeling, 166–168
Models for change, 540
Modernization:
  aging and, 511–522
  affect and, 521
  attitudes toward aging, 518–522
  demography of, 511, 516–518
  future research in aging, 521–522
  theory, 513–514
Monologuing in old age, 548
Mood tone, 617, 621, 623
Moral development, 347–349, 384
Morale, 616, 617, 618, 620
Mother:
  infant interaction, 28, 164–171
  presence and absence of, 171–174
Motion, role of, 95–102
  head movements, 99–103
    perception of occlusions, 102, 103
    visual following, 99–102

  object motion, 95–102
  self-produced *vs.* other-produced body motion, 102, 103
  visual scanning, 99–102
Multilevel Assessment Instrument, 615, 622
Multiple classification, 229, 231
Music, popular, in adolescent, 398, 400, 407, 411

National Center for Health Statistics, 619
Neuronal specificity, 21
Neurosis, 619, 621
Neuroticism, 624, 625
Noncognitive factors, 579
  anxiety, 580
  cautiousness, 580
  motivation, 579
Nonverbal communication, 157
NORC General Survey, 622
Novelty, 79–89. *See also* Recognition memory
Numerosity, 121–123
Nursing home residents, 529

Object:
  definition of, 93–95
  identity, 116–118
  permanence, 111–114
  position, 118–121
Observational learning, 246
  of response classes, 246
Observational-operant research, 527
  behavior patterns, 528
Operant conditioning, 66–68
Operant paradigm, 527–531
  convergent operations, 527
  experimental-operant research, 527, 531
  observational-operant research, 527–531
  operant ecological intervention, 527
  operant psychological principles, 526
  research strategies, 527
  reversibility of behaviors, 527
  work, 527
Organic brain syndrome, 625

Paced, serial addition task (PASAT), 541, 543
Pain, 62
Parental behavior, conditioning of, 14, 15
Parent-child relationships, 485
Parenting:
  adjustment to infants, 157
  behavior, 154
  in evolution, 156
  intuitive, 155
  primary didactic care, 156
Peer:
  influence, adolescent subcultures, 398, 403–406
  interactions in childhood, 336–337

interactions in infancy, 164–171
  sociability, 164–171
  unpredictability, 166–168
Perceived quality of life, 614, 616, 621, 622, 624, 625
Perception, development of, 95–97
  constraints, 95–97
  differentiation, 95–97, 99–102
    convex vs. concave, 99–102
    curved vs. straight, 95–97, 99–102
    different configurations of same elements, 99–102
    rigid vs. elastic, 99–102
  perceptual learning, 95–97
Perceptual development, 79–85
Perceptual shielding, 230
Performance, 218
  definition of, 218
Personality:
  dimensions of:
    extraversion, 603, 606, 609, 611
    masculinity, 603
    neuroticism, 603, 606, 610, 611
    openness, 603, 606, 609, 611
  measurement of, 605
    objective tests, 603, 607
    retrospective accounts, 611
    TAT, 603
  models of, 602, 603, 604, 605, 607, 612
    dimensional continuity, 602–607
    growth/decline, 602, 603
    life stage, 603, 604, 612
    typological, 605
  stability of, 602–611
Philadelphia Geriatric Center Morale Scale, 620, 621
Philosophical understanding, 347–350
Phonetic perception, 38–48
Physical training, 558
Piagetian theory, 218, 219, 225, 344–347, 351, 352, 435, 448, 449, 452
  formal operations, 436, 449, 450
  function, 219
  operations, 219
  stage, 219, 225
  structures, 219
Piano playing, 547
Pivot grammar, 254
Place error, 111–114
Planning and self-control, 295
Plasticity, 524–526, 532–537
  concurrent plasticity, 525
  conditions for, 525
  developmental plasticity, 525
  ecological factors, 526
  extent of, 525
  history-graded influence, 536
  interactionistic models, 527
  non-normative influences, 536

Play:
  face-to-face, 164–177
  father-infant, 164–177
  mother-infant, 164–177
  peer, 164–177
Pleasure, 62
Popularity, in adolescence, 399–400, 403
Positive mental health, 616
Positive reinforcement, 243
Post-formal operations, 443
Practice, 540
Pragmatics, 255
Precursor model, 164, 165
Predictions of future events, 546
Predictive control, 547
Pregnancy, in adolescence, 418–431
Preparedness, 12–14
  maturation, 12–14
Prepared responding, 70
Probability, 529
  base, 529
  conditional, 529
Production deficiency, 271
Prosocial behavior, 316–323
  role-taking, 318
  modeling effects, 322
Prototypes, 109–111
Psychological resources, 457
Psychopathology, 616, 617, 618, 622, 624
Psychophysiological symptoms, 619, 620
Psychosis, 625
Psychotherapy:
  aggression in adolescence, 389
  learning in, 563

Rationalism, 111–114, 220, 223, 224
Reactive control, 547
Recognition memory, 79–89
  facilitation of, 79–87
  forgetting and, 79–87
  intelligence and, 87–90
  origins of, 79–85
  paradigms, 79–81
    habituation-dishabituation, 79–85
    high-amplitude sucking, 79–81
    paired-comparison, 79–85
Reflection-impulsivity:
  age, relation to, 587–588
  definition of, 587
Regression, 450, 451
Rehearsal, 270, 276
Reinforcement, 12–14
Representation, role in memory, 279
Resolution and fortitude, 617
Respiratory occlusion, 75
Response classes, 246
  defining properties, 246
  formation of, 245
  generative, 247

Retirement, 504, 608, 609
  adjustment, 562
Reward, delay of, 293
Rosenberg Self-Esteem Scale, 620, 621, 622
Routes, descriptions of, 548
  mental representations of, 549

Schemas (schemata):
  cognitive, 346
  figurative, 111–114
  operative, 109–111
Schooling, 439, 448–449
  effects on memory, 282
Schools:
  adolescence, 395–417
  drug use, 368–369
Scientific reasoning, 345–347
Script knowledge and memory, 275
Scripts, 111–114
Self-concept, 617
Self-control, 291
Self-efficacy, 531
Self-esteem, 617, 620, 623
Self-instructions, 295
Self-regulation, 290–301
Self-synchrony, 22
Semantics, 254
  syntactic categories, 255
  word meanings, 255
Sensation-seeking:
  and drug use, 363
  and conduct disorder, 384
Sensorimotor sequences, 121–123
Sensory adaptation, 559
Sensory register, 270
Sequential lag analysis, 528
Sex differences, 234, 348–359, 381
  alcohol use, 358–359
  arrest rates, 381
  drug use, 358–359
  moral development, 348–349
Sex-role development, 115, 307
  age changes in children, 313
  adults' influence, 307
  Freudian theory, 305
  gender identity, 311
  gender role behaviors, 305, 311
Sex roles, 463
Sex-role socialization:
  parent attitudes, 306
  parent behaviors, 306
  peer behaviors, 307
  teacher attitudes, 307
  teacher behaviors, 306
Sexual intercourse, non-marital:
  adolescence, 419–421
  associated factors, 420–421
Shapes, 121–123
Sharing, 316–323

Short-term store, 270
Slant:
  gradients of texture, 93–97
  perception of, 93–97
  perspective transformations, 93–97
Social behavior, 243, 615
Social class, 464, 465
Social communication, 151
Social competence, 327
Social development:
  interactions, 185–187
  issues for theory, 180–182
  multiple vs. fixed paths, 187–190
  relationships, 185–187
  social knowledge, 180–182
  themes, 180–182
Social deviance:
  and alcohol use, 363
  and drug use, 362–364
  psychological dimension of, in adolescence, 382–383
  subcultures of, 367, 395–417
Social ease, 622
Social episodes, 531
Social integration of cognition, 451–453
Social interaction, 150, 157–159, 309
  failures, 158
  in preterm infants, 159
  ontogeny of, 157, 158
Socialization, 493, 494
Socialized aggression
  adolescence, 383, 387–388
  correlates of, 386
  family interaction and, 387
  frequency of, 388
  persistence of, 388–389
  prevention of, 390
  treatment of, 389–390
Social learning, 12–14
Social perception, 20
Social relationships, development of:
  acquaintance, 185–187
  adult-child asymmetry, 182–185
  attachment, 187–190
  biological basis, 187–190
  bonding, 190–192
  cognitive structures, 182–185, 205–208
  critical period, 190–192
  deterministic nature, 187–192
  development of self, 182–185
  dimensions, 180–182
  empathy, 182–187
  epigenetic model, 180–182, 187–195
  family, 197–205
  father-child, 180–182, 195–197
  friendship, 185–187, 192–195, 197–199
  functions, 199–205
  habits, 182–185
  interactions, 180–187

interconnections, 187–190, 197–199
love, 185–187
mother-child, 195–199
needs, 199–208
others, 195–199
parental, 197–199
parent-child, 180–182
peer, 192–195, 197–199
sequence, 187–190
sibling, 195–199, 205–208
social matrix, 202–208
social network systems model, 180–182, 190–208
strangers, 185–190
traits, 187–195
types, 180–182, 185–187
Social stimuli, 21
Social systems, 150
Social validity, 251
Sociolinguistics, 255
social context and language, 256
Space, 224, 229, 234
Spatial configurations, 121–123
three-dimensional, 123–126
two-dimensional, 121–123
Speech perception, 22, 32, 48–54
Stability, 450, 451
Stimulus attributes, 121–123
average values, 128–130
componential, 121–123
modal values, 128–130
wholistic, 121–123
Stimulus classes, 248
Stimulus control, 299
Stimulus feature detectors, 21
Story grammars and memory, 274
Strategies:
of control, 541
of performance, 541
Stress, psychological:
adolescence, 403
British, 398, 411
Marxism, sociological, 398, 410
Structure, invariants of, 93–102
detection of, during motion, 93–102
motivation, role of, 97–99
shape constancy, 97–99
Subcultures:
addict, 367–368
adolescence, 367, 395–417
varieties of in adolescence, 399–411
Subjective discomfort, 616
Succorance need, 471

Sucking behavior, 64
Sudden infant death syndrome, 72
Supramodal integration, 28, 29
Supramodal perception, 21
Synchronized events, 130–133
Syntax, 253
Systems theory, 150

Task simplification, 226
Task and situational factors, 226, 233
familiarity, 230
figurative task features, 229
instructions, 231, 233
Telegraphic speech, 254
Television, verbal language of, 257
Temptation, resistance to, 295
Theoretical models, 570
associative approach, 570
contextual approach, 570
information-processing approach, 570
Theory of bounded rationality, 441
Theory of constructive operators, 435
schemes, 437
silent operators, 437
Time use, 615, 622
Training, 227, 231
Trait theories, 461
Transitory mood, 618

Utilization, 232
definition of, 232

Visual preference, 24
Visual reinforcement, 67
Visual scanning, 23
Visual search tasks, 546
Visual system, newborn, 95–97
accommodation, 95–97
acuity, 95–97
binocular vision, 95–97
contrast sensitivity, 95–97
visual evoked potential, 95–97
Voice perception, 22

Walking, changes in old age, 547
Wellbeing, sectors of, 614, 616
Work and health, 505
Working memory, 541, 543, 546
Worries, 622

Zest *vs.* apathy, 617
Zung Depression Scale, 619